Japanese–English Chemical Dictionary

Edited by
Markus Gewehr

Japanese–English Chemical Dictionary

Including a Guide to Japanese Patents and Scientific Literature

Edited by
Markus Gewehr

With Contributions by Irene Schellner and Klaus Hinkelmann

WILEY-VCH Verlag GmbH & Co. KGaA

The Editor

Dr. Markus Gewehr
BASF AG
Global Strategic IP Management Agricultural Products,
APM/B
67117 Limburgerhof
Germany

Cover Illustration:
Photo and design idea by Markus Gewehr.

■ All books published by Wiley-VCH are carefully produced. Nevertheless, authors, editors, and publisher do not warrant the information contained in these books, including this book, to be free of errors. Readers are advised to keep in mind that statements, data, illustrations, procedural details or other items may inadvertently be inaccurate.

Library of Congress Card No.: applied for

British Library Cataloguing-in-Publication Data
A catalogue record for this book is available from the British Library.

Bibliographic information published by the Deutsche Nationalbibliothek
Die Deutsche Nationalbibliothek lists this publication in the Deutsche Nationalbibliografie; detailed bibliographic data are available in the Internet at ⟨http://dnb.d-nb.de⟩

© 2008 WILEY-VCH Verlag GmbH & Co. KGaA, Weinheim

All rights reserved (including those of translation into other languages). No part of this book may be reproduced in any form – by photoprinting, microfilm, or any other means – nor transmitted or translated into a machine language without written permission from the publishers. Registered names, trademarks, etc. used in this book, even when not specifically marked as such, are not to be considered unprotected by law.

Typesetting   Asco Typesetters, Hong Kong
Printing   Strauss GmbH, Mörlenbach
Binding   Litges & Dopf GmbH, Heppenheim
Cover Design   Grafik-Design Schulz, Fußgönheim

Printed in the Federal Republic of Germany
Printed on acid-free paper

ISBN 978-3-527-31293-1

Contents

Preface *XI*

Abbreviations and General Notes *XV*

Contributors *XVII*

I **General Part** *1*

1 **Introduction to the Japanese Language** *3*

1.1   The Japanese Language   *3*
1.1.1   Characteristics of the Language   *3*
1.1.2   Basics and Principles of the Japanese Grammar   *4*
1.1.3   Overview of Particles   *6*
1.1.4   Conjugation and Overview of Morphological Endings and their Use   *9*
1.2   Japanese Writing   *14*
1.2.1   Basic Principles   *14*
1.2.1.1 Types of Characters and their Use   *14*
1.2.1.2 Punctuation Marks   *15*
1.2.2   The *kana* Syllabaries and the Japanese Sound System   *17*
1.2.2.1 Introduction to the *kana* Syllabaries   *17*
1.2.2.2 The Japanese Sound System   *18*
1.2.2.3 Long Vowels   *21*
1.2.2.4 Other Aspects   *23*
1.2.3   *kanji* 漢字   *24*
1.2.3.1 Introduction to *kanji*   *24*
1.2.3.2 Structure and Elements of *kanji*   *26*
1.2.3.3 Classification of *kanji*   *27*
1.2.3.4 *kanji* Combination   *27*
1.2.3.5 Combination of *kanji* and *kana*   *29*
1.2.4   Transcription of Foreign Words into Japanese Phonology   *30*

Japanese-English Chemical Dictionary. Edited by Markus Gewehr
Copyright © 2008 WILEY-VCH Verlag GmbH & Co. KGaA, Weinheim
ISBN: 978-3-527-31293-1

2	**Japanese in Scientific and Technical Publications** *41*
2.1	Scientific and Technical Publications *41*
2.1.1	General Characteristics of Language used in Scientific and Technical Publications *41*
2.1.2	Expression of Specific Terms by Nouns in Chemical Literature *43*
2.1.3	Verbs and Adjectives in Technical Publications *45*
2.2	Frequently used *kanji* *48*
2.2.1	Important *kanji* in Scientific Publications *49*
2.2.2	Important *kanji* Combinations *59*
2.3	Numbers, Symbols and Units *65*
2.3.1	Numbers *65*
2.3.2	Mathematical Symbols and Terminology *69*
2.3.3	Units *70*
2.3.4	Other Symbols and Characters in Scientific and Technical Literature *74*
2.3.4.1	Greek Letters *74*
2.3.4.2	Colors *74*
2.4	Suggestions for Reading Japanese Scientific and Technical Publications *75*
2.4.1	Method for Sentence Analysis *75*
2.4.1.1	Step 1: Separation of Sentences in Subordinate Clauses *76*
2.4.1.2	Steps 2 and 3: Identification of Elements and Phrases within the Subordinate Clauses *77*
2.4.1.3	Step 4: Looking up of Unknown *kanji* and Translation of Single Words *78*
2.4.1.4	Steps 5, 6 and 7: Translation of the Sentence *79*
2.4.2	Relationship between Sentences *79*
2.4.3	Identification of Subordinate Clauses by Conjunctions *80*
2.4.4	Identification of Subordinate and Attribute Clauses by Verb Endings *84*
2.5	Example Translations *87*
2.5.1	Example Translation of Patent Claims *87*
2.5.1.1	JP 2001-342183 *87*
2.5.1.2	JP 2001-220439 *89*
2.5.2	Example Translation of Patent Description *90*
2.5.2.1	Prior Art (JP 2001-213861) *91*
2.5.2.2	Technical Field of the Invention (JP 2001-286762) *92*
2.5.2.3	Embodiment of the Invention (JP 10-114701) *93*
2.5.2.4	Description of Substituents (WO 01/10825) *95*
2.5.2.5	Example (JP 07-145156) *97*
2.5.2.6	Example (JP 2001-276618) *98*
2.6	Tools for Supporting Text Analysis *99*
2.6.1	Printed Support *100*
2.6.2	Identification of *kanji* in Character Dictionaries *102*
2.6.2.1	Identification of *kanji* with Radicals *102*
2.6.2.2	Identification of a *kanji* by Stroke Counting *104*
2.6.3	General Online Support for Japanese Language *105*
2.6.4	Online Support for Analysis of Chemical Literature *107*

3	**Naming of Chemical Compounds**	*113*
3.1	Naming of Elements and Inorganic Compounds	*113*
3.1.1	Elements	*113*
3.1.2	General Aspects of Naming Inorganic Compounds	*119*
3.1.3	Inorganic Acids	*124*
3.1.4	Inorganic Salts and Ores	*135*
3.1.4.1	Naming Simple Salts	*136*
3.1.4.2	Salts of Dibasic Acids	*138*
3.1.4.3	Salts of Tribasic Acids	*138*
3.1.4.4	Complex Salts	*139*
3.1.4.5	Common Names	*141*
3.2	Naming of Organic Compounds	*142*
3.2.1	General	*142*
3.2.2	Substituents	*146*
3.2.2.1	Substituent Classes	*146*
3.2.2.2	Specific Substituents	*147*
3.2.3	Number and Position of Substituents and Functional Groups	*150*
3.3	Overview of Specific Organic Molecules	*153*
3.3.1	Functional Groups and Chemical Classes in Organic Chemistry	*153*
3.3.2	Naming of Heterocycles	*157*
3.3.3	Organic Acids	*165*
3.3.4	Organic Acid Derivatives	*168*
4	**Japanese Patent Documentation**	*171*
	Irene Schellner and Markus Gewehr	
4.1	The Japanese Patent System	*172*
4.1.1	Development of the Japanese Patent System	*172*
4.1.2	The modern Japanese Patent System	*173*
4.1.3	The Japanese Patent Office	*174*
4.2	Special Characteristics of Japanese Patent Documentation	*176*
4.2.1	Document Types and Kind of Document Codes	*176*
4.2.2	INID Codes and Headlines	*178*
4.2.3	Japanese Patent Numbers	*180*
4.2.3.1	Numbering Systems	*180*
4.2.3.2	Japanese Imperial Years	*182*
4.2.3.3	Frequently Encountered Problems	*183*
4.2.4	Special Japanese Classification Systems	*184*
4.2.5	File Index Classification Lists	*185*
4.2.6	File forming Terms Lists	*188*
4.3	Online Sources of Japanese Patent Information	*192*
4.3.1	Patent Information from Patent Offices	*192*
4.3.2	Commercial and other Patent Information Sources	*195*

5 Overview of Japanese Patent Law 203
Klaus Hinkelmann

5.1 Introduction 203
5.1.1 Patentable Inventions 203
5.1.2 Patentability Criteria 204
5.1.2.1 Novelty 204
5.1.2.2 Exceptions from Loss of Novelty – Grace Period 205
5.1.2.3 First to File Principle 205
5.1.2.4 Inventive Step 206
5.1.3 Official Fees 206
5.2 Drafting of Japanese Patent Applications 207
5.2.1 Patent Claims (Section 36(5)(6) JPL) 208
5.2.2 Detailed Explanation of the Invention (Section 36(4) JPL) 210
5.3 Filing of Japanese Patent Applications 213
5.3.1 Direct Japanese Patent Filing 213
5.3.2 Entry into the National Japanese Phase for PCT Applications 215
5.4 Examination of Japanese Patent Applications 215
5.4.1 Substantive Examination upon Request 216
5.4.2 Examination of Japanese Patent Applications 217
5.4.2.1 Examination before the Japanese Patent Office 217
5.4.2.2 Appeal to the Intellectual Property High Court against Rejection Decision of the JPO 219
5.4.2.3 Patent Grant 220
5.4.2.4 Divisional Applications (分割出願 *bunkatsu shutsugan*) 220
5.5 Attack on Patent Applications and Patents 221
5.5.1 Submission of Information Regarding Patentability to the JPO (情報提供 *jouhou teikyou*) 221
5.5.2 Attack on Granted Patents – Invalidation Trial (無効審判 *mukou shimpan*) 222
5.5.3 Correction Trials (訂正審判 *teisei shimpan*) 224
5.6 The Patent Right 225
5.6.1 Term of Patent Right – Patent Term Extension 225
5.6.2 Rights of Inventors and Patent Owners 226
5.6.2.1 Rights of Inventors 226
5.6.2.2 Rights of the Owner of a Patent Application or a Patent 227
5.6.2.3 Rights of a Patentee 228
5.6.3 Licensing of Patent Rights 229
5.6.3.1 *senyou* Licenses 229
5.6.3.2 *tsuujou* Licenses 230
5.6.4 Further aspects of the Patent Right 230
5.6.4.1 Limits of the Patent Right 230
5.6.4.2 Exhaustion of the Patent Right 230
5.6.4.3 Prior User Rights 231
5.6.4.4 Further Limitations (Sections 69(2)(3) JPL) 231
5.6.5 Interpretation of Patent Claims 231

5.6.5.1	Literal Claim Interpretation	*232*
5.6.5.2	Claim Interpretation by the Doctrine of Equivalence	*232*
5.7	Enforcement of Patent Rights	*233*
5.7.1	Remedies for Patent Infringement	*233*
5.7.1.1	Injunction – Permanent and Preliminary	*233*
5.7.1.2	Damages	*234*
5.7.1.3	Recovery of Reputation	*235*
5.7.2	Procedural aspects of the Enforcement before the Courts	*235*
5.7.2.1	Jurisdiction and Standing to Sue	*235*
5.7.2.2	Before any Court Proceedings	*236*
5.7.2.3	Attorneys-at-Law (弁護士 *bengoshi*) and Patent Attorneys (弁理士 *benrishi*)	*236*
5.7.2.4	Court Proceedings	*236*

II	**Japanese–English Dictionary**	*239*
6	**Dictionary Structure and Explanations**	*241*
6.1	General Explanations	*241*
6.2	Dictionary Part I: Scientific Terms beginning with *kana*	*242*
6.3	Dictionary Part II: Scientific Terms beginning with basic *kanji*	*243*
6.4	Dictionary Part III: Further Scientific Terms beginning with *kanji*	*245*
7	**Dictionary Part I: Scientific Terms Beginning with *kana***	*249*
8	**Dictionary Part II: Scientific Terms Beginning with Basic *kanji***	*411*
8.1	Scientific Terms Beginning with *kanji* for Figures and Quantities	*411*
8.2	Scientific Terms Beginning with *kanji* for Chemical Elements	*427*
8.3	Scientific Terms Beginning with Characters Frequently Appearing in the Initial Position of Chemical Terms	*452*
8.4	Scientific Terms Beginning with Characters Representing Important Prefixes for Chemical Words	*509*
9	**Dictionary Part III: Further Scientific Terms Beginning with *kanji***	*531*
9.1	*kanji* without Radicals	*531*
9.2	*kanji* based on Radicals	*540*

III	**Appendices**	*653*
10	**Bibliography**	*655*
10.1	Character Dictionaries	*655*
10.2	Grammar and Related Topics	*655*
10.3	General Japanese–English Dictionaries	*656*

10.4 Scientific Books and Dictionaries *657*
10.5 Further Literature and Information Sources *658*
10.5.1 Online Sources of Japanese Chemical Societies *658*
10.5.2 Online Sources of Authorities and Institutes in Japan *658*

11 **Subject Index** *659*

Preface

For several decades Japan has been one of the most important countries for innovative research and development activities. In particular, in the fields of chemistry, chemical engineering and life sciences, Japanese universities, such as Kyoto University, University of Tokyo, Osaka University, Tokyo Institute of Technology, Tohoku University and Nagoya University, are recognized as sources of reliable scientific knowledge. This contribution has been acknowledged by four recent Nobel prizes for chemistry, namely to Professor Kenichi Fukui (Kyoto University; for theories concerning the course of chemical reactions, 1981), Professor Hideki Shirakawa (Tsukuba University; for the discovery and development of conductive polymers, 2000), Professor Ryoji Noyori (Nagoya University; for his work on chirally catalyzed hydrogenation reactions, 2001) and Koichi Tanaka (Shimazu Corporation; for ionization methods for mass spectrometric analysis of bio-macromolecules, 2002). A huge number of researchers, in universities, private companies and research institutes, is steadily creating scientific knowledge and making it available by publishing in scientific journals as well as in patents. Especially, the number of patent applications filed in Japan increased dramatically in the last part of the last century. Unfortunately, many publications are still written in Japanese, in particular most patents. Although Japanese scientists today also publish their results in international journals, and patents are translated before filing in further countries, there is still a great need to understand and consider documents written in Japanese.

In contrast to the importance of sources of scientific information from Japan are the hurdles in their reading and understanding, if they are written in Japanese. Although there is a permanent exchange of scientists with Japan and contact with Japanese Universities is steadily increasing, there are still few scientists with thorough Japanese language capabilities. In addition, there are still only a few tools available that facilitate Japanese text analysis.

A basis for the understanding of scientific literature from Japan is an elementary knowledge of the Japanese language and a thorough knowledge of the technical field of the documents involved. In the case of patent literature, experience with patent documents and legal phraseology is also necessary. However, linguistic problems are widely considered as the main challenge to understanding Japanese patents. While many researchers may have a basic knowledge of Japanese and understand literary texts, the linguistic features of Japanese – such as missing subjects and plural forms, continuously written text without clear word borders, complicated nesting of attributive and subordinate clauses – lead to hurdles in interpretation. For instance, patents often consist of paragraph-long sentences and it is difficult to identify the relationship between sentence elements. Even the relationship between sentences is sometimes hard to discover. The same problem appears with Japanese characters: a basic knowledge of important *kanji* used in everyday life is easy to obtain, but it is the many rarely used characters that exacerbate the interpretation of scientific texts, together with *kanji* combinations that are unique in their use in a scientific environment. Even simple *katakana* expressions may cause trouble because of the non-standardized transcription of scientific English terms.

For these reasons, this book aims to facilitate analysis of the chemical literature written in Japanese. The main intended audiences are the researcher who has some need for information from Japanese sources, patent attorneys dealing with Japanese patent applications and translators active in the field of chemistry. While scientists from universities and research institutes may focus on journals, most of the other target groups rely on patent information. Because patents contain relevant knowledge, in particular necessary for industrial research, a specific focus is on scientists in industry with some basic knowledge of the Japanese language.

To meet their need for a comprehensive tool for understanding Japanese patents, this book consists of two parts. The main part is the dictionary of scientific terms, covering major aspects of chemistry, such as organic and inorganic chemistry, biochemistry, polymer chemistry and chemical engineering. The focus is on basic organic, inorganic and macromolecular chemistry, in particular words from general chemistry, names of minerals, polymers and organic compounds, expressions for general chemical transformations and reaction types, terms describing physical properties of substances and physicochemical methods. The dictionary is based on a compilation of scientific terms that includes, currently, over 60 000 entries. For this compilation, various sources have been utilized, in particular patents and publications in scientific journals, but also textbooks, scientific dictionaries, internet publications and online dictionaries. The compilation has been established and verified over the last nine years. From this, over 15 000 terms have been selected for inclusion in this book. As criteria, frequency of use, relevance for text understanding, importance for chemistry and diversity were applied. Words are arranged within the dictionary in a way that facilitates their retrieval and translation for non-native Japanese speaking readers. Chapter 6 explains the structure and organizing principles. The dictionary is divided into three parts: Chapter 7 covers terms beginning with *kana*. Words starting with *kanji* are included in Chapters 8 and 9, subdivided into characters with specific importance for chemistry (Chapter 8) and further characters (Chapter 9).

To support people with only a limited understanding of the Japanese language and no professional practice in translation, the book also contains a general part. The goal of the first three chapters is to support text analysis in general. Chapter 1 clarifies basic principles of the Japanese language, such as characteristics of the language, differences to English, basic grammar topics, and overviews two characteristic features of Japanese, namely particles and morphological endings. Owing to the importance of text analysis, this chapter also describes in detail the Japanese way of writing, by explaining *kana*, *kanji* and their phonology (Section 1.2). Chapter 2 discuss the characteristics of scientific documents. By providing detailed lists of frequently used *kanji* and *kanji* combinations, a basic understanding of key terms can be obtained, thereby facilitating navigation within Japanese documents. Readers with little experience may also use this list to build up a basic knowledge of scientific terms. Methods for text analysis are given in Section 2.4, followed by some example translations of different parts of Japanese patents in Section 2.5. The methodical approach that is described facilitates sentence analysis by providing a step-by-step procedure. This is useful in particular for complicated sentences in Japanese patents, which are difficult to understand because of nested subordinate clauses, long listings and complex attributes and attribute clauses. Further tools that might be applied are provided at the end of Chapter 2. While most printed dictionaries are mainly intended for use by native Japanese speaking scientists and, therefore, in general not very useful for foreigners, the internet has recently become an important source for tools that are helpful for text analysis. Some tools, which are recognized as useful, are described; however, further valuable tools may arise soon. Because the names of chemical compounds are considered as key information in chemical texts, Chapter 3 gives detailed descriptions for naming inorganic and organic compounds.

Probably the most important information source from Japan is patents, especially for industrial research. This is because, whereas many Japanese researchers from universities publish their results in international journals in English, most knowledge obtained from company research is only initially available after publication in Japanese patent documents. Because of this need, important aspects of Japanese patents are described in two additional chapters. The Japanese patent system is the topic of Chapter 4, including its historical development and the role of the Japanese patent office. For proper patent document searches, patents numbers, document types and FI- and F-term classification are discussed. The section also contains Japanese terms used for INID codes and as headlines in patents, which helps in navigating within patent documents. Chapter 4 was written largely by Irene Schellner, an expert on Japanese patent documentation within the European patent office in Vienna. Additional information is available in her further publications. Her contribution has been supplemented by Sections 4.2.2 and 4.3. Various facets of Japanese patent law are specified in Chapter 5, such as basic terminology, patentability criteria, rights of patent owners and licensing opportunities. The complete patenting process is described, beginning with filing the patent application and the examination steps up to patent right enforcement. Chapter 5 is written by patent attorney Dr. Klaus Hinkelmann, who has extensive experience of Japanese patent law. More detailed comments can be found in his publications.

Various people have accompanied the development of the chemical dictionary and this book over the last few years and have made valuable contributions, either by discussions of terms and their arrangement within the dictionary, by reviewing the best procedure to analyze Japanese documents, by arguments on general Japanese language issues, by support in analysis of patents or articles from Japan or by encouragement in finalizing the project. I express my gratitude to Yoko Kono, Dr. Alexandra Schichtel, Dr. William K. Moberg, Kuniko Owada, Keiko Wiskamp, Konrad Vester, Dr. Kuniaki Shimbo, Tomoko Oshiro, Dr. Sandra Löhr and Dr. Anke Eberhardt. Special thanks go to my Japanese friends and colleagues who reviewed Japanese terms in the dictionary and language topics in the general part, in particular to Dr. Hiroyuki Sawada (BASF Japan), Dr. Tsuyoshi Kono (Kyoto University) and Professor Fumitake Yoshida (emeritus, Kyoto University).

Mannheim, July 2007 *Markus Gewehr*

Abbreviations and General Notes

AIPPI	International Association for the Protection of Intellectual Property
EDICT	Japanese–English Dictionary Project (Section 2.6.3)
EPIDOS	European Patent Information and Document Service
EPO	European Patent Office
FI	File Index
INID	Internationally agreed numbers for the identification of bibliographic data (Section 4.2.2)
INPADOC	International Patent Documentation Center
IPDL	Industrial Property Digital Library
IPC	International Patent Classification
JP	Country code for national Japanese patents and patent applications
JPL	Japanese Patent Law
JPO	Japanese Patent Office
JST	Japanese Science and Technology Agency
METI	Ministry of Economy, Trade and Industry
MITI	Ministry of International Trade and Industry
N	noun
NA	*na*-type adjective
NCIPI	National Center for Industrial Property Information and Training
neg.	negative
PAJ	Patent abstracts of Japan
PCT	Patent Cooperation Treaty
URL	Uniform Resource Locator (used as uniform resource identifier for internet addresses)
V	verb
VA	i-type adjective
WIPO	World Intellectual Property Organization

Japanese-English Chemical Dictionary. Edited by Markus Gewehr
Copyright © 2008 WILEY-VCH Verlag GmbH & Co. KGaA, Weinheim
ISBN: 978-3-527-31293-1

General Notes

- Readings (pronunciation, phonology) of *kanji* is indicated by italics and usually located immediately after the Japanese word, e.g. 酸素 *sanso*.
- The reading of *hiragana* and *katakana* is explained in Section 1.2.2. Because it is assumed that *kana* reading is acquired after some experience in translation, *ro-maji* is not indicated after all words written in *kana* (except where there is a need for emphasis).

Contributors

For suggestions, recommendations or questions, please contact japanese_engl_chem_dic_project@yahoo.de or one of the contributors.

Dr. Markus Gewehr
BASF Aktiengesellschaft
Global Strategic IP Management Agricultural Products, APM/B
67117 Limburgerhof, Germany
e-mail: markus.gewehr@basf.com

Markus Gewehr studied chemistry at Johannes Gutenberg-University in Mainz, Germany. He received his Ph.D. 1996 in the group of Professor Horst Kunz on the use of enzymatic cleavable protecting groups in the synthesis of peptides and glyocopeptides. In 1990–1991 he studied at Kyoto University with Professor Norio Ise and Professor Hiromi Kitano in the field of macromolecular chemistry and chemical engineering, accompanied by learning the Japanese language. Since then, he is a regularly visitor to Japan and is the author of essays on different topics in anthologies about Japan. The results of his research activities are published in several journals.

After joining BASF Aktiengesellschaft in Ludwigshafen, Germany, in 1997, Markus Gewehr worked on fungicide active ingredient discovery research. By translating patents and other chemical literature from Japan, he gained experience in text analysis of scientific documents written in Japanese. He is involved in over 160 patent applications related to different patent families. Within BASF's agricultural division, he is now responsible for global strategic IP management.

Irene Schellner
Japanese Patent Information
European Patent Office
Rennweg 12
1030 Vienna, Austria
e-mail: ischellner@epo.org
www.european-patent-office.org

Japanese-English Chemical Dictionary. Edited by Markus Gewehr
Copyright © 2008 WILEY-VCH Verlag GmbH & Co. KGaA, Weinheim
ISBN: 978-3-527-31293-1

Having studied Japanese at the University of Vienna, Irene Schellner joined the European Patent Office in 1994 and has been working as a Japanese patent information specialist since then. She carries out searches in Japanese databases for clients, gives training courses in Japanese patent law and patent information matters, and is in charge of the EPO's helpdesk service for European users of Japanese and East Asian patent information. Irene Schellner has published several articles about Japanese patent documentation and specific topics of the Japanese patent system, e.g. in "World patent information" or in documents from the EPO.

Dr. Klaus Hinkelmann
Hinkelmann&Huebner
Kaiserplatz 2
80803 Muenchen, Germany
e-mail: khinkelmann@hinkelmannhuebner.de

Klaus Hinkelmann studied chemistry at Albert-Ludwigs-Universität Freiburg, Germany, and received his Ph.D. in organic electrochemistry. After a post-doctoral stay with Professors Heeger and Wudl at the University of California, Santa Barbara, USA, he joined the IP department of BASF Aktiengesellschaft in Ludwigshafen, Germany, where he worked for more than ten years. From 1995 to 1998, he was East Asia Representative of BASF's IP department in Japan.

Klaus Hinkelmann is a European and German patent attorney and is currently working in his own IP firm in Munich, Germany. He is the author of numerous publication on Japanese IP law, including the German language reference book *Industrial Property Protection in Japan*, published by Carl Heymanns Verlag.

Dr. Hiroyuki Sawada
BASF Agro Ltd.
Roppongi 25 Mori Bldg. 23F
1-4-30, Roppongi, Minato-ku
106-0032 Roppongi, Japan
e-mail: hiroyuki.sawada@basf-agro.co.jp

Hiroyuki Sawada studied chemistry at both Kyoto University and Purdue University in the U.S.A. He was awarded his Ph.D. for organometallic synthesis from Kyoto University in 1986.

Hiroyuki Sawada worked for Ube Industries for 20 years in the chemical business. He then moved to the agrochemical business and worked for American Cyanamid as a licensing manager in Japan for three years. In 2000, he joined the agricultural division of BASF and is currently working as licensing manager in Tokyo. He is also involved in the analysis of Japanese patent applications.

I
General Part

1
Introduction to the Japanese Language

1.1
The Japanese Language[1]

1.1.1
Characteristics of the Language

The Japanese language (日本語 *nihongo*) is spoken by over 127 million people. It has been heavily influenced by the Chinese language over a period of at least 1 500 years. Therefore, Japanese is written with a mix of Chinese characters (漢字 *kanji*) and two syllabaries that are also based on Chinese writing. Much vocabulary has been imported from China or created using the Chinese method of word formation. Today, three sources for words are used in the Japanese language:

- Original Japanese words were already used before the introduction of Chinese writing in the 4th–6th century. With the new possibility of writing down the language, suitable *kanji* had to be found. The pronunciation of *kanji* with the original Japanese sounds is called *on*-reading.
- Chinese words (漢語 *kango*) were introduced while the Japanese adopted Chinese characters and the pronunciation is called *kun*-reading. Chinese-based words comprise as much as 70% of the total vocabulary of the Japanese language and form as much as 30–40% of words used in speech. A small number of words has also been borrowed from Korean and Ainu.
- Words from Western languages have entered Japanese from the 16th century onwards, beginning with Portuguese and followed by borrowing from the Dutch during Japan's isolation in the Edo period. After the reopening in the Meiji restoration in the 19th century, vocabulary from German, French and English was introduced. Today, words from Western languages are transliterated and written in *katakana* (外来語 *gairaigo*).

As consequence, Japanese is written today with a mixture of *kanji*, syllables of two syllabaries (平仮名 *hiragana* and 片仮名 *katakana*), Roman letters and Arabic figures. The bulk of

---

[1] Because some aspects are not relevant for the understanding of scientific publications, they are not described in this book, such as dialects, the honorific system of politeness, origin and characteristics of vocabulary, history of the language, orthographic reforms and the characteristics of standard Japanese forms.

Japanese text is usually written in either *kanji* or *hiragana*. Within a sentence, words are not separated with spaces – this agglutinative nature gives the "word" a different definition from words in English.

Morphologically, Japanese is an agglutinative language, meaning that words are formed by joining morphemes together or adding affixes to the bases of words.

Japanese is characterized by a complex system of honorifics, reflecting the hierarchical structure of Japanese society. The language has an extensive grammatical system as well as its own vocabulary to express politeness and formality. Three levels of honorific speech are distinguished: plain Japanese (also called informal, 砕けた *kudaketa*, 普通 *futsuu*), simple polite form (also called plain formal, 丁寧語 *teineigo*) and the advanced polite form (敬語 *keigo*) that itself has different levels (respectful language 尊敬語 *sonkeigo*, humble language 謙譲語 *kenjougo*). The choice of the right level of honorifics is determined by various factors, including social position, age, job, gender and experiences, and, within the *keigo*, the content of the speech is also relevant. As will be discussed later, in scientific publications the plain form is exclusively used and, because of this, no detailed explanation of the honorific system and the grammatical measures of *teineigo* and *keigo* is given here.

Dozens of dialects are spoken in Japan. They typically differ in terms of pitch accent, inflectional morphology, vocabulary and pronunciation. As Standard Japanese is prevalent nationwide, dialects do not have to be considered while analyzing scientific literature from Japan.

1.1.2
Basics and Principles of the Japanese Grammar

Compared with some other important languages, the Japanese grammar is often characterized as "not too complicated". However, there are many differences to English and other European languages and there are many typical rules that are necessary to consider in the analysis of Japanese publications. In the following, some basics and principles are introduced. A comprehensive description of the Japanese grammar can be found in the books listed in the appendix.

Words in Japanese are classified into two broad categories, which are further divided as follows:
- independent words (自立語 *jiritsugo*) having internal meaning:
  - conjugable words (活用語 *katsuyougo*): verbs (動詞 *doushi*) and i-type adjectives (形容詞 *keiyoushi*);
  - non-conjugable words (非活用語 *hikatsuyougo*): nouns (名詞 *meishi*), pronouns (代名詞 *daimeishi*), na-type adjectives (形容動詞 *keiyoudoushi*), adverbs (副詞 *fukushi*), conjunctions (接続詞 *setsuzokushi*) and interjections (感動詞 *kandoushi*);
- ancillary words (付属語 *fuzokugo*) are modifiers without own meaning: particles (助詞 *joshi*), prenominals (連体詞 *rentaishi*), counter words (助数詞 *josuushi*) and auxiliary verbs (助動詞 *jodoushi*).[2]

[2] There is no common agreement about the definition of the word in the Japanese language. With conjugable words, stems of verbs and adjectives as well as some endings (as "auxiliary verbs") are own words in many grammar books. In this book, a word is the finally conjugated verb or adjective, i.e., the combination of verb- or adjective-stem and ending.

Because of their importance in scientific literature, verbs, i-type adjectives and particles are described in the following subsections. Conjunctions and interjections are discussed with the sentence analysis in Chapter 2 (Section 2.4).

Japanese nouns have neither number nor gender. To indicate more than one, either a suffix is added (e.g., 々, 達 and 等 are possible for a small number of selected nouns) or the quantity is named. Thus 分子 bunshi may mean "one molecule" or "molecules" without any implied preference for singular and plural. In addition, nouns do not inflect to show politeness or respect.

There is only a small number of true adverbs. Most adverbs are derived from other words; therefore, adverbs are often not considered as an independent class of words, but rather a function used by other words.

The most important part of the sentence is the predicate. It contains the conjugable words: verbs, i-type adjectives or the copula. The copula (だ, です) is a special word forming sentences like "A is B". All conjugable words consist of a stem and a changeable ending that expresses tense, negation, politeness, conditional and other functions. The predicate is, therefore, the most important part of the sentence.

There are two principle structures for word order and sentence structure in the Japanese language: The basic word order is "subject–object–verb" whereas most Western languages are classified as subject–verb–object languages. As consequence, each sentence ends in a verb, an adjective or a form of the copula, except if sentence-final particles are present and except in some rhetorical and poetic usage, which is of no relevance in scientific literature. The basic word order is also valid for sub-clauses, which facilitates understanding (Section 2.1.1).

The basic sentence structure is "topic–comment". Hence, the topic comes at the sentence-initial position and is indicated separately from the subject and both do not always coincide. Topic may also be other elements of the clause, such as objects, temporal or spatial information.

The order of the other sentence elements is relatively free, as long as the modifying element stands in front of the modified word (e.g., the adverb precedes the modified verb, the relative clause is in front of the modified noun and genitive nominal precedes the possessed nominal).

Elements of a sentence are not only defined by single words (単語 tango). It is grammatically more reasonable to speak of "phrases" (文節 bunsetsu) as the sentence-forming equivalents of English "words"[2]. A phrase consist of a word that is followed by auxiliary verbs, verb- and adjective-endings, suffixes and particles that modify its meaning and define its grammatical role. In addition, phonologically, the postpositional elements are part of the word they follow, because the pitch accent falls behind the combination. Particles like は, が, の or を indicate the topic of the sentence, define the function of words and regulate the relation of the sentence elements to each other. Section 1.1.3 gives a list of important particles.

Interrogative questions have the same structure as affirmative sentences, but with intonation rising at the end. Sentence-final particle, か or の, may be positioned at the end of the sentence, depending of the formality of the speech.

The subject or object of a sentence need not be stated if it is obvious from the context. Therefore, words are omitted from the sentence; this is preferred to referring to them with pronouns. Although there are pronouns in the Japanese language, these are not used as frequently as pronouns in Western languages. They are only used in situations implying some emphasis and this is correlated with the honorific system. Personal pronouns are deleted unless it is necessary to emphasize who is speaking to whom.

1.1.3
Overview of Particles[3]

Japanese particles (助詞 *joshi*) are used postpositionally and have a wide range of grammatical functions. The most important is the definition of the grammatical role of the term they follow. Particles indicate the topic of the sentence, emphasize words, define the function of words, stress the direction of actions, determine ownership and by this regulate the relation of the sentence elements to each other. They also indicate a question, the speaker's assertiveness and a wide range of emotions. Some Japanese particles work like prepositions in English, but they are unlike prepositions in many ways and Japanese does not have any equivalents of prepositions. Instead, nouns and verbs are used for modifying where English might use prepositions.

Particles are written in *hiragana* even if some have old *kanji* forms. For three common particles, *hiragana* diverge from its pronunciation: は is read "wa", へ is read "e" and を is read "o".

Particles can be classified into ordinary, sentence final and compound particles. Sentence-final particles indicate an interrogative sentence (か or の) and express emotions and emphasis and are of minor importance for the understanding of scientific literature. Compound particles are formed with one particle together with other words that might also be particles. Usually, their understanding can be derived from the single particle and, therefore, they are not discussed here.

For foreigners, the most difficult issue with particles is the distinction and correct use of は and が. Some rough ideas can be used as primers, as given in Table 1.1.

As an example, focus and emphasis are illustrated by the following pair of sentences:

モルホリンは有機塩其だ。= Morpholine is an organic base. ("organic base" is stressed
moruhorin wa yuukienki da.   and there might be other organic bases among the
                                  molecules that are considered)
モルホリンが有機塩其だ。= It is morpholine that is the organic base. ("morpholine"
moruhorin ga yuukienki da.   is stressed in the sense that it is morpholine of all the
                                  organic bases that is considered)

Table 1.2 describes the main functions of important particles. Particles and functions without relevance for scientific literature are not listed.

---

[3] Sometimes, conjugations, interjections and particles are discussed together. In this book, conjugations are discussed in Section 2.4 as they are a suitable instrument for sentence analysis. Some words may be used as particles as well as conjugations. The table only contains the particle function, e.g., the sentence connecting function of *kara* or *demo* is not listed.

Table 1.1 Use of particles *wa* and *ga*.

	Use of は *wa*	Use of が *ga*
Focus of the sentence	No emphasis; or emphasis of the predicate or an object	Emphasis of the subject
Translation into English	With the definite article ("the")	With the indefinite article ("a")
Question	Question for the object	Question for subject or object
Answer to a question	If question has a negative predicate or particle は	If question has affirmative predicate and particle が
Specific verbs and adjectives		• Existential verbs (いる, ある, 住む) • Verbs and adjectives that express abilities, desire and sympathy • The request form (-たい)

Table 1.2 Overview of important particles and their function.

Particle	Following	Function
は	Nouns, particles	• Indication of the topic of the sentence • Contrasting the current topic from other possible topics

が	Nouns	• Indication of the topic with emphasis of the subject
		• Determination of sentence topic in case with second subject
		• Marking of interrogative pronouns in W&H questions (who, where, what, ..., how)
		• Marking of the direct object in case of
		    – specific verbs (いる, ある, 住む)
		    – verbs and adjectives which express abilities, desire and sympathy
		    – the request form -たい

も	Nouns, phrases	• Indication of the topic with the meaning "also", "either ... or" or "neither ... nor" • Marking the object indicating emphasis (translated as, for example, "even") • Modification of interrogatives in terms of generalization, for example, when: いつ → always: いつも
を	Nouns	• Marking of the direct object • Marking of the indirect object for verbs of locomotion
の	Nouns Phrases	• Modification of nouns (combination of two noun phrases resulting in indicating an ownership between nouns) • Sentence-final particle to indicate interrogative sentences

Table 1.2 *(continued)*

Particle	Following	Function
に	Nouns	• Indicating absolute dates that do not depend on the point of time they are mentioned • Marking of the indirect object referring to destination, target person or the goal of an action • Marking the object in passive clauses • Expressing the purpose of an action ("in order to") • Indicating a place of existence
へ	Nouns	• Marking of the indirect object referring to destination, direction, target person or the goal of an action
で	Nouns	• Marking of the location of action (except place of existence) • Indicating the instrument of an action • Referring to a causal relationship ("because of") • Marking of quantitative, temporal or spatial separation and temporal arrival point
と	Nouns	• Exhaustive listing of counted objects • Indicating the partner of activities • Marking of quotations, e.g., of indirect speech, thoughts, naming, expression of opinion and similar wording
から	Nouns	• Indicating the temporal, spatial or personal starting point of or the reason for an action
	Verbs (te-form)	• Indicating a point of time after a given action
まで	Nouns, verbs	• Indicating the temporal or spatial end point of an action • Marking of unexpected amounts ("even") • Together with に (→ までに) indicating a time limit for an action as "by … (a certain time)"
か	Nouns, verbs	• Marking of alternatives ("or") • Modification of interrogatives in terms of generalization, e.g., who: だれ → somebody: だれか
	Phrases	• As sentence-final particle indicating interrogative sentences
でも	Nouns, particles	• Marking of constrictions as "even" • Modification of interrogatives in terms of restriction, e.g., when: いつ → never: いつでも
ばかり, しか, だけ, きり, のみ	ばかり, しか, だけ: nouns, verbs きり, のみ: nouns	• Expressing limitations like "only", "just", "nothing but" or "nobody but"
くらい, ほど, ごろ	Nouns	• Expressing an approximate quantity ("about, around, approximately")
ほど	Verbs, adjectives	• Indicating the extend, upper limit or degree of an action or property
より	Nouns	• Indicating the person or object of comparison
ずつ	Quantifier	• Modifying quantifiers by expressing equal distribution of quantity

Table 1.2 (continued)

Particle	Following	Function
など	Nouns	• Indicating exemplification ("like, for example, and similar, such as")
ごとに	Nouns	• Indicating regular activities in terms of time or place ("every")
って	Nouns, verbs, adjectives	• Indicating a topic in the meaning "speaking of …"
さえ, すら	Nouns	• Marking the object indicating emphasis (translated as, e.g., "even"), さえ with positive emphasis, すら with negative emphasis

1.1.4
Conjugation and Overview of Morphological Endings and their Use

The Japanese predicate contains the only conjugable words, fulfils many functions and is, therefore, considered as the most important sentence element. Apart from a few exceptions, it is constrained to the ends of clauses. It can be classified into:
- verbal predicate with conjugable verbs (the copula may be used);
- adjectival predicate with conjugable i-type adjectives (the copula may be used);
- adjectival predicate with non-conjugable na-type adjectives and the copula and
- noun-type predicate with the copula.

The most important functions of the predicate are the expression of tenses, voice, aspect, politeness and the formation of negative sentences. This is achieved by conjugating verbs, i-type adjectives or the copula. All three types of conjugable words consist of a stem, which expresses the meaning of the verb and which is written with *kanji*, *hiragana* or both. Behind the stem, the ending is attached, which indicates the above-described functions and which is written in *hiragana*. Some endings are also conjugable and, therefore, called "auxiliary verbs".

The copula is used as conjugable word, if the predicate contains non-conjugable words like nouns or na-type adjectives. It is also called the "be-verb" in English. In scientific literature, its most important forms are the plain present forms である and だ. Further conjugated plain forms are given in Table 1.3.[4]

Conjugated verbs[5] consist of the verb stem, a stem extension and the ending. Whereas the ending of all verbs is usually the same (except the shift from -て to -で and -た to -だ for few

---

4) The affirmative polite form of the copula is です. It is not only used as predicate but also to modify a sentence predicate with stative verbs into a more polite form. です is further inflected like group-1-verbs, e.g., in it is affirmative past polite form でした.

5) There is an alternative description of verbs used in most of Japanese grammars by Western linguists. It uses "stem consonants", "reduced stems" and the insertion of additional vowels between stems and endings for group-1-verbs. These terms are not used by Japanese as they think of syllables instead of isolated consonants and vowels. Because it is also a complicated approach with many exceptions, it was decided to use a more systematic description.

Table 1.3 Conjugation of the copula.

	Present tense	Past tense	Gerundive	Presumtion
Affirmative	だ da	だった datta	で de	だろう darou
	である de aru	であった de atta		
Negative	ではない dewa nai	ではなかった dewa nakatta	ではなくて dewa nakute	
	じゃない ja nai	じゃなかった ja nakatta		

types), the stem extension depends of the verb class. Except for the two irregular verbs する *suru* and 来る *kuru*, every Japanese verb is a member of one of the two main verb classes, both of which can be divided into sub-classes:

- The plain present form of group-1-verbs (五段 *godan*) ends with a syllable of the u-column. Because the consonant of the last syllable is different and determines the sub-class (く-, ぐ-, す-, つ-, ぬ-, ぶ-, む-, る- and う-verbs), verbs of this class are also called "consonantal verbs".[6] For all conjugations, a stem extension has to be placed between verb stem and ending. The stem extension is always a syllable of the same row (e.g., -ま-, -み-, -む-, -め-, -も-) but the row depends on the sub-class. There are also some irregular modifications for plain past, gerundive and conditional forms (Table 1.4).

- All group-2-verbs (一段 *ichidan*) consist in their plain present form only of verb stem and the ending -る. Depending on the vowel before, which can be either i or e, they can be divided into *iru*-verbs (上一段 *kamiichidan*, e.g., to discuss: 論じる *ronjiru*) and *eru*-verbs (下一段 *shimoichidan*, e.g., to heat: 温める *atatameru*). They are also called "vocal verbs". In contrast to group-1-verbs, there is no stem extension necessary for many conjugations and some endings are directly attached to the stem. Other stem extensions are not regularly derived from one row of the 50-sound-matrix as it is for group-1-verbs.

6) The only exception of the characteristics are consonantal verbs with the ending う, as they do not end with a consonant (e.g., 会う au, 言う iu). In former times, う was "*wu*" and this still can be seen in some forms, like the negative non-past form (e.g., 会わない *awanai*, 言わない *iwanai*). If the last syllable of a verb is る and if the syllable before る is from the e- or i-column it can't be decided by the plain non-past form whether the verb is group-1- or group-2-type. It either has to be known or it can be identified by other conjugated forms.

Table 1.4 Plain past, gerundive and conditional forms for group-1-verbs.

Sub-class	Example	Plain past		Gerundive	
		Ending	Example	Ending	Example
く-	書く kaku	-いた	書いた kaita	-いて	書いて kaite
ぐ-	泳ぐ oyogu		泳いだ oyoida		泳いで oyoide
す-	話す hanasu	-した	話した hanashita	-して	話して hanashite
ぬ-	死ぬ shinu	-んだ	死んだ shinda	-んで	死んで shinde
ぶ-	遊ぶ asobu		遊んだ asonda		遊んで asonde
む-	住む sumu		住んだ sonda		住んで sonde
つ-	持つ motsu	-った	持った motta	-って	持って motte
る-	掛かる kakaru		掛かった kakatta		掛かって kakatte
う-	言う iu		言った itta		言って itte

The combination of verb stem and stem extension is called "verb base". All endings can be attached to seven different bases, as summarized in Table 1.5.[7] The overview includes some important verb endings and the formation of the group-1-verb 足す tasu and the group-2-verb 見る miru as examples. The irregular verb する suru is included due to its importance for scientific literature (Section 2.1.3). The 連用形 base is used to attach many other additional verbs or other words, such as -すぎる (e.g., too big: 大きすぎる ookisugiru), -かた (e.g., the way of writing: 買いかた kaikata), -そうだ, -始める (e.g., to start to add: 足し始める tashihajimeru), -終わる (e.g., to finish to examine: 調べ終わる shirabeowaru), -安い (e.g., easy to understand: 分かり安い wakariyasui). Other group-1-verbs are modified in a similar way to 足す tasu, with some irregularities for plain past, gerundive and conditional forms, depending on the group-1-sub-class (Table 1.4). Table 1.5 lists the seven bases and some important endings.

Some endings may be conjugated as well, which leads to the formation of agglutinated multiple verbal endings. For instance, the polite ending -ます conjugates as a consonantal verb (e.g., -ます → -ました) and passive and causative endings conjugate as vocal verbs (e.g., -られる → -られられます, -させる → -させられた). The order of agglutinating endings is not arbitrary. For example, the combination of negative and causative forms is only possible by adding a negative ending to the causative form of the verb, e.g., 見る miru → 見させる misaseru → 見させない misasenai (not possible is the opposite approach adding causative ending the negative verb form 見ない minai → 見なさせる minasaseru).

---

7) Two basis have the same verb conjugation but conjugated verbs are used differently, either as sentence terminal verbs (終止形) or as attributes (連体形). The form 連用形 may be derived into two sub-classes, one for which no stem extension is necessary for class-2-verbs and a second for which -ら- and -さ- as stem extension are used.

Table 1.5 Japanese verb bases.

Base name	Form	Ending	Group-1-verbs example: 足す tasu (stem: 足- ta-) Stem extension	Conjugated verb	Group-2-verbs example: 見る miru (stem: 見- mi-) Stem extension	Conjugated verb	する suru Conjugated verb
終止形 shuushikei 連体形 rentaikei	Plain present	(No ending)	-す	足す tasu	-る	見る miru	する suru
連用形 renyoukei	Polite present	-ます	-し-	足します tashimasu	(No stem extension)	見ます mimasu	します shimasu
	Polite present neg.	-ません		足しません tashimasen		見ません mimasen	しません shimasen
	Imperative	-なさい		足しなさい tashinasai		見なさい minasai	しなさい shinasai
	Request form	-たい		足したい tashitai		見たい mitai	したい shitai
	Purpose	-に		足しに tashini		見に mini	しに shini
	Plain past	-た (-だ)		足した tashita		見た mita	した shita
	Gerundive	-て (-で)		足して tashite		見て mite	して shite
	Conditional	-たら (-だら)		足したら tashitara		見たら mitara	したら shitara
未然形 mizenkei	Plain present neg.	-ない	-さ-	足さない tasanai	(No stem extension)	見ない minai	しない shinai
	Plain past neg.	-なかった		足さなかった tasanakatta		見なかった minakatta	しなかった shinakatta
	Passive	-れる		足される tasareru	-ら-	見られる mirareru	される sareru
	Causative	-せる		足させる tasaseru	-さ-	見させる misaseru	させる saseru
仮定形 kateikei	Conditional	-ば	-せ-	足せば taseba	-れ-	見れば mireba	すれば sureba
	Potential	-る		足せる taseru	-られ-	見られる mirareru	できる dekiru
命令形 meireikei	Imperative	(No ending)	-せ-	足せ tase	-ろ	見ろ miro	しろ shiro
音便形 onbinkei	Volitional	-う	-そ-	足そう tasou	-よ-	見よう miyou	しよう shiyou

Table 1.6 Conjugation of i-adjectives and examples for i- and na-type adjectives.

Form	i-adjective ending	Example i-adjective: easily inflammable 燃え安い	Example na-adjective: viscous 粘稠な
Plain present	-い	燃え安い *moeyasui*	粘稠な *nenchou na*
Plain past	-かった	燃え安かった *moeyasukatta*	粘稠だった *nenchou datta*
Plain present negative	-くない	燃え安くない *moeyasukunai*	粘稠ではない *nenchou dewa nai*
Plain past negative	-くなかった	燃え安くなかった *moeyasukunakatta*	粘稠ではなかった *nenchou dewa nakatta*
Gerundive	-くて	燃え安くて *moeyasukute*	粘稠で *nenchou de*
Gerundive negative	-くなくて	燃え安くなくて *moeyasukunakute*	粘稠ではなくて *nenchou dewa nakute*
Conditional	-ければ	燃え安ければ *moeyasukereba*	粘稠なら/ならば *nenchou nara/naraba* 粘稠であれば *nenchou de areba*

There are two types of adjectives in Japanese. na-type adjectives (形容動詞 *keiyoudoushi*) are not conjugable and, therefore, a form of the copula replacing the final な has to be used, indicating tense, politeness and other functions. They have some properties similar to nouns and many na-adjectives are nouns that are affixed with na and, because of this, they are also called quasi-adjectives or adjectival nouns.

i-type adjectives (形容詞 *keiyoushi*) inflect, i.e., they can become, for example, past or negative and because of this they have verbal character. However, they do not have the full range of conjugation seen for verbs. Conjugated i-type adjectives are formed by replacing the final -i with the appropriate ending. The polite form of the copula may follow to make the adjective more polite but this is not applied in scientific publications. In principle, the same bases exist for adjectives as described for verbs above. Some verb endings can be considered as i-type adjectives with the same further conjugation, such as the plain past form -ない or the request form -たい. Table 1.6 gives the conjugation of both types of adjectives and examples.

The two types of adjectives are also used differently in the case of adverbal use. In i-type adjectives, the ending is replaced by -く, and between na-type adjective and verb に is introduced. Nouns can be formed from some i-type adjectives by adding the ending さ to the adjective stem, such as for big: 大きい *ookii* → biggest: 大きさ *ookisa* and high: 高い *takai* → highest: 高さ *takasa*. For a few adjectives, the ending み *mi* is also possible, e.g., for weak: 弱い *yowai* → weakest: 弱み *yowami*.

1.2
Japanese Writing[8]

1.2.1
Basic Principles

1.2.1.1   Types of Characters and their Use

Written Japanese is composed of a mixture of four systems of characters: two phonetic syllabaries, 平仮名 *hiragana* and 片仮名 *katakana*; 漢字 *kanji*, which are characters of Chinese origin; and, rarely, Japanese words written in Roman letters. The bulk of Japanese text is usually written in a combination of *kanji* and *hiragana*, but *katakana* are also common in scientific publications. After a brief introduction, each is described more detailed.

kanji are used in Japanese either as pictograms, ideograms or phonograms. They usually represent a concept or an idea, but are occasionally used simply for their sound. *kanji* are used for nouns, adjectives, the stems of verbs and verbal adjectives and Japanese names. Most have two or more readings.

hiragana syllables have a defined reading but are not connected to a specific meaning or content. They are used for the inflecting endings of verbs and adjectives (送り仮名 *okurigana*), for words with only grammatical function (e.g., particles), for all other Japanese words without *kanji*, such as する or きれい or for *kanji* that have recently became unpopular, e.g., for phosphorous (燐 *rin*): リン or りん. *hiragana* are also used for words for which the *kanji* form is too difficult and not expected to be known to the readers, is not known to the writer, or is too formal for the writing purpose. As 振り仮名 *furigana*, *hiragana* are also used as help to indicate the correct reading of difficult *kanji*.

To express words of foreign origin, the second syllabary *katakana* is used. It is more frequently found in chemical literature than in non-scientific publications, as many technical expressions, especially names of chemical compounds, are borrowed from foreign languages, i.e., English, Latin or German. Depending on the author's philosophy, plant and animal names may also be written in scientific literature with *katakana*, although original Japanese names exist (e.g., wheat コムギ in pesticide chemistry or mouse ネズミ in pharmaceutical publications). *katakana* can also be found for onomatopoetica, or if a Japanese expression shoud be emphasized, especially in signs and advertisements. Words to be emphasized in a sentence are also sometimes written in *katakana*, and, in some Japanese *kanji* dictionaries, *katakana* are used to indicate the *on*-reading of a *kanji*. In addition, *kanji* in scientific literature are sometimes replaced by *katakana* if the *kanji* is expected to be too difficult.

Finally, Latin letters and Arabic figures are also used in scientific literature. Numbers are often written with Arabic forms, especially for specification of quantity and temperature, for code numbers of pharmaceutical or pesticide drugs, and to indicate the position of substituents in the names of chemical compounds. In addition, some English expressions, such as abbreviations (e.g., "DMF" and "tert.") and the oxidation state in the names of inorganic compounds, are usually written with Roman letters. Empirical formulas are always written in the international way. The following sentence illustrates the different characters used in scientific Japanese literature:

---

[8] The following aspects are not described in this book: Systems of encoding *kanji* and *kana*, e.g., unicode, EUC-JP, Shift-JIS, half-width and full-width *katakana*; industrial standards for the writing of *kanji* and *kana*; methods to write individual *kana* and *kanji*, e.g., stroke order.

（２－クロロフェニル）メチルアミン４２.８ｇ（０.３ｍｏｌ）をジクロロメタン
１１８.３ｇに溶解し、無水酢酸３２.２ｇ（０.３１５ｍｏｌ）を２０～４０°Ｃで
五時間かけて滴下した。

| verb conjugation with *hiragana* | noun in *kanji* | chemical compound name with *katakana* | amount with Arabic figures | Latin letters | particle with *hiragana* |

For transliterating Japanese in the Roman alphabet, different methods may be used. The most important system is the Hepburn method, which is also used in this book and is described in Section 1.2.2. Transcribed Japanese is called ローマ字 *ro-maji*. Romanization is also the most common way to input Japanese words into word processors and computers.

All words in Japanese can be written in either *hiragana*, *katakana* or *ro-maji*. The choice of which type of writing to use depends on style, conventions, and, to some extend, the preference of the authors.

There is no fixed rule regarding the direction of writing. Whereas in newspapers, magazines and fiction, Japanese is still written traditionally in columns going from top to bottom, with columns ordered from right to left (縦書き *tategaki* or 縦組み *tategumi*), it is common to write horizontally from left to right in scientific and technical publications (横書き *yokogaki* or 横組み *yokogumi*). Independent from the direction of writing, all characters are placed directly one after each other. No special inter-space is used to separate words, or even sentences. All symbols (*kanji* and *kana*), including the punctuation marks, get the same space, i.e., they are written into an imaginary square of the same size.

1.2.1.2 Punctuation Marks

There are no standardized rules for the use of punctuation marks in the Japanese language. It is left to the author's discretion when and how to use them. The number of marks that can be found in scientific literature is very limited, as grammatical constructions like indirect speech are uncommon. Because of the generally close relationship of chemical literature to the English language, the use of marks and signs has been adapted to the standard of international publications.

The two most important Japanese punctuation marks are the small dot "。" (丸 *maru*) and the drop shaped comma "、" (点 *ten* or 読点 *touten*, "reading point"). *maru* is used to finalize all types of sentences (therefore, it is also called 句点 *kuten* "sentence point"), including interrogative and exclamatory sentences, as there are traditionally no interrogation and exclamation marks in Japanese. *ten* may be inserted, if it helps in understanding complex sentences by grouping parts of the sentence that belong together. There are few cases in which the *ten* is essential for understanding the sentence, such as in following examples:

> 研究者が、昨日研究室へ来た時に塩酸ヒドロキシルアミンを反応
> 混合に添加した。
>
> *kenkyuusha ga, kinou kenkyuushitsu e kita toki ni ensanhidorokishiruamin o hannoukongou ni tenka shita.*
>
> When the researcher yesterday came to the laboratory, he added hydroxylamine hydrochloride to the reaction mixture.

研究者が昨日研究室へ来た時に塩酸ヒドロキシルアミンを反応混合に添加した。

kenkyuusha ga kinou kenkyuushitsu e kita toki ni ensanhidorokishiruamin o hannoukongou ni tenka shita.

When the researcher yesterday came to the laboratory, somebody (e.g., the researcher or another person; subject not named) added hydroxylamine hydrochloride to the reaction mixture.

ten is also used to separate successive numbers (e.g., 二、三時間 *ni, san jikan* = "two, three hours") and to partition four- and multi-digit numbers in triple-digit groups, such as in 五、二九〇キロ (*gosen nihyaku kyuuju kiro*). Table 1.7 gives further punctuation marks that are used in the scientific literature.

Table 1.7 Punctation marks.

Mark	Japanese name	Description
•	中点 *nakaten*, 中黒 *nakaguro*	• Used to separate items in lists of nouns and names, e.g., "fluorine, chlorine, bromine and others" = 弗素・塩素・臭素など *fusso・enso・shuuso nado* • Indication of the beginning of decimal places of fractions (reading: *ten*), e.g., 六・五% • Separation of components of chemical names and terms for clearer understanding, e.g., "in vitro": イン・ビトロ; trimethylacetaldehyde: トリメチル・アセトアルデヒド
～	波形 *namigata*	• Indication of the range between figures like "from … until"
「 」	鉤 *kagi*, 鉤括弧 *kagikakko*	• Quotation marks to indicate beginning and end of citations
『 』	二重鉤 *futaekagi*, 二十鉤括弧 *nijuukagikakko*	• Quotation marks for double quotes (used for citations within citations or when indicating a book title)
…	点線 *tensen*	• Indication of an incomplete sentence, thought or thread
〔 〕	亀甲 *kikkou*	• Used to insert comments into quoted text
【 】	すみつき括弧 *sumitsukikakko*	• Used in headings, e.g., in dictionary definitions
( ) [ ] 〈 〉	括弧 *kakko*, 丸括弧 *marugakko* かぎかっこ *kagikakko* 山括弧 *yamakakko*	• Further types of brackets
《 》 ≪ ≫	二重括弧 *futaekakko*	• Brackets within brackets

1.2.2
The *kana* Syllabaries and the Japanese Sound System

1.2.2.1 Introduction to the *kana* Syllabaries

The Japanese language consists phonetically of syllables. They are represented by the two phonetic syllabaries, 平仮名 *hiragana* and 片仮名 *katakana*. Each sound may be written down by one *kana* character. In addition, the phonology of *kanji* is subject to the same system. Each *kana* is either a vowel (e.g., あ), a consonant followed by a vowel (e.g., き) or the nasal sonorant ん.

When *kanji* were introduced in Japan, each native Japanese word could be written down, but the longer the word, the more characters were necessary (with each character consisting of few to many stroke numbers). To make writing less difficult and time-consuming, the *kanji* were increasingly simplified, resulting in the *hiragana* syllabary (8th century). *hiragana* first gained popularity among women, who were not allowed access to the same level of education as men ("women's writing", 女手 *onnade*). Later, the *katakana* syllabary matured finally from the *kanji* during the early Heian period, when monastery students of Buddhism used simple characters to make notes to explain the pronunciation of difficult *kanji*. Later, *hiragana* was used for unofficial writing because of its flowing style, e.g., for personal letters, while *katakana* was used for official documents.

To express the sounds of the Japanese language for non-Japanese, the syllables are converted into the Latin alphabet. Today, there are several different romanization systems but, mainly, two methods are used for the transcription process.[9] Preferred among foreigners and in scientific literature is the Hepburn system (ヘボン式ローマ字 *hebonshiki ro-maji*). It was developed 1885 by a commission of Japanese and foreign scientists and became popular after publication by the American linguist J.C. Hepburn. Some optimization has taken place since then, and the currently used version is called the "Revised Hepburn system". The method is geared to the English language, following English phonology for consonants, and vowels also to the German and Roman (i.e., Spanish) languages. This book uses the Hepburn system, with only a few exceptions, which are described below.[10]

---

[9] There are other methods besides the Hepburn system for romanization. The *kunreishiki ro-maji* (訓令式ローマ字) was developed by the Japanese authorities. It does not consider the irregular reading of some *kana* characters (し, ち, つ, ふ and for palatalized *kana* derived from し and ち, e.g., しゃ, ちゅ, じょ) and, therefore, looks neater because of greater consistency. In total there are only a few differences from the Hepburn system. It is recommended by the Japanese government and taught to Japanese elementary school students.

The *nihonshiki* (日本式) is the least used of the three main systems. It strictly follows Japanese phonology and the syllabary order and, therefore, is the only method that allows lossless mapping to and from *kana*.

[10] In this book, there is no use of apostrophes (like "*n'i*" for ンい) to distinguish between the syllable combinations of ン with vowels (ンあ n + a, ンい n + i, ンう n + u, ンえ n + e and ンお n + o) and the syllables な, に, ぬ, ね and の.

1.2.2.2 The Japanese Sound System[11]

The "basic" Japanese sounds can be arranged in a 50-sound-matrix (五十音 *gojuuon*) that associates each syllable with one *kana* character.[12] The syllables are traditionally arranged in rows by vowels in the order "a-i-u-e-o" and in columns by consonants in the order "no consonant, k-s-t-n-h-m-y-r-w". Nowadays, it has changed into rows for the consonants and the new arrangement is also used in this book. In recent centuries, linguistic development has led to the loss of some sounds because of their disuse. The number of official approved characters for basic *hiragana* and *katakana* today is 46 each, including the only sound that is not part of the syllable system but later introduced to express nasal sounds, ン. The *kana* を, originally read as *wo*, is only used as a grammatical indicator. The transcription of the syllables with the Hepburn system into the Latin alphabet takes the irregular phonology of the syllables し, ち, つ and ふ into account.

Changes of sounds during the centuries and the appearance of new sounds that were not represented by a basic character of the 50-sound-matrix made it necessary to add auxiliary markers for the existing *kana*, creating new ways of reading. Therefore there are additional derived sounds and characters besides the 50-sound-matrix, called "modified syllables": starting from the syllables of the か-, さ-, た- and は-rows, a marker (濁り点 *nigoriten* or 濁点 *dakuten* = adding two small lines to the top-right corner of the *kana* character) turns an unvoiced consonant into a voiced one, such us k into g or h into b. By this, "clouded sounds" (voiced sounds, 濁音 *dakuon*) are formed, such as ば, じ, で or ご. To *kana* beginning with an h, a small circle (半濁り点 *hannigoriten* or 半濁点 *handakuten*) can also be added, changing the h into a p and forming the "semiclouded sounds" ぱ, ぴ, ぷ, ぺ and ぽ (半濁音 *handakuon*). Figures 1.1 and 1.2 summarize the *hiragana* and *katakana* characters of the 50-sound-matrix, including clouded and semiclouded sounds and their transcription by the Hepburn system.

With clouding, there are two cases in which a sound may be expressed with two *kana*: There are two *kana* pronounced *ji* (*hiragana* じ and ぢ, *katakana* ジ and ヂ) and two *kana* pronounced *zu* (*hiragana* ず and づ, *katakana* ズ and ヅ). These pairs are not interchangeable. Usually, the *kana* characters of the s-row are used, i.e., *ji* is written as じ or ジ, and *zu* is written as ず or ズ.

Sounds that are represented by only one *kana* character are called unpalatalized (直音 *chokuon*). Besides these, there are further derived sounds that are called palatalized (拗音 *youon*). They are formed by combining some syllables of the i-column (き, し, ち, に, ひ, み and り) or the clouded sounds ぎ, び, ぴ or じ followed by a subscript version of the *kana* for ya, yu or yo, e.g., きゃ *kya*, にゅ *nyu*, ひょ *hyo*, みゃ *mya*, りょ *ryo*, gyo, びゃ *bya* and ぴゅ *pyu*. For the syllables し, ち and じ, the "y" is skipped, giving pronunciations such as しゃ *sha*, じゅ *ju* and ちょ *cho*. Figure 1.3 gives all of the palatalized sounds.

---

11) Although the expressions "vowels" and "consonants" are used here, it has to be kept in mind that Japanese do not think of their sounds in the Japanese language in terms of vowels and consonants but in terms of syllables. In addition, whereas Westerners think in terms of letters, Japanese think in terms of *kana*.

12) Besides the *gojuuon* ordering, an old-fashioned *iroha* ordering is sometimes used. The ordering is derived from a Buddhist poem and, therefore, is not useful for foreigners.

あ *a*	い *i*	う *u*	え *e*	お *o*						
か *ka*	き *ki*	く *ku*	け *ke*	こ *ko*	→	が *ga*	ぎ *gi*	ぐ *gu*	げ *ge*	ご *go*
さ *sa*	し *shi*	す *su*	せ *se*	そ *so*	→	ざ *za*	じ *ji*	ず *zu*	ぜ *ze*	ぞ *zo*
た *ta*	ち *chi*	つ *tsu*	て *te*	と *to*	→	だ *da*	ぢ *ji*	づ *zu*	で *de*	ど *do*
な *na*	に *ni*	ぬ *nu*	ね *ne*	の *no*						
は *ha* *(wa)*	ひ *hi*	ふ *fu*	へ *he (e)*	ほ *ho*	→	ば *ba*	び *bi*	ぶ *bu*	べ *be*	ぼ *bo*
					→	ぱ *pa*	ぴ *pi*	ぷ *pu*	ぺ *pe*	ぽ *po*
ま *ma*	み *mi*	む *mu*	め *me*	も *mo*						
や *ya*		ゆ *yu*		よ *yo*						
ら *ra*	り *ri*	る *ru*	れ *re*	ろ *ro*						
わ *wa*				を *(w)o*						
				ん *n*						

Fig. 1.1 *hiragana* Characters: 50-sound-matrix, character *n*, *hiragana* for clouded and semiclouded sounds and their *ro-maji*, transcribed by the Hepburn system.

ア a	イ i	ウ u	エ e	オ o						
カ ka	キ ki	ク ku	ケ ke	コ ko	→	ガ ga	ギ gi	グ gu	ゲ ge	ゴ go
サ sa	シ shi	ス su	セ se	ソ so	→	ザ za	ジ ji	ズ zu	ゼ ze	ゾ zo
タ ta	チ chi	ツ tsu	テ te	ト to	→	ダ da	ヂ ji	ヅ zu	デ de	ド do
ナ na	ニ ni	ヌ nu	ネ ne	ノ no						
ハ ha	ヒ hi	フ fu	ヘ he	ホ ho	→	バ ba	ビ bi	ブ bu	ベ be	ボ bo
					→	パ pa	ピ pi	プ pu	ペ pe	ポ po
マ ma	ミ mi	ム mu	メ me	モ mo						
ヤ ya		ユ yu		ヨ yo						
ラ ra	リ ri	ル ru	レ re	ロ ro						
ワ wa				ヲ (w)o						
				ン n						

Fig. 1.2 *katakana* Characters: 50-sound-matrix, character *n*, *katakana* for clouded and semiclouded sounds and their *ro-maji*, transcribed by the Hepburn system.[13]

13) Because of the similar shape, there are two pairs of *katakana* that can be easily mixed up and, therefore, have to be carefully distinguished: ソ so and ン n; and シ shi and ツ tsu.

きゃ *kya*	きゅ *kyu*	きょ *kyo*	キャ *kya*	キュ *kyu*	キョ *kyo*
ぎゃ *gya*	ぎゅ *gyu*	ぎょ *gyo*	ギャ *gya*	ギュ *gyu*	ギョ *gyo*
しゃ *sha*	しゅ *shu*	しょ *sho*	シャ *sha*	シュ *shu*	ショ *sho*
じゃ *ja*	じゅ *ju*	じょ *jo*	ジャ *ja*	ジュ *ju*	ジョ *jo*
ちゃ *cha*	ちゅ *chu*	ちょ *cho*	チャ *cha*	チュ *chu*	チョ *cho*
にゃ *nya*	にゅ *nyu*	にょ *nyo*	ニャ *nya*	ニュ *nyu*	ニョ *nyo*
ひゃ *hya*	ひゅ *hyu*	ひょ *hyo*	ヒャ *hya*	ヒュ *hyu*	ヒョ *hyo*
びゃ *bya*	びゅ *byu*	びょ *byo*	ビャ *bya*	ビュ *byu*	ビョ *byo*
ぴゃ *pya*	ぴゅ *pyu*	ぴょ *pyo*	ピャ *pya*	ピュ *pyu*	ピョ *pyo*
みゃ *mya*	みゅ *myu*	みょ *myo*	ミャ *mya*	ミュ *myu*	ミョ *myo*
りゃ *rya*	りゅ *ryu*	りょ *ryo*	リャ *rya*	リュ *ryu*	リョ *ryo*

Fig. 1.3 Hepburn system for the transcription of *hiragana* and *katakana*: palatalized sounds.

1.2.2.3 Long Vowels[14]

The writing of long vowels (also called doubled vowels) depends on the origin of the word. For originally Japanese words (*kun*-readings), the same vowel is added to the vowel that should be elongated, such as かあ, みい or ふう.[15] For words of Chinese origin (*on*-

---

14) The expression "long vowel" is only applied for vowels within words that are written by only one *kanji*. There is no vowel elongation with noun- or *kanji* combinations, if the first noun ends with the same vowel the second noun is starting with. Then, a short break in speaking is applied and the rules explained here are not applicable, e.g., in zinc oxide: さんかあえん (酸化亜鉛) *sankaaen* (酸化 *sanka* + 亜鉛 *aen*), dolomite: はくうんせき (白雲石) *hakuunseki* (白 *haku* + 雲 *un* + 石), stereoisomerism: りったいいせい (立体異性) *rittaiisei* (立体 *rittai* + 異性 *isei*). For the same reason, it is not called a long vocal if a Japanese expression is combined with a foreign word (in *katakana*) and two vowels meet, e.g., in aluminium chloride: えんかアルミニウム (塩化アルミニウム) *enkaaruminiumu* (塩化 *enka* + アルミニウム) or in molecular ion: ぶんしイオン (分子イオン) *bunshiion* (分子 *bunshi* + イオン).

15) There are only few words with long a or long i and they are all of Japanese origin. There is no *kanji* including long a or i. In addition, there are very few words of Japanese origin with a long e.

Table 1.8 Expression of long vowels.

Vowel	Chinese origin *on*-reading	Japanese origin *kun*-reading	Western origin (only *katakana*)
a		ああ	アー
i		いい	イー
u	うう	うう	ウー
e	えい	ええ	エー
o	おう	おお	オー

reading), *kana* in the u- and o-columns are doubled by adding u; and *kana* in the e-column have their vowels lengthened by adding i (Table 1.8).

Examples: long u: cast iron: ち<u>ゅう</u>てつ = 鋳鉄 *ch<u>uu</u>tetsu*
electric current: でんり<u>ゅう</u> = 電流 *denr<u>yuu</u>*
long e: protein biosynthesis: たんぱく<u>せい</u>ごう<u>せい</u> = 蛋白生合成 *tanpaku<u>sei</u>gou<u>sei</u>* (Chinese origin)
long o: many: <u>おお</u>い = 多い <u>oo</u>i (Japanese origin)
quantum number: り<u>ょう</u>しすう = 量子数 *r<u>you</u>shisuu* (Chinese origin)

Long vowels in foreign words that are written in *katakana* are formed by attaching a horizontal or vertical (vowel extender mark, 長音 *chouon*) line behind the vowel[16] in the center of the text, with the width of one *kana* character. It is written vertically in vertical text and horizontally in horizontal text as in the following examples:

citral: シトラール *shitora-ru*
strychnine: ストリキニーネ *sutorikini-ne*
safrole: サフロール *safuro-ru*
desiccator: デシケーター *deshike-ta-*
Foucault's pendulum: フーコー振子 *fu-ko-furiko*
Ziegler-Natta catalyst: チーグラーナッタ触媒 *chi-gura-nattashokubai*

If the word in *katakana* is derived from an original Japanese term with an own *kanji*, a long vowel is also expressed with an additional vowel such as for *hiragana*. If Japanese words are written with *kana*, usually *hiragana* are used. However, in scientific publications, *katakana* are often applied, especially for the names of the elements and chemical compounds.

---

[16] The vowel extender mark chouon used in *katakana* is rarely used for *hiragana*. One example of this non-standard use is らーめん, which is also often written in *katakana*. The elongation line always follows the direction of the text: horizontal in left-to-right-writing and vertical in up-to-down-writing. If a Japanese word is written in *katakana*, long vowels are usually written as they would be in *hiragana*, but there are a few exceptions, such as for candle: ローソク (蝋燭 *rousoku*).

Example for long "e":	silicon: ケイ素 = けい素 = 珪素 *keiso*
Examples for long "o"	iodine: ヨウ素 = よう素 = 沃素 *youso*
	kojic acid: コウジ酸 = こうじ酸 = 麹酸 *koujisan*
	sodium borate: ホウ酸ナトリウム = ほう酸ナトリウム = 硼酸ナトリウム *housannatoriumu*
	uric acid: ニョウ酸 = にょう酸 = 尿酸 *nyousan*

Also for words written in *katakana*, an additional vowel may be used instead of a vowel extender mark. Sometimes, this is a sign of an intermission regarding content or a short break in the flow of words, e.g., because the word is composed of two independent parts (this might not be called a real long vowel, as described for *hiragana* above).

Examples:
phenylurethane: フェニルウレタン *feniruuretan* (フェニル *feniru* + ウレタン *uretan*)
fluorescein: フルオレシイン *furuoreshiin*
isooctane: イソオクタン *isookutan* (イソ *iso* + オクタン *okutan*)

For other examples, there is no hint for this kind of composition, though no vowel extender mark is used but an additional vowel, e.g.

sphaerite: スフェエライト *sufeeraito*
squalene: スクウアレン *sukuuaren*
mendelejevite: メンデレエフ石 *mendereefuseki*

There is a variation that is applied in this book in contrast to the usually applied revised Hepburn system: To guarantee a clear, accurate conversion of scientific expressions it was necessary to change the transcription of long vowels: Long vowels are expressed in *hiragana* by adding an additional *kana* of the same or of another vowel after the affected vowel (あ after *a*; い after *i*; お after *o*; う after *u*; お or う after *o*; い or え after *e*). The Hepburn system takes long vowels into account with a macron on top of the vowel: ē, ā, ī, ō and ū. But with a long "o"-sound, an additional お or う may be added. Using the Hepburn system, it cannot be decided which vowel had been added, as both sounds are written with ō. Therefore, in this book, all vowels that operate as elongation vowels are also transferred into Latin characters and no macrons are used. This ensures both compatibility with the Japanese dictionary and distinction between おお (*oo*) and おう (*ou*). This is also important for most computerized systems to properly convert keystrokes on a Roman keyboard into *kana* (if no characters from outside the ASCII character set may be used).

Long vowels in *katakana* are indicated with an elongation line behind the vowel. For the same reason, i.e., to stay close to the Japanese typeface, long vowels will not be expressed with a macron but with an elongation line behind the vowel, similar to the Japanese origin. For the differences between this book and Hepburn see the examples in Table 1.9.

1.2.2.4 Other Aspects
- Doubled consonants (also called long consonants or long consonants) appear in Japanese in front of fricatives or short stops of reading within one word. They are indicated by a

Table 1.9 Different transcription in this book and by Hepburn.

English word	Japanese	In this book	Hepburn system
bromine	臭素	*shuuka*	*shūka*
chain compound	鎖状化合物	*sajoukagoubutsu*	*sajōkagōbutsu*
copper sulfide	硫化銅	*ryuukadou*	*ryūkadou*
protease	プロテアーゼ	*purotea-ze*	*puroteāze*
indole blue	インドールブルー	*indo-ruburu-*	*indōruburū*
coulometer	クーロメーター	*ku-rome-ta-*	*kūrometā*

small *kana* つ or ツ (called 促音 *sokuon*) and by doubling the following consonant in *ro-maji*, e.g.:

in *hiragana*: platinic acid: はっきんさん (白金酸) ha*kk*insan
                dehydration: だっすいそ (脱水素) da*ss*uiso
in *katakana*: nickel chloride: 塩化ニッケル enkani*kk*eru
                ytterbium: イッテルビウム i*tt*erubiumu

Originally, doubled consonants occur only for unvoiced ones (k, s, t, p, n, m). Doubled voiced consonants (gg, dd, bb, jj) can only be found in words with foreign origin (*gairaigo*), such as for Bragg angle: ブラッグの角 *buraggunokaku*, hybrid: ハイブリッド *haiburiddo*, Abbé condenser: あっべ集光器 *abbeshuukouki* and Wheatstone bridge: オイートストーン・ブリッジ *oi-tosuto-nburijji*.

- The nasal sonorant ん, ン cannot be at the beginning of a word. If it is directly followed by a vowel within a word, notably, n may not be merged with the vowel to get a *kana* of the n-row. The Hepburn system (but not this book) uses an apostrophe between n and the vocal in case the "n" is originally ん or ン.[10] For example: crotonaldehyde: クロトンアルデヒド *kuroto*n*arudehido* (Hepburn: *kuroton'arudehido*).
- There are three particles that are written by *hiragana* but pronounced differently: を (*wo*), へ (*he*) and は (*ha*), pronounced o, e and wa. As is common in Japanese language learning materials for foreigners, they are romanized in this book, using their pronunciation.
- A small *katakana ke* ケ (pronounced "ka", a simplified version of the *kanji* 箇), is used to indicate quantity such as period of months (e.g., five months: 五ヶ月 *gogetsu*).

1.2.3
kanji 漢字

1.2.3.1 Introduction to *kanji*[17]

The Japanese originally did not possess characters to write down their spoken language. The development of the current Japanese writing system started with the introduction of Chinese

---

[17] 漢字 *kanji*, literally "Han characters", is the Japanese term for Chinese characters (Hanzi).

characters in the 4th–6th century by Buddhist monks from China via the Korean peninsula. Initially, *kanji* were used as phonetic characters independent of their meaning, but to express similar sounding Japanese syllable sounds. By this, each Japanese word could be written down, but the longer the word, the more characters were necessary (with each character consisting of few to many stroke numbers). Later, *kanji* were also incorporated into the way of writing relating to their meaning and independent of their phonetic properties to express Japanese words with the same or a similar meaning. In this case, they were read with the originally Japanese pronunciation for the word. During the following centuries, the set of Chinese characters used in the Japanese language constantly changed: New characters were introduced, existing *kanji* changed their shapes, characters were simplified, others got lost and new characters were invented.

After the Meiji Restoration at the end of the 19th century, the government, for the first time, regulated the writing system. As part of activities to modernize the country, the writing was facilitated. Many characters were simplified, the total number of *kanji* in circulation was reduced, difficult *kanji* were replaced by easy ones, the number or readings as well as the number of *kanji* taught in school were limited and standardized rules for the way of writing established. The current set of *kanji* in use[18] is the result of some main orthographic reforms until the 1950s. Less important changes have followed since. The official guidelines from the government are recommendations, hence many characters outside the standards are still known and commonly used. A Japanese person with an average education will know about 3000 characters, but the number of *kanji* that are regularly used in publications (fiction and basic technical literature) is estimated as between 6 000 and 7 000.[19]

Because of the above-described ways *kanji* has been adopted into Japanese, a single *kanji* has one or more different readings and may be used to write different morphemes. The decision as to which reading has to be chosen depends on the context and its use in combination with other *kanji* as well as on the intended meaning. Readings are classified[20] as either Chinese-derived *on*-reading (also Sinojapanese reading, 音読み *onyomi*) or Japanese *kun*-reading (also native reading, 訓読み *kunyomi*).[21]

The *on*-reading is based on the original Chinese pronunciation of the *kanji* when it was introduced in Japan. Because some characters were introduced at different times or from different parts of China, they also have multiple *on*-readings, such as for 度: *do*, *taku* and *to*.[22] If

[18] In 1946, the Japanese government identified 1850 *kanji* as "current-use characters" (当用漢字, *touyou kanji*). The current official governmental guideline comprehends the following number of *kanji* for teaching in school and for use in ministerial publications:
- 1006 characters that Japanese children learn in elementary school (教育漢字 *kyouiku kanji*),
- 1945 characters (including *kyouiku kanji*) taught in high school (常用漢字 *jouyou kanji*),
- 983 additional characters (人名用漢字 *jinmeiyou kanji*) for use in people's names.

[19] The 大漢和辞典 *daikanwa jiten*, one of the largest dictionary of *kanji* ever compiled, has about 50 000 entries. The recently published *kanji* compilation *kou kanwa jiten* includes about 20 000 characters in modern Japanese.

[20] To distinguish *on*- and *kun*-readings, some dictionaries use *hiragana* for *kun*-readings and *katakana* for *on*-readings. In other books, *on*-readings are given in capital and *kun*-readings in lower-case letters.

[21] There are *kanji* without *kun*-reading. Other characters have a *kun*-reading only in combination with other *kanji*, e.g., 方 *kata* in combination with verbs (*kun*-reading) or 御 *o* as honorific prefix.

[22] *on*-readings are classified into four types: 呉音 (*goon*) were introduced from Japan during the 5th and 6th centuries from the Wu region (now Shanghai), 漢音 (*kanon*) were originally used in the Chinese standard language of the Tang Dynasty (7th to 9th century), 唐音 (*touon*) is related to the pronunciations of later dynasties and 慣用音 (*kanyouon*) denotes mistaken readings that have become accepted.

a *kanji* is part of a multi-*kanji* compound word, usually the *on*-reading has to be applied, because, beside the *kanji*, these words were also adapted from the Chinese language. This covers words that either did not exist in Japanese or could not be articulated in exactly the same way with native words.

The *kun*-reading is similar to the pronunciation of the native Japanese word for which the Chinese character was chosen because of its close meaning. There might be more than one or, also, no *kun*-reading for one *kanji*. This reading is chosen mainly for isolated *kanji* that are not part of a combination of *kanji*. They often function as simple nouns or inflected adjectives and verbs.

As an aid for correct pronunciation, small *kana* (振り仮名 *furigana*) may be printed next to a *kanji*, e.g., in texts for children and foreign learners, or in newspapers for rare or unusual readings and if characters are not included in the official lists of essential *kanji*.

Example:
        さんか かんげんでンい
  oxidation-reduction potential: 酸化還元電位 *sankakangendeni*

Usually *hiragana* is used, but when it is necessary to distinguish between *kun*- and *on*-reading, the Japanese pronunciations are written in *hiragana*, and the Chinese ones are written in *katakana*.

1.2.3.2 Structure and Elements of *kanji*

Simple *kanji* may be single pictograms, ideograms or phonograms, but most of the characters consist of different elements that might be classified into:
- a phonetic element indicating the sound or reading of a *kanji*, which is usually in the right part of the *kanji*;
- a radical indicating the general area of meaning of the character and dominantly positioned in the left part of the *kanji*;
- further elements modifying the meaning of the radical and determining the overall sense of the character. These elements might also be radicals, but only one is taken for the identification of the character.

The phonetic element exists as an own *kanji*. In more complex characters, it can take every position, but is often in the right, upper right and lower right part of the *kanji*. Usually, it does not contribute to the meaning of the character and only determines its pronunciation. However, the presence of such an element does not necessarily mean the character is read the same, as the reading may also determined by other elements or be totally different. As an example, the *kanji* for five, 五 *go*, is also the phonetic element in the following characters that are all read "go": 語, 梧, 伍, 吾 and 悟. The *kanji* 工 *kou*, meaning "work, manufacturing", is contained in 紅, 江, 控, 紅, 肛, 腔, 紅 and 虹, and all characters can be pronounced *kou*, but may also have additional *on*- and *kun*-readings.

Radicals are a set of characters that are used to classify *kanji* in the Chinese and Japanese language. The historical Chinese system consisted of 214 radicals; modern Japanese dictionaries use only 79 radicals. In opposition to phonetic elements, radicals take the dominant function in determining the *kanji*'s meaning, which is further modified in complex characters with additional radicals and elements. For instance, the radical 金 appears in over 140 characters and the meaning of many of them has some connection to metal, gold, other metallic elements or metal products in older times, like 鉄 (*tetsu*, iron), 鉦 (*shou*, fermium), 鈷 (*ko*, cobalt), 鉛 (*namari*, lead), 鉱 (*kou*, ore), 銭 (*sen*, money), 銀 (*gin*, silver), 銑 (*sen*, pig iron), 銃 (*juu*, gun), 銅 (*dou*, copper), 鋳 (*chuu*, cast metal), 錐 (*sui*, gimlet), 錫 (*suzu*, tin), 錆 (*shou*, rust), 錚 (*sou*, metallic sound), 錨 (*byou*, anchor), 鍵 (*kagi*, key), 鋼 (*hagane*, steel), 鎖 (*kusari*, chain), 鐘 (*shou*, bell). Important radicals that can be found in *kanji* for chemical terms and methods to identify *kanji* by radicals are described in Section 2.6.2.

1.2.3.3 Classification of *kanji*

From the above-described characteristics and possible elements of a *kanji*, four basic ways to systematically order Chinese characters are possible and realized in dictionaries or other publications: Order by category, by reading, with radicals or by stroke counting. To identify characters in dictionaries, mainly the order by radicals is useful. Dictionaries intended to be used by native speakers require a knowledge of the reading. For trained users, the order by stroke counting may also be helpful. Classification by reading, with radicals or by stroke counting, is described in more detail in Sections 2.6.1 and 2.6.2.

Classification of *kanji* by category: Chinese characters are traditionally classified by their structure and function. This classification is not useful for the identification of the meaning as most of the characters belongs to one, and some to more than one, group. There are the following six categories (六書 *rikusho*):

- Pictrographic or diagrammatic characters 象形文字 (*shoukeimoji*), which are simple pictures or sketches of the object they represent, e.g., tree: 木 *ki* or mountain 山 *yama*.
- Ideographic characters and symbols 指事文字 (*shijimoji*) representing an abstract idea of their meaning, e.g., up: 上 *ue*, three: 三 *san*.
- Ideograms 会意文字 (*kaiimoji*), which are combination of pictograms that contribute to the overall meaning of the *kanji*, e.g., wood 森 *mori* (three trees), coal 炭 *tan* (mountain + slope + fire).
- Phonetic-ideographic and radical-phonetic characters 形声文字 (*tenchuumoji*), which consist of a combination of some elements, typically two components, one radical indicating the general area of meaning (semantic context) and one phonetic element showing its sound or reading, e.g., powder 粉 *fun*. This is the largest category with about 85% of the characters.
- Derived characters with extended usage 転注文字 (*tenchuumoji*) – a vaguely defined group of characters, of which the meaning has become extended or the character was used in a meaning derived but different from its original one.
- Phonetic loan characters 仮借文字 (*kashamoji*) have no relationship between their semantic and structure rather a phonetic that is derived from a former meaning.

1.2.3.4 *kanji* Combination

When the Japanese adopted Chinese writing and introduced it to their language, they did not only take over the Chinese characters with their meanings, they also adapted the Chinese method of forming complex new words by joining several characters together. The existing number of *kanji* is a high but limited source for words, but with the method for combining

Table 1.10 Examples of *kanji* combination with two characters.

First character	Second character	*kanji* combination
transform, make into 化 *ka*	+science, learning, study 学 *gaku*	→ chemistry 化学 *kagaku*
original, fundamental 原 *gen*	+child 子 *shi*	→ atom 原子 *genshi*
dividing 分 *bun*	+child 子 *shi*	→ molecule 分子 *bunshi*
positive 陽 *you*	+child 子 *shi*	→ proton 陽子 *youshi*
odor 臭 *shuu*	+element 素 *so*	→ bromine 臭素 *shuuso*
take away 脱 *datsu*	+water 水 *sui*	→ dehydration 脱水 *dassui*
acid 酸 *san*	+transform make into 化 *ka*	→ oxidation 酸化 *sanka*
firefly 蛍 *kei*	+light 光 *kou*	→ fluorescence 蛍光 *keikou*
turn, change 転 *ten*	+place 位 *i*	→ rearrangement 転位 *teni*
red 赤 *seki*	+external 外 *gai*	→ infrared 赤外 *sekigai*
agriculture 農 *nou*	+medicine 薬 *yaku*	→ agricultural chemicals 農薬 *nouyaku*
try out 試 *shi*	+effect 験 *ken*	→ experiment 試験 *shiken*
egg 蛋 *tan*	+white 白 *paku*	→ protein 蛋白 *tanpaku*
steam 蒸 *jou*	+departure, emit 発 *hatsu*	→ evaporation 蒸発 *jouhatsu*
color 色 *shiki*	+element 素 *so*	→ pigment, dye 色素 *shikiso*
many 多 *ta*	+form, shape 形 *kei*	→ polymorph 多形 *takei*

kanji, Japanese can express all kinds of complex contents. If new words emerged, e.g., by technological developments, *kanji* combinations could be created.

Words formed by combining several *kanji* and pronounced with sounds adopted from Chinese are called 熟語 *jukugo*. Most combinations are 漢語 *kango*, which are read with their *on*-readings.[23] In many cases, the meaning of a *kanji* combination can be derived from the meaning of the single characters, such as for the examples given in Table 1.10. This also works for terms consisting of more than two *kanji* (Table 1.11).

kanji or *kanji* groups can not be arbitrarily combined but the *kanji* order is relevant for the meaning. Sometimes the combination works only in one order, in other cases different meanings result. For instance, two terms can be formed by combining "chemistry" 化学 and "industry" 工業: chemical industry 化学工業 *kagakukougyou* and industrial chemistry 工業化学 *kougyoukagaku*.

In some cases, a scientific term can be written with two different *kanji* (or, in other words, there are two *kanji* with the same pronunciation expressing the same content), such as for 形 and 型 in following examples:

---

23) Very few *kanji* combinations are read with the *kun*-readings of the individual characters contained in them. Also, mixed *on*- and *kun*-readings are rare. Some combinations have irregular readings that are not related to the readings of the individual characters. Occasionally, there are combinations that can be read in two ways.

reduced form: 還元形 and 還元型 *kangengata*
boat form, boat conformation: 舟形 and 舟型 *funegata*

With the *kanji* combination, specific syllables may come together, which complicates the reading of the composition. In certain cases phonetic modifications are applied, facilitating the reading, i.e., the deletion of syllables and the insertion of a double consonant with ツ (especially if the first *kanji* ends with i or u of a syllable with a voiceless consonant and the reading of the second *kanji* begins with the same voiceless consonant). In addition, changes within the consonants system occur from voiceless to voiced consonants, e.g., from h to b or p.

Examples: 脱 *datsu* + 水 *sui* → dehydration: 脱水 *dassui*
          測 *soku* + 光 *kou* → photometry: 測光 *sokkou*
          物 *butsu* + 体 *tai* → substance, object: 物体 *buttai*
          石 *seki* + 灰 *kai* → lime: 石灰 *sekkai*
          熱 *netsu* + 湯 *tou* → boiling water: 熱湯 *nettou*
          接 *setsu* + 着 *chaku* → adhesion: 接着 *secchaku*
          弗 *futsu* + 素 *so* → fluorine: 弗素 *fusso*
          六 *roku* + 百 *hyaku* → six hundred: 六百 *roppyaku*
          三 *san* + 千 *sen* → three thousand: 三千 *sanzen*

Table 1.11 Examples of *kanji* combination with more than two characters.

First part	Second part	*kanji* combination
chlorine 塩素 *enso*	+ treat, deal with 処理 *shori*	→ chlorination 塩素処理 *ensoshori*
tall 高 *kou*	+ molecule 分子 *bunshi*	→ macromolecule 高分子 *koubunshi*
research 研究 *kenkyuu*	+ room 室 *shitsu*	→ research laboratory 研究室 *kenkyuushitsu*
living creature 生物 *seibutsu*	+ engineering 工学 *kougaku*	→ biotechnology 生物工学 *seibutsukougaku*
atom 原子 *genshi*	+ nucleus 核 *kaku* + chemistry 化学 *kagaku*	→ nuclear chemistry 原子核化学 *genshikakukagaku*
add 加 *katsu*	+ water 水 *sui* + decomposition 分解 *bunkai*	→ hydrolysis 加水分解 *kasuibunkai*
hydrogen 水素 *suiso*	+ linkage 結合 *ketsugou*	→ hydrogen bond 水素結合 *suisoketsugou*

1.2.3.5 Combination of *kanji* and *kana*

Nouns may not only be formed by combining several *kanji* but also by joining *kanji* and *kana*, especially with *katakana* in scientific terms. The *kanji* part can consist of a single character as well of a *kanji* combination. In more complex terms, several *katakana* may also occur between *kanji* or some *kanji* are placed between two *katakana* expressions.

Examples: palmitic acid ($CH_3(CH_2)_{14}COOH$): パルミチン酸 *parumichinsan*
            Raman effect: ラマン効果 *ramankouka*

magnesium chloride (MgCl$_2$):	塩化マグネシウム	*enkamaguneshiumu*
hydrogen cyanide (HCN):	シアン化水素	*shiankasuiso*
Boltzmann constant:	ボルツマン定数	*borutsumanteisuu*
fluorescent screen:	螢光スクリーン	*keikousukuri-n*
phenyl isocyanate (C$_6$H$_5$NCO):	イソシアン酸フェニル *isoshiansanfeniru*	
phthalic anhydride (C$_8$H$_4$O$_3$):	無水フタル酸	*musuifutarusan*

1.2.4
Transcription of Foreign Words into Japanese Phonology

If terms from foreign languages are intended to be written with Japanese characters, they have to be expressed by the existing syllables. The sounds are transcribed with the syllable that is as close as possible to the phonology of the foreign sound. One of the most characteristic differences between western languages and Japanese is the missing Japanese equivalent for consecutive consonants. To solve this problem, a syllable of the u-row (sometimes also of i- or o-row) is used and the ending vowel is spoken weakly or not at all. Consonants at the end of a foreign word are transcribed in a similar way, mainly ending with a syllable of the u-row or, if the foreign expressions ends with "t" or "d", by ト or ド.

Examples:			
	magnesium:	マグネシウム	*maguneshiumu*
	benzyl alcohol:	ベンジルアルコール	*benjiruaruko-ru*
	dextrin:	デキストリン	*dekisutorin*
	redox reaction:	レドックス反応	*redokkusuhannou*
	leucopterin:	ロイコプテリン	*roikoputerin*
	lactose:	ラクトース	*rakuto-su*
	methylate:	メチラート	*mechira-to*
	polyamide:	ポリアミド	*poriamido*

Especially within the scientific and technical language, many words have to be expressed with sounds that are not represented by the syllables in the 50-sound-matrix or the above-described variations (clouded sounds) and combinations (palatalized sounds). Because of this, and parallel to the increasing importance of literature from abroad during the last few decades, new combinations of syllables have been developed. They are written like the palatalized sounds with small characters of vowels, such as ディ in ディアステレオメル *diasutereomeru* (diastereomer), フォ in フォスゲン *fosugen* (phosgene, COCl$_2$) or ウェ in カルウェオール *karuweo-ru* (carveol, C$_{10}$H$_{16}$O), or with small characters of the y-row, such as デュ in デュロキノン *dyurokinon* (duroquinone, C$_{10}$H$_{12}$O$_2$). In this way, some gaps in the sound system concerning existing combinations of consonants and vowels were also closed. For instance, "tsu" is written as ツ, but "tsa", "tsi", "tse" and "tso" could not be expressed. The same applies with "wa", which is written as ワ, and the missing characters for "wi", "wu", "we" and "wo". For the sounds in the t-row, タ, テ and ト exist, but there was no way to write "ti" and "tu" because the two other characters in the row (チ and ツ) were read differently.

Even a new clouded character ヴ was introduced for foreign sounds starting with "v". ヴ can be combined with small characters of vowels, e.g., ヴァ in vanadium: ヴァナジウム *vanajiumu* and ヴォ in flavone フラヴォン *furavon*. As all new characters and character combinations are only used to support the transcription of foreign words, they can only be found for *katakana* (Fig. 1.4). Because there are no strict rules regarding the use of *kana* to express foreign sounds, other combinations can also be found.

	new a-sounds	new i-sounds	new u-sounds	new e-sounds	new o-sounds
combinations with the character ヴ	ヴァ va ヴャ vya	ヴィ vi	ヴ vu ヴュ vyu	ヴェ ve	ヴォ vo ヴョ vyo
new combinations in the vowel-row		ウィ wi	ウゥ wu˙	イェ ye ウェ we	ウォ wo
new combinations in the k-row	クァ kwa グァ gwa	クィ kwi グィ gwi	クゥ kwu˙ グゥ gwu˙	クェ kwe グェ gwe˙	クォ kwo グォ gwo˙
new combinations in the s-row				シェ she ジェ je	
new combinations in the t-row		ティ ti	テゥ tu˙ トゥ tu テュ tyu		
	ツァ tsa	ツィ tsi		ツェ tse チェ che	ツォ tso
	デャ dya	ディ di	デゥ du˙ ドゥ du デュ dyu		デョ dyo
new combinations in the h-row	ファ fa フャ fya	フィ fi ヒィ hyi˙	フュ fyu	フェ fe ヒェ hye	フォ fo フョ fyo

Fig. 1.4 Expression of foreign sounds with new *katakana* and *katakana* combinations. New combinations marked with an asterisk are basically possible but not commonly accepted. More new combinations can be found on the internet (e.g., at http://www.roomazi.org and http://www.wul.waseda.ac.jp/opac/aboutw/roman/).

The small *tsu*, which is used in Japanese words to indicate double consonants, may be used in places that have no equivalent in native sounds, e.g., to express "x", such as in redox potential: レドックス電位 *redokkusudeni*. It is often used for the sound "ch", which is common in German names (e.g., Esbach's reagent: エスバッハ試薬 *esubahhashiyaku*).

The sounds represented by "r" and "l" in English are unknown in the Japanese language. The Japanese sounds expressed by *kana* of the r-row (ラリルレロ) start with a short roll of a trilled "r" which is a quick flap of the tongue against the gum. However, these *kana* are usually used to transcribe both, r- and l-sounds of foreign terms.

It has to be kept in mind that there is no standardized or even generally accepted form of romanization for the new *kana* combinations and the use of the small ツ. Even with the increased number of sounds that can be expressed by using *katakana*, it is still not always easy to carry out a proper transcription. In the end, it is the understanding of the author and his personal phonetic perception of a word. This leads to different possibilities for its transformation into the Japanese syllable system. Table 1.12 gives some ideas of possibilities for expressing foreign sounds for which no traditional Japanese syllable in the 50-sound-matrix exists, either by known *katakana* or with the above-described new *katakana* combinations. Because the transcription is based on the sounds and not on the writing, the chosen *katakana* syllable can not always be easily foreseen by the isolated letter in the English word, and the sounds before and behind have to be taken into account.

As can be seen, for many sounds there is more than one possibility for transferring them into Japanese writing. Because many scientific words are not often used, they do not have a

Table 1.12 Examples for the expression of foreign sounds with *katakana* syllables.

Foreign sound		Transcription	Examples
a-sounds	ca	カ, キャ	carrier: キャリヤー *kyariya-*
i-sounds	shi, schi, dschi, si, sy, ci, cy, zi, zy, thi, thy, ty	シ, チ, ジ	shikimic acid: シキミ酸 *shikimisan*; cresyl: クレシル其 *kureshiruki*
	ci, ti, ty	チ, ティ	mesitylene: メシチレン *meshichiren*; myristic acid: ミリスティン酸 *mirisutinsan*
e-sounds	je, ye	イエ	meyerhofferite: マイエルホッフェル石 *maieruhofferuseki*
	she	セ, シェ	shellac: セラック *serakku*; echellite: エシェル石 *esheruseki*; shaker: シェーカー *she-ka-*
	eu	syllable o-column + イ	leucine: ロイシン *roishin*

Table 1.12 (continued)

Foreign sound		Transcription	Examples
German umlaute	ä	syllable from e-column	jäneckite: エーネカイト *e-nekaito*; Gräbe-Ullmann synthesis: グレーベウールマン合成 *gurebeu-rumangousei*
	ö	syllable from e-column	Mössbauer spectrum: メスバウアースペクトル *mesubaua-supekutoru*; trögerite: トレゲル石 *toregeruseki*; Brönsted acid: ブレーンステズ酸 *bure-nsutezusan*
	ü	syllable from i-column with small ユ; syllable from u-column	hüttenbergite: ヒュッテンベルグ石 *hyuttenberuguseki*; Geiger–Müller counter: ガイガーミュラー計数管 *gaiga-myura-keisuukan*; Büchner funnel: ブッヒナー吸引漏斗 *buhhina-kyuuinrouto*
d-sounds	di	ディ, ジ	diastereomer: ディアステレオメル *diasutereomeru*; acridine: アクリジン *akurijin*
	du	ズ, シュ, ジュ, チュ	dulcin: ズルチン *zuruchin*; duralumin: ジュラルミン *jurarumin*; isodurene: イソチュロール *isodyuro-ru*
hu		フ	humic acid: フミン酸 *huminsan*
l-sounds	la	ラ, ロ, レ	riboflavine: リボフラビン *ribofurabin*; laser: レーザー *re-za-*; lawsonite: ローソン石 *ro-sonseki*
	le	レ, ラ	lens: レンズ *renzu*; roller mill: ローラーミル *ro-ra-miru*
	li	リ	linolenic acid: リノレン酸 *rinorensan*
	lo	ロ	linalool: リナロール *rinaro-ru*
	lu	ル	luminol: ルミノール *rumino-ru*
f- and ph-sounds	fa, pha	ファ, ハ	farnesol: ファルネゾール *farunezo-ru*; rhodophane: ロドファン *rodofan*
	fi, phi	フィ, ヒ, ファ	olefin: オレフィン *orefin*; fiber: ファイバ *faiba*
	fe, phe	フェ, ヘ	phenyl: フェニル *feniru*
	fo, pho	フォ, ホ	iodoform: ヨードフォルム *yo-doforumu*; foil: ホイル *hoiru*

Table 1.12 *(continued)*

Foreign sound		Transcription	Examples
pf-sounds	pfe	ペ, フェ	cupferron: クペロン *kuperon* or クッフェロン *kufferon*; kaempferol: ケンフェロール *kenfero-ru*;
	pfu, pf + consonant	プ, フ	Schoellkopf's acid: シェールコップの酸 *she-rukoppunosan*; hoepfnerite: フェフナー石 *fefuna-seki*
q-sounds	qua	クァ	equatorial bond: エクアトリアル結合 *ekuatoriaruketsugou*; squalane: スクアラン *sukuaran*
	que	クェ, クエ, クウェ, ケ	sequencing: シークエンシング *shi-kuenshingu*; quebrachite: クウェブラキト *kuweburakito*; quercitol: ケルシット *kerushitto*
	qui	クィ, クイ, キ	equilenine: エクイレニン *ekuirenin*; orthoquinone: オルトキノン *orutokinon*
	quo	クォ, クオ, コ	aliquot: アリコート *ariko-to*
ch at word end	-ach	ッハ	Wallach reaction: ヴァラッハの反応 *varahhanohannou*; weisbachite: ワイスバッハ石 *waisubahhaseki*
	-ich	ッヒ	Mannich reaction: マンニッヒ反応 *mannihhihannou*; Ehrlich's reagent: エールリッヒ試薬 *e-rurihhishiyaku*
	-och	ッホ	Koch's acid: コッホ酸 *kohhosan*
t-sounds	ti	ティ, チ	antimony: アンチモン *anchimon*; elastin: エラスティン *erasutin*
	tu	ツ, チュ	galacturonic acid: ガラクツロン酸 *garakutsuronsan*; spatula: スパチュラ *supachura*
	voiceless th	s- and t-rows	methyl: メチル *mechiru*; threonine: スレオニン *sureonin*; ether: エーテル *e-teru*
	voiced th	z-row	rutherfordium: ラザホージウム *razaho-jiumu*
v-sound	va	ワ, バ, ヴァ, ウア	vaseline: ワゼリン *wazerin*; mevalonic acid: メバロン酸 *mebaronsan*; rivanol: リヴァノール *rivano-ru*
	ve	ベ, ヴェ, ウエ, バ	veratrole: ベラトロール *beratoro-ru*; flaveanic acid: フラヴェアン酸 *furaveansan*; receiver: レシーバー *reshi-ba-*

Table 1.12 *(continued)*

Foreign sound		Transcription	Examples
	vi	ビ, ヴィ, ウイ	flavine: フラビン *furabin*; pyruvic acid: ピルヴィン酸 *piruvinsan*; virus: ウイルス *uirusu*
	vo	ボ, ヴォ, ウオ	flavone: フラボン *furabon*; volt: ヴォルト *voruto*
	vu	ブ, ヴ	levulinic acid: レブリン酸 *reburinsan*; levuline: レヴリン *revurin*
w-sounds	we	ベ, ウェ, ウエ, ワエ	veronal: ヴェロナール *verona-ru*; websterite: ウエブスター石 *uebusuta-seki*
	wi	ビ, ウィ, ヴィ, ワイ	wittichenite: ウィチヘン鉱 *wichihenkou*; Neuwied green: ノイヴィードグリーン *noivi-doguri-n*; wine yeast: ワイン酵母 *wainkoubo*
	wo	ボ, ウォ, ウ, ワオ	wolfram: ウォルフラム *worufuramu*; Wood's metal: ウッド金 *uddokin*
	wu	ブ, ヴュ, フ, ウ, ウゥ, ワウ	wurtzite: ウルツ鉱 *urutsukou*; Wurtz synthesis: ヴュルツの合成法 *vyurutsunogouseihou*
x-sounds	xa	クサ, キサ	siloxane: シロクサン *shirokusan*; rhodoxanthin: ロドキサンチン *rodokisanchin*
	xe	クセ, キセ	hexene: ヘキセン *hekisen*; siloxene: シロクセン *shirokusen*
	xi, xy	クシ, キシ	lipoxygenase: リポキシゲナーゼ *ripokishigena-ze*; aloxite: アロクシット *arokushitto*
	xo	クソ, キソ	lyxose: リキソーゼ *rikiso-ze*; hexose: ヘクソーゼ *hekuso-ze*
	xu, -x	クス, キス	dextrin: デキストリン *dekisutorin*; redox reaction: レドックス反応 *redokkusuhannou*

commonly accepted way of being written and the choice of the *katakana* syllables depends on the opinion of the author. Consequently, different spellings for many foreign scientific and technical terms can be found in different publications. The number of different spellings is even increased because, in English, different expressions may, sometimes, also be used and terms are also adopted from other languages, in particular from German, such many element names, e.g., sodium: ナトリウム (from the German "Natrium"), potassium: カリウム (from "Kalium") and chromium: クロム (from "Chrom"). The following examples demonstrate the variety of possible translations.

benzene:	ベンゼン, ベンゾール	
virus:	ヴァイラス, ヴィールス, ウイルス, ビールス	
spectrum:	スペクトル, スペクトラム	
molar:	モーラル, モル	
millimeter:	ミリメーター, ミリメートル	
Cannizzaro reaction:	カニザロー反応 *kanizaro-hannou*	
	カニッツァーロ反応 *kanissha-rohannou*	
	カニシャロ反応 *kanisharohannou*	
	カニッシャロ反応 *kanissharohannou*	

The following examples show the critical areas that cause most of the different transcriptions that can be found for chemical terms:

- Sounds that begin with "f…" or "v…" in English can be expressed with the new characters ヴ, ファ, フィ, フェ and フォ but, usually, the written form with an alternative transcription, i.e., with syllables of the h- and b-rows, can also be found in the literature, such as for:

"va"	valine:	ヴァリン or バリン
	ovalbumin:	オヴァルブミン or オボアルブミン
"vi"	vinyl:	ヴィニル or ビニル
	flavine:	フラヴィン or フラビン
"ve"	biliverdin:	ビリヴェルヂン or ビリベルジン
"vo"	sulfonic acid:	スルフォン酸 *surufonsan* or スルホン酸 *suruhonsan*
	Avogadro's number	アヴォガドロ数 *avogadorosuu* or アボガドロ数 *abogadorosuu*
	flavone:	フラヴォン or フラボン
	formaldehyde:	フォルムアルデヒド or ホルムアルデヒド
	phospholipase:	フォスフォリパーゼ or ホスフォリパーゼ
	chloroform:	クロロフォルム or クロロホルム

- If the names of chemical compounds end with -ose, e.g., sugars or some plastics, the written form may end with ゼ or by ス, e.g.:

arabinose:	アラビノーゼ or アラビノース
hexose:	ヘキソーゼ or ヘキソース
viscose:	ビスコース or ビスコーゼ
cellulose:	セルロース or セルローゼ

- For the sounds "lo" and "ro", in some cases, the two transcription possibilities ロ or ル are used equivalently, e.g.:

acrylonitrile	アクリルニトリル or アクリロニトリル
chlorobenzene:	クロロベンゾール or クロルベンゾール

fluorosilicic acid: フルオルケイ酸 *furuorukeisan* or フルオロケイ酸 *furuorokeisan*

- English terms having an e-sound at the beginning may be translated with a *kana* of the a- or e-column, e.g., for berkelium: バーケリウム or ベーケリウム.
- It is not always clear if a vowel is long or short; therefore, some words can be found with or without a vowel extender mark (長音 *chouon*) or with an additional character for the same vowel:

  Examples: austenite: オーステナイト or オウステナイト
             iodal: ヨーダール or ヨーダル
             lipoxygenase: リポオキシゲナーゼ or リポキシゲナーゼ
             isoxazole: イソオキサゾール or イソキサゾール

- A consonant and a following vowel in a foreign term may be phonetically combined to one syllable, e.g., s + a = サ. Alternatively, the consonant is expressed by a syllable of the u-row and the vowel stays independent (e.g., s + a = スア). The second way is used if the term is a word composition and the border is just between the consonant and the syllable. For instance, the term "propyl alcohol", which consists of two parts, could not reasonably be transcribed as "プロピラルコール" because the word "alcohol" can hardly be recognized. Instead, プロピルアルコール is used. As the pronunciation of both transcriptions is different, it is usually clear which method has to be applied. However, for some English terms two Japanese expressions can be found, even if one, for the reason described above, does not properly reflect the English reading (Table 1.13).
- For foreign words that contain "xa" or "xo", often, two Japanese forms can be seen: one in which the "x..." is transcribed with キ and another with ク, as in following examples:

  cyclohexane: シクロヘキサン or シクロヘクサン
  oxonium salt: オキソニウム塩 or オクソニウム塩
  exopeptidase: エキソペプチダーゼ or エクソペプチダーゼ
  oxaminic acid: オキサミド酸 *okisamidosan* or オクザミド酸 *okuzamidosan*
  dioxan: ジオキサン or ジオクザン
  hexanol: ヘキサノール or ヘクサノール

Table 1.13 Examples of different transcriptions.

	Combined syllable	Consonant and vowel separated
chloramine:	クロラミン *kuroramin*	クロルアミン *kuroruamin*
rosaniline:	ローザニリン *ro-zanirin*	ローズアニリン *ro-zuanirin*
butaldehyde:	ブチラルデヒド *buchirarudehido*	ブチルアルデヒド *buchiruarudehido*
benzamide:	ベンザミド *benzamido*	ベンズアミド *benzuamido*

- As the pronunciation of a foreign word may not be always clear, voiceless and voiced syllables can be found in same expressions, e.g.:

sarcosine:	サルコシン or ザルコシン
farnesol:	ファルネソール or ファルネゾール
oxalacetic acid:	オキサロ酢酸 *okisarosakusan* or オキザロ酢酸 *okizarosakusan*
resorcinol:	レソルシノール or レゾルシノール

- The transformation of i- and similar sounds is the source of most of the non-uniform appearance of scientific terms, because there are many i-sounds in chemical expressions and also many similar possibilities to transcribe. Typical similarities that occur are ディ vs. ジ, ティ vs. チ, ジ vs. チ, ジ vs. ギ or シャ vs. ジヤ to give some examples. They are used, for instance, in the following words:

stigmasterol:	スチグマステロール or スティグマステロール
benzidine:	ベンジジン or ベンチジン
abscisic acid:	アブシシン酸 *abushishinsan* or アブシジン酸 *abushijinsan*
butyne:	ブチン or ブテイン
dysprosium:	ジスプロシウム or ディスプロシウム
elastin:	エラスチン or エラスティン
brucine:	ブルシン or ブルチン
fluorescein:	フルオレシイン or フルオレセイン
gibberellin:	ギベレリン or ジベレリン
anthraquinone:	アントラキノン or アントラヒノン
potential:	ポテンシャル or ポテンジヤル
amidine:	アミジン or アミディン

In addition, other vowels or consonants and whole syllables may be transcribed with different characters. As with most of the scientific terms, there is no clear spelling in the Japanese language – it is up to the authors to find a proper transformation of foreign expressions into Japanese.

Examples of different vowels:

leucoaniline:	リュコアニリン or ロイコアニリン
alkylate:	アルキラート or アルキレート
phenetidine:	フェニチジン or フェネチジン
burette:	ビュレット or ブレット

Examples of different consonants or complete syllables:

benzhydrol	ベンズヒドロール or ベンツヒドロール
Brönsted acid:	ブレンステット酸 *burensutettosan* or ブレーンステズ酸 *bure-nsutezusan*

proinsulin:	プロインスリン or プロインシュリン
cellulose:	セルロース or シェルロース
sodium amalgam:	ナトリウムアマルガム or ナトリウムアムルガム
cupferron:	クッフェロン or クペロン
duralumin:	ジュラルミン or デュラルミン
carbene:	カルベン or カーベン

In some cases, the transcription of different foreign terms leads to the same Japanese word and it has to be decided by context which expression is intended by the author. Examples of this problem are:

ベンジル *benjiru*	benzyl ($-CH_2C_6H_5$) or benzil ($C_6H_5COCOC_6H_5$)
アスタチン *asutachin*	astatine (At, element 85) or astacine ($C_{40}H_{48}O_4$)
フレーム *fure-mu*	flame or frame
カプリン酸 *kapurinsan* and カプリル酸 *kapurirusan*	capric acid [$CH_3(CH_2)_8COOH$] or caprylic acid [$CH_3(CH_2)_6COOH$]
コラミン *koramin*	colamine ($H_2NCH_2CH_2OH$) or coramin ($C_{10}H_{14}ON_2$)

2
Japanese in Scientific and Technical Publications

2.1
Scientific and Technical Publications

2.1.1
General Characteristics of Language used in Scientific and Technical Publications

The Japanese language in scientific publications is characterized by a clear focus on the description of the technical topic. No sophisticated style or means to express politeness, such as polite verb endings, prefixes for polite nouns (を and ご) or vocabulary showing respect and humility, are used. The level of language is restricted to informal speech and verb endings such as the *masu*-form are not used. In patents, usually only the present tense is applied. In general, the past tense is rarely found in chemical literature. Besides the plain present form from Table 1.5, a further frequently used form is the te-form. It is used to express successively occurring actions or simultaneous existing states.

As in English publications, a passive style is more often applied in scientific texts than in fictional literature. This is realized in Japanese by passive verb forms and passive clauses in which, usually, the subject of the active sentence is not mentioned. Expressions that can be found, are for example:

is expressed:	表される *arawasareru*
is substituted:	置換された *chikan sareta*
is limited	限られている *kagirarete iru*
is used, but ... :	持ちいられているが *mochiirarete iru ga*
an example is given:	例は挙げられる *rei wa agerareru*
a catalyst is added by the chemist:	触媒はその化学者により付加された *shokubai wa sono kagakusha ni yori fuka sareta*

Passive verbs are formed by adding the ending れる after the verb stem and an extension as explained in Table 1.5. The resulting passive verbs are conjugable like group-2-verbs (see Section 1.1.4). For instance, the verb "to substitute" has the following passive forms:

affirmative passive verb form:	置換される *chikan sareru*
negative passive verb form:	置換されない *chikan sarenai*

Japanese-English Chemical Dictionary. Edited by Markus Gewehr
Copyright © 2008 WILEY-VCH Verlag GmbH & Co. KGaA, Weinheim
ISBN: 978-3-527-31293-1

past tense of passive verb form:	置換された	*chikan sareta*
negative past tense:	置換されなかった	*chikan sarenakatta*
te-form of passive verb form:	置換されて	*chikan sarete*

Personal pronouns are rarely used in Japanese; in scientific literature they are used even less. Therefore, the person who carries out an action is usually not named.

Sentences are often long, consisting of many embedded subordinate clauses, and contain many more characters than sentences in fiction, newspapers and journals. Also, in comparison with scientific publications written in English, much more information is included within individual sentences in Japanese. For instance, most patent claims only consist of one sentence.

Scientific texts use a reduced vocabulary compared with fiction, especially for verbs and adjectives. In contrast, significantly more nouns are used than in other Japanese texts. In addition, a higher rate of loanwords from foreign languages can be found. Even if original Japanese words written with Chinese characters exist, parallel equivalent foreign terms, transcribed from English or German languages and written by *katakana*, are used.

Examples:	accumulator:	蓄電地 *chikudenchi* or アキュムレーター
	alcohol dehydrogenase:	アルコール脱水素酵素 *aruko-rudassuisokouso* or アルコールデヒドロゲナーゼ
	anion:	陰イオン *inion* or アニオン
	anode:	陽極 *youkyoku* or アノード
	antagonist:	拮抗体 *kikkoutai* or アンタゴニスト
	metal:	金属 *kinzoku* or メタル
	hydroxylation:	水酸化 *suisanka* or ヒドロキシル化 *hidorokishiruka*
	autoclave:	高圧釜 *kouatsugama* or オートクレーブ

Because of the huge number of technical terms that can not be expressed with the traditional vocabulary existing in Japan or introduced from China in the 4th–6th century, it has been permanently necessary to generate new words. Nowadays, Japanese technical terms are either combinations of simple words, and written with series of *kanji*, or words that remain independent and are combined using the particle の. For some expressions, both methods can be found, depending on the style of the author, for example:

temperature scale:	温度目盛 *ondomemori* or 温度の目盛 *ondo no memori*
heat loss:	熱損失 *netsusonshitsu* or 熱の損失 *netsu no sonshitsu*
light velocity:	光速度 *kousokudo* or 光の速度 *hikari no sokudo*
Liebig condenser:	リービッヒ冷却器 *ri-bihhireikyakuki* or リービッヒの冷却器 *ri-bihhi no reikyakuki*
element of symmetry:	対称要素 *taishouyouso* or 対称の要素 *taishou no youso*

Another feature of scientific literature is the presence of dots to separate long words that are taken from foreign languages into individual parts. Dots facilitate understanding, but their application depends on the style of the author.

Examples: methylphenyl ketone: メチル・フェニルケトン
Haber–Bosch process: ハーバー・ボッシュ法
in vitro: イン・ビトロ
perillaaldehyde: ペリラ・アルデヒド
Jena glass: イェナ・ガラス

2.1.2
Expression of Specific Terms by Nouns in Chemical Literature

Nouns are the most important word class in scientific literature. Even many terms that are formed by verbs and adjectives in Western languages are derived from nouns in Japanese (Section 2.1.3). In chemical literature, many expressions are formed using specific Japanese words, resulting in the frequent appearance of certain *kanji*. Important Chinese characters are listed in Section 2.2. One of the nouns with the broadest use is *koto* 事, which describes many concrete things and general matters, e.g., 物事 *monogoto* (things), 事例 *jirei* (example), 事実 *jijitsu* or 事項 *jikou* (facts), 仕事 *shigoto* (work) and 事理 *jiri* (reason). The terms for most of the substances in chemistry, i.e., chemical compounds, products, molecules and materials, contain one of the nouns listed in Table 2.1. They are attached to chemical terms, e.g., *kishaku* (稀釈 dilution) + *zai* (剤) gives *kishakuzai* (稀釈剤 diluent).

These nouns are sometimes exchangeable, as for antihistamine: 坑ヒスタミン薬 *kouhisutaminyaku* or 坑ヒスタミン剤 *kouhisutaminzai*. The form or shape of an object is indicated using the nouns *jou* 状 or *kata*. *kata* can be written by two different *kanji*, 形 or 型. Examples are ロート状の *ro-tojou no* (funnel-shaped), 鎖状分子 *sajoubunshi* (chain molecule), 酸化形 *sankagata* (oxidized form), 舟形 *funegata* (boat conformation), 還元型 *kangengata* (reduced form).

For the description of states and changes of states, e.g., chemical transformations and reaction types, which appear frequently in chemical literature, a certain number of nouns are applied. Vocabulary for states in English usually end with -ness or -ity. In Japanese, one of the following nouns can be found at the final position in corresponding words:

sei 性 (character, nature; e.g., insolubility: 不溶性 *fuyousei*, optical activity:
光学活性 *kougakukassei*, adhesiveness: 粘着性 *nenchakusei*);
do 度 (degree; e.g., volatility 揮発度 *kihatsudo*)
joutai 状態 (state, condition; e.g., transition state: 転移状態 *tenijoutai*,
singlet state: 一重項状態 *ichijuukoujoutai*);

For many examples, both, *sei* and *do* can be found with the same meaning, like for acidity: 酸度 *sando*, 酸性 *sansei* and 酸性度 *sanseido*; and for viscosity: 粘度 *nendo* and 粘性 *nensei*. *sei* is one of the nouns often found in adjective-type expressions (Section 2.1.3). Of limited use is the noun *ritsu*, which originally means coefficient, rate or portion, but is also used in chemical literature to describe states. For instance in 透磁率 *toujiritsu* (permeability) and 誘導率 *yuudouritsu* (inductivity).

For small objects in chemistry and physics, the noun for child is used to indicate small objects, for instance in 電子 *denshi* (electron), 陽子 *youshi* (proton), 原子 *genshi* (atom), 分子 *bunshi* (molecule), 量子 *ryoushi* (quantum) and 光子 *koushi* (photon).

Changes of physical or chemical properties are expressed with the noun *ka* 化, like in 脱イ

Table 2.1 *kanji* in expressions for chemical compounds.

English	Japanese	Examples
body, object	体 *karada, tai*	glucoside: 配糖体 *haitoutai*, powder: 粉体 *funtai*, metal complex: 金属錯体 *kinzokusakutai*
preparation, medicine	剤 *zai*	dye: 着色剤 *chakushokuzai*, inhibitor: 禁止剤 *kinshizai*
chemical, medicine	薬 *kusuri*	toxicant: 毒薬 *dokuyaku*, electrophilic reagent: 求電子試薬 *kyuudenshishiyaku*
thing, object, matter	物 *mono, butsu*	additive: 添加物 *tenkabutsu*, suspended matter: 浮遊物 *fuyuubutsu*
chemical compound	化合物 *kagoubutsu*	homopolar compound: 同極化合物 *doukyokukagoubutsu*, unsaturated compound: 不飽和化合物 *fuhouwakagoubutsu*
product	生成物 *seiseibutsu*	decomposition product: 分解生成物 *bunkaiseiseibutsu*
substance, material	物質 *busshitsu*	product: 生産物 *seisanbutsu*, anhydride: 無水物 *musuibutsu*, distillate: 留出物 *ryuushutsubutsu*
product	産物 *sanbutsu*	byproduct: 副産物 *fukusanbutsu*, mineral: 鉱産物 *kousanbutsu*
material	料 *ryou*	raw material 原料 *genryou*, dyestuff: 着色料 *chakushokuryou*
element, base	素 *moto, so*	chlorine: 塩素 *enso*, fibrin: 繊維素 *seniso*
material	材 *zai*	raw material: 原材料 *genzairyou*

オン化 *datsuionka* (deionization), 変化 *henka* (change), 非局在化 *hikyokuzaika* (delocalization), ジアゾ化 *jiazoka* (diazotization), 転化 *tenka* (conversion) and マイクロカプセル化 *maikurokapuseruka* (micro-encapsulation). It is generally applicable to form much vocabulary describing chemical reactions, transformations and other changes that often end in English with -ation by combining with Japanese, Sinojapanese and foreign nouns.

Examples:
    chlorine 塩素 *enso* → chlorination 塩素化 *ensoka*
    liquid state 液状 *ekijou* → 液状 liquefaction: 液化 *ekika*
    chelate キレート → chelation キレート化 *kire-toka*
    to get rid of, to retreat: → degeneration 退化 *taika*
        退く退, ける *shirizoku, nokeru*
    deuterium 重水素 *juusuiso* → deuteration 重水素化 *juusuisoka*
    resin 樹脂 *jushi* → resinification 樹脂化 *jushika*
    dimer 二量体 *niryoutai* → dimerization 二量体化 *niryoutaika*
    enol form エノール形 *eno-rugata* → enolization エノール化 *eno-ruka*
    gel ゲル → gelation, gelling ゲル化 *geruka*
    coal 炭 *tan* → carbonization 炭化 *tanka*

The noun *ka* also appears in many chemical names that are derived from a word describing a transformation, such as for fluorination 弗化 *fukka* → sodium fluoride 弗化ナトリウム *fukkanatoriumu* and sulfurization 硫化 *ryuuka* → hydrogen sulfide 硫化水素 *ryuukasuiso*. General changes can also be expressed with the nouns *keisei* or *seisei*, both meaning "formation", for instance in 環形成 *kankeisei* (ring formation), 雑種形成 *zasshukeisei* (hybridization) and 結晶生成 *kesshouseisei* (crystallization; however, respective terms with 化 are more common: 環化, 雑種化 and 結晶化).

For changes within a molecule, such as substitution of a atom or a chemical group, a second possibile way of expressing changes is the noun *chikan* (置換 "substitution, replacement"), e.g., in ハロゲン置換 *harogenchikan* (halogenation), 求核置換 *kyuukakuchikan*, (nucleophilic substitution), アシル置換 *ashiruchikan* (acylation) and アミノ基置換 *amin okichikan* (amination). The noun also appears in words related to this, e.g., 置換基 *chikanki* (substituent) or 置換生成物 *chikanseiseibutsu* (substitution product).

Some nouns are important for the description of various types of machines, apparatuses and measuring instruments, e.g., *ki* and *kei*. *ki* can be expressed by two Chinese characters, 機 and 器 and is used, for example, in 遠心分離機 *enshinbunriki* (centrifuge), 押出機 *oshidashiki* (extruder), 集光器 *shuukouki* (condenser) and 質量分析器 *shitsuryoubunsekiki* (mass spectrograph). For many machines, both can be applied with the same meaning, e.g., for extractor: 抽出機 or 抽出器 *chuushutsuki* or filter press: 圧濾機 or 圧濾器 *atsuroki*. *kei* is found in many gages, such as 分光計 *bunkoukei* (spectrometer), ガス温度計 *gasuondokei* (gas thermometer) and アルコール計 *aruko-rukei* (alcoholometer).

2.1.3
Verbs and Adjectives in Technical Publications

Although verbs and adjectives are not the dominant word types in scientific literature, there are some characteristics related to the description of technical topics. Two verbs predominante in scientific texts written in Japanese. The first is である, which expresses existence. It is one plain form of the copula and used in sentences like "A is B". Its conjugated forms are given in Section 1.1.4. Because it is a sentence-final form, it is positioned at the end of main clauses as well as of subordinate clauses and in attribute clauses in front of nouns. である can be considered as consisting of the te-form of the copula で and the auxiliary verb ある. ある can be used after the each verb's te-form and expresses that an action has been carried out and the system is still in that state. As the agent is not important, or unpersonalized under a scientific point of view, "he" is usually omitted. Similarly, the verb いる may be used after the te-form. Whereas Verb-てある is usually translated as "has been done" (and it is obvious that somebody is responsible for the activity), the focus of Verb-ている is the state and it is not implied that someone was active.

The second important verb in technical literature is する. It basically means "to do, to perform, to make". There are many expressions formed with する and even some idiomatic phrases, e.g., to express a decision on something (ことにする, ことにしている) or someone's act of making sure that someone will do something (ようにする). The most characteristic property in technical texts is its combination with nouns. If する is attached to a noun it changes that preceding noun into a verbal phrase. Traditionally, many Chinese nouns can be used in this combination, forming many verbs, e.g., 勉強する *benkyou suru* (to study), 掃除する *souji suru* (to clean), 用意する *youi suru* (to prepare) and 説明する *setsumei suru* (to explain). Alternatively, the Sinojapanese compound itself can be used as the direct object of する. For instance, "to learn chemistry" can be expressed with 化学を勉強する *kagaku o benkyou suru*. If the direct object 化学 is connected to the Sinojapanese noun 勉強 by the particle の, a noun phrase is created that can be used as a direct object of *suru* as 化学の勉強をする *kagaku no benkyou o suru*, also meaning "to learn chemistry".

Because the noun-type vocabulary is disproportionately high in natural sciences, many activities in chemistry are also described by the combination of noun and する.

Examples:
- oxidation: 酸化 *sanka* → to oxidize: 酸化する *sanka suru*
- crystal: 結晶 *kesshou* → to crystallize: 結晶する *kesshou suru*
- evaporation: 蒸発 *jouhatsu* → to evaporate: 蒸発する *jouhatsu suru*
- distillation: 蒸留 *jouryuu* → to distil: 蒸留する *jouryuu suru*
- polymerization: 重合 *juugou* → to polymerize: 重合する *juugou suru*

Even longer word compositions can be verbalized, e.g., perhydrogenization: 過酸化水素化 *kasankasuisoka* → to perhydrogenize: 過酸化水素化する *kasankasuisoka suru*. In contemporary scientific Japanese, it is very common to use *suru* also with English loanwords or terms in which one part is derived from a Western language, for instance:

- sulfonation: スルフォン化 → to sulfonate: スルフォン化する
- ester formation: エステル化 → to esterify: エステル化する
- recycle: リサイクル → to recycle: リサイクルする
- annealing, tempering: アニーリング → to anneal: アニーリングする

Adjectives are even less important for scientific literature because there are very few true adjectives. The descriptive of properties of nouns is mainly undertaken by other nouns with a typical way of formation in the Japanese language. Nouns can be widely connected using the particle の. As described in Section 2.1.2, it is broadly used in technical language because of the necessity to form new terms with a limited number of basic words. One way to translate noun relationships with の into English is to use adjective constructions, as in the following examples:

- inorganic: 無機の *muki no*
- intermolecular: 分子間の *bunshikan no*
- amorphous: 無定形の *muteikei no*
- enzymatic: 酵素の *kouso no*

In scientific Japanese both are possible, i.e., to form words by simple combination of nouns or by the connection with の, and there is no clear preference for one method. For example, "hydrophobic site" can be translated as 疎水性部位 *sosuiseibui* or 疎水性の部位 *sosui no seibui*. If terms become too complex because of too many *kanji* or if one part is emphasized, forms with の are sometimes favored. In many of these noun compositions, some specific Japanese words can be found that are frequently used in the description of states and properties by nouns (Section 2.1.2), especially 性 *sei*, 状 *jou* and 質 *shitsu*. If terms end with one of these words, many adjective-like items can be formed.

Examples with 性 *sei*:
    acidic: 酸性の *sansei no*
    basic: 塩基性の *enkisei no*
    hydrophilic: 親水性の *shinsuisei no*
    explosive: 爆発性の *bakuhatsusei no*

Examples with 状 *jou* (condition, appearance, shape, form):
    liquid: 液状の *ekijou no*
    ethereal: エーテル状の *e-terujou no*
    fibrous: 繊維状の *senijou no*

Examples with 質 *shitsu* (quality, nature):
    colloidal: 膠質の *koushitsu no*
    calcareous, limy: 石灰質の *sekkaishitsu no*
    heterogeneous: 異質の *ishitsu no*

This formation can also be found for nouns derived from foreign languages, e.g., サイクリックの (cyclic) or 等エンタルピーの *touentarupi- no* (isenthalpic). The number is low because English adjectives can directly be transcribed into Japanese, which is realized for few examples, e.g., モーラル (molar), リキッド (liquid), アキラル (achiral), イソタクチック (isotactic) and ピリオディック (periodic). Because of these different possibilities

Table 2.2 Examples for the formation and use of na-type adjectives.

Noun	na-type adjective			Examples for direct combination with nouns	
chemistry	化学 *kagaku*	→	chemical	化学的な *kagakuteki na*	chemical stability	化学的安定性 *kagakutekianteisei*
machine	機械 *kikai*	→	mechanical	機械的な *kikaiteki na*	mechanical energy	機械的エネルギー *kikaitekienerugi-*
nucleophilicity	求核性 *kyuukakusei*	→	nucleophilic	求核的な *kyuukakuteki na*	nucleophilic substitution	求核的置換 *kyuukakutekichikan*
microscope	顕微鏡 *kenbikyou*	→	microscopic	顕微鏡的な *kenbikyouteki na*	microcrystalline	顕微鏡的結晶の *kenbikyoutekikesshou no*
enzyme	酵素 *kouso*	→	enzymatic	酵素的な *kousoteki na*	enzymatic degradation	酵素的分解 *kousotekibunkai*

of formation, various Japanese terms may be used for some English adjectives, such as for crystalline: 結晶性の *kesshousei no*, 結晶質の *kesshoushitsu no* or クリスタリン.

One of the few possibilities for creating true adjectives from nouns is to attach 的 *teki* at the end of nouns. This method leads to na-type adjectives that are themselves not conjugable and, therefore, need the copula for conjugation (Section 1.1.4). Examples are 有機的な *yuukiteki na* (organic), 立体特異的な *rittaitokuiteki na* (stereospecific), 熱力学的な *netsurikigakuteki na* (thermodynamic) and 等方的な *touhouteki na* (isotropic). Used adnominal as affirmative present tense form, terms formed with 的 can also be directly combined with other nouns. 的 can be attached to various nouns that themselves are not necessarily able to describe states or properties (for examples see Table 2.2).

2.2
Frequently used *kanji*

Besides the 6000–7000 Chinese characters that are commonly used in newspapers and fiction, an additional number of *kanji* are specific for scientific usage. Although the total number of different characters used in chemical publications may be over 10 000, there is a manageable number of *kanji* that can be very frequently found. Knowledge of them is essential and facilitates analysis of chemical literature.

The following characters have been selected – classified because of their position within *kanji* combination. They represent about 2–3% of the commonly used *kanji*. Some further characters were not chosen because they always occur frequently (e.g., characters for numbers), independent of scientific texts and are, therefore, considered as known. The characters in Tables 2.3–2.5 are ordered by radicals. To help readers to find the *kanji* in other dictionaries, the last columns contain the identification numbers of the characters in the two standard *kanji* dictionaries.[1]

Table 2.6 lists some of the most important *kanji* combination expressing essential chemical and scientific terms. As some characters occur only in specific combinations with other *kanji*, they are not listed in Tables 2.3–2.5. For instance, 酢 *saku* is most commonly used in combinations with 酸 *san* (acetic acid and its derivatives, e.g., zinc acetate 酢酸亜鉛 *sakusanaen*), 析 *seki* in 分析 *bunseki* and 解析 *kaiseki*, 酵 *kou* in 酵素 *kouso* (enzyme, e.g., reductase 還元酵素 *kangenkouso*), and 応 *ou* almost only in 反応 *hannou* (reaction). There are some *kanji* that seem to belong to a combination, because they occur frequently one after another, such as 酸塩. This is purely accidental, and as they do not build a unit regarding content they are not listed here.

Some characters have a similar meaning and the same reading as they consist of the same sound-determining radical and, therefore, they may have even a similar shape. In certain

---

1) Because many other dictionaries refer to one of the two standard *kanji* dictionaries, their *kanji* identification numbers are useful. The first number (written in italics) is the "descriptor" from the *Japanese Character Dictionary* (by M. Spahn and W. Hadamitzky, Nichigai Associates Inc., Tokyo, 1989), consisting of the character's radical number, the number of strokes in its non-radical part, and, separated by a decimal point, its sequential number within the group of characters having the same radical and same residual stroke-count. The second and third numbers are taken from the *Japanese-English Character Dictionary* (by A.N. Nelson, Charles E. Tuttle Co., Inc., Tokyo, 1970). The second number is the successive number of all *kanji*. In brackets, the number of the radical followed by the stroke count of the non-radical part is given.

combinations with other *kanji*, they can be replaced, giving the same meaning; both *kanji* combinations can be found in publications. In combination with other characters, the meaning is different, although also the reading is the same. One example is the pair 型 and 形, both of which read *kei*. Together with 同 *dou* and 造 *zou* they form the same expressions, e.g., 同型 or 同形 *doukei* (isomorphism), 造形 or 造型 *zoukei* (molding). With 原 *gen*, the meaning of the combinations is not equal 原型 *genkei* (prototype, model) and 原形 *genkei* (original form). Another example is the pair 留 and 溜 *ryuu*, e.g., in 蒸留 or 蒸溜 *jouryuu* (distillation) and in 乾留 or 乾溜 *kanryuu* (carbonization).

2.2.1
Important *kanji* in Scientific Publications

(a) Table 2.3 compiles *kanji* that are exclusively (as prefixes) or dominantly used at the first position of chemical terms in Japanese.[2]

Table 2.3 Frequent *kanji* at the beginning of technical terms.

kanji	ro-maji[a]	English	Example	Hadamitsky, Nelson
不	FU	un-, anti-, not	unsaturated solution 不飽和溶液 *fuhouwayoueki* inactive substance 不活性物質 *fukasseibusshitsu*	0a4.2 17 (1/3)
中	CHUU	middle	neutrality 中性 *chuusei* intermediate 中間物 *chuukanbutsu*	0a4.40 81 (1/3)
半	HAN	half, semi-	semipermeability 半透性 *hantousei* hemihydrate 半水化物 *hansuikabutsu*	0a5.24 132 (1/4)
未	MI	un-, not yet	incomplete 未完 *mikan* unreacted 未反応 *mihannou*	0a5.27 179 (1/4)
多	TA, ou(i)	many, poly-, multi-	polysilicic acid 多珪酸 *takeisan* multilayer 多分子膜 *tabunshimaku*	0a6.5 1169 (36/3)
再	SAI	again, twice, re-	recrystallization 再結晶 *saikesshou* verification, reexamination 再検査 *saikensa*	0a6.26 35 (1/5)
亜	A	rank next, -sub, -ous	sulfurous acid H_2SO_3 亜硫酸 *aryuusan* zinc 亜鉛 *aen*	0a7.14 43 (1/6)
非	HI	non-, un-	nonmetal 非金属 *hikinzoku* asymmetry 非対称 *hitaishou*	0a8.1 5080 (175/0)

[2] Depending on the context, the meaning and the intended interpretation, some of the listed characters may have additional readings, other meanings or may be used on other positions of *kanji* combinations in Japanese terms. If a *kanji* is classified to be dominantly at the beginning or end of a *kanji* combination, the combination is a term with defined meaning. Other *kanji* may be in front of the combination or may follow, forming more complex expressions. Because of this, prefixes and suffixes can also be in the middle of a multi character term.

Table 2.3 *(continued)*

kanji	ro-maji[a]	English	Example	Hadamitsky, Nelson
重	JUU	di- (see also Table 2.5)	dichromic acid H$_2$Cr$_2$O$_7$ 重クロム酸 *juukuromusan* sodium bicarbonate 重曹 *juusou*	0a9.18 224 (4/8)
付	FU, *tsu(keru)*	attach, affix	cycloaddition 付加環化 *fukakanka* protonation プロトン付加 *purotonfuka*	2a3.6 363 (9/3)
全	ZEN	all, total, complete	overall yield 全収率 *zenshuuritsu* total hardness 全硬度 *zenkoudo*	2a4.16 384 (9/4)
次	JI, SHI, *tsugi*	next	hypochlorous acid 次亜塩素酸 *jiaensosan*	2b4.1 638 (15/4)
防	BOU	anti-, -proof, -resistant	waterproof 防水 *bousui* antiseptic 防腐剤 *boufuzai*	2d4.1 4980 (170/4)
陰	IN, ON	negative	anion 陰イオン *inion* electronegative element 陰性元素 *inseigenso*	2d8.7 4491 (154/4)
陽	YOU	positive	proton 陽子 *youshi* anodic oxidation 陽極酸化 *youkyokusanka*	2d9.5 5012 (170/9)
正	SEI, SHOU	positive, ortho-, regular	orthoantimonic(V) acid H$_3$SbO$_4$ 正アンチモン酸 *seianchimonsan* positive electrode 正極 *seikyoku*	2m3.3 27 (1/4)
比	HI, *kura(beru)*	compare, specific (see also Table 2.4)	specific viscosity 比粘度 *hinendo* colorimetry 比色分析 *hishokubunseki* reference cell 比較セル *hikakuseru*	2m3.5 2470 (81/0)
反	HAN	anti-, opposite	inversion 反転 *hanten* antiferromagnetism 反強磁性 *hankyoujisei*	2p2.2 817 (27/2)
過	KA, *su(giru)*	excess, per-	peroxide 過酸化物 *kasankabutsu* overheating 過食 *kashoku*	2q9.18 4723 (162/9)
混	KON, *ma(zeru)*	mix, include	miscibility 混和性 *konwasei* mixed ether 混成エーテル *konseie-teru*	3a8.14 2604 (85/8)
清	SEI	pure, clear	pure water 清水 *seisui* cleaner 清浄剤 *seijouzai*	3a8.18 2605 (85/4)
減	GEN	decreasing, reduce	reduced pressure 減圧 *genatsu* depolarization 減極 *genkyoku*	3a9.37 2637 (85/9)
濃	NOU	undiluted, dark, thick	concentration 濃度 *noudo*, 濃縮 *noushuku* concentrated hydrochloric acid 濃塩酸 *nouensan*	3a13.7 2711 (85/13)
超	CHOU	hyper-, ultra-	ultrasonic waves 超音波 *chouonpa* hyperconjugation 超共役 *choukyouyaku*	3b9.18 4543 (156/5)

Table 2.3 *(continued)*

kanji	ro-maji[a]	English	Example	Hadamitsky, Nelson
単	TAN	single, mono-, uni-	monomer 単量体 *tanryoutai* single bond 単結合 *tanketsugou*	3n6.2 139 (3/8)
脱	DATSU	take away	dehydration 脱水 *dassui* dehalogenation reaction 脱ハロゲン反応 *datsuharogenhannou*	4b7.8 3775 (130/7)
無	MU, na(i), na(shi)	un-, anti-, not, -free, -less	colorless 無色 *mushoku* inorganic chemistry 無機化学 *mukikagaku* acetic anhydride 無水酢酸 *musuisakusan*	4d8.8 2773 (86/8)
特	TOKU	special	stereospecific 立体特異的 *rittaitokuiteki* characteristic frequency 特性振動数 *tokuseishindousuu*	4g6.1 2860 (93/6)
自	JI, SHI, ono(zukara)	self, spontaneously, naturally	self-inductance 自己インダクタンス *jikoindakutansu* autocatalysis 自触媒作用 *jishokubaisayou* natural science 自然科学 *shizenkagaku*	5c1.1 3841 (132/0)
補	HO	assist, supplement	coenzyme 補酵素 *hokouso* adjuvant 補助剤 *hojozai*	5e7.1 4242 (145/7)
異	I, koto(naru)	difference, be different	isomer 異性体 *iseitai* anisotropy 異方性 *ihousei* heteropolar bond 異極結合 *ikyokuketsugou*	5f6.7 3008 (102/6)
等	TOU, hito(shii)	equal, iso-, homo-	homogeneity 等質 *toushitsu* isotropy 等方性 *touhousei*	6f6.9 3396 (118/6)

[a] *on*-readings are written with capitalized letters. Non-capitalized letters are *kun*-readings (for an explanation of *on*- and *kun*-readings see Section 1.2.3).

(b) Table 2.4 lists Chinese characters that are often placed at the end of *kanji* combination representing chemical terms. Some of them are solely used as suffixes.[2)]

Table 2.4 Frequent *kanji* at the end of technical terms.

kanji	ro-maji[a]	English	Example	Hadamitsky, Nelson
事	JI, koto	thing, matter	conflagration 火事 *kaji* construction 工事 *kouji*	0a8.15 272 (6/7)

Table 2.4 *(continued)*

kanji	ro-maji[a]	English	Example	Hadamitsky, Nelson
業	GYOU	industry, business	chemical industry 化学工業 *kagakukougyou* agriculture 農業 *nougyou*	0a13.3 143 (4/12)
位	I, *kurai*	grade, place	electric potential 電位 *deni* rearrangement 転位 *teni*	2a5.1 401 (9/5)
価	KA	value, valence	atomic valence 原子価 *genshika* polyvalent alcohol 多価アルコール *takaaruko-ru*	2a6.3 422 (9/6)
子	SHI, SU	child (small things)	atom 原子 *genshi* molecule 分子 *bunshi*	2c0.1 1264 (39/0)
剤	ZAI	medicine, preparation	solvent 溶剤 *youzai* detergent 洗剤 *senzai*	2f8.6 5424 (210/2)
率	RITSU	rate, value, coefficient	viscosity coefficient 粘性率 *nenseiritsu* flow rate 移動率 *idouritsu* percentage 百分率 *hyakubunritsu*	2j9.1 319 (8/9)
比	HI	ratio (see also Table 2.3)	mixing ratio 混合比 *kongouhi* flow rate 移動比 *idouhi*	2m3.5 2470 (81/0)
用	YOU	use, for use in	hydration 水和作用 *suiwasayou* substitute 代用 *daiyou*	2r3.1 2993 (101/0)
型	KEI, *kata*	form, model, type (see 形)	polymorphism 遺伝子多型 *idenshitagata* atomic model 原子模型 *genshimokei* racemic form ラセミ型 *rasemigata*	3b6.11 1077 (32/6)
器	KI	apparatus	filter 濾過器 *rokaki* manometer 検圧器 *kenatsuki*	3d12.13 994 (30/12)
形	KEI, GYOU, *kata*, *nari*	form, shape, figure (see 型)	crystal form 結晶形 *kesshougata* linear molecule 直線形分子 *chokusenkeibunshi* amorphous sulfur 無定形硫黄 *muteikeiiou*	3j4.1 1589 (59/4)
度	DO	degree, -times, extend, measure, limit	flow rate 流動速度 *ryuudousokudo* activity 活動度 *katsudoudo* degree of cross-linking 橋かけ度 *hashikakedo*	3q6.1 1511 (53/6)
的	TEKI	-ic (adjectiv forming suffix)	systematic 系統的な *keitouteki* exothermal 発熱的 *hatsunetsuteki* stereospecific 立体特異的 *rittaitokuiteki*	4c4.12 3097 (106/3)
然	NEN, ZEN	as, like such	natural 天然 *tennen*, 自然 *shizen* suddenly, unexpected 突然 *totsuzen*	4d8.10 2770 (86/8)
物	BUTSU, *mono*	thing, object, matter	hydrate 水化物 *suikabutsu* mineral 鉱物 *koubutsu* dry substance 乾燥物 *kansoubutsu*	4g4.2 2857 (93/4)

Table 2.4 *(continued)*

kanji	ro-maji[a]	English	Example	Hadamitsky, Nelson
性	SEI	-ness, -ity	toxicity 毒性 *dokusei* viscosity 粘性 *nensei*	4k5.4 1666 (61/5)
式	SHIKI	formula, type, system	chemical formula 化学式 *kagakushiki* cyclic compound 環式化合物 *kanshikikagoubutsu*	4n3.2 1556 (56/3)
産	SAN	produce	industry 産業 *sangyou* mineral 鉱産物 *kousanbutsu*	5b6.4 3354 (117/6)
料	RYOU	material	fertilizer 肥料 *hiryou* pigment 顔料 *ganryou*	6b4.4 3468 (119/4)
計	KEI, *haka(ru)*	measure, gauge	thermometer 温度計 *ondokei* vacuum gauge 真空計 *shinkuukei*	7a2.1 4312 (149/2)

[a] *on*-readings are written with capitalized letters. Non-capitalized letters are *kun*-readings (for an explanation of *on*- and *kun*-readings see Section 1.2.3).

(c) Table 2.5 lists further characters that can be found at any position within *kanji* combinations with similar probability.

Table 2.5 Frequent *kanji* at any position in chemical terms.

kanji	ro-maji[a]	English	Example	Hadamitsky, Nelson
内	NAI, DAI, *uchi*	inside, among	intramolecular 分子内の *bunshinai no* internal pressure 内圧 *naiatsu*	0a4.23 82 (2/3)
生	SEI, SHOU, *i(kiru)*, *nama*, *ki*	birth, life, pure, unprocessed	biochemistry 生化学 *seikagaku* heat generation 熱発生 *neppatsusei* de novo synthesis 新生合成 *shinseigousei*	0a5.29 2991 (100/0)
気	KI, KE	spirit, air, atmosphere, flavour	electricity 電気 *denki* magnetism 磁気 *jiki* atmospheric pressure 空気圧 *kuukiatsu*	0a6.8 2480 (84/0)
発	HATSU	discharge, emit	formation 発生 *hassei* luminescence 発光現象 *hakkougenshou*	0a9.5 3902 (140/4)
重	JUU, CHOU, *omo(i)*	heavy, weight (see also Table 2.3)	specific gravity 比重 *hijuu* polymerization 重合 *juugou*	0a9.18 224 (4/8)

Table 2.5 (continued)

kanji	ro-maji[a]	English	Example	Hadamitsky, Nelson
化	KA	transform, make into, -ization	sulfurization 硫化 ryuuka biochemistry 生化学 seikagaku invert sugar 転化糖 tenkatou	2a2.6 350 (9/2)
合	GOU	together	bond energy 結合エネルギー ketsugouenerugi- condensation 縮合 shukugou	2a4.18 383 (9/4)
体	TAI, TEI, karada	body, object, form, condition	volume 体積 taiseki isomer 同分体 doubuntai polymer 重合体 juugoutai	2a5.6 405 (9/5)
低	TEI	low	low-molecular-weight compound 低分子量化合物 teibunshiryoukagoubutsu	2a5.15 406 (9/5)
保	HO, HOU, tamo(tsu)	preserve, maintain	preservative 保恒剤 hokouzai retention time 保持時間 hojijikan protecting group 保護基 hogoki	2a7.11 455 (9/7)
冷	REI, tsume(tai)	cold	cooling 冷却 reikyaku refrigerator 冷却装置 reikyakusouchi	2b5.3 642 (15/5)
凝	GYOU, ko(ru)	get stiff, be absorbed in	adhesion 凝集 gyoushuu anticoagulant 抗凝剤 kougyouzai	2b14.1 652 (15/14)
力	RYOKU, RIKI, chikara	power, force, strength	intermolecular force 分子間力 bunshikanryoku gravity 重力 juuryoku	2g0.1 715 (19/0)
加	KA, kuwa(eru)	add, increase	heating 加熱 kanetsu hydrolysis 加水分解 kasuibunkai protonation プロトン付加 purotonfuka	2g3.1 716 (19/3)
変	HEN, ka(waru)	change, strange	transformation, conversion 変換 henkan isobaric change 等圧変化 touatsuhenka	2j7.3 306 (8/7)
高	KOU, taka	high, amount	high pressure 高圧 kouatsu polymer chemistry 高分子化学 koubunshikagaku	2j8.6 5248 (189/0)
止	SHI, to(meru), ya(meru)	stop, limited	termination reaction 停止反応 teishihannou inhibitor 防止剤 boushizai	2m2.2 2429 (77/0)
点	TEN	point	melting point, melting temperature 融点 yuuten	2m7.2 804 (25/15)
色	SHOKU, iro	color	hemoglobin 血色素 kesshikiso achromatism 色消し irokesi	2n4.1 3889 (139/0)
分	BUN	dividing, portion	molecule 分子 bunshi analysis 分析 bunseki	2o2.1 578 (12/2)

Table 2.5 (continued)

kanji	ro-maji[a]	English	Example	Hadamitsky, Nelson
着	CHAKU	attach	adhesive 粘着物 nenchakubutsu adsorption 吸着 kyuuchaku	2o10.1 3665 (123/6)
圧	ATSU	pressure	compression 圧迫性 appakusei isobar 等圧線 touatsusen	2p3.1 818 (27/3)
原	GEN	original, fundamental	element 原素 genso atom 原子 genshi crude ore, raw ore 原鉱 genkou	2p8.1 825 (27/8)
同	DOU, ona(ji)	the same	isomer 同分体 doubuntai isotope 同性元素 douseigenso	2r4.2 619 (13/4)
水	SUI, mizu	water	carbohydrate 炭水化物 tansuikabutsu sodium hydroxide (NaOH) 水酸化ナトリウム suisankanatoriumu hydrogenation 水素化 suisoka	3a0.1 2482 (85/0)
沈	CHIN, shizu(mu)	to sink	precipitation 沈澱 chinden sediment 沈積物 chinsekibutsu	3a4.9 2209
油	YU, YUU, abura	oil	mineral oil 石油 sekiyu lipophilic group 親油基 shinyuki	3a5.6 2508 (85/4)
法	HOU	method, law	analytical method 分析法 bunsekihou law 法則 housoku	3a5.20 2535 (85/5)
活	KATSU, i(kiru)	life, activity, to be alive	activity 活動度 katsudoudo inert gas 不活性ガス fukasseigasu	3a6.16 2552 (85/6)
流	RYUU, RU, naga(reru)	flow, current	reflux 還流 kanryuu viscous liquid 粘性流体 nenseiryuutai	3a7.10 2576 (85/7)
液	EKI	liquid, fluid	suspension 懸濁液 kendakueki Fehling's solution フェーリング液 fe-ringueki	3a8.29 2599 (85/8)
温	ON, atata(kai)	warm	temperature 温度 ondo isothermal 等温線 touonsen	3a9.21 2634 (85/9)
溶	YOU, to(keru), to(kasu)	melt, dissolve	solvent 溶剤 youzai dissolved oxygen 溶存酸素 youzonsanso	3a10.15 2659 (85/10)
基	KI	radical, basis, group	functional group 官能基 kannouki biradical 二端遊離基 nitanyuuriki	3b8.12 1098 (32/8)
塩	EN, shio	salt	ferrite 亜鉄酸塩 atetsusanen lactate 乳酸塩 nyuusanen	3b10.4 1125 (32/10)

Table 2.5 (continued)

kanji	ro-maji[a]	English	Example	Hadamitsky, Nelson
換	KAN, ka(eru)	substitute	substitution, displacement 置換 chikan cation exchange 陽イオン交換 youionkoukan bromination 臭素置換 shuusochikan	3c9.15 1964 (64/9)
蒸	JOU, mu(su)	steam	evaporation 蒸着 jouchaku water vapor 水蒸気 suijouki	3k9.19 4002 (140/9)
薬	YAKU, kusuri	chemical, medicine	agricultural chemical, pesticide 農薬 nouyaku adjuvant 補助薬 hojoyaku drug tolerance 薬剤抵抗性 yakuzaiteikousei	3k13.15 4074 (140/13)
空	KUU, sora, aki	sky, empty	vacuum 真空 shinkuu air 空気 kuuki	3m5.12 3317 (116/3)
光	KOU, hikari	light	photo reaction 光反応 kouhannou phosphorescence 燐光 rinkou	3n3.2 1358 (42/3)
学	GAKU	science, learning, study	engineering 工学 kougaku experimental chemistry 実験化学 jikkenkagaku	3n4.2 1271 (39/5)
炭	TAN, sumi	carbon, coal, charcoal	carbon monoxide 一酸化炭素 issankatanso saccharide 炭水化物 tansuikabutsu	3o6.5 1418 (46/6)
固	KO, kata(i)	solid, hard	a solid 固体 kotai freezing point 凝固点 gyoukoten	3s5.2 1036 (31/5)
核	KAKU	nucleus, core	hydrogen nucleus 水素原子核 suisogenshikaku nucleophilic reaction 求核反応 kyuukakuhannou	4a6.22 2254 (75/6)
極	KYOKU	pole, end, extremely	polar molecule 極性分子 kyokuseibunshi dipole 双極子 soukyokushi, 二重極 nijuukyoku	4a8.11 2305 (75/8)
脂	SHI, abura, yani	fat, resin, gum, tar	resin 樹脂 jushi lipid 脂質 shishitsu degreasing agent 脱脂剤 dasshizai	4b6.7 3766 (130/6)
乾	KAN	dry	dryness 乾性 kansei dry cell 乾電池 kandenchi	4c7.14 784 (24/9)
晶	SHOU	crystal	crystallization 晶化 shouka crystallography 結晶学 kesshougaku liquid crystal 液晶 ekishou	4c8.6 2137 (72/8)
量	RYOU	quantity, size, measure	quantitative analysis 定量分析 teiryoubunseki tetramer 四量体 yonryoutai	4c8.9 2141 (72/8)
火	KA	fire	fire resistance 耐火性 taikasei ignition temperature 発火温度 hakkaondo	4d0.1 2743 (86/0)

Table 2.5 (continued)

kanji	ro-maji[a]	English	Example	Hadamitsky, Nelson
熱	NETSU	heat	heating 加熱 *kanetsu* thermodynamic 熱力学的 *netsurikigakuteki*	4d11.4 2797 (86/11)
燐	RIN	phosphorus	sodium hypochlorite NaOCl 次亜塩素酸ナトリウム *jiaensosannatoriumu* phosphorescence 燐光 *rinkou*	4d13.4 2807 (86/12)
珪	KEI	silicon	calcium silicate $CaSiO_3$ 珪酸カルシウム *keisankarushiumu* quartz sand 珪砂 *keisha*	4f6.4 2937 (96/6)
理	RI	reason truth, logic	theory 理論 *riron* physics 物理学 *butsurigaku*	4f7.1 2942 (96/7)
環	KAN, wa, tamaki	ring, circle	cyclization 環化 *kanka* cyclic ester 環状エステル *kanjouesuteru*	4f13.1 2970 (96/13)
解	KAI, GE, to(ku)	untie, release	dissociation 解離 *kairi* depolymerization 解重合 *kaijuugou*	4g9.1 4306 (148/6)
数	SUU, SU, kazu	number	constant 恒数 *kousuu* logarithm 対数 *taisuu* frequency 振動数 *shindousuu*	4i9.1 2057 (66/9)
成	SEI, JOU, na(ru), na(su)	become, consist of	formation, generation 生成 *seisei* hybridization 雑種形成 *zasshukeisei* crystallization 結晶生成 *kesshouseisei*	4n2.1 1799 (62/2)
石	SEKI, ishi	stone	limestone ($CaCO_3$) 石灰石 *sekkaiseki* diamond 金剛石 *kongouseki*	5a0.1 3176 (112/0)
硫	RYUU	sulfur	sulfuric acid (H_2SO_4) 硫酸 *ryuusan*	5a7.3 3191 (112/7)
磁	JI	magnetism, porcelain	magnetite (Fe_3O_4) 磁鉄鉱 *jitekkou* electromagnetic field 電磁場 *denjiba* porcelain funnel 磁製ロート *jiseiro-to*	5a9.6 3209 (112/9)
立	RITSU, RYUU, ta(tsu)	stand, rise	isolation 孤立化 *koritsuka* stereochemistry 立体化学 *rittaikagaku*	5b0.1 3343 (117/0)
製	SEI	make, manufacture	product 製品 *seihin* crude metal 粗製金属 *soseikinzoku*	5e8.9 4249 (145/8)
複	FUKU	double, multiple, again	duplication 重複 *choufuku* bicyclic compound 複環式化合物 *fukukanshikikagoubutsu*	5e9.3 4255 (145/9)

Table 2.5 (continued)

kanji	ro-maji[a]	English	Example	Hadamitsky, Nelson
界	KAI	limit, boundary, circle	surface chemistry 界面化学 kaimenkagaku boundary layer 境界層 kyoukaisou critical temperature 臨界温度 rinkaiondo	5f4.7 2998 (102/4)
系	KEI	system, line, series	orthorhombic system 斜方晶系 shahoushoukei three-component mixture 三元分系 sangenbunkei Lyman series ライマン系列 raimankeiretsu	6a1.1 195 (4/6)
素	SO	element	hydrogen 水素 suiso toxin 毒素 dokuso	6a4.12 3511 (120/4)
結	KETSU	bind	bond, linkage 結合 ketsugou crystal 結晶 kesshou	6a6.5 3540 (120/6)
線	SEN	line	absorption line 吸収線 kyuushuusen multiplet 多重線 tajuusen calibration curve 検量線 kenryousen	6a9.7 3580 (120/9)
粘	NEN	stick, persist	mucin 粘素 nenso viscosity 粘性 nensei	6b5.4 3472 (119/5)
糖	TOU	sugar	fructose 果糖 katou lipopolysaccharide リポ多糖 ripotatou	6b10.3 3485 (119/10)
質	SHITSU	quality, nature	protein 蛋白質 tanpakushitsu heterogeneity 異質性 ishitsusei	7b8.7 4518 (154/8)
転	TEN	turn, change	rearrangement 転位 teni rotational energy 回転エネルギー kaitenenerugi-	7c4.3 4615 (159/4)
酸	SAN	acid	oxygen 酸素 sanso nitric acid (HNO_3) 硝酸 shousan	7e7.2 4789 (164/7)
金	KIN, KON, kana, kane	gold, metal	auric chloride ($AuCl_3$) 塩化第二金 enkadainikin alloy 合金 goukin nonmetal 非金属 hikinzoku	8a0.1 4815 (167/0)
鉄	TETSU, kurogane	iron	steel 鉄鋼 tekkou ferromagnetism 鉄磁性 tetsujisei iron(III) oxide (Fe_2O_3) 酸化第二鉄 sankadainitetsu	8a5.6 4844 (167/5)
鉛	EN, namari	lead	minium 鉛丹 entan zinc acetate [$Zn(CH_3COO)_2$] 酢酸亜鉛 sakusanaen	8a5.14 4842 (167/5)
鉱	KOU	ore	hematite [α-Fe_2O_3] 赤鉄鉱 sekitekkou chromite クロム鉄鉱 kuromutekkou	8a5.15 4843 (167/5)
銀	GIN, shirogane	silver	silver dichromate [$Ag_2Cr_2O_7$] 重クロム酸銀 juukuromusangin mercury 水銀 suigin	8a6.3 4855 (167/6)

Table 2.5 (continued)

kanji	ro-maji[a]	English	Example	Hadamitsky, Nelson
銅	DOU, aka	copper	copper(I) chloride (CuCl) 塩化第一銅 enkadaiichidou brass 黄銅 oudou	8a6.12 4853 (167/6)
錫	SHAKU, suzu	tin	stannane 水素化錫 suisokasuzu tin foil 錫箔 suzuhaku	8a8.7 4873 (167/8)
電	DEN	electricity	ionization 電離 denri electrolysis 電解 denkai	8d5.2 5050 (173/5)

[a] on-readings are written with capitalized letters. Non-capitalized letters are kun-readings (for an explanation of on- and kun-readings see Section 1.2.3).

2.2.2
Important kanji Combinations

Table 2.6 lists approximately 100 of the most important combinations of two or more kanji.[3]

Table 2.6 Important kanji combination in chemical literature.

kanji	ro-maji	English	Examples
活性	kassei	activity	active oxygen 活性酸素 kasseisanso activation energy 活性化エネルギー kasseikaenerugi-
異性	isei	isomerism	isomer 異性体 iseitai rotamer 回転異性体 kaiteniseitai
酸性	sansei	acidity	acid resistance 耐酸性 taisansei hydrogencarbonate 酸性炭酸塩 sanseitansanen
極性	kyokusei	polarity	polar bond 極性結合 kyokuseiketsugou non-polar solvent 無極性溶媒 mukyokuseiyoubai
水性	suisei	aqueous	hydrophobic 疎水性の sosuiseino water resistance 耐水性 taisuisei
水素	suiso	hydrogen	hydrogen fluoride (HF) 弗化水素 fukkasuiso tritium 三重水素 sanjuusuiso

[3] The characters are roughly ordered by analogy and similarity. The combination of kanji is described in Section 1.2.3.

Table 2.6 *(continued)*

kanji	ro-maji	English	Examples
元素	genso	element	radioactive element 放射性元素 *houshaseigenso* elementary analysis 元素分析 *gensobunseki*
酵素	kouso	enzyme	lipase 脂肪分解酵素 *shiboubunkaikouso* enzyme inhibition 酵素阻害 *kousosogai*
炭素	tanso	carbon	carbon monoxide (CO) 一酸化炭素 *issankatanso* asymmetric carbon atom 非対称炭素原子 *hitaishoutansogenshi*
酸素	sanso	oxygen	liquid oxygen 液体酸素 *ekitaisanso*
窒素	chisso	nitrogen	nitrogen assimilation 窒素固定 *chissokotei* nitrogen peroxide (N_2O_4) 過酸化窒素 *kasankachisso*
弗素	fusso	fluorine	hydrofluoric acid 弗酸 *futsusan*
塩素	enso	chlorine	chlorous acid ($HClO_2$) 亜塩素酸 *aensosan*
臭素	shuuso	bromine	sodium bromate ($NaBrO_3$) 臭素酸ナトリウム *shuusosannatoriumu* debromination 脱臭素 *dasshuuso*
沃素	youso	iodine	silver periodate ($AgIO_6$) 過沃素酸銀 *kayousosangin*
珪素	keiso	silicon	silicon dioxide (SiO_2) 二酸化珪素 *nisankakeiso*
硼素	houso	boron	trimethylborane [$B(CH_3)_3$] トリメチル硼素 *torimechiruhouso*
砒素	hiso	arsenic	arsenic poisoning 砒素中毒 *hisochuudoku*
酸化	sanka	oxidation	catalytic oxidation 接触酸化 *sesshokusanka* mercury oxide 酸化水銀 *sankasuigin*
硫化	ryuuka	sulfurization, sulfide	copper sulfide 硫化銅 *ryuukadou*
臭化	shuuka	bromide	silver bromide (AgBr) 臭化銀 *shuukagin*
沃化	youka	iodide, iodination	silver iodide (AgI) 沃化銀 *youkagin*
弗化	fukka	fluoride	hydrogen fluoride 弗化水素 *fukkasuiso*
塩化	enka	chloride	silver chloride (AgCl) 塩化銀 *enkagin* carbon tetrachloride (CCl_4) 四塩化炭素 *shienkatanso*
炭化	tanka	carbonization, in carbon compounds (e.g., carbides)	silver acetylide (Ag_2C_2) 炭化銀 *tankagin* hydrocarbon 炭化水素 *tankasuiso*
炭酸	tansan	carbonic acid	sodium bicarbonate ($NaHCO_3$) 重炭酸ナトリウム *juutansannatoriumu* dry ice 雪状炭酸 *yukijoutansan*

Table 2.6 *(continued)*

kanji	ro-maji	English	Examples
硝酸	shousan	nitric acid	silver nitrate (AgNO$_3$) 硝酸銀 *shousangin* nitrocellulose 硝酸繊維素 *shousanseniso*
硫酸	ryuusan	sulfuric acid	ammonium sulfate [(NH$_4$)$_2$SO$_4$] 硫酸アンモニウム *ryuusananmoniumu* fuming sulfuric acid 発煙硫酸 *hatsuenryuusan*
酢酸	sakusan	acetic acid	copper acetate 酢酸銅 *sakusandou* phenylacetic acid [C$_6$H$_5$CH$_2$COOH] フェニル酢酸 *fenirusakusan*
燐酸	rinsan	phosphoric acid	zinc phosphate 燐酸亜鉛 *rinsanaen*
原子	genshi	atom	carbon atom 炭素原子 *tansogenshi* diatomic molecule 二原子分子 *nigenshibunshi*
量子	ryoushi	quantum	volume 量感 *ryoukan* light quantum 光量子 *kouryoushi*
分子	bunshi	molecule	linear molecule 線状分子 *senjoubunshi* polymer 高分子 *koubunshi*
格子	koushi	lattice	lattice constant 格子定数 *koushiteisuu*
分光	bunkou	light diffraction	spectral analysis 分光分析 *bunkoubunseki* infrared spectroscopy 赤外分光法 *sekigaibunkouhou*
化合物	kagoubutsu	chemical compound	organic compound 有機化合物 *yuukikagoubutsu* cyclic compound 環状化合物 *kanjoukagoubutsu*
合成	gousei	synthesis, synthetic	asymmetric synthesis 不斉合成 *fuseigousei* artificial fiber 合成繊維 *gouseiseni*
結合	ketsugou	bond, linkage, combination	double bond 二重結合 *nijuuketsugou* antibonding orbital 反結合軌道 *hanketsugoukidou* disulfide bridge ジスルフィド結合 *jisurufidoketsugou*
混合	kongou	mixture	entropy of mixing 混合エントロピー *kongouentoropi-*
重合	juugou	polymerization	polymer 重合体 *juugoutai* depolymerization 解重合 *kaijuugou*
物質	busshitsu	matter, substance	chemical 化学物質 *kagakubusshitsu* inorganic compound 無機物質 *mukibusshitsu*
化学	kagaku	chemistry	polymer chemistry 高分子化学 *koubunshikagaku* chemical engineering 工業化学 *kougyoukagaku*
光学	kougaku	optics	optical isomer 光学異性体 *kougakuiseitai* photoelectron spectroscopy 光電子分光学 *koudenshibunkougaku*

Table 2.6 (continued)

kanji	ro-maji	English	Examples
分解	bunkai	decomposition, breakdown, disintegration	hydrolysis 加水分解 kasuibunkai ketone cleavage ケトン分解 ketonbunretsu enzymatic degradation 酵素学的分解 kousogakutekibunkai
分析	bunseki	analysis	crystal structure analysis 結晶構造分析 kesshoukouzoubunseki mass spectrometry 質量分析 shitsuryoubunseki spectral analysis スペクトル分析 supekutorubunseki
解析	kaiseki	analysis	X-ray analysis X線解析 Xsenkaiseki conformational analysis 配座解析 haizakaiseki
置換	chikan	substitution	substituent 置換分 chikanbun, 置換基 chikanki
蒸留 蒸溜	jouryuu	distillation	distillation column 蒸溜搭 jouryuutou vacuum distillation 真空蒸留 shinkuujouryuu
反応	hannou	reaction	exothermic reaction 発熱反応 hatsunetsuhannou reaction mechanism 反応機構 hannoukikou Mannich reaction マンニッヒ反応 mannihhihannou
蒸気	jouki	vapor	vapor phase 蒸気相 joukisou steam bath 水蒸気浴 suijoukiyoku
空気	kuuki	air	atmospheric pressure 空気圧 kuukiatsu liquid air 液状空気 ekijoukuuki
気体	kitai	gas	inert gas 不働気体 fudoukitai
電気	denki	electricity	electrolysis 電気分解 denkibunkai electronegativity 電気的陰性 denkitekiinsei
電子	denshi	electron	electron shell 電子殻 denshikaku electrophilic substitution 求電子置換 kyuudenshichikan
電流	denryuu	electric current	amperemeter 電流計 denryuukei
電位	deni	electric potential	oxidation potential 酸化電位 sankadeni
電極	denkyoku	electrode, pole	graphite electrode 黒鉛電極 kokuendenkyoku
電池	denchi	battery, element, electric cell	galvanic cell ガルヴァー二電池 garuva-nidenchi
金属	kinzoku	metal	noble metal 貴金属 kikinzoku organometallic compound 有機金属化合物 yuukikinzokukagoubutsu
水銀	suigin	mercury	mercury chloride 塩化水銀 enkasuigin calomel 角水銀鉱 kakusuiginkou
白金	hakkin	platinum	platinum electrode 白金電極 hakkindenkyoku hexachloroplatinate(IV) ヘクサクロロ第二白金酸塩 hekusakurorodainihakkinsanen

Table 2.6 (continued)

kanji	ro-maji	English	Examples
亜鉛	aen	zinc	zinc bromide (ZnBr$_2$) 臭化亜鉛 shuukaaen
塩基	enki	base	nucleobase 核酸塩基 kakusanenki Lewis base ルイス塩基 ruisuenki
硫黄	iou	sulfur	sulfur bacterium 硫黄細菌 iousaikin
加硫	karyuu	curing, vulcanization	vulcanizer 加硫機 karyuuki soft rubber 軟質加硫ゴム nanshitsukaryuugomu
試薬	shiyaku	reagent	Schweizer's reagent シュヴァイツアー試薬 shuvaitsua-shiyaku nucleophile 求核試薬 kyuukakushiyaku
触媒	shokubai	catalyst	catalytic hydrogenation 触媒水素化 shokubaisuisoka phase-transfer catalyst 相関移動触媒 soukanidoushokubai
結晶	kesshou	crystal	crystallization 結晶化 kesshouka single crystal 単結晶 tankesshou
樹脂	jushi	resin	polystyrene resin ポリスチロール樹脂 porisuchiro-rujushi cation-exchange resin 陽イオン交換樹脂 youionkoukanjushi
染料	senryou	dye, dyestuffs	diazo dye ジアゾ染料 jiazosenryou
繊維	seni	fiber	synthetic fiber 合成繊維 gouseiseni fibrin 繊維素 seniso
蛋白	tanpaku	protein	lipoprotein リポ蛋白室 ripotanpakushitsu protease 蛋白酵素 tanpakukouso
脂肪	shibou	fat	fatty acid 脂肪酸 shibousan lipase 脂肪分解酵素 shiboubunkaikouso
生物	seibutsu	biological compound, living creature	organism 生物体 seibutsutai bioorganic chemistry 生物有機化学 seibutsuyuukikagaku
指示薬	shijiyaku	indicator	fluorescence indicator 螢光指示薬 keikoushijiyaku
還元	kangen	reduction	Rosenmund reduction ローゼンムンドの還元法 ro-zenmundonokangenhou reductase 還元酵素 kangenkouso
滴定	tekitei	titration	iodimetry 沃素滴定 yousotekitei conductometric titration 伝導滴定 dendoutekitei
飽和	houwa	saturation	supersaturated solution 過飽和溶液 kahouwayoueki unsaturated compound 不飽和化合物 fuhouwakagoubutsu
吸収	kyuushuu	absorption	absorber 吸収体 kyuushuutai
実験	jikken	experiment, test	empirical formula 実験式 jikkenshiki laboratory bench 実験台 jikkendai

Table 2.6 *(continued)*

kanji	ro-maji	English	Examples
試験	*shiken*		Fehling's test フェーリング試験法 *fe-ringushikenhou* laboratory 試験室 *shikenshitsu*
対称	*taishou*	symmetry	asymmetric carbon atom 非対称炭素原子 *hitaishoutansogenshi* symmetry axis 対称軸 *taishoujiku*
構造	*kouzou*	structure	structural formula 構造式 *kouzoushiki* molecular structure 分子構造 *bunshikouzou*
軌道	*kidou*	orbital	bonding orbital 結合軌道 *ketsugoukidou* frontier orbital フロンティア軌道 *furontiakidou*
無水	*musui*	anhydrous, dry	acetic acid anhydride 無水酢酸 *musuisakusan* absolute alcohol 無水アルコール *musuiaruko-ru*
乾燥	*kansou*	dry	desiccator 乾燥器 *kansouki* vacuum drying 真空乾燥 *shinkuukansou*
有機	*yuuki*	organic	organic chemistry 有機化学 *yuukikagaku*
細胞	*saibou*	cell (biology)	intercellular 細胞間の *saiboukan no* cytology 細胞学 *saibougaku*
真空	*shinkuu*	vacuum	high-vacuum 高真空 *koushinkuu* vacuum distillation 真空蒸留 *shinkuujouryuu*
質量	*shitsuryou*	mass	molar weight モル質量 *morushitsuryou* relative molecular mass 相対分子質量 *soutaibunshishitsuryou*
効果	*kouka*	effect	synergistic effect 相乗効果 *soujoukouka* nuclear Overhauser effect 核オーバーハウザー効果 *kakuo-ba-hauza-kouka*
原理	*genri*	principle	Franck–Condon principle フランクコンドン原理 *furankukondongenri* basic concept, fundamental idea 基本原理 *kihongenri*
係数	*keisuu*	coefficient	extinction coefficient 級光係数 *kyuukoukeisuu*
定数	*teisuu*	constant	gas constant 気体定数 *kitaiteisuu*
速度	*sokudo*	velocity	diffusion rate 拡散速度 *kakusansokudo* reaction kinetics 反応速度論 *hannousokudoron*
標準	*hyoujun*	standard	standard solution 標準溶液 *hyoujunyoueki* normal element 標準電池 *hyoujundenchi*
状態	*joutai*	state, condition	crystalline state 結晶状態 *kesshoujoutai* phase diagram 状態図式 *joutaizushiki*
測定	*sokutei*	measure	spectrometry 分光測定 *bunkousokutei* immunoassay 免疫学的測定法 *menekigakutekisokuteihou*

Table 2.6 (continued)

kanji	ro-maji	English	Examples
平衡	heikou	equilibrium	redox equilibrium 酸化還元平衡 sankakangenheikou
温度	ondo	temperature	ignition temperature 発火温度 hakkaondo thermometer 温度計 ondokei
放射	housha	radiation, emission, discharge	thermal radiation 熱放射 netsuhousha radioactive 放射性の houshaseino
研究	kenkyuu	research	chemical research 化学研究 kagakukenkyuu experimental study 実験的研究 jikkentekikenkyuu
法則	housoku	law, rule	Markovnikov's rule マルコーヴニコフの法則 maruko-vunikofunohousoku

2.3
Numbers, Symbols and Units

2.3.1
Numbers[4]

In technical Japanese, numbers are usually written with Arabic characters if they represent date and quantity. In the names of chemical compounds, either *kanji* or *kana* are applied to indicate number and position of substituents[5] (Table 2.7). For numbers above ten, the *kanji* for the power is usually added, e.g., for 4567: 四千五百六十七 *yonsengohyakurokujuunana*. In some cases, the Chinese characters simply replace the corresponding Arabic numeral, e.g., for 4567: 四五六七. For numbers up to ten, the Japanese system of counting is also applied. It uses the same characters but with additional *kana* つ, resulting in a different pronunciation, e.g., for five: 五 *go* (Sinojapanese number) and 五つ *itsutsu* (Japanese number).

English counts sets of 1 000, while Japanese counts sets of 10 000. The Japanese sets are, therefore, divided into groups of four: 一, 十, 百 and 千, for instance 一万 (ten thousand), 十万 (one hundred thousand), 百万 (one million) and 千万 (ten million).

---

[4] Concerning pronunciation, the numbers 4, 7 and 9 have different readings that depend on whether these are used alone (四 *shi*, 七 *shichi*) or in combinations (14: 十四 *juuyon*, 27: 二十七 *nijuunana*). There are some phonetic changes for certain combination of *kanji* (Section 1.2.2), such as for 600 六百 *roppyaku* or 3000: 三千 *sanzen*.

Phonetic changes are also found in combination of numbers with counters, if the counter starts with one of the consonants h, f, k, s, p or t, such as in 一個 *ikko*, 三本 *sanbon* and 六分 *roppun*.

[5] See also Section 3.2 for a detailed explanation of the use of Japanese characters to indicate the number and position of substituents.

Table 2.7 Numbers.

Number	kanji	katakana
0	零 rei	ゼロ
1, first, mono, primary, single	一 ichi, 単 tan, 単一 tanitsu	モノ
2, second, di, secondary, both, double, pair	二 ni, 両 ryou, 双 sou	ジ (ダイ)
3, tri, tertiary	三 san	トリ
4, tetra, quaternary	四 shi, yon	テトラ
5, penta	五 go	ペント
6, hexa	六 roku	ヘクサ
7, hepta	七 shichi, nana	ヘプタ
8, octa	八 hachi	オクタ
9, nona	九 kyuu, ku	ノナ
10, deca	十 juu	デカ
11, undeca	十一 juuichi	ウンデカ
12, dodeca	十二 juuni	ドデカ
20	二十 nijuu	
100	百 hyaku	
1000	千 sen	
10 000	万 man	
100 000 000	億 oku	
1 000 000 000 000	兆 chou	
infinite	無限大 mugendai	
poly	多 ta	
multi	多重 tajuu	

If figures are used for the number and position of substituents in names of organic molecules, the number of substituents are expressed with the transcribed English expression in *katakana*, e.g., for トリメチルアミン [trimethylamine, N(CH$_3$)$_3$]. The Japanese expression for two ("di- …") is usually ジ, e.g., in ジクロル酢酸 *jikurorusakusan* (dichloroacetic acid, Cl$_2$CHCOOH), and in a few cases also ダイ, e.g., for ダイアミンブラック (diamine black). If the chemical term is written with *kanji*, such as for inorganic compounds or general expressions, the number is also represented by a *kanji*, like for 三炭糖 *santantou* (triose) or 三水化物 *sansuikabutsu* (trihydrate). In cases where the substituent may be expressed by either a transcribed English expression or a Japanese *kanji*, the same rule is applicable, e.g., for "dinitro": with *kanji* in 二硝酸塩 *nishousanen* and with *kana* in ジニトロベンゾール (dinitrobenzene). Because of this, sometimes *kana* and *kanji* can be found for numbers in molecules, e.g., for tetrachloromethane, carbon tetrachloride (CCl$_4$): 四塩化炭素 *shienkatanso* or テトラクロルメタン. Examples are given in Table 2.8.

Whereas Japanese numbers are usually used alone (like 一つ *hitotsu*, 二つ *futatsu*,…, 十 *too*), *kanji* expressing numbers are combined with counters, if they are not used in names of chemical compounds or terms like those above. Counters are characteristic words for the object to be counted and may be represented by a *kanji* or *kana*. In scientific texts, also Arabic numbers are sometimes combined with Japanese counters. The counter is directly attached

Table 2.8 Examples for the use of numbers.

Number	Examples	
one, mono	nitrogen monoxide:	一酸化窒素 *issankachisso*
	monocarboxylic acid:	モノカルボン酸 *monokarubonsan*
	monochloride:	一塩化物 *ichienkabutsu*
	monomolecular reaction:	一分子反応 *ichibunshihannou*
	adenosine monophosphate:	アデノシン一燐酸 *adenoshinichirinsan*
two, di	copper(II) chloride, cupric chloride:	塩化第二銅 *enkadainidou*
	second-order reaction:	二次反応 *nijihannou*
	dimer:	二量体 *niryoutai*
	dichloroacetic acid:	ジクロル酢酸 *jikurorusakusan*
	secondary alcohol:	第二アルコール *dainiaruko-ru*
three, tri	adenosine triphosphate:	アデノシン三燐酸 *adenoshinsanrinsan*
	trichloromethane:	トリクロルメタン *torikurorumetan*
	trialkylphosphate:	燐酸トリアルキル *rinsantoriarukiru*
	trihexose:	三六炭糖 *sanrokutantou*

behind the number. As examples, flat objects, like metal sheets, foils, lenses and plates are counted with 枚 *mai*:

Examples:	four tin foils:	四枚の錫箔 *yonmai no suzuhaku*
	two metal sheets:	板金を二枚 *itagane o nimai*
	five filter plates:	濾板の5枚 *rokahan no 5 mai*

The expression of quantity might be placed either in front of or behind the object. When it is in front, counter and object are connected with の like two noun phrases. This word order [number + counter の object] is the usual one in scientific publications. If the quantity follows the object, the object is marked with the particle that indicates the function of the object, e.g., を, は or が (for use of particles see Section 1.1.3). Table 2.9 lists some important counters in technical literature.

Examples:	ten carbon atoms:	十個の炭素原子 *juuko no tansogenshi*
	10^5 morphine molecules:	十の五乗個のモルフィンの分子 *juu no gojuuko no morufin no bunshi*
	three light rays:	三本の光線 *sanbon no kousen*
	two set of equations:	二組の方程式 *nikumi no houteishiki*

To express approximate numbers, various words exist that are placed either before the number or after it. In front of the numbers, やく, ほぼ and およそ are used. ぐらい, くらい, ほど and ごろ are positioned after the number. The meaning may be slightly different, depending on the context the words are used in. Some may be used only in a specific context,

Table 2.9 Selected important counters in technical literature.

	Counter	ro-maji	Use	Remarks
Objects	部	bu	printed materials	books, journals
	個	ko	small round objects	electrons, atoms, molecules, cells, balls
	本	hon	long and thin objects	bottles, rods, wires
	枚	mai	thin and flat objects	sheets, foils, plates
	冊	satsu	printed matter	books, magazines, journals
	台	dai	machines, electronic devices	computer, spectrometer
Abstract matters	番	ban	number, ordinals	
	号	gou	number, ordinals	
	組	kumi	group, class, set	
	倍	bai	times, multiples	twice: 二倍 nibai, five times: 五倍 gobai
	回	kai	times	
	度	do	times, degrees	
	分	bun	part, segment, degree, rate	
Time	年間	nenkan	years	
	ヶ月	kagetsu	months	
	週間	shuukan	weeks	
	日	nichi	days	
	時間	jikan	hours	
	分間	funkan	minutes	
	秒間	byoukan	seconds	

such as ごろ only for time. Depending on their use, other words can also be used to express approximate numbers, like ばかり if placed after a term for time or quantity, which usually means "only".

If two *kanji* representing numbers are used one after the other but separated by a comma, they have a similar meaning, e.g., 二、三時間 *ni, sanjikan* ("two or three hours") or 四、五百倍 *yon, gohyakubai* ("four or five times"). Only consecutive numbers may be used and only up to 七、八. Ordinal numbers are created by either adding the suffix 目 *me* to the Japanese number or to the combination of Chinese number and counter. Alternatively, the prefix 第 *dai* may be used in front of Chinese numbers, such as in following examples for flat objects:

the first:	一つ目 *hitotsume*	一枚目 *ichimaime*	第一 *daiichi*
the second:	二つ目 *futatsume*	二枚目 *nimaime*	第二 *daini*
the third:	三つ目 *mitsume*	三枚目 *sanmaime*	第三 *daisan*

The multiple of numbers may be formed by adding either adding 第 *dai* or じゅう, like for five times: 五第 *godai* and 五じゅう *gojuu*. There are other terms that are used before or after figures to modify time and quantity, like:

exactly:	丁度 *choudo*, e.g., exactly 100 g: 丁度百グラム *choudo hyaku guramu*
exactly:	だけ
every:	毎 *mai*, e.g., every hour: 毎時 *maiji*
every, at an interval of:	ごとに, e.g., every two hours: 二時間ごとに *nijikan goto ni*
each, at a time:	ずつ
in the order of:	程度 *teido*, e.g., in the order of 10 000: 万程度 *manteido*
more than:	から *kara*, e.g., more than 3 g: 三グラムから *sanguramukara*
more than:	以上 *ijou*, e.g., more than hours: 二時間以上 *nijikanijou*
less than:	以下 *ika*, e.g., less than 30 seconds: 三十秒以下 *sanjuubyouika*

2.3.2 Mathematical Symbols and Terminology

The four basic arithmetic operations and their results are:

to add:	たす	sum:	和 *wa*
to subtract:	ひく	difference:	差 *sa*
to multiply:	かける	product:	積 *seki*
to divide:	わる	quotient:	商 *shou*

Equations are formed with the terms for mathematical symbols given in Table 2.10.

In contrast to other languages, the denominator in fractions is read before the numerator, e.g., for $\frac{3}{4}$: 四分の三 *yon bun no san*. Examples for equations:

$5 + 4 = 9$	五プラス四イコール九 *go purasu yon iko-ru kyuu*
$10 - 8 = 2$	十ひく八は二 *juu hiku hachi wa ni*
$8 \div 2 = 4$	八分の二イコール四 *hachi bun no ni iko-ru yon*
$3 \neq 5$	三イコールならず五 *san iko-ru narazu go*
$10 > 9$	十大なり九 *juu dainari kyuu*
$\sqrt{4} = 4^{1/2}$	ルート四イコール四の分の一乗 *ru-to yon iko-ru yon no bun no ichi jou*
$\int_0^{90°} \cos^2(x)\,dx$ $= \frac{1}{2}\sin(x)\cdot\cos(x)$ $+ \frac{1}{2}x + c$	インテグラル零度から九十度までコサイン二乗 x dx イコール二分の一サイン x かけるコサイン x プラス二分の x プラス c *integuraru reidou kara kyuujuudou made kosain nijou xdx ikoru ni bun no ichi sain x kakeru kosain x purasu ni bun no x purasu c*

Table 2.10 Mathematical symbols.

Symbol	English	Japanese
+	plus	プラス
−	minus	マイナス
×	multiply	かける
÷	divided by	分の, ぶんの
x^n	power of	乗 *jou*
$\sqrt{\ }$	root	ルート, 乗紺 *joukon*
e	exponent	エクスポーネンシャル
log	logarithm	ログ
ln	natural logarithm	ログナチュラル
n!	factorial	nの階乗 *n no kaijou*
sin	sine	正弦 *seigen*, サイン
cos	cosine	余弦 *yogen*, コサイン
tan	tangent	正接 *seisetsu*, タンジェント
cot	cotangent	余接 *yosetsu*, コタンジェント
∫	integral	インテグラル
=	equal	イコール, は
≠, < >	unequal	イコールならず
>	greater than	大なり
<	less than	小なり
≥	greater than or equal	大なるイコール
≤	less than or equal	小なりイコール

For two and three, powers and roots may be expressed by the terms mentioned above, but also with special expressions, for example:

10^2 十の自乗 *juu no jijou* 十の平方 *juu no heihou*
10^3 十の三乗 *juu no sanjou* 十の立方 *juu no rippou*
$\sqrt[3]{x} = x^{1/2}$ xの自乗根 *x no jijoukon* xの平方根 *x no heihoukon*
$\sqrt[2]{x} = x^{1/3}$ xの三乗根 *x no sanjoukon* xの立方根 *x no rippoukon*

The two possibilities for each trigonometric functions are used differently: with the expression derived from English, the value is placed behind the operation, whereas with *kanji*, the value is connected with の, e.g., for sine 90: 九十の正弦 *kyuujuu no seigen* or サイン九十 *sain kyuujuu*.

2.3.3
Units

Most scientific units are named with transcribed English terms, derived mainly from the international accepted system of units (SI) (Table 2.11). For some units, there are original

Table 2.11 Modern units used in chemical publications.

Measure	Japanese	Abbr.	Unit
Chemical and physical units:			
amount of substance	物質量 busshitsuryou	mole	モル
length	距離 kyori	m	メートル
area	面積 menseki	m^2	平方メートル heihoume-toru
volume	体積 taiseki	m^3	立方メートル rippoume-toru
		L	リットル
weight	質量 shitsuryou	g	グラム
angle	角度 kakudo	°	度 do
		rad	ラジアン
time	時間 jikan	h	時 ji, アワー
		min	分 bun, ミニッツ
		s	秒 byou, セコンド
velocity	速度 sokudo	v	
temperature	温度 ondo	°C	度シー doshi-
		K	ケルビン
pressure	圧力 atsuryoku	Pa	パスカル
		torr	トル
frequency	周波数 shuuhasuu	Hz	ヘルツ
force	力 chikara	N	ニュートン
energy	エネルギー	J	ジュール
kinematic viscosity	動粘性率 dounenseiritsu	St	ストークス
dipole moment	双極子モーメント soukyokushimo-mento	D	デバイ
viscosity	粘性率 nenseiritsu	P	ポアズ
radioactivity	放射能 houshanou	Ci	キュリー
exposure to ionizing radiation	X線の照射線量 xsennoshoushasenryou	R	レントゲン
Electronic and magnetic units:			
voltage, (electric potential)	電圧 denatsu (電位 deni)	V	ボルト
electric current	電流 denryuu	A	アンペア
impedance, resistance	インピーダンス, 抵抗 teikou	Ω	オーム
electric charge	電荷 denka	C	クーロン
electric power	仕事率 shigotoritsu, 電力 denryoku	W	ワット
magnetic field density	磁束密度 jisokumitsudo	G	ガウス
capacitance	電気容量 denkiyouryou	F	ファラッド

Table 2.11 *(continued)*

Measure	Japanese	Abbr.	Unit
magnetic flux	磁束 *jisoku*	Wb	ウエーバー, ウェーバー
magnetic field	磁場 *jiba*	Oe	エルステッド
inductance	透磁率 *toujiritsu*	H	ヘンリー
conductance	コンダクタンス	S	ジーメンス

Table 2.12 Existing *kanji* for units.

Measure	Unit	Japanese	
Linear measures	millimeter	粍	*mirime-toru, miri*
	centimeter	糎	*senchime-toru, senchi*
	decimeter	粉	*deshime-toru (fun)*
	meter	米	*me-toru*
	decameter	籵	*dekame-toru*
	hectometer	粨	*hekutome-toru*
	kilometer	粁	*kirome-toru*
Volume measures	milliliter	竓	*mirirrittoru*
	centiliter	竰	*senchirittoru*
	deciliter	竕	*deshirittoru*
	liter	立	*rittoru*
	decaliter	竍	*dekarittoru*
	hectoliter	竡	*hekutorittoru*
	kiloliter	竏	*irorittoru*
Weight measures	milligram	瓱	*miriguramu*
	centigram	甅	*senchiguramu*
	decigram	瓰	*deshiguramu*
	gram	瓦	*guramu*
	decagram	瓧	*dekaguramu*
	hectogram	瓸	*hekutoguramu*
	kilogram	瓩	*kiroguramu*

Japanese words, e.g., for time (e.g., hour, minute). For some units, also other less common readings may be found.

Powers in units are written and read with the following *kana* and used together with the unit as in English, e.g., dl: デシリットル, cm: センチメートル, ms: ミリセコンド, μg: マイクログラム, km: キロメートル, MV: メガボルト, GHz: ギガヘルツ.

deci	10^{-1}	デシ	kilo	10^3	キロ
centi	10^{-2}	センチ	mega	10^6	メガ
milli	10^{-3}	ミリ	giga	10^9	ギガ
μ	10^{-6}	マイクロ	tera	10^{12}	テラ
nano	10^{-9}	ナノ			
pico	10^{-12}	ピコ			

For linear, weight and volume measures, *kanji* also exist that are pronounced as the transcribed English expression. They are rarely used. Besides a main radical indicating the kind of measure, a second element represents the powers. For deca, hecto and kilo, the radical correlates to the numbers ten, hundred and thousand (Table 2.12).

Although the metric system is official in Japan, in older publication the old units may still be found (Table 2.13).

Table 2.13 Traditional Japanese units.

Measure		Japanese unit		English equivalent
Linear measures	一厘	1 *rin*		0.303 millimeter
	一分	1 *bu*	= 10 *rin*	3.03 millimeters
	一寸	1 *sun*	= 10 *bu*	3.03 centimeters
	一尺	1 *shaku*	= 10 *sun*	30.3 centimeters
	一丈	1 *jou*	= 10 *shaku*	3.03 meters
	一間	1 *ken* or		
	一尋	1 *hiro*	= 6 *shaku*	1.82 meters
	一丁	1 *chou*	= 60 *ken*	109 meters
	一里	1 *ri*	= 36 *cho*	3.93 kilometers
Volume measures	一勺	1 *shaku*		0.018 liter
	一合	1 *gou*	10 *shaku*	0.18 liter
	一升	1 *shou*	= 10 *gou*	1.8 liter
	一斗	1 *to*	= 10 *shou*	18 liters
	一石	1 *koku*	= 10 *to*	180 liters
Weight measures	一匁	1 *monme*		3.75 grams
	一百目	1 *hyakume*	= 100 *monme*	375 grams
	一斤	1 *kin*	= 160 *monme*	0.6 kilograms
	一貫	1 *kann* or		
		1 *kamme*	= 1000 *monme*	3.75 kilograms
	一担	1 *bikoru*	= 100 *kin*	60 kilograms

2.3.4
Other Symbols and Characters in Scientific and Technical Literature

2.3.4.1 Greek Letters
Greek letters are represented by *katakana* given in Table 2.14.

Table 2.14 Transcription of Greek letters (older or less common terms in bracket).

Greek	Japanese
α, A	アルファ
β, B	ベータ（ヴィタ）
γ, Γ	ガンマ（ガマ）
δ, Δ	デルタ
ε, E	イプシロン（エプシロン）
ζ, Z	ゼータ、シェータ（ジタ）
η, H	イータ（イタ）
ϑ, Θ	シータ、テータ（シタ）
ι, I	イオタ（ヨタ）
κ, K	カッパ（カパ）
λ, Λ	ラムダ
μ, M	ミュー（ミ）
ν, N	ニュー（ニ）
ξ, Ξ	クサイ、グザイ、クシー（クシ）
o, O	オミクロン
π, Π	パイ（ピ）
ρ, P	ロー（ロ）
σ, Σ	シグマ
τ, T	タウ
υ, Y	ウプシロン（イプシロン）
φ, Φ	ファイ（フィ）
χ, X	カイ（ヒ、キー）
ψ, Ψ	プサイ（プシー、プシ）
ω, Ω	オメガ

2.3.4.2 Colors
Colors are frequently used in chemical names as well as in the description of scientific phenomenon. They can be either expressed with *kanji* representing the original Japanese word or by transcription of English expressions (Table 2.15).

Less important are grey: 灰色 *haiiro* (also グレー or グレイ), brown: 茶色 *chairo* (also 褐色 *kasshoku* or ブラウン), orange: オレンジ色 *orenjiiro*, gold: 金色 *kiniro*, silver: 銀色 *giniro* and pink: 桃色 *momoiro*.

Table 2.15 Expressions of colors and similar states.

Color	by *kanji*	by *katakana*	Examples
transparent	透明 *toumei*		opacity 不透明 *futoumei*
white	白い *shiroi*	ホワイト	white lead ore 白鉛鉱 *hakuenkou*; protein 蛋白 *tanpaku*; white gold ホワイトゴールド
black	黒い *kuroi*	ブラック	carbon black 炭素黒 *tansoguro*; graphite 黒鉛 *kokuen*; diamine black ダイアミンブラック
red	赤い *akai*	レッド	red iron ore, hematite 赤鉄鉱 *sekitekkou*; infrared 赤外 *sekigai*; redshift レッドシフト
yellow	黄色 *kiiro*	エロー	brass 黄銅 *oudou*; naphthol yellow ナフトールエロー
green	緑色 *midoriiro*	グリーン	smaragd, emerald 緑玉石 *ryokugyokuseki*; chlorophyll 葉緑素 *youryokuso*; malachite green マラカイトグリーン
blue	青い *aoi*	ブルー, ブリュー, ブラウ	hydrogen cyanide (German: Blausäure) 青酸 *seisan*; lapis lazuli, azure stone 編青石 *henseiseki*; cobalt blue コバルトブルー
purple, violet	紫色 *murasakiiro*	ヴァイオレッ, バイオレッテオ, パープル	ultraviolet radiation 紫外線 *shigaisen*; crystal violet クリスタルヴァイオレット

2.4
Suggestions for Reading Japanese Scientific and Technical Publications

2.4.1
Method for Sentence Analysis

The only commonly used punctuation mark is "。" (丸 *maru*), which is positioned at the end of a sentence. Therefore, the analysis of scientific publications has to be focused, and achieved, on the sentence level. The following steps are proposed for the analysis of individual sentences:

Step 1: Separation of sentences in subordinate clauses.
Step 2: Identification of elements within the subordinate clauses.
Step 3: Division of phrases within the elements.
Step 4: Looking up of unknown *kanji* and translation of single words.

Step 5: Determination of the meaning of phrases and sentence elements.
Step 6: Connection of sentence elements and translation of subordinate clauses.
Step 7: Connection of main, subordinate and attribute clauses and translation of the complete sentence.

These steps may be used for all languages and seem not to be characteristic for the analysis of Japanese texts, but the procedure within the individual steps strongly depends on the features of the Japanese language.

2.4.1.1 Step 1: Separation of Sentences in Subordinate Clauses

One of the most difficult steps in text analysis is the identification of subordinate clauses within a given sentence. There are various possibilities of nesting independent and dependent subordinate clauses. For understanding scientific texts, distinction can be made between:
- sentence connection with equal subordinate clauses,
- main and dependent subordinate clauses,
- embedded sentences in which one element is represented by a subordinate claus,
- embedded sentences with main clauses in which elements are modified with attribute clauses.

The easiest way to separate a complex sentence into subordinate clauses would be with a comma. As the use of the Japanese comma strongly depends on the writing style of the author, it is not necessarily present. More helpful are conjunctions that connect sentences in various ways. Conjunctions are nouns, particles or specific verb endings. They are usually positioned at the end of the subordinate clause that is in front of the main clause and form the first two types of above classified nesting sentences. Depending on the conjunction, the subordinate clause can only be applied together with a following main clause (for most of the conjunctions) or may also be used independently (e.g., with から in the case of an answer to a previous given question). Most important noun and particle conjunctions are listed in Section 2.4.3. There are some additional conjunctions that connect sentences in the same sense, but the individual clauses usually remain independent and are separated by a dot. These conjunctions are taken into account in Section 2.4.2. Conjunctions that do not connect sentences but words (e.g., と, や, とか, -たり) are not useful in step 1 and not listed in the table.

For embedded sentences in which one element is represented by a subordinate clause, topic, subject and object subordinate clauses are possible. They are present if a predicate is nominalized with の or こと (Section 2.4.4) and followed by the necessary particle, e.g., the subject marker が in case of subject subordinate clauses or the topic marker は in case of topic subordinate clauses. They can be identified because only a few verb forms are nominalizable.

The second type of embedded sentences contains attribute clauses. They are placed in front of a noun and can be characterized because of their sentence-final predicate form. A simple attribute clause may exist only of an i-type adjective in front of the noun. More complex attribute sentences may contain several objects that are also modified with attribute clauses. They are described in Section 2.4.4.

2.4.1.2 Steps 2 and 3: Identification of Elements and Phrases within the Subordinate Clauses

Each sentence, plain clauses as well as subordinate clauses, consists of different elements. These can be subject, objects and predicate; or topic and comments. Each element can be either a single phrase or consist of a few phrases. A phrase is considered as the combination of a single word (e.g., a noun, a verb or an adjective) with auxiliary verbs, verb- and adjective endings, suffixes and particles that modify its meaning and define its grammatical role. If an element consist of two phrases, each phrase has its own particle, and at the end of the element an additional particle is used.

Elements form significant coherence and are, therefore, are worth separating. The most important separation is between topic and comment. The topic can be easily identified because it is marked with は. The topic of a sentence is what the sentence is about and can be translated as "Concerning ...", "Talking about ...", "Referring to ..." or "As for ...". In general, any noun phrase can be topicalized. Besides the topic marker は, the topic can, therefore, consist of one as well as of many additional words, e.g.:

硫酸は ... *ryuusan wa* ...
= Concerning sulfuric acid ... (noun + topic marker)

この分子は ... *kono bunshi wa* ...
= Referring to this molecule ... (demonstrative *kono* + noun + topic marker)

古い分光写真器は ... = *furui bunkoushashinki wa* ...
= As for the old spectrograph ... (adjective + noun + topic marker)

発泡剤を使うのは ... *happouzai o tsukau no wa* ...
= Speaking about using the foaming agent ... (noun + object marking particle + verb + nominalizing particle + topic marker)

Once a topic is known, it does not need to be repeated unless another topic is introduced. The remaining elements of the sentence form its comment.

A second method of separation refers to the word order. As described before, Japanese is a subject–object–verb language. However, if a topic is defined, it is placed at the sentence-initial position. The subject and any object can be the topic of the sentence marked with は. The topic is followed by the subject, which is marked with が, and the objects. Concerning the objects, various particles are possible for the determination of their function but, at this point, particles are only used to identify individual objects and words for further translation. The sentence structure can be summarized as in Fig. 2.1.

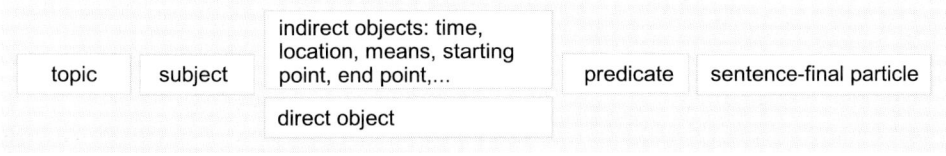

Fig. 2.1 Basic structure of Japanese sentences.

2.4.1.3 Step 4: Looking up of Unknown *kanji* and Translation of Single Words

Elements consist of phrases and phrases of single words marked with particles. Hence, after classifying phrases, the next step is the translation of words. Words written in *hiragana* and *katakana* can be read and looked up in a dictionary. Notably, words in Japanese dictionaries are not ordered by the Latin alphabet but by Japanese syllables according to the order in the 50-sound-matrix. The syllable order in Japanese dictionaries is explained in Section 2.6.1. The main problem is, most probably, unknown *kanji*. Methods for identification of the Chinese characters are described in Section 2.6.2. The meaning of *kanji* combination is often but not always derived from the meaning of individual *kanji*. For proper translation and understanding of scientific literature it is, therefore, reasonable to use a specific technical dictionary.

Apart from the interpretation of *kanji*, the only word types that cause an additional special attention are conjugable words. i.e., verbs and i-type adjectives. They consist not only of a

Table 2.16 Examples for modal verbs.

Function	Modal verb	Verb base to be attached at
Constraint	-なければならない, -なくてはいけない	未然形 *mizenkei*
"The point is reached, that"	-なければならなくなる, -なければいけなくなる	
Permission	-てもいい	連用形 *renyoukei*
Prohibition	-てはならない, -てはいけない	
Degree in terms of "too much"	-すぎる	連用形 *renyoukei*
Begin of an activity	-始める -*hajimeru*, -だす	
End of an activity	-終わる -*owaru*	
Easiness	-安い -*yasui*, -にくい- -ずらい	
Examples for modal verbs with limited use:		
To carry out an action upwards	-上げる -*ageru*	連用形 *renyoukei*
To do something together	-会う -*au*	
To carry out an action again	-返す -*kaesu*	
To be going to	掛ける -*kakeru*	
To carry out an action until the end	切る -*kiru*	
To carry out an action exhaustively	込む *komu*	
To carry out an action towards a goal	付ける *tsukeru*	

stem representing their basic meaning but also of endings that modify significantly their sense. Because of the agglutinative character of the language, some endings may be again conjugated and multiple verbal endings are formed. If a phrase is analyzed, the complete conjugated verb or adjective has to be taken into account, including all verb and adjective endings. Important basic endings for verbs and adjectives are summarized in Section 1.1.4. Beyond the listed endings, many verbs can be modified by adding other, specific verbs or adjectives. Dependent verbs and adjectives that are attached are called modal verbs. Some of them may be used with all other verbs, like 始める *hajimeru* or 安い *yasui*, while others are only used in specific combinations. The meaning of the finally conjugated verbs may not always be easily recognizable regarding the individual parts. Table 2.16 contains some further endings and combinations.

2.4.1.4 Steps 5, 6 and 7: Translation of the Sentence

Knowing Chinese characters and translation of words, the function of singles phrases and sentence elements, which were separated in steps 2 and 3, can now be determined using the marking particles. Topic and subject marker have already been discussed at step 3 (Section 2.4.1.2). Now, the direct and indirect objects have to be characterized and the meaning within the subordinate clauses determined. As for the objects, major elements within a sentence can be direct object (marked with を), absolute date (marked with に), means (marked with で), location (marked with に or で), direction (marked with へ or に), starting point (marked with から) and end point (marked with まで). A complete overview of all important particles for the identification of the function of individual sentence elements is given in Section 1.1.3.

With the translation of sentence elements, the understanding of subordinate clauses can be conducted. The sense of subordinate clauses can be connected by using the relationships identified in step 1 and conjunctions that are listed in Section 2.4.3. Special focus has to be made on attributes and attribute sentences, because embedded attributes consisting of attribute clauses within attribute clauses and the right correlation of sentence elements like subjects may alter the overall meaning of the sentence (Section 2.4.4).

2.4.2
Relationship between Sentences

If individual sentences are translated, they have to be considered within their direct environment. In step 1, conjunctions within sentences were used to identify subordinate clauses. In addition, there are conjunctions at the beginning of clauses that have to be considered while determining the sense of sentences. Table 2.17 summarizes important conjunctions that are in the initial position of a sentence and create a relationship to the previous sentence. Because the Japanese language tries to avoid any irrelevant information within a sentence, the meaning of a clause can, sometimes, only be understood if the previous text is taken into account. For instance, if the topic is mentioned and clear, it is not repeated in subsequent sentences and there will be no topic unless another one is presented. There is a general strong tendency in Japanese to minimize effort in presenting information. Generally, elements that can be understood from the context are omitted. However, sentences have to be grammatically correct and, therefore, the predicate can never be deleted.

Table 2.17 Conjunctions connecting independent sentences.

Function	Conjunction	Examples for English
consecutive actions and states	そして formal: それでは, では informal: それじゃ, じゃ, じゃあ それ なら	and; and then and; then and; then; in that case
alternative actions and states	それとも	or; either ... or ...
contrastive sentence connectors	でも, しかし, だが, だけど	but
cause and result	それで[a] だから[a]	therefore; because of that; that's why therefore; because of that; that's why

[a] Initial position in the sentence, which describes the effect; the reason is described in the previous sentence.

2.4.3
Identification of Subordinate Clauses by Conjunctions

Table 2.18 gives important conjunctions, along with their function for the identification of subordinate clauses, as described for step 1 in Section 2.4.1.1.

Table 2.18 Conjunctions within sentences.

Function	Conjunction	Examples for English	Formation	Remarks
consecutive actions and states	し	and; not only ... but also ...; in addition	V-u shi V-ta shi VA-i shi VA-katta shi N/NA da shi N/NA datta shi	• at the end of the first subordinate clause • can be repeated more than once in the sentence
	それから	and; and then; after that	V-te sore kara V-連用形 (V-renyoukei) sore kara VA-ku sore kara VA-kute sore kara NA de sore kara	• beginning of second subordinate clause • separation of clauses by dot possible

Table 2.18 (continued)

Function	Conjunction	Examples for English	Formation	Remarks
alternative actions and states	か	or; either … or	V-*u ka* V-*ta ka* VA-*i ka* VA-*katta ka* NA/N *ka* NA/N *datta ka*	
contrastive sentence connectors	が	but	after all verb and adjective forms	• end of first subordinate clause
	かわりに	but; instead of; although	V-*u kawari ni* V-*ta kawari ni* VA-*i kawari ni* VA-*katta kawari ni* NA *na kawari ni* NA *datta kawari ni* N *no kawari ni*	
	のに	although; but; even though; in contrast to	V-*u noni* V-*ta noni* VA-*i noni* VA-*katta noni* NA/N *na noni* NA/N *datta noni*	• end of first subordinate clause
	けれども, けれど, けど	although	V-*u keredomo* V-*ta keredomo* VA-*i keredomo* VA-*katta keredomo* NA/N *da keredomo* NA/N *datta keredomo*	• end of first subordinate clause
	ても	although; even if	V-*te mo* VA-*kute mo* NA/N *de mo*	• end of first subordinate clause
temporal connection of sentences	前に *mae ni*	before	V-*u mae ni*	• end of first subordinate clause • subordinate clause (expressing later action) positioned before main clause (describing earlier action) • also only with nouns possible, but no sentence connection (before, in front of)

Table 2.18 *(continued)*

Function	Conjunction	Examples for English	Formation	Remarks
	後で *ato de*	after	V-*ta ato de*	• end of first subordinate clause • subordinate clause (describing earlier action) positioned before main clause (expressing later action) • also only with nouns possible, but no sentence connection
	から	after	V-*te kara*	• end of first subordinate clause • subordinate clause (describing earlier action) positioned before main clause (expressing later action)
	間 (に) *aida (ni)*	during; while	V-*u aida (ni)* V-*te iru aida (ni)* VA-*i aida (ni)* NA *na aida (ni)*	• end of first subordinate clause • also only with nouns possible, but no sentence connection • non-temporal use of *aida* ("between")
	うちに	during; while	V-*u uchi ni* V-*te iru uchi ni* VA-*i uchi ni* NA *na uchi ni*	• end of first subordinate clause • also only with nouns possible, but no sentence connection
	時, とき *toki*	when; during; right after	V-*ta toki* VA-*katta toki* NA *datta toki*	• noun • end of first subordinate clause • also only with nouns possible, but no sentence connection • predicate before *to* must be past; if non-past, than conditional sentence connection
	と	when	V-*ta to* VA-*katta to* NA/N *datta to*	• end of first subordinate clause • predicate before *to* must be past; if non-past, then conditional sentence connection
cause and result	から	because; since	V-*u kara* V-*ta kara* VA-*i kara* VA-*katta kara* NA/N *da kara* NA/N *datta kara*	• end of first subordinate clause • first: subordinate clause describing reason, second: main clause expressing result

Table 2.18 *(continued)*

Function	Conjunction	Examples for English	Formation	Remarks
	ので	because; since	V-*u node* V-*ta node* VA-*i node* VA-*katta node* NA/N *na node* NA/N *datta node*	• end of first subordinate clause • first: subordinate clause describing reason, second: main clause expressing result
	ために	because of; in order to	V-*u tame ni* V-*ta tame ni* VA-*i tame ni* VA-*katta tame ni* NA *na tame ni* N *no tame ni* NA/N *datta tame ni*	• end of first subordinate clause • first: subordinate clause describing purpose, second: main clause expressing result
conditional	なら	if; if it is the case that	V-*u no nara* V-*ta no nara* VA-*i no nara* VA-*katta no nara* NA/N *nara* NA/N *datta kara*	• end of first subordinate clause • *nara* is the simplified conditional form of the copula • first: subordinate clause describing condition, second: main clause expressing result
	と	if	V-*u to* VA-*i to* NA/N *da to*	• end of first subordinate clause • first: subordinate clause describing condition, second: main clause expressing result • predicate before *to* must be non-past; if past, than temporal sentence connection
	時, とき *toki*	if; when; always when; at the time when; right before	V-*u toki* VA-*i toki* NA *na toki*	• end of first subordinate clause • also only with nouns possible, but no sentence connection • predicate before *toki* must be non-past; if past, than temporal sentence connection
	は	if	V-*te wa* VA-*kute wa* NA/N *de wa*	• end of first subordinate clause • used in fixed expressions, e.g., prohibition (-*te wa ikenai*)

2.4.4
Identification of Subordinate and Attribute Clauses by Verb Endings

Besides conjunctions, subordinate clauses may be connected using specific verb endings. The most important one is the gerundive-form (also called te-form). All verbs, i-type adjectives and the copula possess te-forms characterized by the final syllables て or で (Section 1.1.4). The te-form fulfils several functions. Its basic characteristic is the expression of a temporal sequence. This means that, after the action or state that is described by the te-form, further actions or states follow. Hence, most of the uses can roughly be translated with "... and then ...", but to reflect the correct intention of the sentence, other translations are usually preferred. For instance, the commonly used combination with the verb いる is understood as "be doing ...". The te-form can be found in short combinations, like some fixed expressions

Table 2.19 Conjunctions within sentences.

Function	Form	English	Base for formation	Remarks
two actions occur sequentially	て	and	連用形 renyoukei	
two states exist parallel	て	and	連用形 renyoukei	
cause and result	て	because	連用形 renyoukei	the first subordinate clause describes the reason or the cause of the activity or state expressed in the second clause
means of an activity	て	by	連用形 renyoukei	the first subordinate clause describes the means by which the activity expressed in the second clause is carried out
manner of an activity	て	by	連用形 renyoukei	the first subordinate clause describes the manner in which the activity expressed in the second clause is carried out
conditional	ば	if	仮定形 kateikei	the first subordinate clause describes the condition
conditional or temporal connection of sentences	たら	if; when; after	連用形 renyoukei	the first subordinate clause describes the condition or an action that is carried out before the action expressed in the main clause
temporal connection of sentences	ながら	while; during	連用形 renyoukei	the first subordinate clause describes an action that is carried out concurrently or simultaneously with the action expressed in the main clause
cause and result	に	in order to	連用形 renyoukei	the first subordinate clause describes the reason or the cause of the activity or state expressed in the second clause

(e.g., "must not do": -てはいけない) as well as for the connecting of subordinate sentences. In linking sentences, the last element of the predicate of a clause is the te-form followed by elements and predicate of the second subordinate sentence.

Besides the te-form, there are some other verb endings used in this manner, e.g., the conditional endings ば and たら. Table 2.19 gives important verb endings that are used in connecting subordinate sentences. The second clause is always the main clause.

A different type of verb and adjective endings is used to identify subordinate clauses that are used either to replace a sentence element or to modify nouns. The common property of these type of constructions is their noun-type character, i.e., the subordinate clause ends with a sentence-final predicate form followed by a noun, either the noun that is intended to modify or a particle that turns the clause into a noun-type expression. This later process is called nominalization. It is used, in general, if a verbal or adjectival term is intended to be used as a noun-type sentence element (e.g., the topic requires a noun-type expression to be marked with は). Once a sentence has become a noun phrase, it can be used anywhere where a regular noun can be used, e.g., as subject or direct object. The two nominalizing particles are の and こと. The choice depends on the type of verb or adjective and the context. They are placed immediately after verb, i-type adjective or the copula. The four plain affirmative/negative and present/past forms are possible in front of の or こと, as, for example, for the verb する, the adjective 難しい *muzukashii* and the copula in combination with の: するの, しないの, したの, しなかったの, 難しいの, 難しくないの, 難しかったの, 難しくなかったの, ではないの, だったの and ではなかったの. There are also some idiomatic phrases with *koto*, e.g., the expression of capabilities (こと が できる), decision (こと に する), result (こと に なる), occasional events and experiences (こと が ある).

The same verb and adjective endings are necessary if nouns are modified with attribute clauses. Attribute clauses have the conventional word order and may contain almost all sentence elements as ordinary clauses but do not contain a topic. The sentence element marked with は is, therefore, always part of the main clause. A subject in an attribute sentence has to be marked with either が or の. If there is only one element marked with が within the whole sentence, it might either be the subject of the main clause or the subject of the attribute sentence and the correct translation has to be derived from the context.

If more than one attribute is present within a sentence, they can modify the same or different nouns. Two attributes modify the same noun, if there is no noun following the first sentence-final predicate form (Fig. 2.2).

An example is the expression 長鎖デ不飽和の脂肪酸 *chousa de fuhouwa no shibousan* (long-chain unsaturated fatty acid), in which the noun 脂肪酸 *shibousan* (fatty acid) is modified by the two noun-type attributes 長鎖 *chousa* (long chain) and 不飽和 *fuhouwa* (unsaturation). Similar expressions exists with two verb-type attributes, two adjective attributes or mixed attributes containing different word types. For instance, the expression 新しくて

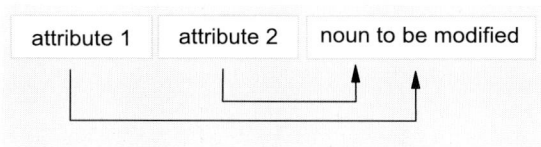

Fig. 2.2 Two attributes modify the same noun.

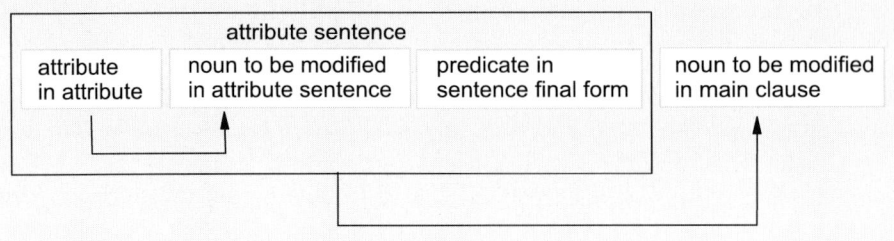

Fig. 2.3 Two attributes modify different nouns.

便利な分析法 *atarashikute benri na bunsekihou* (new and convenient analytical method) contains i- and na-type adjectives. The expression 環状の粘稠な高分子 *kanjou no nenchou na koubunshi* (cyclic viscous macromolecule) consists of a noun-type attribute as well as a na-adjective to modify the noun 高分子. In principle, multiple attribute modifications are also possible, such as 使った立体特異的で定量的な酸化 *tsukatta rittaitokuiteki de teiryouteki na sanka* (the applied quantitative stereospecific oxidation).

In addition to this, various attributes and attribute clauses can be present in a sentence, modifying different nouns. They either modify different nouns in the main clause, or nouns within an already existing attribute sentence (Fig. 2.3).

Nouns that are modified by an attribute or attribute sentence are used within sentences in their usual function. The following sentence gives examples of attribute clauses that modify the direct objects of the main clause:

> 私たちは自分たちの新しい反応を使って、京都大学の研究者が
> 最初に天然物から見つけた化合物を合成した。
> *watashitachi wa jibuntachi no atarashii hannou o tsukatte, kyouto daigaku no
>   kenkyuusha ga saisho ni tennenbutsu kara mitsuketa kagoubutsu o gousei shita.*
> We synthesized, by our new reaction, the molecule, that researchers from
>   Kyoto University first found in a natural product.

The main sentence is 私たちは化合物を合成した。*watashitachi wa kagoubutsu o gousei shita* (We synthesized a molecule.). The direct object 化合物 *kagoubutsu* is modified by two attribute sentences. The first attribute sentence is 自分たちの新しい反応を使って *jibuntachi no atarashii hannou o tsukatte* and can be translated with "… using our new reaction …". It contains the noun 反応 *hannou*, which follows two of its attributes: the noun-type 自分たちの *jibuntachi no* (our own) and the i-type adjective 新しい *atarashii* (new). The second attribute sentence is 京都大学の研究者が最初に天然物から見つけた *kyouto daigaku no kenkyuusha ga saisho ni tennenbutsu kara mitsuketa*. It contains the subject 京都大学の研究者 *kyouto daigaku no kenkyuusha* (researcher from Kyoto University), which is followed by the subject marker が. The subject consist of two phrases, the noun 研究者 *kenkyuusha* (researcher) and a noun-type attribute 京都大学の *kyouto daigaku no*. Both sentences can be recognized as attributes because of their verb endings, namely te-form in 使って and simple past form in 見つけた. Figure 2.4 illustrates the overall sentence structure. For further examples see the translations in Section 2.5.

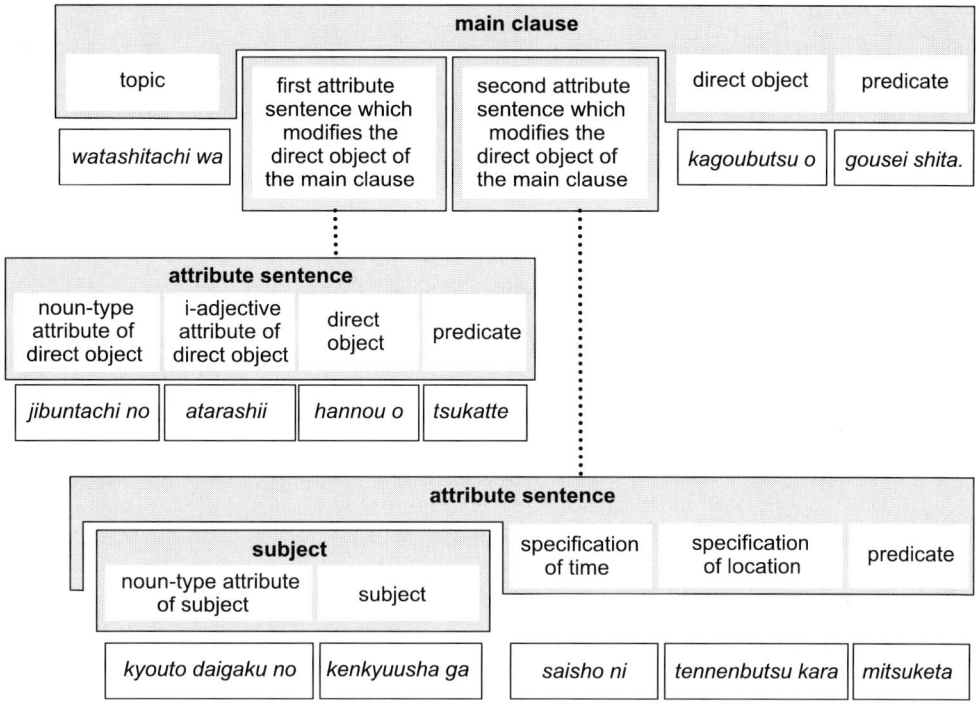

Fig. 2.4 Sentence structure of an example sentence containing two attribute sentences.

2.5
Example Translations

2.5.1
Example Translations of Patent Claims

As examples for the translation of patent claims, some claims have been chosen from the two Japanese patent applications JP 2001-342183 [title: 4-(1-fluoroethyl)thiazole-5-carboxylic acid amide derivatives and agricultural and horticultural pest control agents] and JP 2001-220439 [title: A continuous method for the production of poly(alkylene ether) glycol diester].

2.5.1.1 JP 2001-342183
【特許請求の範囲】
【請求項1】 次式（1）：
【化1】

(1)

で示される４－（１－フルオロエチル）チアゾール－５－カルボン酸アミド誘導体。なら、式中のR、R^{1}及びAは、次の通りである。Rは、置換又は非置換のフェニル基、置換又は非置換のフェノキシ、置換又は非置換のチエニル基を表す。R^{1}は、炭素原子数１～４個のアルキル基を表す。Aは、直接結合、直鎖及び分岐状の炭素原子数１～６個のアルキレン基を表す。

【請求項２】 次式（２）：
【化２】

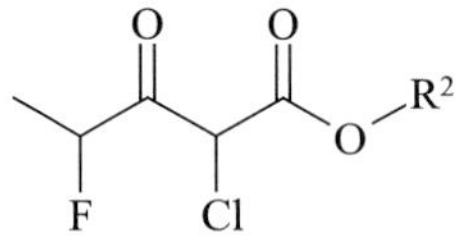

$$\text{(structure of 2-chloro-4-fluoropentanoic acid ester)} \qquad (2)$$

で示される２－クロロ－４－フルオロペンタン酸エステル化合物。なら、式中のR^{2}は、炭素原子数１～４個のアルキル基を表す。

【請求項３】 次式（３）：
【化３】

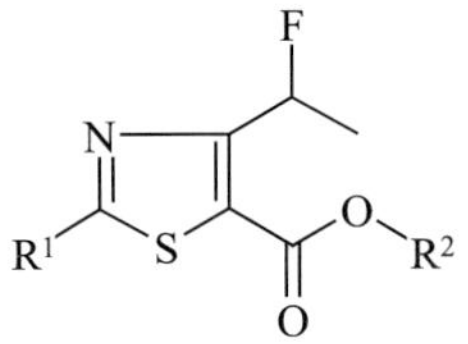

$$\text{(structure of thiazole-5-carboxylate)} \qquad (3)$$

で示される４－（１－フルオロエチル）チアゾール－５－カルボン酸エステル化合物。なら、式中のR^{1}は、請求項１の記載と同義であり；R^{2}は、前記の記載と同義である。

【請求項５】 請求項１に記載の式（１）で示される４－（１－フルオロエチル）チアゾール－５－カルボン酸アミド誘導体を有効成分とする農園芸用の有害生物防除剤。

transcription into *ro-maji*:
[tokkyoseikyuu no hani]
[seikyuukou 1] jishiki (1):
[ka 1]
de shimesareru 4-(1-furuoroechiru) chiazo-ru-5-karubonsanamidoyuudoutai (誘導体 derivate). *nara, shikichuu* (式中 in the formula) *no R, R^1 oyobi A wa, tsugi no touri de aru. R wa, chikanmata* (置換 substitution) *wa hichikan* (非置換 unsubstituted) *no feniruki* (フェニル基 phenyl group), *chikanmata wa hichikan no fenokishi* (フェノキシ phenoxy), *chikanmata wa hichikan no chieniruki* (チエニル基 thienyl group) *o arawasu. R^1 wa, tansogenshisuu* (炭素原子数 number of carbon atoms) *1~4 ko* (個 counter for objects) *no arukiruki o arawasu. A wa, chokusetsuketsugou* (直接 direct; 結合 bond), *chokusaoyobi* (直鎖及び straight chain) *bunkijou* (分岐 divergence) *no tansogenshisuu 1~6 ko no arukirenki o arawasu.*

[seikyuukou 2] jishiki (2):
[ka 2]
de shimesareru 2-kuroro-4-furuoropentansanesuterukagoubutsu. nara, shikichuu no R^2 wa, tanso-genshisuu 1~4 ko no arukiruki o arawasu.

[seikyuukou 3] jishiki (3):
[ka 3]
de shimesareru 4-(1-furuoroechiru)chiazo-ru-5-karubonsanesuterukagoubutsu. nara, shikichuu no R1 wa, seikyuukou 1 no kisai (記載 report) *to dougi* (同義 the same meaning) *de ari; R2 wa, zenki* (前記 the above mentioned) *no kisai to dougi de aru.*

[seikyuukou 5] *seikyuukou 1 ni kisai no shiki (1) de shimesareru 4-(1-furuoroechiru)chiazo-ru-5-karubonsanamidoyuudoutai oyuukouseibun* (有効成分 active ingredient) *to suru nouengeiyou* (farm, plantation; arts, skills; to use) *no yuugaiseibutsujozai* (有害 harmful; 生物 biological; 防 protect; 除 remove; 剤 agent).

translation:
[scope of the patent claims]
[claim 1]
chemical formula 1
4-(1-fluoroethyl)thiazole-5-carboxylic acid amide derivatives as shown by formula (1). R, R^1 and A in the formula are as follows. R represents either a substituted or unsubstituted phenyl group, a substituted or unsubstituted phenoxy or a substituted or unsubstituted thienyl group. R^1 expresses an alkyl group with 1 to 4 carbon atoms. A represents a direct bond, a straight chain or a branched alkylene group with 1 to 6 carbon atoms.

[claim 2]
chemical formula 2
2-chloro-4-fluoropentanoic acid ester compounds as shown by formula (2). R^2 in the formula represents an alkyl group with 1 to 4 carbon atoms.

[claim 3]
chemical formula 3
4-(1-fluoroethyl)thiazole-5-carboxylic acid ester compounds as shown by formula (3). R, R^1 and A in the formula are as follows. Wherein, R^1 has the same meaning as described in claim 1 and R^2 has the same meaning as mentioned above.

[claim 5]
Agricultural and horticultural pest control agents comprising 4-(1-fluoroethyl)thiazole-5-carboxylic acid amide derivatives according to the formula (1) shown in claim 1 as active ingredients.

2.5.1.2 JP 2001-220439
【請求項1】撹拌装置を備えた反応器において、環状エーテル及びカルボン酸無水物を０．０１～３ｍｍの粒子径を有する固体酸触媒の存在下で連続的に反応させ、ポリアルキレンエーテルグリコールジエステルを連続的に製造する際に、反応器内に設置した円筒型フィルターを通した反応液を抜き出すことを特徴とするポリアルキレンエーテルグリコールジエステロの連続製造法。

【請求項2】円筒型フィルター平織りの金属メッシュで且つ目開きが少なくとも０．０１mmである請求項１に記載の連続製造法。

【請求項3】円筒型フィルターの濾過速度ガ０．１〜３ｍ／ｈｒである請求項１又は２に記載の連続製造法。

transcription into *ro-maji*:
[seikyuukou 1] kakuhansouchi (撹拌装置 stirring equipment) *o sonaeta* (備える to possess) *hannouki* (反応器 reactor) *ni oite, kanjou* (環状 cyclic) *e-teru oyobi karubonsanmusuibutsu* (カルボン酸 carboxylic acid; 無水物 anhydride) *o 0.01~3 mm no ryuushikei* (粒子径 particle size) *o yuu suru kotaisanshokubai* (体酸 solid; 触媒 catalyst) *no sonzaika de renzokuteki* (連続的 continuous) *ni hannou* (反応 reaction) *sase, poriarukirene-teruguriko-rujiesuteru* (ポリアルキレン polyalkylene; エーテル ether; グリコール glycol; ジエステル diester) *o renzoku-teki ni seizou suru* (製造する to produce) *sai ni, hannoukinai* (反応器内 in the reactor) *ni secchi shita* (設置する to establish, to equip) *entoukata firuta-* (円筒 cylinder; 型 shape; フィルター filter) *o tsuu shita hannoueki* (反応液 reaction solution) *o nukidasu* (抜き出す to extract, to leave out, to remove) *koto o tokuchou* (特徴 distinctive feature) *to suru poriarukirene-teruguriko-rujiesutero no renzokuseizouhou* (連続 continuous; 製造 production; 法 method).

[seikyuukou 2] entoukatafiruta-ga hiraori (平織り plain fabrics) *no kinzokumesshu* (金属 metal; メッシュ mesh) *de, katsu mehiraki* (目開き opening) *ga sukunaku* (少ない few, little) *tomo 0.01 mm de aru seikyuukou 1 ni kisai* (記載 report) *no renzokuseizouhou*.

[seikyuukou 3] entoukatafiruta- no rokasokudo (濾過 filtration; 速度 velocity) *ga 0.1 to 3 m/hr de aru seikyuukou 1 mata wa 2 ni kisai no renzokuseizouhou*.

translation:
[claim 1] A method for the continuous production of poly(alkylene ether) glycol diester characterized by reacting a cyclic ether and a carboxylic acid anhydride and a solid acid catalyst having a particle size of 0.01 to 3 mm in a reactor equipped with a stirring apparatus and draining off the reaction solution through a cylindrical filter within the reactor.

[claim 2] A method for the continuous production according to claim 1 where the cylindrical filter is a plain weave metal mesh and the mesh openings are at least 0.01 mm.

[claim 3] A method for the continuous production according to claim 1 or 2 where the filtration rate of the cylindrical filter is 0.1 to 3 m h^{-1}.

2.5.2
Example Translation of Patent Description

The description of the patent consists of various parts, of which exemplary parts are translated: prior art from JP 2001-213861 (title: A method for production of ε-caprolactam.), technical field of the invention from JP 2001-286762 (title: Method for regenerating heteropoly acid type catalyst and method for producing methacrylic acid.), embodiment of the invention from JP 10-114701 (title: Method for producing acetic acid), description of substituents from WO 01/10825 (title: Carbamate derivatives and agricultural and horticultural bactericides), formulation examples from JP 07-145156 (title: Thiazolecarboxamide derivatives and horticultural fungicide containing this as active component) and synthesis example from JP 2001-276618.

2.5.2.1 Prior Art (JP 2001-213861)
【０００２】
【従来の技術】ε－カプロラクタンハ、６－ナイロン第の原料そして用いられる非常に重要な基幹化学原料である。その工業的製造方法としては、工業的に容易に得られるベンゼンを出発原料とした後述のようないくつかの製造方法が採用されている。工業的に最も広く行われている製造方法は、ベンゼンを完全水素化してシクロヘキサンにしこれを空気酸化してシクロヘキサノールとシクロヘキサノンの混合物を得、シクロヘキサノールはさらに脱水素してシクロヘキサノンにした後、別途アンモニアを空気酸化さらに水素化して製造したヒドロキシルアミン塩と反応させてシクロヘキサノンオキシムを得、これを硫酸の触媒を用いで、液相下で転位反応を行い、を得るものである。しかしながら、この方法では工程数ガ多く、シクロヘキサンの空気酸化工程において、選択率を向上させるために転化率を３～１０％程度に低く抑える必要があるため生産性が低く、未反応シクロヘキサンのリサイクルのために多量のエネルギーを必要とする上、選択率も７３～８３％程度とあまり高くない。かつ、カルボン酸、エステル第の多量の副生成物が生成し、これを除去するためにアルカリによる処理が必要である第、操作が煩雑である。

transcription into *ro-maji*:
[0002]
[*juurai no gijutsu*] ε-*kapurorakutan wa, 6-nairondai no genryou* (原料 raw material) *soshite mochiirareru* (用いる to use) *hijou* (非常 very, extremely) *ni juyou na* (重要 important) *kikanka-gakugenryou* (基幹 fundamental; 化学 chemistry; 原料 raw material) *de aru. sono kougyouteki-kiseizouhouhou* (工業的 industrial; 製造方法 manufacturing method) *toshite wa, kougyouteki ni youi* (容易 easy, simple) *ni erareru* (得る to gain, to obtain) *benzen o shuppatsugenryou* (出発原料 starting raw material) *toshita koujutsu* (後述 discussed below) *no you na ikutsuka no seizouhouhou* (製造方法 manufacturing process) *ga saiyou sarete iru* (採用する to adopt). *kougyouteki ni* (工業的 industrial) *itomo* (最も extremely) *hiroku* (広い wide) *iwarete iru seizouhouhou wa* (製造方法 manufacturing method), *benzen o kanzensuisoka* (完全 complete; 水素化 hydrogenation) *shite shikurohekisan* (シクロヘキサン cyclohexane) *ni shi kore o kuukisanka* (空気酸化 air oxidation) *shite shikurohekisano-ru* (シクロヘキサノール cyclohexanol) *to shikurohekisanon* (シクロヘキサノン cyclohexanone) *no kongoubutsu* (混合物 mixture) *o toku* (得 profit, advantage), *shikurohekisano-ru wa sara ni* (さらに further, then) *dassuiso* (脱水素 dehydration) *shite shikurohekisanon ni shita ato, betto* (別途 special) *anmonia o kuukisanka sara ni suisioka* (水素化 hydrogenation) *shite seizou shita hidorokishiruaminen* (ヒドロキシルアミン塩 hydroxylamine salt) *to hannou* (反応 reaction) *sasete shikurohekisanonokishimu* (シクロヘキサノンオキシム cyclohexanone oxime) *o toku, kore o ryuusan* (硫酸 sulfuric acid) *no shokubai* (触媒 catalyst) *o mochiide* (用いる to use), *ekisouka de* (液相 liquid phase) *tenihannou* (転位 rearrangement) *o okonai* (行う to carry out), *o eru mo no de aru. shikashi* (しかし but) *nagara, kono houhou* (方法 method) *de wa kouteisuu ga ooku* (多い many), *shikurohekisan no kuukisankakoutei ni oite, sentakuritsu* (選択率 selectivity) *o*

koujou (向上 improvement) *saseru tame ni tenkaritsu* (転化率 inversion ratio) *o 3~10% teido* (程度 extend, degree) *ni hikuku* (低い low) *osaeru* (抑える to hold down, to suppress, to control) *hitsuyou* (必要 necessity) *ga aru tame seisansei* (生産性 productivity) *ga hikuku, mihannou* (未反応 unreacted) *shikurohekisan no risaikuru* (リサイクル recycling) *no tame ni taryou* (多量 large quantity) *no enerugi-* (エネルギー energy) *o hitsuyou to suru kami, sentakuritsu mo 73~83% teido to amari takakunai* (高い high). *katsu, karubonsan* (カルボン酸 carboxylic acid), *esuterudai no taryou no fukuseiseibutsu* (副生成物 byproduct) *ga seisei* (生成 formation) *shi, kore o jokyo suru* (除去する to remove) *tame ni arukari ni yoru shori* (処理 treatment) *ga hitsuyou de aru dai, sousa* (操作 operation, handling, control) *ga hanzatsu* (煩雑 complicated) *de aru*.

translation:
[0002]
[prior art] ε-Caprolactam, which is used as raw material for nylon-6, is a very important fundamental chemical raw material. Using benzene as a starting raw material, which can be easily obtained on an industrial scale, various production methods such as those described below, can be used for the industrial production of this material. The most widely used production method involves the complete hydrogenation of benzene giving cyclohexane, which is then oxidized by air to form a mixture of cyclohexanol and cyclohexanone; after dehydration of cyclohexanole, cyclohexanone is reacted with a hydroxylamine salt which has been produced by oxidation of ammonia with air and hydrogenation, giving cyclohexanone oxime; this is rearranged in the liquid phase with sulfuric acid catalyst. However, among these methods, there are many processes for the oxidation of cyclohexane by air, but the productivity is low because the conversion ratio must be limited to a low degree of 3 to 10% in order to improve the selectivity, and a large quantity of energy is necessary to recycle unreacted cyclohexane with also a not high selectivity of 73 to 83%. Furthermore, large quantities of carboxylic acids and esters are formed as by-products, and in order to remove them, complicated alkali treatment is necessary.

2.5.2.2 Technical Field of the Invention (JP 2001-286762)
【発明の詳細な説明】
【０００１】
【発明の属する技術分野】本発明はヘテロポリ酸率触媒の再生方法、およびメタクリル酸の製造方法に関する。詳しくは、接触気相酸化反応に長期間使用するなどして活性劣化したヘテロポリ酸率触媒を再生する方法、およびこの再生されたヘテロポリ酸率触媒の存在下にメタクロレイン、イソブチルアルデヒドおよび/またはイソ酪酸を気相酸化または気相酸化脱水素してメタクリル酸を製造する方法に関する。

transcription into *ro-maji*:
[hatsumei no shousai na setsumei]
[0001]
[hatsumei no zoku suru gijutsubunya] honhatsumei wa heteroporisanritsushokubai (ヘテロポリ酸 heteropoly acid; 率 type; 触媒 catalyst) *no saiseihouhou* (再生 recycling; 方法 method) *oyobi metakurirusan* (メタクリル酸 methacrylic acid $H_2C=C(CH_3)COOH$) *no seizouhouhou*

(製造方法 manufacturing process) *ni kansuru. kuwashiku wa, sesshokukisousankahannou* (接触 contact; 気相 vapour pahse; 酸化反応 oxidation reaction) *ni choukikanshiyou suru* (長期 long-term; 使用する to use) *nado shite kasseirekka* (活性 activity; 劣化 degradation) *shita heteroporisanritsushokubai o saisei suru houhou, oyobi kono saisei sareta heteroporisanritsushokubai no sonzaika ni metakurorein, isobuchiruarudehido oyobi/mata wa isorakusan* [イソ酪酸 isobutyric acid $(CH_3)_2CHCOOH$] *o kisousanka mata wa kisousankadassuiso* (気相酸化 gas-phase oxidation; 脱水素 dehydration) *shite metakurirusan o seizou suru* (製造する to produce) *houhou ni kansuru.*

translation:
[detailed description of the invention]
[0001]
[technical field of the invention] The present invention is related to a method for recycling a heteropoly acid type catalyst and a manufacturing process for methacrylic acid. More detailed, it is related to a method for recycling a heteropoly acid type catalyst whose activity has been degraded, by long-term use in a gas-phase oxidative contact reaction for instance; and a method with this recycled heteropoly acid type catalyst, for producing methacrylic acid by gas-phase oxidation or gas-phase oxidative dehydration of methacrolein, isobutyraldehyde and/or isobutyric acid.

2.5.2.3 Embodiment of the Invention (JP 10-114701)
【発明の実施の形態】
【０００８】本発明における酢酸製造の触媒としては元素周期律表の第８各金属を使用できるが、特にロジウムが高い活性を有しているため好ましい。ロジウムの使用形態としては、反応条件に可溶性であって、反応系中ロジウムカルボニル錯体種を成形し得るものであればどのようなものでもかまわない。工業的にはヨウ化ロジウムなどが挙げられるが、これに限定されるものではない。反応液中のロジウム濃度は１００〜１０,０００ｐｐｍ、好ましくは２００〜３,０００ｐｍである。

【０００９】本発明において酢酸を製造する際、ヨウ素化合物を反応率にに添加することが好ましい。ヨウ素化合物としてはヨウ化メチルとヨウ化リチウムが特に好ましく。特開昭６０－１４９５４２号で公知のようにヨウ化リチウムとヨウ化メチルと混合して使用することが最も好ましい。ヨウ化メチルは反応率に直接添加できるし、反応率でヨウ化メチルが生ずるような方法で添加してもかまわない。反応液中のヨウ化メチルの濃度が高いと、ギ酸メチルの異性化反応が促進されるが、ヨウ化メチルは回収して、反応器に再循環させる必要があり、循環工程の設備規模、エネルギーの使用量から経済的に最も有利な反応液中のヨウ化メチルの濃度は５〜２０ｗｔ％範囲である。

transcription into *ro-maji*:
[hatsumei no jisshi no keitai]
[0008] honhatsumei ni okeru sakusanseizou (酢酸 acetic acid; 製造 production) *no shokubai* (触媒 catalyst) *toshite wa gensoshuukiritsuhyou* (元素周期律表 periodic table of the elements) *no dai 8 kakukinzoku* (各金属 each metal) *o shiyou dekiru ga, toku ni rojiumu* (ロジウム rho-

dium) *ga takai kassei* (活性 activity) *o yuu shite iru tame konomashii. rojiumu no shiyoukeitai* (形態 form) *toshite wa, hannoujouken* (反応 reaction; 条件 condition) *ni kayousei* (可溶性 solubility) *de atte, hannoukeichuu* (反応系 reaction system; 中 inside) *rojiumukarubonirusakutaishu* (ロジウムカルボニル錯体 rhodiumcarbonyl complex; 種 species) *o seikei* (成形 moulding) *shi eru mono de areba dono you na mono de mo kamawanai* (-てもかまわない don't mind, don't care). *kougyouteki* (工業的 industrial) *ni wa youkarojiumu* (ヨウ化ロジウム rhodium iodide) *nado ga agerareru* (挙げる to raise, to perform) *ga, kore ni gentei* (限定 limit) *sareru mono de wa nai. hannouekichuu no rojiumunoudo* (濃度 concentration) *wa 100~10,000 ppm, konomashiku wa 200~3,000 ppm de aru*.

[0009] honhatsumei ni oite sakusan o seizou suru sai, yousokagoubutsu (ヨウ素化合物 iodine compound) *o hannouritsu ni tenka suru* (添加する to add) *koto ga konomashii. yousokagoubutsu toshite wa youkamechiru* (ヨウ化メチル methyl iodide) *to youkarichiumu* (ヨウ化リチウム lithium iodide) *ga toku ni konomashiku. tokkaishou 60-149542 gou de kouchi* (公知 public knowledge) *no you ni youkarichiumu to youkamechiru to kongou* (混合 mixture) *shite shiyou suru* (使用する to use) *koto ga mottomo konomashii. youkamechiru wa hannouritsu ni chokusetsutenka* (直接 direct, 添加 addition) *dekiru shi, hannouritsu de youkamechiru ga shouzuru you na houhou* (方法 method) *de tenka shite mo kamawanai. hannouekichuu no youkamechiru no noudo ga takai to, gissanmechiru* (ギ酸メチル methyl formate) *no iseikahannou* (isomerization reaction) *ga sokushin* (促進 promotion) *sareru ga, youkamechiru wa kaishuu shite* (回収する to recover), *hannouki* (反応器 reactor) *ni saijunkan saseru* (再循環する to recycle) *hitsuyou* (要 necessity) *ga ari, junkankoutei* (循環 circulation; 工程 process) *no setsubikibo* (設備 equipment; 規模 scale), *enerugi- no shiyouryou kara keizaiteki* (経済的 economic) *ni mottomo yuuri* (有利 advantageous) *na hannouekichuu no youkamechiru no noudo wa 5~20 wt% hani* (範囲 range) *de aru*.

translation:
[embodiment of the invention]
[0008] In accordance with the present invention, each group eight element of the periodic table of elements can be used as catalyst for the production of acetic acid and, because of the high activity, rhodium is preferred. Rhodium may be used in any form as long as it is soluble under the reaction conditions and can form rhodium carbonyl complex species in the reaction system. Rhodium iodide can be used in industry without any limitations. The concentration of rhodium in the reaction solution is 100 to 10 000 ppm, and preferably 200 to 3000 ppm.

[0009] According to the present invention, an iodine compound is preferably added to the reaction system during production of acetic acid. As iodine compounds, methyl iodide and lithium iodide are preferred. As it is public knowledge from JP 60-149542, using a mixture of methyl iodide and lithium iodide is preferred. Methyl iodide can be directly added to the reaction system, or by a different method, methyl iodide is formed in the reaction system. If the concentration of methyl iodide in the reaction solution is high, the isomerization reaction of methyl formate is promoted, but it is necessary to recover the methyl iodide and to recycle it into the reactor. Because of the scale of equipment in the circulation process and the used energy, the most economic and advantageous methyl iodide concentration in the reaction solution is in the range of 5 to 20%.

2.5.2.4 Description of Substituents (WO 01/10825)

発明の開示
即ち、本発明は、般式［Ｉ］。

$$\text{R}^1-\text{G}-\overset{\text{O}}{\underset{}{\text{C}}}-\underset{\text{R}^3}{\overset{\text{R}^2}{\text{N}}}-\text{CH}-\text{C}_6\text{H}_3(\text{X}_n)-\underset{\text{Q}}{\text{C}}=\text{N}-\text{O}-\text{Y}$$

［Ｉ］

式中、Ｘはハロゲン原子、Ｃ１～Ｃ６アルキル基、Ｃ１～Ｃ６アルコキシ基、Ｃ１～Ｃ６ハロアルキシ基又はＣ１～Ｃ６ハロアルコキシ基を表し、ｎは０又は１から４の整数を表し、Ｒ¹はＣ１～Ｃ６アルキル基を表し、Ｒ²は水素原子、Ｃ１～Ｃ６アルキル基、Ｃ２～Ｃ６アルケニル基、Ｃ２～Ｃ６アルキニル基、Ｃ１～Ｃ６アルコキシ基、Ｃ１～Ｃ６アルコキシＣ１～Ｃ６アルキル基、Ｃ１～Ｃ６アルキルカルボニル基、Ｃ１～Ｃ６アルコキシカルボニル基、Ｃ１～Ｃ６アルキルカルボニルＣ１～Ｃ６アルキル基又は置換されてもよいベンジル基を表し、Ｒ³は水素原子又はＣ１～Ｃ６アルキル基を表し、Ｇは酸素原子、硫黄原子又は－ＮＲ⁴－基［Ｒ⁴は水素原子もしくはＣ１～Ｃ６アルキル基を表す。］を表し、Ｙは水素原子、Ｃ１～Ｃ１０アルキル基（該基は同一か又は異なる一個以上の、ハロゲン原子、シアノ基、ニトロ基、ヒドロキシ基、Ｃ３～Ｃ６シクロアルキル基、Ｃ１～Ｃ６アルコキシ基、アミノ基、モノＣ１～Ｃ６アルキルアミノ基、ジーＣ１～Ｃ６アルキルアミノ基、Ｃ１～Ｃ６アルキルチオ基、Ｃ１～Ｃ６アルキルスルフィニル基、Ｃ１～Ｃ６アルコキシイミノ基又はＣ（Ｏ）ＮＲ⁵Ｒ⁶（ここで、Ｒ⁵、Ｒ⁶は同一か又は異なる水素原子又はＣ１～Ｃ６アルキル基を示す）で置換されてもよい）、Ｃ２～Ｃ２０アルケニル基を表し、Ｑは水素原子、ハロアルキル基、シアノ基又はフェニル基（該基には一個以上の、ハロゲン原子、シアノ基、ニトロ基、ヒドロキシ基又はＣ１～Ｃ４アルコキシカルボニル基が置換されてもよい）を表すで示されるカーバメート誘導体及びこれらを有効成分とする農園芸用殺菌剤である。本明細書に記載された記号及び用語について説明する。ハロゲン原子とはフッ素原子、塩素原子、臭素原子又はヨウ素原子である。Ｃ１～Ｃ６アルキル基とは、直鎖又は分岐鎖状のアルキル基を示し、例えばメチル、エチル、ｎ－プロピル、ｉｓｏ－プロピル、ｎ－ブチル、ｉｓｏ－ブチル、ｓｅｃ－ブチル、ｔｅｒｔ－ブチル、ｎ－ペンチル、ｉｓｏ－ペンチル、ネオペンチル、ｎ－ヘキシル、１，１－ジメチルプロピル、１，１－ジメチルブチル等の基を挙げることができる。

transcription into *ro-maji*:
hatsumei (発明 invention) *no kaiji* (開示 publication)
sunawachi, honhatsumei (本発明 present invention) *wa, hanshiki* (般式 general formula) *[1]*.
shikijuu, X wa harogengenshi (ハロゲン原子 halogen atom), $C_1 \sim C_6$ *arukiruki* (アルキル基 alkyl group), $C_1 \sim C_6$ *arukokishiki*, $C_1 \sim C_6$ *haroarukishiki mata wa* $C_1 \sim C_6$ *haroarukokishiki o arawashi, n wa 0 mata wa 1 kara 4 no seisuu* (整数 integer) *o arawashi, R^1 wa* $C_1 \sim C_6$ *arukiruki o arawashi, R^2 wa suisogenshi* (水素原子 hydrogen atom), $C_1 \sim C_6$ *arukiruki,* $C_2 \sim C_6$ *arukeni-*

ruki, $C_2 \sim C_6$ arukiniruki, $C_1 \sim C_6$ arukokishiki, $C_1 \sim C_6$ arukokishi $C_1 \sim C_6$ arukiruki, $C_1 \sim C_6$ arukirukarubōniruki, $C_1 \sim C_6$ arukokishikarubōniruki, $C_1 \sim C_6$ arukirukarubōniru $C_1 \sim C_6$ arukiruki mata wa chikan (置換 *substitution*) *sarete mo yoi benjiruki* (ベンジル基 *benzyl group*) *o arawashi, R^3 wa suisogenshi mata wa $C_1 \sim C_6$ arukiruki o arawashi, G wa sansogenshi* (酸素原子 *oxygen atom*), *iougenshi* (硫黄原子 *sulfur atom*) *mata wa –NR^4- ki [R^4 wa suisogenshi moshiku wa $C_1 \sim C_6$ arukiruki o arawasu.] o arawashi, Y wa suisogenshi, $C_1 \sim C_{10}$ arukiruki (gaiki wa douitsu* (同一 *the same, identical*) *ka mata wa kotonaru* (異なる *to differ*) *ichiko* (一個 *counter for object*) *ijou* (以上 *more than, above, over, or more*) *no, harogengenshi, shianoki, nitoroki* (ニトロ基 *nitro group*), *hidorokishiki* (ヒドロキシ基 *hydroxyl group*), $C_3 \sim C_6$ *shikuroarukiruki* (シクロアルキル基 *cycloalkyl group*), $C_1 \sim C_6$ *arukokishiki, aminoki, mono $C_1 \sim C_6$ arukiruaminoki, ji-$C_1 \sim C_6$ arukiruaminoki, $C_1 \sim C_6$ arukiruchioki, $C_1 \sim C_6$ arukirusurufiniruki, $C_1 \sim C_6$ arukokishiiminoki mata wa $C(O)NR^5R^6$ (koko de, R5, R6 wa douitsu ka mata wa kotonaru suisogenshi mata wa $C_1 \sim C_6$ arukiruki o shimesu) de chikan sarete mo yoi), $C_2 \sim C_{20}$ arukeniruki o arawashi, Q wa suisogenshi, haroarukiruki, shianoki* (シアノ基 *cyano group*) *mata wa feniruki (gaiki ni wa ichiko ijou no, harogengenshi, shianoki, nitoroki, hidorokishiki mata wa $C_1 \sim C_4$ arukokishikarubōniruki ga chikan sarete* (置換する *to substitute*) *mo yoi) o arawasu de shimesareru ka-bame-toyuudoutaioyobi* (カーバメート *carbamate;* 誘導体 *chemical derivative) korera o yuukouseibun* (有効 *effective;* 成分 *composition, ingredient*) *to suru nouengeiyousakkinzai* (農園 *plantation;* 殺菌剤 *germicide*) *de aru. honmeisaisho* (本明細書 *detailed statement*) *ni kisaisareta kigou oyobi yougo* (用語 *term, terminology*) *ni tsuite setsumei suru* (説明する *to explain*). *harogengenshi to* (と *marking of quotations*) *wa fussogenshi, ensogenshi, shuusogenshi mata wa yousogenshi de aru. $C_1 \sim C_6$ arukiruki to wa, chokusa* (直鎖 *straight chain*) *mata wa bunkisajou* (分岐鎖 *branched chain;* 状 *form*) *no arukiruki o shimeshi, tatoeba* (例えば *for example*) *mechiru, echiru, n-puropiru, iso-pujropiru, n-buchiru, iso-buchiru, sec-buchiru, tert-buchiru, n-penchiru, iso-penchiru, neopenchiru* (ネオペンチル *neopentyl*), *n-hekishiru, 1,1-jimechirupuropiru, 1,1-jimechirubuchiru tou* (等 *etc.*) *no ki o ageru* (挙げる *to perform*) *koto ga dekiru.*

translation:
Publication of the invention
Namely, the present invention is related to the general formula [1].
In the formula, X represents a halogen atom, C_1–C_6-alkyl group, C_1–C_6-alkoxy group, C_1–C_6-haloalkyl group and C_1–C_6-haloalkoxy group, and n represents zero and one to four integer, and R^1 represents a C_1–C_6-alkyl group, and R^2 represents a hydrogen atom, C_1–C_6-alkyl group, C_2–C_6-alkenyl group, C_2–C_6-alkinyl group, C_1–C_6-alkoxy group, C_1–C_6-alkoxy-C_1–C_6-alkyl group, C_1–C_6-alkylcarbonyl group, C_1–C_6-alkoxycarbonyl group, C_1–C_6-alkylcarbonyl-C_1–C_6-alkyl group and benzyl group, which can also be substituted, R^3 represents a hydrogen atom and a C_1–C_6-alkyl group, and G represents an oxygen atom, a sulfur atom and a -NR^4- group [R^4 represents a hydrogen atom and also a C_1–C_6-alkyl group], Y represents a hydrogen atom, a C_1–C_{10}-alkyl group (which can be the same or different, and can also be substituted by one or more of the following groups: halogen atom, cyano group, nitro group, hydroxyl group, C_3–C_6-cycloalkyl group, C_1–C_6-alkoxy group, amino group, mono-C_1–C_6-alkylamino group, di-C_1–C_6-alkylamino group, C_3–C_6-alkylthio group, C_3–C_6-alkylsulfinyl group, C_1–C_6-alkoxyimino group and $C(O)NR^5R^6$ (in here, R^5, R^6 which can be the same or different, are hydrogen atoms or C_1–C_6-alkyl groups), C_2–C_{20}-alkenyl group, and Q represents a hydrogen atom, a haloalkyl group, a cyano group and a phenyl group (which can be

substituted by one or more of the following groups: halogen atom, cyano group, nitro group, hydroxyl group and C_1–C_4-alkoxycarbonyl group) and the carbamate derivative is an active ingredient in the use as germicide in agriculture. In the following detailed statements, used numbers and terms are explained. Halogen atom means fluorine atom, chlorine atom, bromine atom and iodine atom. A C_1–C_6-alkyl group can be a straight chain and a branched chain alkyl group, for example methyl, ethyl, n-propyl, iso-propyl, n-butyl, iso-butyl, sec-butyl, tert-butyl, n-pentyl, iso-pentyl, neopentyl, n-hexyl, 1,1-dimethylpropyl, 1,1-dimethylbutyl etc.

2.5.2.5 Example (JP 07-145156)

【００３７】製剤例１粉剤
化合物番号１の化合物３部、ケイソウ土２０部、白土３０部およびタルク４７部を均一に粉砕混合して粉剤１００部を得た。

【００３８】製剤例２水和剤
化合物番号２の化合物３０部、ケイソウ土４７部、白土２０部、リグニンスリホン酸ナトリウム１部およびアルキルベンゼンスルホン酸ナトリウム２部を均一に粉砕混合して水和剤１００部を得た。

【００３９】製剤例３乳剤
化合物番号１の化合物１０部、シクロヘキサン１０部、キシレン５０部およびソルボール〔東邦化学製界面活性剤〕２０部を均一に溶解混合し、乳剤１００部を得た。

【００４２】製剤例６フロワブル剤
化合物番号１の化合物４０部、カルボキシメチルセルロース３部、リグニンスルホン酸ナトリウム２部、ジオクチルスルホサクシネートナトリウム塩１部および水５４部をサンドグラインダーで湿式粉砕し、フロワブル剤１００部を得た。

transcription into *ro-maji*:
[0037] seizairei 1 (製剤 formulation) *funzai* (粉剤 powder)
kagoubutsubangou (化合物 compound 番号 number) *1 no kagoubutsu 3 bu* (部 part), *keisoudo 20 bu, hakudo* (白土 clay, China clay) *30 bu oyobi taruku* (タルク talc) *47 bu o kinitsu* (均一 uniform) *ni funsaikongou shite* (粉砕混合する to pulverise and mixe) *funzai 100 bu o eta.*

[0038] seizairei 2 suiwazai
kagoubutsubangou 2 no kagoubutsu 30 bu, keisoudo 47 bu, hakudo 20 bu, riguninsurihonsannatoriumu (リグニンスリホン酸ナトリウム sodium ligninsulfonate) *1 bu oyobi arukirubenzensuruhonsannatoriumu 2 bu o kinitsu ni funsaikongou shite suiwazai 100 bu o eta.*

[0039] seizairei 3 nyuuzai (乳剤 emulsion)
kagoubutsubangou 1 no kagoubutsu 10 bu, shikurohekisan (シクロヘキサン cyclohexane) *10 bu, kishiren* (キシレン xylene) *50 bu oyobi sorubo-ru [touhoukagakuseikaimenkasseizai]* (界面活性 surface activity) *20 bu o kinitsu ni youkaikongou shi, nyuuzai 100 bu o eta.*

[0042] seizairei 6 furowaburuzai
kagoubutsubangou 1 no kagoubutsu 40 bu, karubokishimechiruseruro-su (カルボキシメチルセルロース carboxymethylcellulose) *3 bu, riguninsuruhonsannatoriumu 2 bu, jiokuchirusuruhosakushine-tonatoriumuen* (ジオクチルスルホサクシネートナトリウム

塩 sodium dioctylsulfosuccinate salt) *1 bu oyobi sui 54 bu o sandogurainda-* (サンドグラインダー sand grinder) *de shitsushikifunsai* (湿 moisture, 粉剤 powder) *shi, furowaburuzai 100 bu o eta.*

translation:
[0037] Formulation example 1 powder
Three parts of compound number 1, two parts of diatomaceous earth, 30 parts of China clay and 47 parts of talc were uniformly pulverized and mixed to give ten parts of the powder.

[0038] Formulation example 2 water-dispersible powder
30 parts of compound number 2, 47 parts of diatomaceous earth, 20 parts of China clay, one part of sodium ligninsulfonate and two parts of sodium alkylbenzenesulfonate were uniformly pulverized and mixed to give 100 parts of water-dispersible powder.

[0039] Formulation example 3 emulsion
Ten parts of compound number 1, ten parts of cyclohexane, 50 parts of xylene and 20 parts of sorbol [Touhou Chemicals surfactant] were homogeneously dissolved and mixed to give 100 parts of emulsion.

[0042] Formulation example 6, flowable formulation
40 parts of compound number 1, three parts of carboxymethylcellulose, two parts of sodium ligninsulfonate, one part of sodium dioctylsulfosuccinate and 54 parts of water were powdered under wet condition in a sand grinder to give 100 parts of flowable formulation.

2.5.2.6 Example (JP 2001-276618)
【実施例】
【０１０３６】触媒の調整
仕込み組成式$Mo_{01}V_{0.31}Nb_{0.07}Sb_{0.20}B_{1.0}O_n$／４６．２ｗｔ％－$SiO_2$で示される酸化物触媒を次のようにして調整した。水１３３００ｇにヘプタアモリブデン酸アンモニウムを２０１４．２ｇ、メタバナジン酸アンモニウムを４１３．８ｇ、三酸化二アンチモンを３３２．５ｇ、オルトホウ酸７０８．２ｇを加え、攪拌しながら９０℃で２時間３０分間加熱した後、約７０℃まで冷却して混合液A－１を得た。

【０１０３７】得られた混合液A－１にSiO_2として３０．６ｗｔ％を含有するシリカゾル７８４３．１ｇを添加した。更にH_2O_2として１５ｗｔ％含有する過酸化水素水５１６．８ｇを添加し、５０℃で１時間攪拌を続けた。次にニオブ含有液を１２１３．９ｇ添加して原料調分液を得た。得られた原料調分液を、遠心式噴霧乾燥器に供給して乾燥し、微小球状の乾燥紛体を得た。乾燥機のは入口温度は２１０℃、そして出口温度は１２０℃であった。

transcription into *ro-maji*:
[jisshirei]
[0036] shokubai (触媒 catalyst) *no chousei* (調整 preparation)
shikomi (仕込む to bring up) *soseishiki* (組成式 composition formula) *ga* $Mo_1V_{0.31}Nb_{0.07}Sb_{0.20}B_{1.0}O_n$/46.2 wt%-$SiO_2$ *de shimesareru sankabutsushokubai o tsugi* (次 next) *no you ni shite chousei shita. mizu 13300 g ni heputamoribudensananmoniumu* (ヘプタアモリ

ブデン酸アンモニウム ammonium heptamolybdate $(NH_4)_6Mo_7O_{24} \cdot 4H_2O$) *o 2014.2 g, metabanajinsananmoniumu* (メタバナジン酸アンモニウム ammonium metavanadate NH_4VO_3) *o 413.8 g, sansankanianchimon* (三酸化二アンチモン antimonous oxide Sb_2O_3) *o 332.5 g, orutohousan* (オルトホウ酸 orthoboric acid H_3BO_3) *708.2 g o kuwae, kakuhan* (攪拌 stirring) *shinagra 90 °C 2 jikan 30 bunkankanetsu* (加熱 heating) *shita ato, yaku* (約 approximately) *70 °C made reikyaku* (冷却 cooling) *shite kongoueki* (混合 mixing; 液 liquid) *A-1 o eta.*

[0037] erareta kongoueki A-1 ni SiO$_2$ to shite 30.6 wt% o ganyuu suru (含有する to contain) *shirikazoru* (シリカゾル silica sol) *7843.1 g o tenka shita* (添加する to add). *sara ni H2O2 to shite 15 wt% o ganyuu suru kasankasuisosui* (hydrogen peroxide H_2O_2) *516.8 g o tenka shi, 50 °C de 1 jikankakuhan o tsuzuketa* (続ける to continue). *tsugi ni niobu ganyuueki o 1213.9 g tenka shite genryouchoubuneki* (原料 raw material) *o eta. erareta genryouchoubuneki o, enshin-shikifunmukansouki* (遠心 centrifugal; 式 type; 噴霧 spraying; 乾燥器 dryer) *ni kyoukyuu shite* (供給する to supply) *kansou* (乾燥 drying) *shi, bishoukyuujou* (微小 minute; 球状 spherical shape) *no kansoufuntai* (乾燥 drying; 紛体 powder) *o eta. kansouki* (乾燥機 dryer) *no iriguchiondo* (入口温度 entrance temperature) *wa 210 °C, soshite deguchiondo* (出口温度 exit temperature) *wa 120 °C de atta.*

translation:
[Actually performed example]
[0036] Preparation of the catalyst
The oxidation catalyst with the composition formula $Mo_1V_{0.31}Nb_{0.07}Sb_{0.20}B_{1.0}O_n/46.2$ wt%-SiO_2 was prepared as follows. To 13300 g water, 2014.2 g ammonium heptamolybdate, 413.8 g ammonium metavanadate, 332.5 g antimonous oxide and 708.2 g orthoboric acid were added; during agitation, it was heated at 90 °C for 2 h 30 min, and afterwards cooled to approximately 70 °C to get the composition liquid A-1.

[0037] To the obtained composition liquid A-1, 7843.1 g silica sol, containing 30.6 wt% SiO_2 were added. Furthermore 516.8 g hydrogen peroxide, containing 15 wt% H_2O_2, was added, and stirring continued for one hour. Next, by adding 1213.9 g of the niobium containing solution, the starting solution was obtained. The obtained starting solution was dried with a centrifugal type spraying dryer and a minute dry powder with spherical shape was obtained. The entrance temperature of the drying machine was 210 °C and the exit temperature 120 °C.

2.6
Tools for Supporting Text Analysis

There are three types of useful tools for analysis of scientific literature from Japan. Most importantly are printed dictionaries, which have a long tradition and provide reliable information. Many books are available, including some on specific scientific topics. The identification of words is described below. How to find characters within specific *kanji* dictionaries is also explained. In addition to printed dictionaries, electronic dictionaries have gained in popularity very rapidly since the early 1990s because of their speed, portability, and unique *kanji* and word searching features. Various models are available, from Canon, Sharp and other companies. Because the range of products is still changing rapidly, they are not described here. Updated information about their features is available on the company's web sites. As a third support possibility, online tools have become more popular in recent years. Most of them are non-commercial and based on private initiatives. Important web tools for general language support as well as for scientific purposes are described below.

2.6.1
Printed Support

Much printed literature is available that might be used for text analysis, covering general topics, in particular Japanese–English dictionaries and books about Japanese grammar written in English. There are also dictionaries that cover chemical and other scientific terms. As most of them are intended for use by native Japanese speakers, they are only of limited use for foreigners. The main challenge is the identification of Chinese characters within dictionaries as the knowledge of *kun-* and *on-*readings is essential. To find the meaning or reading of single characters, a different type of dictionary has to be consulted. There are some *kanji* dictionaries that have developed a system of arranging characters in ways that facilitate the identification without knowing the correct reading. They give also English translations of common vocabulary and some scientific expressions. With more specific terms, word translation requires a two-step process, beginning with the search for the reading, followed by identification of the entire word within a dictionary. However, even experienced translators can be stumped by unfamiliar characters, and looking up and cross-referencing *kanji* can be time-consuming.

Among *kanji* dictionaries, two books have set the benchmark: *The Modern Reader's Japanese-English Character Dictionary* by A.N. Nelson and the *Japanese Character Dictionary* by M. Spahn and W. Hadamitsky. Identification of characters within these books is described in Section 2.6.2. Concerning Japanese–English word dictionaries, a list with books covering general vocabulary as well as chemical terms is provided in Chapter 10. In Japanese–English dictionaries, words are usually not ordered according to the Latin alphabet (with some exceptions for foreigners) but by using the Japanese sound system, which is described in Section 1.2.2. Although the 50-sound-matrix is the basis of the Japanese syllabaries, the characters are arranged in dictionaries not separately from the additional clouded, semiclouded and palatalized *kana*. Each syllable from the 50-sound-matrix is followed by its derived syllables like the following compilation, e.g., か is followed by its clouded derivation が, ほ is followed by its clouded and semiclouded derivation ぼ and ぽ, and に is followed by the palatalized *kana* にゃ, にゅ and にょ. A distinction is made between *hiragana* and *katakana*. Figure 2.5 gives an overview of the order of syllables in Japanese dictionaries.

In addition, *kanji*, as their reading also refers to the syllable alphabet, are ordered in a dictionary using the same method together with words written in *kana*, resulting in a mixture of characters that makes it difficult for inexperienced readers to find the searched word.

A vowel extender mark (long *katakana* vowel) is considered as the separate vowel that is expressed by the line. Accordant characters follow directly behind the vowel they extend, such as:

... キ ... クウ クー グウ グー ... ケ ...

2.6 Tools for Supporting Text Analysis

Fig. 2.5 Order of syllables in Japanese dictionaries (with *katakana* as an example).

The small つ *tsu* for duplicated consonants is treated like an ordinary つ *tsu* and is ranked right behind つ, i.e.,

...　ち　つ　っ　て　...

The following extract from a dictionary demonstrates the word order: ..., ひ, び, ぴ, ひあし, ピアス, ひあそび, ひいあぼあさん, ビーカー, ひいき, ピーク, びいしき, ヒース, ビーズ, ヒーター, ビーだま, ..., ひちょう, ぴちん, ひつ, ぴっ, ひつあつ, ひつう, ひっか, ..., ひやく, びやく, ひゃくがい, びゃくしん, ひやけ, ひやす, ひゃっか

じてん, ひゃっぽう, ひやとい, ひやり, いゆ, ヒューズ, ビューティ, びゅん, びょん ぴょん, ひら, ビラ, ...

English–Japanese dictionaries are also very helpful in verifying a translation, e.g., *Dictionary of Chemical Terms* by Kitiro Hashimoto (三共出版株式会社, Tokyo 1975). The book covers over 80 000 words from English, German and Latin. It includes general terms as well as names of chemical compounds. *Sanyo's Tri-Lingual Glossary of Chemical Terms*, compiled by Hiroshi Yamada, has a similar scope. Some information about general topics on analysis of Japanese patents can be found in *Japanese Patent Translation Handbook* (issued by the American Translators Association) and *Basic Technical Japanese* by E. Daub, R. Bird and N. Inoue. Examples of useful books, including Japanese grammar books, further *kanji* dictionaries as well as general and scientific word dictionaries are given in Chapter 10, which also lists all bibliographic data of the books mentioned above.

2.6.2
Identification of *kanji* in Character Dictionaries

Until few years ago, printed character dictionaries were the only tool for looking up *kanji*. In recent years, very useful online *kanji* dictionaries have also been developed. They are, in principle, based on the same *kanji* order and provide similar search options, i.e., the identification using radicals or stroke counting. Online dictionaries are described in Section 2.6.4. The two most used standard dictionary books suitable for foreigners are *The Modern Reader's Japanese-English Character Dictionary* by A.N. Nelson and the *Japanese Character Dictionary* by M. Spahn and W. Hadamitsky. Both contain several thousand single *kanji* as well as *kanji* combinations. Identification is based on the radical system. Whereas Nelson uses the traditional 214 radicals, Spahn/Hadamitsky reduced the number to 79 to improve lucidity and to enhance the logical arrangement.

2.6.2.1 Identification of *kanji* with Radicals

As described in Section 1.2.3, *kanji* consist of radicals. Lists of either 214 or 79 radicals are used to classify all *kanji* and may be applied for the identification of characters. If the *kanji* consists of more than one radical, the traditional one has to be identified, which is used for the character's classification. Fortunately, only a few *kanji* have an unmanageable number of radicals.

To determine under which radical a *kanji* is listed in a dictionary, modern books (like the two standard dictionaries mentioned above) use easy prioritization criteria. The position of the radical is the most important factor, followed by the radical's stroke number. There are some easy rules that aim to speed up the identification process:

- If the entire character is itself a radical, a variant of a radical or consists only of one radical, then that is the radical.
- When both sides of a *kanji* are radicals, the left-hand radical is preferred to the one at the right-hand side. This rule is also to be applied if two radicals have the same horizontal position within the *kanji* – then the radical has to be taken whose leftmost point is farther left.
- When top and bottom of a *kanji* are radicals, the upper radical is preferred to the one at the bottom. If two radical have the same vertical position, the one with the higher highest point must be taken.

Concerning the position, there is the following basic order (the position of the radical is indicated in gray). For "others", different rules are applied, depending on the dictionary used.

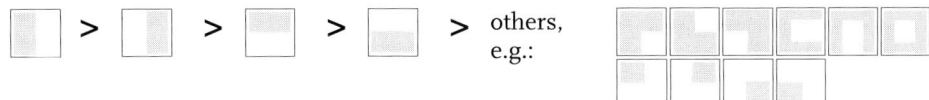

About half of all *kanji* divide themselves into a left- and right-hand side, with the left-hand side being the radical. The left-before-right- and top-before-bottom-rules help to identify about 75% of the characters. Identification of the radical is sometimes complicated, because some radicals appear in different shapes. If the radical cannot be decided by the position, then the element that consists of more strokes is the radical for the character's classification.

Radicals can be ordered within a dictionary by three criteria:
- by the historical radical number, like in Nelson's dictionary;
- by the stroke number of the modern 79 radicals, like in the *Japanese Character Dictionary*;
- by the position of the radical in most of its *kanji* (some radicals may appear at different positions within the characters, as described above, but usually one position is favored). For example, the following radicals occur mainly at the indicated position:

木	リ	雨	心	广	辶	匚	門	囗	气
柱	別	電	忠	広	近	医	開	図	気
chuu	betsu	den	chuu	kou	kin	i	kai	zu	ki

All *kanji* with the same determining radical are listed together in *kanji* dictionaries. Within a radical class, characters are again classified, according to the number of strokes of the remaining elements. Only a few possible characters match these two characteristics and the *kanji* can now be identified.

Because of their meaning, typical radicals in *kanji* of chemical and technical terms are "fire" 火, "water" 水,"rain" 雨, "stone" 石, "metal" 金, "earth" 土 and "strength" 力. In their case, the radicals are also independent *kanji* (or, in other words, the *kanji* consist only of one radical). There are other radicals without their own *kanji*, such as "water" 氵 (as left part of *kanji*) and "ice" 冫. Radicals may also change their shape or even have a different number of strokes, depending on their position within the characters. Table 2.20 shows examples of their occurrence in more complex characters and their position within these characters.[6]

---

6) (a) Other radicals may occur more frequently in *kanji* of chemical terms but the radicals for the examples were chosen because of the meaning related to natural science. (b) The *Japanese Character Dictionary* (Spahn/Hadamitzky) applies the modern 79 radical set. Some of the deleted radicals were assigned to the remaining radicals. *kanji* that are not covered by the 79 radical set have to be classified with other radicals they consist of. (c) The historical radical number is used in Nelson's Dictionary.

Table 2.20 Selection of important radicals for chemical literature.

Radical	Number in Hadamitsky	Historical radical number					Other position
火 fire	4d	86	燐 rin	–	炎 en	炙 sha	毯 tan
氵, 水 water	3a	85	液 eki	–	沓 tou	泉 sen	–
土 earth	3b	32	塔 tou	–	走 sou	基 ki	在 zai
石 stone	5a	112	硫 ryuu	–	磊 rai	砦 toride	–
金 metal	8a	167	鉄 tetsu	–	–	鏨 tagane	–
冫 ice	2b	15	冷 rei	–	–	–	弱 jaku
酉 liquid	7e	164	酸 san	–	–	醤 shou	–
力 strength	2g	19	加 ka	動 dou	脅 kyou	勇 yuu	–
立 stand	5b	117	竦 shou	–	童 dou	竪 ju	競 kyou

2.6.2.2 Identification of a *kanji* by Stroke Counting

The way in which a character is written with a pencil or brush is the basis of counting the strokes. A single stroke is any line that is drawn with the pencil in continuous contact with the paper, independent of the curse of the line, which may have cusps and curves.

One difficulty in the identification of a *kanji* by stroke counting is that the printed version of a *kanji* usually looks different to the hand written one. For the printed *kanji*, many single strokes look like separate strokes. If the reader is not familiar with writing *kanji*, some experience is needed to be able to count the stroke number of the printed characters. The following examples show that the stroke number may be estimated wrongly when starting with the printed character:

kanji:	乙	乃	与	弔	双	比	逐	滅	凝	臓	巌	鑑
reading:	otsu	nai	yo	chou	futa	hi	chiku	metsu	gyou	zou	gan	kan
stroke number:	1	2	3	4	4	4	10	13	16	19	20	23

In addition, although the stroke number is standardized, different dictionaries may count variably, e.g., for 子 *ji*, which is counted with two or three strokes. The biggest hurdle is the number of *kanji* with the same stroke number. There are only a few characters with small and high numbers of strokes. For instance, among the 2000 *kanji* in O'Neill's dictionary there are two with one stroke, 14 with two strokes and 33 with three strokes. The number of *kanji* increases rapidly with the stroke number up to more than 200 *kanji* for twelve strokes, and then decreases again (Fig. 2.6). Few characters have more than 20 strokes (21 strokes: four *kanji*, 22 strokes: two *kanji*, 23 strokes: one *kanji*).

As most *kanji* have stroke numbers between seven and thirteen, many characters have to be checked for the searched one. If the stroke number was counted wrongly, a second or third

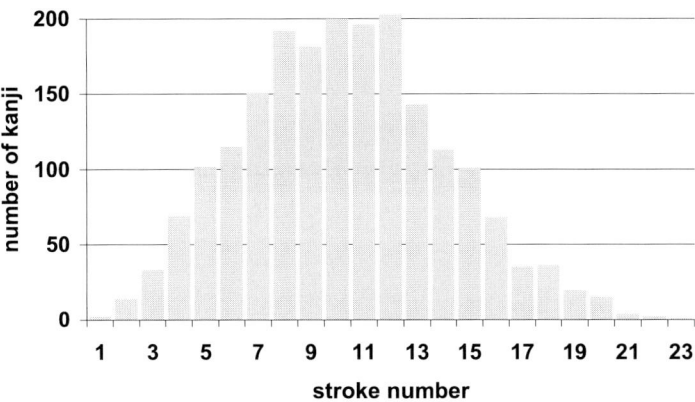

Fig. 2.6 Number of *kanji* with same stroke number.

attempt may have to be made, resulting in having to read up to more than 500 *kanji* to find one specific character. As this method is obviously not effective and is very time-consuming, it is rarely used. Usually, the possibility of identifying a character by stroke counting is given as an additional option within the dictionaries besides the radical system.

2.6.3
General Online Support for Japanese Language

In recent years, the number of online resources providing support for identification of characters, translation of texts and general understanding of the Japanese language has snowballed. Search engines, electronic dictionaries and online forums have also greatly improved their quality and usability, giving translators, nowadays, the opportunity to analyze Japanese documents much faster. However, most internet sites focus on the general language. For specific subjects, such as chemistry, printed media are still indispensable.

A great pioneering work was carried out by James William Breen at Monash University, Australia. In 1991, he started to compile a database ("EDICT") with Japanese terms and their English translations in machine-readable form. The scope and quality were steadily increased and further databases were compiled. Numerous people provided materials or other support to develop the dictionary, demonstrating the power of a virtual collaboration that is facilitated by the internet. Jim Breen's dictionary files are the benchmark of Japanese–English online dictionary information. Currently, the online *kanji* dictionary ("KanjiDic") includes 6355 characters, his online dictionary has about 60 000 regular words and the dictionary of Japanese proper names over 160 000 entries. All databases are freely available and utilized by many other websites. Since 2000, all files are owned by The Electronic Dictionary Research and Development Group at Monash University. URLs for free download and further illustration are given in Table 2.21. Jim Breen's personal internet page is a helpful source for additional information about the Japanese language and translation resources.

Because of its public availability, many websites provide access to EDICT and KanjiDic data and search options within the database files. Table 2.21 lists some interfaces. Among these,

Table 2.21 Internet sources based on J.W. Breen's activities.

Topic	Name, scope	URL
Jim Breen's personal web sites	homepage with further links relevant for text translations	http://www.csse.monash.edu.au/~jwb/
	file archive	http://www.csse.monash.edu.au/~jwb/j_monftp.html
	WWWJDIC Translation Aid	http://www.csse.monash.edu.au/~jwb/wwwjtrans.html
Dictionary files	EDICT (Japanese-English Dictionary Project): basic dictionary file (related: EDICT2, EDICT_SUB)	http://www.csse.monash.edu.au/~jwb/edict.html
	JMDict (Japanese Multilingual Dictionary): multilingual lexical database with Japanese as pivot language and translations into English, German, France, Russia and other languages	http://www.csse.monash.edu.au/~jwb/j_jmdict.html
	KANJIDIC, KANJIDIC2, KANJD212: *kanji* database files, data sets consisting of *kanji*, readings, codes (e.g., JIS code, unicode), indices (e.g., Spahn/Hadamitsky, Nelson, Halpern), radical number, stroke-count, grade	http://www.csse.monash.edu.au/~jwb/kanjidic.html, http://www.csse.monash.edu.au/~jwb/kanjidic2/index.html
	ENAMDICT/JMnedict: Japanese proper names (place-names, surnames, given names, company names, product names)	http://www.csse.monash.edu.au/~jwb/enamdict_doc.html
Interfaces	WWWJDIC	http://www.csse.monash.edu.au/~jwb/wwwjdic.html
	漢字ディービー KanjiDB	http://www.whiteknightlogic.net/kanjidb/
	Jeffrey's Japanese–English Dictionary Gateway	http://dict.regex.info/cgi-bin/j-e
	Denshi Jisho Dictionary	http://www.jisho.org/
	FOKS	http://www.foks.info/index.html
	Animelab Dictionary	http://www.animelab.com/anime.manga/dictionary/
	JTango Dictionary	http://www.glpwd.com/jtango-web/search.action
	Risu Dictionary	http://dict.risukun.com/
	Dima Chirkin – English–Japanese Dictionary	http://icecube.berkeley.edu/~dima/stuff/japanese/
	Local Translation online dictionary	http://www.online-dictionary.biz/english/japanese

recommendable interfaces are Jim Breen's own interface "WWWJDIC", Martin Thorne's "KanjiDB" and Kim Ahlström's "Denshi Jisho Dictionary". In addition to simple Japanese to English word translations, they provide various tools for *kanji* identification, word and text translation. Characters may be searched by reading, English translation, stroke count, radical numbers or codes. Searches can be narrowed by simultaneous application of more than one search criteria. Identification by stroke count or radical is, in principle, the same as when searching in printed *kanji* dictionaries (Section 2.6.2). Also very useful is the "multi-radical search" in WWWJDIC (similar to "compositional *kanji* search" in KanjiDB and *"kanji* by radical" in Denshi Jisho), which considers all radicals included in the searched *kanji*. These three tools also list all *kanji* combination containing the search character. "FOKS" consist of an "intelligent search tool" that allows one to lookup of dictionary entries even with the predictably incorrect reading. Jeffrey's Japanese–English Dictionary Gateway provides further useful information and links.

There are some other online dictionaries worth mentioning (URLs are given in Table 2.22): Eijiro is the free online version of a huge Japanese–English dictionary sold on CD-ROM. It contains about 1.7 million entries. The idea of the jeKai project is to create an open and free online dictionary to which everybody can contribute. However, the number of entries is still limited. Some online dictionaries are based on printed dictionaries, such as "WebDictionary" by Sanseido and NTT's information project "goo". The best German online dictionary is *wadoku jiten* by Ulrich Apel. It contains 235 000 entries.

Recently, tools have become available that offer real time translation of Japanese web pages. After starting Popjishyo or Rikai and entering the URL of a Japanese page, reading and English translation of characters, which are met by the cursor, are given automatically.

If general topics on the Japanese language are searched, the internet is an inexhaustible source of information. Table 2.22 summarizes some useful links.

Finally, some translator's associations also provide information about the Japanese language and offer translation services. See for instance: Japan Association of Translators (JAT, http://www.jat.org/), Japanese Translation Federation (http://www.jtf.jp), Institute of Translation Interpreting (UK, http://www.iti-jnet.org.uk/), American Translators Association, Japanese Language Division (http://www.ata-divisions.org/JLD/index.htm) and The Kanji Foundry (UK, http://www.thekanjifoundry.com/index.html). An interesting tool is *honyaku* (http://honyakuhome.org/), providing a mailing list to discuss specific translation issues among the translator community.

2.6.4
Online Support for Analysis of Chemical Literature

The world wide web provides various information and tools that facilitate the analysis of scientific literature. For some specific subjects, excellent online pages are available that are mainly based on academic projects or the initiatives of private persons. However, there is still no comprehensive source covering all aspects of chemistry and related topics. Among the most useful web sites are some projects that focus on the names and nomenclature of chemical compounds. For instance, the Japanese Science and Technology Agency (JST) provides a chemical substance database with data on approx. 2 million organic compounds. They are searchable by name, molecular formula and molecular weight. Sadahiro Tomoyuki operates

Table 2.22 Internet sources for character and text analysis (in addition to sources listed in Table 2.21).

Topic	Name	URL
Word dictionaries	Eijiro	http://www.alc.co.jp/
	je 海 (jeKai)	http://www.jekai.org
	和独辞典 (WaDoku jiten)	http://www.wadoku.de/
	三省堂 WebDictionary	http://www.sanseido.net/
	goo 辞書 (goo jisho)	http://dictionary.goo.ne.jp/
	KOD (Kenkyusha online dictionary)	http://kod.kenkyusha.co.jp/service/
	Infoseek	http://dictionary.www.infoseek.co.jp/
	yahoo dictionary	http://dic.yahoo.co.jp/
	excite	http://www.excite.co.jp/world/english/
	RNN	http://www.rnnnews.jp/
	Bigblobe	http://search.biglobe.ne.jp/dic/
	Livedoor	http://dic.livedoor.com/
	So-net	http://so-net.dictionary.goo.ne.jp
	Webster's Online Dictionary	http://www.websters-online-dictionary.org/
	Altavista Babelfish	http://babelfish.altavista.com/tr
	Systran	http://www.systransoft.com/index.html
kanji dictionaries	彩雅 Saiga Japanese kanji dictionary	http://www.saiga-jp.com/kanji_dictionary.html
	kanji networks	http://www.kanjinetworks.com/
	Japanisch-Deutsches kanji-Lexikon	http://www.bibiko.de/kanji/index.html
Webpage translation	Pop 辞書 (popjisho)	http://www.popjisyo.com/WebHint/Portal.aspx
	Rikai	http://www.rikai.com/perl/Home.pl
General information about Japanese language	Wikipedia	http://ja.wikipedia.org/wiki; http://en.wikipedia.org/wiki, e.g., http://en.wikipedia.org/wiki/Japanese_grammar, http://en.wikipedia.org/wiki/Japanese_language
	A logical Japanese grammar	http://homepage3.nifty.com/jgrammar/
	Nihongo resources	http://www.nihongoresources.com
	ジェイグラム jGram	http://www.jgram.org/
	日本語と日本の文化 Japanese language and culture	http://www.epochrypha.com/japanese/#links
	Japanese language	http://japanese.about.com/
	A guide to Japanese grammar	http://www.guidetojapanese.org

2.6 Tools for Supporting Text Analysis | 109

Table 2.23 Online sources for chemistry related topics.

Information provider/Name	URL	Details
Wikipedia	http://ja.wikipedia.org/wiki	General chemical topics, such as chemistry (http://ja.wikipedia.org/wiki/%E5%8C%96%E5%AD%A6), organic chemistry (http://ja.wikipedia.org/wiki/%E6%9C%89%E6%A9%9F%E5%8C%96%E5%AD%A6), inorganic chemistry (http://ja.wikipedia.org/wiki/%E7%84%A1%E6%A9%9F%E5%8C%96%E5%AD%A6), name reactions (http://ja.wikipedia.org/wiki/%E5%8C%96%E5%AD%A6%E5%8F%8D%E5%BF%9C%E3%81%AE%E4%B8%80%E8%A6%A7) or sulfur compounds (http://ja.wikipedia.org/wiki/%E7%A1%AB%E9%BB%84); many further topics are covered
Lifescience Dictionary Project	http://wwwsoc.nii.ac.jp/lsdproject/en/index.html	Free dictionary of scientific terms related to medicine, biochemistry, bioorganic and medical chemistry
Nomenclature	http://homepage1.nifty.com/nomenclator/index.htm	Naming of organic substances (http://homepage1.nifty.com/nomenclator/jiyaku/jklite.htm), names and nomenclature rules (e.g., for carbonic acid derivatives: http://homepage1.nifty.com/nomenclator/triv/carbonic.htm)
Bizstyle	http://www.kw-guide.jp	General scientific explanations, chemical terms at http://www.kw-guide.jp/science
JST	http://nikkajiweb.jst.go.jp/nikkaji_web/pages/top.html	Database with information on approx. 2 million organic chemicals, and search options
National Institute of Health Sciences (NIHS), Japan	http://www.nihs.go.jp/ICSC/	Compound database with search options and detailed information on approx. 5 000 substances
various providers	http://www.saglasie.com/tr/chemical/; http://hit-hit-web.hp.infoseek.co.jp/aiueo2.html	Chemical substance databases with names of up to 25 000 compounds, and various search options

Table 2.23 (continued)

Information provider/Name	URL	Details
鉱物たちの庭 koubutsutachi noniwa	http://www.ne.jp/asahi/lapis/fluorite/name/sidxlist.html	Mineral name list
Medical English Dictionary Online	http://www.medo.jp/0.htm	Lists of medical terms
Showa	http://www.st.rim.or.jp/~shw/	Reagents and product lists of suppliers of chemicals (e.g., http://www.st.rim.or.jp/~shw/cat_dex.html)
Panasonic	http://panasonic.co.jp/eco/suppliers/data/lst_csg3f_j.pdf	Names of chemical compounds (e.g., see http://panasonic.co.jp/eco/suppliers/data/lst_csg3f_j.pdf#search=%22%E5%BC%97%E5%8C%96%E3%82%AB%E3%83%AA%E3%82%A6%E3%83%A0%20%22potassium%20fluoride%22%22)
JPO	http://www.jpo.go.jp/shiryou/	Lists with chemicals and their Japanese and English names
Ministry of Health, Labour and Welfare	http://www.mhlw.go.jp/	Health related chemicals list, e.g., http://www.mhlw.go.jp/topics/yunyu/tetsuzuki-fains/

a machine-generating naming tool for organic compounds ("nomenclator"). He also offers lists of IUPAC rules for Japanese nomenclature of organic compounds. Further useful tools for name searches on chemicals are offered by the National Institute of Health Sciences and some private pages. General chemical terms can be searched using kw-guide, which covers approx. 10 000 records. For URLs see Table 2.23.

In addition, there are many websites that do not intend to support text analysis and translations, but which may be used to verify names of chemical compounds. Among these are, for instance, product lists of companies selling chemicals (e.g., many names of commercial chemicals can be found in Showa's reagent lists) and compilations of official authorities regarding health or medical issues (e.g., the Japanese Ministry of Health, Labour and Welfare). Even lists of the JPO, intended to classify patent applications, may be utilized to check scientific terms.

For the field of life science topics and medical chemistry, some internet tools have been developed. The Lifescience Dictionary Project of Professor Shuji Kaneko of Kyoto University is analyzing English texts of medical journals, resulting in the compilation of Japanese–English dictionaries of life-science related scientific terms. The free online database covers some 40 000 records.

Finally, general internet tools also allow the identification or verification of scientific terms and facilitate text analysis. Free online encyclopedias, in particular Wikipedia, might be the first choice, featuring detailed descriptions on specific topics. In addition, enquiries using key words in search engines, such as google or yahoo, may also lead to new online sources of valuable information.

3
Naming of Chemical Compounds

3.1
Naming of Elements and Inorganic Compounds

3.1.1
Elements

The Japanese name for elements, 元素 *genso*, consists of the *kanji* for "origin" 元 *gen* and for "beginning" 素 *so*. Some element's name include the latter *kanji* of element, e.g., some non-metals such as hydrogen: 水素 *suiso* and nitrogen: 窒素 *chisso*; and halogens such as chlorine: 塩素 *enso* and iodine: 沃素 *youso*. Some other elements, especially metals, which have long been known and also have some importance in everyday life, possess their own *kanji*, like gold: 金 *kin*, copper: 銅 *dou*, iron: 鉄 *tetsu* or mercury: 水銀 *suigin*.

The names of these elements derive from different properties. The *kanji* may be determined by the element's appearance in nature, like for hydrogen, 水素 *suiso*, the "element which is part of water" or for chlorine, 塩素 *enso*, the "salt element" because chlorine is part of table salt (sodium chloride) and other important salts. As oxygen is the main component of most acids, its name 酸素 *san* consists of the *kanji* for acid and element. Some other element names are derived from their appearance. The name for mercury, 水銀 *suigin*, is composed of the two *kanji* for water and silver, because mercury seems like liquid silver, an impression, which is also responsible for its Latin name, hydrargyrum ("water silver"). Also because of its appearance, sulfur and platinum own *kanji* for colors in their names: For sulfur, 硫黄 *iou*, it is the color of the elemental rhombic or monocline sulfur, yellow (黄 *ou*, *o*, *ki*). The *kanji* for platinum, 白金 *hakkin*, can be translated as "white gold", either because of its similar, but whiter, brighter and matt, color or because it is also found in rivers and was first discovered by people originally looking for gold. Because of the jewel-like brilliance of quartz SiO_2, the *kanji* for silicon, 珪 *kei*, possesses the radical for jewel 王 as well as for earth 土, giving the entire character the meaning like "jewels out of earth". Because of their metallic properties, other element names consist of the radical for metal 金, e.g., lead: 鉛 *en* or tin: 錫 *suzu*. The similarity to metals and gold can also be found in the *kanji* for silver and copper. Both elements belong together with gold in the same subgroup and have similar properties as gold, e.g., their natural deposit and their use for manufacturing of coins: 銅 (*dou*, copper) and 銀 *gin* (silver). The second component of 銅 is the *kanji* for "same" 同 *dou* (which also determines the spelling of 銅); therefore, the entire *kanji* can be seen as "same than gold".

Japanese-English Chemical Dictionary. Edited by Markus Gewehr
Copyright © 2008 WILEY-VCH Verlag GmbH & Co. KGaA, Weinheim
ISBN: 978-3-527-31293-1

Another mechanism for the formation of names can be found for the *kanji* of carbon, bromine and iodine: The first character of the name for carbon, 炭素 *tanso*, consists of radicals for mountain 山, slope 厂 and fire 火, indicating on charcoal, its deposit at the mountain slope and its use to make fire. Bromine got its name 臭素 *shuuso*, which contains the *kanji* for odor (臭 *shuu*), because of its characteristic, stifling smell. Because of the phonetic similarity to the English name, the *kanji* 沃 *you* is used in the name of iodine, 沃素 *youso*.

Even if *kanji* exist, sometimes *kana* are used, expressing the same spelling like the corresponding *kanji*, especially in newer publications, such as for boron (硼素 or ホウ素 *houso*), tin (錫 or スズ *suzu*), phosphorous (燐 or リン *rin*) or iodine (沃素 or ヨウ素 *youso*).

Besides the above-described element names, most elements have names derived from the English terms, which are transcribed into the Japanese syllable alphabet, e.g., for magnesium: マグネシウム, argon: アルゴン or europium: ユーロピウム. In addition, some names are adopted from the German language into Japanese, like sodium: ナトリウム, potassium: カリウム, antimony: アンチモン, manganese: マンガン and others. For some of them, nowadays, two words can be found, one older transcribed from German and a newer one transcribed from English, e.g., for uranium: ウラン and ウラニウム or for selenium: セレン and セレニウム. Table 3.1 summarizes the Japanese names for chemical elements.

Table 3.1 The elements.

English element name	Symbol	Atomic number	Atomic weight	Japanese
actinium	Ac	89	227.0278	アクチニウム
aluminium	Al	13	26.98154	アルミニウム
americium	Am	95	243[a]	アメリシウム
antimony	Sb	51	121.75	アンチモン
argon	Ar	18	39.948	アルゴン
arsenic	As	33	74.9216	砒素 *hiso*, ひ素 *hiso*
astatine	At	85	210[a]	アスタチン
barium	Ba	56	137.33	バリウム
berkelium	Bk	97	247[a]	バークリウム, バーケリウム, ベーケリウム, ベークリウム
beryllium	Be	4	9.01218	ベリリウム
bismuth	Bi	83	208.9804	ビスマス
bohrium	Bh	107	262[a]	ボーリウム
boron	B	5	10.81	硼素 *houso*, ホウ素 *houso*
bromine	Br	35	79.904	臭素 *shuuso*
cadmium	Cd	48	112.41	カドミウム
calcium	Ca	20	40.08	カルシウム

Table 3.1 (continued)

English element name	Symbol	Atomic number	Atomic weight	Japanese
californium	Cf	98	251[a]	カリホルニウム, カリフォルニウム
carbon	C	6	12.011	炭素 tanso, カーボン
cerium	Ce	58	140.12	セリウム
cesium	Cs	55	132.9054	セシウム
chlorine	Cl	17	35.453	塩素 enso
chromium	Cr	24	51.996	クロム
cobalt	Co	27	58.9332	コバルト
copper	Cu	29	63.546	銅 dou
curium	Cm	96	247[a]	キュリウム
darmstadtium (former eka-platinum, ununnilium)	Ds (Eka-Pt, Uun)	110	269[a]	ダルムスタチウム, ウンウンニリウム
dubnium	Db	105	262[a]	ドブニウム
dysprosium	Dy	66	162.50	ジスプロシウム, ディスプロシウム
einsteinium	Es	99	252[a]	アインスタイニウム
erbium	Er	68	167.26	エルビウム
europium	Eu	63	151.96	ユウロピウム, ユーロピウム, ヨオロピウム
fermium	Fm	100	257[a]	フェルミウム
fluorine	F	9	18.998403	弗素 fusso, フッ素 fusso
francium	Fr	87	223[a]	フランシウム
gadolinium	Gd	64	157.25	ガドリニウム
gallium	Ga	31	69.72	ガリウム
germanium	Ge	32	72.61	ゲルマニウム
gold	Au	79	196.9665	金 kin
hafnium	Hf	72	178.49	ハフニウム
hassium	Hs	108	265[a]	ハッシウム
helium	He	2	4.00260	ヘリウム
holmium	Ho	67	164.9304	ホルミウム
hydrogen	H	1	1.0079	水素 suiso
indium	In	49	114.82	インジウム
iodine	I	53	126.9045	沃素 youso, ヨウ素 youso, ヨード
iridium	Ir	77	192.22	イリジウム
iron	Fe	26	55.847	鉄 tetsu
krypton	Kr	36	83.80	クリプトン

Table 3.1 (continued)

English element name	Symbol	Atomic number	Atomic weight	Japanese
lanthanum	La	57	138.9055	ランタン
lawrencium	Lr	103	260.11	ローレンシウム
lead	Pb	82	207.2	鉛 namari
lithium	Li	3	6.941	リチウム
lutetium	Lu	71	174.967	ルテチウム
magnesium	Mg	12	24.305	マグネシウム
manganese	Mn	25	54.9380	マンガン
meitnerium (former eka-iridium)	Mt (Eka-Ir)	109	266[a]	マイトネリウム
mendelevium	Md	101	258[a]	メンデレビウム
mercury	Hg	80	200.59	水銀 suigin
molybdenum	Mo	42	95.94	モリブデン 水鉛 suien
neodymium	Nd	60	144.24	ネオジム
neon	Ne	10	20.179	ネオン
neptunium	Np	93	237[a]	ネプシニウム
nickel	Ni	28	58.69	ニッケル
niobium	Nb	41	92.9064	ニオブ
nitrogen	N	7	14.0067	窒素 chisso
nobelium	No	102	259[a]	ノーベリウム
osmium	Os	76	190.2	オスミウム
oxygen	O	8	15.9994	酸素 sanso
palladium	Pd	46	106.42	パラジウム
phosphorus	P	15	30.97376	燐, リン rin
platinum	Pt	78	195.08	白金 hakkin
plutonium	Pu	94	244[a]	プルトニウム
polonium	Po	84	210[a]	ポロニウム
potassium	K	19	39.0983	カリウム
praseodymium	Pr	59	140.9077	プラセオジム
promethium	Pm	61	145[a]	プロメチウム
protactinium	Pa	91	231.0359	プロトアクチニウム
radium	Ra	88	226[a]	ラジウム
radon	Rn	86	222[a]	ラドン
rhenium	Re	75	186.207	レニウム

Table 3.1 *(continued)*

English element name	Symbol	Atomic number	Atomic weight	Japanese
rhodium	Rh	45	102.9055	ロジウム
roentgenium (former eka-gold, unununium)	Rg (Eka-Au, Uuu)	111	272	レントゲニウム, ウンウンウニウム
rubidium	Rb	37	85.4678	ルビジウム
ruthenium	Ru	44	101.07	ルテニウム
rutherfordium	Rf	104	261[a]	ラザホージウム
samarium	Sm	62	150.36	サマリウム
scandium	Sc	21	44.9559	スカンジウム
seaborgium	Sg	106	263[a]	シーボーギウム, ジーボーギウム
selenium	Se	34	78.96	セレン, セレニウム
silicon	Si	14	28.0855	珪素 *keiso*, ケイ素 *keiso*
silver	Ag	47	107.868	銀 *gin*
sodium	Na	11	22.98977	ナトリウム
strontium	Sr	38	87.62	ストロンチウム
sulfur	S	16	32.066	硫黄 *iou*, サルファ
tantalum	Ta	73	180.9479	タンタル
technetium	Tc	43	98	テクネチウム
tellurium	Te	52	127.60	テルル
terbium	Tb	65	158.9254	テルビウム
thallium	Tl	81	204.383	タリウム
thorium	Th	90	232.0381	トリウム
thulium	Tm	69	168.9342	ツリウム
tin	Sn	50	118.69	錫, スズ *suzu*
titanium	Ti	22	47.88	チタン, チタニウム
tungsten	W	74	183.85	タングステン, ウォルフラム
ununhexium, eka-polonium[b]	Uuh, Eka-Po	116		ウンウンヘクシウム, ウンウンヘキシウム
ununbium, eka-mercury	Uub, Eka-Hg	112	277[a]	ウンウンビウム
ununpentium, eka-bismuth[b]	Uup, Eka-Bi	115		ウンウンペンチウム
ununquadium, eka-lead	Uuq, Eka-Pb	114		ウンウンクアジウム

Table 3.1 (continued)

English element name	Symbol	Atomic number	Atomic weight	Japanese
ununseptium, eka-astat[c]	Uus, Eka-At	117		ウンウンセプチウム
ununtrium, eka-thallium[c]	Uut, Eka-Tl	113		ウンウントリウム
uranium	U	92	238.0289	ウラン, ウラニウム
vanadium	V	23	50.9415	バナジウム, ヴァナジウム
xenon	Xe	54	131.29	キセノン, ゼノン
ytterbium	Yb	70	173.04	イッテルビウム
yttrium	Y	39	88.9059	イットリウム
zinc	Zn	30	65.39	亜鉛 *aen*
zirconium	Zr	40	91.22	ジルコニウム, ジルコン

[a] Atomic weight of most long-living isotope.
[b] Unconfirmed.
[c] Undiscovered.

In the Periodic Table of Elements (元素の周期表 *genso no shuukihyou*), the elements are classified by the criteria given in Table 3.2.

Table 3.2 Classification of elements.

Criteria	Japanese classification	English translation
metallic character:	金属元素 *kinzokugenso*	metallic element
	半金属元素 *hankinzokugenso*	semimetal element
	非金属元素 *hikinzokugenso*	nonmetallic element
	人工元素 *jinkougenso*	artificial element
consistence:	常温で固体 *jouon de kotai*	solid under normal temperature
	常温で液体 *jouon de ekitai*	liquid under normal temperature
	常温で気体 *jouon de kitai*	gaseous under normal temperature
group character:	アルカリ金属 *arukarikinzoku*	alkali metal
	アルカリ土類金属 *arukaridoruikinzoku*	alkaline earth metal
	ハロゲン *harogen*	halogen
	希ガス *kigasu*	noble gas
	遷移元素 *senigenso*	transition element

3.1.2
General Aspects of Naming Inorganic Compounds

The names of inorganic compounds are built by either using existing Japanese expressions, by transcription of English common names or by mixtures of both methods.

Japanese names exist for many traditionally known compounds, especially many acids, minerals and ores, like for sulfuric acid (H_2SO_4): 硫酸 *ryuusan*, fluorspar (CaF_2): 螢石 *keiseki*, chalcopyrite ($CuFeS_2$): 黄銅鉱 *oudoukou*, opal ($SiO_2 \cdot nH_2O$): 蛋白石 *tanpakuseki*, common salt (NaCl): 食塩 *shokuen*, minium (read lead, Pb_3O_4): 鉛丹 *entan* and hematite (red iron ore, α-Fe_2O_3): 赤鉄鉱 *sekitekkou*. For many ores, *kanji* representing one or more elements of which the compound consists are used, e.g., copper 銅 in azurite $2CuCO_3 \cdot Cu(OH)_2$: 藍銅鉱 *randoukou* or iron 鉄 in pyrite FeS_2: 黄鉄鉱 *outekkou* and limonite $2Fe_2O_3 \cdot 3H_2O$: 褐鉄鉱 *kattekkou*.

In salt and acid names, the element's *kanji* are usually combined with either 素 *so* ("element"), 化 *ka* ("transform, make into") or no additional *kanji* (Table 3.3). 素 *so* is used if the character appears in the basic element name (as with oxygen: 酸素 *sanso*) and if the element represents the cation or the electropositive atom in the compound, like the halogens in oxygen acids and oxides, or carbon in oxides and halogenides. If the element name consists only of one *kanji*, such as for gold 金, phosphorus 燐, lead 鉛 and silver 銀 or a derivation of them (mercury 水銀, platinum 白金 and tin 亜鉛) or if the name consists only of *kana*, no additional *kanji* is attached. 化 *ka* is used if the element represents the electronegative atom or anion, e.g., the halogens in halogenides, carbon in acetylides and sulfur in sulfides. The two important exceptions are hydrogen in hydrides, which are named 水素化 *suisoka*, and sulfur as electropositive atom by 硫黄 *iou*, e.g., in sulfur dioxide SO_2: 二酸化硫黄 *nisankaiou*. Occasionally, other exceptions from this rule may be found, e.g., borane BH_3 ボラン, which has covalent boron–hydrogen bonds but is also sometimes called 硼化水素 *houkasuiso* (derived from "boron hydride").

As a result, simple[1] binary compounds consisting of atoms of only two elements have names in Japanese that correspond to their chemical formulas, as they do in English. The order is the reverse of the chemical formula, beginning with the anion or electronegative element and followed by the cation or electropositive element, e.g.:

hydrogen chloride (HCl):	塩化水素 *enkasuiso*
silver bromide (AgBr):	臭化銀 *shuukagin*
carbon tetrachloride (CCl_4):	四塩化炭素 *shienkatanso*
bromine chloride (BrCl):	塩化臭素 *enkashuuso*
mercuric iodide (HgI_2):	沃化水銀 *youkasuigin*
hydrocarbon:	炭化水素 *tankasuiso*

If two atoms of different elements share the negative partial charge, the *kanji* of both elements are attached and followed by 化 *ka*. As usual, 素 *so* is attached to the electropositive

---

[1] Simple compound means that either only one compound is possible (e.g., NaCl: 塩化ナトリウム *enkanatoriumu*), only one compound of more possible is formed or it is not intended to indicate one specific compound (e.g., copper chloride: 塩化銅 *enkadou*, including CuCl and $CuCl_2$).

3 Naming of Chemical Compounds

Table 3.3 Examples of chemical names with 素 and 化.

Element	Combination with 素 *so*	Combination with 化 *ka*
Hydrogen	hydrogen peroxide (H_2O_2): 過酸化水素 *kasankasuiso*; sodium hydrogencarbonate ($NaHCO_3$): 炭酸水素ナトリウム *tansansuisonatoriumu*	lithium aluminium hydride ($LiAlH_4$): 水素化アルミニウムリチウム *suisokaaruminiumurichiumu*
Boron	tetrafluoroboric acid (HBF_4): 弗化硼素酸 *fukkahousosan*	selenium(II) boride (Se_3B_2): 硼化セレン *houkaseren*
Carbon	carbon disulfide (CS_2): 二硫化炭素 *niryuukatanso*	sodium carbide (Na_2C_2): 炭化ナトリウム *tankanatoriumu*
Silicon	silicon tetrabromide ($SiBr_4$): 四臭化珪素 *shishuukakeiso*	silicide: 珪化物 *keikabutsu*
Nitrogen	nitrogen monoxide (NO): 一酸化窒素 *issankachisso*	lithium nitride (Li_3N): 窒化リチウム *chikkarichiumu*
Arsenic	arsenic(v) sulfide (As_2S_5): 五硫化砒素 *goryuukahiso*	iron arsenide ($FeAs$): 砒化鉄 *hikatetsu*
Oxygen	dioxygen difluoride (O_2F_2): 二弗化二酸素 *nifukkanisanso*	arsenic trioxide (AsO_3): 三酸化砒素 *sansankahiso*
Sulfur	sulfur chloride (S_2Cl_2): 塩化硫黄 *enkaiou*	hydrogen sulfide (H_2S): 硫化水素 *ryuukasuiso*
Chlorine	ammonium chlorate (NH_4ClO_3): 塩素酸アンモニウム *ensosananmoniumu*	iron trichloride ($FeCl_3$): 塩化第二鉄 *enkadainitetsu*
Bromine	hypobromous acid ($HBrO$): 亜臭素酸 *jiashuusosan*	carbon tetrabromide (CBr_4): 四臭化炭素 *shishuukatanso*
Iodine	silver periodate ($AgIO_6$): 沃素酸銀 *kayousosangin*	silver iodide (AgI): 沃化銀 *youkagin*

atom, as in carbon oxysulfide COS: 酸硫化炭素 *sanryuukatanso*. In a few cases the *kanji* order is arbitrarily in English, in which case the characters may also be combined in both ways. After the first element in the compound name, 化 *ka* is added, after the second, 窒 *so* is attached. Therefore, for nitrogen silicide = silicon nitride (Si_3N_4) the possible Japanese names are 珪化窒素 *keikachisso* and 窒化珪素 *chikkakeiso*.

For some minerals, the transcription of the English name is used without any further *kanji*, like for polyargyrite ($Ag_{24}Sb_2S_{15}$): ポリアージライト, magnesia (MgO): マグネシア, pyrochroite [$Mn(OH)_2$]: パイロクロイト and tachydrite ($CaCl_2 \cdot 2MgCl_2 \cdot 12H_2O$): タキハイドライト.

Besides this, as most of element names are derived from English and expressed with *katakana* in Japanese, most of inorganic compound names are also mixtures of *kanji* and *kana*

Table 3.4 Examples for names of oxides.

Oxide	Japanaese name	Example
monoxide	一酸化 *issanka*	osmium monoxide (OsO): 一酸化オスミウム *issankaosumiumu*
sesquioxide	三二酸化 *sannisanka*	osmium sesquioxide (Os_2O_3): 三二酸化オスミウム *sannisankaosumiumu*
dioxide	二酸化 *nisanka*	osmium dioxide (OsO_2): 二酸化オスミウム *nisankaosumiumu*
trioxide	三酸化 *sansanka*	phosphorous trioxide (P_2O_3 or P_4O_6): 三酸化二燐 *sansankarin*
tetroxide	四酸化 *shisanka*	osmium tetroxide (OsO_4): 四酸化オスミウム *shisankaosumiumu*

relating to the elements the compounds are built of. As described above, the part forming the anion or the electronegative element is placed before the cation, like palladium hydride (Pd_2H): 水素化パラジウム *suisokaparajiumu*, sodium sulfate (Na_2SO_4): 硫酸ナトリウム *ryuusannatoriumu*, potash alum [$KAl(SO_4)_2 \cdot 12H_2O$]: カリ明礬 *karimyouban*, magnesium chloride ($MgCl_2$): 塩化マグネシウム *enkamaguneshiumu* and calcium oxide (CaO): 酸化カルシウム *sankakarushium*. The same principle is used with functional groups having *katakana* names, e.g., calcium cyanide [$Ca(CN)_2$]: シアン化カルシウム and sodium azide (NaN_3): アジ化ナトリウム.

Some *kanji* can be often found in inorganic substance names, indicating the type of compound: For instance, acids are named by adding the *kanji* for acid (酸 *san*) to *kanji* or *kana* of the central element, like for sulfuric acid H_2SO_4 硫酸 *ryuusan* and chromic acid H_2CrO_4 クロム酸 *kuromusan*. Most mineral names end with one of the following *kanji*: 石 *seki* (stone), 鉱 *kou* (ore), 岩 *gan* (rock) or 玉 *gyoku* (jewel). Further *kanji* are used to indicate oxidation number, salts (塩 *en*), compounds with lower oxidation numbers (e.g., with 亜 *a* in: nitric acid HNO_3 硝酸 *shousan* → nitrous acid HNO_2 亜硝酸 *ashousan* or with 次亜 *jia*, e.g., hypochlorous acid HClO 次亜塩素酸 *jiaensosan*), peroxides and peracids [with 過 *ka*, e.g., perchloric acid ($HClO_4$): 過塩素酸 *kaensosan* or barium peroxide (BaO_2) 過酸化バリウム *kasankabariumu*] and orthoacids (with 正 *sei*, e.g., orthophosphoric acid: 正燐酸 *seirinsan*). The naming of acids and salts is described in detail below.

Oxides are named with the Japanese expression for oxide (酸化 *sanka*). The number of oxygen atoms within an inorganic oxide can be indicated by different methods. *kanji* for the number can be used in front of 酸化, as with the examples in Table 3.4.

For some oxides it is common to also specify the number of other atoms. If less than one equivalent of oxygen is within the compound, it is indicated with 亜 *a* (see also names of acids below). In addition, some oxides may be interpreted as acid anhydrites. In conclusion, several possible names exist for some compounds. For instance, oxides of nitrogen may be named as described in Table 3.5.

Table 3.5 Names of nitrogen oxides.

Oxide	Formula	Japanese name
dinitrogen monoxide, nitrogen(I) oxide, nitrous oxide	N_2O	亜酸化窒素 *asankachisso*, 一酸化二窒素 *issankanichisso* 酸化窒素（Ⅰ） *sankachisso(I)*
nitrogen monoxide, nitrogen(II) oxide, nitric oxide	NO	一酸化窒素 *issankachisso* 酸化窒素（Ⅱ） *sankachisso(II)*
dinitrogen trioxide, nitrogen sesquioxide nitrogen(III) oxide, nitrous acid anhydride	N_2O_3	三酸化二窒素 *sansankanichisso* 無水亜硝酸 *musuiashousan* 酸化窒素（Ⅲ） *sankachisso(III)*
nitrogen dioxide, nitrogen(IV) oxide	NO_2	二酸化窒素 *nisankachisso* 酸化窒素（Ⅳ） *sankachisso(IV)*
dinitrogen tetraoxide dinitrogen(IV) oxide, nitrogen peroxide	N_2O_4	四酸化二窒素 *shisankanichisso* 酸化二窒素（Ⅳ） *sankanichisso(IV)* 過酸化窒素 *kasankachisso*
dinitrogen pentoxide, nitrogen(V) oxide, nitric acid anhydride	N_2O_5	五酸化二窒素 *gosankanichisso* 無水窒素 *musuishousan* 酸化窒素（Ⅴ） *sankachisso(V)*
nitrogen trioxide	NO_3	三酸化窒素 *sansankachisso*

The oxidation number may also be used as in English. It is added in brackets behind the *kanji* for the metal, such as in the following examples:

copper(I) oxide (cuprous oxide, Cu_2O):	酸化銅（Ⅰ） *sankadou(I)*
nickel monoxide [nickelous oxide, nickel(II) oxide, NiO]:	酸化ニッケル（Ⅱ） *sankanikkeru(II)*
iron(III) oxide (iron sesquioxide, ferric oxide, Fe_2O_3):	酸化鉄（Ⅲ） *sankatetsu(III)*

Finally, the prefix for ordinals 第 *dai* indicates the oxide in the manner of "the first, second,... oxide of ...". To identify the correct formula, the number of possible oxides has to be known.

Examples:	iron monoxide [iron(II) oxide, ferrous oxide, FeO]:	酸化第一鉄 *sankadaiichitetsu*
	mercurous oxide [mercury(I) oxide, Hg_2O]:	酸化第一水銀 *sankadaiichisuigin*
	mercuric oxide [mercury(II) oxide, HgO]:	酸化第二水銀 *sankadainisuigin*

The following tin oxides show the use of the different possibilities for naming oxides:

  tin monoxide [stannous oxide, 一酸化錫 *issankasuzu* or
   tin(II) oxide, SnO]:   酸化錫（II）*sankasuzu(II)* or
             酸化第一錫 *sankadaiichisuzu*

  tin dioxide [stannic oxide,   二酸化錫 *nisankasuzu* or
   tin(IV) oxide, SnO$_2$]:   酸化スズ（IV）*sankasuzu(IV)*
            or 酸化第二スズ *sankadainisuzu*

In a similar way, the names of inorganic hydroxides, sulfides and halogenides are formed. The term for hydroxides is 水酸化 *suisanka* and for sulfides it is 硫化 *ryuuka*.

  sodium hydroxide (NaOH):   水酸化ナトリウム
              suisankanatoriumu

  carbon monosulfide (CS):   一硫化炭素 *ichiryuukatanso*
  antimony trisulfide [antimony(III) 三硫化アンチモン
   sulfide, Sb$_2$S$_3$]:   *sanryuukaanchimon*
  ferrous sulfide [iron(II) sulfide, FeS]: 硫化第一鉄 *ryuukadaiichitetsu*
  platinum disulfide (PtS$_2$):   硫化第二白金 *ryuukadainihakkin*

Halides may be 弗化 (or フッ化) *fukka* (fluorides), 塩化 *enka* (chlorides), 臭化 *shuuka* (bromides) and 沃化 (or ヨウ化) *youka* (iodides). Similar to oxides, the number of halogen atoms can be expressed with either the prefix for ordinals 第 *dai* or with kanji for numbers, e.g.:

  phosphorous trichloride (PCl$_3$): 三塩化燐 *sanenkarin*
  phosphorus pentachloride (PCl$_5$): 五塩化燐 *goenkarin*
  osmium hexafluoride (OsF$_6$):   六弗化オスミウム
              rokufukkaosumiumu
  sodium aurochloride (NaAuCl$_2$): 第一金塩化水素酸ナトリウム
              daiichikinenkasuisosannatoriumu
  sodium aurichloride (sodium   第二金塩化水素酸ナトリウム
   chloroaurate, NaAuCl$_4$):   *dainikinenkasuisosannatoriumu*

For oxyhalogenides and sulfohalogenides, the halogen is named in the usual way (e.g., 弗化 *fukka*), whereas the oxygen, sulfur or the rest of the molecule is expressed with the transcription of the English word into *kana*.

 Examples: phosphorus oxychloride,   オキシ塩化燐 *okishienkarin* or
       phosphoryl chloride (POCl$_3$): 塩化ホスホリル *enkahosuhoriru*
       thionyl chloride (SOCl$_2$):   塩化チオニル *enkachioniru*
       phosphorus thiochloride (PSCl$_3$): スルフォ塩化燐 *surufoenkarin*

For other inorganic salts, see also Section 3.1.4. For more complex inorganic compounds, such as metal carbonyls, the transcription of the English names are used for type and number of the ligands:

Table 3.6 Zinc and lead derivatives with same Japanese names.

Zinc derivatives		Lead derivatives	
zincic acid (H_2ZnO_2)	亜鉛酸 *aensan*	plumbous acid, plumbous hydroxide, lead(II) hydroxide [$Pb(OH)_2$]	亜・鉛酸, 亜ナマリ酸 *anamarisan* 水酸化鉛（II） *suisankanamari(II)*
zincate (M_2ZnO_2)	亜鉛酸塩 *aensanen*	plumbite (M_2PbO_2)	亜・鉛酸塩, 亜ナマリ酸塩 *anamarisanen*
sodium hydrogenzincate ($NaHZnO_2$)	亜鉛酸水素ナトリウム *aensansuisonatoriumu*	sodium hydrogenplumbite ($NaHPbO_2$)	亜・鉛酸水素ナトリウム, 亜ナマリ酸水素ナトリウム *anamarisansuisonatoriumu*
magnesium zincate ($MgZnO_2$)	亜鉛酸マグネシウム *aensanmaguneshium*	magnesium plumbite ($MgZnO_2$)	亜・鉛酸マグネシウム, 亜ナマリ酸マグネシウム *anamarisanmaguneshiumu*

nickel tetracarbonyl [$Ni(CO)_4$]: ニッケルテトラカルボニル
iron pentacarbonyl [$Fe(CO)_5$]: 鉄ペンタカーボニル *tetsupentaka-boniru*
iron tetranitrosyl [$Fe(NO)_4$]: 鉄テトラニトロシル *tetsutetoranitoroshiru*

There is a danger of confusion concerning the *kanji* 鉛, which is used in the names of lead (鉛 *namari, en*) as well as of zinc (亜鉛 *aen*). The mix-up may occur between zinc compounds and plumbous acid $Pb(OH)_2$ or its derivatives, as the names of plumbous acid derivatives also contain 亜鉛 *aen*. To resolve this problem, a dot (中点 *nakaten*) may be used in case of plumbous acid or its derivatives or *kana* are used instead of the *kanji*, as with the examples in Table 3.6.

3.1.3
Inorganic Acids

Inorganic acid names consist of the combination of element name and the *kanji* for acid, 酸 *san*, like for iodic acid (HIO_3): 沃素酸 *yousosan*, ferric acid (H_2FeO_4): 鉄酸 *tetsusan* or manganic acid (H_2MnO_4): マンガン酸 *mangansan*. As described above, *kana* are sometimes used instead of *kanji* for the element names, such as for phosphoric acid (H_3PO_4): 燐酸 or リン酸 *rinsan* and arsenic acid (H_3AsO_4): 砒酸 or ひ酸 *hisan*. Derivatives of the main acid of an element are indicated with further *kanji* or *kana* in front of the element name (Table 3.7).

Concerning their oxidation numbers, inorganic acids are named using prefixes explained in Fig. 3.1. If acids with two atoms of the central element are known (e.g., disulfuric acid), the same prefixes are used.

Table 3.7 *kanji* and *kana* in names of inorganic acids.

Acid type	Japanese	Example
Acids with lower oxidation number than the main acid (ending "-ous" in English)	亜 *a*	selenious acid (H_2SeO_3): 亜セレン酸 *aserensan*
Hypoacids (acids with two oxidation numbers lower than the main acid)	次亜 *jia*	hypochlorous acid (HClO): 次亜塩素酸 *jiaensosan*
Peracids (peroxo acids)	過 *ka* ペルオキシ ペルオキソ	permanganic acid ($HmnO_4$): 過マンガン酸 *kamangansan* peroxysulfuric acid (H_2SO_5): ペルオキシ硫酸 *peruokishiryuusan*
Pyroacids	ピロ	pyrophosphoric acid ($H_4P_2O_7$): ピロ燐酸 *pirorinsan*
Orthoacids	正 *sei* or オルト	orthosilicic acid (H_4SiO_4): 正珪酸 *seikeisan*, orthotelluric acid (H_6TeO_6): オルトテルル酸 *orutoterurusan*
Metaacids	メタ	metaboric acid (HBO_2): メタ硼酸 *metahousan*
Paraacids	パラ	parasorbic acid ($C_6H_8O_2$): パラソルビン酸 *parasorubinsan*
Thio acids	チオ	thiosulfuric acid ($H_2S_2O_3$): チオ硫酸 *chioryuusan*
Poly acids	ポリ or 多 *ta*	polysilicic acid: ポリ珪酸（ポリケイ酸）*porikeisan*, 多珪酸（多ケイ酸）*takeisan*

acid with two lower oxidation numbers than main acid	acid with lower oxidation number than main acid	main acid	peroxo acid
hypo-...-ous 次亜 ... 酸 e.g. HClO, H_2SO_2, $H_4P_2O_4$	...-ous 亜 ... 酸 e.g. $HClO_2$, H_2SO_3, $H_4P_2O_5$	...-ic ... 酸 e.g. $HClO_3$, H_2SO_4, $H_4P_2O_7$	peroxo-...-ic 過 ... 酸 e.g. $HClO_5$, H_2SO_5, $H_4P_2O_8$

Fig. 3.1 Prefixes in names of inorganic acids.

Tables 3.8–3.19 compile the names of selected important inorganic acids, including some acidic compounds that are not stable in the free form, e.g., sulphuric(II) acid (H_2SO_2), or acids that are only hypothetical, such as manganic acid (H_2MnO_4). Because of a similar naming of certain salts in Japanese, amphoteric hydroxides [e.g., $Cu(OH)_2$] or oxides (e.g., MoO_3) that form salts under basic conditions (e.g., by dissolving in concentrated alkalki hydroxide solutions) and other salt-forming inorganic molecules (like titan dioxide, which forms titanates by melting together with other metal salts, i.e., alkali oxides and hydroxides) are added to the table. However, not all names are commonly used, e.g., 臭化水素酸 is favored over 臭酸.

Table 3.8 Group VII compounds.

Compound name	Formula	Japanese name
hydrofluoric acid (aqueous hydrogen fluoride)	aq. HF	弗化水素酸 *fukkasuisosan*, フッ化水素酸 *fukkasuisosan*, 弗酸 *fussan*, フツ酸 *fussan*
hypofluorous acid	HFO	次亜フッ素酸, 次亜弗素酸 *jiafussosan*
hydrochloric acid (aqueous hydrogen chloride)	aq. HCl	塩酸 *ensan*, 塩酸水素酸 *enkasuisosan*
hypochlorous acid, chloric(I)acid	HClO	次亜塩素酸 *jiaensosan*
chlorous acid, chloric(III) acid	$HClO_2$	亜塩素酸 *aensosan*
chloric acid, chloric(V) acid	$HClO_3$	塩素酸 *ensosan*
perchloric acid, chloric(VII) acid	$HClO_4$	過塩素酸 *kaensosan*
hydrobromic acid (aqueous hydrogen bromide)	aq. HBr	臭化水素酸 *shuukasuisosan*, 臭酸 *shuusan*
hypobromous acid, bromic(I) acid	HBrO	次亜臭素酸 *jiashuusosan*
bromous acid, bromic(III) acid	$HBrO_2$	亜臭素酸 *ashuusosan*
bromic acid, bromic(V) acid	$HBrO_3$	臭素酸 *shuusosan*
perbromic acid, bromic(VII) acid	$HBrO_4$	過臭素酸 *kashuusosan*
hydroiodic acid (aqueous hydrogen iodide)	aq. HI	沃化水素酸 *youkasuisosan*, ヨウ化水素酸 *youkasuisosan*, ヨウ酸 *yousan*, 沃酸 *yousan*
hypoiodous acid, iodic(I) acid	HIO	次亜ヨウ素酸 *jiayousosan*, 次亜沃素酸 *jiayousosan*
iodous acid, iodic(III) acid	HIO_2	亜ヨウ素酸 *ayousosan*, 亜沃素酸 *ayousosan*
iodic acid, iodic(V) acid	HIO_3	ヨウ素酸 *yousosan*, 沃素酸 *yousosan*
periodic acid	HIO_4	過ヨウ素酸 *kayousosan*, 過沃素酸 *kayousosan*, 過ヨード酸 *kayo-dosan*
orthoperiodic acid	H_5IO_6	オルト過ヨウ素酸 *orutokayousosan*, オルト過沃素酸 *orutokayousosan*

3.1 Naming of Elements and Inorganic Compounds

Table 3.9 Group VI compounds (for oxoacids of sulfur see Tables 3.10–3.13).

Compound name	Formula	Japanese name
water	H_2O	水 mizu
hydrogen peroxide	H_2O_2	過酸化水素 kasankasuiso
hydrogen sulfide	H_2S	硫化水素 ryuukasuiso
polysulfane, hydrogen polysulfide	H_2S_n	多硫化水素 taryuukasuiso
hydrogen selenide	H_2Se	セレン化水素 serenkasuiso
selenious acid	H_2SeO_3	亜セレン酸 aserensan
selenic acid	H_2SeO_4	セレン酸 serensan
hydrogen telluride	H_2Te	テルル化水素 terurukasuiso
tellurous acid	H_2TeO_3	亜テルル酸 aterurusan
telluric acid	H_2TeO_4	テルル酸 terurusan
orthotelluric acid	H_6TeO_6	オルトテルル酸 orutoterurusan

Table 3.10 Monosulfuric acids.

H_2SO_2	H_2SO_3	H_2SO_4	H_2SO_5
HO–S–OH	O=S(OH)(OH)	O=S(=O)(OH)(OH)	O=S(=O)(OH)(OOH)
sulfoxylic acid (hyposulfurous acid) スルホキシル酸	sulfurous acid 亜硫酸	sulfuric acid 硫酸	peroxysulfuric acid, Caro's acid 過硫酸, カロ酸, ペルオキソ一硫酸

Table 3.11 Disulfuric acids.

$H_2S_2O_4$	$H_2S_2O_5$	$H_2S_2O_6$	$H_2S_2O_7$	$H_2S_2O_8$
dithionous acid 亜ジチオン酸	disulfurous acid 二亜硫酸, ピロ亜硫酸	dithionic acid ジチオン酸	disulfuric acid 二硫酸	peroxodisulfuric acid Marshall's acid ペルオキソ二硫酸, マーシャルの酸

Table 3.12 Thiosulfuric acids.

$H_2S_2O_2$	$H_2S_2O_3$	$H_2S_3O_6$
thiosulfurous acid チオ亜硫酸	thiosulfuric acid チオ硫酸	trithionic acid トリチオン酸

Table 3.13 Polysulfuric acids.

$H_2S_3O_{10}$	$H_2S_4O_{13}$
trisulfuric acid トリ硫酸	tetrasulfuric acid テトラ硫酸
$H_2S_nO_6$	$H_2S_nO_{3n+1}$
polythionic acid ポリチオン酸	polysulfuric acid ポリ硫酸

Table 3.14 Group V compounds (for oxoacids of phosphorus see Tables 3.15–3.17).

Compound name	Formula	Japanese name
ammonia	NH_3	アンモニア
hydrazine, diamide, diazane	N_2H_4	ヒドラジン, ジアミド
azoimide, hydrazoic acid, hydrogen azide	HN_3	窒化水素酸 *chikkasuisosan*, ヒドラゾ酸 *hidorazosan*
hyponitrous acid	$H_2N_2O_2$	次亜硝酸 *jiashousan*
nitrous acid	HNO_2	亜硝酸 *ashousan*
nitric acid	HNO_3	硝酸 *shousan*
peroxynitric acid	HNO_4	ペルオキシ硝酸 *peruokishishousan*

Table 3.14 (continued)

Compound name	Formula	Japanese name
hydroxylamine	H_3NO	ヒドロキシルアミン
orthonitric acid	H_3NO_4	オルト硝酸 *orutoshousan*
phosphine, phosphane, phosphorous trihydride	PH_3	燐化水素 *rinkasuiso*, フォスフィン
polyphosphoric acid	$H_{n+2}P_nO_{3n+1}$	ポリ燐酸 *poririnsan*
arsane, arsenic hydride, arsine	AsH_3	砒化水素 *hikasuiso*, 水素化砒素 *hikasuiso*, アーシン, アルシン
arsenious acid	H_3AsO_3	亜砒酸 *ahisan*, 亜ひ酸 *ahisan*
arsenic acid	H_3AsO_4	砒酸 *hisan*, ひ酸 *hisan*
stibane, antimony hydride	SbH_3	アンチモン化水素 *anchimonkasuiso*, スティビン
antimonous acid	$H_3SbO_3 = Sb(OH)_3$	亜アンチモン酸 *aanchimonsan*
antimonic acid	H_3SbO_4[a]	アンチモン酸 *anchimonsan*

[a] Only as $H_3SbO_4 \cdot 2H_2O$ [$HSb(OH)_6$].

Table 3.15 Monophosphoric acids and their tautomeric forms.

H_3PO	H_3PO_2	H_3PO_3	H_3PO_4	H_3PO_5
H—P(OH)(H)	H—P(=O)(OH)(H)	H—P(=O)(OH)(OH)	HO—P(=O)(OH)(OH)	HO—P(=O)(OOH)(OH)
phosphinous acid 亜ホスフィン酸, 亜フォスフィン酸	phosphinic acid ホスフィン酸, フォスフィン酸	phosphonic acid ホスホン酸, フォスフォン酸	phosphoric acid, orthophosphoric acid 燐酸, リン酸, オルト燐酸, オルトリン酸, 正燐酸, 正リン酸	peroxophosphoric acid 過燐酸過リン酸[a], ペルオキソ燐酸, ペルオキソリン酸
H—P(=O)(H)(H)	H—P(OH)(OH)	HO—P(OH)(OH)		
phosphine oxide ホスフィンオキシド	phosphonous acid 次亜燐酸, 次亜リン酸, 亜ホスホン酸	phosphorous acid 亜燐酸, 亜リン酸		

[a] 燐酸 is sometimes used as abbreviation for "superphosphate" [$Ca(H_2PO_4)_2 + 2CaSO_4$ fertilizer], which is correctly named 過燐酸石灰.

Table 3.16 Diphosphoric acids, their tautomeric and isomeric forms.

$H_4P_2O_2$	$H_4P_2O_3$	$H_4P_2O_4$	$H_4P_2O_5$	$H_4P_2O_6$	$H_4P_2O_7$	$H_4P_2O_8$
hypodiphosphonous acid 次二亜ホスホン酸[a]	diphosphonous acid 二亜ホスホン酸[a], 二燐酸 (I), 二リン酸 (I)[a]	hypodiphosphonic acid 次ジホスホン酸	diphosphonic acid ジホスホン酸	hypodiphosphonic acid, diphosphonic(IV) acid 二燐酸 (IV), 二リン酸 (IV), 次燐酸, 次リン酸	diphosphoric acid, pyrophosphoric acid ピロ燐酸, ピロリン酸, 二燐酸, 二リン酸	peroxodiphosphoric acid 過二燐酸, 過二リン酸
		hypodiphosphorous acid 二燐酸 (II), 二リン酸 (II)	diphosphorous acid 亜二燐酸, 二亜リン酸	diphosphorous acid 亜二燐酸 (III, V), 二リン酸 (III, V)		
				diphosphoric(II,IV) acid 二燐酸 (II, IV), 二リン酸 (II, IV)		
				isohypophosphoric acid, diphosphoric(III,V) acid 二燐酸 (III, V), 二リン酸 (III, V), イソ次燐酸, イソ次リン酸		

[a] Names not commonly accepted.

Table 3.17 Oligo- and polyphosphoric acids.

Orthooligo phosphoric acids	$H_5P_3O_{10}$ triphosphoric acid 三燐酸, 三リン酸, トリ燐酸, トリリン酸	$H_6P_4O_{13}$ tetraphosphoric acid テトラ燐酸, テトラリン酸	$H_{n+2}P_nO_{3n+1}$ polyphosphoric acid ポリ燐酸, ポリリン酸
Metaoligo phosphoric acids	$H_3P_3O_9$ trimetaphosphoric acid トリメタ燐酸, トリメタリン酸	$H_4P_4O_{12}$ tetrametaphosphoric acid テトラメタ燐酸, テトラメタリン酸	$(HPO_3)_n$ metaphosphoric acid メタ燐酸, メタリン酸

Table 3.18 Group IV and III compounds.

Compound name	Formula	Japanese name
carbonic acid	H_2CO_3	炭酸 tansan
orthocarbonic acid	H_4CO_4	オルト炭酸 orutotansan
metasilicic acid, silantriole	$H_4SiO_3 = H_3Si(OH)$	メタ珪酸 metakeisan, メタケイ酸 metakeisan
silicic acid, orthosilicic acid	H_4SiO_4	ケイ酸 keisan, 珪酸 keisan, 正珪酸 seikeisan, 正ケイ酸 seikeisan, オルト珪酸 orutokeisan, オルトケイ酸 orutokeisan
disilicic acid	$H_2Si_2O_5$	二ケイ酸 nikeisan, 二珪酸 nikeisan, ジケイ酸 jikeisan, ジ珪酸 jikeisan
trisilicic acid	$H_8Si_3O_{10}$	トリ珪酸 torikeisan, トリケイ酸 torikeisan, 三珪酸 sankeisan, 三ケイ酸 sankeisan
metatrisilicic acid	$H_4Si_3O_8$	メタトリ珪酸 metatorikeisan, メタトリケイ酸 metatorikeisan, メタ三珪酸 metasankeisan, メタ三ケイ酸 metasankeisan
tetrasilicic acid	$H_{10}Si_4O_{13}$	テトラケイ酸 tetorakeisan, テトラ珪酸 tetorakeisan, 四珪酸 yonkeisan, 四ケイ酸 yonkeisan
metatetrasilicic acid	$H_6Si_4O_{11}$	メタテトラケイ酸 metatetorashikeisan, メタテトラ珪酸 metatetorashikeisan, メタ四珪酸 metayonkeisan, メタ四ケイ酸 metayonkeisan
polysilicic acid	$[H_2SiO_3]_x$, $[H_6Si_4O_{11}]_x$, $[H_2Si_2O_5]_x$	ポリ珪酸 porikeisan, ポリケイ酸 porikeisan, 多珪酸 takeisan, 多ケイ酸 takeisan
germanous acid	H_2GeO_2	亜ゲルマニウム酸 agerumaniumusan
metagermanic acid	H_2GeO_3	メタゲルマニウム酸 metagerumaniumusan
germanic acid	H_4GeO_4	ゲルマニウム酸 gerumaniumusan
metastannic acid	H_2SnO_3	メタ錫酸 metasuzusan, メタスズ酸 metasuzusan
orthostannic acid	H_4SnO_4	オルト錫酸 orutosuzusan, オルトスズ酸 orutosuzusan, 正錫酸 seisuzusan, 正スズ酸 seisuzusan
stannic acid	$H_2SnO_3/H_2Sn(OH)_6$	錫酸 suzusan, スズ酸 suzusan
plumbous acid plumbic(II) acid	$H_2PbO_2 = Pb(OH)_2$	亜・鉛酸 aensan
metaplumbic acid	H_2PbO_3	メタ鉛酸 metaensan
orthoplumbic acid	$H_4PbO_4 = Pb(OH)_4$	正鉛酸 seiensan
metaboric acid	HBO_2	メタ硼酸 metahousan, メタホウ酸 metahousan

3.1 Naming of Elements and Inorganic Compounds

Table 3.18 (continued)

Compound name	Formula	Japanese name
boric acid, boric(III) acid	$H_3BO_3 = B(OH)_3$	硼酸 housan, ホウ酸 housan
tetraboric acid pyroboric acid	$H_2B_4O_7 = B_4O_5(OH)_2$	四硼酸 yonhousan, テトラホウ酸 tetorahousan
pentaboric acid	$H_3B_5O_9 = B_5O_6(OH)_3$	五硼酸 gohousan, ペンタホウ酸 pentahousan
aluminum hydroxide	$Al(OH)_3$	水酸化アルミニウム suisankaaruminiumu
salts: aluminate	$MAlO_2$	アルミン酸塩 aruminsanen

Table 3.19 Subgroup element compounds.

Compound name	Formula	Japanese name
copper(II) hydroxide salts: cuprate(II), cuprite	$Cu(OH)_2$ $M_2[Cu(OH)_4]$	水酸化第二銅 suisankadainidou 銅酸塩 dousanen
gold(III) hydroxide, "auric acid" salts: aurate	$H_3AuO_3 = Au(OH)_3$ $M[Au(OH)_4]$	水酸化金 suisankakin, 金酸 kinsan 金酸塩 kinsanen
zinc hydroxide, "zincic acid" salts: zincate	$H_2ZnO_2 = Zn(OH)_2$ $M_2[Zn(OH)_4]$	水酸化亜鉛 suisankaaen, 亜鉛酸 aensan 亜鉛酸塩 aensanen
Scandium hydroxide salts: scandate	$Sc(OH)_3$ $MScO_2, M_3ScO_3,$ $M_3[Sc(OH)_6]$	水酸化スカンジウム suisankasukanjiumu スカンジウム酸塩 sukanjiumusanen
titanic acid	H_2TiO_3/H_4TiO_4	チタン酸 chitansan
metatitanic acid	H_2TiO_3	メタチタン酸 metachitansan
orthotitanic acid	H_4TiO_4	オルトチタン酸 orutochitansan
orthozirconic acid	$H_4ZrO_4 = Zr(OH)_4$	オルトジルコン酸 orutojirukonsan
metazirconic acid	H_2ZrO_3	メタジルコン酸
vanadic acid	$(HVO_3)_n/H_3VO_4/H_4V_2O_7$	ヴァナジン酸 vanajinsan, バナジン酸 banajinsan
orthovanadic acid	H_3VO_4	オルトバナジン酸 orutobanajinsan, オルトヴァナジン酸 orutovanajinsan
metavanadic acid	$(HVO_3)_n$	メタヴァナジン酸 metavanajinsan, メタバナジン酸 metabanajinsan

Table 3.19 *(continued)*

Compound name	Formula	Japanese name
chromous acid	$HCrO_2$	亜クロム酸 *akuromusan*
chromium hydroxide	$Cr(OH)_3$	水酸化クロム *suisankakuromu*
chromic acid	H_2CrO_4	クロム酸 *kuromusan*
dichromic acid	$H_2Cr_2O_7$	重クロム酸 *juukuromusan*, ジクロム酸 *jikuromusan*, 二クロム酸 *nikuromusan*
trichromic acid	$H_2Cr_3O_{10}$	トリクロム酸 *torikuromusan*
tetrachromic acid	$H_2Cr_4O_{13}$	テトラクロム酸 *tetorakuromusan*
polychromic acid	$H_2[Cr_nO_{3n+1}]$	ポリクロム酸 *porikuromusan*
perchromic(vi) acid	H_2CrO_6	過クロム酸（VI）*kakuromusan(VI)*
perchromic(v) acid	H_3CrO_8	過クロム酸（V）*kakuromusan(V)*
molybdenum trioxide, "molybdic acid"	MoO_3 H_2MoO_4	三酸化モリブデン *sansankamoribuden*, モリブデン酸 *moribudensan*
tungsten trioxide, "tungstic acid", "orthotungstic acid"	WO_3 H_2WO_4	三酸化タングステン *sansankatangusuten*, タングステン酸 *tangusutensan*, ウォルフラム酸 *worufuramusan*, オルトウォルフラム酸 *orutoworufuramusan*, オルトタングステン酸 *orutotangusutensan*
manganous acid	H_2MnO_3	亜マンガン酸 *amangansan*
hypomanganic acid, manganic(v) acid	H_3MnO_4	ヒポマンガン酸 *hipomangansan*
manganic acid	H_2MnO_4	マンガン酸 *mangansan*
permanganic acid	$HmnO_4$	過マンガン酸 *kamangansan*
rhenous acid	H_2ReO_3	亜レニウム酸 *areniumusan*
rhenic acid	H_2ReO_4	レニウム酸 *reniumusan*
perrhenic acid	$HreO_4$	過レニウム酸 *kareniumusan*
ferrous hydroxide	$Fe(OH)_2$	水酸化第一鉄 *suisankadaiichitetsu*
ferric hydroxide salts: ferrite, ferrate(iii)	$Fe(OH)_3$ $MFeO_2$	水酸化第二鉄 *suisankadainitetsu* 亜鉄酸塩 *atetsusanen*
ferric(iv) acid	H_4FeO_4	鉄（IV）酸 *tetsu(IV)san*
ferric(v) acid	H_3FeO_4	鉄（V）酸 *tetsu(V)san*
ferric acid, ferric(vi) acid	H_2FeO_4	鉄酸 *tetsusan*
ruthenic(vi) acid	H_2RuO_4	ルテニウム（VI）酸 *ruteniumu(VI)san*

Table 3.19 (continued)

Compound name	Formula	Japanese name
ruthenic(VII) acid	$HRuO_4$	ルテニウム（VII）酸 *ruteniumu(VII)san*
perruthenic acid	$HRuO_4$	過ルテニウム酸 *karuteniumusan*
osmic acid	H_2OsO_4	オスミウム酸 *osumiumusan*
platinic acid	H_2PtO_3	白金酸 *hakkinsan*

Notably, Japanese nomenclature may not always be consistent, in particular for less common compounds, and the correct meaning has to be recognized by the context. For instance, 亜ホスフィン酸 is mainly used for phosphinous acid [$H_2P(OH)$], but can also be found for phosphenous acid [$PO(OH)$]. Occasionally, for some phosphoric acids, the names of the tautomers are used, as this was also not clear in English in the past. For instance, 次亜リン酸 ("hypophosphorous acid") is phosphonous acid, but sometimes is also used for phosphinic acid. Another example is prefixes, e.g., 三 before sulfuric acid may be interpreted as trisulfuric acid ($H_2S_3O_{10}$, also called トリ硫酸) or as three sulfate anions in salts of sulfuric acid, e.g., in 三硫酸二鉄 [$Fe_2(SO_4)_3$].

For some acids, various names are used. Some of the names are more common than others. For instance, persulfuric acid (H_2SO_5) is most commonly called 過硫酸 *karyuusan* or ペルオキソ硫酸 *peruokisoryuusan*. Its trivial name is カロー酸 *karo-san*. The term "peroxy" may also be transcribed differently, so that ペルオキシ硫酸 *peruokishiryuusan* or just オキシ硫酸 *okishiryuusan* are also possible. Finally, to differentiate from peroxodisulfuric acid ($H_2S_2O_8$), persulfuric acid may be named ペルオキソ一硫酸, ペルオキソモノ硫酸, ペルオキシ一硫酸 and ペルオキシモノ硫酸.

3.1.4
Inorganic Salts and Ores

Generic salt names are built by adding the *kanji* 塩 *en* after the acid's name, such as for nitrates 硝酸塩 *shousanen* (from nitric acid: 硝酸 *shousan*) or for hydrogen selenides セレン化水素塩 *serenkasuisoen*. For specific salts and if the indication of the oxidation number is not necessary, the name of the acid is simply followed by the name of the metal cation, as it is described above, e.g.:

silver chloride (AgCl):	塩化銀 *enkagin*
potassium nitrate (KNO_3):	硝酸カリウム *shousankariumu*
magnesium bromide ($MgBr_2$):	臭化マグネシウム *shuukamaguneshiumu*
aluminium hydroxide [$Al(OH)_3$]:	水酸化アルミニウム *suisankaaruminiumu*

3.1.4.1 Naming Simple Salts

There are three ways to indicate the oxidation number of inorganic salts:

- First, according to English names, the oxidation number of the metal cation is added in brackets at the end of the name; this is mainly used for oxygen acids.

Examples:	mercury(I) chloride (mercurous chloride, Hg_2Cl_2):	塩化水銀（I）*enkasuigin (I)*
	iron(II) sulfate (ferrous sulfate, $FeSO_4$):	硫酸鉄（II）*ryuusantetsu(II)*
	arsenic(III) fluoride (arsenic trifluoride, AsF_3):	弗化砒素（III）*fukkahiso(III)*
	lead(II) acetate [$Pb(CH_3COO)_2$]:	酢酸鉛（II）*sakusanen (II)*

- The second possibility is to add the number of acid residue parts in front of the acid's name. This applies if the acid residue consists of only one atom, like oxides, halogenides and sulfides. Examples of salts with just one metal cation are:

	platinum(II) oxide (platinum monoxide, PtO):	一酸化白金 *issankahakkin*
	rhodium dioxide (RhO_2):	二酸化ロジウム *nisankarojiumu*
	carbon disulfide (CS_2):	二硫化炭素 *niryuukatanso*
	samarium trichloride ($SmCl_3$):	三塩化サマリウム *sanenkasamariumu*
	phosphorus pentoxide (P_2O_5):	五酸化燐 *gosankarin*

If the salt contains more than one cation atom, their number is also added to the compound name, either together with the number of acid residues or in front of the cation's name.

Examples:	sulfur sesquioxide [sulfur(III) oxide, S_2O_3]:	三二酸化硫酸 *sannisankaryuusan*
	titanium sesquisulfide (Ti_2S_3):	三二硫化チタン *sanniryuukachitan*
	trinickel tetroxide [nickel(II,III) oxide, Ni_3O_4]:	四三酸化ニッケル *shisansankanikkeru*
	dinitrogen tetroxide (N_2O_4):	四酸化二窒素 *shisankanichisso*
	manganese sesquioxide (Mn_2O_3):	三二酸化マンガン *sannisankamangan*

For some oxygen acids, this method can not be applied, as a specific derivative is named. For example, 二硫酸錫 *niryuusansuzu* does not mean tin(IV) sulfate, $Sn(SO_4)_2$, but tin disulfate (SnS_2O_7 because it is derived from disulfuric acid (pyrosulfuric acid, $H_2S_2O_7$) 二硫酸 *niryuusan*.

- Finally, salt names can be formed with the prefix for ordinals 第 *dai* in combination with the figure in front of the metal cation with the meaning of, for example, "the first sulfate of …" or "the second chloride of …". Because this is not directly derived from the oxidation numbers, the common oxidation numbers of the metals have to be known.

Table 3.20 Metal names in salts using 第.

Metal	第一	第二
tin	Sn(II): 第一錫	Sn(IV): 第二錫
lead	Pb(II): 第一鉛	Pb(IV): 第二鉛
phosphorus	P(III): 第一燐	P(V): 第二燐
arsenic	As(III): 第一砒素	As(V): 第二砒素
antimony	Sb(III): 第一アンチモン	Sb(V): 第二アンチモン
copper	Cu(I): 第一銅	Cu(II): 第二銅
silver	Ag(I): 第一銀	Ag(II): 第二銀
gold	Au(I): 第一金	Au(III): 第二金
zinc	Zn(II): 第一亜鉛	Zn(II): 第二亜鉛
mercury	Hg(I): 第一水銀	Hg(II): 第二水銀
titanium	Ti(II): 第一チタン	Ti(IV): 第二チタン
chromium	Cr(II): 第一クロム	Cr(III): 第二クロム
manganese	Mn(II): 第一マンガン	Mn(III): 第二マンガン
iron	Fe(II): 第一鉄	Fe(III): 第二鉄
cobalt	Co(II): 第一コバルト	Co(III): 第二コバルト
nickel	Ni(II): 第一ニッケル	Ni(III): 第二ニッケル
palladium	Pd(II): 第一パラジウム	Pd(IV): 第二パラジウム #
platinum	Pt(II): 第一白金	Pt(IV): 第二白金

Examples are given in Table 3.20. 第三 *daisan* is used rarely and is only common for tertiary salts of phosphoric acids (see below).

This way of naming can be compared with the English pairs of names ending with either "-ous" or "-ic", such as cuprous (Cu^+) and cupric (Cu^{2+}) or platinous (Pt^{2+}) and platinic (Pt^{4+}). In the case of "-ous", 第一 *daiichi* is used. 第二 *daini* indicates "-ic".

Examples:

for platinum:	platinous sulfide (PtS):	硫化第一白金 *ryuukadaiichihakkin*
	platinum disulfide (PtS$_2$):	硫化第二白金 *ryuukadainihakkin*
for mercury:	mercury(I) iodide (mercurous iodide, Hg$_2$I$_2$):	沃化第一水銀 *youkadaiichisuigin*
	mercury(II) iodide (mercuric iodide, HgI$_2$):	沃化第二水銀 *youkadainisuigin*
for tin:	tin(II) chromate (stannous chromate, SnCrO$_4$):	クロム酸第一錫 *kuromusandaiichisuzu*
	tin(IV) chromate [stannic chromate, Sn(CrO$_4$)$_2$]:	クロム酸第二錫 *kuromusandainisuzu*

Because of these possibilities for nomenclature, different names can be found for inorganic salts in Japanese publications. For instance, iron oxides are named with 酸化鉄（II）, 酸化第一鉄 or 一酸化鉄 for iron(II) oxide (ferrous oxide, iron monoxide, FeO) and with 酸化鉄（III）, 酸化第二鉄 or 三二酸化鉄 *sannisankatetsu* for iron(III) oxide (ferric oxide, iron sesquioxide, Fe_2O_3). In addition, as described above, for some elements either *kanji* or *kana* may be used, resulting in different written names, as for the following two examples:

$$\text{iodine pentafluoride (IF}_5\text{):} \quad \text{五弗化沃素, 五フッ化ヨウ素, 五ふっかよう素 } \textit{gofukkayouso}$$

$$\text{phosphorous bromides (PBr}_3/\text{PBr}_5\text{):} \quad \text{臭化燐, シュウ化リン, しゅう化りん } \textit{shuukarin}$$

3.1.4.2 Salts of Dibasic Acids

For dibasic acids such as sulfuric acid H_2SO_4 the number of hydrogen atoms in salts has to be indicated. The following methods are possible:

- The most common and modern way is to include 水素 *suiso* (hydrogen) between the acid's name and the name of the metal cation. For example, for sulfuric acid (H_2SO_4) 硫酸 *ryuusan* two potassium salts are possible:

  potassium hydrogensulfate ($KHSO_4$): 硫酸水素カリウム *ryuusansuisokariumu*

  potassium sulfate (K_2SO_4): 硫酸カリウム *ryuusankariumu*

- For hydrogen salts of some acids, 重 *juu* ("heavy", in this context: "bi-") may be added in front of the acid's name. For example, sodium salts of carbonic acid (H_2CO_3) 炭酸 *tansan* are sodium hydrogencarbonate (sodium bicarbonate, $NaHCO_3$): 重炭酸ナトリウム *juutansannatoriumu* and sodium carbonate (Na_2CO_3): 炭酸ナトリウム *tansannatoriumu*. This is only possible for selected acids, because, in other cases, the name refers to different acids. For example, for chromic acid H_2CrO_4, 重 modifies the name into dichromic acid ($H_2Cr_2O_7$): 重クロム酸銀 *juukuromusangin*. Therefore, 重クロム酸銀 *juukuromusangin* means silver dichromate ($Ag_2Cr_2O_7$) and not silver hydrogenchromate ($AgHCrO_4$).
- Hydrogen salts can also be named with 酸性 *sansei* ("acidic") in front of the acid's name, such as for ammonium salts of sulfurous acid (H_2SO_3) 亜硫酸 *aryuusan*: ammonium hydrogensulfite [$(NH_4)HSO_3$]: 酸性亜硫酸アンモニウム *sanseiaryuusananmoniumu* and ammonium sulfite [$(NH_4)_2SO_3$]: 亜硫酸アンモニウム *aryuusananmoniumu*.

3.1.4.3 Salts of Tribasic Acids

For tribasic acids like phosphoric acid H_3PO_4, the following methods are applied:
- With the number of hydrogen atoms: Behind the acid's name, 水素 *suiso* is added for one hydrogen atom or with the additional *kanji* 二 *ni* for two hydrogen atoms.

Table 3.21 Names of sodium and magnesium salts of phosphoric acid.

	Sodium salts	Magnesium salts
$M^I H_2 PO_4$	sodium dihydrogenphosphate (NaH_2PO_4): 燐酸二水素ナトリウム *rinsannisuisonatoriumu* 燐酸二水素一ナトリウム *rinsannisuisoichinatoriumu*	magnesium dihydrogenphosphate, magnesium biphosphate [$Mg(H_2PO_4)_2$]: 燐酸二水素マグネシウム *rinsannisuisomaguneshiumu* 燐酸二水素一マグネシウム *rinsannisuisoichimaguneshiumu*
$M^I_2 HPO_4$	disodium hydrogenphosphate, dibasic sodium phosphate (Na_2HPO_4): 燐酸水素二ナトリウム *rinsansuisoninatoriumu* 第二燐酸ナトリウム *dainirinsannatoriumu*	magnesium hydrogenphosphate ($MgHPO_4$): 燐酸水素二マグネシウム *rinsansuisonimaguneshiumu* 第二燐酸マグネシウム *dainirinsanmaguneshiumu*
$M^I_3 PO_4$	sodium phosphate, trisodium phosphate, tribasic sodium phosphate (Na_3PO_4): 燐酸ナトリウム *rinsannatoriumu*[a] 燐酸三ナトリウム *rinsansannatoriumu* 第三燐酸ナトリウム *daisanrinsannatoriumu*	magnesium phosphate [$Mg_3(PO_4)_2$]: 燐酸マグネシウム *rinsanmaguneshiumu*[a] 燐酸三マグネシウム *rinsansanmaguneshiumu* 第三燐酸マグネシウム *daisanrinsanmaguneshiumu*

[a] 燐酸ナトリウム and 燐酸マグネシウム are also used as general terms covering all kinds of sodium and magnesium phosphates.

- The prefix for ordinals 第 *dai* can be used in front of the name in the meaning of 第一 *daiichi* for "primary salt", i.e., the dihydrogen salt ($H_2PO_4^-$), of 第二 *daini* for "secondary salt", i.e., the monohydrogen salt (HPO_4^{2-}), or of 第三 *daisan* for "tertiary salt" (PO_4^{3-}).
- The number of metal atoms can be indicated with the corresponding *kanji* in front of the metal's name, e.g., with 三 *san* in the case of trisodium sulfate.

For the possible sodium and magnesium salts of phosphoric acid (H_3PO_4) 燐酸 *rinsan*, the names given in Table 3.21 can be found.

3.1.4.4 Complex Salts

For complex salts, like hexacyanoferrates(III), [$Fe(CN)_6$]$^{3-}$, or tetrachloroplatinates(II), [$PtCl_4$]$^{2-}$, most names start with the number and denotation of the ligand, which is usually written in *kana*. This is followed by the names of the central atom or ion and the counter cation. The oxidation number of the central atom may be specified with 第一 and 第二 in front of its name as explained above or with the number in brackets behind its name.

Examples for chloroplatinates:

    tetrachloroplatinic(II) acid (H_2PtCl_4): テトラクロロ第一白金酸 *tetorakurorodaiichihakkinsan*

    → ammonium tetrachloroplatinate(II) ($(NH_4)_2PtCl_4$): テトラクロロ第一白金酸アンモニウム *tetorakurorodaiichihakkinsananmoniumu* テトラクロロ白金（II）酸アンモニウム *tetorakurorohakkin(II)sananmoniumu*

    hexachloroplatinic(IV) acid (H_2PtCl_6): ヘキサクロロ第二白金酸 *hekusakurorodainihakkinsan*

    → ammonium hexachloroplatinate(IV) ($(NH_4)_2PtCl_6$): ヘキサクロロ第二白金酸アンモニウム *hekusakurorodainihakkinsananmoniumu* ヘキサクロロ白金（IV）酸アンモニウム *hekisakurorohakkin(IV)sananmoniumu*

If the oxidation status of the central ion and the number of ligand groups is clear, the names can also be abbreviated, such as to:

    ammonium chloroplatinite, $(NH_4)_2[PtCl_4]$: 塩化第一白金酸アンモニウム *enkadaiichihakkinsananmoniumu*

    ammonium chloroplatinate(IV), $(NH_4)_2[PtCl_6]$: 塩化第二白金酸アンモニウム *enkadainihakkinsananmoniumu*

The names of two cyanoferrates are ヘキサシアノ鉄（II）酸カリウム *hekisashianotetsu(II)sankariumu* for $K_4[Fe(CN)_6]$ [potassium hexacyanoferrate(II), potassium ferrocyanide, yellow prussiate of potash] and ヘキサシアノ鉄（III）酸カリウム *hekisashianotetsu(III)sankariumu* for $K_3[Fe(CN)_6]$ [potassium hexacyanoferrate(III), potassium ferricyanide, red prussiate of potash].

    Alternatively, the name order can be changed and the name starts with the specification of the central atom, followed by names of the ligand, which end with 化 *ka*, and finally of the counter ion. Therefore, for cyanaurates, for example, the following names are possible:

    ammonium aurocyanide (ammonium cyanaurite, ammonium dicyano-auriate, $(NH_4)[Au(CN)_2]$): シアン化第一金酸アンモニウム *shiankadaiichikinsananmoniumu* or 第一金シアン化アンモニウム *daiichikinshiankaanmoniumu* or ジシアノ第二金酸アンモニウム *jishianodainikinsananmoniumu*

	ammonium auricyanide (ammonium cyanaurate, ammonium tetracyanoaurate, $(NH_4)[Au(CN)_4]$):	シアン化第二金酸アンモニウム *shiankadainikinsananmoniumu* or 第二金シアン化アンモニウム *dainikinshiankaanmoniumu* or テトラシアノ第二金酸アンモニウム *tetorashianodainikinsananmoniumu*
Further examples:	ammonium fluorotitanate(IV), $(NH_4)_2[TiF_6]$:	弗化チタン（IV）酸アンモニウム *fukkachitan(IV)sananmoniumu*
	potassium tetrarhodanocobaltite, $K_2[Co(SCN)_4]$:	テトラロダノ第一コバルト酸 カリウム *tetorarodanodaiichikobarutosankariumu*
	ammonium fluorosilicate, $(NH_4)_2[SiF_6]$:	珪弗化アンモニウム *keifukkaanmoniumu*
	sodium tetrachloropalladite, $Na_2[PdCl_4]$:	テトラクロロ第一パラジウム酸 ナトリウム *tetorakurorodaiichiparajiumusannatoriumu*
	potassium chloroiridate, $K_2[IrCl_6]$:	ヘクサクロロイリジウム（IV）酸カリウム *hekusakuroroirijiumu(IV)sankariumu*

3.1.4.5 Common Names

Besides the systematic names, many minerals and ores also have common names. Apart from a few exceptions, they all end with one of the following *kanji*: 石 *seki* (stone), 鉱 *kou* (ore), 岩 *gan* (rock) or 玉 *gyoku* (jewel). The proper name is expressed either with *kana*, which are transcribed from English, or with own *kanji*. Examples for names derived from the English common names are ワイス石 *waisuseki* (weissite, Cu_2Te), パラゴナイト岩 *paragonaitogan* (palagonite), マイエルホッフェル石 *maieruhofferuseki* (meyerhofferite, CaB_4O_7) and メロネス鉱 *meronesukou* (melonite, $NiTe_2$). Mineral names consisting of *kanji* may refer to the elements they are built of, such as 硝石 *shouseki*, meaning "nitrogen stone" (saltpeter, KNO_3), 臭銀鉱 *shuuginkou*, meaning "bromine-silver-ore" (bromyrite, $AgBr$), 硫銅銀鉱 *ryuudouginkou*, meaning "sulfur-copper-silver-ore" (strohmeyerite, $CuAgS$) and 水鉛鉛鉱 *suienenkou*, meaning "molybdenum-lead-ore" (wulfenite, $PbMoO_4$). Others mineral names are abstract, sometimes derived from an existing English common name, e.g., 天河石 *tenkaseki* (lit. "milky way stone" for amazonite, $KAlSi_2O_3$), 褐鉄鉱 *kattekkou* (lit. "brown iron ore" for climonite, $2Fe_2O_3 \cdot 3H_2O$), 螢石 *keiseki* (lit "firefly stone" for fluorite, CaF_2), 氷晶石 *hyoushouseki* (lit. "ice crystal stone" for cryolite, Na_3AlF_6), 凍石 *touseki* (lit. "frozen stone" for steatite, $Mg_3(OH)_2[Si_4O_{10}]$), 鋼玉 *kougyoku* (lit. "steel jewel" for corundum, Al_2O_3), 血石 *ketsuseki* (lit. "blood stone" for heliotrope, SiO_2), 赤鉄鉱 *sekitekkou* (lit. "red iron ore" for hematite, $\alpha\text{-}Fe_2O_3$), 黒銅鉱 *kokudoukou* (lit. "black lead ore" for melaconit, CuO) and 光線石 *kousenseki* (lit. "light beam stone" for abichite, $Cu_3(OH)_3AsO_4$). Finally, for some minerals, names by both methods are used, such as for acanthite (Ag_2S): アカンサイト (derived from the English common name) and 流銀鉱 *ryuuginkou* (meaning "sulfur-silver-ore").

3.2
Naming of Organic Compounds

3.2.1
General

There are different possibilities for naming organic substances in Japanese. Basically, Japanese organic nomenclature follows the English method of naming. A few compounds that have some connection with everyday life have original Japanese names. They are usually written with *kanji*, such as, for instance, most of the basic organic acids, e.g., acetic acid (酢酸 *sakusan*), tartaric acid (酒石酸 *shusekisan*) and lactic acid (乳酸 *nyuusan*). For a full list of the names of acids see Section 3.3.3. Even if original Japanese common names with their own *kanji* exist, compounds may also named by *kana*. Depending on the philosophy of the authors, the use of either *kanji* in general or of *kana* in general is preferred. In a few cases, the German language is the origin of the Japanese names of organic compounds, e.g., malic acid [HOOCCH(OH)CH$_2$COOH]: 林檎酸 or リンゴ酸 *ringosan* (derived from "Apfelsäure"), fumaric acid (HOOCCH=CHCOOH): フマル酸 *fumarusan* (derived from "Fumarsäure") and formic acid (HCOOH): 蟻酸 or ギ酸 *gisan* (derived from "Ameisensäure").

Most organic compounds are named by *kana*. Either the English common name or a systematic name is directly transcribed into *kana*. Therefore, all name parts and substituents and also the word order follows the English nomenclature, except for a few name parts for which the Chinese characters are always used (e.g., for acid: 酸 *san*, or for halogens if they are not considered as substituent but as electronegative element, e.g., ethyl bromide: 臭化エチル *shuukaechiru* or ブロムエタン). If two common names are used in English, they can usually also be found in Japanese publications, such as with sylvic acid/abietic acid (C$_{20}$H$_{30}$O$_2$): シルヴィン酸 *shiruvinsan* or アビエチン酸 *abiechinsan*. As there are various possibilities in English to name a specific chemical with systematic elements, for instance by taking a different unit as the core element of the compound, the same possibilities exist in Japanese. The number of names is even higher due to various possibilities of transcription of English expressions into the Japanese syllables, like long or short vocals, substituents (e.g., nitro: ニトロ or ナイトロ) or prefixes (e.g., pyro: ピロ or パイロ) (see also Sections 3.1.2 and 1.2). For instance, for "cellulose", シェルロース, セルロース, セルローゼ and 繊維素 can be found. All the following transcriptions mean cortisone (C$_{21}$H$_{28}$O$_5$): コーチゾン, コルチゾン, コチゾン, コルチソン, コーチソン and コチソン, but only コーチゾン and コルチゾン should be used.

The opposite problem may also occur: There are some similar sounding Japanese names that do not belong to the same English expression and may be confused, such as:

allyl: アリル	↔	aryl: アリール
benzine: ベンジン	↔	benzyne: ベンザイン
silicone: シリコーン	↔	silicon: シリコン

In less common word compositions, the "wrong" transcriptions can also be found, such as ビアリル for biaryl (usually transcribed with ビアリール). For some English terms, no distinctive Japanese transcriptions exist, which may lead easily to confusion. For instance, the following *kana* terms may be translated into two different words:

ベンジル	→ benzyl ($C_6H_5CH_2$-)	or benzil ($C_{14}H_{10}O_2$)
ホスホラン	→ phospholane (C_4H_8P)	or phosphorane (PR_5)
ニトリル	→ nitrile (RCN)	or nitryl (NO_2^+)
ピロリジン	→ pyrrolidine (C_4H_9N)	or pyrrolizine (C_7H_7N)
コリン	→ choline ($C_5H_{14}NOCl$)	or corrin ($C_{19}H_{22}N_4$)
デシル	→ decyl (n-$C_{10}H_{21}$-)	or desyl ($C_6H_5COCH(C_6H_5)$-)

Although the word order usually follows the English nomenclature, the opposite order is used for esters, carboxylic acid salts and some simple compounds, e.g.:

methyl chloride (CH_3Cl): 塩化メチル *enkamechiru*
ethyl acetate ($CH_3COOC_2H_5$): 酢酸エチル *sakusanechiru*
sodium benzoate (C_6H_5COONa): 安息酸ナトリウム *ansokusannatoriumu*

Hence, to form the name of salts or organic compounds, the name of the acid is placed in front of the organic base, e.g.:

nicotine ($C_{10}H_{14}N_2$): → nicotine hydrochloride ($C_{10}H_{14}N_2 \cdot HCl$):
ニコチン　　　　　　　　塩酸ニコチン *ensannikochin*
morphine ($C_{17}H_{19}NO_3$): → morphine acetate ($C_{18}H_{23}NO_5$):
モルフィン　　　　　　　酢酸モルフィン *sakusanmorufin*

Plural names may be formed with 類 *rui* at the end of the names, e.g., terpenes = テルペン類 *terupenrui*. Because many Japanese scientists went to study chemistry in Germany, the German language is also a source of transcribed organic names, as has been described already for element names. Important examples for organic chemistry are ベンゾール (benzene), トルオール (toluene) and スチロール (styrene).

For the alkyl series and the corresponding alkenyles, the names are compiled in Table 3.22. Similar to this, the names of other derivatives are formed, like for alkinyles or alcohols. To indicate isomeric and similar derivatives, prefixes are used in similar a way to English. Some are written in Latin letters, like cis, trans, sec or tert. Transcribed forms for cis: シス and trans: トランス can be found only in a few examples (Table 3.23).

Some authors prefer to insert dots between different parts of the molecule names, to enhance the clearness of chemical names, such as in following examples:

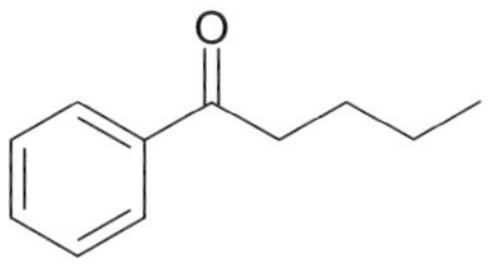

butylphenyl ketone:
ブチル・フェニルケトン

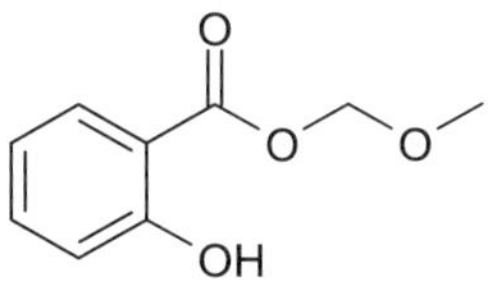

methoxymethyl salicylate:
サリチル酸・メトキシメチル・エステル

Table 3.22 Japanese names for alkyl, alkenyl, alkene and alkyne radicals.

Number carbon atoms	Alkene series	Formula	kana	Alkyne series	Formula	kana
1	methane	CH_4	メタン	methyl	$-CH_3$	メチル
2	ethane	C_2H_6	エタン	ethyl	$-C_2H_5$	エチル
3	propane	C_3H_8	プロパン	propyl	$-C_3H_7$	プロピル
4	butane	C_4H_{10}	ブタン	butyl	$-C_4H_9$	ブチル
5	pentane	C_5H_{12}	ペンタン	pentyl	$-C_5H_{11}$	ペンシル
6	hexane	C_6H_{14}	ヘキサン (ヘクサン)	hexyl	$-C_6H_{13}$	ヘキシル
7	heptane	C_7H_{16}	ヘプタン	heptyl	$-C_7H_{15}$	ヘプチル
8	octane	C_8H_{18}	オクタン	octyl	$-C_8H_{17}$	オクチル
9	nonane	C_9H_{20}	ノナン	nonyl	$-C_9H_{19}$	ノニル
10	decane	$C_{10}H_{22}$	デカン	decyl	$-C_{10}H_{21}$	デシル
11	undecane	$C_{11}H_{24}$	ウンデカン	undecyl	$-C_{11}H_{23}$	ウンデシル
12	dodecane	$C_{12}H_{26}$	ドデカン	dodecyl	$-C_{12}H_{25}$	ドデシル
13	tridecane	$C_{13}H_{28}$	トリデカン	tridecyl	$-C_{13}H_{27}$	トリデシル
14	tetradecane	$C_{14}H_{30}$	テトラデカン	tetradecyl	$-C_{14}H_{29}$	テトラデシル
15	pentadecane	$C_{15}H_{32}$	ペンタデカン	pentadecyl	$-C_{15}H_{31}$	ペンタデシル
16	hexadecane	$C_{16}H_{34}$	ヘキサデカン	hexadecyl	$-C_{16}H_{25}$	ヘキサデシル
17	heptadecane	$C_{17}H_{36}$	ヘプタデカン	heptadecyl	$-C_{17}H_{35}$	ヘプタデシル
18	octadecane	$C_{18}H_{38}$	オクタデカン	octadecyl	$-C_{18}H_{37}$	オクタデシル
19	nonadecane	$C_{19}H_{40}$	ノナデカン	nonadecyl	$-C_{19}H_{39}$	ノナデシル
20	icosane, eicosane	$C_{20}H_{42}$	イコサン, エイコサン	icosyl, eicosyl	$-C_{20}H_{41}$	イコシル, エイコシル
21	henicosane	$C_{21}H_{44}$	ヘンイコサン	henicosyl	$-C_{21}H_{43}$	ヘンイコシル
30	triacontane	$C_{30}H_{62}$	トリアコンタン	triacontyl	$-C_{30}H_{61}$	トリアコンチル
40	tetracontane	$C_{40}H_{82}$	テトラコンタン	tetracontyl	$-C_{40}H_{81}$	テトラコンチル
100	hectane	$C_{100}H_{202}$	ヘクタン	hectyl	$-C_{100}H_{201}$	ヘクチル

Table 3.22 (continued)

Number carbon atoms	Alkene series	Formula	kana	Alkyne series	Formula	kana
2	ethene, ethylene	C_2H_4	エテン, エチレン	ethyne, acetylene	C_2H_2	エチン, アセチレン
3	propene, propylene	C_3H_6	プロペン, プロピレン	propyne	C_3H_4	プロピン
4	butene, butylene	C_4H_8	ブテン, ブチレン	butyne	C_4H_6	ブチン
5	pentene	C_5H_{10}	ペンテン	pentyne	C_5H_8	ペンチン
6	hexene	C_6H_{12}	ヘキセン	hexyne	C_6H_{10}	ヘキシン
7	heptene	C_7H_{14}	ヘプテン	heptyne	C_7H_{12}	ヘプチン
8	octene	C_8H_{16}	オクテン	octyne	C_8H_{14}	オクチン
9	nonene	C_9H_{18}	ノネン	nonyne	C_9H_{16}	ノニン
10	decene	$C_{10}H_{20}$	デセン	decyne	$C_{10}H_{18}$	デシン
11	undecene	$C_{11}H_{22}$	ウンデセン	undecyne	$C_{11}H_{20}$	ウンデシン
12	dodecene	$C_{12}H_{24}$	ドデセン	dodecyne	$C_{12}H_{22}$	ドデシン
13	tridecene	$C_{13}H_{26}$	トリデセン	tridecyne	$C_{13}H_{24}$	トリデシン
14	tetradecene	$C_{14}H_{28}$	テトラデセン	tetradecyne	$C_{14}H_{26}$	テトラデシン
15	pentadecene	$C_{15}H_{30}$	ペンタデセン	pentadecyne	$C_{15}H_{28}$	ペンタデシン
16	hexadecene	$C_{16}H_{32}$	ヘキサデセン	hexadecyne	$C_{16}H_{30}$	ヘキサデシン
17	heptadecene	$C_{17}H_{34}$	ヘプタデセン	heptadecyne	$C_{17}H_{32}$	ヘプタデシン
18	octadecene	$C_{18}H_{36}$	オクタデセン	octadecyne	$C_{18}H_{34}$	オクタデシン
19	nonadecene	$C_{19}H_{38}$	ノナデセン	nonadecyne	$C_{19}H_{36}$	ノナデシン
20	icosene, eicosene	$C_{20}H_{40}$	イコセン, エイコセン	icosyne, eicosyne	$C_{20}H_{38}$	イコシン, エイコシン
21	henicosene	$C_{21}H_{42}$	ヘンイコセン	henicosyne	$C_{21}H_{40}$	ヘンイコシン
30	triacontene	$C_{30}H_{60}$	トリアコンテン	triacontyne	$C_{30}H_{58}$	トリアコンチン
40	tetracontene	$C_{40}H_{80}$	テトラコンテン	tetracontyne	$C_{40}H_{78}$	テトラコンチン
100	hectene	$C_{100}H_{200}$	ヘクテン	hectyne	$C_{100}H_{198}$	ヘクチン

Table 3.23 Naming of alkyl derivatives.

Derivative	Japanese name	Example
isomeric derivatives:	イソ	isopropyl, -CH(CH$_3$)$_2$: イソプロピル
		isovaleric acid [(CH$_3$)$_2$CHCH$_2$COOH]: イソ吉草酸 *isokissousan*
	ｉｓｏ-	iso-propyl, -CH(CH$_3$)$_2$: ｉｓｏ-プロピル
	ｎ-	n-pentyl, -(CH$_2$)$_4$CH$_3$: ｎ-ペンチル
	ｓｅｃ-	sec-butyl, -CH(CH$_3$)CH$_2$CH$_3$: ｓｅｃ-ブチル
	ｔｅｒｔ-	tert-butyl, -C(CH$_3$)$_3$: ｔｅｒｔ-ブチル
cyclic derivatives:	シクロ	cyclobutane (C$_4$H$_8$): シクロブタン
		cyclooctatetraene (C$_8$H$_8$): シクロオクタテトラエン
cis and trans derivatives	ｃｉｓ	cis-coniferin (C$_{16}$H$_{22}$O$_8$): ｃｉｓ-コニフェリン
	ｔｒａｎｓ	trans-cinnamic acid (C$_9$H$_8$O$_2$): ｔｒａｎｓ-桂皮酸 *trans-keihisan*

In rare cases, complex names are separated for easier reading by the insertion of the particle の, especially if a specific substituent or functional group should be pointed out, such as "the ethyl ester of ...'' or "the bromide of ...''.

Examples: methyl acetate: 酢酸のメチルエステル
                                                  sakusannomechiruesuteru

4,4′,5,5′,6,6′-hexahydroxydiphenic   4, 4′, 5, 5′, 6, 6′-ヘクサヒドロオキ
acid dilactone (ellagic acid):   シジフェン酸のジラクトン

3.2.2
Substituents

3.2.2.1 Substituent Classes

To cover a broad chemical structure, it is usual to use generic formulas and descriptions in claims of patents. As a result, substituents are usually described by using ranges, especially for organic groups. This is realized in a similar way to English by adding the range in front of the concerned group, such as:

C_1–C_8 alkyl: C_1–C_8 アルキル基
C_1–C_3 halogenalkyl: ハロC_1–C_3 アルキル基,
                               C_1–C_3 ハロゲン化アルキル基,
                               C_1–C_3 ハロアルキル基
phenyl C_1–C_2 alkyl: フェニルC_1–C_2 アルキル基
C_1–C_6 alkoxy C_1–C_6 alkyl: C_1–C_6 アルコキシC_1–C_6 アルキル基
C_1–C_4 alkylsulfonyl C_1–C_4 アルキルスルホニルC_1–C_3 ア
  C_1–C_3 alkoxy: コキシ基

Table 3.24 Terms for generic substituents.

Substituent class	kana
alkyl	アルキル基
cycloalkyl	シクロアルキル基
alkenyl	アルケニル基
cycloalkenyl	シクロ アルケニル基
alkynyl	アルキニル基
haloalkyl	ハロアルキル基, ハロゲン化アルキル基
alkoxy	アルコキシ基
haloalkoxy	ハロアルコキシ基, ハロゲン化アルコキシ基
alkylthio	アルキルチオ基
haloalkylthio	ハロアルキルチオ基
alkylsulfinyl	アルキルスルフィニル基
alkylsulfonyl	アルキルスルホニル基, アルキルスルフォニル基
alkylamino	アルキルアミン基
dialkylamino	ジアルキルアミン基
aryl	アリール基
heteroaryl	ヘテロアリール基
alkylcarbonyl	アルキルカルボニル基
acyl	アシル基
acyloxy	アシルオキシ基
alkoxycarbonyl	アルコキシカルボニル基
halogen	ハロゲン基

In older patents, expressions like 炭素数3から8まで (*tansosuu 3 kara 8 made*, 3 to 8 carbon atoms) can also be found. Table 3.24 compiles typical examples for generic substituents.

3.2.2.2 Specific Substituents

Although there is an original Japanese expression for an element (like for chlorine or bromine), for substituents in the names of chemical compounds the transcription of the English term is used (except for hydrogen). For instance, fluorine as a substitutent in organic molecules is named as フルオロ *furuoro* or フルオル *furuoru*, but not with 弗化. In some cases, as already described, more than one transcription into the Japanese syllables is possible. Table 3.25 gives the most important specific substituents for organic compounds.

Table 3.25 Terms for important substituents in organic compound names.

Substituent	Formula	kana
hydrogen	-H	水素原子 *suisogenshi*
methyl	-CH$_3$	メチル
difluoromethyl	-CHF$_2$	ジフルオロメチル, ジフルオルメチル
trifluoromethyl	-CF$_3$	トリフルオロメチル, トリフルオルメチル
ethyl	-C$_2$H$_5$	エチル
propyl	-C$_3$H$_7$	プロピル
isopropyl	-CH(CH$_3$)$_2$	イソプロピル
cyclopropyl	-C$_3$H$_5$	シクロプロピル
butyl	-C$_4$H$_9$	ブチル
isobutyl	-CH$_2$CH(CH$_3$)$_2$	イソブチル
pentyl	-C$_5$H$_{11}$	ペンチル
neopentyl	-CH$_2$C(CH$_3$)$_3$	ネオペンチル
cyclopentyl	-C$_5$H$_9$	シクロペンチル
hexyl	-C$_6$H$_{13}$	ヘキシル
cyclohexyl	-C$_6$H$_{11}$	シクロヘキシル
heptyl	-C$_7$H$_{15}$	ヘプチル
octyl	-C$_8$H$_{17}$	オクチル
nonyl	-C$_9$H$_{19}$	ノニル
decyl	-C$_{10}$H$_{21}$	デシル
undecyl	-C$_{11}$H$_{23}$	ウンデシル
dodecyl	-C$_{12}$H$_{25}$	ドデシル
vinyl	-CH=CH$_2$	ビニル
propenyl	-C$_3$H$_5$	プロペニル
allyl	-CH$_2$CH=CH$_2$	アリル
propargyl	-CH$_2$C≡CH	プロパルギル
phenyl	-C$_6$H$_5$	フェニル
benzyl	-CH$_2$C$_6$H$_5$	ベンジル
tolyl	-C$_6$H$_4$(CH$_3$)	トリル
trityl	-C(C$_6$H$_5$)$_3$	トリチル
phenetyl	-CH$_2$CH$_2$C$_6$H$_5$	フェネチル
styryl	-CH=CHC$_6$H$_5$	スチリル
cinnamyl	-CH$_2$CH=CHC$_6$H$_5$	シンナミル
phenacyl	-CH$_2$COC$_6$H$_5$	フェナシル
xylyl	-CH$_2$C$_6$H$_4$(CH$_3$)	キシリル
mesityl	-C$_6$H$_2$(CH$_3$)$_3$	メシチル
picryl	-C$_6$H$_2$(NO$_3$)$_3$	ピクリル

3.2 Naming of Organic Compounds

Table 3.25 (continued)

Substituent	Formula	*kana*
naphthyl	-$C_{10}H_7$	ナフチル
furyl	-C_4H_3O	フリル
thienyl	-C_4H_3S	チエニル
pyridyl	-C_5H_4N	ピリジル
piperidyl	-$C_5H_{10}N$	ピペリジル
formyl	-CHO	フォルミル, ホルミル
acetyl	-$COCH_3$	アセチル
carboxyl	-COOH	カルボキシ
oxal-, oxalyl	-COCOOH	オキサロ, オキザロ, オクザル, オキサリル, オクザリル
cyano	-CN	シアノ
fluoro	-F	フルオロ, フルオル
chloro	-Cl	クロロ, クロル
bromo	-Br	ブロム, ブロモ
iodo	-I	ヨード
hydroxy	-OH	ヒドロキシ, ヒドロオキシ
methoxy	-OCH_3	メトキシ
trifluoromethoxy	-OCF_3	トリフルオロメトキシ, トリフルオルメトキシ
ethoxy	-OC_2H_5	エトキシ
propoxy	-OC_3H_7	プロポキシ
butoxy	-OC_4H_9	ブトキシ
phenoxy	-OC_6H_5	フェノキシ
acetoxy	-$OOCCH_3$	アセトキシ
pivaloyl	-$COC(CH_3)_3$	ピバロイル
benzoyl	-COC_6H_5	ベンゾイル
cinnamoyl	-$COCH=CHC_6H_5$	シンナモイル
amino	-NH_2	アミノ
methylamino	-$NHCH_3$	メチルアミノ
dimethylamino	-$N(CH_3)_2$	ジメチルアミノ
anilino	-NHC_6H_5	アニリノ
nitro	-NO_2	ニトロ, (ナイトロ)
nitroso	-NO	ニトロソ
mercapto	-SH	メルカプト
methylsulfonyl (mesyl)	-SO_2CH_3	メチルスルフォニル, メシル
toluenesulfonyl (tosyl)	-$SO_2C_6H_4CH_3$	トシル

3.2.3
Number and Position of Substituents and Functional Groups

In most cases, for the number of substituents in the names of organic chemicals, expressions transcribed from English are used, like mono: モノ, di: ジ, tri: トリ, tetra: テトラ, penta: ペンタ, hexa: ヘキサ, hepta: ヘプタ or octa: オクタ.

Examples: dimethyl malonate: マロン酸ジメチルエステル
maronsanjimechiruesuteru
tribromobenzene: トリブロムベンゾール
tetrabromoethylene: テトラブロムエチレン
pentachloroethane: ペンタクロルエタン
hexamethylenediamine: ヘキサメチレンジアミン

In a few cases, *kanji* are used for the number of substituents of building blocks (as it is in the normal way of naming inorganic compounds), mainly for natural compounds, in biochemistry or if inorganic parts such as acid residues are involved.

Examples: adenosine monophosphate: アデノシン一燐酸 *adenoshinichirinsan*
trisaccharide: 三糖類 *santourui*
pentaerythritol tetranitrate: 四硝酸ペンタエリスリット
shishousanpentaerisuritto

There are also examples for which both possibilities can be found:

tetrachloromethane 四塩化炭素 *shienkatanso* or テトラクロ
(carbon tetrachloride, CCl₄): ルメタン

To specify the position of the substituents, functional groups or multiple bonds, the same methods as in English are applied for alkyl chains, cyclic hydrocarbons, phenylic compounds, heterocylces and other organic compounds. Commonly used are figures according to the same rules as in English. The following examples demonstrate the use:

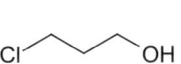

3-chloro-propyl alcohol
(trimethylene chlorohydrin)
3-クロル-1-プロパノール
(トリメチレンクロルヒドリン)

2-hydroxy-2,4,6-cycloheptatrienone
(tropolone)
2-ヒドロキシ-シクロヘプタ-2, 4, 6-トリエン-1-オン (トロポロン)

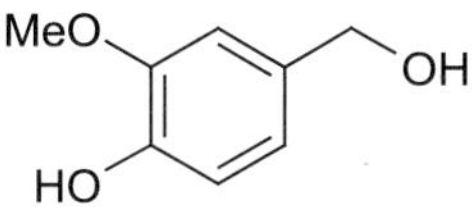

4-hydroxy-3-methoxy-benzyl alcohol
(vanillyl alcohol)
4-ヒドロキシ-3-メトキシ-ベンジルアルコール(バニリルアルコール)

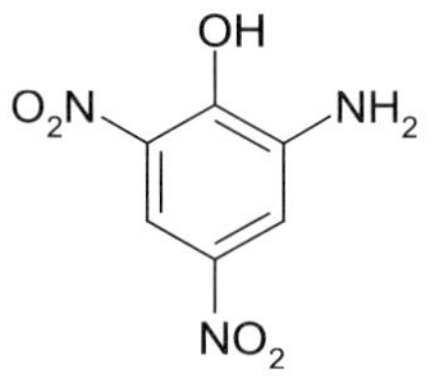

2-amino-4,6-dinitrophenol
(picramic acid)
2-アミノ-4, 6-ジニトロフェノール
(ピクラミン酸)

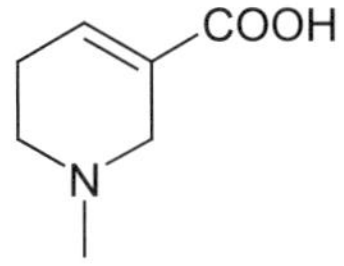

1-methyl-1,2,5,6-tetrahydronicotinic acid
(arecaidine)
1-メチル-1, 2, 5, 6-テトラヒドロ-ニコチン酸(アレカイジン)

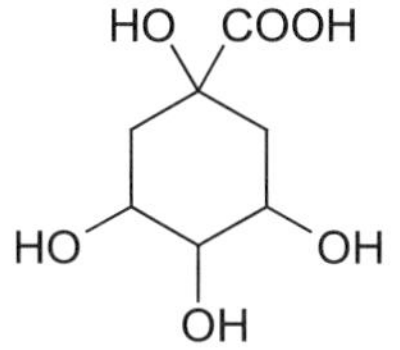

1,3,4,5-tetrahydroxycyclohexan-1-carboxylic acid (quinic acid)
1, 3, 4, 5-テトラオキシ-シクロヘキサン-1-カルボン酸(キナ酸)

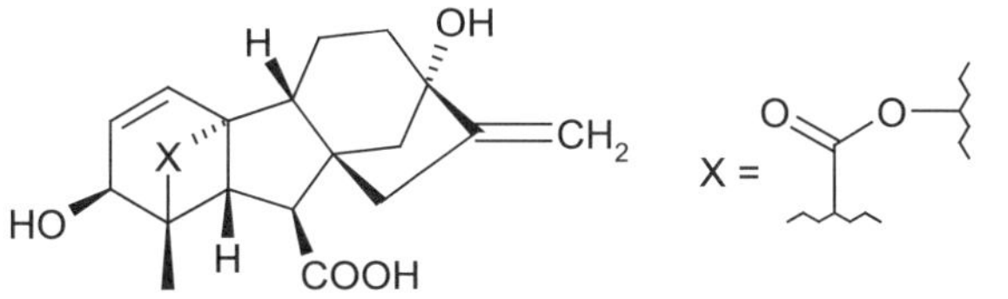

$2\beta,7$-dihydroxy-1β-methyl-8-methylidene-13-oxo-$4a\alpha,1\alpha$-(epoxymethano)-$4b\beta$-gibb-3-ene-10β-carboxylic acid (gibberellic acid)
$2\beta, 7$-ジヒドロキシ-1β-メチル-8-メチリデン-13-オキソ-$4a\alpha, 1\alpha$-(エポキシメタノ)-$4b\beta$-ジバ-3-エン-10β-カルボン酸(ジベレリン酸)

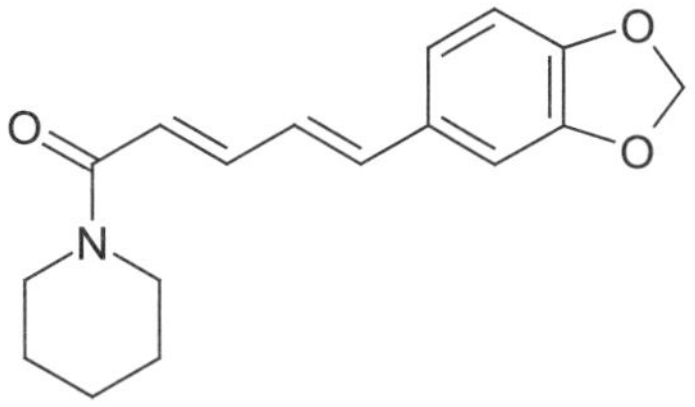

1-[(2E,4E)-5-(1,3-benzodioxol-5-yl)penta-2,4-dienoyl]piperidine (piperine)
1-[(2E, 4E)-5-(1, 3-ベンゾジオキソール-5-イル)ペンタ-2, 4-ジエノイル]ピペリジン (ピペリン)

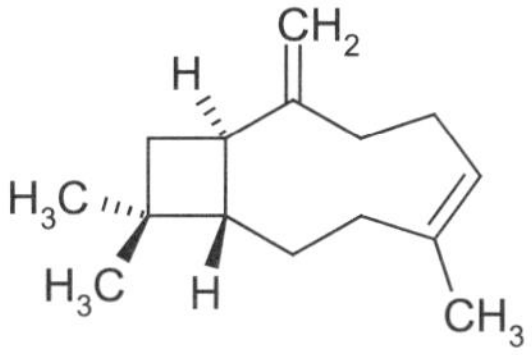

(E)-(1R,9S)-4,11,11-trimethyl-8-methylidenebicyclo[7.2.0]undec-4-ene (caryophyllene)
(E)-(1R, 9S)-4, 11, 11-トリメチル-8-メチリデンビシクロ[7. 2. 0]ウンデカ-4-エン(カリオフィレン)

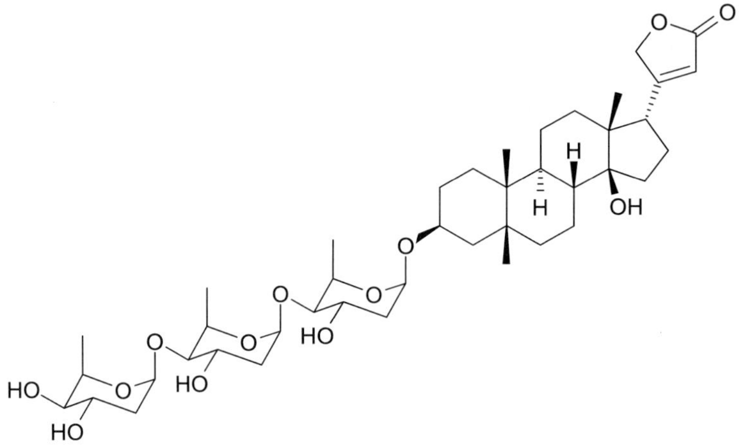

3β-[(2,6-dideoxy-β-D-*ribo*-hexopyranosyl-(1 → 4)-2,6-dideoxy-β-D-*ribo*-hexopyranosyl-(1 → 4)-2,6-dideoxy-β-D-*ribo*-hexopyranosyl)oxy]-14-hydroxy-5β-card-20(22)-enolide (digitoxin)
3β-[(2,6-ジデオキシ-β-D-リボ-ヘキソピラノシル-(1 → 4)-2,6-ジデオキシ-β-D-リボ-ヘキソピラノシル-(1 → 4)-2,6-ジデオキシ-β-D-リボ-ヘキソピラノシル)オキシ]-14-ヒドロキシ-5β-カルダ-20(22)-エノリド（ジギトキシン）

The use of further parameters to indicate the position of a substituent is also as common, as in English. If the substituent is attached to a heteroatom, the element symbol of the heteroatom may be used to indicate the position of the substituent, e.g., *N*-dimethylformamide [HCON(CH$_3$)$_2$]: N-ジメチルホルムアミド.

To indicate the position of a substituent relative to a given substituent on a phenyl ring with ortho, meta and para, either the Latin initial or the transcription by *katakana* (オルト, メタ and パラ) can be used. The more complex the chemical compound, the greater the preference for the simpler one-letter alternative. Examples:

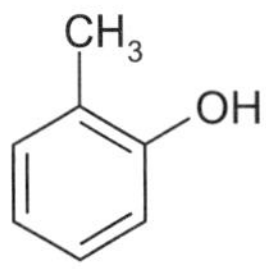

ortho-cresol
オルトクレゾール

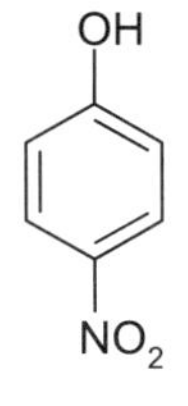

para-nitrophenol
パラニトロフェノール

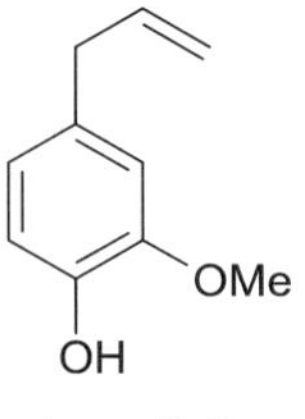

p-hydroxy-*m*-methoxyallylbenzene (eugenol)
p-オキシ-m-メトキシ-アリルベンゼン

Greek letters are adopted directly, such as for α-hydroxyacetophenone (phenacyl alcohol): α-オキシ・アセトフェノン and β-picoline: β-ピコリン.

3.3
Overview of Specific Organic Molecules

3.3.1
Functional Groups and Chemical Classes in Organic Chemistry

Tables 3.26–3.32 give an overview of the main chemical classes and functional groups in organic chemistry.

Table 3.26 Japanese names of hydrocarbons.

Chemical group	Formula	Japanese names and examples
alkane	C_nH_{2n+2}	アルカン, e.g., propane, $CH_3CH_2CH_3$: プロパン
alkene, olefin	C_nH_{2n}	アルケン, オレフィン, e.g., propene, $CH_3CH=CH_2$: プロパン
alkyne	C_nH_{2n-2}	アルキン, e.g., propyne, $H_3CC\equiv CH$: プロピン
cycloalkane	C_nH_{2n}	シクロアルカン, e.g., cyclobutane, C_4H_8: シクロブタン
cycloalkene, cycloolefin	C_nH_{2n-2}	シクロアルケン, シクロオレフィン, e.g., cyclohexene, C_6H_{10}: シクロヘキセン

Table 3.27 Japanese names of oxygen-containing hydrocarbons.

Chemical group	Formula	Japanese names and examples
alcohol	ROH	アルコール, e.g., ethyl alcohol, ethanol (C_2H_5OH): エチルアルコール, エタノール
phenol	C_6H_5OH	フェノール
ether	ROR'	エーテル, e.g., diethyl ether, ethyl ether ($C_2H_5OC_2H_5$): ジエチルエーテル, エチルエーテル
aldehyde	RCHO	アルデヒド, e.g., salicylaldehyde, $C_6H_4(OH)COOH$: サリチルアルデヒド
ketone	RCOR'	ケトン, e.g., phenyl methyl ketone, $C_6H_5COCH_3$: フェニルメチルケトン
hemiacetal, semiacetal	RCH(OH)(OR')	ヘミアセタール, セミアセタール
acetal	RCH(OR')(OR'')	アセタール
ketal, ketone acetal	RC(OR')(OR'')R'''	ケタール
carboxylic acid	RCOOH	カルボン酸 *karubonsan*, e.g., propenoic acid, $H_2C=CHCOOH$: プロペン酸 *puropensan*
benzoquinone	$C_6H_4O_2$	ベンゾキノン, e.g., tetrachloro-*p*-benzoquinone, $C_6Cl_4O_2$: テトラクロル-p-ベンゾキノン

Table 3.28 Japanese names of carboxylic acid derivatives.

Chemical group	Formula	Japanese names and examples
carboxylic acid ester, carboxylate	RCOOR′	カルボン酸エステル *karubonsanesuteru*, e.g., ethyl acrylate ($H_2C=CHCOOC_2H_5$): アクリル酸エチル *akurirusanechiru*
carboxylic acid anhydride	RCOOOCR′	無水カルボン酸 *musuikarubonsan*, e.g., succinic anhydride ($C_4H_4O_3$): 無水コハク酸 *musuikohakusan*
carboxylic acid chloride	RCOCl	カルボン酸クロライド *enkakarubonsankuroraido*, e.g., propionyl chloride (CH_3CH_2COCl): プロピオン酸クロライド *puropionsankuroraido*
carboxamide	$RCONH_2$	カルボン酸アミド *karubonsanamido*, e.g., stearic acid amide, stearamide, $H_3(CH_2)_{16}CONH_2$: ステアリン酸アミド *sutearinsanamido*, ステアリルアミド
carboxylic acid anilide	$RCONHC_6H_5$	カルボン酸アニリド *karubonsananirido*, e.g., acetic acid anilide ($CH_3CONHC_6H_5$): 酢酸アニリド *sakusananirido*
carboxylic acid hydrazide	$RCONHNH_2$	カルボン酸ヒドラジド *karubonsanhidorajido*, e.g., levulinic acid hydrazide ($CH_3COCH_2CH_2CONHNH_2$): レブリン酸ヒドラジド *reburinsanhidorajido*
carboxylic acid azide	$RCON_3$	カルボン酸アジド *karubonsanajido*, e.g., vanillic acid azide, $C_6H_3(OH)(OCH_3)CON_3$: バニリン酸アジド *banirinsanajido*
orthoester, orhocarboxylic ester	$RC(OR')_3$	オルトカルボン酸エステル *orutokarubonsanesuteru*, オルトエステル *orutoesuteru*, e.g., triethyl orthoacetate, $CH_3C(OC_2H_5)_3$: オルト酢酸トリエチル *orutosakusantoriechiru*
thiocarboxylic acid	RC(=S)OH	チオカルボン酸 *chiokarubonsan*
imido ester	RC(=NH)OR′	イミドエステル, イミノエステル
amidine	RC(=NH)NH2	アミジン
lactone		ラクトン, e.g., butyrolactone $C_4H_6O_2$: ブチロラクトン
nitrile	RCN	ニトリル, e.g., acrylonitrile ($H_2C=CHCN$): アクリロニトリル

3.3 Overview of Specific Organic Molecules

Table 3.29 Japanese names of carbonic acid derivatives.

Chemical group	Formula	Japanese names and examples
urea, carbamide	RNHCONHR′	尿素 *nyouso*, e.g., N,N′-diphenylurea ($C_6H_5NHCONHC_6H_5$): N,N′-ジフェニル尿素 N,N′-*jifenirunyouso*
urethane, carbamate	ROCONH$_2$, ROCONHR′	ウレタン, カルバミン酸エステル *karubaminsanesuteru*
semicarbazide, carbamoyl hydrazide	RNHCONHNHR′	セミカルバジド, カルバモイルヒドロジド
guanidine, imidourea, carbamidine	RNHC(=NH)NHR′	グアニジン, e.g., nitroguanidine, HN=C(NH$_2$)NHNO$_2$: ニトログアニジン
xanthogenic acid	ROC(=S)SH	キサントゲン酸 *kisantogensan*
xanthate, xanthogenate, dithiocarbonate	ROC(=S)SR′	キサントゲン酸エステル *kisantogensanesuteru*
thiourea	RNHC(=S)NHR′	チオ尿素 *chionyouso*
isocyanate, carbimide	RN=C=O	イソシアナート, e.g., = phenyl isocyanate ($C_6H_5N=C=O$): イソシアン酸フェニル *isoshiansanfeniru*, フェニルイソシアナート
cyanic acid ester	ROCN	シアン酸エステル *shiansanesuteru*, e.g., methyl cyanate (CH$_3$OCN): シアン酸メチル *shiansanmechiru*
cyanamide	RNHCN	シアナミド
carbodiimide	RN=C=NR′	カルボジイミド, e.g., = dicyclohexylcarbodiimide ($C_6H_{11}N=C=NC_6H_{11}$): シクロヘキシルカルボジイミド
thiocyanic acid ester	RSCN	チオシアン酸エステル *chioshiansanesuteru*, e.g., methyl thiocyanate (CH$_3$SCN): チオシアン酸メチル *chioshiansanmechiru*
isothiocyanic acid ester, thiocarbimide	RN=C=S	イソチオシアン酸エステル *isochioshiansanesuteru*, チオカルビミド

Table 3.30 Japanese names of nitrogen-containing hydrocarbons.

Chemical group	Formula	Japanese names and examples
amine	$R-NH_2$	アミン, e.g., dimethylamine, $NH(CH_3)_2$: ジメチルアミン
aniline	C_6H_5NHR	アニリン, e.g., N,N-dimethylaniline $C_6H_5N(CH_3)_2$ N,N-ジメチルアニリン
imine	$RC(=NH)R'$	イミン
oxime	$RC(=NOH)R'$	オキシム
hydrazine	$RNHNHR'$	ヒドラジン, e.g., phenylhydrazine $(C_6H_5NHNH_2)$: フェニルヒドラジン
azo compound	$RN=NR'$	アゾ化合物 *azokagoubutsu*, e.g., triaminoazobenzene (vesuvine), $H_2NC_6H_4-N=N-C_6H_3(NH_2)_2$: トリアミノアゾベンゼン
azide	RN_3	アジド, e.g., phenyl azide $(C_6H_5N_3)$: フェニルアジド
isocyanide, isonitrile	RNC	イソシアニド, イソニトリル *isonitoriru*
diazo compound	$RCHN_2$	ジアゾ化合物 *jiazokagoubutsu*, e.g., diazomethane (CH_2N_2): ジアゾメタン
nitrene	RN	ナイトレン

Table 3.31 Japanese names of sulfur-containing hydrocarbons.

Chemical group	Formula	Japanese names and examples
thiol, mercaptan, thioalcohol, hydrosulfide	$R-SH$	チオール, メルカプタン, e.g., ethyl mercaptane (C_2H_5SH): エチルメルカプタン, エタンチオール
thioether	$R-S-R'$	チオエーテル
sulfoxide	$R-S(=O)-R'$	スルホキシド, e.g., dimethyl sulfoxide, $(CH_3)_2SO$: ジメチルスルホキシド
sulfone	$R-SO_2-R'$	スルホン, スルフォン, e.g., diethyl sulfone, $SO_2(C_2H_5)_2$: ジエチルスルホン
sulfonic acid	$R-SO_3H$	スルホン酸 *suruhonsan*, スルフォン酸 *surufonsan*, e.g., ethanesulfonic acid $(C_2H_5SO_3H)$: タンスルホン酸 *etansuruhonsan*
sulfinic acid	$R-SO_2H$	スルフィン酸 *surufinsan*, e.g., ethylsulfinic acid $(C_2H_5SO_2H)$: エチルスルフィン酸 *echirusurufinsan*
sulfenic acid	$R-SOH$	スルフェン酸 *surufensan*
sulfonyl chloride	$R-SO_2Cl$	スルホニルクロライド, e.g., toluenesulfonyl chloride, $C_6H_4(CH_3)SO_2Cl$: トルエンスルホニルクロライド

Table 3.32 Japanese names of phosphorus-containing hydrocarbons.

Chemical group	Formula	Japanese names and examples
phosphine	PR_3	ホスフィン, フォスフィン, e.g., tribenzylphosphine, $P(CH_2C_6H_5)_3$: トリベンジルホスフィン
phosphonic acid	$RPO(OH)_2$	フォスフォン酸 *fosufonsan*, ホスホン酸 *hosuhonsan*, e.g., methylphosphonic acid, $CH_3PO(OH)_2$: メチルフォスフォン酸 *mechirufosufonsan*
phosphorous acid ester	$P(OR)_3$	亜リン酸エステル *arinsanesuteru*, 亜燐酸エステル *arinsanesuteru*, e.g., phosphorous acid trimethyl ester, $P(OCH_3)_3$: 亜燐酸トリメチル *arinsantorimechiru*
phosphinic acid	$R_2PO(OH)$	ホスフィン酸 *hosufinsan*, フォスフィン酸 *fosufinsan*, e.g., phenylphosphinic acid, $C_6H_5PO(OH)$: フェニルフォスフィン酸 *fenirufosufinsan*
phosphine oxide	$R_3P=O$	ホスフィンオキシド, e.g., triphenylphosphine oxide, $(C_6H_5)_3PO$: トリフェニルホスフィンオキシド
phosphoric acid ester	$PO(OR)_3$	リン酸エステル *rinsanesuteru*, 燐酸エステル *rinsanesuteru*, e.g., phosphoric acid diethyl ester, $HPO_2(OC_2H_5)_2$: 燐酸ジエチルエステル *rinsanjiechiruesuteru*

3.3.2
Naming of Heterocycles

All common names of heterocyclic compounds are derived from the English expressions by transcription into the Japanese syllables. If more English names exist then, usually, all transcriptions can also be found in Japanese publications. Figure 3.2 summarizes the names of selected important heterocycles. The number and position of substituents tend towards the commonly used rules, as in the examples given in Table 3.33. If a substituent's name is attached behind the name of the heterocycle, the position of the substiutent is placed, in most cases, before the heterocycle, e.g., for pyrazole-4-carboxylic acid: 4-ピラゾールカルボン酸.

If the heterocycle itself is used as a substituent, the end of the heterocycle's name is modified according to the English "yl" ending by adding イル or some similar ending, depending on the last letter of the English name, such as for pyridyl: ピリジル or imidazolyl: イミダゾーリル. The position at which the heterocycle is attached to the rest of the molecule can be named by adding the number in front of the heterocycle or before the *iru*-ending, similar to the English naming, e.g., histidine [2-amino-3-(3*H*-imidazol-4-yl)propanoic acid]: 2-アミノ-3-(3*H*-イミダゾール-4-イル)プロパン酸（ヒスチジン）and for nicotine [3-(1-methylpyrrolidin-2-yl)pyridine]: 3-(1-メチルピロリジン-2-イル)ピリジン（ニコチン）.

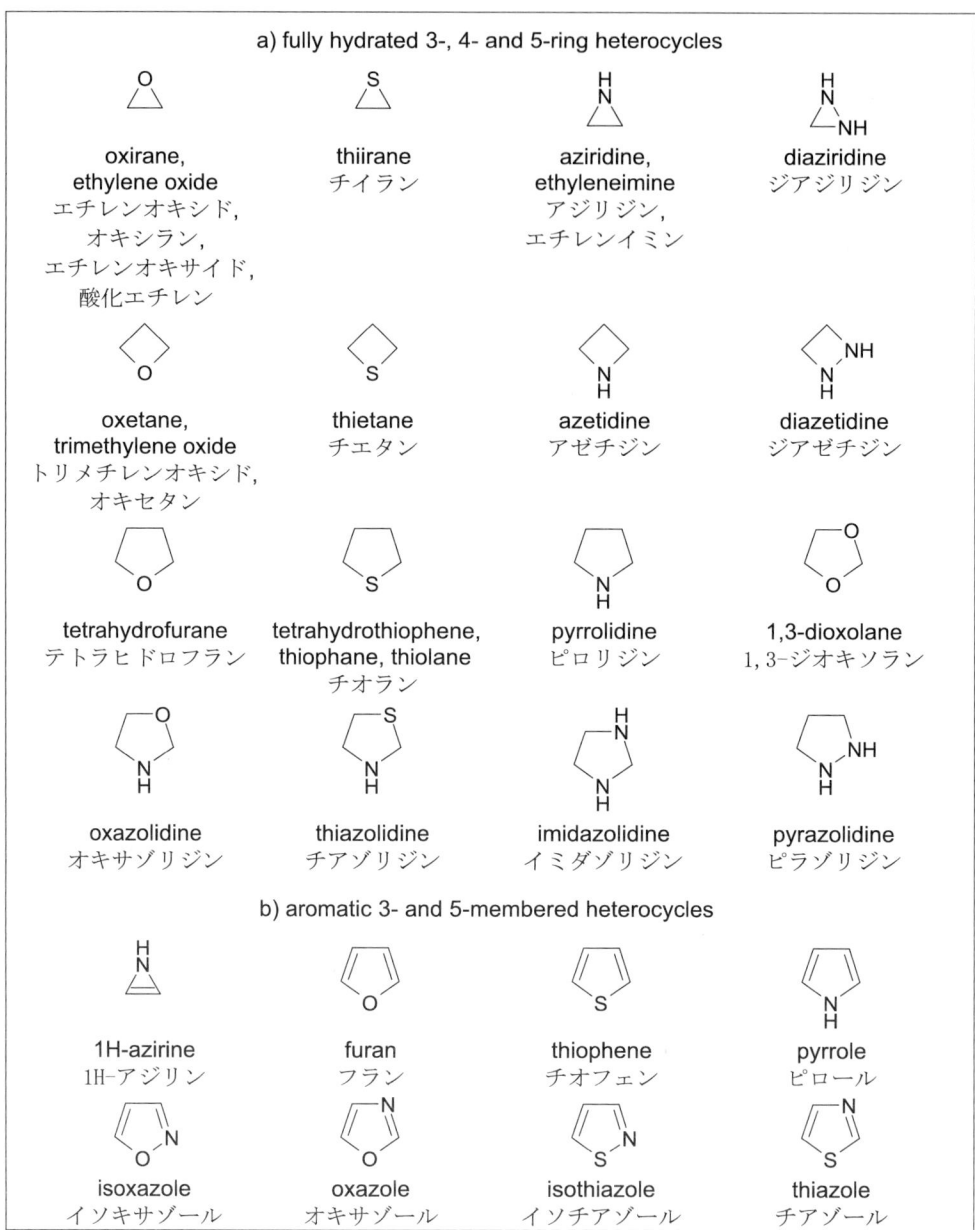

Fig. 3.2 Japanese names of selected heterocycles.

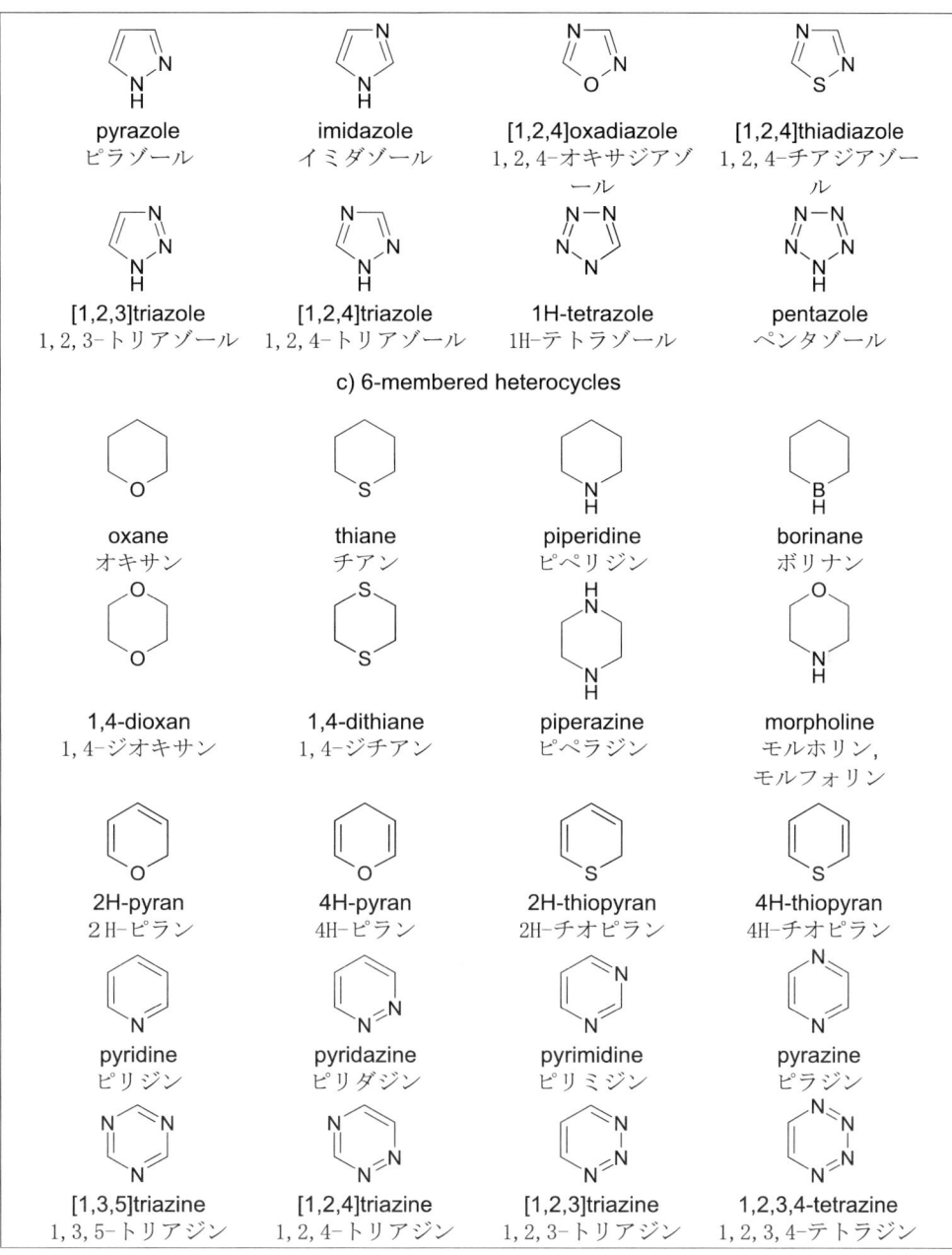

Fig. 3.2 *(continued)*

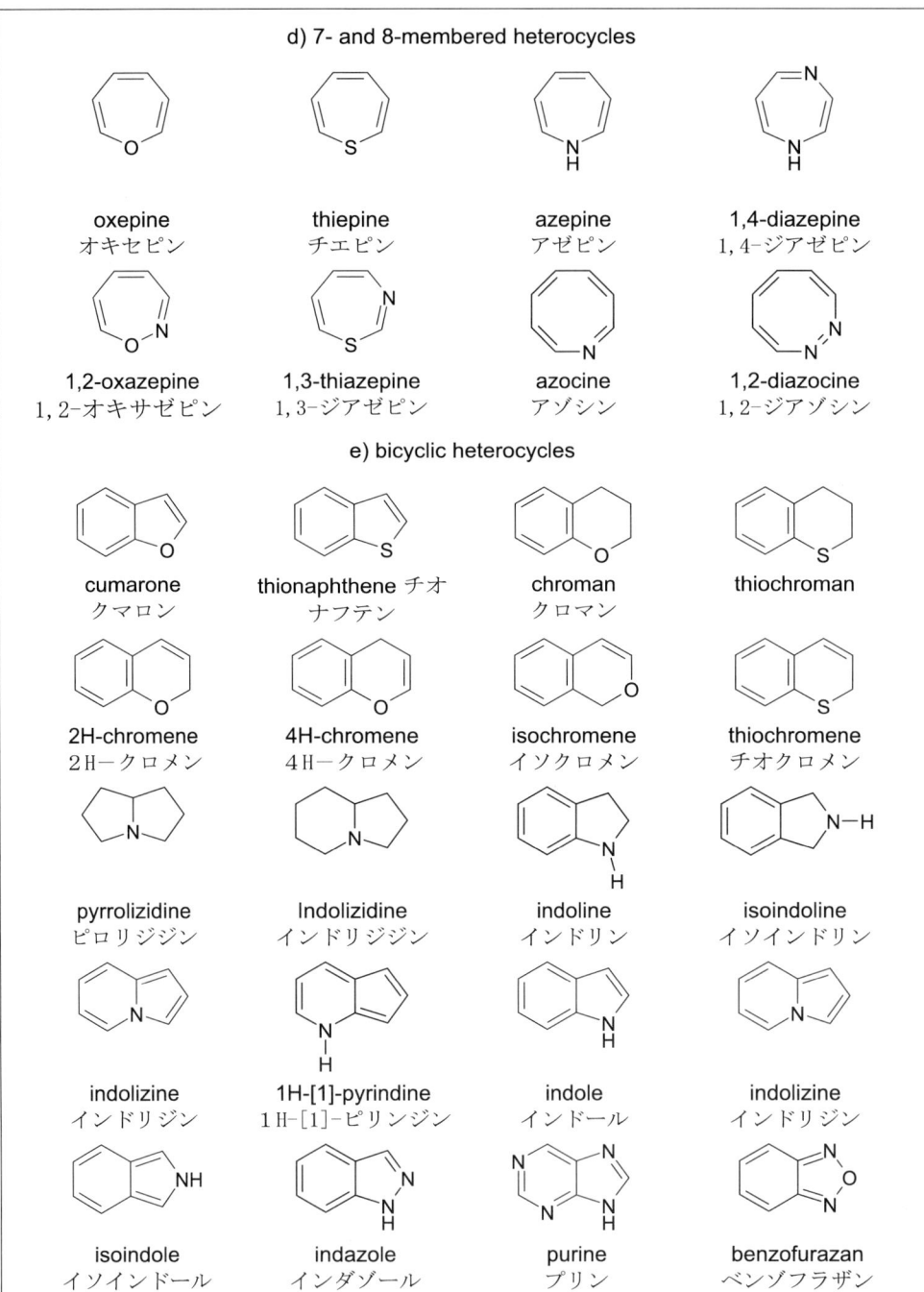

Fig. 3.2 (continued)

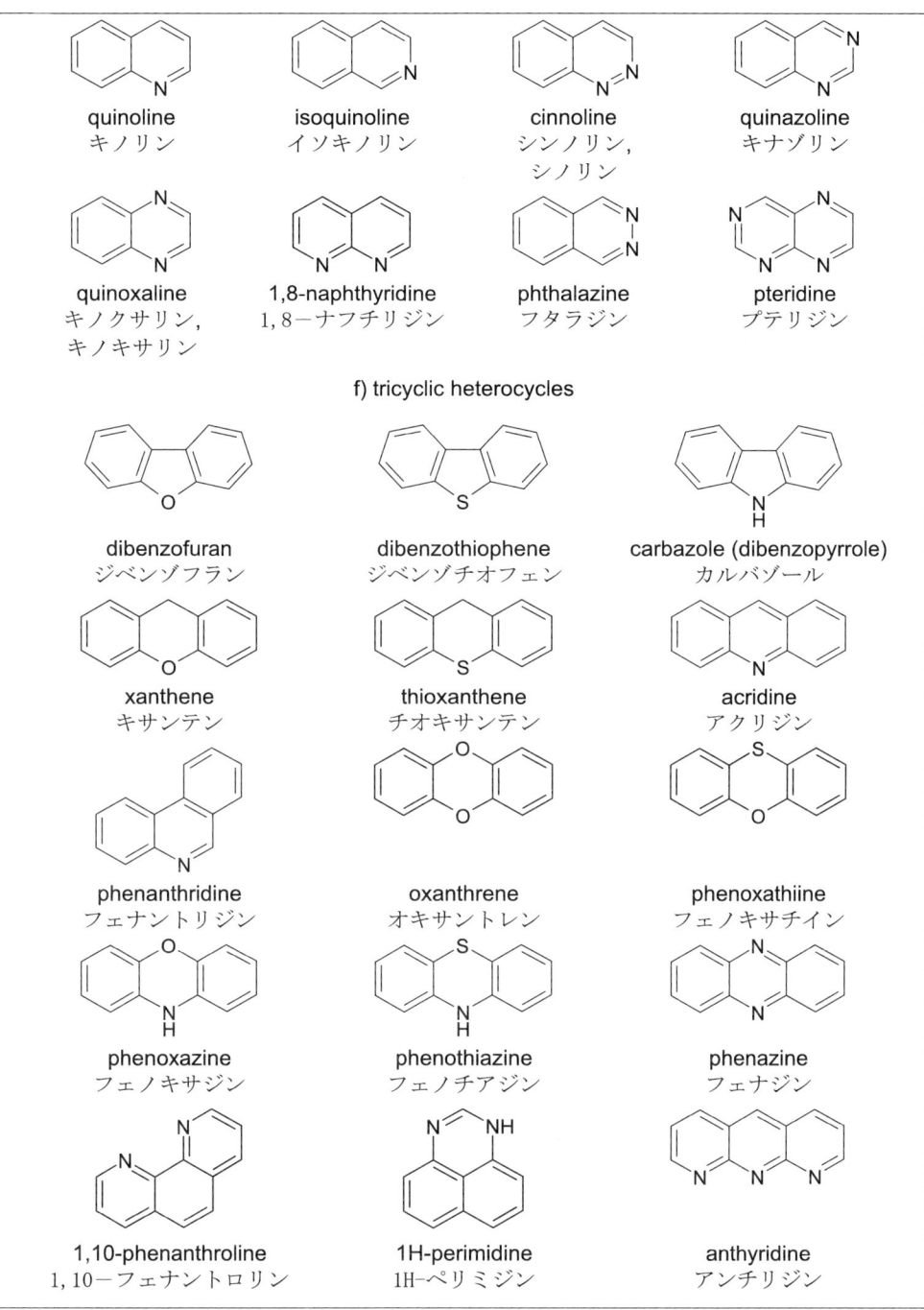

Fig. 3.2 (continued)

Table 3.33 Examples for heterocyclic nomenclature.

Common name	Systematic term	Japanese name
α-picoline	2-methylpyridine	2-メチルピリジン（2-ピコリン）
uvic acid	2,5-dimethyl-3-furanecarboxylic acid	2,5-ジメチル-3-フランカルボン酸（ウヴィク酸）
guanine	2-aminopurin-6-one	2-アミノ-6-オキシ-プリン（グアニン）
δ-valerolactam	piperidin-2-one	ピペリジン-2-オン（δ-バレロラクタム）
β-collidine	3-ethyl-4-methylpyridine	3-エチル-4-メチルピリジン（β-コリジン）
berberonic acid	pyridine-2,4,5-tricarboxylic acid	ピリジン-2,4,5-トリカルボン酸（ベルベロン酸）
muscazone	α-amino-2-oxo-2,3-dihydro-1,3-oxazole-5-acetic acid	α-アミノ-2-オキソ-2,3-ジヒドロ-1,3-オキサゾール-5-酢酸（ムスカゾン）
ibotenic acid	α-amino-3-oxo-2,3-dihydroisoxazole-5-acetic acid	α-アミノ-3-オキソ-2,3-ジヒドロイソオキサゾール-5-酢酸（イボテン酸）
phenolphthalein	3,3-bis(4-hydroxyphenyl)-1,3-dihydroisobenzofuran-1-one	3,3-ビス-(4-ヒドロキシフェニル)-1,3-ジヒドロイソベンゾフラン-1-オン（フェノールフタレイン）

Table 3.34 Stem terms for the Hantzsch–Widman system.

Ring size	Unsaturated rings	Saturated rings without nitrogen	Saturated rings containing nitrogen
3	-irene: イレン -irine: イリン (for rings containing nitrogen)	-irane: イラン	-iridine: イリジン
4	-ete: エト	-etane: エタネ	-etidine: エチジン
5	-ole: オール	-olane: オラン	-olidine: オリジン
6	-ine: イン -inine: イニン (for rings containing B, P, As, Sb, halogens)	-ane: アン (for rings containing O, S, Se, Te, Bi, Hg) -inane: イナン (for rings containing B, P, As, Sb, Si, Ge, Sn, Pb, halogens)	-inane: イナン
7	-epine: エピン	-epane: エパン	-epane: エパン
8	-ocine: オシン	-ocane: オカン	-ocane: オカン
9	-onine: オニン	-onane: オナン	-onane: オナン
10	-ecine: エシン	-ecane: エカン	-ecane: エカン

3-[(4-amino-2-methylpyrimidin-5-yl)methyl]-5-(2-hydroxyethyl)-4-methyl-1,3-thiazol-3-ium chloride (vitamin B_1; thiamin hydrochloride)
3-[(4-アミノ-2-メチルピリミジン-5-イル)メチル]-5-(2-ヒドロキシエチル)-4-メチル-1,3-チアゾール-3-イウムクロリド(ビタミンB_1; チアミンクロリド)

Besides common names, IUPAC recommendations for the nomenclature of heterocycles are, nowadays, broadly applied. Heteromonocycles are named by the Hantzsch–Widman system, which combines prefixes representing the heteroatoms with endings that indicate ring size. Important prefixes are (with declining priority in the Hantzsch–Widman system), for example, オキサ (for oxygen), チア (for sulfur), セレナ (for selenium), アザ (for nitrogen) and フォスファ (for phosphorus). Table 3.34 shows the Japanese equivalents for such endings.

For their application and for further rules see the chemical literature. The position of heteroatoms within the heterocycle is specified according to IUPAC rules, as, for example, in 2,6-naphthyridine: 2, 6-ナフチリジン and the following examples:

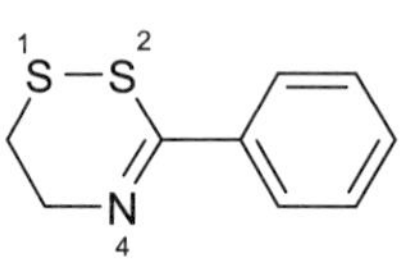

3-phenyl-5,6-dihydro-
[1,2,4]dithiazine
3-フェニル-5, 6-ジヒドロ-1, 2, 4-ジチアジン

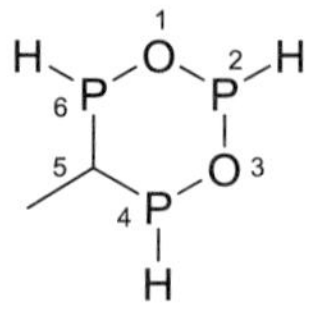

5-methyl-[1,3,2,4,6]-
dioxatriphosphinane
5-メチル-1, 3, 2, 4, 6-ジオキサトリホスフィナン

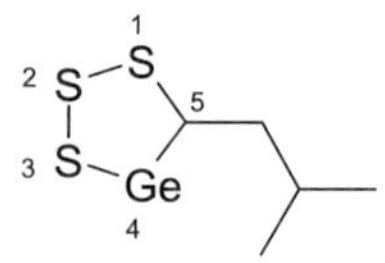

5-isobutyl-[1,2,3,4]-
trithiagermolane
5-イソブチル-1, 2, 3, 4-トリチアゲルモラン

Heterocycles with condensed rings are also named according to IUPAC recommendations, such as in the following examples. The cycle that is mentioned at the beginning of the name, changes its name, such as ベンゾ (benzo-), フロ (furo-), チエノ (thieno-), ピリド (pyrido-), ピリミド (pyrimido-), イミダゾ (imidazo-), オキサゾロ (oxazolo-), キノ (quino-), イソキノ (isoquino-), キノリノ (quinolino-), イソキノリノ (isoquinolino-), ベンゾイミダゾ (benzimidazo-) and ベンゾオキサゾロ (benzoxazolo-).

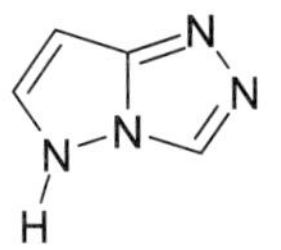

5H-pyrazolo[5,1-c][1,2,4]-
triazole
5H-ピラゾーロ [5, 1-c]
[1, 2, 4]-トリアゾール

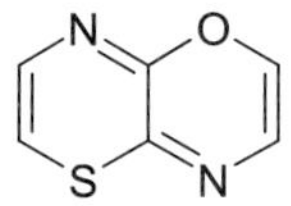

1,4-thiazino[3,2-b]-1,4-oxazine
1, 4-チアジノ [3, 2-b]-
1, 4-オキサジン

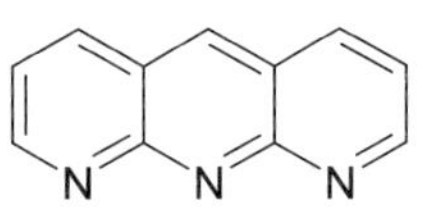

pyrido[2,3-b][1,8]
naphthyridine
ピリド [2, 3-b] [1, 8]
ナフチリジン;
1,8,9-triaza-anthracene
1, 8, 9-トリアザアントラセン

3.3.3
Organic Acids

The names of important organic acids are given in Tables 3.35–3.39. For some acids, especially in new publications, *katakana* or *hiragana* are used instead of *kanji*, such as for oxalic acid HOOC-COOH: シュウ酸 instead of 蓚酸, both pronounced *shuusan*.

Table 3.35 Japanese names of saturated unbranched monoacids (including saturated fatty acids).

Acid	IUPAC nomenclature	Japanese common names
formic acid (HCOOH)	メタン酸 *metansan*	蟻酸, ギ酸 *gisan*
acetic acid (CH_3COOH)	エタン酸 *etansan*	酢酸 *sakusan*
glacial acetic acid (CH_3COOH)		氷酢酸 *hyousakusan*
propionic acid, n-propanoic acid (CH_3CH_2COOH)	n-プロパン酸 *n-puropansan*	プロピオン酸 *puropionsan*
butyric acid, n-butanoic acid [$CH_3(CH_2)_2COOH$]	n-ブタン酸 *n-butansan*	酪酸 *rakusan*, ブチル酸 *buchirusan*
valeric acid, n-pentanoic acid [$CH_3(CH_2)_3COOH$]	n-ペンタン酸 *n-pentansan*	吉草酸 *kissousan*, バレリアン酸 *bareriansan*, ヴァレリアン酸 *vareriansan*
caproic acid, n-hexanoic acid [$CH_3(CH_2)_4COOH$]	n-ヘキサン酸 *n-hekisansan*	カプロン酸 *kapuronsan*
oenanthic acid, n-heptanoic acid [$CH_3(CH_2)_5COOH$]	n-ヘプタン酸 *n-heputansan*	エナント酸 *enanchirusan*
caprylic acid, n-octanoic acid [$CH_3(CH_2)_6COOH$]	n-オクタン酸 *n-okutansan*	カプリル酸 *kapurirusan*
pelargonic acid, nonanoic acid [$CH_3(CH_2)_7COOH$]	n-ノナン酸 *n-nonansan*	ペラルゴン酸 *perarugonsan*
capric acid, n-decanoic acid [$CH_3(CH_2)_8COOH$]	n-デカン酸 *n-dekansan*	カプリン酸 *kapurinsan*
n-undecanoic acid, n-undecylic acid [$CH_3(CH_2)_9COOH$]	n-ウンデカン酸 *n-undekansan*	n-ウンデシル酸 *n-undeshirusan*
lauric acid, n-dodecanoic acid [$CH_3(CH_2)_{10}COOH$]	n-ドデカン酸 *n-dodekansan*	ラウリン酸 *raurinsan* n-ドデシル酸 *n-dodeshirusan*
n-tridecanoic acid [$CH_3(CH_2)_{11}COOH$]	n-トリデカン酸 *n-toridekansan*	n-トリデシル酸 *n-torideshirusan*

Table 3.35 *(continued)*

Acid	IUPAC nomenclature	Japanese common names
myristic acid, n-tetradecanoic acid [$CH_3(CH_2)_{12}COOH$]	n-テトラデカン酸 *n-tetoradekansan*	ミリスチン酸 *mirisuchinsan*
n-pentadecanoic acid [$CH_3(CH_2)_{13}COOH$]	n-ペンタデカン酸 *n-pentadekansan*	n-ペンタデシル酸 *n-pentadeshirusan*
palmitic acid, n-hexadecanoic acid [$CH_3(CH_2)_{14}COOH$]	n-ヘキサデカン酸 *n-hekisadekansan*	パルミチン酸 *parumichinsan*
margaric acid, n-heptadecanoic acid [$CH_3(CH_2)_{15}COOH$]	n-ヘプタデカン酸 *n-heputadekansan*	マルガリン酸 *marugarinsan*
stearic acid, n-octadecanoic acid [$CH_3(CH_2)_{16}COOH$]	n-オクタデカン酸 *n-okutadekansan*	ステアリン酸 *sutearinsan*
tuberculostearic acid, nonadecanoic acid [$C_{18}H_{37}COOH$]	ノナデカン酸 *nonadekansan*	ツベルクロステアリン酸 *tsuberukurosutearinsan*
arachic acid, arachidic acid, n-eicosanoic acid [$CH_3(CH_2)_{18}COOH$]	n-エイコサン酸 *n-eikosansan*	アラキジン酸 *arakijinsan*, アラキン酸 *arakinsan*
behenic acid, n-docosanoic acid [$CH_3(CH_2)_{20}COOH$]	n-ドコサン酸 *n-dokosansan*	ベヘン酸 *behensan*
lignoceric acid, n-tetracosanoic acid [$CH_3(CH_2)_{22}COOH$]	n-テトラコサン酸 *n-tetorakosansan*	リグノセリン酸 *rigunoserinsan*
cerotic acid, n-hexacosanoic acid [$CH_3(CH_2)_{24}COOH$]	n-ヘキサコサン酸 *n-hekisakosansan*	セロチン酸 *serochinsan*
montanic acid, n-octacosanoic acid [$CH_3(CH_2)_{26}COOH$]	n-オクタコサン酸 *n-okutakosansan*	モンタン酸 *montansan*
melissic acid, n-triacontanoic acid [$CH_3(CH_2)_{28}COOH$]	n-トリアコンタン酸 *n-toriakontansan*	メリシン酸 *merishinsan*, メリッシン酸 *merisshinsan*
lacceric acid, n-dotriacontanoic acid [$CH_3(CH_2)_{30}COOH$]	n-ドトリアコンタン酸 *n-dotoriakontansan*	ラッセル酸 *rasserusan*
ceromelissic acid, n-tritriacontanoic acid [$CH_3(CH_2)_{31}COOH$]	n-トリトリアコンタン酸 *n-toritoriakontansan*	セロメリッシン酸 *seromerisshinsan*
geddic acid, n-tetratriacontanoic acid [$CH_3(CH_2)_{32}COOH$]	n-テトラトリアコンタン酸 *n-tetoratoriakontansan*	ゲダ酸 *gedasan*
ceroplastic acid, n-pentatriacontanoic acid [$CH_3(CH_2)_{33}COOH$]	n-ペンタトリアコンタン酸 *n-pentatoriakontansan*	セロプラスチン酸 *seropurasuchinsan*

Table 3.36 Japanese names of saturated diacids and hydroxy diacids.

Acid	Japanese name
oxalic acid (HOOCCOOH)	蓚酸, シュウ酸 shuusan
malonic acid (HOOCCH$_2$COOH)	マロン酸 maronsan
succinic acid [HOOC(CH$_2$)$_2$COOH]	琥珀酸, コハク酸 kohakusan
glutaric acid [HOOC(CH$_2$)$_3$COOH]	グルタル酸 gurutarusan
adipic acid [HOOC(CH$_2$)$_4$COOH]	アジピン酸 ajipinsan
pimelic acid [HOOC(CH$_2$)$_5$COOH]	ピメリン酸 pimerinsan
suberic acid [HOOC(CH$_2$)$_6$COOH]	スベリン酸 suberinsan
lepargylic acid, azelaic acid [HOOC(CH$_2$)$_7$COOH]	レパルギル酸 reparugirusan, アゼライン酸 azerainsan
sebacic acid [HOOC(CH$_2$)$_8$COOH]	セバシン酸 sebashinsan
malic acid [HOOCCH(OH)CH$_2$COOH]	林檎酸 ringosan, リンゴ酸 ringosan
tartaric acid [HOOC(CHOH)$_2$COOH]	酒石酸 shusekisan
mucic acid [HOOC(CHOH)$_4$COOH]	粘液酸 nenekisan, ムチン酸 muchinsan
saccaric acid [HOOC(CHOH)$_4$COOH]	糖酸 tousan, サッカリン酸 sakkarinsan

Table 3.37 Japanese names of unsaturated and aromatic acids.

Acid	Japanese name
acrylic acid (CH$_2$=CHCOOH)	アクリル酸 akurirusan
methacrylic acid [CH$_2$=C(CH$_3$)COOH]	メタクリル酸 metakurirusan
crotonic acid (trans-CH$_3$CH=CHCOOH)	クロトン酸 kurotonsan
isocrotonic acid (cis-CH$_3$CH=CHCOOH)	イソクロトン酸 isokurotonsan
cinnamic acid (C$_6$H$_5$CH=CHCOOH)	桂皮酸, ケイ皮酸 keihisan
maleic acid (cis-HOOCCH=CHCOOH)	マレイン酸 mareinsan
fumaric acid (trans-HOOCCH=CHCOOH)	フマル酸 fumarusan
oleic acid [CH$_3$(CH$_2$)$_7$CH=CH(CH$_2$)$_7$COOH]	油酸 yusan, オレイン酸 oreinsan
cetoleic acid [H$_3$C(CH$_2$)$_9$CH=CH(CH$_2$)$_9$COOH]	鯨油酸 geiyusan
benzoic acid (C$_6$H$_5$COOH)	安息香酸 ansokukousan, 安息酸 ansokusan
toluic acid [C$_6$H$_4$(CH$_3$)COOH]	トルイル酸 toruirusan
naphthoic acid [C$_{10}$H$_7$COOH]	ナフトエ酸 nafutoesan
phthalic acid [C$_6$H$_4$(COOH)$_2$]	フタル酸 futarusan
picric acid [HOC$_6$H$_2$(NO$_2$)$_3$]	ピクリン酸 pikurinsan

Table 3.38 Japanese names of formic acid derivatives.

Acid	Japanese name
hydrogen cyanide, hydrocyanic acid (HCN)	シアン化水素 shiankasuiso, 青酸 seisan, シアン化水素酸 shiankasuisosan
carbamic acid (H_2NCOOH)	カルバミン酸 karubaminsan
cyanic acid (NC-OH)	シアン酸 shiansan
isoyanic acid (HN=C=O)	イソシアン酸 isoshiansan
fulminic acid (HCNO)	雷酸 raisan
thiocyanic acid, rhodanic acid (HSCN)	ロダン酸 rodansan, チオシアン酸 chioshiansan

Table 3.39 Japanese names of further acids.

Acid	Japanese name
camphoric acid, 1,2,2-trimethylcyclopentane-1,3-dicarboxylic acid ($C_{10}H_{16}O_4$)	樟脳酸 shounousan, ショウノウ酸 shounousan
folic acid, pteroylglutamic acid ($C_{19}H_{19}N_7O_6$)	葉酸 yousan
gallic acid (3,4,5-trihydroxybenzoic acid) [$C_6H_2(OH)_3COOH$]	没食子酸 bosshokushisan
hippuric acid ($C_6H_5CONHCH_2COOH$)	馬尿酸 banyousan
inosinic acid ($C_{10}H_{13}N_4PO_8$)	イノシン酸 inoshinsan
kainic acid, digenic acid ($C_{10}H_{15}NO_4$)	カイニン酸 kaininsan, 海人酸 kaininsan
kojic acid ($C_6H_6O_4$)	麹酸 koujisan, コウジ酸 koujisan
lactic acid [$CH_3CH(OH)COOH$]	乳酸 nyuusan
pivalic acid [$CH_3C(CH_3)_2COOH$]	ピバル酸 pibarusan
uric acid (2,6,8-trioxypurine) ($C_5H_4N_4O_3$)	尿酸 nyousan

3.3.4
Organic Acid Derivatives

Salts of organic acids are named by adding the name of the metal cation at the end of the acid name.

Examples: acetic acid (CH_3COOH): 酢酸 sakusan → magnesium acetate [$(CH_3COO)_2Mg$]: 酢酸マグネシウム sakusanmaguneshiumu;

tartaric acid [$HOOC(CHOH)_2COOH$]: 酒石酸 shusekisan → sodium tartrate [$NaOOC(CHOH)_2COONa$]: 酒石酸ナトリウム shusekisannatoriumu;

formic acid (HCOOH): 蟻酸 gisan → platinum formate [$(HCO)_2Pt$]: 蟻酸白金 gisanhakkin

If the *kanji* 塩 *en* us added to the name of the acid, the general salt is meant, e.g., acetic acid (CH$_3$COOH) 酢酸 *sakusan* → acetate (CH$_3$COM) 酢酸塩 *sakusanen*. In addition, for esters, the name of the alcohol residue is attached to the acid name. Furthermore, エステル may be added. For example, valeric acid CH$_3$(CH$_2$)$_3$COOH 吉草酸 *kissousan* → ethyl valerate, valeric acid ethyl ester CH$_3$(CH$_2$)$_3$COOC$_2$H$_5$: 吉草酸エチル *kissousanechiru* or 吉草酸エチルエステル *kissousanechiruesuteru*.

For organic acid halogenides, the halogenide function is indicated by either *kana* at the end of the name or *kanji* at the beginning. In the first case, the basic acid may be named with its *kanji* or a *kana* term for the acid residue, e.g., 乳酸ブロミド *yuusanburomido* [for lactic acid bromide, CH$_3$CH(OH)COBr] or プロピオニルフルオリド (for propionic acid fluoride, C$_2$H$_5$COF). If the *kanji* for the halogenide is used, it has to be combined with a *kana* expression of the acid residue (-yric or -oic), such as 弗化ヘキサノイル *fukkahekisanoiru* (for hexanoic acid fluoride, C$_5$H$_{11}$COF). The combination with the acid's name in the last example would not refer to the acid halogenide, but to an acid with a halogenated alkyl chain. 弗化酪酸 *fukkarakusan* or 弗化ブチル酸 *fukkabuchirusan* mean fluorobutyric acid (C$_3$H$_6$F-COOH) but not butyric acid fluoride. For some less common or more complex compounds, the whole English name is transcribed into Japanese; therefore, the halogenide is positioned at the end of the name in *katakana*, e.g., toluenesulfonic acid, C$_6$H$_4$(CH$_3$)SO$_3$H, トルエンスルホン酸 *toruensuruhonsan* → toluenesulfonyl chloride, C$_6$H$_4$(CH$_3$)SO$_2$Cl, トルオルスルフォクロリド.

In the names of halogenated acid halogenides, the halogen substituent is always indicated at the beginning of the name while the halogenide function is described at its end, e.g., in pentafluorobenzoyl bromide (C$_6$H$_5$COBr): ペンタフルオルベンゾイルブロマイド. For malonic acid derivatives, the following names are conceivable, even though they are not all commonly used:

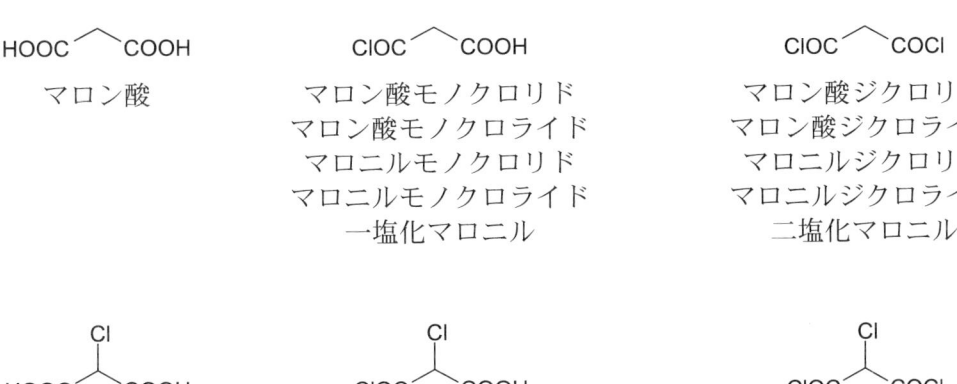

Cl Cl HOOC—C—COOH	Cl Cl ClOC—C—COOH	Cl Cl ClOC—C—COCl
二塩化マロン酸 ジクロロマロン酸	二塩化マロン酸モノクロリド 二塩化マロン酸モノクロライド 二塩化マロニルモノクロリド 二塩化マロニルモノクロライド ジクロロマロニルモノクロリド ジクロロマロニルモノクロライド	二塩化マロン酸ジクロリド 二塩化マロン酸ジクロライド 二塩化マロニルジクロリド 二塩化マロニルジクロライド ジクロロマロニルジクロリド ジクロロマロニルジクロライド

For other acid derivatives, *kanji* or *katakana* are added to the beginning of the acid name: 無水 *musui* for anhydrides (e.g., maleic acid マレイン酸 *mareinsan* → maleic acid anhydride 無水マレイン酸 *musuimareinsan*), 過 *ka* for peracids (e.g., perbenzoic acid C_6H_5COOOH: 過安息香酸 *kaansokukousan*), メタ for meta acids and オルト or 正 *sei* for ortho acids (e.g., orthoformic acid オルト蟻酸 *orutogisan* or orthobenzoic acid 正安息香酸 *seiansokukousan*).

4
Japanese Patent Documentation

During the 1970s, the Japan Patent Office (JPO), 特許庁 *tokkyochou*, was receiving over 100 000 patent applications a year. By the early 1980s the number had doubled and the impressive upwards trend in patent filings continued throughout that decade and well into the 1990s. Since 1985 the annual number of patent applications filed in Japan has consistently been well in excess of 350 000, stabilizing after 1998 at a level of just over 400 000 filings a year (Fig. 4.1).

Japanese patent applicants are also highly active internationally: in 2005, the number of PCT applications originating from Japan increased by 23% over the previous year. Japan was the second biggest PCT filing country in 2005, with some 25 000 PCT filings, or 18.8% of all PCT applications that year.

The above figures illustrate a strong pro-patent orientation in Japan and the strategic value that Japan places on protecting inventions with patents. For the patent searcher these figures mean that a substantial part of worldwide patent literature is to be found in Japan. In today's global economy, no one can afford to ignore such a significant component of prior art documentation.

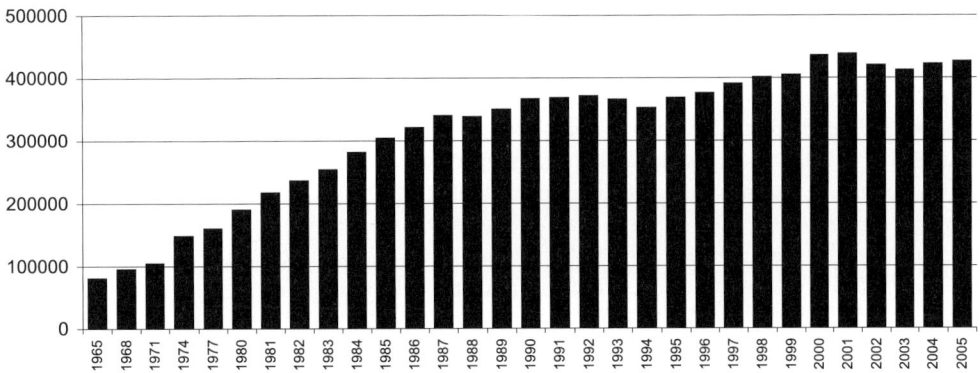

Fig. 4.1 Total number of patent applications in Japan.

Japanese-English Chemical Dictionary. Edited by Markus Gewehr
Copyright © 2008 WILEY-VCH Verlag GmbH & Co. KGaA, Weinheim
ISBN: 978-3-527-31293-1

4.1
The Japanese Patent System

4.1.1
Development of the Japanese Patent System

Looking at the high numbers of patent applications in Japan, it may surprise the reader that compared with Western countries Japan was rather late in adopting a patent system. The reasons for this are historical, as will now be explained.

Throughout the more than 250 years of isolation policy in Japan (1603–1867) and the closing of the country to foreign influences, seen as a threat to the stability of its feudal society by the military "Shogunate" government, inventions and new developments were discouraged. At a time when many countries in Europe already had enacted patent laws and had patent systems in place, a "Ban on Novelty" (新規御法度 shinki gohatto) was decreed in Japan in 1721. By issuing this decree, the Shogunate government wanted to protect and promote existing commercial and industrial guilds, prohibiting the introduction and consumption of new goods and items.

The Shogunate ended in 1868. The Shogun was forced to resign and the young Emperor Meiji was restored as the new head of the Japanese government. After the Meiji Restoration, Japan embarked on a rapid and determined course of modernization. Seriously lagging behind the West in technology and industrial progress, Japan sent envoys to America and Europe to discover what made Western countries great powers.

A book by the famous Japanese author Fukuzawa Yukichi (1835–1901) introduced the idea of patents to the Japanese people. Fukuzawa had been a member of the first ever Japanese delegation to the United States in 1860, and had visited several countries in Europe. He published a famous book in 1867 about *Conditions in the West* (西洋事情 *seiyou jijou*), describing how the protection of inventors' rights would encourage further creative activities and thus lead to the nation's development. Japanese government delegations also paid visits to the United States' patent office and returned home with large volumes of materials on the patent system, which subsequently influenced the drafting of Japan's own patent law.

In 1886, the Japanese government sent Takahashi Korekiyo (1854–1936), who a year later became the first Commissioner of Patents, on a mission to study the patent systems in Europe and the United States. When asked during his visit to the United States' patent office why the people of Japan wanted a patent system, Takahashi is reported to have replied: "We have looked about us to see what nations are the greatest, so that we can be like them. We said: 'What is it that makes the United States such a great nation?', and we investigated and found that it was patents and so we will have patents."

In the fourth year of the Meiji era, 1871, an experimental patent system was implemented, but abandoned a year later. In 1884, the first trade mark bylaws were passed. It was, however, on 18 April 1885 that the first substantial patent law in Japan was established by the "Patent Monopoly Act" (専売特許条例 *senbai tokkyo jourei*) – 18 April has been celebrated as "Invention Day" in Japan since 1954. The "Patent Act" (特許条例 *tokkyo jourei*) replaced the Patent Monopoly Act in 1888. The same year saw the enactment of a design law, and a utility model law followed in 1905.

The first patents under the Patent Monopoly Act were granted in August 1885. The first patent was granted to Hotta Mizumatsu for an anticorrosive paint and painting method. Hotta was a lacquer ware craftsman and his invention was effective in protecting ships' keels from corrosion (his patent, number "1", is archived in the Japan Patent Office's "Industrial Property Digital Library", from where it can be retrieved using the kind of document code "C" and the patent number "1"). Altogether, some 400 patent applications were filed in 1885, of which around 100 were granted.

Shortly after the public announcement of the Patent Monopoly Act, Japan entered the industrial revolution, centering on yarn-making and spinning industries, which eventually led to Japan's rise as a world power. In 1891, Toyoda Sakichi, the father of the founder of Toyota Industries, obtained his first patent for a wooden hand loom. Toyoda further developed his invention, creating a wooden power loom as well as the world's first automatic loom. Toyoda Sakichi was granted a total of 84 patents and 35 utility models.

Even though much of early Japanese patent law was modeled on the patent systems found in Western countries – in particular it copied many features of the United States' system – there were some special features that reflected Japanese needs. For example, under the Patent Monopoly Act, and subsequently the Patent Act, patents were not granted to foreigners, and patent protection could not be obtained for fashion or food products, or medicines; patents not exploited within three years could be revoked and severe sanctions were imposed for infringement, going as far as sentences of hard labor. When Japan became a signatory of the Paris Convention in 1899, the Patent Act was replaced by the "Patent Law" (特許法 *tokkyohou*), bringing legislation into line with the Convention, and extending patent protection to foreigners. Also in 1899, the "Patent Agent Registration Regulations" were passed and by the end of the year some 130 patent agents had registered. Before this, it had been possible for anybody to represent a patent applicant before the Japanese patent office.

4.1.2
The Modern Japanese Patent System

During the early part of the 20th century, several major reforms of the Japanese patent system reflected a strong influence of German patent law. One example was the introduction of utility models in 1905. Japan completely revised its Patent Law in 1921, replacing the "first-to-invent" principle, which had been in force until that time, by "first-to-file". 1921 also saw the introduction of an opposition procedure. While no patent protection was available for medicine, food and chemical products, protection was possible for processes relating to their manufacture.

Japan entered a decade of high economic growth in 1955, during which there were calls for amendments to the industrial property laws that had not undergone any fundamental changes since 1921. Thus, another substantial revision to the industrial property laws took place in 1959, marking the starting point of the modern industrial property system in Japan. The most important changes in the patent law concerned the definition of invention and novelty, including the "distribution in publications overseas" as one of the reasons for lack of novelty, as well as a grace period for disclosure.

The patent law amendment of 1971 brought the deferred examination system (search and examination only upon request within seven years from the application date), and obligatory publication of the application ("laid-open publications") 18 months after the application date (or priority date). The introduction of the laid-open publication established the practice of publishing the full text of patent applications in the "Patent Gazettes" and of including abstracts with the patent applications.

1976 saw the introduction of the multiple claim system, and of patent protection for medicines and chemical substances (as opposed to methods for their production), which had not been possible under the previous version of the patent law. Japan was a founding member of the Patent Co-operation Treaty (PCT), making it possible for Japan to be designated by this route from 1978 onwards. The standard "JP" country code for Japan replaced the previous "JA" code in the same year. In 1985, Japan adopted the international priority system and in 1988 introduced the possibility for patent term extensions for chemical, in particular pharmaceutical and agrochemical, patents.

In the 1990s, several amendments to the patent and utility model laws aimed to harmonize the Japanese system with international practice. Substantive examination for utility models was abolished by the 1993 amendment to the utility model law, leading to a drastic decrease in application numbers for utility models.

The patent law was amended in 1994 to comply with the Agreement on Trade Related Intellectual Property Rights (TRIPS) of the World Trade Organization (WTO). The amendments, which became effective in July 1995, included the term of a patent right (now defined as "20 years from application date", as opposed to the previous definition of "15 years from examined publication date but not in excess of 20 years from application date"), the possibility of filing patent applications in English, and the restoration of a lapsed patent right (if the applicant can prove that there were reasons beyond his control for failing to pay annual fees within the legal time limit). The pre-grant opposition system was replaced by a post-grant opposition system in 1996, and accelerated examination was introduced. To streamline the patent examination and granting procedure, the time limit for filing requests for examination was shortened from seven to three years for applications filed on 1 October 2001 or later. (Owing to the reduction of the examination request period to three years, the number of requests for examination received at the JPO from 2003 to 2004 increased by 26 percent.)

A more recent major change to the Japanese patent law occurred in 2004, when the opposition system was abolished, leaving invalidation proceedings as the only way of challenging granted patents in Japan.

In an attempt to render utility models more attractive, major changes were made to the utility model law in 2005. These include a longer term for utility models (extended to ten years from six) and the possibility of filing a patent application based on (and claiming the priority of) a registered utility model within three years of the utility model's application date.

4.1.3
The Japanese Patent Office

The predecessor of the Japanese patent office was the "Trademark Registration Office", which had been established in 1884 and a year later – with the official announcement of the "Patent Monopoly Act" – renamed into "Patent Monopoly Office". Takahashi Korekiyo was the first Director General of the Patent Monopoly Office. In 1886, the office's name was changed to "Patent Monopoly Bureau", which later became simply the "Patent Bureau". When it was first founded, the "Patent Monopoly Bureau" only had three judges, one examiner and one assistant examiner. By 1899, more than 1500 patent applications were coming in each year and the patent office's staff had grown to five judges, 15 examiners and 20 assistant examiners. After the Second World War, the Bureau became the "Patent Standards Bureau" and was attached to the Ministry of Commerce and Industry. When the Ministry of Commerce and Industry was changed into the Ministry of International Trade and Industry (MITI) in 1949, the Patent Bureau became the "Patent Office". Today, the Japan Patent Office (JPO), 特許庁 *tokkyochou*, belongs to the Ministry of Economy, Trade and Industry (METI). It receives over 400 000 patent applications annually and has more than 2500 staff, about 1400 of whom are examiners and a further 400 appeal examiners (numbers as of March 2005).

The Japanese patent office defines its role as: to grant industrial property rights, to draft plans for industrial property policies, to participate in international exchange and cooperation, to review the Japanese industrial property system, and to disseminate information on industrial property rights. On its homepage, the JPO states:[1]

> *The aim of the industrial property system is to contribute to the nation's industrial development through adequate protection and effective utilization of inventions and other forms of intellectual creations. To help promote science and technology, the IP system is expected to play an increasingly important role in Japan in the 21st century.*

In 1984, the JPO committed itself to the vision of a "paperless system" and in 1990 was the first patent office in the world to accept patent filings in electronic form. While, initially, electronic applications could be submitted to the JPO on floppy disk or via dedicated terminals, it was possible from 1998 on to file applications via personal computers. An internet-based online filing system was introduced in October 2005. By this time, some 97% of all patent filings at the JPO were being received in electronic form.

Within the framework of the paperless system, the JPO started in 1993 to publish its official gazettes for unexamined patent and utility model applications on CD-ROM, and from 1994 examined patent and utility models were also published in this way. Since January 2006, the JPO has published its gazettes (for registered utility models) via the internet (http://www.publication.jpo.go.jp/utility/do/usr/topmenu?lang=e). Further services, such as free access to patent documents and machine-translation of patent applications, are described in Section 4.3.1.

In 1983, the Japanese Patent Office, together with the European Patent Office and the United States Patent and Trademark Office, set up the "Trilateral Cooperation" to harmonize industrial property administration, to further develop the protection of industrial property rights, and to fully exploit the potential in patent searches, the examination process and the utilization of electronic tools. Today, the Trilateral Offices process the greater part, i.e., over 85%, of all patent applications filed worldwide, including PCT applications. Further information is available at http://www.trilateral.net/. A similar service is available from the United States Patent and Trademark Office (http://www.uspto.gov/patft/index.html).

---

1) Internet address of the Japan Patent Office:
http://www.jpo.go.jp/shoukai_e/index.htm.

4.2
Special Characteristics of Japanese Patent Documentation

4.2.1
Document Types and Kind of Document Codes

The earliest Japanese patents (from 1885 to 1922) were only published once they were granted. These oldest publications of Japanese patents are usually given the kind of document code "C" and use a continuous number sequence starting from the number 1. In 1921, an opposition system was introduced in Japan, and from 1922 onwards patent documents were published for opposition before grant. From 1922 to 1971, Japanese patent documents were only published once, after substantive examination and before grant. In 1971, Japan introduced the early publication system, whereby patent applications were laid-open to public inspection within 18 months of the filing date. This "laid-open" publication is called 公開 *koukai* in Japanese. The *koukai* stage of a patent application is assigned the kind of document code "A". This code can be found on the first pages of Japanese laid-open publications as well as in many databases.

Before 1996, when the pre-grant opposition system was still in force, the second publication stage following the *koukai* stage was called 広告 *koukoku*. This was the publication of the examined specification, but not yet the granted patent document, and came with the kind of document code "B". Following its publication as an examined case, the document was open for a period of three months (pre-grant opposition system), during which time third parties could oppose the final grant of the patent. If no opposition occurred during this time, the patent was granted and an entry in the patent register was effected. The entry in the register was given a special registration number, in a continuous running sequence. The registration number also came with a kind of document code, "C". Importantly, only a few databases collected the information on this third document type or made these C-codes available for searching.

Since the introduction of a post-grant opposition system in 1996, the second stage of publication following the *koukai* has been the granted patent specification. Called 登録 *touroku* in Japanese, it replaced the older *koukoku* publication and carries the "B2" kind of document code. *touroku* publications have a running number sequence starting at 2500001. Usually, there is a time delay of some two to three months between the day of grant and the publication of the granted patent. The new six-month opposition period used to start on the day of publication of the *toroku* document, but the opposition system was abolished on 1 January 2004, leaving invalidation suits as the only way of challenging granted patents in Japan.

With the adoption of accelerated examination in 1996, a new type of *touroku* document came about: where the applicant requests accelerated examination, a decision of grant and the publication of the granted patent can occur even before the early *koukai* publication stage. In such cases, the publication of the granted patent is issued without a previous laid-open or *koukai* publication and is given the kind code "B1". The reader should note that this may prove problematic for prior art searches that rely on English abstracts for Japanese patent documents, as English abstracts are only produced for Japanese unexamined patent applications (A-documents), not for granted documents. The corresponding A-publication, and even

```
(19) 日本国特許庁(JP)        (12) 公 表 特 許 公 報(A)        (11) 特許出願公表番号
                                                              特表2006-513849
                                                                (P2006-513849A)
                                                      (43) 公表日  平成18年4月27日(2006.4.27)
```

Fig. 4.2 Example of a Japanese *kouhyou* publication (PCT application having entered the national phase in Japan).

more so the English abstract for a "B1" document, usually appears with a considerable time delay. There are even instances of "B1" documents without corresponding laid-open publications or English abstracts, and it is, therefore, easy to miss them in monitoring or prior art searches.

Another special aspect of Japanese patent documentation concerns international applications (PCT) entering the national phase in Japan. The first publications of such PCT applications were issued in 1979. To distinguish these documents from domestic Japanese patent filings, the Japanese patent office decided to use a specific number series starting with 500001. Once these PCT applications have entered the national phase in Japan and been laid-open to public inspection, they are assigned the usual "A" code to indicate the first publication stage. But they can be easily recognized by the special number series of 500000. In Western databases these publications, which are called 公表 *kouhyou* in Japan, often receive a "T" code to indicate the Japanese translation of a PCT international application (Fig. 4.2).

A second type of international application is those originating from Japan and designating Japan. When these PCT applications enter the national phase in Japan and are laid-open to the public, they are a simple re-publication of the original (Japanese) PCT publication and bear a publication number in the original PCT publication number format (with the WO country code preceding the number) on their front pages. In Japan, these publications carry the kind code "A1" and are referred to as 再公表 *saikouhyou* (Fig. 4.3).

In both of the above cases of PCT international applications, the documents receive a regular *touroku* number and kind code (B2), once they are granted and published. At grant stage, they can, therefore, no longer be distinguished from other granted patents in Japan. Table 4.1 summarizes all former and current document codes.

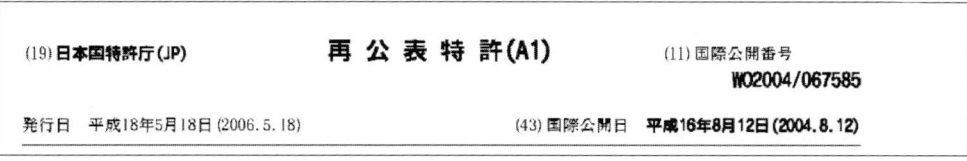

Fig. 4.3 Example of a Japanese *saikouhyou* publication (Japanese re-publication of a PCT application with Japanese priority).

Table 4.1 Codes of Japanese patent documents.[a]

Code	Document type	Currently used
A1	PCT publication from Japan, entering the national phase in Japan (*saikouhyou*)	Yes
A2	Laid-open publication of unexamined document 18 months after filing (*koukai*, first publication stage); also: non-Japanese PCT application having entered the national phase in Japan and been laid-open to the public (*kouhyou*, number starting with 500001)	Yes
B	Publication of the examined patent application but not yet granted (*koukoku*, former second publication stage)	No (used until 1996)
B1	In case of request for accelerated examination: publication of the granted patent (*touroku*), without A2 *koukai* publication (may be faster than *koukai*; used since 1996)	Yes
B2	Publication of the granted patent (number starting at 2500001), also for former PCT applications (*touroku*; used since 1996)	Yes
B4	Publication of examined patent application	No (used until 1996, currently used only in INPADOC[b])
C	Registration number for publication of granted patents	No (used 1885–1992, currently used only in INPADOC[b])
T1	Translation of international PCT application	No (used only in INPADOC[b])
T2	Publication of unexamined patent application based on an international PCT application	Yes
T3	Publication of examined patent application based on an international PCT application	No (used only in INPADOC[b])
T4	Publication of granted patent application based on an international PCT application	No (used only in INPADOC[b])

[a] Sources: http://www.european-patent-office.org/inpadoc/stats/kdcodes_0634.txt;
http://www.delphion.com/help/kindcodes; http://pk2id.delhi.nic.in/faq/kdcode3.html;
http://www.thomsonscientific.com/support/patents/patinf/patentfaqs/jplaw.
[b] See http://www.european-patent-office.org/inpadoc/stats/kdcodes_0634.txt.

4.2.2
INID Codes and Headlines

Japanese language can be found in national Japanese patents (JP) published by 日本国特許庁 (the Japanese Patent Office), and in international applications (PCT) published under the Patent cooperation treaty. Table 4.2 summarizes the terms for the categories, giving general information on the patent in the order as they appear on the first page of JP and PCT patent publications.

Table 4.2 Japanese expressions on the first pages of JP and PCT applications.

Information	INID code[a]	Japanese	Translation
publishing organization	(19)	日本国特許庁 nihonkokutokkyochou	Japanese Patent Office
	–	世界知的所有権機関、国際事務局 sekaichitekishoyuukenkikan, kokusaijimukyoku	World Intellectual Property Organization, International Bureau
type of publication	(12)	公開特許公報 koukaitokkyokouhou	publication of unexamined patent application
	–	国際出願公開番号 kokusaishutsugankoukabangou	International application published under the PCT
publication number	(11)	特許出願公開番号 tokkyoshutsugankoukaibangou 特開 tokkai	publication number of patent application publication
	(10)	国際公開番号 kokusaikoukaibangou	international publication number
date of publication	(43)	公開日 koukaibi	publication day
	(43)	国際公開日 kokusaikoukaibi	international publication day
international patent classification	(51)	国際特許分類 kokusaitokkyobunrui	international patent classification
	–	識別記号 shikibetsukigou	identification code
	–	テーマコード (参考) te-mako-do (sankou)	theme code (reference)
	–	審査請求 shinsaseikyuu	request for examination
	–	請求 seikyuu	requested
	–	未請求 miseikyuu	not yet requested
	–	請求項の数 seikyuukou no kazu	number of claims
	–	全 … 頁 zen … peiji	total number of pages
application number	(21)	出願番号 shutsuganbangou	application number
		国際出願番号 kokusaishutsuganbangou	international application number
date of application	(22)	出願日 shutsuganbi	filing date
		国際出願日 kokusaishutsuganbi	international filing date
publication language	(26)	日本語 nihongo	Japanese
priority details	(30)	優先権データ yuusenkende-ta	priority data
	(31)	優先権主張番号 yuusenkenshuchoubangou	priority application number
	(32)	優先日 yuusenbi	priority date
	(33)	優先権主張国 yuusenkenshuchoukoku	country of priority (country in which priority application was filed)
names of involved people	(71)	出願人 shutsugannin	applicant
		米国を除く全ての指定国について beikoku o nozoku subete no shiteikoku ni tsuite	for all designated states except US

Table 4.2 (continued)

Information	INID code[a]	Japanese	Translation
	(72)	発明者 hatsumeisha	inventor
		米国についてのみ beikoku ni tsuite nomi	for US only
	(74)	代理人 dairinin	agent, attorney or representative
designated states	(81)	指定国 shiteikoku	designated states
	(81)	国内 kokunai	national
	(84)	広域 kouiki	regional
		ヨーロッパ特許 yo-roppatokkyo	European patent
		ユーラシア特許 yu-rashiatokkyo	Eurasian patent
title	(54)	【発明の名称】hatsumei no meishou	title of the invention
abstract	(57)	【要約】youyaku	abstract, summary
		【課題】kadai	subject, problem, theme, topic
		【解決手段】kaiketsushudan	means for solution
		【効果】kouka	effect

[a] Internationally agreed numbers for the identification of bibliographic data (INID Codes) are defined by the World Intellectual Property Organization (WIPO) and identify bibliographic elements in patents specifications. For further information and standards see http://www.wipo.int/index.html.en. INID codes in this table represent only a selection.

Navigation within the JP patent is facilitated because headlines and top lines are given in brackets 【~】. Table 4.3 gives important and frequently used headlines in claims and description of JP publications.

4.2.3
Japanese Patent Numbers

4.2.3.1 Numbering Systems

Before the changes to the Japanese patent law in 1996, all numbers assigned to a Japanese patent document up to grant had exactly the same number format, with each number consisting of a two-digit year part (Japanese imperial years) followed by up to six digits for the serial number part. This number format was used for application and publication numbers (unexamined and examined) alike. The only exception to this general formatting rule was the registration number, a seven-digit serial number without year indication that was assigned to granted patents. Importantly, this registration number was only used for an entry in the patent register but was not directly linked to an actual publication. Only a few Western database services offered these registration numbers for searching. In total, a Japanese patent docu-

Table 4.3 Headlines in claims and description of JP patents.

Headline	ro-maji	English
【特許請求の範囲】	tokkyoseikyuu no hani	scope of the patent claims
【請求項1】	seikyuukou1	claim 1
【請求項2】	seikyuukou2	claim 2
【請求項一】	seikyuukouichi	claim 1
【請求項二】	seikyuukouni	claim 2
【明細書】	meisaisho	specification
【発明の詳細な説明】	hatsumei no shousai na setsumei	detailed description of the invention
【発明の属する技術的分野】	hatsumei no zoku suru gijutsutekibunya	technical field of the invention
【発明の属する技術分野】	hatsumei no zoku suru gijutsubunya	technical field of the invention
【産業上の利用分野】	sangyoujou no riyoubunya	industrial field of application
【従来の技術】	juurai no gijutsu	prior art
【発明が解決しようとする課題】	hatsumei ga kaiketsu shiyoutosuru kadai	problem to be resolved by the invention
【課題を解決するための手段】	kadai o kaiketsu suru tame no shudan	means for resolving the problem
【発明の実施の形態】	hatsumei no jisshi no keitai	embodiment of the invention (practical form of the invention)
【実施例一】	jisshirei ichi	example of execution 1
【実施例1】	jisshirei1	example of execution 1
【比較例】	hikakurei	comparative example
【参考例】	sankourei	reference example
【発明の効果】	hatsumei no kouka	effect of the invention
【性能試験】	seinoushiken	performance test
【表一】	hyou ichi	table 1
【図面の簡単な説明】	zumen no kantannasetsumei	explanation of drawings

ment under the old Japanese patent law was assigned four different numbers: application number, unexamined publication number, examined publication number, and registration number (Table 4.4).

With the change to the post-grant opposition system in 1996, the examined publication was dropped and, instead, patent documents were published after grant. A new number format for the publication number of the granted patent was introduced. This new number format consists of seven digits and does not include a year part. To distinguish this new grant

Table 4.4 Examples of Japanese patent documents numbers.

Document type	Examples of document numbers (number system before 1996)	Examples of document numbers (number system since 1996)
Application number	特願昭５５－０１３１４４	特願２００２－３２２０４１
Publication number unexamined documents	特開昭５６－１１１５４２	特開２００４－１５５６９３
Publication number examined documents	昭５９－００８４６５	–
Registration number (grant number)	１２３４５６７	１２３４５６７

number from the old seven-digit registration number, it was decided to start the new grant number series at 2500000. The first granted patent published in this way carried the number 2500001 and was published on 29 May 1996.

The JPO has used four-digit Western years since 2000, instead of the two-digit Japanese year for unexamined publication numbers. Thus, Japanese laid-open publications now have ten-digit numbers consisting of four digits for the Western year and up to six digits for the serial number. Grant numbers are seven digit serial numbers without a year indication. Under the present Japanese patent law, a Japanese patent document only has three different numbers assigned: application number, unexamined publication number and grant number (Table 4.4).

4.2.3.2 Japanese Imperial Years

Historically, Japan has used its own calendar system based on the years of the Emperor's reign. These "imperial years" form the basis of the dating system for official documents in Japan, including patents. Since 2000, however, the JPO has been using four-digit Western years on its patent publications, but, for application numbers, the imperial years are still in use. Since the searcher will very often be confronted with documents published before 2000, it is worth taking a closer look at the Japanese imperial years and dating system.

Since the introduction of a patent system in Japan, there have only been four emperors. Each is referred to by the name given to the era of his reign, rather than his personal name. The first emperor in modern Japan, Matsuhito, was on the throne from 1868 to 1912 and his era is referred to as the 明治 *meiji* period. On the earliest patent publications in Japan, the years are indicated as 明治一八年 (*meiji* year 18 = 1885), 明治一九年 (*meiji* year 19 = 1886), and so on. Upon his death in 1912, Emperor Matsuhito was succeeded by Yoshihito, whose reign is known as the 大正 *taishou* period.

On the more modern patent documents, the searcher is likely to come across two imperial periods, 昭和 *showa* (Emperor Hirohito, 1926–1989) and 平成 *heisei* (Emperor Akihito, 1989 to present). Usually, on patent publications the years as such are written in Arabic numbers

Table 4.5 Imperial periods and examples for Japanese year's names.

Imperial period	Japanese name	Examples English	Japanese	
1868–1912	明治 *meiji*			
1912–1926	大正 *taishou*			
1926–1989	昭和 *shouwa*	1926	昭和一年	*shouwa 1*
		1988	昭和６３年	*shouwa 63*
		1989	昭和６４年	*shouwa 64*
			平成１年	*heisei 1*
1989 to date	平成 *heisei*	1990	平成２年	*heisei 2*
		2007	平成１９年	*heisei 19*

but preceded by one Japanese *kanji* character or a combination of two characters indicating the *shouwa* era or *heisei* era (before the 1950s, the numbers on patent documents would also be written in Japanese *kanji* characters, making even dates and numbers quite hard to read for the non-Japanese speaker). In some database entries or document number quotes, the year part of the patent document number may be preceded by the Western transliteration rather than the *kanji* characters. In such cases, the letters "SHO" or "S" indicate the *shouwa* imperial years, and "HEI" or "H" stand for the *heisei* years.

A new imperial period starts when the emperor accedes to the throne. The first year of the new emperor's reign is counted from the day of his accession through to the end of the year (31 December). In 1989, when Hirohito died and was succeeded by his son Akihito, the new era of *heisei* began. Thus, the final year of *shouwa* (*shouwa* year 64) was at the same time the first year of *heisei* (*heisei* year one). At the changeover from *showa* to *heisei*, the patent publications issued early that year still carried the *shouwa* imperial era 昭和６４年 (*shouwa* year 64), with the new *heisei* imperial era 平成１年 (*heisei* year 1, or 平成元年 *heisei gannen*) appearing on the patent publications published after April 1989. Table 4.5 gives an overview on the imperial periods and some examples.

The era name stands at the beginning of the full application and publication data. The year is followed by month and day. After each number, Chinese characters indicate year, month and day, like in 平成１８年１１月２５日, which is November 25, 2006.

4.2.3.3 Frequently Encountered Problems

The special characteristics of Japanese patent documentation described above can lead to confusion and problems for the searcher. Some of the most frequently encountered problems are:

- Same numbering system for different industrial property rights: The same document numbering system is applied for all types of industrial property rights in Japan, which means that without knowing the type of right or kind-of-document code it is virtually impossible to deduce whether one is looking for a patent, utility model, trade mark or design right.

- Different numbers for different publication stages: Furthermore, each stage in the grant process of a Japanese industrial property right is assigned a different number, with most of these numbers having the same format (exception is the seven-digit registration number). Since the numbers are random, it is impossible to establish number concordances intellectually. This means that any searcher needs to have additional information, not just the document number, to identify the correct Japanese document. Ideally, this additional information will be the publication stage or the kind-of-document code. If this information is not available, it is difficult or impossible to find the correct document. It is recommended to try to obtain additional information such as the applicant's name, an IPC class, publication dates, etc. and cross-check any documents retrieved against this information.
- No uniform numbering system on different database services: Traditionally, different database providers have taken different approaches and used a different logic to handle Japanese document numbers. In some cases, the Japanese imperial years get translated into Western years, sometimes the Japanese one-digit imperial years are given as one-digit numbers, in other instances as two-digit numbers with a leading zero.

 The JPO's "Industrial Property Digital Library" is one example of a services available that offers Japanese document number search using Western years or using Japanese imperial years. Western years have to have four digits, e.g., 2000-123456. However, when searching this number with Japanese years, one has to be careful about the correct input, as it is different on the "PAJ" service (12-123456) and on the "Patent&Utility Model Gazette DB" service (H12-123456).

Knowledge of the number formatting rules applied by different database producers is helpful for efficient searches in Japanese patent documentation.

4.2.4
Special Japanese Classification Systems

Two distinctive and useful features of Japanese patent documentation are the "File Index (FI)" and "File Forming Terms (F-terms)" classification systems developed by the Japanese patent office.

Up to the 1970s, the JPO had used a system for classifying patent documents called the "Japanese Patent Classification (JPC)", which consisted of rather broad classes comparable to the classification applied by the United States patent office. In the early 1970s, the Japanese classification was phased out in favor of the International Patent Classification (IPC). In 1978, IPC^2 was officially introduced at the JPO. However, with the rising numbers of patent applications in Japan, the IPC soon became insufficient for the JPO's examiners' search needs.

Still working largely with paper files, the JPO examiners started to subdivide these IPC-based files using three-digit numbers and alphabetic symbols to discriminate between subdivided files. This practice resulted in a new classification scheme, the "File Index" or "FI". With the advent of the JPO's "paperless project" and the introduction of a computerized search system in 1985, a new search index was developed for more efficient prior art searches and for coping with the ever-increasing rise in patent applications and emerging new technologies. The new search index was called "File Forming Terms" or "F-terms" and allowed JPO examiners to search prior art documentation from multiple technical viewpoints.

FI and F-terms are assigned to Japanese patent and utility model documents in addition to IPC classes. FI classes have been published on the first pages of Japanese patent applications since 1992. Since 2000, the first pages have also carried F-terms. The complete backfile documentation at the JPO gets reclassified whenever changes in FI classes or F-terms are necessary. Thus, using FI or F-terms in searching gives users access to the complete documentation on Japanese patents and utility models.

Initially, FI classification and F-terms were only available for searches in Japanese, within the Japanese-language services on the JPO's "Industrial Property Digital Library (IPDL)". In 2001, English versions of the FI and F-term classification scheme were made available to the public. Search facilities in English using FI and F-terms were installed as part of the IPDL's English language services,[2] as well as a reference service, called the "Patent Map Guidance"[3], which includes a searchable concordance list that allows users to determine relevant FI and F-terms on the basis of a known IPC class.

Even though FI and F-terms may not be the solution to every search problem, they can in some technical fields offer good alternatives to keyword searches, giving the searcher access to the complete Japanese patent documentation without the "noise" generated by keyword searches.

4.2.5
File Index Classification Lists

As described in Section 4.2.4, the File Index (FI) classification is an internal system of the Japanese Patent Office based on the IPC classification.[4] Whereas the IPC system has some

---

2) http://www.ipdl.ncipi.go.jp/homepg_e.ipdl.
3) http://www5.ipdl.ncipi.go.jp/pmgs1/pmgs1/pmgs_E.
4) The International Patent Classification (IPC) is a hierarchical system created in 1971 under the Strasbourg Agreement, in which the whole area of technology is divided into a range of sections, classes, subclasses and groups. The Classification is updated on a regular basis to improve the system and to take account of technical developments. The IPC system is considered as indispensable for the retrieval of patent documents in the search for establishing the novelty of an invention or determining the state of the art in a particular area of technology. For further information about the IPC see http://www.wipo.int/classifications/ipc/en/, http://www.depatisnet.de/ipc/init.do or http://en.wikipedia.org/wiki/International_Patent_Classification. The current edition of the IPC (IPC8) entered into force on January 1, 2006. From this edition, the classification has been divided into "core" and "advanced" levels. Whereas the core level will be updated every three years, the advanced level provides more detailed classification and is updated every three to four months. With IPC8, patent offices get the choice between a simpler to implement but more general classification using the core classifications or a more detailed but more complex to maintain advanced classification. See http://www.wipo.int/classifications/ipc/ipc8/ or http://www.wipo.int/classifications/ipc/en/other/guide/guide_ipc8.pdf.

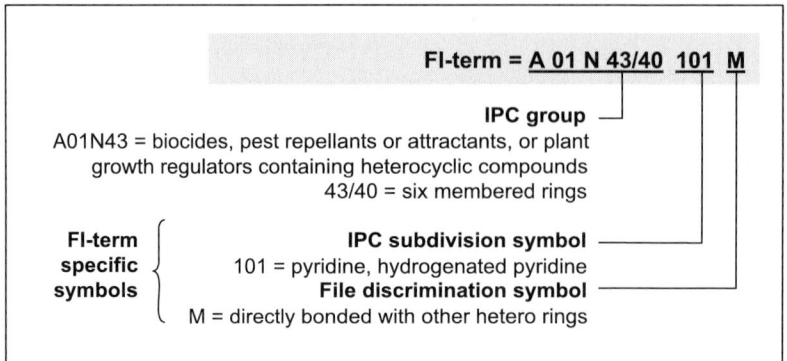

Fig. 4.4 Example for FI class.

70 000 classes, the JPO's classification consists of over 170 000 classes. In the example given in Fig. 4.4, the FI class "A01N43/40 101M" corresponds to the IPC class "A01N43/40".

Each FI classification term consists of a symbol such as "A01N43/40101M". The first letter is the section symbol. The IPC and FI systems divide technology into eight sections, ranging from A through H. The most important sections for chemistry are A ("Human Necessities", including pharmaceutical and agricultural chemistry), C ("Chemistry, Metallurgy", including inorganic chemistry, organic chemistry, macromolecular compounds, biochemistry, alloys, combinatorial chemistry and chemistry of natural compounds such as sugars), D ("Textiles, Paper", including chemical treatment of fibers), and G ("Physics", including nuclear chemistry). Each section is further divided into classes and subclasses, represented by a two-digit number and a letter, respectively. This is followed by a one- to three-digit "group" number, an oblique stroke and a number of at least two digits representing a "main group" or "subgroup". In addition to the IPC symbol, a FI terms consists of a three-digit subdivision number (e.g., "101" in C12N9/00 101 or an one letter file discrimination symbol (e.g., "A" in C01D3/04A), or of both (e.g., "010D" in A01N43/82 010D). Table 4.6 gives some examples of FI terms.

A list of all FI terms can be found on the homepage of the National Center for Industrial Property Information and Training, www.ipdl.ncipi.go.jp/homepg_e.ipdl. FI term searches can be carried out from the same homepage, using the Industrial Property Digital Library database at http://www4.ipdl.ncipi.go.jp/Tokujitu/tjftermena.ipdl?N0000=114. The use of the search function is described at http://www.ipdl.ncipi.go.jp/HELP/pmgs_en/database/inqhelp.html.

However, if a certain IPC class is to be identified before the patent search, it is recommended to use at first the service provided by the WIPO at http://www.wipo.int/classifications/ipc/ipc8/ or the European patent office at http://v3.espacenet.com/eclasrch. Because of a more favorable graphic illustrations and additional explanations, it facilitates the identification of a IPC class in case of unknown hierarchy (e.g., for A01N43/34 and A01N43/40 which consist of the same code level but different hierarchy, see Table 4.6). For some FI terms, no English translations but figures are provided as explanations for certain chemical substituents or functional groups, e.g., for A01N43/40 101 G in Table 4.6.

Table 4.6 Examples of FI-terms.

Field	Symbol	Explanation
Agricultural chemistry	A	Human necessities
	A01	Agriculture, forestry, animal husbandry, hunting [...]
	A01N	Preservation of bodies of humans or animals or plants [...]; biocides, e.g., as disinfectants, as pesticides, as herbicides
	A01N43	Biocides, pest repellants or attractants, or plant growth regulators containing heterocyclic compounds
	A01N43/34	... having rings with one nitrogen atom as the only ring hetero atom
	A01N43/40	... six membered rings
	A01N43/40 101	... pyridine; hydrogenated pyridine
	A01N43/40 101 G, A01N43/40 101 H, A01N43/40 101 J, A01N43/40 101 K	... see figure:
	A01N43/40 101 M	... directly bonded with other hetero rings
Inorganic chemistry	C	Chemistry; metallurgy
	C01	Inorganic chemistry
	C01F	Compounds of the metals beryllium, magnesium, aluminum, calcium [...]
	C01F7/00	Compounds of aluminum
	C01F7/02	Aluminium oxide, aluminum hydroxide
	C01F7/02D	Processing methods
	C01F7/02G	... preparation methods of pellets, fragments or other moulds
	C01F7/02H	... by splitting a liquid substance into small droplets and solidifying them
	C01F7/02J	... in liquid media

Table 4.6 (continued)

Field	Symbol	Explanation
Macro-molecular chemistry	C	Chemistry; metallurgy
	C08	Organic macromolecular compounds; their preparation or chemical working-up [...]
	C08F	Macromolecular compounds obtained by reactions only involving carbon-to-carbon unsaturated bonds
	C08F2/00	Process of polymerization
	C08F2/00A	By adding the monomer or polymer catalyst
	C08F2/12	... polymerization in non-solvents
	C08F2/16	... aqueous medium
	C08F2/22	... emulsion polymerization
	C08F2/24	... with the aid of emulsifying agents
	C08F2/24 A	Emulsifier containing a polymerized unsaturated radical

4.2.6
File Forming Terms Lists

File Forming Terms (F-Terms) are a characteristic feature of the JPO designed for more efficient prior art searches in patent examinations (Section 4.2.4.) It is a completely independent classification system and applied to patent documents in addition to the FI classification described in Section 4.2.5. The F-term classification system is based on 2800 themes that are further split into term codes assigned according to various technical viewpoints, such as materials, operations, products or purposes. Chemical viewpoints are, for example, chemical classes, use of chemicals, preparation processes, apparatuses and laboratory equipment. Currently, there are some 350 000 terms.

F-terms are applied retrospectively to the first Japanese patents in 1885 but do not exist for all Japanese patent documents and the coverage depends on the field of technology. While the IPC classifies documents mostly from a single technical viewpoint, the F-term classification works from multiple viewpoints. Therefore, the combination of IPC and F-terms effectively narrows down relevant documents in prior art searches.

Assignment of F-terms is largely based on the FI classification. Every theme code has a corresponding FI coverage. An F-term consists of a five-digit theme code and a two-digit term code (technical viewpoint symbol), followed by further two-digit subdivisions (Fig. 4.5). Extensions may be added at the end of the F-term to explain from which part of the patent the F-term arises (e.g., "A" means that the F-term is derived from the claims and "B" means that the term is derived from the experimental section).

Theme Codes start with a two-digit theme group. The most important themes for chemistry are within the groups summarized in Table 4.7. Further relevant main themes related to chemical compounds and synthesis are distributed among other theme groups, such as

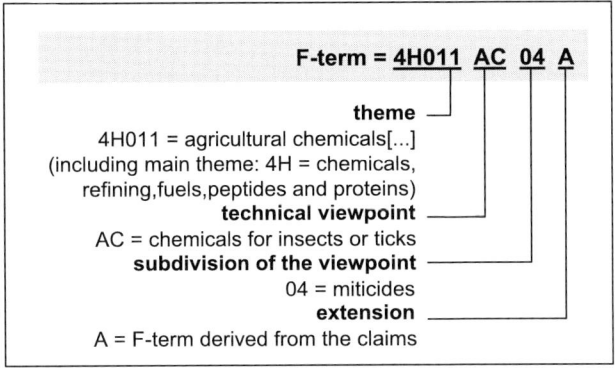

Fig. 4.5 An example of F-term classification.

2E191 (fire-extinguishing agents), 4B050 (enzymes and modification thereof), 4D039 (neutralization), 4D055 (crystallizing), 4L031 (chemical treatment of fibers), 4L0235 (artificial filaments), 4L037 (inorganic fibers), 4M104 (electrodes for semiconductors), 4D077 (emulsifier, dispersant, foaming agent, wetting agent), 5G303 and 5G305 (insulating materials).

Instruments, measurement technologies, analytical methods, laboratory equipment and specific processes are described in other themes, e.g., 2B230 (chemical treatments for wood), 2F014 (measurement of the levels of liquids or fluidic solids), 2F043 (volume, volumetric flow rate, liquid level test and calibration), 2F056 (measuring of temperature or quantity of heat), 2G004 (concentration cells, measuring oxygen concentration), 2G020 (spectrometry), 2G054 (investigating or analyzing materials by the use of chemical reactions), 2G063 (investigating and analyzing materials by the use of chromatography), 4D005 (coagulation and sedimentation treatment), 4D012 (separation of gases by adsorption), 4D014 (separation of liquids from each other), 4D025 (treatment of water by ion exchange), 4E001 (arc welding and cutting), 4F070 (processes for treating macromolecular substances), 4F073 (treatment of macromolecular formations), 4F207 (extrusion molding of plastics) and certain themes related to batteries, electrodes and cells in the 5H main theme (e.g., 5H018, 5H024, 5H025, 5H026, 5H032 and 5H050).

Theme codes and corresponding term codes are explained in F-term charts and are available on the JPO's internet homepage at http://www5.ipdl.ncipi.go.jp/pmgs1/pmgs1/!frame_E?hs=1&gb=2&dep=1&sec=2C&cls=&scls=&mgrp=&idx=&sgrp=&sf=&bs=&dt=0&wrd=&nm=. As a result of a cooperation project with the EPO, most of these charts have been translated into English. Figure 4.6 shows an example for a F-term given in Table 4.7; and Table 4.8 lists exemples of some subdivisions of the main theme given in Fig. 4.5, "agricultural chemicals and associated chemicals". F term searches can be carried out from the JPO's homepage, using the Industrial Property Digital Library database at http://www4.ipdl.ncipi.go.jp/Tokujitu/tjftermena.ipdl?N0000=114. The use of the search function is described at http://www.ipdl.ncipi.go.jp/HELP/pmgs_en/database/inqhelp.html.

Table 4.7 Theme groups of F-terms relevant for chemical patents.

Theme group	Examples for complete main theme
4C aromatics, heteroaromatics, pharmaceuticals, medical, dental, ophthamological, devices and supplies	4C023: heterocyclic compounds containing sulfur atoms; 4C031: quinoline compounds; 4C033: thiazole and isothiazole compounds; 4C048: epoxy compounds; 4C050: nitrogen-containing condensed heterocyclic rings; 4C051: lactam manufacturing methods; 4C054: hydrogenated pyridines; 4C057: saccharide compounds; 4C091: steroid compounds; 4C201: therapeutic activity of compounds or pharmaceuticals
4G glass, ceramics, inorganics, colloids, catalysts, physical and chemical processes	4G042: oxygen, ozone, oxides; 4G043: sulfur, nitrogen and their compounds; 4G047: inorganic compounds of heavy metals; 4G049: ammonia, cyanogen and their compounds; 4G065: colloid chemistry; 4G067: ionic exchange; 4G073: silicates, zeolites and molecular sieves; 4G075: physical and chemical processes and apparatuses; 4G077: crystals; 4G140: hydrogen, water and hydrids
4H chemicals, refining, fuels, peptides and proteins	4H006: esters [...]; 4H007: manufacturing of synthetic gases [...]; 4H011: agricultural chemicals [...]; 4H045: peptides and proteins; 4H056: dyes
4J polymers, resins, paints, inks	4J001: polyamides; 4J011: polymerization methods; 4J026: graft and block polymers; 4J036: epoxy resins
4K metallurgy	4K020: manufacture of alloys [...]; 4K025: electrochemical [...] coating; 4K026: chemical treatment of metals

4C036	Nitrogen- or sulfur-containing heterocyclic ring compounds with rings of six or more members
	C07D279/00–293/12

	AA00	AA01	AA02	AA03		AA05	AA06	AA07	AA08	AA09	
AA	THIAZINE COMPOUNDS	. Condensation with a carbocyclic ring or ring system	.. The thiazine ring is condensed with a carbocyclic ring or ring system wherein the carbocyclic ring is a benzene ring	... Atom bonded directly to the benzene ring		. Hydrocarbyl bonded directly to thiazine ring carbon	. Substituted hydrocarbyl bonded directly to thiazine ring carbon	. Hetero atom bonded directly to thiazine ring carbon	.. Nitrogen atom bonded directly to thiazine ring carbon	. Carboxylic acid or functional derivative thereof bonded directly to thiazine	
		AA11	AA12	AA13	AA14	AA15	AA16	AA17	AA18	AA19	AA20
		. Hydrocarbyl bonded directly to thiazine ring nitrogen	. Substituted hydrocarbyl bonded directly to thiazine ring nitrogen	.. Carbonyl bonded directly to thiazine ring nitrogen	. Nitrogen substituted hydrocarbyl bonded directly to thiazine	. Thiazine ring sulfur is bonded directly to oxygen	. Substituting groups other than the aforementioned are directly bonded to the thiazine ring	. Claims of chemical substances	. Claims of chemical compound manufacturing methods	.. Claims wherein heterocyclic rings are manufactured	. Claims with pharmaceutical applications

	AB00	AB01	AB02	AB03	AB04	AB05	AB06	AB07	AB08	AB09	AB10
AB	COMPOUND HAVING A SEVEN- OR MORE MEMBERED HETEROCYCLIC RING WITH ONE NITROGEN AND ONE SULFUR AS THE ONLY RING HETERO ATOMS (E.G., THIAZEPINS)	. Ring hetero atoms are in positions 1 and 8	. Ring hetero atoms are in positions 1 and 9	. Ring hetero atoms are in positions 1 and 10	. The heterocyclic ring is condensed with a carbocyclic ring or ring system	. The carbocyclic ring is a benzene ring	. Hetero atoms attached to six-membered aromatic ring carbon atoms	. Hydrocarbyl bonded directly to the ring carbon of the heterocyclic ring	. Substituted hydrocarbyl bonded directly to the ring carbon of the heterocyclic ring	. Hetero atom bonded directly to ring carbon of the heterocyclic ring	. The hetero atom is oxygen
		AB11	AB12	AB13	AB14	AB15	AB16	AB17	AB18	AB19	AB20
		. Carboxylic acid or functional derivative thereof bonded directly to ring carbon of the heterocyclic ring	. Hydrocarbyl group bonded to the ring nitrogen of the ring	. Substituted hydrocarbyl group bonded directly to ring nitrogen of the heterocyclic ring	. Nitrogen-substituted hydrocarbyl group bonded directly to ring nitrogen of the heterocyclic ring	. Ring sulfur is bonded directly to oxygen	. Substituting groups other than the aforementioned are directly bonded to the heterocyclic ring	. Claims of chemical substances	. Claims of chemical compound manufacturing methods	.. Claims wherein heterocyclic rings are manufactured	. Claims of pharmaceutical applications

	AC00	AC01	AC02	AC03	AC04	AC05	AC06	AC07		AC09	AC10
AC	COMPOUNDS THAT HAVE A HETEROCYCLIC RING WITH ONE NITROGEN ATOM AND ONE SULFUR ATOM ONLY AS RING HETERO ATOMS	. Ring hetero atoms are in positions 1 and 8	. Ring hetero atoms are in positions 1 and 9	. Ring hetero atoms are in positions 1 and 10	. Heterocyclic ring has four or fewer members	. Heterocyclic ring has five members	. Heterocyclic ring has six members	. Heterocyclic ring has seven or more members		. One of the heterocyclic rings is condensed with a carbocyclic ring or ring system	. The carbocyclic ring is a benzene ring
		AC11	AC12	AC13	AC14		AC16	AC17	AC18	AC19	AC20
		. Hydrocarbyl or substituted hydrocarbyl bonded directly to ring carbon of one of the hetero atoms	. Hydrocarbyl group bonded directly to ring nitrogen or of the ring	. Carbon chain hydrocarbyl or substituted hydrocarbyl bonded directly to ring nitrogen of one of the heterocyclic rings	. Ring sulfur is bonded directly to oxygen		. Substituting groups other than the aforementioned are directly bonded to the heterocyclic ring	. Claims of chemical substances	. Claims of chemical compound manufacturing methods	.. Claims wherein heterocyclic rings are manufactured	. Claims of pharmaceutical applications

Fig. 4.6 Example of part of an F-term chart derived from main theme 4C036. (Source: http://www5.ipdl.ncipi.go.jp/pmgs1/pmgs1/pmgs_E.)

Table 4.8 Examples for F-terms derived from main theme 4H011.

Technical viewpoint			Examples for subdivisions of the viewpoint
AA	chemicals for microorganisms, e.g., bacteria or fungi	AA01	germicide or bacteriostatic agent
		AA03	antifungal agents
		AA04	antiviral agents
AB	chemicals for plants	AB01	herbicides, e.g., for field [...]
		AB02	herbicides for wet paddies
		AB03	growth regulators
AC	chemicals for insects or ticks	AC02	insecticides for cockroaches, mosquitoes, or flies
		AC04	miticides
		AC07	insect or tick attractants, e.g., pheromones
BA	roles of compounds in mixtures	BA01	combinations of active ingredients and inactive ingredients
		BA03	active ingredients and ingredients for reducing toxicity or chemical harmfulness
BB	specification of active ingredients	BB02	halogenated hydrocarbons
		BB03	compounds containing organic oxygen or organic sulfur
		BB05	aldehydes, ketones or oximes
		BB08	oxygen- or sulfur-containing heterocyclic rings
CB	object of conservation	CB02	mammalian
		CB08	cells
		CB09	insects
		CB10	plants
DA	shade of drug product	DA01	solid
		DA02	granular or powdery [...]
		DA03	pellet, tablet or pill
DE	application method [...]	DE05	mosquito coil
		DE15	spraying

4.3
Online Sources of Japanese Patent Information

4.3.1
Patent Information from Patent Offices

A convenient access to various kinds of patent information is provided by official patent offices and organizations around the world. Related to Japanese patent information, one of the most exhaustive source is the Japanese Patent Office (JPO), 特許庁 *tokkyochou*, which is described in Section 4.1.3. By means of the "National Center for Industrial Property Information and Training" (NCIPI), the JPO provides access to Japanese industrial property informa-

Fig. 4.7 English internet page of the NCIPI with various search functions.

tion based on the official gazettes.[5] In March 1999, a free-of-charge internet search service was launched, called the "Industrial Property Digital Library" (IPDL, http://www.ipdl.ncipi.go.jp/homepg.ipdl). Besides original Japanese data, it provides some access in English to all types of Japanese industrial property rights on its English homepage at http://www.ipdl.ncipi.go.jp/homepg_e.ipdl (Fig. 4.7). The data are limited to JP patent applications filed in Japan; information on international PCT patents written in Japanese is not covered.

Search functions are within the "patent&utility model". English abstracts, legal status information and complete patent documents can be searched in the "Patent&Utility Model Concordance" or the "PAJ" submenus. The PAJ search module provides English and Japanese patent abstracts. They can be searched by numbers (application, publication or grant numbers) at http://www19.ipdl.ncipi.go.jp/PA1/cgi-bin/PA1INDEX or by text searches (e.g., applicant, title of invention or abstract) and narrowed by application or publication date and IPC classes at http://www19.ipdl.ncipi.go.jp/PA1/cgi-bin/PA1INDEX. The correct input format of numbers and text is described using the help function at http://www19.ipdl.ncipi.go.jp/PA1/html/help/index.html or http://www.ipdl.ncipi.go.jp/HELP/tokujitu/bansaku_en/help_index.html. Figure 4.8 displays the research result for publication number 07-145156. Applying the "Japanese" button gives access to the complete original Japanese document.

---

5) To access the Japanese language services with Western browser software, a special character set or encoding may be required.

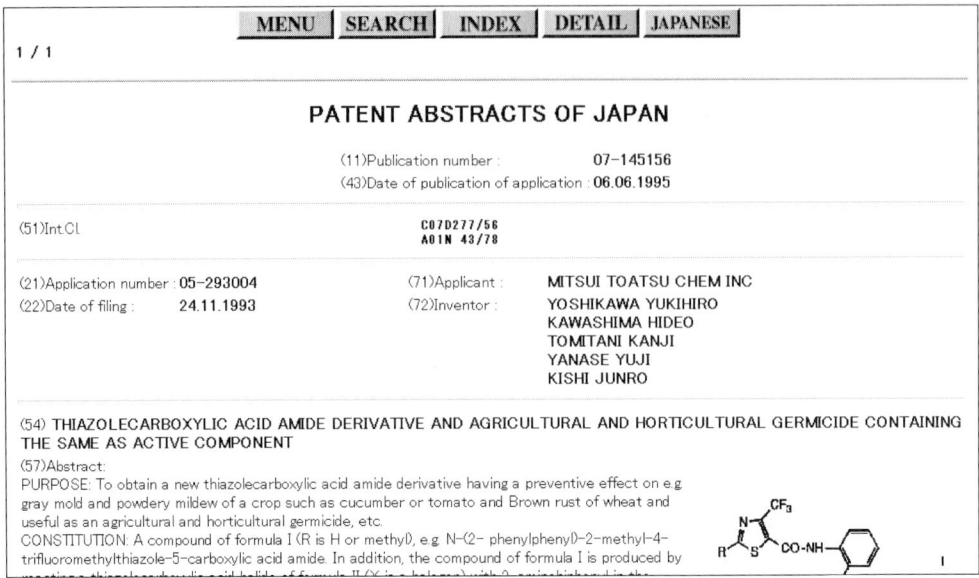

Fig. 4.8 Exemplary search result using the PAJ search module.

For more recent patent and utility model publications (1993/1994 onwards), a free machine-translation engine is also available. This very helpful tool can be initiated with the "Detail" button. The automatic translation process is started individually for each patent document part, e.g., for claims, detailed description, prior art or examples. Even if the translation is, to some extent, inaccurate, and the translation of text in figures and some tables is not possible, this tool provides the most convenient and rapid access to patent information to non-Japanese speaking scientists.

Using the PAJ service, patent abstracts published since 1976 onwards can be retrieved. PAJ from January 1993 onwards includes the legal status information. While JPO's IPDL database includes JP documents and bibliographic data without any time lag, English abstracts are usually available with three to four months delay.

FI- and F-terms (Section 4.2.4) can be used in patent document searches in the submenu "FI/F-term Search". Access to extended FI- and F-term lists as well as their general descriptions is given in the submenu "Patent Map Guidance".

Besides the JPO, other patent offices also provide access to certain Japanese patent documentation. The European patent office maintains "esp@cenet", a free internet database with various search functions (http://ep.espacenet.com). For Japanese patents, JP as well as PCT documents, English abstracts, bibliographic data, the full original Japanese documents and some legal status information are given. Abstracts for JP patents are available for documents from 1976. International PCT patents and patent applications that are written in Japanese can also be searched at the internet presence of the World Intellectual Property Organization (WIPO) at http://www.wipo.int/pctdb/en/. Full documents, English abstracts and bibliographic information are provided.

The most comprehensive set of databases on patent bibliography is hosted from the EPO in its Vienna subsidiary: The EPIDOS-INPADOC (EPIDOS: "European Patent Information and Document Service", INPADOC: "International Patent Documentation Center") contains documents and bibliography of all patents filed and granted in over 70 countries since 1968, representing over 33 million references and covering approx. 95% of all patents published worldwide since 1973. Up to 40 000 documents are added each week to one of its databases. The goals and scope of the EPIDOS-INPADOC databases are described at http://www.european-patent-office.org/inpadoc/general.htm. Japanese records are from PATOLIS (see below). Legal status information is provided by 22 patent offices and can be requested by the PRS ("Patent Register Service") of EPIDOS-INPADOC. The tool provides all significant steps in the lifetime of an invention, starting with its filing to the end of term of the patent, including data such as examination requests, revocation, change of owner and granting. For Japanese patents, data are available since 1996 for PCT applications with some gaps. Legal status information of JP documents is currently not included in the PRS. EPIDOS-INPADOC databases can be accessed online using the espacenet of the EPO (see above) or on various hosts, e.g., JAPIO, STN, Questel and Dialog (Section 4.3.2). Content and coverage of the different tools are listed in certain downloadable files at http://www.european-patent-office.org/inpadoc/statistics_dwld.htm.

In addition to patent documentation, the JPO and EPO offer further services on intellectual property management and provide information on Japanese patents. For instance, the EPO have established a Q&A section on the Japanese patent system at http://patentinfo.european-patent-office.org/prod_serv/far_east/faq/japan/index.en.php. Basic information on the Japanese patent law, fees for patent filing and maintenance in Japan is provided by the JPO at http://www.jpo.go.jp/quick_e/index_tokkyo.htm. A collection of useful links related to patents, e.g., patent databases and patent information provider, can be found at http://www.european-patent-office.org/onlinelinks/d/. EPO offers additional services related to Japanese patents, such as document delivery, legal status searches (http://www.european-patent-office.org/jpinfo/search), patent document translation (http://www.european-patent-office.org/jpinfo/translation) and seminars and training courses on certain topics related to the Japanese patent system (http://www.european-patent-office.org/jpinfo/training).

4.3.2
Commercial and other Patent Information Sources

Besides the sources described in Chapter 3, there are various possibilities for obtaining information on Japanese patents on the world wide web. Services beyond the free data provided by JPO or EPA usually require subscription. Three kinds of services are available: direct access to patent databases, information providers, which give user access to databases (search functions to be applied by the user), and professional search services. Table 4.9 summarizes selected internet services, their benefits and limitations (for limitations of INPADOC data see Section 4.3.1).

In addition to patent data, the internet has become an exhaustive source for all kinds of information and services related to intellectual property. Table 4.10 lists some examples.

Table 4.9 Selected sources of Japanese patent information.

Information source	URL	Source type and provided information	Features
The British Library (UK)	http://www.bl.uk/collections/patents.html	• patent information provider, some databases only accessible in British library on CD-ROM • professional patent search service	• some services free of charge, for professional searches subscription required
Delphion (Thomson, USA)	http://www.delphion.com/	• patent information provider • broad search and output options • legal status information • full text download	• utilization of various databases, e.g., JAPIO, INPADOC, Derwent world patent index • subscription required
Derwent World Patent Index (Thomson, USA)	http://scientific.thomson.com/products/dwpi/	• patent database with broad search options • English patent abstracts, various classifications, full bibliographic data	• worldwide most comprehensive database of patent documents (currently covering 14.5 million patent records, complete agrochemical, pharmaceutical, chemical patents since 1970) • own classification, index, English abstracts • counterpart patent publications collected in same record (patent families) • subscription required
Dialog (Thomson, USA)	http://www.dialog.com	• patent documents, bibliographic information	• most comprehensive information collection from scientific and technical literature and patents • access to more than 600 databases, including some databases relevant for Japanese patents, e.g., JAPIO, Derwent world patent index • subscription required

Table 4.9 (continued)

Information source	URL	Source type and provided information	Features
Europat (Europatent Gesellschaft für europäischen Patentdienst, Germany)	http://www.wps.de/	• full text delivery • patent family, bibliographic and legal status information	• subscription required
Gongwell (USA)	http://www.gsishop.com/	• patent information provider • patent searches and patent translation	• specialized in patents from Asian countries, including Japan • subscription required
Google patent search (USA)	http://www.google.com/patents	• full text delivery • overview on selected patent with bibliographic information, claims, drawings and citations	• free of charge • based on data from US patent office • search for JP patents requires existing equivalent US patent • advanced searches (number title, inventor, assignee, classification, issue date, filing date) at http://www.google.com/advanced_patent_search
IP Newsflash (Rolf Claessen, Germany)	http://www.ipnewsflash.com/	• patent family and legal status information, IPC classes, some bibliographic information	• free of charge • data taken from INPADOC
JAPIO database (Japan)	http://www.japio.or.jp	• patent database with broad search options • some bibliographic data, classification codes, English titles and abstracts	• hosted by The Japan Patent Information Organization, the biggest information provider on patent information in Japan currently covering approx. 8 million patent records • subscription required
LexisNexis (USA)	http://www.lexisnexis.com	• patent document delivery • machine translation of Japanese patents	• subscription required

Table 4.9 (continued)

Information source	URL	Source type and provided information	Features
MicroPatent (Thomson, USA)	http://www.micropat.com/static/index.htm	• patent information provider • broad search options including full text search • legal status information • professional search services • front-page bibliographic database	• world's largest, commercial collection of searchable full-text data, featuring granted patents and published applications • utilization of various databases, e.g., INPADOC • subscription required
Paterra (USA)	http://www.paterra.com/	• patent database • machine translation of Japanese patents • patent number search • English patent abstracts, IPC classification, F-terms, few bibliographic data • general explanation of F-terms	• own proprietary Japanese-English dictionary with more than 340 000 entries and authentic translations • patent search: free; machine translation: subscription required
Patolis (Patolis Corporation, Japan)	http://www.patolis.co.jp/en/index.html	• patent database with broad search options • full JP documents, English abstracts, bibliographic data, legal status • complete list of extended pharmaceutical patents in Japan	• most comprehensive industrial Property retrieval system in Japan • subscription required • formerly part of JAPIO • English service terminated September 2006. Access to legal status data by EPO or by Questel-Orbit
Paton (TU Ilmenau, Germany)	http://www.paton.de	• patent information provider • search services, e.g., prior art search • patent delivery service • collection of web links related to patent information	• in German • subscription required

Table 4.9 (continued)

Information source	URL	Source type and provided information	Features
Patro (Patro Information, Japan)	http://www.patro.co.jp/e/index.html	• patent information provider • Translation of Japanese patents by humans • professional patent search service	• subscription required
Patscan (Patex, Canada)	http://www.patscan.ca//	• patent information provider	• subscription required • utilization of various databases, e.g., JAPIO, INPADOC, Derwent world patent index
Patware/PATWWW (Incom, Germany)	http://www.incom-ips.de, http://www.patwww.com	• patent information provider • full text delivery • patent family, bibliographic and legal status information	• utilization of various databases, e.g., INPADOC • subscription required
Questel (France)	http://www.questel.orbit.com	• patent information provider • own patent search or professional search services • delivery of patent documents • Markush search option	• utilization of various databases, e.g., JAPIO, INPADOC, Derwent world patent index • subscription required • after termination of English Patolis service, Questel provides exclusive access to data in established Patolis-e-format
SciFinder (CAS, American Chemical Society, USA)	http://info.cas.org/SCIFINDER/	• information provider, access to various kinds of scientific information including patents • broad search options • chemical substructure search	• subscription required • world's largest collection of biochemical, chemical and chemical engineering information • further services from CAS
SIP (Germany)	https://www.patentfamily.de/(zlz0cs55zwltuyeom3e2i145)/index.aspx?lang=en&ab=an	• patent information provider • full text download, simple legal status, patent family observation	• free patent search, for other services subscription required

Table 4.9 (continued)

Information source	URL	Source type and provided information	Features
STN (Germany/USA/Japan)	http://www.stn-international.de/	• information provider, access to various kinds of scientific information, including patents • broad search options and output options	• information network of FIZ Karlsruhe (Germany), CAS (USA) and The Science and Technology Agency (Japan) • utilization of various databases, e.g., JAPIO, MARPAT, Derwent World Patent Index, INPADOC • subscription required
Univentio (The Netherlands)	http://www.univentio.com	• patent information provider • full text searches possible • bibliographic information, full patent text delivery	• subscription required • formerly Univentio, acquired by LexisNexis (USA), see http://www.lexisnexis.com and above

Table 4.10 Selected internet sources for general IP information and services in Japan.

Information/ service provider	URL	Provided information and services
Japanese Institute of Invention and Innovation	http://www.jiii.or.jp/english/e.htm	Support of R&D activities and utilization of intellectual property
Japan Patent Attorneys Association (Japan)	http://www.jpaa.or.jp/english/index.html	Facilitation of contact between companies and Japanese patent attorneys
IPCC (Industrial Property Cooperation Center, Japan)	http://www.ipcc.or.jp/	Utilization of intellectual property (in Japanese)
Patent application processing center (Japan)	http://www.papc.or.jp/	Utilization of intellectual property (in Japanese)
Institute of Intellectual Property (Japan)	http://www.iip.or.jp/e/index.html	Research on IP topics, support of IP projects, various databases (e.g., patent applicants, document numbers, law suit examples)
Global IP Group (Japan)	http://www.shinjyu.com/	General IP news from Japan; law suit examples; assistance in field of IP law

Much further information is available from online sources. Useful links to general information about the Japanese patent system or related topics can be found, for example, at following web pages: http://www.ipmenu.com, http://www.patentreference.com, http://www.european-patent-office.org/onlinelinks/e, http://www.paton.tu-ilmenau.de/links, http://jguide.stanford.edu/index.html, http://www.okuyama.com, http://www.indiana.edu/~cheminfo/ca_cps.html and http://sciencelinks.jp.

5
Overview of Japanese Patent Law

5.1
Introduction

Inventions can be protected in Japan in particular through patents and utility models. In contrast to a patent, protection as a utility model requires a particular shape of the invented object. Moreover, processes are also not protectable as utility models. Patents are thus the intellectual property rights of choice for inventions of chemical or biotechnological products. The following overview deals, thus, only with the Japanese Patent Law (in the following also abbreviated as "JPL"). The paragraphs cited refer to the JPL as of May 1, 2006, an English translation of which is included in the book *Japanese Laws Relating to Industrial Property*, AIPPI Japan, 2006.

This chapter should be read while keeping in mind that the purpose of the Japanese Patent Law is to encourage inventions by promoting their protection and utilization so as to contribute to the development of industry (Section 1 JPL). The interests of industry are thus considered to be more important than the rights of individual inventors.

5.1.1
Patentable Inventions

According to the JPL, an invention means the highly advanced creation of technical ideas by which a law of nature is utilized (§2(1) JPL). Inventions belong to the categories product, process or process of manufacturing a product. Any person who has made an invention that is industrially applicable can obtain a patent therefore (§29(1) JPL). There is no explicit enumeration of non-patentable subject matter in the JPL as is the case, for example, in the German Patent Act or the European Patent Convention. However, inventions liable to contravene public order, morality or public health shall not be patented (§32 JPL).

According to the practice before the Japanese Patent Office, as outlined in the Examination Guidelines (see www.jpo.go.jp/infoe), methods for the treatment of the human body by means of surgery or therapy or methods of diagnosis practiced on the human body are not patentable. This exemption is not applicable to animals.

The patentability of medical processes, and in particular the extension of patentable medical subject matter has been a hot topic in Japan over the years. As a result, the Examination Guidelines with regard to medicinal inventions[1] have been revised with effect of July 1, 2005. As a result, processes for the processing of biologic material that is taken from one patient and which is then returned to the same person are considered as technically applicable and thus patentable. Specifically, a method for preparing a medicament (such as a blood preparation, vaccine, recombinant preparation) or a medical device (e.g., artificial replacements or substitutes for parts of the human body such as an artificial bone or a cultured skin cell sheet) that are prepared using a human derived material as a raw material should be regarded as patentable subject matter, even if the method is directed to a processing of human body derived materials, assuming that they are to be injected or transplanted back into the same person for therapeutic reasons.[2] Moreover, "a method of operating a medical device" is no longer considered to be directed to "a method of operating on, treating or diagnosing a human".

The patentability of software has undergone considerable changes over the years. According to the Examination Guidelines, computer readable storage media on which a computer program exists are patentable since April 1, 1997. Computer programs as such have been patentable since January 10, 2001. Accordingly, business methods are patentable, in principle, and are treated as a kind of software invention.

5.1.2
Patentability Criteria

An invention, to be patentable, must fulfill the requirements of novelty and inventive step. Moreover, the invention must be described in the description in a manner sufficiently clear and complete for the invention to be carried out by a person having ordinary skill in the art to which the invention pertains. The latter requirement is discussed in more detail under "III. Drafting of a Japanese Patent Application" (Section 5.2).

5.1.2.1 Novelty
The following inventions are not patentable because of lack of novelty (§29(1) JPL):
- Inventions that were publicly known in Japan or elsewhere prior to the filing of the patent application;
- Inventions that were publicly worked in Japan or elsewhere prior to the filing of the patent application;
- Inventions that were described in a distributed publication or made available to the public through electric telecommunication lines in Japan or elsewhere prior to the filing of the patent application.

These provisions are applicable to patent applications filed from January 1, 2000 on. Before that date, public knowledge or public working was considered to be novelty destroying only if

---

[1] "The Examination Guidelines for Medicinal Invention" as the third Examination Guidelines in Part VII: Examination Guidelines for Inventions in Specific Fields, which follow the Computer Software-Related Invention and Biological Inventions.

[2] Cf. Yuasa and Hara *IP News*, July 2005, Vol. 17: "Establishment of the Examination Guidelines in Japan for Medicinal Invention", Shinobu Fukusho, pages 3–8; "Recent Revision of the Examination Guidelines for Industrially Applicable Inventions", R. Izumiya, K. Totsuka, T. Ito, pages 9–13.

it occurred in Japan. By the same effective date the rendering of public availability through electric communication lines was introduced as a novelty defeating act.

Novelty can also be taken away by means of older patent or utility model applications that have not been published before the application date of the patent application (Section 29bis JPL). With respect to an examination of novelty, the older patent/utility model application is considered based upon the whole contents of its disclosure. The older patent/utility model application is, however, not considered with respect to an examination of inventive step. Notably, an older patent or utility model application has no effect on the patent application if, on the application date of the patent application, both applications are derived from the same applicant or the same inventors.

5.1.2.2 Exceptions from Loss of Novelty – Grace Period

The JPL prescribes several kinds of disclosures of the invention that is the subject of a patent application that are not considered to be novelty destroying (Section 30 JPL). Namely, the following exceptions to lack of novelty exist:

- Novelty destroying disclosures that are due to the fact that a person having the right to obtain a patent has conducted an experiment, has made a presentation in a printed publication, has made a presentation through electric telecommunication lines, or has made a presentation in writing at a study meeting held by a scientific body designated by the Commissioner of the Patent Office.
- Novelty destroying acts committed against the will of the person having the right to obtain a patent.
- The person having the right to obtain a patent has exhibited the invention at an exhibition held by the Government or by any local public entity or one that is designated by the Commissioner of the JPO, or at an international exhibition held in the territory of a country party to the Paris Convention or of a Member of the World Trade Organization by its government, etc. or by a person authorized thereby etc. (cf. Section 30(3) JPL).

The aforementioned novelty destroying acts can be overcome only when the applicant files the Japanese patent application (not the priority patent application) within six months from the date of the novelty destroying disclosure. The exemption to loss of novelty is moreover only applicable if the applicant submits a written statement to that effect to the Commissioner of the JPO simultaneously with the patent application and if they submit to the JPO within 30 days of the filing of the patent application a document proving that the invention has fallen under one of these cases.

Loss of novelty can thus not be avoided if the invention was published before the patent application by a third person with the agreement of the inventor, i.e., when the subject of the invention was made accessible to third parties voluntarily. This will be, for example, the case if the inventor has allowed third parties to publish his invention before the patent application.

5.1.2.3 First to File Principle

Where two or more patent applications relating to the same invention are filed on different dates, only the first applicant may obtain a patent for the invention (Section 39(1) JPL). In particular, where two or more patent applications relating to the same invention are filed on the same date, only one such applicant, agreed upon after mutual consultation among all the applicants, may obtain a patent for the invention. If no agreement is reached or no consultation is possible, none of the applicants shall obtain a patent for the invention (Section 39(2) JPL).

The effect of an older application under Section 39 JPL will, however, depend upon its status. Namely, where a patent application or a utility model application is abandoned, withdrawn or dismissed, or where an examiner's decision or trial decision that a patent application is to be refused has become final and conclusive, such application shall be deemed never to have been made. For the purposes of Section 39 JPL, the applications are compared with respect to their patent claims ("prior claim approach").

5.1.2.4 Inventive Step

To obtain a patent, the requirement of an inventive step must be fulfilled (Section 29(2) JPL):

> *Where an invention could easily have been made, prior to the filing of the patent application, by a person with ordinary skill in the art to which the invention pertains, on the basis of an invention or inventions referred to in any of [novelty defeating disclosures of Section 29(1)], a patent shall not be granted.*

The Japanese term used to describe an inventive step is 進歩性 *shimposei* (progressiveness). This term is, however, not to be found in the JPL. This is illustrative of the fact that surprising advantageous effects are of comparatively high importance for satisfying the requirement of the inventive step. The so-called "problem–solution approach" that is applied by the European Patent Office is of minor importance.

The JPO appears to require a larger inventive step standard than the U.S. Patent & Trademark Office and the European Patent Office. The Examination Guidelines state that

> *the reasoning can be made from various and extensive aspects. The examiner evaluates whether a claimed invention falls under a selection of an optimal material, a workshop modification of design, a mere juxtaposition of features on the basis of prior art inventions or whether the contents of cited prior art inventions disclose a cause or a motivation for a person skilled in the art to arrive at the claimed invention ...,*

The JPO often denies the existence of an inventive step based on the argument that the invention *could* easily have been made. In these cases it does not consider it necessary to demonstrate that the skilled person *would* have made the invention on the basis of the prior art (as is the principle applied by the European Patent Office). As of December 2006, the inventive step standard to be applied by the JPO is a widely discussed and researched topic.

5.1.3
Official Fees

The JPO requires the payment of certain official fees for handling a patent application as well as for its maintenance (annual fees). The fee structure of the JPO has been amended drastically with effect from April 1, 2004. As a result, the filing fee was reduced considerably, while

Table 5.1 Official fees for patent applications.

Activity	Fee for patent application filed after April 1, 2004	Fee for patent application filed before April 1, 2004
Filing of a patent application	¥16 000	¥21 000
Requesting examination	¥168 000 + ¥4 000 for each patent claim	¥84 300 + ¥2 000 for each patent claim
Reduction of examination fee for PCT applications (if JPA is international search authority)	40% reduction	80% reduction
Reduction of examination fee for PCT applications (if JPA is not international search authority)	10% reduction	20% reduction
Maintenance fees for 1–3 years	¥2600 + ¥200 for each patent claim	¥13 000 + ¥1100 for each patent claim
Maintenance fees for 4–6 years	¥8100 + ¥600 for each patent claim	¥20 300 + ¥1600 for each patent claim
Maintenance fees for 7–9 years	¥24 300 + ¥1900 for each patent claim	¥40 600 + ¥3 200 for each patent claim
Maintenance fees for 10–25 years	¥81 200 + ¥6400 for each patent claim	¥81 200 + ¥6 400 for each patent claim

the examination fee was doubled. Moreover, the fees that are to be paid for the maintenance of Japanese patents have also been reduced. The examination fees are different for PCT applications, and the reduction depends on whether the JPA is the international search authority. Table 5.1 provides an overview on official fees.

Effective April 1, 2004 a system of reimbursement of the examination fee was introduced: 50% of the examination fee is reimbursed upon request, which is to be filed within six months after the patent application has been abandoned or withdrawn if no official action has been issued until that time. To maintain Japanese patents the annual fees given in Table 5.1 have to be paid from the recordal of the patent right onwards (i.e., not from the application date).

5.2
Drafting of Japanese Patent Applications

A Japanese patent application must include a description of the invention, one or more patent claims and, if applicable, one or more drawings and an abstract. The description shall state the title of the invention, a brief explanation of the drawings and a detailed explanation of the invention.

5.2.1
Patent Claims (Section 36(5)(6) JPL)

In view of the limited possibilities for amending the claims under Japanese patent practice it is useful to draft a sufficient number of claims. In most cases around 10 claims is sufficient. Care should, however, be taken not to file too many independent claims, to avoid a finding of lack of unity of invention under Section 37 JPL.

In the patent claim(s) there shall be set forth by statements separated on a claim by claim basis, all matters that an applicant for patent considers necessary in defining an invention for which a patent is sought. The invention claimed in one claim can be the same as an invention claimed in another claim; i.e, an overlap in the subject matter of claims is admissible.

The claims shall set forth the invention(s) for which a patent is sought and must be described in the detailed explanation of the invention. The claims must be clear and concise, and comply with an ordinance of METI. This requirement is not fulfilled if the invention is expressed differently in the detailed description of the invention and the claims. It would also be inadmissible to indicate numerical ranges in the claims that are neither described nor implied in the detailed description. Multiple dependency of claims is allowable, in contrast to the U.S.A. A dependent claim may refer to another claim of a different claim category. "Means-plus-function"-claims are admissible.

Claims should, in principle, be directed to a product, process for manufacture and other processes. The examination practice is, however, less strict and claims directed to a specific use are now admissible. Notably, use claims in Japan are often in the format of product claims (product A ... for the use ...).

For claims that are directed to alloys, pharmaceuticals, catalysts, compositions, Japanese examiners often require the field of use to be indicated in the claim.

Disclaimers are in principle admissible.

The invention must be clearly identifiable on the basis of the claims. The following claim would be considered unclear because the total amount of the maximum amount of component A, and the minimum amounts for components B and C, respectively would exceed 100 wt.-%: "Alloy, consisting of 40 to 60 wt.-% A, 30 to 50 wt.-% B and 20 to 30 wt.-% C."

The following claim is admissible only if the technical definition and the X-laboratory test method are described in the detailed explanation of the invention and this information does belong to the state of the art at the time of filing: "Adhesive composition, containing a component Y with a viscosity of a to b pas/sec as measured by the X-laboratory test method."

A claim might be unclear if non-technical features such as, for example, trademarks are used. In such a case, it must be clear to the person skilled in the art that a specific quality, composition and structure of the product were maintained at least for a specific term until the date of the patent application. Moreover, the claim category must be clear. The following claims are not admissible:
- Method or device, comprising ...
- Method and device, comprising ...
- A claim where it cannot be understood whether it is directed to a product or a process, because the claim describes only operations, characteristics, purposes or effects of things, e.g., "anticancerogenic effect of chemical substance A".

Expressions like 方式 *houshiki* (system) (e.g., 電話方式 *denwa houshiki*, telephone system) belong to the product category. 使用 *shiyou* and 利用 *riyou* (use) are interpreted as method for the application of things and thus as "process". "Use of compound X as insecticide" is interpreted as "method for the use of substance X as insecticide". "Use of compound X for the manufacture of a medicament for the therapeutical application Y" is considered to be a "Process for the use of substance X for the manufacture of a medicament for the therapeutical application Y".

A claimed invention is deemed to be unclear if features are indicated as alternatives but the alternatives have no similar characteristics or function. The group of inventions that arise from the indication of alternatives should thus belong to the same technical field or one of the following conditions should be fulfilled:

- The group of inventions (e.g., chemical compounds) have in common a technical problem that was not solved until the application date of the group of inventions.
- The group of inventions has in common a major part of the features constituting the invention.

Moreover, a claim directed to chemical compounds, wherein an intermediate product and a final product are described as alternatives, is admissible only if the intermediate product as such is already a final product and fulfils the aforementioned requirements.

A claim "Substance, obtainable by processing in an organic solvent" is inadmissible, since it is unclear here what is meant by "processing". When only one example is given in the patent description, but no practical method and no definition, the expression "processing" has too many meanings and the invention remains unclear.

Claims should not relate to drawings or the patent description if the scope of the claims becomes unclear, as is, for example, the case with "an automatic drilling machine as shown in Fig. 1". Drawings are often ambiguous and may be interpreted in different ways. The same applies if a part of the description is referred to that is not clearly indicated. This is, however, not applicable if the invention may be indicated clearly and concisely in a claim such as the following, wherein a particular relationship among the components in an alloy is indicated that can be defined by referring to a drawing as clearly as with a numerical or other literal expression:

"Heat resistant Fe-Cr-Al alloy for electric heating, consisting of Fe, Cr and Al with the composition range which is limited by points A, B, C and D in Figure 1, and impurities of less than x wt.-%."

The requirements for claims in specific fields of technology, e.g., gene technology, are described in considerable detail in the Examination Guidelines of the JPO[3] ["Biological Inventions", "Guidelines for describing taxonomic characters", and "Examples of Examination on the Inventions related to Genes (DNA fragments, full-length cDNAs, and Single Nucleotide

---

[3] Cf. Examination Guidelines of JPO: "Biological Inventions", "Guidelines for describing taxonomic characters", and "Examples of examination on the inventions related to genes (DNA Fragments, full-length cDNAs, and Single Nucleotide Polymorphisms)", accessible via the homepage of JPO (www.jpo.go.jp); cf. *AIPPI Journal*, (July 2000), Vol. 25(4), pages 178–180.

Polymorphisms)"]. New plant varieties can be patented. In a patent application for plants as such, the plant needs to be defined in the claims:

"Plant, belonging to the eggplant family, with a T-DNS, in which toxine genes, which contain the base sequence ATGACT..., are incorporated."
"Water melon with 33 chromosomes in somatic cells, obtained by cross breeding of tetraploidic water melons, which have been obtained by ploidic treatment of diploidic water melons."

The inventions of pharmaceutically active compounds can be protected broadly if the application in the medical field has been found for the first time: "Compound A for the use in medicaments". In case of the invention of second or further medical indications, these can be claimed by specifying the indication in the claim:

"A pharmaceutical composition for the treatment of illness X, comprising compound Y as active ingredient."

Section 36(5)(ii) JPL is violated in case of a chemical invention if the name or the structural formula of the chemical substance is not indicated. However, if this were not possible, the chemical compound might also be specified by means of its physical or chemical characteristics. In case this were also not possible, the method of manufacture might also be included.

Product-by-Process claims are thus admissible. A product-by-process claim covers each product obtainable by the indicated process, independent of whether it is obtained via the indicated or another process.[4] There is no difference in the interpretation of "obtainable by" and "obtained by".

The unity of invention principle[5] of §37 JPL must be complied with: "Where there are two or more inventions, they may be the subject of a patent application in the same request only when they have the technical relationship as provided for in an ordinance of METI. §25(8) of the Patent Law Enforcement Guidelines" (to be referred to in the following as "Enforcement Guidelines") provide moreover: The technical relationship prescribed by METI in Section 37 JPL means a technical relationship, where, by having the same or corresponding special technical features, two or more inventions are linked as to form a single general inventive concept. The special technical features define the contribution an invention makes over the prior art. The existence of the technical relationship shall be determined irrespective of whether two or more inventions are claimed in separate claims or as alternatives within a single claim.

5.2.2
Detailed Explanation of the Invention (Section 36(4) JPL)

The detailed explanation of the invention must describe the invention in a manner sufficiently clear and complete for the invention to be carried out by a person having ordinary

---

4) Cf. Klaus Hinkelmann, *AIPPI Journal*, (2000), Vol. 25(1), pages 39–50; "Product-by-Process-Claims in Japan" by Yoshiro Hashimoto, *AIPPI Journal*, (2001), Vol. 26(5), pages 294–299.

5) See also "The Revisions to Requirement of Unity of Invention of the Patent Law – Practical Implications", by Satoshi Kabasawa, *AIPPI Journal*, (2005), Vol. 30(4), pages 217–224.

skill in the art to which the invention pertains, as provided for in an ordinance of the Ministry of Economy, Trade and Industry (METI). Related prior art should be discussed in the description. There are, however, no sanctions if the latter is not complied with in the patent application as filed.

The detailed explanation of the invention shall describe the invention in a manner sufficiently clear and complete for the invention to be carried out by a person having ordinary skill in the art to which the invention pertains, as provided for in an ordinance of METI (Section 36(4) JPL). According to §24bis Enforcement Guidelines, the description shall contain a description of the problem to be solved and of its solution and other matters that are necessary for the person skilled in the art to understand the invention. The use of SI units is mandatory (§3 Enforcement Guidelines).[6] There is no duty in Japan to disclose the best mode of the invention.

As the applicant is required to indicate in the description the pertinent state of the art known to him (Section 36(4)(ii) JPL), the examiner should issue a notice of rejection demanding the provision of such information if this requirement is not dealt with. In case the applicant does not submit comments or an amendment, the patent application will be rejected. A granted patent will, however, not be deemed unenforceable if a violation of this obligation is recognized thereafter.

If the invention is directed to a particular use that solves a problem by utilizing a particular feature of a chemical compound (e.g., in the field of pharmaceuticals or agrochemicals), the use needs also to be described in detail in the description. Test data should be supplied for such use inventions. If a chemical compound as such were to be claimed, test data may be omitted. Advantageous effects should be described as they might be important for any discussion of inventive step.

In technical areas where it is in general difficult to predict the effects of an invention (e.g., chemical compounds), normally one or more representative working embodiments should be described that allow the person skilled in the art to carry out the invention without difficulties.

The listing of numerous chemical compounds without any physical or chemical data, i.e., without an actual proof of their synthesis provides no advantages to support broad patent claims. However, such a listing of compounds might be novelty destroying state of the art that might prevent their patenting by third parties. There will be no prior art effect if the technical knowledge at the application date was insufficient to allow the manufacture of a chemical compound. A chemical intermediate, e.g., a monomer for a polymer, will be considered industrially applicable if a way to a final product and a utility of the final product is disclosed.

An insufficient description of the invention exists if the skilled person must conduct numerous or complicated experiments that exceed the extent that might be expected from him to find a way to work the invention.

The applicant must describe at least one way for the working of the invention. This implies, for the invention of a product, that it must be possible to manufacture and use the product in accordance with the description.

---

6) Cf. T. Nakata, "New Measurement Law enforced on October 1, 1999", Yuasa and Hara *IP News*, Vol. 6 (December 1999), pages 3–4.

If a product is characterized by its functions, characteristics etc., and these are neither standard nor commonly used by the skilled person, the detailed description must contain a definition of these functions etc. or the method for their determination, such that the invention may be performed by the skilled person.

For the drafting of patent applications it should be kept in mind that the JPA does not allow incorporation by reference, i.e., it is not possible to incorporate the contents of publications, patent applications etc. by simply referring to them.

A way to work the claimed invention must not be described for all embodiments or alternatives. However, if the examiner has good reasons as to why the skilled person is not able to extend the specific way of working described explicitly in the detailed description (or a working example) to the whole scope of the patent claim, the invention might be considered as insufficiently disclosed and thus not patentable.

Functional claims and thus claims (inventions) that are characterized by means of parameters are admissible under Section 36(5) JPL. These claims may afford a great scope of protection, but it may often be difficult to draft a description that sufficiently supports claims that are defined by means of parameters. The thus defined invention might be vague and the description might contain no executable actions.

The definition of a parameter must be sufficiently clear. For example, if the melt viscosity of a polymer is indicated as being from 10^2 to 10^3 centipoise, both the method of measurement and the conditions employed should be described. The measuring instrument should be identified according to its trade name, model number and producer, since even minor differences might yield different results. The method should be described as exact as possible to enable a person skilled in the art who repeats the measurement to arrive at the same result.

As the causal relationship between parameter and effect is indirect (the specific constitution is omitted) it is recommendable to describe a large number of examples. It is important to describe a mechanism that creates a correlation between the parameter range claimed and the effect observed, to strengthen the weak indirect connection between parameter and effect.

Patent applications regarding microorganisms require the indication of taxonomic data. A deposition of a microorganism must be effected before the filing of the priority patent application. The patent application must be accompanied by a copy of a receipt or a certificate from the JPO that confirms the deposition (§27bis(1) Enforcement Guidelines). This requirement may only be avoided if the microorganism was publicly available before the application date, as it would fulfill the requirement under §27bis(2) Enforcement Guidelines "that a person with ordinary knowledge on the field of the filed invention may obtain the microorganism without difficulties". Amino acid and nucleotide sequences must be filed in computer readable form. Submission through a recording medium is not required.

Japanese disclosure requirements are especially strict in the field of pharmaceutical inventions.[7] For an invention of a pharmaceutical, numerical pharmacological test data regarding specific compounds must already be included in the application as originally filed. At the time of a Japanese patent application such data may be included. These data will then enjoy at least the priority of the direct filing in Japan. In the case of the entry of the national Japanese phase for a PCT application such a supplementation will not be possible. A pharma-

---

[7] Tokyo High Court, October 30, 1998; Case No. 1996 Gyo-Ke 201.

ceutical patent application must not contain acute toxicity data. However, such data in the application might be used to support a finding of inventive step.

§24 Enforcement Guidelines prescribes the following order for claims, description, figures and summary in a patent application:
1. Title
2. Claims
3. Detailed explanation of the invention
4. Short explanation of the drawings
5. Summary
6. Figures

5.3
Filing of Japanese Patent Applications

A patent application or entry into the national Japanese phase can be filed before the JPO online, via diskette or in paper form. Japanese patent applications can, however, not be filed by telefax. Online filing is preferred by the JPO. High data conversion fees need not be paid (as is the case with paper filings) if the application is filed online or via diskette. By using online filing, the patent application will be completed by means of the software provided, at no charge by the JPO, and filed online. An application number will be accorded immediately.

A patent application can in principle be filed on the basis of a utility model (UM) application or registration (Sections 46, 46bis JPL) within three years from the date of filing of the UM application. Moreover, patent applications can be converted into an UM (Section 10 Japanese UM Law) or design application (Section 13 Japanese Design Law) within 9.5 years from the filing date, or within 30 days from the transmission of reasons for rejection.

5.3.1
Direct Japanese Patent Filing

A direct Japanese patent application can be filed either in Japanese or English. Notably, the JPL allows filing in a foreign language (Section 36bis JPL). However, according to §25(4) Enforcement Guidelines, the only allowed foreign language at present is English. A Japanese translation must be filed within 14 months from the filing date of an English language application (or, if applicable, from the filing date of a priority application) (Section 36bis(2) JPL). Otherwise, the patent application will be deemed withdrawn (Section 36bis(3) JPL). The Japanese translation can be corrected on the basis of the English language application (§17bis(2) JPL). However, a foreign priority application in accordance with the Paris Convention cannot be a basis for correcting the Japanese translation.

A Japanese inventor is not obliged to file a patent application first in Japan. A foreign resident, i.e., a person who has neither his domicile nor residence (nor, in the case of a legal entity, its establishment) in Japan may not, except where prescribed by Cabinet Order, proceed before the JPO or institute a suit against any measure taken by an administrative agency in accordance with the JPL or an order or ordinance thereunder, except through his representative with respect to his patent who has his domicile or residence in Japan (Section 8 JPL). The so-called patent administrator shall represent the principal in all procedures and in a suit instituted against measures taken by an administrative agency in accordance with the JPL etc. This is, however, not applicable where a resident abroad restricts the scope of power of attorney of his patent administrator.

Application Requirements: Any person desiring a patent shall submit a request to the Commissioner of the JPO stating the following:
- the name and the domicile or residence of the applicant for the patent;
- the name and the domicile or residence of the inventor.

The request shall be accompanied by the description, one or more patent claims and, if applicable, one or more drawings and the abstract. To obtain a date of application the following documents must be furnished:
- request with name and nationality of the applicant
- description
- drawings (if applicable)
- priority claim (name of country and date)
- name of the representative

Abstract, address of applicant, name and address of inventor, application number of priority application, and power of attorney may be submitted later.

The priority of an earlier patent application must be claimed upon filing and within one year after the filing of the first patent or utility model application (priority application). Moreover, a legalized copy of the priority document must be furnished within 16 months after the filing or priority date (the earlier date is applicable).

In view of the electronic exchange of priority documents between the JPO, the European Patent Office and the U.S. Patent and Trademark Office, this requirement is waived, if the priority of a European or U.S. patent application is claimed.

The Power of Attorney (general or specific) must be filed only upon request of the JPO. However, other procedures (filing an appeal or an invalidity action) require the provision of a Power of Attorney. The submission of a deed of assignment, nationality certificate (in the case of an individual person) or company certificate (in the case of a company) is in general not required. The documents may be filed in a foreign language, they must, however, be accompanied by a Japanese translation. Formal drawings can be supplemented if the application was filed with informal drawings. There is no communication in the case of missing or wrong drawings.

For patent applications that require for the confirmation of their working the deposition of a microorganism, of animal or plant cells, animal embryos, etc., a copy of a confirmation of an international depository in accordance with the Budapest Treaty or a confirmation of deposition by the National Institute of Bioscience and Human Technology (NIBH) is filed together with the patent application (Section 36(4) JPL; §27(2)(i) Enforcement Guidelines). The access number must be given in the description. Late submission will lead to rejection of the application.

If the right to obtain a patent belongs to several inventors, the patent application must be filed by all of them (Section 38 JPL) since it will be rejected otherwise (Section 49(1) JPL). If one of the joint owners does not want to file a patent application, the patent application may not be filed by the others. 審判 *shimpan* proceedings (e.g., an appeal against a rejection decision) before the JPO must be initiated by all joint owners (Section 132(3) JPL).

In Japan it is very difficult to amend the indication of the inventors. The inventors must be known at the time of filing. Affidavits must be submitted for any amendments. These complications are a principal flaw that might indicate an unlawful transfer of the right to the patent and thus weaken the patent. Incorrect naming of the inventors may even be a reason of invalidity if the applicant is neither the inventor nor the owner of the rights of all inventors.

5.3.2
Entry into the National Japanese Phase for PCT Applications

Entry into the national Japanese phase can be effected within 30 months after the first priority date, irrespective of whether an international preliminary examination was requested (§184(4) JPL). The following requirements before the JPO must be fulfilled within these 30 months (Section 184^{ter}ff.. JPL; in particular Section 184^{quinquies} JPL):

- Request to enter the national Japanese phase.
- Copy of the international publication, including cover page, patent description, claims, abstract and drawings; or:
  Copy of PCT request, patent description, claims, abstract and, if applicable, drawings in the version originally filed with the PCT.
- If applicable, a copy of the amendment in accordance with Article 19(1) PCT (Sections 184^{septies} JPL) and/or according to Article 34(2)(b) PCT (Section 184^{octies} JPL). A communication pursuant to Art. 20 PCT and 36(3)(a) PCT, respectively, to JPO renders the amendment also efficient.
- Payment of the national fee.

A Japanese translation of a foreign language PCT application must be filed within two months after the filing of a request for entry into the national Japanese phase. The submission of a Japanese translation is, thus, possible until up to 32 months after the first priority date. The Japanese translation may be corrected on the basis of the text of the foreign language PCT application. The international preliminary examination report is not binding on the Japanese patent examiner.

5.4
Examination of Japanese Patent Applications

A patent application will in general be laid open (i.e., published) after 18 months from the filing date (Section 64(1) JPL). The applicant can request an earlier laying open of the application; this request cannot be withdrawn (Section 64bis JPL). After the laying open, a third party can request with the JPO a file inspection.

The JPO will first conduct a formal examination. In the case of formal deficiencies (non-payment of fees, missing naming of inventors) it may invite a correction within a prescribed time limit (Section 17(3) JPL). If drawings are not filed in time, the application date will not be shifted to the date of filing the drawings as it is in Germany.

5.4.1
Substantive Examination upon Request

The merits of a Japanese patent application is examined by an Examiner upon request (審査請求 *shinsa seikyuu*), which can be filed with the JPO by anybody (Section 48bis JPL). The term for filing a request for examination is 7 years from the filing date for patent applications whose application date is before October 1, 2001 (Section 48ter(1) JPL), and 3 years from the application date for patent applications filed from October 1, 2001 on. In the case of a divisional application pursuant to Section 44(1) JPL examination may still be requested within 30 days after filing (Section 48ter(2) JPL). If a request for examination is not filed in time, the patent application is deemed withdrawn after the expiration of the term for requesting examination (Section 48ter(4) JPL). A restitution in integrum in the term for requesting examination is not available.

The number of official communications in examination and the possibilities to effect amendments to the description, drawings and claims are very limited in the Japanese examination proceedings. It is thus recommended to file at the latest time, upon requesting examination, an optimized set of claims that considers examination results from foreign countries and/or prior art that might have meanwhile come to the attention of the applicant.

Request for a speedy examination: The JPO provides two procedures for a rapid examination of selected inventions. The "preferential examination" (優先審査 *yuusen shinsa*) in accordance with §48^{sexies} JPL may be requested by the applicant as well as by a third party who is in dispute with the applicant (however, not by licensees), whereas the "accelerated examination" (早期審査 *souki shinsa*) may be requested from the applicant or his licensee. Although only the "preferential examination" is expressly prescribed in the JPL (cf. Section 48^{sexies} JPL) and the "accelerated examination" has been developed in the practice of the JPO,[8] the latter is used nearly exclusively since the "preferential examination" has certain disadvantages.

"Preferential examination" may be requested by the applicant only when the invention is used in Japan by a third party. The patent application must be laid open. To support the request, the demandant (applicant, third party) must submit according to §31ter Enforcement Guidelines documents regarding the working of the invention with the JPO. These documents will become accessible, which is usually not what the demandant desires.

The "accelerated examination" of the patent application does not require the previous laying open of the application. To file the request, the existence of an equivalent foreign patent application to the Japanese patent application is sufficient. One of the following two requirements must be fulfilled:
- The patent applicant or his licensee are working the invention or intend to do so within 2 years.
- The patent applicant has filed a patent application outside Japan or with an intergovernmental organization like the WIPO.

---

8) Cf. *Practices in Examination and Appeals under 1994-Revised Patent Law in Japan*, Japan Patent Office, AIPPI Japan, 1996, Chapter VI. Guidelines for accelerated Examination System and Accelerated Appeal Examination System (pages 239–284).

For the request, the applicant/licensee must submit an "explanation regarding the circumstances of the accelerated examination". For case (a) details regarding the working and for case (b) details on at least one parallel foreign patent application should be presented. A prior art search has to be submitted in both cases, which has been made either by the applicant, a patent office or an intergovernmental organization.

For completeness, it is noted here that a separate request for accelerated appeal proceedings may be filed against a decision by the examiner to reject a patent application. Similarly, an explanation regarding the circumstances of the accelerated examination and the reasons for the urgency of the patent grant must be provided.

5.4.2
Examination of Japanese Patent Applications

5.4.2.1   Examination before the Japanese Patent Office
First Instance (Examiner): The application is examined by one examiner, who is usually provided with the search results of another institution close to the JPO. However, the examiner can also carry out an (additional) examination on their own. The examiner examines as to whether a patent can be granted on the claimed invention (Section 48bis JPL). The examination will be conducted separately for each patent claim. The applicant will thus know which claims might be rejected and which might probably be granted. The examiner is not bound by any statements of the applicant in the description regarding the state of the art. The first office action can be either a notice of grant or a notification of reasons for refusal.

The examiner must reject the patent application if it falls under one of the points mentioned in Section 49 JPL; in particular, non-fulfillment of the patentability criteria (e.g., lack of statutory subject matter; lack of novelty or inventive step; insufficiency of disclosure). The decision shall be in writing and shall state the reasons for rejection (Section 52 JPL).

The JPL provides three types of rejection communications: a "notice of reasons for rejection" (拒絶理由通知 kyozetsu riyuu tsuuchi), possibly a second communication with a "last notice of reasons for rejection" (最後の拒絶理由通知 saigo no kyozetsu riyuu tsuuchi) and the final rejection (拒絶査定 kyozetsu satei). These communications are further distinguished by the possibilities for amendments by the applicant (Section 17bis JPL). The comprehensive possibilities to effect amendments after the first communication, forwarded pursuant to Section 50 JPL (Section $17^{bis}(1)(i)$ JPL), are reduced after the issuance of the second communication with a last notice of reasons for rejection (Section $17^{bis}(1)(ii)$ JPL). After issuance of the final rejection the extent to which the claims might be amended is limited pursuant to Sections $17^{bis}(4)$ and $17^{bis}(5)$ JPL.

When examiners intend to render a decision that an application is to be rejected, they shall notify the applicant for the patent of the reasons for rejection and also give the applicant with an opportunity to submit a statement of the arguments, designating an adequate time limit (Section 50 JPL). The term designated for responding to the official action is 3 months for patent applicants living abroad, which may be prolonged, once, by three months upon the payment of an administration fee (60 days for applicants residing in Japan, which cannot be extended). The term begins with the transmission of the official action. This first official action ("Notice of Reasons for Rejection": 拒絶理由通知書 kyozetsu riyuu tsuuchisho) should contain the opinion of the examiner regarding the patentability of all claims and should contain all reasons for rejection. The examiner should propose suitable amendments or a division of the application if these measures help the applicant to respond to the official action.

Objections by the examiner against the workability of the invention are to be explained in detail.

In response to an official action pursuant to Section 50 JPL, the applicant may submit comments (意見書 *ikensho*) and/or an amendment (補正 *hosei*).

In the comments, the applicant may undertake to rebut the rejection reasons put forward by the examiner. The approach to be taken will depend on the type of rejection reason and the type of evidence presented. In Japan, broad claims may be supported, for example, by submitting further working examples and/or the indication of publications etc.

In the amendment, the applicant may amend the description, including patent claims and/or the drawings. Possibilities for effecting amendments exist in particular in response to the first office action (cf. Section 17bis(1)(i) JPL). A scope-shifting amendment, i.e., an amendment such that the invention recited in each claim that has been already examined in an official action is changed to an invention that does not meet the requirement of unity in relation to the already examined invention, is not allowed since an amendment to the JPL effective 2007. Since further amendments are only available to a limited extent (cf. Section 17bis(4) JPL), a careful response to the first official action is of utmost importance. In response to the first official communication the whole disclosure in the detailed description of the invention can be used to effect amendments. Even a broadening of claims is possible. However, after the expiration of the term for filing a response to the first official communication only minor amendments to the claims are available, by specifying, for example, in more detail a feature already present in the claims. A full recourse to the detailed description is no longer available.

The last notice with reasons for rejection is in principle a communication that indicates only such reasons for rejection that have become necessary by the amendments in response to the first non-concluding communication. An amendment in response to a final communication is admissible, if it is a claim amendment by "correcting mistakes in the description" or the "clarification of ambiguous description" pursuant to Sections 17bis(4)(iii) and 17bis(4)(iv) or an amendment of the detailed description of the invention.

The second official action may be a final rejection, if the reasons of rejection were caused by the amendments carried out by the applicant in response to the first official communication. This is the case, for example, if the applicant added new features to the claims to avoid objections of lack of novelty and/or inventive step without, however, removing these objections. If the objections raised in the second official communication have not been mentioned in the first official communication and have not been caused by amendments of the applicant in response to the first office action, then the second office action may not be a final rejection. An example is the presentation of new reasons for rejection because of prior art that was newly found by the examiner.

In general, the second official communication is final. The applicant or a representative should thus exploit all possibilities for an effective communication with the examiner. As auxiliary requests and oral proceedings are not available under the Japanese patent examination system, it is important to elucidate the attitude of the examiner with regard to any potential amendments to the claims etc. and to respond correspondingly. For example, the examiner might be contacted by phone after the first official communication to obtain comments on it. Possible arguments and amendments of the applicant might be brought to the attention of the examiner by phone or telefax in advance and discussed with him, before an official argument and amendment is submitted on the pending official communication.

The examiner will consider the examination guidelines of the JPO although they are not legally binding. The "Examination Guidelines for Patent and Utility Model" can be found on the homepage of JPO (www.jpo.go.jp). If the examiner is of the opinion that the reasons for rejecting the application continue to exist despite the amendments carried out and the arguments put forward by the patent applicant, they shall issue a decision to reject the patent application.

Second Instance (Appeal to the Trial Board): An appeal against the rejection decision of the examiner (拒絶査定不服審判 *kyozetsu satei fufuku shimpan*) may be filed with the Trial Board (審判部 *shimpan bu*) within the JPO (Section 121 JPL). The appeal is to be filed within 30 days (automatic prolongation by 60 days for applicants residing abroad) upon transmittal of the decision. In the case of joint owners of the right to obtain a patent, the appeal must be made jointly by all the joint owners (Section 132 JPL).

Information required for the demand of an appeal pursuant to Sections 131(1) and (3) JPL (including the appeal reasons) and documents may be supplemented upon invitation by the JPO within 30 days pursuant to Section 133(1) JPL.

If the applicant has effected amendments to the claims or the description within 30 days after an appeal has been raised, the appeal will be examined firstly by the examiner of the first instance (Section 162 JPL), who examines as to whether a redress can be given on account of the arguments and amendments presented by the applicant (*zenchi* examination). The examination will start again, whereby a new reason of rejection might be asserted by the examiner (Section 163 JPL).

The purpose of the *zenchi* examination is to reexamine the correctness of the examination of the patent application. The examiner shall determine whether the reasons for rejection have been removed by the amended description, claims and drawings, whereby the arguments in the appeal brief should also be considered. The *zenchi* will determine whether additional reasons for rejecting the application exist.

If the reasons of rejection have been removed by an admissible amendment and if no new reasons for rejecting the application have been found, a decision of grant will be given.

If the examiner is of the opinion that the reasons of rejection continue to exist, they shall make a report to the Commissioner of JPO on the result of the examination without rendering a decision with respect to the demand for a trial (Section 164(3) JPL). All of the reasons for the rejection decision of the examiner not removed should be expressly stated and explained in the *zenchi* report, if applicable together with later found reasons of rejection. If no redress is available, the Trial Board (審判部 *shimpan bu*) will conduct written appeal proceedings. In cases where the applicant did not perform any amendments to the description or the patent claims, the appeal will go directly without an examination by the examiner of first instance to the corresponding *shimpan bu*.

5.4.2.2 Appeal to the Intellectual Property High Court against Rejection Decision of the JPO

A further appeal may be raised against rejection decisions of the Trial Board of the JPO to the Intellectual Property High Court in Tokyo (§ 178(1) JPL). The term to be observed is 30 days (plus 90 days for applicants residing abroad) beginning with the transmission of the rejection decision. If the timely filing of the complaint is sufficient the appeal brief may be filed later. The competence of the court is limited to revoke the decision of the JPO or to reject the appeal. The court may, however, not rule on the grant of a patent.

Basically, the Intellectual Property High Court may still make a hearing of evidence in appeal proceedings against a decision of the JPO that had not been considered in the proceedings before the JPO, and arrive at a revocation of the decision based on this material. If the court revokes a decision of the JPO and orders a new decision on the request to grant a patent on the application, the JPO is bound thereto in accordance with Section 181(2) JPL. In principle, a further appeal might be raised to the Supreme Court with respect to a violation of matters of law. This is, however, extremely rare.

5.4.2.3 Patent Grant

When the examiner finds no reasons for refusal with respect to the patent application, they shall render a decision that a patent is to be granted (Section 51 JPL). The examiner is bound by their decision. The annual fee for each year from the first to the third year shall be paid in a lump sum within 30 days from the date of transmittal of the examiner's decision or the trial decision that the patent is to be granted (Section 108(1) JPL). There is no separate communication from the JPO to effect this payment. The establishment of a patent right shall be registered when this payment has been effected or exemption or deferment of such payment has been granted (Section §66(2) JPL).

Translation errors in the description cannot be corrected if the essential content of the description is going to be amended. The priority document may only be used for proving the priority date, not for correcting translation errors. For example, the erroneous translation of the term "polyvinylacetal" in a German priority application to "polyvinyl acetate" could not be corrected.[9] Likewise, it was not possible to correct the wrong translation of "boron" to "bromine".[10]

5.4.2.4 Divisional Applications (分割出願 bunkatsu shutsugan)

The compact examination proceedings before the JPO have increased the attractiveness of divisional applications (Section 44 JPL).[11] Upon the filing of a divisional application, all options for amendments to the claims are preserved. This is contingent, however, upon a timely filing of a divisional application. Until recently divisional applications could only be filed as long as the basic patent application could be amended. This has led to situations where no opportunity for filing a divisional application remained. For example, upon a notice of grant a divisional application could not be filed. Moreover, after the issuance of a final rejection, an appeal had to be filed first to allow the filing of a divisional application. In 2006, however, the JPL was amended to allow in the first instance the filing of a divisional application within 30 days after a notice of patent grant or the transmittal of a final rejection by the examiner. Once an application is pending in appeal proceedings against an examiner's decision of rejection, it is not possible to file a divisional application when an applicant receives an examiner's decision to grant a patent as a result of the *zenchi* examination or an appeal decision by the *shim-*

---

9) "Charge Carrier Foil", Tokyo High Court, June 27, 1978, Case No. 1977 Gyo-Ke 46; cf. *IIC*, Vol. 10 (1979), pages 758–761.

10) Tokyo High Court, March 24, 1983, Case No. 1981 Gyo-Ke 82.

11) M. Motsenbocker, *Mitteilungen* 1996, pages 151–155.

pan bu to grant a patent as a result of the appeal examination (Section 44 (1) (ii) JPL). Moreover, one or more divisional applications can be filed to counter an objection by the examiner under the requirement of unity of invention pursuant to Section 37 JPL or upon initiative of the applicant.

5.5
Attack on Patent Applications and Patents

It is possible to fight against a patent application before patent grant as well as after patent grant. In Japan, the opposition system was abolished effective January 1, 2004. The only possibilities for attack are, thus, the submission of patentability relevant information to the JPO and the filing of an invalidation demand.

5.5.1
Submission of Information Regarding Patentability to the JPO (情報提供 *jouhou teikyou*)

Before patent grant, i.e., as long as the patent application is pending, a third person may submit with the JPO information that it deems to be relevant for the patentability of the applied invention. According to §13bis Enforcement Guidelines, anybody may submit to pending patent applications (i.e., not to granted patents) information regarding the patentability. The information may relate, for example, to novelty (Sections 29(1), 29bis JPL), inventive step (Section 29(2) JPL) or the first-to-file principle (Section 39(1)–(4) JPL). The patent relevant information may also relate to the addition of new material upon amendments by the applicant to the patent application (Sections 17(2)(3), 17bis(3) JPL), the addition of new material to the Japanese translation of a foreign language patent application, which extends beyond the scope of the original foreign documents (Section 49(v) JPL), and description requirements regarding the detailed explanation of the invention and the claims (Section 36(4) and (6) JPL, except for Section 36(6)(iv) JPL, which concerns formal requirements for the drafting of claims).

In addition to publications or patent documents, other documents and drawings may be submitted, even in the form of electronic data. However, things that are different from documents such as, for example, video movies that explain the operation of a device are not admissible. The JPO will not examine evidence by hearing witnesses, investigating materials, or questioning of the parties concerned.

The patent relevant information may thus relate to public knowledge or working of the invention before the priority date. If the working of an invention is not possible on the basis of the detailed description, experimental reports may be submitted by a third person that demonstrate that the invention is not disclosed to such an extent as to allow the person skilled in the art to work the invention. Moreover, information can be provided if the patent claims do not conform with Section 36(6)(ii) JPL, because the features used are not usually used by the skilled person and the definitions or test or measuring methods may not be understood by the person skilled in the art.

No information can, however, be submitted with respect to a possible violation of public order and morality (Section 32 JPL). Moreover, no information can be provided if the applicant for patent who is not the inventor has not succeeded to the right to obtain a patent for the invention concerned (Section 49(vii) JPL) or if there were a violation of the requirement of unity of invention (Section 37 JPL).

The applicant will be informed by the JPO that a third party submitted patent relevant information, but not of its content, about which the applicant can learn by means of a file inspection. The applicant can comment on this information only in cases where the official examination report refers to it. The patent examiner and informant may not contact each other regarding the clarification of information, explanations to patentability etc. The informant may, if they wish, be informed about the utilization of the patent relevant information.

5.5.2
Attack on Granted Patents – Invalidation Trial (無効審判 *mukou shimpan*)

The only means to attack a granted patent is an invalidation trial pursuant to Section 123 JPL before the JPO. Other provisions of the JPL regarding trial (審判 *shimpan*) procedures are applicable. The second instance would be the Intellectual Property High Court in Tokyo. In extremely rare cases a further appeal may be raised to the Supreme Court of Japan.

In principle, anybody may demand a patent invalidation trial. A specific interest must not be shown except for two cases. Firstly, if a patent has been granted on a patent application filed by a person who is not the inventor and who has not succeeded to the right to obtain a patent for the invention concerned. Secondly, if not all joint owners of the patent application filed it with the JPO.

A patent invalidation trial may be demanded even after the extinguishment of the patent right. Where a patent invalidation trial has been demanded, the JPO will notify the exclusive licensee with respect to the patent right and other persons who have any registered rights relating to the patent accordingly. In principle, the reasons for invalidating a patent relate to the non-fulfillment of the patentability criteria set out above (in particular, lack of novelty or of inventive step, or it might not comply with the first-to-file principle, violate the principle of public order and morality, was filed against the will of the true inventor, etc.).

Where a trial decision that a patent is to be invalidated has become final and conclusive, the patent right shall be deemed never to have existed (Section 125 JPL). Any person who has an interest in the result of the invalidation proceedings may intervene in the trial to assist one of the parties, until the conclusion of the trial (Section 148 JPL). The intervenor may continue the invalidation proceedings even after the demand for the trial has been withdrawn by the original party. The request for invalidating a patent is to be submitted in writing to the JPO stating the following:

- the name and the domicile or residence of the demandant and his representative;
- an identification of the trial case;
- the relief sought in the demand and the grounds therefore.

The grounds shall concretely specify the fact constituting the grounds for invalidating the patent and state its relationship with the evidence for each fact required to be proven. When a correction trial is demanded, the corrected description, patent claim(s) or drawing(s) shall be attached to the written demand.

An amendment to the written demand shall not change the gist thereof. The amendment may be admitted, however, by the JPO by a ruling when there is clearly no apprehension that the amendment does cause unreasonable delay in the trial and was necessitated by the amendments of the patentee.

Namely, the patentee (defendant in a patent invalidation trial) may demand a correction of the description, the patent claims or drawings (Section 134bis JPL). In particular, the patentee may restrict patent claims, correct errors or incorrect translation, clarify any ambiguous statement in the description or claims.

Accordingly, the demandant may amend his grounds, if the amendment to the grounds for the demand was necessitated by the amendments the patentee had carried out.

The JPO may examine even invalidation grounds not pleaded by a party or an intervenor if it deems a demand for a correction by the patentee not admissible (Section 134bis JPL). In this case, when not allowing said demand for correction on the relevant ground, the JPO shall notify the parties and intervenors of the result of the trial examination and give them an opportunity to state their opinion thereon, designating an adequate time limit.

In general only supporting evidence may be supplied by the demandant. For example, if the demandant puts forward that the claimed product is identical with a product A publicly known before the priority date, further evidence can be submitted that proves the sales period for product A. In addition, witnesses may be questioned.

The correction of the name of the patentee or the indication of a further joint patentee not mentioned in the initial invalidation request are admissible amendments of an invalidation request.

A demand for an invalidation trial may be withdrawn before a trial decision becomes final and conclusive. However, after a written reply under Section 134 JPL has been submitted by the patentee, the demand for an invalidation trial may only be withdrawn with the consent of the adverse party. When a demand for a patent invalidation trial has been made with regard to two or more claims covered by a patent that has two or more claims, the demand may be withdrawn for any of the claims.

When the case is ready for the rendering of a trial decision, the JPO (*shimpan* division) shall notify the parties and the intervenors of the conclusion of the trial examination and render the decision within 20 days from this notification, unless the case is complicated or where there are unavoidable circumstances (Section 156 JPL). The trial decision shall be in writing and state the number of the trial, the name and the domicile or residence of the parties and the intervenors as well as of representatives; an identification of the trial case; the conclusions of the trial decision and the reasons therefore and the date of the trial decision. The decision will be transmitted to the parties, the intervenors and persons whose demand to intervene has been refused.

The *shimpan* division will declare each patent claim either invalid or it will reject the invalidation demand with respect to this claim. When a final and conclusive trial decision in a patent invalidation trial or an extended registration trial has been registered, no one may demand a trial on the basis of the same facts and the same evidence (Section 167 JPL). However, as long as new evidence is presented, anybody may demand a new invalidation trial on the basis of the same invalidation reasons.

The patent invalidation trial shall be conducted by oral trial examination; however, the JPO may decide to conduct the trial by documentary examination on a motion by a party or an intervenor or ex officio (Section 145 JPL). The oral examination shall be conducted in public unless public order or morality are liable to be injured thereby.

If the parties involved are not satisfied with JPO's decision on the invalidation demand, they may appeal to the Intellectual Property High Court (IP High Court, established on April 1, 2005). The term to be observed is 30 days (plus 90 days for residents abroad). The JPO will not be a party before the IP High Court.

Where the IP High Court finds for the plaintiff in the court action, it shall annul the trial decision or ruling (Section 181(1) JPL). Where the IP High Court recognizes that it is reasonable to order a further examination to be carried out in an action instituted against the trial decision with respect to the patent invalidation trial under Section 178(1) for invalidating that patent, and when the patentee has made a demand for a correction trial or intended to do so, it may annul the trial decision by a ruling for remitting the case to the trial examiner (Section 181(2) JPL). Before rendering this ruling the IP High Court shall hear the opinion of the party.

The IP High Court reexamines the lawfulness of the invalidity decision. The subject is thus limited to the invalidation reasons asserted before the JPO. However, new evidence relating to these invalidation reasons may be presented before the IP High Court.

In principle, a further appeal can be raised against a decision by the IP High Court to the Supreme Court. The Supreme Court will only decide on matters of law. The decision must either be legally incorrect or violate the Japanese Constitution. In practice, only a few IP cases are handled by the Supreme Court.

5.5.3
Correction Trials (訂正審判 *teisei shimpan*)

A granted patented may only be amended through a correction trial pursuant to Section 126 JPL before the JPO, unless an invalidation trial with regard to that patent is pending before the JPO since a granted patent may also be amended in an invalidation trial. Accordingly, no correction to a patent can be made before the IP High Court.

The patentee may demand a correction trial making a correction of the description, patent claim(s) or drawing(s) attached to the request. However, such correction is limited to the restriction of claim(s); the correction of errors in the description or of incorrect translation; and the clarification of an unambiguous description (Section 126 JPL). The correction of the description, patent claim(s) or drawing(s) may not substantially enlarge or modify the claim(s).

A correction trial may not be demanded during the period between the time when the patent invalidation trial has come to be pending before the JPO and the time when the trial decision has become final and conclusive. This is, however, not applicable within 90 days from the date when an action against the trial decision was instituted with respect to a patent invalidation trial.

A correction trial may be demanded even after the extinguishment of the patent right, unless the patent has been invalidated on a patent invalidation trial. Where there is an exclusive license, a pledge or a non-exclusive licensee, the patentee may demand a correction trial only with the consent of such person (Section 127 JPL).

Where a trial decision that the description, patent claim(s) or drawing(s) attached to the request are to be corrected has become final and conclusive, the patent application, the laying open of the application, the examiner's decision or the trial decision that the patent is to be granted, or the registration of establishment of the patent right shall be deemed to have been made on the basis of the corrected description, patent claim(s) or drawing(s) (Section 128 JPL).

5.6
The Patent Right

5.6.1
Term of Patent Right – Patent Term Extension

The term of the patent right shall be 20 years from the filing date of the patent application (Section 67 JPL). The term may be extended, upon application for registration of an extension, by a period up to five years if there was a period in which it was not possible to work the patented invention, because a considerable period of time was required on account of the necessity of obtaining an approval or other disposition that is governed by provisions in laws intended to ensure safety, etc. in the working of the patented invention. Essentially, this provision is applicable to pharmaceutical and agrochemical inventions. Extension of the term of the patent right must be applied for with the JPO. The application for term extension must state the following:
- the name and the domicile or residence of the applicant;
- the patent number;
- the term of the extension applied for (up to five years);
- particulars of the disposition as provided for in the Cabinet Order.

The application shall be accompanied by materials that give reasons for the extension, as provided for in an ordinance of METI. The application for registration of extension of the term of a patent right shall be made within the time limit prescribed by Cabinet Order counting from the date of obtaining the disposition provided for in the Cabinet Order. After the expiration of the term of a patent right the application cannot be made. Where a patent is owned jointly, each of the joint owners may not, except jointly with the other owners, apply for registration of an extension of the term of a patent right.

Where an application for registration of an extension of the term of a patent right is filed, the term of the patent right shall be deemed to have been extended. However, this provision shall not apply when the examiner's decision that the application is to be refused has become final and conclusive or when an extension of the term of the patent right has been registered.

It may, however, occur that it is impossible to obtain the disposition as provided for in the Cabinet Order by the day before six months prior to the expiration date of the term of a patent right of 20 years. In such case, a person desiring to apply for registration of an extension of the term of a patent right shall submit by that day to the Commissioner of the JPO a document stating the following matter:
- the name and the domicile or residence of the person desiring the application;
- the patent number;
- the disposition as provided for in Cabinet Order referred to in Section 67(2) JPL.

Where the document required to be submitted is not submitted, an application for the registration of an extension of the term of a patent right may not be made after six months prior to the date of expiration of the term of a patent right as provided for in Section 67(1) JPL.

After an extension has been granted, the patent is effective in the extended period only for the specific aspect to which the request for term extension relates, i.e., in particular only for the corresponding chemical compounds, the corresponding indication, etc.

5.6.2
Rights of Inventors and Patent Owners

5.6.2.1 Rights of Inventors

The inventor has, according to Section 36(1) JPL, the right to be named in the request for a patent application. A co-inventor can be a person that did not contribute to the mental concept of the invention but who contributed to its practical realization.[12] In the case where the right to obtain a patent is owned jointly, the patent may be applied for only jointly by all the joint owners (Section 38 JPL), i.e., it would mean a serious flaw if a patent were to be applied for by a person who did not obtain the right to apply for a patent by all the inventors. This would be a reason for invalidation that might lead to the revocation of a patent if invalidation proceedings were to be instituted.

The means by which the true inventor can protect themselves against the filing of a patent application for their invention by a third party who unlawfully appropriated the invention are limited. If the unlawful applicant refuses to transfer the application to the true inventor, and if a civil lawsuit regarding the transfer of the patent application or the patent granted thereon is unsuccessful, the true inventor can only try to destroy the patent right by means of invalidation proceedings before the JPO. The true inventor will, essentially, not be in a position to prevent any limitation or the withdrawal/abandonment of the patent application/patent by the unlawful applicant.

Inventors can either fully dispose of their inventions and the patent rights granted thereon (free inventors) or their rights may be restricted when they are employed inventors in private companies or work at national or private universities.

Inventions of employed inventors: The legal basis for the handling of inventions of employed inventors is Section 35 JPL,[13] whose current version came into effect on April 1,

---

12) "Mah-jong", Tokyo High Court, April 27, 1976; Case No. Showa 47 Gyo-Ke25; cf. IIC, Vol. 9 (1978) pages 48–51.

13) Section 35 JPL:
(1) An employer, a legal entity or a state or local public entity (hereinafter referred to as the "employer, etc.") shall have a non-exclusive license on the patent right concerned where an employee, an executive officer of a legal entity or a national or local public official (hereinafter referred to as the "employee, etc.") has obtained a patent for an invention which by reason of its nature falls within the scope of the business of the employer, etc. and an act or acts resulting in the invention were part of the present or past duties of the employee, etc. performed on behalf of the employer, etc. (hereinafter referred to as an "employee's invention") or where a successor in title to the right to obtain a patent for an employee's invention has obtained a patent therefore.

(2) In the case of an employee's invention made by an employee, etc. which is not an employee's invention, any contractual provision, service regulation or other stipulation providing in advance that the right to obtain a patent or the patent right shall pass to the employer, etc. or that he shall have an exclusive license on such invention shall be null and void.

2005. There was much litigation in Japan in the past between (former) employees and their (former) employers with regard to the amount of remuneration for the inventions made by the employees. Until a basic decision was rendered in 1999, the remuneration for employee's inventions was very low. Triggered by that decision, numerous court cases were filed by inventors. As a result, the courts awarded increasing amounts of remuneration up to the huge amount of ¥20 billion in the "Blue LED Case" that was awarded by the Tokyo District Court on June 30, 2004 to the inventor, Mr. Shuji Nakamura. However, in the appeal instance the amount was reduced in a settlement mediated by the Tokyo High Court to ¥608 million for all patent rights of Mr. Nakamura. The Tokyo High Court referred in the reasoning regarding the settlement to Section 1 JPL, according to which it is the purpose of the JPL to encourage inventions by promoting their protection and utilization so as to contribute to the development of industry.[14] The main focus should thus not be the interests of the inventor. It is thus expected that the remuneration awarded by Japanese courts will be lower in the future.

An invention is considered to be created because of the duty against the employer and thus an employee invention, if the time and place of the creation and completion of the invention cannot be determined, but if the invention belongs to the service duties of the employee and was made during the time of employment.

The transfer of the rights to the invention does not incur the obligation for the employer to file a patent application. The employer may also keep the invention secret and use it as internal know-how of the company. Moreover, the employer is not obliged to allow the employee to file foreign patent applications, if the employer has no own interest. Moreover, the employee is not entitled to information and cooperation in the patent prosecution. In Japan there is neither a guideline for remuneration of employee inventions nor a board of arbitration for these questions associated with the JPO as is the case in Germany.

As regards inventions at universities, it is noted that although Section 35 JPL is applicable to state employees in general; its application to university professors and other university members is still in dispute.

5.6.2.2 Rights of the Owner of a Patent Application or a Patent

With regard to the rights of a patent applicant or owner of a patent right it has to be differentiated whether the right is a patent application or a granted patent.

(3) The employee, etc. shall have the right to a reasonable remuneration when he has enabled the right to obtain a patent or the patent right with respect to an employee's invention to pass to the employer, etc. or has given the employer, etc. an exclusive right to such invention in accordance with the contract, service regulation or other stipulations.

(4) The payment of the remuneration in the preceding subsection, as provided for in the contract, service regulation or other stipulation, shall not be considered to be unreasonable, in view of the situation under which a negotiation is carried out between employer, etc. and employee, etc. in the course of establishing the criteria for determining the remuneration, the situation under which the criteria established are disclosed, and the situation under which the views of employee, etc. are heard for calculating the amount of the remuneration, etc.

(5) Where there is no stipulation with respect to remuneration referred to in the preceding subsection or where the payment of the remuneration determined thereby shall be considered to be unreasonable, the amount of the remuneration referred to in subsection (3) shall be determined, taking into consideration the amount of profits that the employer, etc. will make from the invention, the burden assumed and contribution made by the employer, etc. in connection with the invention, and the treatment upon the employee, etc. and other circumstances.

14) *AIPPI Journal*, (November 2005), Vol. 30(6), pages 311–314.

After the laying open of a patent application, the applicant is entitled to claim the payment of compensation, in a sum of money equivalent to what they would be entitled to receive for the working of the invention if the invention were patented, against a person who has commercially worked the invention (Section 65 JPL). To this end, the applicant has to give a warning with a written statement setting forth the contents of the invention claimed in the patent application. Even in the absence of the warning, the same shall apply to a person who commercially worked the invention before the registration of the establishment of the patent right, knowing that the invention was claimed in the patent application laid open for public inspection.

For international (PCT) patent applications designating Japan and being filed in a foreign language, the claim to compensation arises only after publication of the Japanese translation of the international application (Section 184^{decies} JPL).

The right to claim compensation may not be exercised until after establishment of the patent right is registered. The exercise of the right to claim the compensation does not preclude the exercise of the patent right. However, where a patent application has been abandoned, withdrawn or dismissed after the laying open of the patent application, or where the decision that the patent issued thereon is to be invalidated has become final and conclusive, the right to compensation shall be deemed never to have arisen.

The right to obtain a patent may be transferred, but it may not be the subject of a pledge; a joint owner of the right to obtain a patent may not assign their share without the consent of all the other joint owners (Section 33 JPL).

5.6.2.3 Rights of a Patentee

A patentee shall have an exclusive right to commercially work the patented invention. However, where the patent right is the subject of an exclusive license, this provision shall not apply to the extent that the exclusive licensee possesses exclusively the right to work the patented invention (Section 68 JPL). "Working" of an invention in JPL means the following acts:

- in the case of an invention of a product (including a program, etc. – hereinafter referred to as "product"), acts of manufacturing, using, assigning, etc. (meaning assigning and leasing; and including, where the product is a program, etc., its providing through electric telecommunication lines), exporting or importing or offering for assignment, etc. (including displaying for the purpose of assignment, etc. – hereinafter referred to as "assignment, etc.") of, the product;
- in the case of an invention of a process, acts of using the process;
- in the case of an invention of a process of manufacturing a product, acts of using, assigning, etc., or importing or offering for assignment, etc. of, the product manufactured by the process, in addition to the acts mentioned in the preceding paragraph.

Protection of a process thus extends to the product manufactured directly by the patented process.

The following needs to be observed in case of several applicants/right owners of a patent right (Section 73 JPL): Each of the joint owners of a patent right may neither transfer their share nor establish a pledge upon it without the consent of all the other joint owners. Moreover, each of the joint owners may, except as otherwise prescribed by contract, work the patented invention without the consent of the other joint owners. Each of the joint owners may grant neither an exclusive license nor a non-exclusive license without the consent of all the other joint owners.

5.6.3
Licensing of Patent Rights[15]

Licenses in patent rights are regulated in Sections 77–99 JPL. It is essential to differentiate between two kinds of licenses, namely exclusive licenses (専用実施権 *senyou jisshiken*; specific right to work the invention; in the following *senyou* license) and non-exclusive licenses (通常実施権 *tsuujou jisshiken*; ordinary right to work the invention; in the following *tsuujou* license). Upon concluding a *senyou* license, the licensee is awarded with essentially all rights that belong to the patentee. In contrast, the *tsuujou* license as simple license is essentially an agreement pursuant to the law of obligations that is only effective among the contract partners. If the ordinary license is, however, registered with the JPO it may be hold against all later owners of the protective right or a *senyou* licensee. Guidelines for patent and know-how licensing agreements (under the Antimonopoly Act), Fair Trade Commission, 1998, are published in English at www.jftc.gojp/e-page/guideli/patent99.htm.

Patent applications can also be licensed. The corresponding license contract ends, however, automatically once the application has been rejected by the JPO.[16] An exclusive license contract relating to a patent application finally rejected will not be null and void;[17] the payment obligations only end on the day on which the patent application was rejected. Know-how can only be the subject of a license contract if it can be worked.[18] Otherwise, the license contract can be terminated.

License contracts end by the expiration of the term stipulated in the contract or by termination. In the case of expiration of the stipulated term, the courts might examine whether a non-prolongation might lead to a considerable damage for the licensee with regard to any investments made. Termination of a license contract by the licensor because of nonworking of the licensed subject is only possible if it is explicitly allowed for in the contract. Revocation of the patent rights upon which the license contract is based is a reason for termination even in the absence of any explicit stipulation. Sublicenses automatically end together with the license contract.[19]

5.6.3.1 *senyou* Licenses
A 専用 *senyou* license requires the following:
- The licensor must transfer an undivided interest in at least part of the patent right. The *senyou* license can be divided temporarily, geographically or with respect to its technical field. Joint *senyou* licensees have under Section 77(5) JPL the same rights and duties as joint patentees under Section 74 JPL. They can only jointly transfer or execute patent rights.

---

15) Christopher Heath, Chapter 6, "Technology Transfer in Japan", in *Legal Rules of Technology Transfer in Asia*, Eds. Christopher Heath Kung-Chung Liu, Kluwer Law International, 2002, pages 99–137.
16) Supreme Court of Japan, October 19, 1993; GRUR Int. 1995, 341.
17) Tokyo High Court, July 20, 1997, 868 Hanrei Jihô 46.
18) "Mangala", Kobe District Court, September 25, 1985; 575 Hanrei Times 52.
19) "Hummel", Osaka District Court, December 9, 1987, 1268 Hanrei Jihô 130.

- The licensor has no right to work the invention on his own.
- The license must be registered with the JPO. If a later concluded license agreement were to be registered, the *senyou* licensee cannot cancel this later concluded license agreement. Instead, the licensor may be sued because of breach of contract.

A *senyou* license may be transferred together with the corresponding business unit to which it belongs or in the case of a universal succession (merger, inheritance). Otherwise, the agreement of the patent owner is required for any transfer. A *senyou* license does not prevent the licensee instituting invalidity proceedings against the patent.

5.6.3.2 *tsuujou* Licenses

There are several kinds of ordinary (通常 *tsuujou*) licenses. Most common is a non-exclusive license based upon a contract with the right owner. A *tsuujou* licensee has in general no independent standing to sue patent infringers before the courts. In the case of an exclusive *tsuujou* license (i.e., a non-registered exclusive license), the licensee may, however, institute court proceedings to claim damages (no injunction may be requested). In principle, compulsory licenses are available under the JPL (see Sections 83(1), 92, 93 JPL). There is, however, no practical relevance.

5.6.4
Further Aspects of the Patent Right

5.6.4.1 Limits of the Patent Right

The effects of the patent right shall not extend to the working of the patent right for the purpose of experiment or research (Section 69(1) JPL). This means that the utilization of a patented invention for research purposes at universities, state or private institutes or the industry cannot be prevented by the patentee or exclusive licensee. Moreover, experiments utilizing the patented invention, for example, for education purposes in schools may also not be prohibited.

It was long disputed in Japan whether the exemption of Section 69(1) JPL meant that clinical trials or field experiments that are conducted during the term of a patent right may be carried out by a third person with the purpose of obtaining an official permission for the production and sale of patented pharmaceuticals or agrochemicals. These discussions were essentially concluded by the ruling of the Supreme Court of Japan in the case "Ono Pharmaceutical Co. Ltd. vs. Kyoto Pharmaceutical Co." of April 16, 1999. The Supreme Court ruled that the conducting of clinical trials during the patent term does not constitute patent infringement. It was decisive for the court that if such clinical trials were not to be allowed during the patent term the third party could, as a result, not utilize the patented invention for a considerable time even after the termination of the patent term. This would be in violation of the principles of the patent system.

5.6.4.2 Exhaustion of the Patent Right

As soon as the patented product or the product obtained directly through the patented process has been put on the market in Japan by the patentee or with his consent, the patent right will be exhausted; no rights remain with the right owner with respect to ensuing use actions like, for example, sale or rental ("national exhaustion"). However, the refill of patented ink cartridges is an act of infringement both regarding the product patent of the ink tank and the method patent for the manufacture of ink tanks.[20] The new use of a disposable camera by a third party who substituted the film and sold it again was found to constitute patent infringement.[21]

The import into Japan of products that are protected by a Japanese patent is admissible, if these products were put on the market outside of Japan by the patentee or with his consent.[22] If the patentee clarifies, however, and indicates it on the products covered by the patent, that Japan is to be excluded from the sales and use region, parallel import is unlawful.

5.6.4.3 Prior User Rights

A right to a working of the patented invention in the sense of a non-exclusive license is given to a person who, without knowledge of the contents of an invention claimed in a patent application, made the invention by himself or has learnt the invention from another person who has made the invention without knowing thereof, and has been commercially working the invention or has been making preparations therefore in Japan actually at the time of filing of the patent application (Section 79 JPL). This non-exclusive license is limited to the extent of the patent that is being worked or for which preparations for working are being made and to the purpose of such working or the preparations therefore.

5.6.4.4 Further Limitations (Sections 69(2)(3) JPL)

The patent right is furthermore limited in that the effects of the patent right do not extend to the following:
- vessels or aircraft merely passing through Japan or machines, instruments, equipment or other accessories used therein;
- products existing in Japan prior to the filing of the patent application.

Moreover, the effects of the patent right for inventions of medicines (namely, products used for the diagnosis, cure, medical treatment or prevention of human disease) to be manufactured by mixing two or more medicines or for inventions of processes for manufacturing medicines by mixing two or more medicines shall not extend to acts of preparing medicines in accordance with the prescriptions of physicians or dentists or to medicines prepared in accordance with the prescriptions of physicians or dentists.

5.6.5
Interpretation of Patent Claims[23]

The scope of protection the patentee enjoys from the patent is reflected by the "technical scope of a patented invention" (特許発明の技術的範囲 *tokkyo hatsumei no gijutsuteki*

---

[20] "Canon Inc. vs. Recycle Assist Co., Ltd."; IP High Court, January 31, 2006.
[21] Tokyo District Court, August 31, 2000; cf. Toshiko Takenaka, in *CASRIP Newsletter* (Univ. of Washington, Seattle, WA) (Autumn 2000), page 9.
[22] Supreme Court of Japan, July 1, 1997.
[23] Cf. Toshiko Takenaka, *Interpreting Patent claims: the United States, Germany, and Japan*, IIC Studies 1995, Vol. 17, pages 193ff.

hanii), which shall be determined on the basis of the patent claims (Section 70 JPL). The meaning of a term or terms of the patent claim(s) shall be interpreted in the light of the description and the drawing(s). No statements of the abstract attached to the request shall be taken into account for this purpose. As long as the wording of a claim is clear, the invention is determined by this wording and the description of the invention is not to be considered. Only under particular circumstances, when the technical meaning of the claims cannot be understood without ambiguity or in case of obvious clerical errors, is the inclusion of the patent description admissible.[24]

The technical scope of a patented invention may be interpreted in accordance with the wording of the claim ("literal interpretation") or by means of the doctrine of equivalence ("equivalent interpretation") which will lead to an interpretation beyond the literal meaning of the claims. In rare cases the claim interpretation remains below the literal meaning of the claims.

5.6.5.1 Literal Claim Interpretation

To obtain the scope of protection according to the wording of a claim, often the expressions used in a patent claim need to be interpreted. To this end, the gist (要旨 *youshi*) of the invention, the object underlying the invention (目的 *mokuteki*), any file wrapper estoppel (禁反言 *kinhangen*), the state of the art, referral to other applications or patent, opinions etc., will be taken into account.

In case the gist of the invention is described in the description more broadly than is apparent from the patent claims, this may not lead to a broadened interpretation of the patent claims. The reverse case is however possible. A too narrow description of the object underlying an invention might be used to narrow the scope of protection for a broad claim. It may thus be advisable, not to include too many objects into the description.

The patentee might have made statements that affect the scope of protection. In such a case, the courts may use such statements to remove a part from the technical scope of the invention, even if such a statement did not result in an amendment to the claims or description.

To interpret the wording of a claim, recourse was often taken by the courts to the state of the art. This usually led to a restriction of the claims. The reason behind was the fact that the courts were obliged to consider granted patents valid, even if reasons for their invalidity existed. However, since the decision of the Supreme Court of Japan of April 11, 2000, the invalidity of a patent right is considered in general by an infringement court. The court can now reject an infringement suit on the basis that there are reasons to invalidate the patent.

Japanese courts interpret functional claims similarly to structural claims. There is no limitation to the expressly disclosed embodiment in the patent description.

5.6.5.2 Claim Interpretation by the Doctrine of Equivalence

The extension of the scope of protection beyond the literal meaning of the claims has long been practiced in Japan, although the term "doctrine of equivalence" was not referred to in most cases. The application of the doctrine was, however, essentially limited to the me-

---

24) Supreme Court of Japan, March 8, 1991.

chanical field. The first decision regarding the application of the doctrine of equivalence in the field of chemistry/biotechnology by a Japanese High Court was rendered on March 29, 1996 in the "Genentech vs. Sumitomo Pharmaceuticals" case.

The first comprehensive decision regarding the doctrine of equivalence was rendered by the Supreme Court of Japan on February 24, 1998 in the "Ball Spline Bearing" case. The Supreme Court outlined for the first time clearly basic conditions for applying the doctrine of equivalence in Japanese patent infringement litigation.

In principle, when differences can be found in the claimed features in comparison with the contested embodiment (infringing product or process), this embodiment can be said to be outside the scope of protection. Nevertheless, even when any difference can be found between the features of the claim and the features of the contested embodiment, the contested embodiment would fall within the technical scope of protection of the patent as an equivalent form to the claimed invention if all of the following requirements are met:

- The difference does not relate to the essential part of the patented invention.
- The object of the patented invention can be achieved in the same manner, and the function and result of the patented invention can also be achieved by the contested embodiment, even when the feature of the claim that is different from the contested embodiment is replaced with the corresponding feature of the contested embodiment.
- The person skilled in the art could easily arrive at this replacement at the time of manufacturing this contested embodiment.
- The contested embodiment is not identical with the prior art at the filing date of the patent application for the protected invention, or the contested embodiment is not obvious to a person skilled in the art in view of this prior art.
- The contested embodiment was not intentionally excluded from the technical scope of the invention by the applicant in the patent examination proceedings.

5.7
Enforcement of Patent Rights

5.7.1
Remedies for Patent Infringement

5.7.1.1 Injunction – Permanent and Preliminary

A patentee or an exclusive licensee may require a person who is infringing or is likely to infringe the patent right or exclusive license to discontinue or refrain from such infringement (Section 100 JPL). In addition, the destruction of articles by which an act of infringement was committed (including articles manufactured by an act of infringement in the case of a patented invention of a process of manufacture), the removal of the facilities used for the infringing act, or other measures necessary to prevent the infringement, may be demanded.

Indirect patent infringement is also unlawful (Section 101 JPL). The following acts are thus deemed to be an infringement of a patent right or exclusive license:

- In the case of a patent for an invention of product, acts of manufacturing, assigning, etc., or importing or offering for assignment, etc. of, in the course of trade, things to be used exclusively for the manufacture of the product.
- In the case of a patent for an invention of product, acts of manufacturing, assigning, etc., or importing or offering for assignment, etc. of, in the course of trade, articles to be used for the manufacture of the product (excluding those which are generally distributed in Japan) and indispensable for solving the problems through the invention concerned, knowing that the invention is a patented invention and that the articles are to be used for the working of the invention.
- In the case of a patent for an invention of a process, acts of manufacturing, assigning, etc., or importing or offering for assignment, etc. of, in the course of trade, things to be used exclusively for the working of such invention.
- In the case of a patent for an invention of a process, acts of manufacturing, assigning, etc., or importing or offering for assignment, etc. of, in the course of trade, articles to be used for the use of such process (excluding those which are generally distributed in Japan) and indispensable for solving the problems through the invention concerned, knowing that the invention is a patented invention and that the articles are to be used for the working of the invention.

In the case of a patent for an invention of a process of manufacturing a product, where such a product was not publicly known in Japan prior to the filing of the patent application concerned, any identical product shall be presumed to have been manufactured by that process (Section 104 JPL).

The patentee or exclusive licensee will generally seek a permanent injunction. However, very often it is important to stop immediately infringing activities. This can be achieved by filing with a court a request for a preliminary injunction (仮処分 *karishobun*). A preliminary injunction will, however, be granted in a patent infringement case only if the legal situation is not complicated and if there are no significant doubts in the validity of the patent. This occurs in particular with technically simple cases.

5.7.1.2 Damages

The patentee or exclusive licensee may also claim damages for patent infringement from a person who has intentionally or negligently infringed the patent right, as compensation for damage caused to him by the infringement. The JPL provides several methods for the determination of the amount of damages (Section 102 JPL):

- The sum of money which results when the profit of the right owner per unit of the infringing articles is multiplied by the number of articles which the patentee or exclusive licensee could have sold in the absence of the infringement may be estimated as the amount of damage suffered by the patentee or exclusive licensee within a limit not exceeding an amount attainable depending on the working capability of the patentee or exclusive licensee. Where there is any circumstance that prevents the patentee or exclusive licensee from selling part or the whole of the number of articles, a sum equivalent to this number of articles shall be deducted.

- The profits gained by the infringer through the infringement.
- An amount of money he would be entitled to receive for the working of the patented invention ("license analogy").
- The preceding subsection shall not preclude a claim to damages exceeding the amount referred to therein. In such a case, where there has been neither willfulness nor gross negligence on the part of the person who has infringed the patent right or the exclusive license, the court may take this into consideration when awarding damages.

A person who infringes a patent right or exclusive license of another person shall be presumed to have been negligent as far as the act of infringement is concerned (Section 103 JPL).

The application of the above criteria by the Japanese courts has led in the past years to huge amounts of damage awards. For example, in the two Pachisuro cases handed down by the Tokyo District Court on March 19, 2002, the damages awarded amounted to ¥8.4 billion (i.e., about €65 million). The court applied the first of the above-mentioned methods and ruled, interestingly, that the working capacity of the patentee refers to the working capacity during the term of the patent, and not to the working capacity at the time of infringement.

5.7.1.3 Recovery of Reputation

Upon the request of a patentee or an exclusive licensee, the court may, in lieu of damages or in addition thereto, order a person who has injured the business reputation of the patentee or exclusive licensee by infringing the patent right or exclusive license, whether intentionally or negligently, to take the measures necessary for the recovery of the business reputation (Section 106 JPL).

5.7.2
Procedural Aspects of the Enforcement before the Courts

5.7.2.1 Jurisdiction and Standing to Sue

Patent infringement litigation is to be started either before the Tokyo District Court or the Osaka District Court, depending in general on the location of the presumed infringer or the infringing act. If either party is not satisfied with the outcome, an appeal may be raised to the Intellectual Property High Court in Tokyo, which will re-examine the case. In very rare cases, a further appeal may be raised to the Supreme Court of Japan, which will deal, however, only with legal matters.

In general the litigation will be started either by the patentee or the exclusive licensee. A non-exclusive licensee has, usually, no standing to sue. In a recent Supreme Court decision, it was clarified that a patentee shall not lose their right to seek injunctive relief even when they have granted an exclusive, registered license (専用実施権 *senyou jisshiken*) to a third party.[25]

The owner of a non-exclusive license, whether registered or not, has in general no standing to sue against alleged infringers. Sometimes a suit for damages is, however, admitted.

---

25) Supreme Court of Japan, June 17, 2005; Case No.
   Hei 16 (Ju) 997; cf. S. Tanaka, N. Itai, in *AIPPI Journal*, Vol. 30(6) (2005).

5.7.2.2 Before any Court Proceedings

In general, the patentee or the exclusive licensee will transmit a warning letter to the alleged infringer. This might avoid litigation or the infringer might at least no longer contend that he acted in good faith. However, if a preliminary injunction is sought against the patent infringing activities, a warning letter might be disadvantageous in that it can indicate a lack of urgency. Although Section 187 JPL prescribes the marking of patented products, the lack of doing so has been, so far, irrelevant in patent infringement proceedings.

The JPO can give, upon request by the patentee, licensee or a party accused of patent infringement, an opinion (判定 *hantei*) as to whether a particular product or process comes under the technical scope of the patented invention (Section 71 JPL). The existence of an actual dispute is not necessary and the product or process to be put forward for comparison must not really exist. The opinion will usually depend on documents presented by the parties. The trial examiners may, however, hold oral proceedings for the presentation of evidence or arguments. A *hantei* is an official opinion of the JPO and should thus be carefully considered by the courts. There is, however, no binding effect in that the *hantei* is considered to be one expert opinion among others possible. No appeal is available against a *hantei*.

5.7.2.3 Attorneys-at-Law (弁護士 *bengoshi*) and Patent Attorneys (弁理士 *benrishi*)

Infringement litigation will be handled in general through the cooperation of an attorney-at-law with a patent attorney. Representation by a patent attorney is not admissible. In Japan, there is, however, no obligation to be represented by an attorney.

As regards patent matters, a patent attorney in particular represents clients in all proceedings before the JPO and the corresponding appeals before the Intellectual Property High Court. In patent infringement matters, a patent attorney can cooperate with an attorney-at-law. Recently, a patent attorney can now appear independently before the courts as long as the representation of a party by an attorney-at-law is clear.

5.7.2.4 Court Proceedings

The ways court proceedings are handled have changed considerably over the years. Patent infringement suits proceed now much more efficiently than before and enable an efficient enforcement of patent rights. In Japan there is no "pre-trial discovery" as in the U.S.A. Obtaining and securing evidence might thus be difficult, if the infringing products are not directly available.

Documentary evidence and witness evidence are in addition to the questioning of the parties the most common means to obtain evidence. To clarify complicated issues, experts will often be involved. Until recently, the possibility for the patentee to obtain evidence from the other party was very restricted. In recent years the Japanese Law of Civil Procedure (JLCP) has, however, been revised completely. Many amendments to the JLCP as well as the JPL allow now an easier obtainment of evidence and allow the streamlining of the court proceedings, as illustrated by the introduction of the "principle of concentrated taking of evidence" of Section 182 JLCP on January 1, 1998. The following details illustrate the present situation:

In Japan, the examination and decision on an asserted patent infringement belongs to the realm of the civil courts while any decision relating to the validity of the patent right is to be made by the JPO. Accordingly, a court handling a patent infringement matter cannot rule on the validity of a patent. Moreover, until recently it had also to assume the validity of the patent. In 2000, the Supreme Court of Japan ruled, however, that a court handling a patent infringement matter should also examine whether there are obvious reasons for invalidating the patent. If this were the case the infringement suit were to be rejected because of an abuse of right (権利の乱用 *kenri no ranyou*) pursuant to Section 1a of the Japanese Civil Law. Thereafter, many courts have applied this ruling. Finally, effective April 1, 2005, Section 104ter JPL was implemented:

- In a litigation relating to the infringement of patent right or exclusive license, the patentee or the exclusive licensee shall not exercise the right to the other party, where the patent is considered to be invalidated on a patent invalidation trial.
- With respect to the means of attack or defense under the preceding subsection, the court may, upon motion or ex officio, render a ruling of dismissal, if the submission involving such measures is considered to have been made for the purpose of causing unreasonable delay in the trial.

The infringement court has several means to assure speedy and efficient court proceedings: In a litigation directed to the infringement of a patent right or exclusive license, where denying the material allegation made by a patentee or an exclusive licensee to the effect that an act of infringement is committed with reference to an article or process, the other party shall clarify his relevant act in a concrete manner. However, this provision shall not apply when the other party has an adequate reason for preventing him from disclosing the same (Section 104bis JPL).

Where the court orders, upon the request from a party, an expert opinion to be given with respect to the matters necessary for the proof of the damages caused by the infringement, the other party shall explain to the expert the matters necessary for the expert opinion to be given (Section 105bis JPL).

Where it is recognized that the damage was caused in a litigation relating to the infringement of a patent right or exclusive license, the court may award the reasonable amount of damages, based on the entire purport of the oral argument and the result of the taking of evidence when it is extremely difficult to prove facts necessary for the proof of damages from the nature of such relevant facts (Section 105ter JPL).

The Japanese Constitution prescribes public proceedings. This might be a problem in patent infringement proceedings that involve the disclosure of precious know-how. The presumed infringer risks losing precious know-how and the right owner might be faced with a strong defense by the presumed infringer who asserts a danger of losing precious know-how.

However, recent amendments to the JPL have improved the situation. The court may now issue a secrecy order, upon the request of a party, not to use the trade secret for the purpose other than for prosecuting the litigation or not to disclose the trade secret to a person other than those ordered (Sections 105^{quater}, 105^{quinquies} JPL).

Moreover, as a further means to protect trade secrets, the inspection of court records may be limited (Section 105^{sexies} JPL). Finally, the public can be excluded under particular circumstances from the court proceedings (Section 105^{septies} JPL). To examine the matters concerned privately, the court shall declare to such effect together with the reason therefore before causing the public to withdraw from the courtroom. When the examination of the matters concerned is terminated, the court shall cause the public to be admitted to the courtroom again.

As a rough estimate the court proceedings in the first instance before the Osaka District Court or the Tokyo District Court and the Intellectual Property High Court will each last about one year. In principle, a further appeal may be raised to the Supreme Court, which will, however, take up the case only if there were a serious violation of the law or if a basic decision is required. There must be either a violation of the Japanese constitution or a severe violation of laws or administrative prescriptions. Appeals to the Supreme Court in patent litigation matters are, thus, seldom.

II
Japanese–English Dictionary

6
Dictionary Structure and Explanations

6.1
General Explanations

The dictionary of chemical terms is based on a compilation of scientific words that includes over 60 000 Japanese entries and English translations. Various sources for chemical terms were utilized, in particular patents and publications in scientific journals. In addition, textbooks, scientific dictionaries and printed dictionaries with general Japanese words as well as internet publications and online dictionaries were used to complete and verify the compilation. The collection contains terms from chemistry, i.e., organic and inorganic chemistry, physical chemistry, biochemistry and polymer science, and also from surroundeding fields, covering aspects from medicinal chemistry, physics, mineralogy and chemical engineering. Based on this compilation, a selection of approximately 15 300 terms was made for this book. As criteria, frequency of use, relevance for text understanding, importance for chemistry and diversity were applied. The selection focuses on basic organic, inorganic and macromolecular chemistry, in particular words from general chemistry, names of minerals, polymers and organic compounds, expressions for general chemical transformations and reaction types, terms describing physical properties of substances and physicochemical methods. In addition, some essential terms from life sciences and engineering have been added.

All expressions were taken from one of the above-mentioned sources. However, not all words represent the most accurate scientific term as there is some tolerance in their use in literature from Japan. Additionally, terms are used to describe slightly differing phenomenon. Consequently, it is helpful to incorporate the whole text passage in the translation to identify the exact meaning. It is also recommended that care be taken in translating expressions from English to Japanese as this is not the objective of this dictionary and in many cases a different term may specify more precisely the English meaning.

To facilitate the location searched for terms, words are arranged in three parts within this dictionary. The first part includes all expressions beginning with *kana*. As knowledge of reading *hiragana* and *katakana* is presumed, terms can easily be identified by following the conventional Latin alphabet. Words beginning with *kanji* are listed in the second and third part of the dictionary. Two *kanji* dictionaries are useful, based on the experience that some characters are more easily recognized than others after some time of text translation. For chemists, these are *kanji* representing figures, chemical elements and some specific chemical meanings, such as 化, 熱 and 液. Therefore, a small number of characters at the beginning of chemical terms form almost half of all dictionary entries. As they will be easily recognized after a short period of translation experience, it is a time saving process to look only at the short list of characters provided in Section 6.3. As most translators certainly know one of the standard character dictionaries, the established system of *kanji* identification by radicals and stroke counts is applied for all other characters in the third part of the dictionary.

Very similar words were not always selected when they can easily be translated from related terms, such as less important salt names (e.g. 過沃素酸 periodic acid is included, but not 過沃素酸塩 periodates). Some expressions can be written in *hiragana*, *katakana* and *kanji*, for instance the element names of phosphorus (燐, リン or りん), boron (硼, ホウ or ほう), silicon (珪, ケイ or けい) and iodine (沃, ヨウ or よう). In the dictionary, depending on clarity, either both *kana* and *kanji* terms are listed (either in different parts or within one line) or only one alternative is given. In all dictionary parts, the reading of *kanji* and *kana* is indicated immediately behind the Japanese characters in italicized letters. As described in Section 1.2, the Hepburn system is applied to transcribe into ローマ字. The reading of all *kana* characters is given in Section 1.2.2. For chemical compounds, the dictionary also contains the chemical formulas, either as empirical formula or with semi-structural elements, the latter ones for organic compounds are used to point out functional groups.

6.2
Dictionary Part I: Scientific Terms beginning with *kana*

The first part of the dictionary displays scientific terms beginning with *kana* according to their reading. While the word order in dictionaries published in Japan follows the Japanese sound systems (the succession of characters in the *kana* syllabary is given in Fig. 2.5; see Section 2.6.1), the word order in this book is based on the Latin alphabet (Fig. 6.1) to facilitate the finding for non-native Japanese speaking users. There is no difference if the word is written with *katakana* or *hiragana*. Long vocals represented by a vowel extender mark line (Section 1.2.2) are handled like short vocals. The small *kana* ッ in germate consonants is considered as the consonant it represents, such as "k" in ッカ.

a	ア							
b	バ	ベ	ビ	ボ	ブ	ビャ	ビョ	ビュ
c	チャ	チェ	チ	チョ	チュ			
d	ダ	デ	ディ	ド	デュ			
e, f	エ	ファ	フェ	フィ	フォ	フ		
g	ガ	ゲ	ギ	ゴ	グ	ギャ	ギョ	ギュ
h	ハ	ヘ	ヒ	ホ	ヒャ	ヒョ	ヒョ	
i, j	イ	ジャ	ジ	ジョ	ジュ			
k	カ	ケ	キ	コ	ク	キャ	キョ	キュ
m	マ	メ	ミ	モ	ム	ミャ	ミョ	ミュ
n	ナ	ネ	ニ	ヌ	ニャ	ニョ	ニュ	
o	オ							
p	パ	ペ	ピ	ポ	プ	ピャ	ピョ	ピュ
r	ラ	レ	リ	ロ	ル	リャ	リョ	リュ
s, t	サ	セ	シャ	シ	シェ	ショ	シュ	ソ
	ス	タ	テ	ティ	ト	ツ		
u, v	ウ	ヴァ	ヴェ	ヴィ	ヴォ	ヴ		
w	ワ	ウェ	ウィ	ウォ				
y, z	ヤ	ヨ	ユ	ザ	ゼ	ゾ	ズ	

Fig. 6.1 Order of syllables in Chapter 7 (with *katakana* as an example).

6.3
Dictionary Part II: Scientific Terms beginning with Basic *kanji*

A huge number of scientific terms begin with few different characters. To facilitate the identification of terms within the dictionary, words beginning with a small set of *kanji* are included in the second part of the dictionary. Among these characters are *kanji* that are easy to remember and identify (i.e., characters for numbers) as well as *kanji* which are important for chemical terms, such as characters for elements and frequently occurring prefixes. Figure 6.2

1. Characters for figures and quantities

一	二	三	四	五	六
1	2	3	4	5	6

七	八	九	十
7	8	9	10

半	単	原	多
half, semi	one, single	mono	poly

2. *kanji* for chemical elements (ordered by periodic table group and atomic number)

硼	炭	珪	錫	鉛	窒
boron (^{5}B)	carbon (^{6}C)	silicon (^{14}Si)	tin (^{50}Sn)	lead (^{46}Pb)	nitrogen (^{7}N)

燐	砒	酸	硫	弗	塩
phosphorus (^{15}P)	arsenic (^{33}As)	oxygen (^{8}O)	sulfur (^{16}S)	fluorine (^{9}F)	chlorine (^{17}Cl)

臭	沃	鉄	銅	銀	金
bromine (^{35}Br)	iodine (^{53}I)	iron (^{26}Fe)	copper (^{29}Cu)	silver (^{47}Ag)	gold (^{79}Au)

3. Characters frequently appearing in the initial position of chemical terms (ordered by stroke number)

化	水	中	分	石	生
change, -ization	water	neutral, center, inside	part, degree, minute, rate, change	stone	bio-, life

平	立	気	気	光	芳
stereo, three dimensional	stand up	exist, occur	spirit	light, ray	aromatic

放	重	核	配	流	粒
emit, release	heavy	nucleus	ligand, coordination	flow, current	particle, grain, drop

液	硝	酢	結	電	溶
liquid, fluid	nitrate	sour, vinegar	bond, linkage	electricity	melt, dissolve

蒸	燃	熱
steam, heat	burn, glow	heat, temperature

Fig. 6.2 Order of syllables in Chapter 8 (with *katakana* as an example).

6.4 Dictionary Part III: Further Scientific Terms beginning with kanji | 245

4. *kanji* for important prefixes for chemical terms (ordered by stroke number)					
不	同	亜	非	過	脱
anti-, non-	iso-, equal	next, -ous	non-	per-	de-, remove, get rid of
陰	陽	等	超	無	
negative	positive	iso-, homo, class	ultra-, super-	non-	

Fig. 6.2 *(continued)*

provides the *kanji* order of all characters in the initial position of scientific terms from the second part of the dictionary.

To facilitate the identification of terms where large numbers of entries begin with the same *kanji*, there is a further subdivision for the following four characters:

- entries beginning with 炭 are subdivided into four classes:
  1. terms beginning with 炭化
  2. terms beginning with 炭酸
  3. terms beginning with 炭素
  4. terms beginning with 炭水
- entries beginning with 酸 are subdivided into two classes:
  1. all terms beginning with 酸 except 酸化
  2. terms beginning with 酸化
- entries beginning with 塩 are subdivided into two classes:
  1. all terms beginning with 塩 except 塩化
  2. terms beginning with 塩化
- entries beginning with 電 are subdivided into four classes:
  1. all terms beginning with 電 except 電気, 電子 and 電解
  2. terms beginning with 電気
  3. terms beginning with 電子
  4. terms beginning with 電解

6.4
Dictionary Part III: Further Scientific Terms beginning with *kanji*

Scientific terms that do not begin with one of the basic characters, which are described in Section 6.3, are given in the third part of the dictionary. They are arranged according to a system based on radicals and stroke counts as introduced by Spahn and Hadamitzky in *Japanese Character Dictionary*. For a general description of radicals and stroke counts see Section 2.6.2.

kanji that do not contain any radicals are listed in Section 9.1, according to Fig. 6.3, in the order of their stroke numbers.

2 strokes	入 2a						
3 strokes	工 3a	寸 3b	大 3c	丸 3d			
4 strokes	元 4a	予 4b	互 4c	太 4d	天 4e	内 4f	毛 4g
5 strokes	包 5a	凸 5b	凹 5c	必 5d	斥 5e	左 5f	出 5g
	本 5h	末 5i	未 5j	失 5k	弁 5l	甘 5m	母 5n
6 strokes	両 6a	朱 6b	再 6c	曲 6d			
7 strokes	束 7a	寿 7b					
8 strokes	長 8a	表 8b	画 8c	果 8d	毒 8e	事 8f	
9 and more strokes	発 9a	射 10a	残 10b	疎 11a	野 11b		

Fig. 6.3 Characters without radicals in order of their appearance in Section 9.1.

Spahn and Hadamitsky used 79 radicals to arrange all other Chinese characters. Based on this radical set, each character in Section 9.2 is classified according to the radicals it contains. Because *kanji* usually contain two or more radicals, the main radical has to be determined. Some general rules for the identification of the main radical are explained in Section 2.6.2. Radicals are put in order by the number of strokes, beginning with two strokes. Figure 6.4 lists all radicals, including some variants, in the right order as they are used in the dictionary. For readers that are familiar with the *Japanese Character Dictionary*, the same labeling is used for the radicals. *kanji* having the same radical are arranged in increasing order of number of strokes in the residual part of the character. As an example, the main radical of the character 磁 is 石, having 5 strokes and the label "5a". The number of strokes in the residue is nine. Having found words beginning with a specific *kanji*, all terms with this character are ordered by their reading. However, because there is only a small number for each initial *kanji*, all words can easily be overlooked.

2 strokes	亻,人 2a	冫 2b	孑,子 2c	阝 2d	卩 2e	刂,刀 2f	力 2g
	又 2h	冖 2i	亠 2j	十 2k	卜 2m	儿,𠂉 2n	丶,八 2o
	厂 2p	辶,廴 2q	冂 2r	几 2s	匚 2t		
3 strokes	氵,水 3a	土 3b	扌,手 3c	卩 3d	女 3e	巾 3f	犭,犬 3g
	弓 3h	彳 3i	彡 3j	艹 3k	宀 3m	尚,小 3n	山 3o
	士 3p	广 3q	尸 3r	口 3s			
4 strokes	木 4a	月 4b	日 4c	火,灬 4d	礻,示 4e	王 4f	牛 4g
	方 4h	攵 4i	欠 4j	忄,心 4k	戸 4m	戈,弋 4n	
5 strokes	石 5a	立 5b	目 5c	禾 5d	衤 5e	田 5f	罒 5g
	皿 5h	疒 5i					
6 strokes	糸 6a	米 6b	舟 6c	虫 6d	耳 6e	竹 6f	
7 strokes	言 7a	貝 7b	車 7c	𧾷,足 7d	西 7e		
8 strokes	金 8a	食 8b	隹 8c	雨 8d	門,鬥 8e		
9 and more strokes	頁 9a	馬 10a	魚 11a	鳥 11b			

Fig. 6.4 Radicals and their variants for identification of *kanj* in Section 9.2.

7
Dictionary Part I: Scientific Terms Beginning with *kana*

For explanations see Chapter 6.

ア a

アバメクチン *abamekuchin*
abamectin $C_{48}H_{72}O_{14}/C_{47}H_{70}O_{14}$ (4:1)

アバウト *abauto*
approximately, about, roughly

アッベ屈折計 *abbekussetsukei*
Abbé refractometer

アッベ集光器 *abbeshuukouki*
Abbé condenser

アビエチン *abiechin*
coniferin $C_{16}H_{22}O_8$

アビエチン酸 *abiechinsan*
abietic acid $C_{20}H_{30}O_2$

アビジン *abijin*
avidin

アボガドロ数 *abogadorosuu*
Avogadro's number

アボガドロ定数 *abogadoroteisuu*
Avogadro's constant

アブシジン酸, アブシシン酸
abushijinsan, abushishinsan
abscisic acid $C_{15}H_{20}O_4$

アーチ *a-chi*
arch

アダム鉱 *adamukou*
adamite $Zn_2(AsO_4)(OH)$

アダプタ, アダプター *adaputa, adaputa-*
adapter

アダリン *adarin*
adaline $C_{13}H_{23}NO$ (natural product) or $BrC(C_2H_5)_2CONHCONH_2$ (tradename)

アデナーゼ *adena-ze*
adenase

アデニン *adenin*
adenine $C_5H_5N_5$

アデニル酸 *adenirusan*
adenylic acid

アーデンヌ石 *a-dennuseki*
ardennite $Mn_4Al_4H_5VSi_4O_{23}$

アデノシン *adenoshin*
adenosine $C_{10}H_{13}N_5O_4$

アデノシン一リン酸, アデノシン一燐酸
adenoshinichirinsan
adenosine monophosphate $C_{10}H_{14}N_5O_7P$

アデノシン二リン酸, アデノシン二燐酸
adenoshinnirinsan
adenosine diphosphate $C_{10}H_{15}N_5O_{10}P_2$

アデノシン三リン酸, アデノシン三燐酸
adenoshinsanrinsan
adenosine triphosphate $C_{10}H_{16}N_5O_{13}P_3$

Japanese-English Chemical Dictionary. Edited by Markus Gewehr
Copyright © 2008 WILEY-VCH Verlag GmbH & Co. KGaA, Weinheim
ISBN: 978-3-527-31293-1

アデンシナーゼ adenshina-ze
  adenosinase

アデライト, アデル石 aderaito, aderuseki
  adelite $CaMgHAsO_5$

アドニット adonitto
  adonite $C_5H_7(OH)_5$

アドレナリン adorenarin
  adrenaline
  $C_6H_3(OH)_2CH(OH)CH_2NHCH_3$

アフィニティー標識 afiniti-hyoushiki
  affinity label

アフィニティークロマトグラフィー
  afiniti-kuromatogurafi-
  affinity chromatography

アフラトキシン afuratokishin
  aflatoxin

アフターバーナー afuta-ba-na-
  afterburner

アフターバーニング afuta-ba-ningu
  afterburning

アフタークーラー afuta-ku-ra-
  after-cooler

アフタークロム afuta-kuromu
  afterchroming

アガチン agachin
  agathin $C_6H_4(OH)CONHN(CH_3)C_6H_5$

アガロースゲル agaro-sugeru
  agarose gel

アゴニスト agonisuto
  agonist

アグファカラー法 agufakara-hou
  Agfacolor process

アグノライド agunoraido
  agnolite $H_2Mn_8(SiO_3)_4$

アグリコン agurikon
  aglycon

アグリコラ石 agurikoraseki
  agricolite $2Bi_2O_3 \cdot 3SiO_2$

アグリン agurin
  agurine $C_7H_7O_2N_4Na \cdot CH_3COONa$

アグロメレーション aguromere-shon
  agglomeration

アグルチネーション aguruchine-shon
  agglutination

アグルコン agurukon
  aglycon

アイヒベルグ石 aihiberuguseki
  eichbergite $(Cu,Fe)_2S \cdot 3(Bi,Sb)_2S_3$

アイコサン aikosan
  eicosane $C_{20}H_{42}$

アインホルン反応 ainhorunhannou
  Einhorn reaction

アインシュタインの係数
  ainshutainnokeisuu
  Einstein coefficient

アインスタイニウム ainsutainiumu
  einsteinium (Es, element 99)

アイソラータ、アイソラーター
  aisora-ta, aisora-ta-
  insulator

アイソレージョン aisore-jon
  isolation

アイソタイプ aisotaipu
  isotype

アイソトニック aisotonikku
  isotonic

アイソトープ aisoto-pu
  isotope

アイソザイム aisozaimu
  isoenzyme, isozyme

アジド ajido
  azide

アジ化バリウム ajikabariumu
  barium azide $Ba(N_3)_2$

アジ化物 ajikabutsu
  azide

アジ化物イオン ajikabutsuion
  azide ion

アジ化フェニル ajikafeniru
  phenyl azide $C_6H_5N_3$

アジ化銀 ajikagin
  silver azide AgN_3

アジ化合物 ajikagoubutsu
  azide

アジ化鉛　*ajikanamari*
　lead azide $Pb(N_3)_2$

アジ化ナトリウム　*ajikanatoriumu*
　sodium azide NaN_3

アジ化水素　*ajikasuiso*
　hydrazoic acid, hydrogen azide HN_3

アジミノ化合物　*ajiminokagoubutsu*
　azimino compound -NH-N=N-

アジン　*ajin*
　azine

アジピン酸　*ajipinsan*
　adipic acid $HOOC(CH_2)_4COOH$

アジポイン　*ajipoin*
　adipoin $C_6H_{10}O_2$

アジポニトリル　*ajiponitoriru*
　adiponitrile $NC(CH_2)_4CN$

アジリジン　*ajirijin*
　aziridine C_2H_5N

アジュバント　*ajubanto*
　adjuvant

アカンサイト　*akansaito*
　acanthite Ag_2S

アキラル　*akiraru*
　achiral

アキシアル結合　*akishiaruketsugou*
　axial bond

アコイン　*akoin*
　acoine

アコニチン　*akonichin*
　aconitine $C_{34}H_{47}NO_{11}$

アコニット酸　*akonittosan*
　aconitic acid
　$HOOCCH=C(COOH)CH_2COOH$

アクアマリン　*akuamarin*
　aquamarine $Be_3Al_2Si_6O_{18}$

アクア錯体　*akuasakutai*
　aqua complex

アクチビン　*akuchibin*
　aktivin

アクチン　*akuchin*
　actin

アクチニド　*akuchinido*
　actinide

アクチニウム　*akuchiniumu*
　actinium (Ac, element 89)

アクチニウム系列　*akuchiniumukeiretsu*
　actinium series

アクチノイド　*akuchinoido*
　actinoid

アクチノマイシン　*akuchinomaishin*
　actinomycin

アクチノメーター　*akuchinome-ta-*
　actinometer

アクチノン　*akuchinon*
　actinon

アーク放電　*a-kuhouden*
　arc discharge

アーク放電管　*a-kuhoudenkan*
　arc-discharge tube

アーク加熱　*a-kukanetsu*
　arc heating

アーク光源　*a-kukougen*
　arc source

アクリフラビン　*akurifurabin*
　acriflavine $C_{14}H_{14}N_3Cl$

アクリジン　*akurijin*
　acridine $C_{13}H_9N$

アクリジンエロー　*akurijinero-*
　acridine yellow $C_{13}H_5N(CH_3)_2(NH_2)_2$

アクリジンオレンジ　*akurijinorenji*
　acridine orange $C_{13}H_7N(N(CH_3)_2)_2$

アクリロニトリル　*akurironitoriru*
　acrylonitrile $H_2C=CHCN$

アクリロニトリルブタジエンゴム
　akurironitorirubutajiengomu
　acrylonitrile-butadiene rubber

アクリルアミド　*akuriruamido*
　acrylamide $H_2C=CHCONH_2$

アクリルアルデヒド　*akuriruarudehido*
　acrylic aldehyde, acrolein $H_2C=CHCHO$

アクリルゴム　*akurirugomu*
　acrylic rubber

アクリル樹脂　*akurirujushi*
acrylate resin

アクリルニトリル　*akurirunitoriru*
acrylonitrile $H_2C=CHCN$

アクリルポリマー　*akuriruporima-*
polyacrylate $[-CH_2CH(COOR)-]_n$

アクリル酸　*akurirusan*
acrylic acid $H_2C=CHCOOH$

アクリル酸ブチル　*akurirusanbuchiru*
butyl acrylate

アクリル酸エチル　*akurirusanechiru*
ethyl acrylate $H_2C=CHCOOC_2H_5$

アクリル酸塩　*akurirusanen*
acrylate $H_2C=CHCOOM$

アクリル酸エステル　*akurirusanesuteru*
acrylic ester $H_2C=CHCOOR$

アクリル酸コポリマー
akurirusankoporima-
acrylic acid copolymer

アクリル酸メチル　*akurirusanmechiru*
methyl acrylate $H_2C=CHCOOCH_3$

アクリル繊維　*akuriruseni*
acrylic fiber

アーク炉　*a-kuro*
electric arc furnace

アクロレイン　*akurorein*
acrolein $H_2C=CHCHO$

アクロソーム　*akuroso-mu*
acrosome

アークスペクトル　*a-kusupekutoru*
arc spectrum

アクティノライト　*akutinoraito*
actinolite $Ca_2(Mg,Fe^{2+})_5Si_8O_{22}(OH)_2$

アクトミオシン　*akutomioshin*
actomyosin

アーク溶接機　*a-kuyousetsuki*
arc-welding machine

アキュムレーション　*akyumure-shon*
accumulation

アキュムレータ　*akyumure-ta*
accumulator

あまに油　*amaniyu*
linseed oil

アマリン　*amarin*
amarine $C_{32}H_{46}O_8$

アマルガム　*amarugamu*
amalgam

アマルガム電極　*amarugamudenkyoku*
amalgam electrode

アマルガム化　*amarugamuka*
amalgamation

アメリシウム　*amerishiumu*
americium (Am, element 95)

アミダーゼ　*amida-ze*
amidase

アミディン　*amidin*
amidine $RC(=NH)NH_2$

アミド　*amido*
amide $RCONH_2$

アミド化　*amidoka*
amidation

アミド結合　*amidoketsugou*
amide bond

アミド基　*amidoki*
amido group

アミド基置換　*amidokichikan*
amidation

アミドキシム　*amidokishimu*
amidoxime $RC(=NOH)NH_2$

アミドール　*amido-ru*
amitol $C_2H_4N_4$

アミド態窒素　*amidotaichisso*
amide nitrogen

アミグダリン　*amigudarin*
amygdaline $C_{20}H_{27}NO_{11}$

アミジン　*amijin*
amidine $RC(=NH)NH_2$

アミン　*amin*
amine

アミノ安息香酸エチル
aminoansokukousanechiru
ethyl aminobenzoate $C_6H_4(NH_2)COOC_2H_5$

アミノアシル　*aminoashiru*
　aminoacyl

アミノチアゾール　*aminochiazo-ru*
　thiazolylamine $C_3H_4N_2S$

アミノグリコシド　*aminogurikoshido*
　aminoglycoside

アミノ化　*aminoka*
　amination

アミノ化合物　*aminokagoubutsu*
　amino compound

アミノ基　*aminoki*
　amino group

アミノ基置換　*aminokichikan*
　amination

アミンオキシド　*aminokishido*
　amine oxide RNOH

アミノ基転移　*aminokiteni*
　transamination

アミノ交換反応　*aminokoukanhannou*
　transamination

アミノ末端　*aminomattan*
　amino terminal end

アミノペプチダーゼ　*aminopepuchida-ze*
　aminopeptidase

アミノプラスト　*aminopurasuto*
　aminoplast

アミノリシス　*aminorishisu*
　aminolysis

アミノ酢酸　*aminosakusan*
　amino acetic acid, glycine H_2NCH_2COOH

アミノ酸　*aminosan*
　amino acid

アミノ酸分析　*aminosanbunseki*
　amino acid analysis

アミノ酸置換　*aminosanchikan*
　amino acid replacement

アミノ酸配列　*aminosanhairetsu*
　amino acid sequence

アミノ酸発酵　*aminosanhakkou*
　amino acid fermentation

アミノ酸シーケンサー　*aminosanshi-kensa-*
　amino acid sequencer

アミノ酸側鎖　*aminosansokusa*
　amino acid side chain

アミノ態窒素　*aminotaichisso*
　amino nitrogen

アミノ糖　*aminotou*
　amino sugar $C_6H_{11}O_5NH_2$

アミラーゼ　*amira-ze*
　amylase

アミレン　*amiren*
　amylene, pentene C_5H_{10}

アミリン　*amirin*
　amyrine $C_{30}H_{50}O$

アミロフォスファターゼ　*amirofosufata-ze*
　amylo-phosphatase

アミログラフ　*amirogurafu*
　amylograph

アミロ法　*amirohou*
　amylo process

アミロイド　*amiroido*
　amyloid

アミロイン　*amiroin*
　amyloin

アミロペクチン　*amiropekuchin*
　amylopectin

アミロプラスト　*amiropurasuto*
　amyloplast

アミロール　*amiro-ru*
　amyrol $C_{15}H_{25}OH$

アミロース, アミローゼ　*amiro-su, amiro-ze*
　amylose

アミル　*amiru*
　pentyl, amyl $C_5H_{11}-$

アミルアルコール　*amiruaruko-ru*
　amyl alcohol $CH_3(CH_2)_4OH$

アミルフェノール　*amirufeno-ru*
　amylphenol $HOC_6H_4C(CH_3)_2C_2H_5$

アミトラズ *amitorazu*
amitraz $C_{19}H_{23}N_3$

アミトロール *amitoro-ru*
amitrole $C_2H_4N_4$

アモルファス *amorufasu*
amorphous

アモルファス金属 *amorufasukinzoku*
amorphous metal

アムニオス酸 *amuniosusan*
amniotic acid $C_4H_6N_4O_3$

アナバシン *anabashin*
anabasine $C_{10}H_{14}N_2$

アナフィラキシー *anafirakishi-*
anaphylaxis

アナ異性 *anaisei*
ana-isomerism

アナペ石 *anapeseki*
tamanite $3(Ca,Fe)O \cdot P_2O_5 \cdot 4H_2O$

アナライザー *anaraiza-*
analyser

アナログ *anarogu*
analog

アナログ計器 *anarogukeiki*
analogue instrument

アンバーグリス *anba-gurisu*
ambergris

アンベライト *anberaito*
amberite $Ba(NO_3)_2$

アンブライト *anburaito*
ambrite $C_{40}H_{66}O_5$

アンチフェブリン *anchifeburin*
antifebrine, N-phenylacetamide $CH_3CONHC_6H_5$

アンチフンギン *anchifungin*
antifungin $Mg_3(BO_3)_2$

アンチ形 *anchigata*
anti-form

アンチ位 *anchii*
anti-position

アンチコロイン *anchikoroin*
antikoroin $ZnSO_4, MgSO_4$

アンチモン *anchimon*
antimony (Sb, element 51)

アンチモン白 *anchimonhaku*
antimony white Sb_2O_3

アンチモン華 *anchimonka*
antimony bloom SbO_3

アンチモン化銀 *anchimonkagin*
silver antimonide Ag_3Sb

アンチモン化水素 *anchimonkasuiso*
stibane, antimony hydride SbH_3

アンチモン酸塩 *anchimonsanen*
antimonate $M^I Sb(OH)_6, M^I SbO_2$

アンチモン酸カリウム *anchimonsankariumu*
potassium antimonate $KSbO_2$

アンチモン酸ナトリウム *anchimonsannatoriumu*
sodium antimonate $NaSbO_2$

アンチノック剤 *anchinokkuzai*
antiknock agent

アンチピリン *anchipirin*
antipyrine $C_{11}H_{12}N_2O$

アンドロステロン *andorosuteron*
androsterone $C_{19}H_{30}O_2$

アンドル石 *andoruseki*
andorite $Ag_2S \cdot 2PbS \cdot 3Sb_2S_3$

アネモメータ *anemome-ta*
anemometer

アネシン *aneshin*
anesin $Cl_3CC(CH_3)_2OH$

アネステジン *anesutejin*
anaesthesine $H_2NC_6H_4COOC_2H_5$

アネトール *aneto-ru*
anethole, anise camphor $C_6H_4(OCH_3)CH=CHCH_3$

アンフェタミン *anfetamin*
amphetamine $C_6H_5CH_2CH(NH_2)CH_3$

アンフィー *anfi-*
amphi-

アンフィ位 *anfii*
amphi-position

アンフォトロピン *anfotoropin*
amphotropine $((CH_2)_6N_4)_2C_8H_{14}(COOH)_2$

アンゲリカ酸 *angerikasan*
angelic acid $(Z)\text{-}CH_3CH=C(CH_3)COOH$

アンゲリカ油 *angerikayu*
angelica oil

アンゲリシン *angerishin*
angelicin $C_{11}H_6O_3$

アンハイドライト *anhaidoraito*
anhydrite

アンハロニン *anharonin*
anhalonine $C_{12}H_{15}NO_3$

アンヒドロン *anhidoron*
anhydrone $Mg(ClO_4)_2$

アンヒドロ糖 *anhidorotou*
anhydro-sugar

アニオン *anion*
anion

アニオン重合 *anionjuugou*
anionic polymerization

アニオン界面活性剤 *anionkaimenkasseizai*
anionic surfactant

アニオノイド *anionoido*
anionoid

アニオノイド試薬 *anionoidoshiyaku*
anionoid reagent

アニオンラジカル *anionrajikaru*
anion radical

アニオン酸 *anionsan*
anion acid

アニリド *anirido*
anilide

アニリン *anirin*
aniline $C_6H_5NH_2$

アニリンブラック *anirinburakku*
aniline black $(C_6H_5N)_x$

アニリンブルー *anirinburu-*
aniline blue $((C_6H_5NHC_6H_4)_2=C=C_6H_4=NHC_6H_5)Cl$

アニリン塩酸塩 *anirinensanen*
aniline hydrochloride $C_6H_5NH_3Cl$

アニリンエロー *anirinero-*
aniline yellow $C_6H_5N=NC_6H_4NH_2$

アニリンフォルムアルデヒド樹脂 *anirinforumuarudehidojushi*
aniline formaldehyde resin

アニーリング *ani-ringu*
annealing, tempering

アニリン樹脂 *anirinjushi*
aniline resin

アニリンレッド *anirinreddo*
aniline red $C_{20}H_{20}N_3Cl$

アニリン染料 *anirinsenryou*
aniline dye

アニリンヴィオレット *anirinvioretto*
aniline violet $C_{27}H_{24}N_4$

アニシジン *anishijin*
anisidine C_7H_9NO

アニソール *aniso-ru*
anisole $C_6H_5OCH_3$

アニスアルデヒド *anisuarudehido*
anisaldehyde $C_6H_4(OCH_3)CHO$

アニスアルコーロ *anisuaruko-ru*
anisyl alcohol $C_6H_4(OCH_3)CH_2OH$

アニス酸 *anisusan*
anisic acid $C_6H_4(OCH_3)COOH$

アニス油 *anisuyu*
anise oil

アンジオテンシン *anjiotenshin*
angiotensin

アンモニア *anmonia*
ammonia NH_3

アンモニアガス *anmoniagasu*
ammonia gas

アンモニア合成 *anmoniagousei*
synthesis of ammonia

アンモニアソーダ法 *anmoniaso-dahou*
ammonia soda process

アンモニアスチル *anmoniasuchiru*
ammonia still

アンモニア水　*anmoniasui*
liquid ammonia, ammonia water NH_4OH

アンモニア態窒素　*anmoniataichisso*
ammonia nitrogen

アンモニウム塩　*anmoniumuen*
ammonium salt

アンモニウムイオン　*anmoniumuion*
ammonium ion

アンモニウム明礬　*anmoniumumyouban*
ammonium alum $NH_4Al(SO_4)_2 \cdot 12H_2O$

アンヌレン　*annuren*
annulene

アノード　*ano-do*
anode

アノード液　*ano-doeki*
anolyte

アノード不動態化　*ano-dofudoutaika*
anodic passivation

アノード酸化　*ano-dosanka*
anodic oxidation

アノジニン　*anojinin*
anodynine $C_{11}H_{12}N_2O$

アノマー　*anoma-*
anomer

アノマー化　*anoma-ka*
anomerization

アノン　*anon*
anon, cyclohexanone $C_6H_{10}O$

アノール　*ano-ru*
anol, cyclohexanol $C_6H_{11}OH$

アーノルド塩基　*a-norudoenki*
Arnold's base

アーノルド試薬　*a-norudoshiyaku*
Arnold's reagent

アンペア　*anpea*
ampere

アンペロメトリー　*anperometori-*
amperometry

アンペールの法則　*anpe-runohousoku*
Ampere's law

アンピシリン　*anpishirin*
ampicillin $C_{16}H_{19}N_3O_4S$

アンプル　*anpuru*
ampulla

アンスラセン　*ansurasen*
anthracene $C_{14}H_{10}$

アンスラセンブルー　*ansurasenburu-*
anthracene blue $C_{17}H_9NO_4$

アンスラセンエロー　*ansurasenero-*
anthracene yellow $C_{10}H_6O_4Br_2$

アンタゴニスト　*antagonisuto*
antagonist

アンタゴニズム　*antagonizumu*
antagonism

アンタラ形　*antaragata*
antarafacial

アーントアイシュタート合成　*a-ntoaishuta-togousei*
Arndt-Eistert synthesis

アントフィルライト　*antofiruraito*
anthophyllite

アントラガロール　*antoragaro-ru*
anthragallol $C_{14}H_8O_5$

アントラキノン　*antorakinon*
anthraquinone $C_{14}H_8O_2$

アントラキノン染料　*antorakinonsenryou*
anthraquinone dye

アントラニル酸　*antoranirusan*
anthranilic acid $C_6H_4(NH_2)COOH$

アントラノール　*antorano-ru*
anthranol $C_{14}H_9OH$

アントラー石　*antora-seki*
antlerite $CuSO_4 \cdot 2Cu(OH)_2$

アントラセン　*antorasen*
anthracene $C_{14}H_{10}$

アントラセン染料　*antorasensenryou*
anthracene dye

アントロン　*antoron*
anthrone $C_{14}H_{10}O$

アントロール　*antoro-ru*
anthrol $C_{14}H_9OH$

アントシアニジン　*antoshianijin*
　anthocyanidin

アントシアニン　*antoshianin*
　anthocyanin

アオカビ　*aokabi*
　penicillium, blue mold

アパタイト　*apataito*
　apatite $Ca_5[(F,OH,Cl)(PO_4)_3]$

アピゲニン　*apigenin*
　apigenine $C_{15}H_{10}O_5$

アピイン　*apiin*
　apiin $C_{26}H_{28}O_{14}$

アピオール　*apio-ru*
　apiole $C_{12}H_{14}O_4$

アピオーゼ　*apio-ze*
　apiose $C_5H_{10}O_5$

アポ酵素　*apokouso*
　apoenzyme

アポタンパク質　*apotanpakushitsu*
　apoprotein

アポ蛋白質　*apotanpakushitsu*
　apoprotein

アラビン　*arabin*
　arabin $C_{10}H_{18}O_9$

アラビノース, アラビノーゼ
　arabino-su, arabino-ze
　arabinose $C_5H_{10}O_5$

アラビトール　*arabito-ru*
　arabitol $C_5H_{12}O_5$

アラビット　*arabitto*
　arabit $C_5H_7(OH)_5$

アラビヤゴム　*arabiyagomu*
　gum arabic

アラボン酸　*arabonsan*
　arabonic acid $HOCH_2(CHOH)_3COOH$

アライン　*arain*
　aryne

アラキドン酸　*arakidonsan*
　arachidonic acid $C_{19}H_{31}COOH$

アラキジン酸, アラキン酸
　arakijinsan, arakinsan
　arachic acid $CH_3(CH_2)_{18}COOH$

アラクロール　*arakuro-ru*
　alachlor $C_{14}H_{20}NO_2Cl$

アラモス石　*aramosuseki*
　alamosite $PbSiO_3$

アラニン　*aranin*
　alanine $H_2NCH(CH_3)COOH$

アラントイン　*arantoin*
　allantoin $C_4H_6N_4O_3$

あられ石　*arareishi*
　aragonite $CaCO_3$

アレキサンドル石　*arekisandoruseki*
　alexandrite $BeAl_2O_4$

アレキシン　*arekishin*
　alexine $C_8H_{15}NO_4$

アレクサンドライト, アレクサンドル石
　arekusandoraito, arekusandoruseki
　alexandrite $BeAl_2O_4$

アレモント石　*aremontoseki*
　allemontite $SbAs_3$

アレン　*aren*
　allene $H_2C=C=CH_2$

アレーン　*are-n*
　arene

アレンブロト塩　*arenburotoen*
　salt of alembroth $2NH_4Cl \cdot HgCl_2 \cdot H_2O$

アレニウスの式　*areniusunoshiki*
　Arrhenius equation

アレロパシー　*areropashi-*
　allelopathy

アレロトロピー　*arerotoropi-*
　allelotropism

アレロトロープ　*arerotoro-pu*
　allelotropic

アレルゲン　*arerugen*
　allergen

アレルギー　*arerugi-*
　allergy

アリレン　*ariren*
　arylen

アリルアミン　*ariruamin*
　allylamine $H_2C=CHCH_2NH_2$

アリルアルコール　*ariruaruko-ru*
allyl alcohol $H_2C=CHCH_2OH$

アリルグリシジルエーテル
arirugurishijirue-teru
allyl glycidyl ether
$H_2C=CHCH_2OCH_2(C_2H_3O)$

アリル位　*arirui*
allylic position

アリル基　*ariruki*
allyl group

アリール基　*ari-ruki*
aryl group

アリヴァル　*arivaru*
alival $CH_3CH(OH)CH_2OH$

アリザリン　*arizarin*
alizarine $C_{14}H_8O_4$

アリザリンブラック　*arizarinburakku*
alizarine black $C_{10}H_4O_2(OH)_2$

アリザリンブルー　*arizarinburu-*
alizarine blue $C_{17}H_7NO_2(OH)_2$

アリザリン染料　*arizarinsenryou*
alizarine dyes

アロエモジン　*aroemojin*
aloemodin $C_{14}H_4O_2(OH)_3CH_3$

アロファン酸　*arofansan*
allophanic acid $H_2NCONHCOOH$

アロファン石　*arofanseki*
allophane $Al_2O_3 \cdot 2SiO_2 \cdot H_2O$

アロイル基　*aroiruki*
aroyl radical

アロイソロイシン　*aroisoroishin*
alloisoleucine
$CH_3CH_2CH(CH_3)CH(NH_2)COOH$

アロコラン酸　*arokoransan*
allocholanic acid $C_{24}H_{40}O_2$

アロクロイト　*arokuroito*
allochroite $Ca_3(Fe,Al)_2(SiO_4)_3$

アロクサン　*arokusan*
alloxan, mesoxalylurea $C_4H_2N_2O_4$

アロクシット　*arokushitto*
aloxite Al_2O_3

アロメリズム　*aromerizumu*
allomerism

アロモルフィズム　*aromorufizumu*
allomorphism

アロパラジウム　*aroparajiumu*
allopalladium

アロサン　*arosan*
allosan $C_{15}H_{23}OOCNHCONH_2$

アロステリック効果　*arosuterikkukouka*
allosteric effect

アローゼ　*aro-ze*
allose $C_6H_{12}O_6$

アルブミン　*arubumin*
albumin

アルブミノイド　*arubuminoido*
albuminoid

アルデヒド　*arudehido*
aldehyde RCHO

アルデヒド樹脂　*arudehidojushi*
aldehyde resin

アルデヒド基　*arudehidoki*
aldehyde group

アルティミン　*arudimin*
aldimine $RCH=NH$

アルドヘキソーゼ　*arudohekiso-ze*
aldohexose $HOCH_2(CHOH)_4CHO$

アルドキシム　*arudokishimu*
aldoxime $RCH=NOH$

アルドン酸　*arudonsan*
aldonic acid $HOOC(CHOH)nCH_2OH$

アルドペントーゼ　*arudopento-ze*
aldopentose $HOCH_2(CHOH)_3CHO$

アルドラーゼ　*arudora-ze*
aldolase

アルドリン　*arudorin*
aldrin $C_{12}H_8Cl_6$

アルドール　*arudo-ru*
aldol $CH_3CH(OH)CH_2CHO$

アルドール反応　*arudo-ruhannou*
aldol reaction

アルドール縮合 *arudo-rushukugou*
aldol condensation, aldolization

アルドース, アルドーゼ
arudo-su, arudo-ze
aldose

アルドステロン *arudosuteron*
aldosterone $C_{21}H_{28}O_5$

アルドテトロース *arudotetoro-ze*
aldotetrose $HOCH_2(CHOH)_2CHO$

アルフォール *arufo-ru*
alphol $HOC_6H_4COOC_{10}H_7$

アルギナーゼ *arugina-ze*
arginase

アルギニン *aruginin*
arginine
$H_2NC(=NH)NH(CH_2)_3CH(NH_2)COOH$

アルギン酸 *aruginsan*
alginic acid $(C_6H_8O_6)_n$

アルギン酸カリウム *aruginsankariuimu*
potassium alginate $(C_6H_7O_6K)_n$

アルギン酸カルシウム
aruginsankarushiumu
calcium alginate $((C_6H_7O_6)_2Ca)_n$

アルギン酸ナトリウム
aruginsannatoriumu
sodium alginate $(C_6H_7O_6Na)_n$

アルゴン *arugon*
argon (Ar, element 18)

アルゴリズム *arugorizumu*
algorithm

アルホイル *aruhoiru*
aluminium foil

アルカン *arukan*
alkane C_nH_{2n+2}

アルカンジカルボン酸
arukanjikarubonsan
alkane-dicarboxylic acid $C_nH_{2n}(COOH)_2$

アルカンカルボン酸 *arukankarubonsan*
alkane-carboxylic acid $C_nH_{2n+1}COOH$

アルカノール *arukano-ru*
alkanol $C_nH_{2n+1}OH$

アルカリ *arukari*
alkali

アルカリブリュー *arukariburyu-*
alkali blue $C_{32}H_{27}N_3O_3S$

アルカリ蓄電池 *arukarichikudenchi*
alkaline battery

アルカリ電池 *arukaridenchi*
alkaline cell

アルカリ土類 *arukaridorui*
alkaline earth

アルカリ土類金属 *arukaridoruikinzoku*
alkaline earth metal

アルカリ塩 *arukarien*
alkali salt

アルカリ金属 *arukarikinzoku*
alkali metal

アルカリ性 *arukarisei*
basic, alkaline

アルカリ性加水分解
arukarisei kasuibunkai
alkaline hydrolysis

アルカリ性ケン化 *arukarisei kenka*
alkaline saponification

アルカリ性の *arukariseino*
alkaline

アルカリ滴定 *arukaritekitei*
alkalimetry

アルカリ溶液 *arukariyoueki*
lye

アルカロイド *arukaroido*
alkaloid

アルケン *aruken*
alkene C_nH_{2n}

アルケニン *arukenin*
alkenyne

アルケニル基 *arukeniruki*
alkenyl group

アルケノール *arukeno-ru*
alkenol $C_nH_{2n-1}OH$

アルキド樹脂 *arukidojushi*
alkyd resin

アルキメデスの原理 *arukimedesunogenri*
Archimedes' principle

アルキン *arukin*
alkyne C_nH_{2n-2}

アルキニル基 *arukiniruki*
alkynyl group

アルキノール *arukino-ru*
alkynol $C_nH_{2n-3}OH$

アルキラート *arukira-to*
alkylate

アルキレン *arukiren*
alkylene

アルキレングリコール *arukirengoriko-ru*
alkylene glycol $C_nH_{2n}(OH)_2$

アルキレート *arukire-to*
alkylate

アルキル *arukiru*
alkyl

アルキルアルミニウム
arukiruaruminiumu
aluminum trialkyl R_3Al

アルキルベンゼンスルホン酸
arukirubenzensuruhonsan
alkylbenzene sulfonic acid

アルキル化 *arukiruka*
alkylation

アルキル化剤 *arukirukazai*
alkylation reagent

アルキル基 *arukiruki*
alkyl group

アルキル交換反応 *arukirukoukanhannou*
transalkylation

アルキルマグネシウム
arukirumaguneshiumu
magnesium alkyl R_2Mg

アルキル水銀 *arukirusuigin*
alkyl mercury R_2Hg

アルキルスルフィニル基
arukirusurufiniruki
alkylsulfinyl group RSO-

アルキルスルフィン酸 *arukirusurufinsan*
alkyl sulfinic acid RSO_2H

アルキルスルホニル基
arukirusurufoniruki
alkylsulfonyl group RSO_2-

アルキルスルフォン酸 *arukirusurufonsan*
alkyl sulfonic acid RSO_3H

アルキルスルホニル基
arukirusuruhoniruki
alkylsulfonyl group RSO_2-

アルコキシド *arukokishido*
alkoxide

アルコキシ基, アルコキシル基
arukokishiki, arukokishiruki
alkoxy group

アルコラート *arukora-to*
alcoholate

アルコラーゼ *arukora-ze*
alcoholase

アルコーリシス *aruko-rishisu*
alcoholysis

アルコール *aruko-ru*
alcohol

アルコール脱水素酵素
aruko-rudassuisokouso
alcohol dehydrogenase

アルコールデヒドロゲナーゼ
aruko-rudehidorogena-ze
alcohol dehydrogenase

アルコール発酵 *aruko-ruhakkou*
alcoholic fermentation

アルコール温度計 *aruko-ruondokei*
alcohol thermometer

アルマンジャイト *arumanjaito*
armangite $Mn_3(AsO_3)_2$

アルミ箔 *arumihaku*
aluminium foil

アルミナ *arumina*
alumina, aluminium oxide Al_2O_3

アルミニウム *aruminiumu*
aluminium (Al, element 13)

アルミニウム化合物
aruminiumukagoubutsu
aluminum compounds

アルミノケイ酸ナトリウム
aruminokeisannatoriumu
sodium aluminium silicate
$Na_2Al_2Si_3O_{10} \cdot 2H_2O$

アルミノール　*arumino-ru*
　aluminol $Al_2(C_{10}H_5(OH)(SO_3)_2)_3$

アルミン酸亜鉛　*aruminsanaen*
　zinc aluminate $ZnAl_2O_4$

アルミン酸塩　*aruminsanen*
　aluminate M^IAlO_2

アルミン酸マグネシウム
　aruminsanmaguneshiumu
　magnesium aluminate, spinel $MgAl_2O_4$

アルミン酸ナトリウム
　aruminsannatoriumu
　sodium aluminate $NaAlO_2$

アルミサッシ　*arumisasshi*
　aluminium sash

アルモール　*arumo-ru*
　aluminol, alumol $Al_2(C_{10}H_5(OH)(SO_3)_2)_3$

アルンドアイステルト合成
　arundoaisuterutogousei
　Arndt-Eistert synthesis

アルニチン　*arunichin*
　arnicine $C_{20}H_{30}O_4$

アルノゲン　*arunogen*
　alunogen $Al_2(SO_4)_3 \cdot 18H_2O$

アリル　*aruru*
　allyl $H_2C=CHCH_2-$

アリル樹脂　*arurujushi*
　allyl resin

アリル転位　*aruruteni*
　allylic rearrangement

アルサニル酸　*arusanirusan*
　arsanilic acid $NH_2C_6H_4AsO(OH)_2$

アルシン　*arushin*
　arsane AsH_3

アルソン酸　*arusonsan*
　arsonic acid $RAsO(OH)_2$

アルソール　*aruso-ru*
　alsol $Al_2(C_2H_3O_2)(C_4H_4O_6)(OH)_2$

アルタイ石　*arutaiseki*
　altaite $PbTe$

アルトローゼ　*arutoro-ze*
　altrose $HOCH_2(CHOH)_4CHO$

アルトトリオーゼ　*arutotorio-ze*
　aldotriose $HOCH_2CH(OH)CHO$

アサロン　*asaron*
　asarone $C_{12}H_{16}O_3$

アセチン　*asechin*
　glycerol monoacetate, acetin
　$CH_3COOCH_2CH(OH)CH_2OH$

アセチレン　*asechiren*
　acetylene C_2H_2

アセチレン銀　*asechirengin*
　silver acetylide Ag_2C_2

アセチレンカルボン酸
　asechirenkarubonsan
　acetylene carboxylic acid, propargylic acid
　$HCCCOOH$

アセチリド　*asechirido*
　acetylide

アセチルアセトン　*asechiruaseton*
　acetylacetone $CH_3COCH_2COCH_3$

アセチル価　*asechiruka*
　acetyl value

アセチル化　*asechiruka*
　acetylation

アセチル化剤　*asechirukazai*
　acetylating agent

アセチル基　*asechiruki*
　acetyl group

アセチル基転移酵素　*asechirukitenikouso*
　acetyl transferase

アセチルコリン　*asechirukorin*
　acetylcholine

アセチルサリチル酸　*asechirusarichirusan*
　acetylsalicylic acid $C_9H_8O_4$

アセフェート　*asefe-to*
　acephate $CH_3CONHPO(OCH_3)(SCH_3)$

アセナフチレン　*asenafuchiren*
　acenaphthylene $C_{12}H_8$

アセナフテン　*asenafuten*
　acenaphthene $C_{12}H_{10}$

アセプトール　*aseputo-ru*
　aseptol $C_6H_4(OH)SO_3H$

アセタール　*aseta-ru*
　acetal $RCH(OR')_2$

アセテートフィルム　*asete-tofirumu*
　acetate film

アセテート繊維　*asete-toseni*
　acetate fiber

アセトアミド　*asetoamido*
　acetamide CH_3CONH_2

アセトアミノフェン　*asetoaminofen*
　acetaminophen $C_6H_4(OH)NHCOCH_3$

アセトアニリド　*asetoanirido*
　acetanilide $CH_3CONHC_6H_5$

アセトアルデヒド　*asetoarudehido*
　acetaldehyde CH_3CHO

アセトアルドール　*asetoarudo-ru*
　acetaldol $CH_3CH(OH)CH_2CHO$

アセトフェノン　*asetofenon*
　acetophenone, phenylmethyl ketone $C_6H_5COCH_3$

アセトイン　*asetoin*
　acetoin $CH_3COCH(OH)CH_3$

アセトン　*aseton*
　acetone CH_3COCH_3

アセトナール　*asetona-ru*
　acetonal $Al(OH)_2(C_2H_3O_2)_5Na$

アセトンアルコール　*asetonaruko-ru*
　aceton alcohol, acetol CH_3COCH_2OH

アセトニルアセトン　*asetoniruaseton*
　acetonylacetone $CH_3CO(CH_2)_2COCH_3$

アセトニトリル　*asetonitoriru*
　acetonitrile CH_3CN

アセトンジカルボン酸　*asetonjikarubonsan*
　acetone dicarboxylic acid $HOOCCH_2COCH_2COOH$

アセトンカルボン酸　*asetonkarubonsan*
　acetoacetic acid, acetone carboxylic acid CH_3COCH_2COOH

アセトンシアンヒドリン　*asetonshianhidorin*
　acetone cyanohydrin $(CH_3)_2C(OH)CN$

アセトン糖　*asetontou*
　acetone sugar

アセトリシス　*asetorishisu*
　acetolysis

アセトール　*aseto-ru*
　aceton alcohol, acetol CH_3COCH_2OH

アセト酢酸　*asetosakusan*
　acetoacetic acid, acetone carboxylic acid CH_3COCH_2COOH

アセト酢酸エチル, アセト酢酸エステル　*asetosakusanechiru, asetosakusanesuteru*
　acetoacetic ester, ethyl acetoacetate $CH_3COCH_2COOC_2H_5$

アシッド　*ashiddo*
　acid

アシッドレッド　*ashiddoreddo*
　acid red $C_{29}H_{19}O_4N_4S(SO_3Na)_2$

アーシン　*a-shin*
　arsane AsH_3

アシニトロ　*ashinitoro*
　aci-nitro

アシラーゼ　*ashira-ze*
　acylase

アシロイン　*ashiroin*
　acyloin

アシロイン縮合　*ashiroinshukugou*
　acyloin condensation

アシル置換　*ashiruchikan*
　acylation

アシル化　*ashiruka*
　acylation

アシル化剤　*ashirukazai*
　acylating agent

アシル基　*ashiruki*
　acyl group $RC(O)-$

アシル基転移酵素　*ashirukitenikouso*
　acyltransferase

アシルオキシ　*ashiruokishi*
　acyloxy

アシ式　*ashishiki*
　aci-form

アッセイ　*assei*
　assay

アスベスト　*asubesuto*
　asbestos

アスファルト　*asufaruto*
　asphalt

アスコルビン酸　*asukorubinsan*
　ascorbic acid, vitamin C $C_6H_8O_6$

アスコルビン酸ナトリウム *asukorubinsannatoriumu*
sodium ascorbate $C_6H_7O_6Na$

アスパラギン *asuparagin*
asparagine $H_2NCOCH_2CH(NH_2)COOH$

アスパラギン酸 *asuparaginsan*
aspartic acid $HOOCCH_2CH(NH_2)COOH$

アスパルギルス *asuparugirusu*
aspergillus

アスパルターゼ *asuparuta-ze*
aspartase

アスパルテーム *asuparute-mu*
aspartame

アスペドスペルミン *asupedosuperumin*
aspidospermine $C_{22}H_{30}N_2O_2$

アスペルギルス *asuperugirusu*
aspergillus

アスピレーター *asupire-ta-*
aspirator

アスピリン *asupirin*
aspirin, acetylsalicylic acid $C_6H_4(OCOCH_3)COOH$

アスタチン *asutachin*
astacine

アスタチン *asutachin*
astatine (At, element 85)

アスタキサンチン *asutakisanchin*
astaxanthin $C_{40}H_{52}O_4$

アタクチック *atakuchikku*
atactic

アーティファクト *a-tifakuto*
artifact

アトミック *atomikku*
atomic

アトム *atomu*
atom

アトパイト *atopaito*
atopite $Ca_2Sb_2O_7$

アトラジン *atorajin*
atrazine $C_8H_{14}N_5Cl$

アトロパ異性 *atoropaisei*
atropisomerism

アトロパミン *atoropamin*
atropamin $C_{17}H_{21}NO_2$

アトロパ酸 *atoropasan*
atropic acid $H_2C=C(C_6H_5)COOH$

アトロピン *atoropin*
atropine $C_{17}H_{23}NO_3$

アトロピン硫酸塩 *atoropinryuusanen*
atropine sulfate $(C_{17}H_{23}NO_3)_2 \cdot H_2SO_4 \cdot H_2O$

アウエル合金 *auerugoukin*
auer metal

アウゲライト, アウゲル石 *augeraito*
augelite $Al_2(OH)_3(PO_4)$

アウキシン *aukishin*
auxin $C_{10}H_9NO_2$

アウクビン *aukubin*
aucubin $C_{15}H_{22}O_9$

アヴァテイン *avatein*
avertin CBr_3CH_2OH

アヴォガドロの法則 *avogadoronohousoku*
Avogadro's law

アヴォガドロ数 *avogadorosuu*
Avogadro's number

アヴォガドロ定数 *avogadoroteisuu*
Avogadro's constant

アワー *awa-*
hour

アワルア鉱 *awaruakou*
awaruite Ni_2F

アザフリン *azafurin*
azafrin $C_{27}H_{38}O_4$

アザ芳香族化合物 *azahoukouzokukagoubutsu*
azaaromatics

アゼライン酸 *azerainsan*
azelaic acid $HOOC(CH_2)_7COOH$

アゾベンゼン, アゾベンゾール *azobenzen, azobenzo-ru*
azobenzene $C_6H_5N=NC_6H_5$

アゾフェニン *azofenin*
azophenine $C_{30}H_{24}N_4$

アゾ顔料　*azoganryou*
　azo pigment

アゾ化合物　*azokagoubutsu*
　azo compound

アゾ基　*azoki*
　azo group -N=N-

アゾキシ化合物　*azokishikagoubutsu*
　azoxy compound

アゾリトミン　*azoritomin*
　azolitmine, litmus

アゾール　*azo-ru*
　azole C_4H_5N

アゾ染料　*azosenryou*
　azo dye

アゾトバクター　*azotobakuta-*
　azotobacter

アズレン　*azuren*
　azulene $C_{10}H_8$

アズロール　*azuro-ru*
　adurol $C_6H_3(OH)_2Cl$

バ ba

バブルキャップ　*babburukyappu*
　bubble cap

バビットメタル　*babittometaru*
　Babbitt metal

バッチ　*bacchi*
　batch

バッチ処理　*bacchishori*
　batch process

バーチ還元　*ba-chikangen*
　Birch reduction

バッデレイ石　*baddereiseki*
　baddeleyit ZrO_2

バーデン酸　*ba-densan*
　badische acid $H_2NC_{10}H_6SO_3H$

バイデル石　*baideruseki*
　beidellite $Al_2O_3 \cdot 3SiO_2 \cdot nH_2O$

バイエライト　*baieraito*
　bayerite $Al(OH)_3$

バイカレイン　*baikarein*
　baicalein $C_{15}H_{10}O_5$

バイカリン　*baikarin*
　baicalin $C_{21}H_{18}O_{11}$

バイメタル　*baimetaru*
　bimetal

バイメタル温度計　*baimetaruondokei*
　bimetal thermometer

バインダー　*bainda-*
　binder

バイオアッセイ　*baioassei*
　bioassay

バイオディーゼル　*baiodi-zeru*
　biodiesel

バイオエンジニアリング
　baioenjiniaringu
　bioengineering

バイオエレクトロニクス
　baioerekutoronikusu
　bioelectronics

バイオハザード　*baiohaza-do*
　biohazard

バイオマス　*baiomasu*
　biomass

バイオメカニクス　*baiomekanikusu*
　biomechanics

バイオミメティクス　*baiomimetikusu*
　biomimetics

バイオポリマー　*baioporima-*
　biopolymer

バイオリアクター　*baioriakuta-*
　bioreactor

バイオサイエンス　*baiosaiensu*
　bioscience

バイオテクノロジー　*baiotekunoroji-*
　biotechnology

バイルシュタインの試験
　bairushutainnoshiken
　Beilstein's test

バイルトン石　*bairutonseki*
　bayldonite $4(Pb,Cu)O \cdot As_2O_5 \cdot 2H_2O$

バイヤービリガー酸化 *baiya-biriga-sanka*
Baeyer-Villiger oxidation

バイヤー法 *baiya-hou*
Bayer process

バイヤー酸 *baiya-san*
Baeyer's acid $HOC_{10}H_6SO_3H$

バーケリウム *ba-keriumu*
berkelium (Bk, element 97)

バケットコンベヤー *bakettokonbeya-*
bucket conveyor

バックグラウンド *bakkuguraundo*
background, underground, grounding, foundation

バックグラウンド放射線 *bakkuguraundohoushasen*
background radiation

バックグラウンド信号 *bakkuguraundoshingou*
background signal

バックミキシング *bakkumikishingu*
back mixing

バックラッシュ *bakkurasshu*
backlash

バークリウム *ba-kuriumu*
berkelium (Bk, element 97)

バクテリア汚染 *bakuteriaosen*
bacterial contamination

バクテリオファージ *bakuteriofa-ji*
bacteriophage

バーナー *ba-na-*
burner

バナジン酸 *banajinsan*
vanadic acid H_3VO_4

バナジン酸塩 *banajinsanen*
vanadate $M^I{}_3VO_4$

バナジウム *banajiumu*
vanadium (V, element 23)

バンバーガー転位 *banba-ga-teni*
Bamberger rearrangement

バンベルゲルゴールドシュミット合成法 *banberugerugo-rudoshumittogouseihou*
Bamberger-Goldschmidt synthesis

バンド構造 *bandokouzou*
band structure

バンドスペクトル *bandosupekutoru*
band spectrum

バニリン *banirin*
vanillin $C_6H_3(OH)(OCH_3)CHO$

ばらつき *baratsuki*
dispersion

ばら油 *barayu*
rose oil

バレロラクタム *barerorakutamu*
valerolactam C_5H_9NO

バレロラクトン *barerorakuton*
valerolactone $C_5H_8O_2$

バレル *bareru*
barrel

バレルアルデヒド *bareruarudehido*
valeraldehyde $CH_3(CH_2)_3CHO$

バリン *barin*
valine $(CH_3)_2CHCH(NH_2)COOH$

バリッスシア石 *barissushiaseki*
variscite $AlPO_4 \cdot 2H_2O$

バリタ *barita*
barite $BaSO_4$

バリウム *bariumu*
barium (Ba, element 56)

バール *ba-ru*
bar

バルバロイン *barubaroin*
barbaloin $C_{17}H_{20}O_7$

バルビチュル酸 *barubichurusan*
barbituric acid $C_4H_4N_2O_3$

バルビタール *barubita-ru*
barbital $C_8H_{12}N_2O_3$

バルビツル酸 *barubitsurusan*
barbituric acid $C_4H_4N_2O_3$

バラク *baruku*
bulk

バルマー系列 *baruma-keiretsu*
Balmer series

バルサム *barusamu*
balsam

バソプレッシン　*basopuresshin*
vasopressin $C_{46}H_{65}N_{15}O_{12}S_2$

バッセー法　*basse-hou*
Basset process

バースチング　*ba-suchingu*
bursting

バース石　*ba-suseki*
barthite $3ZnO \cdot CuO \cdot 3As_2O_5 \cdot 2H_2O$

バーストサイズ　*ba-sutosaizu*
burst size

バーストする　*ba-sutosuru*
to burst

バーゼ　*ba-ze*
base

べ be

ベベリン　*beberin*
bebeerine $C_{36}H_{38}N_2O_6$

ベヘン酸　*behensan*
behenic acid $CH_3(CH_2)_{20}COOH$

ベーケリウム　*be-keriumu*
berkelium (Bk, element 97)

ベーキング　*be-kingu*
baking

ベーキングペウダー　*be-kingupeuda-*
baking powder

ベッケ線　*bekkesen*
Becke line

ベックマン温度計　*bekkumanondokei*
Beckmann thermometer

ベックマン転位　*bekkumanteni*
Beckmann rearrangement

ベークライト　*be-kuraito*
bakelite

ベークリウム　*be-kuriumu*
berkelium (Bk, element 97)

ベクター, ベクトル　*bekuta-, bekutoru*
vector

ベメント石　*bementoseki*
bementite $2MnSiO_8 \cdot H_2O$

ベンチジン　*benchijin*
benzidine $H_2NC_6H_4C_6H_4NH_2$

ベンチジン転位　*benchijinteni*
benzidine conversion

ベンチュリ管　*benchurikan*
venturi tube

ベンガラ　*bengara*
Prussian red Fe_2O_3

ベンジジン　*benjijin*
benzidine $H_2NC_6H_4C_6H_4NH_2$

ベンジン　*benjin*
benzine, gasoline, petrol

ベンジリデン　*benjiriden*
benzylidene $C_6H_5CH=$

ベンジル　*benjiru*
benzil $C_6H_5COCOC_6H_5$; benzyl $C_6H_5CH_2$-

ベンジルアミン　*benjiruamin*
benzylamine $C_6H_5CH_2NH_2$

ベンジルアルコール　*benjiruaruko-ru*
benzyl alcohol $C_6H_5CH_2OH$

ベンジルエーテル　*benjirue-teru*
benzyl ether $(C_6H_5CH_2)_2O$

ベンジル酸　*benjirusan*
benzilic acid $C(OH)(C_6H_5)_2COOH$

ベンジル酸転位　*benjirusanteni*
benzilic acid rearrangement

ベンジルセルローズ　*benjiruseruro-zu*
benzylcellulose

ベントナイト　*bentonaito*
bentonite

ベンツヒドロール　*bentsuhidoro-ru*
benzhydrol $C_6H_5CH(OH)C_6H_5$

ベンザイン　*benzain*
benzyne C_6H_4

ベンザミド　*benzamido*
benzamide $C_6H_5CONH_2$

ベンザル　*benzaru*
benzylidene $C_6H_5CH=$

ベンザルデヒド　*benzarudehido*
benzaldehyde C_6H_5CHO

ベンゼン　*benzen*
benzene C_6H_6

ベンゼンチオール　*benzenchio-ru*
  phenylmercaptan C_6H_5SH

ベンゼン核　*benzenkaku*
  benzene nucleus

ベンゼノイド　*benzenoido*
  benzenoid

ベンゼンスルホン酸　*benzensuruhonsan*
  benzene sulfonic acid $C_6H_5SO_3H$

ベンゼン誘導体　*benzenyuudoutai*
  benzene derivatives

ベンゾアントロン　*benzoantoron*
  benzanthrone $C_{17}H_{10}O$

ベンズアルデヒド　*benzoarudehido*
  benzaldehyde C_6H_5CHO

ベンゾチアゾール　*benzochiazo-ru*
  benzothiazole C_7H_5NS

ベンゾフェノン　*benzofenon*
  benzophenone $C_6H_5COC_6H_5$

ベンゾフラビン　*benzofurabin*
  benzoflavine $C_{21}H_{19}N_2 \cdot HCl$

ベンゾフラン　*benzofuran*
  benzofurane C_8H_6O

ベンゾイン　*benzoin*
  benzoin $C_6H_5COCH(OH)C_6H_5$

ベンゾイン縮合　*benzoinshukugou*
  benzoin condensation

ベンゾイル化　*benzoiruka*
  benzoylation

ベンゾキノン　*benzokinon*
  benzoquinone $C_6H_4O_2$

ベンゾニトリル　*benzonitoriru*
  benzonitrile C_6H_5CN

ベンゾピレン　*benzopiren*
  benzopyrene $C_{20}H_{12}$

ベンゾール　*benzo-ru*
  benzene C_6H_6

ベンゾールヘキサクロイド　*benzo-ruhekusakuroido*
  benzene hexachloride $C_6H_6Cl_6$

ベンゾール核　*benzo-rukaku*
  benzene nucleus

ベンゾール環　*benzo-rukan*
  benzene ring

ベンゾールスルフォン酸　*benzo-rusurufonsan*
  benzene sulfonic acid $C_6H_5SO_3H$

ベンゾトリクロリド　*benzotorikurorido*
  benzotrichloride $C_6H_5CCl_3$

ベンゾザリン　*benzozarin*
  benzosalin $C_6H_5COOC_6H_4COOCH_3$

ベンゾゾール　*benzozo-ru*
  benzosol $C_6H_5COOC_6H_4OCH_3$

ベンズアミド　*benzuamido*
  benzamide $C_6H_5CONH_2$

ベンズアニリド　*benzuanirido*
  benzanilide $C_6H_5CONHC_6H_5$

ベンズヒドロール　*benzuhidoro-ru*
  benzhydrol $C_6H_5CH(OH)C_6H_5$

ベラトロール　*beratoro-ru*
  veratrole $C_6H_4(OCH_3)_2$

ベラウン岩　*beraungan*
  berannite $2FePO_4 \cdot Fe(OH)_3 \cdot 2,5H_2O$

ベレンス　*berensu*
  Berlin blue $Fe^{3+}{}_4[Fe^{2+}(CN)_6]_3$

ベリフスチン　*berifusuchin*
  bilifuscin $C_{16}H_{20}N_2O_4$

ベリリウム　*beririumu*
  beryllium (Be, element 4)

ベリリウム酸塩　*beririumusanen*
  beryllate $M^I{}_2BeO_2$

ベリロナイト　*berironaito*
  beryllonite $NaBePO_4$

ベリット　*beritto*
  belite $2CaO \cdot SiO_2C_2S$

ベロノサイト　*beronosaito*
  belonosite $MgMoO_4$

ベロウソフジャボチンスキー反応　*berousofujabochinsuki-hannou*
  Belousov-Zhabotinsky reaction

ベルベリン　*beruberin*
  berberine $C_{20}H_{18}NO_4+$

ベルチェリウス石　*berucheriususeki*
  berzeliite $(Ca,Na_2)(Mg,Mn)_2(AsO_4)_3$

ベルガミオール　*berugamio-ru*
　bergamiol $CH_3COOC_{10}H_{17}$

ベルグマン系列　*berugumankeiretsu*
　Bergmann series

ベルヌーイの定理　*berunu-inoteiri*
　Bernoulli's principle

ベルリン青　*berurinsei*
　Berlin blue, iron blue $FeIII_4[FeII(CN)_6]_3$

ベルト　*beruto*
　belt

ベルトラン石　*berutoranseki*
　bertrandite $2Be_2SiO_4 \cdot H_2O$

ベシュタウ岩　*beshutaugan*
　beschtauite Na_2O

ベッセマー法　*bessema-hou*
　Bessemer process

ベースマテリアル　*be-sumateriaru*
　raw material

ベスゼリーイ石　*besuzeri-iseki*
　veszelyite $7(Zn,Cu)O \cdot (P,As)_2O_5 \cdot 9H_2O$

ベタイン　*betain*
　betaine $C_5H_{11}NO_2$

ベータ線　*be-tasen*
　beta radiation

ベトール　*beto-ru*
　betol $C_6H_4(OH)COOC_{10}H_7$

ベテュロール　*betyuro-ru*
　betulol $C_{30}H_{50}O_2$

ビ bi

ビアセチル　*biasechiru*
　biacetyl $(CH_3CO)_2$

ビチューメン　*bichu-men*
　bitumen

ビフェニル　*bifeniru*
　biphenyl $C_{12}H_{10}$

ビーゲル石　*bi-geruseki*
　beegerite $6PbS \cdot Bi_2S_3$

ビグアニド　*biguanido*
　biguanide $H_2NC(=NH)NHC(=NH)NH_2$

ビーカー　*bi-ka-*
　beaker

ビキシン　*bikishin*
　bixine $C_{25}H_{30}O_4$

ビーム　*bi-mu*
　beam

ビニル　*biniru*
　vinyl $H_2C=CH-$

ビニルアルコール　*biniruaruko-ru*
　vinyl alcohol $H_2C=CHOH$

ビニルアセチレン　*biniruasechiren*
　vinylacetylene $HCCCH=CH_2$

ビニル樹脂　*binirujushi*
　vinyl resin

ビントシェドラーグリーン　*bintoshedora-guri-n*
　Bindschedler green $NH(C_6H_4N(CH_3)_2)_2$

ビオチン　*biochin*
　biotin $C_{10}H_{16}N_2O_3S$

ビオフレネルの法則　*biofurenerunohousoku*
　Biot-Fresnel's law

ビオステリン　*biosuterin*
　biosterin $C_{22}H_{44}O_2$

ビピリジン　*bipirijin*
　bipyridine $C_{10}H_8N_2$

ビピリジル　*bipirijiru*
　bipyridyl

ビーライト　*bi-raito*
　belite $2CaO \cdot SiO_2C_2S$

ビラジカル　*birajikaru*
　biradical

ビリベルジン　*biriberujin*
　biliverdin $C_{33}H_{34}N_4O_6$

ビリルビン　*birirubin*
　bilirubin $C_{33}H_{36}N_4O_6$

ビリヴェルヂン　*biriverudin*
　biliverdin $C_{33}H_{34}N_4O_6$

ビルドアップ法　*birudoappuhou*
　build-up process

ビルケランドエイデ法　*birukerandoeidehou*
　Birkeland-Eyde's process

ビール酵母　*bi-rukoubo*
　brewer's yeast

ビールス　*bi-rusu*
　virus

ビショッフ石　*bishoffuseki*
　bischofite $MgCl_2 \cdot 6H_2O$

ビスフェノール　*bisufeno-ru*
　bisphenol

ビスコース　*bisuko-su*
　viscose

ビスマス　*bisumasu*
　bismuth Bi (element 83)

ビタミン　*bitamin*
　vitamin

ビート糖　*bi-totou*
　beet sugar, root sugar

ビウレット　*biuretto*
　biuret $H_2NCONHCONH_2$

ビウレット反応　*biurettohannou*
　biuret reaction

ビーズ　*bi-zu*
　pearl, bead

ボ bo

ボーア原子模型　*bo-agenshimokei*
　Bohr atomic model

ボーア半径　*bo-ahankei*
　Bohr radius

ボーア磁子　*bo-ajishi*
　Bohr magneton

ボヘミヤガラス　*bohemiyagarasu*
　bohemian glas $K_2O \cdot CaO \cdot 6SiO_2$

ボイラー　*boira-*
　boiler

ボイルマリオットの法則　*boirumariottonohousoku*
　Boyle-Mariotte's law

ボーキサイト　*bo-kisaito*
　bauxite

ボーメ度　*bo-medo*
　Baumé degree

ボーメ比重計　*bo-mehijuukei*
　Baumé's hydrometer

ボンディング　*bondingu*
　bonding

ボンド　*bondo*
　bond, linkage

ボンゼン石　*bonzenseki*
　vonsenite $3(Fe,Mg)O \cdot B_2O_3 \cdot FeO \cdot Fe_2O_3$

ボラン　*boran*
　borane, boron hydride BH_3/B_4H_{10}

ボラゾール　*borazo-ru*
　borazine $B_3N_3H_6$

ボーリウム　*bo-riumu*
　bohrium (Bh, element 107)

ボロマグネサイト　*boromagunesaito*
　boromagnesite $Mg_5B_4O_{11} \cdot 2,5H_2O$

ボロメーター　*borome-ta-*
　bolometer

ボロヴェルチン　*boroveruchin*
　borovertin $C_6H_{12}N_4 \cdot 3HBO_2$

ボルドー液　*borudo-eki*
　Bordeaux mixture

ボールクレー　*bo-rukure-*
　ball clay

ボールミル　*bo-rumiru*
　ball mill

ボルナン　*borunan*
　camphane, bornane $C_{10}H_{18}$

ボルネーン　*borune-n*
　bornene $C_{10}H_{16}$

ボルネオール　*boruneo-ru*
　borneol $C_{10}H_{17}OH$

ボルネオ樟脳　*boruneoshounou*
　borneo camphor $C_{10}H_{17}OH$

ボルンオッペンハイマー近似　*borunhoppenhaima-kinzo*
　Born-Oppenheimer approximation

ボルニルアルコール　*boruniruaruko-ru*
　borneol $C_{10}H_{17}OH$

ボルタンメトリー　*borutanmetori-*
　voltammetry

ボルト　*boruto*
　volt

ボルツマン分布　*borutsumanbunpu*
　Boltzmann distribution

ボルツマン因子　*borutsumaninshi*
　Boltzmann factor

ボルツマン定数　*borutsumanteisuu*
　Boltzmann constant

ボッシュ法　*bosshuhou*
　Bosch process

ボースアインシュタイン統計
　bo-suainshutaintoukei
　Bose-Einstein statistics

ボース粒子　*bo-suryuushi*
　boson

ボーウェナイト　*bo-wenaito*
　bowenite $H_4Mg_3Si_2O_9$

ブ bu

ブーボウブランの反応
　bu-bouburannohannou
　Bouveault-Blanc reduction

ブチン　*buchin*
　butine C_4H_6

ブチニル　*buchiniru*
　butinyl C_4H_5-

ブチラルデヒド　*buchirarudehido*
　butyraldehyde $CH_3(CH_2)_2CHO$

ブチリン　*buchirin*
　butyrin $C_3H_5(C_3H_7COO)_3$

ブチロラクタム　*buchirorakutamu*
　butyrolactam C_4H_7NO

ブチロラクトン　*buchirorakuton*
　butyrolactone $C_4H_6O_2$

ブチル　*buchiru*
　butyl C_4H_9-

ブチルアルデヒド　*buchiruarudehido*
　butyraldehyde $CH_3(CH_2)_2CHO$

ブチルアルコール　*buchiruaruko-ru*
　butanol C_4H_9OH

ぶどう糖　*budoutou*
　dextrose, grape sugar $C_6H_{12}O_6$

ブフォタリン　*bufotarin*
　bufotalin $C_{26}H_{36}O_6$

ブフォテニン　*bufotenin*
　bufotenine $C_{12}H_{16}N_2O$

ブフォトキシン　*bufotokishin*
　bufotocin $C_{40}H_{60}N_4O_{10} \cdot H_2O$

ブフナー漏斗　*bufuna-routo*
　Büchner funnel

ブキシン　*bukishin*
　buxine $C_{18}H_{21}NO_3$

ブナゴム　*bunagomu*
　Buna rubber

ブンセン電池　*bunsendenchi*
　Bunsen cell

ブンゼンバーナー　*bunzenba-na*
　Bunsen burner

ブンゼンロスコーの法則
　bunzenhosuko-nohousoku
　Bunsen-Roscoe law

ブラッグの角　*buraggunokaku*
　Bragg angle

ブラジキニン　*burajikinin*
　bradykinin

ブラジレイン　*burajirein*
　brazilein $C_{16}H_{12}O_5$

ブラジリン　*burajirin*
　brazilin $C_{16}H_{14}O_5 \cdot 1,5H_2O$

ブラジル酸　*burajirusan*
　brasilic acid $C_{12}H_{12}O_6$

ブラジウス流　*burajiusuryuu*
　Blasius flow

ブラッケト系列　*burakketokeinetsu*
　Brackett series

ブランド石　*burandoseki*
　branditite $Ca_2MnAs_2O_8 \cdot 2H_2O$

ブラン反応　*buranhannou*
　Blanc reaction

ブランケット　*buranketto*
　blanket

ブランネル石　*buranneruseki*
　brannerite (UO,TiO,UO$_2$)TiO$_3$

ブラシカステリン　*burashikasuterin*
　brassicasterol C$_{28}$H$_{46}$O

ブラシン酸　*burashinsan*
　brassidic acid C$_8$H$_{17}$CH=CHC$_{11}$H$_{22}$COOH

ブラシル酸　*burashirusan*
　brassylic acid HOOC(CH$_2$)$_{11}$COOH

ブラストサイジン　*burasutosaijin*
　blasticidin C$_{17}$H$_{26}$N$_8$O$_5$

ブラウン管　*buraunkan*
　Braun tube

ブラウン運動　*buraunundou*
　Brownian motion

ブレード石　*bure-doseki*
　warthite MgSO$_4$・Na$_2$SO$_4$・4H$_2$O

ブレーキ　*bure-ki*
　brake

ブレーキ液　*bure-kieki*
　brake fluid

ブレンダー　*burenda-*
　blender

ブレンステット塩基　*burensutettoenki*
　Brönsted base

ブレンステット酸, ブレーンステズ酸
　burensutettosan, bure-nsutezusan
　Brönsted acid

ブレット　*buretto*
　burette

ブリックス度　*burikkusudo*
　Brix degree

ブリネル硬度　*burinerukoudo*
　Brinell hardness

ブリネル数　*burinerusuu*
　Brinell number

ブリン法　*burinhou*
　Brin process

ブリリアントエロ　*buririantoero*
　brilliant yellow C$_{26}$H$_{18}$N$_4$O$_8$S$_2$Na$_2$

ブリリアントグリーン　*buririantoguri-n*
　brilliant green, emerald green
　C$_{27}$H$_{34}$N$_2$O$_4$S/C$_{27}$H$_{33}$N$_2$Cl

ブリスター　*burisuta-*
　blistering

ブリスタリング　*burisutaringu*
　blistering

ブローガス　*buro-gasu*
　blow gas

ブロッキング　*burokkingu*
　blocking

ブロック重合　*burokkujuugou*
　block polymerization

ブロックコポリマー　*burokkukoporima-*
　block copolymer

ブロック共重合体　*burokkukyoujuugoutai*
　block copolymer

ブロックポリマー　*burokkuporima-*
　block polymer

ブロック炉　*burokkuro*
　block furnace

ブロマリン　*buromarin*
　bromalin C$_6$H$_{12}$N$_4$・C$_2$H$_5$Br

ブロマール　*buroma-ru*
　bromal Br$_3$CCHO

ブロメライト　*buromeraito*
　bromellit BeO

ブロメトン　*burometon*
　brometone HOC(CH$_3$)$_2$CBr$_3$

ブロモフルオロメタン
　buromofuruorometan
　bromofluoromethane CH$_2$BrF

ブロモホルム　*buromohorumu*
　bromoform CHBr$_3$

ブロモジフルオロプロパン
　buromojifuruoropuropan
　bromodifluoropropane C$_3$H$_5$F$_2$Br

ブロモジクロロメタン
　buromojikurorometan
　dichlorobromomethane CHBrCl$_2$

ブロモクロロメタン
　buromokurorometan
　chlorobromomethane CH$_3$BrCl

ブロモール　buromo-ru
　bromol $C_6H_3OBr_3$

ブロモ酢酸エチル　buromosakusanechiru
　ethyl bromoacetate $BrCH_2COOC_2H_5$

ブロモトリフルオロメタン；ハロン-1301　buromotorifuruorometan
　bromotrifluoromethane CBr_3F

ブロムアロイン　buromuaroin
　bromaloin $C_{17}H_{15}Br_3O_7$

ブロムアセトン　buromuaseton
　bromoacetone CH_3COCH_2Br

ブロムベンゾール　buromubenzo-ru
　bromobenzene C_6H_5Br

ブロムフェノール　buromufeno-ru
　bromophenol BrC_6H_4OH

ブロムメタン　buromumetan
　methyl bromide CH_3Br

ブロム酢酸　buromusakusan
　bromoacetic acid $BrCH_2COOH$

ブローネル酸　buro-nerusan
　Broenner's acid $H_2NC_{10}H_6SO_3H$

ブルチン　buruchin
　brucine $C_{23}H_{26}N_2O_4$

ブルーミング　buru-mingu
　blooming

ブルーシフト　buru-shifuto
　blue shift

ブルシン　burushin
　brucine $C_{23}H_{26}N_2O_4$

ブリュウスターの法則　buryuusuta-nohousoku
　Brewster's law

ブースター　bu-suta-
　booster

ブタジエン　butajien
　butadiene $H_2C=CHCH=CH_2$

ブタジエンゴム　butajiengomu
　butadiene rubber

ブタン　butan
　butane C_4H_{10}

ブタンジオール　butanjio-ru
　butanediol $HOCH_2CH_2CH_2CH_2OH$

ブタノール　butano-ru
　butanol C_4H_9OH

ブテイン　butein
　butyne C_4H_6

ブテン　buten
　butene C_4H_8

ブテニル　buteniru
　butenyl C_4H_7-

ブテシン　buteshin
　butesin $H_2NC_6H_4COOC_4H_9$

ヒュ byu

ビュレット　byuretto
　burette

チャ cha

チャー　cha-
　char

チャビベトール　chabibeto-ru
　chavibetol $H_2C=CHCH_2C_6H_3(OH)(OCH_3)$

チャビコール　chabiko-ru
　chavicol $H_2C=CHCH_2C_6H_4OH$

チャーチ石　cha-chiseki
　churchite $(Ce,Ca)PO_4 \cdot 2H_2O$

チャージ　cha-ji
　charge

チャート　cha-to
　chart

チェ che

チェック　chekku
　check

チェーン　che-n
　chain

チ chi

チアジン　*chiajin*
 thiazine C_4H_5NS

チアミド, チオアミド　*chiamido, chioamido*
 thioamide $RCSNH_2$

チアミン　*chiamin*
 thiamine $C_{12}H_{17}ClN_4OS$

チアン　*chian*
 thiane $C_5H_{10}S$

チアントレン　*chiantoren*
 thianthrene $C_{12}H_8S_2$

チアゾリジン　*chiazorijin*
 thiazolidine C_3H_7NS

チアゾール　*chiazo-ru*
 thiazole C_3H_3NS

チベトン　*chibeton*
 civetone $C_{17}H_{30}O$

チビアト石　*chibiatoseki*
 chiviatite $PbS \cdot 2Bi_2S_3$

チチバビン反応　*chichibabinhannou*
 Chichibabin reaction

チエタン　*chietan*
 thietane C_3H_6S

チーグラーナッタ触媒
 chi-gura-nattashokubai
 Ziegler-Natta catalyst

チキソトロピー　*chikisotoropi-*
 thixotropy

チクロヘキセン　*chikurohekisen*
 cyclohexene C_6H_{10}

チマーゼ　*chima-ze*
 zymase

チミジン　*chimijin*
 thymidine $C_{10}H_{14}N_2O_5$

チミン　*chimin*
 thymine $C_5H_6N_2O_2$

チミルアミン　*chimiruamin*
 thymylamine $C_6H_3(NH_2)(CH_3)C_3H_7$

チモーゲン　*chimo-gen*
 zymogen

チモヘクソーゼ　*chimohekuso-ze*
 zymohexose

チモヒドロキノン　*chimohidorokinon*
 thymohydroquinone $C_6H_2(OH)_2(CH_3)C_3H_7$

チモキノン　*chimokinon*
 thymoquinone $C_6H_2O_2(CH_3)C_3H_7$

チモール　*chimo-ru*
 thymol, thyme camphor
 $C_6H_3(CH_3)(OH)C_3H_7$

チモールブルー　*chimo-ruburu-*
 thymol blue $C_{27}H_{29}O_5S$

チンダル効果　*chindarugenshou*
 Tyndall effect

チンキ、チンキ剤　*chinki, chinkizai*
 tincture

チオアンチモン酸　*chioanchimonsan*
 thioantimonic acid H_3SbS_4

チオアルデヒド　*chioarudehido*
 thioaldehyde $RCHS$

チオアルコール　*chioaruko-ru*
 thioalcohol RSH

チオ亜硫酸　*chioaryuusan*
 thiosulfurous acid $H_2S_2O_2$

チオアセタール　*chioaseta-ru*
 thioacetal

チオアセトアミド　*chioasetoamido*
 thioacetamide CH_3CSNH_2

チオアセトアニリド　*chioasetoanirido*
 thioacetanilide $CH_3CSNHC_6H_5$

チオエーテル　*chioe-teru*
 thioether

チオフェン　*chiofen*
 thiophene C_4H_4S

チオフェニン　*chiofenin*
 thiophenine C_4H_5NS

チオフェンカルボン酸
 chiofenkarubonsan
 thiophenecarboxylic acid C_4H_3SCOOH

チオフェノール　*chiofeno-ru*
 thiophenol C_6H_5SH

チオフェンスルフォン酸　*chiofensurufonsan*
　thiophenesulfonic acid $C_4H_3S(SO_3H)$

チオフォスゲン　*chiofosugen*
　thiophosgene $CSCl_2$

チオフラン　*chiofuran*
　thiophene, thiofuran C_4H_4S

チオフタリド　*chiofutarido*
　thiophthalide C_8H_6OS

チオフテン　*chiofuten*
　thiophthene $C_6H_4S_2$

チオ蟻酸　*chiogisan*
　thioformic acid $HCOSH$

チオグリコール　*chioguriko-ru*
　thioglycol $HSCH_2CH_2OH$

チオグリコール酸　*chioguriko-rusan*
　thioglycolic acid $HSCH_2COOH$

チオインドキシル　*chioindokishiru*
　thioindoxyl C_8H_6OS

チオジアジン　*chiojiajin*
　thiodiazine $C_3H_3N_2S$

チオジアゾール　*chiojiazo-ru*
　thiodiazole $C_2H_2N_2S$

チオカルバミド　*chiokarubamido*
　thiocarbamide $SC(NH_2)_2$

チオカルバミド酸　*chiokarubamidosan*
　thiocarbamic acid CH_3NS_2

チオカルバニリド　*chiokarubanirido*
　thiocarbanilide $(C_6H_5NH)_2CS$

チオカルビミド　*chiokarubimido*
　isothiocyanic acid $HNCS$

チオケトン　*chioketon*
　thioketone

チオキサントン　*chiokisanton*
　thioxanthon $C_{13}H_8OS$

チオキセン　*chiokisen*
　thioxene C_6H_8S

チオクマロン　*chiokumaron*
　thiocoumarone C_8H_6S

チオクール　*chioku-ru*
　thiocol $C_6H_3(OH)(OCH_3)(SO_3K)$

チオクサンテン　*chiokusanten*
　thioxanthene $C_{13}H_{10}S$

チオクト酸　*chiokutosan*
　thioctic acid $C_8H_{14}O_2S_2$

チオナフテン　*chionafuten*
　thionaphthene C_8H_6S

チオ尿素　*chionyouso*
　thiourea $SC(NH_2)_2$

チオピラン　*chiopiran*
　thiopyran C_5H_6S

チオピロン　*chiopiron*
　thiopyron C_5H_4OS

チオラン　*chioran*
　thiophane, thiolane C_4H_8S

チオ燐酸, チオリン酸　*chiorinsan*
　thiophosphoric acid $SP(OH)_3$

チオール　*chio-ru*
　thiol RSH

チオール酸　*chio-rusan*
　thiolic acid $RCOSH$

チオ硫酸　*chioryuusan*
　thiosulfuric acid $H_2S_2O_3$

チオ硫酸　*chioryuusan*
　trithionic acid $H_2S_3O_6$

チオ硫酸アンモニウム　*chioryuusananmoniumu*
　ammonium thiosulfate $(NH_4)_2S_2O_3$

チオ硫酸塩　*chioryuusanen*
　thiosulfate $M^I{}_2S_2O_3$

チオ硫酸銀　*chioryuusangin*
　silver thiosulfate $Ag_2S_2O_3$

チオ硫酸ナトリウム　*chioryuusannatoriumu*
　sodium thiosulfate $Na_2S_2O_3$
　$(Na_2S_2O_3 \cdot 5H_2O)$

チオ酢酸　*chiosakusan*
　thioacetic acid CH_3COSH

チオ酸　*chiosan*
　thio acid

チオセミカルバジド　*chiosemikarubajido*
　thiosemicarbazide $H_2N\text{-}NH\text{-}CS\text{-}NH_2$

チオシアン　*chioshian*
　thiocyanogen, sulfocyanogen $(SCN)_2$

チオシアン価 *chioshianka*
  thiocyanogen value

チオシアン酸 *chioshiansan*
  hydrogen thiocyanate, rhodanic acid HSCN

チオシアン酸アンモニウム
  chioshiansananmoniumu
  ammonium rhodanate NH_4NCS

チオシアン酸アルキル
  chioshiansanarukiru
  alkyl thiocyanate NCSR

チオシアン酸第一銅
  chioshiansandaiichidou
  copper(I) thiocyanate CuSCN

チオシアン酸第二水銀
  chioshiansandainisuigin
  mercury(II) thiocyanate $Hg(SCN)_2$

チオシアン酸塩 *chioshiansanen*
  thiocyanate M^ISCN

チオシアン酸銀 *chioshiansangin*
  silver thiocyanate AgSCN

チオシアン酸カリウム
  chioshiansankariumu
  potassium thiocyanate, potassium rhodanide KSCN

チオシアン酸ナトリウム
  chioshiansannatoriumu
  sodium thiocyanate, sodium rhodanide NaSCN

チオ炭酸 *chiotansan*
  thiocarbonic acid H_2CS_3

チオトレン *chiotoren*
  thiotolene C_5H_6S

チオウラシル *chiourashiru*
  thiouracil C_4H_4NOS

チラミン *chiramin*
  tyramine $C_6H_4(OH)CH_2CH_2NH_2$

チリー石 *chiri-seki*
  chileite $Fe_2O_3 \cdot H_2O$

チリ硝石, チリー硝石
  chirishouseki, chiri-shouseki
  Chile salpeter, sodium nitrate $NaNO_3$

チログロブリン *chiroguroburin*
  thyroglobulin

チロジナーゼ *chirojina-ze*
  tyrosinase

チロキシン *chirokishin*
  thyroxine $C_{15}H_{11}NO_4I_4$

チロル石 *chiroruseki*
  tyrolite $Cu_3(AsO_4)_2 \cdot 2Cu(OH)_2 \cdot 7H_2O$

チロシン *chiroshin*
  tyrosine $H_2NCH(CH_2C_6H_4OH)COOH$

チロトキシコン *chirotokishikon*
  tyrotoxicon $C_6H_6N_2O$

チル石 *chiruseki*
  tyrite, fergusonite $Y(Nb,Ta)O_4$

チタン *chitan*
  titanium (Ti, element 22)

チタンエロー *chitanero-*
  titan yellow $C_{28}H_{19}N_5S_2(SO_3Na)_2$

チタン顔料 *chitanganryou*
  titanium pigment

チタニウム *chitaniumu*
  titanium (Ti, element 22)

チタン酸 *chitansan*
  titanic acid H_2TiO_3/H_4TiO_4

チタン酸バリウム *chitansanbariumu*
  barium titanate $BaTiO_3$

チタン酸塩 *chitansanen*
  titanate $M^I_2TiO_3/M^I_4TiO_4/M^I_2Ti_2O_5$

チタン酸鉛 *chitansanen*
  lead titanate $TiPbO_3$

チタン鉄鉱 *chitantekkou*
  ilmenite, titanic iron ore $FeTiO_3$

チトクロム, チトクローム
  chitokuromu, chitokuro-mu
  cytochrome

チョ cho

チョーキング *cho-kingu*
  chalking

チョーク *cho-ku*
  chalk, calcium carbonate $CaCO_3$

チョークコイル　*cho-kukoiru*
　choke coil

チョールムーグラ酸　*cho-rumu-gurasan*
　chaulmoogric acid $C_5H_7(CH_2)_{12}COOH$

ちょうじ油　*choujiyu*
　clove oil

チュ chu

チューブリン　*chu-burin*
　tubulin

チュガエフ試薬　*chugaefushiyaku*
　Tschugaeff's reagent, dimethylglyoxime $CH_3C(=NOH)C(=NOH)CH_3$

ダ da

ダチセチン　*dachisechin*
　datiscetin $C_{15}H_{10}O_6$

ダチシン　*dachishin*
　datiscin $C_{27}H_{30}O_{15}$

ダフネチン　*dafunechin*
　daphnetin $C_9H_6O_4$

ダフニン　*dafunin*
　daphnine $C_{15}H_{16}O_9 \cdot 2H_2O$

ダイアグラム　*daiaguramu*
　diagram

ダイアミンブラックBH　*daiaminburakku BH*
　diamine black BH $C_{32}H_{21}N_6O_8S_2Na$

ダイアミングリーンG　*daiaminguri-n G*
　diamine green G $C_{34}H_{21}N_8O_{12}S_2Na_3$

ダイアモンド　*daiamondo*
　diamond

ダイアニルエローR　*daianiruero- R*
　dianil yellow R $C_{37}H_{24}N_7O_4S_4Na$

ダイアセチル　*daiasechiru*
　diacetyl $(CH_3CO)_2$

ダイジェスター　*daijesuta-*
　digester

ダイナマイト　*dainamaito*
　dynamite

ダイナミックス　*dainamikkusu*
　dynamics

ダイオード　*daio-do*
　diode

ダイヤモンド　*daiyamondo*
　diamond

ダームスタチウム　*da-musutachiumu*
　darmstadtium (former ununnilium, eka-platinum) (Ds, former Uun, Eka-Pt, element 110)

ダニエル電池　*danierudenchi*
　Daniell cell

ダーリン　*da-rin*
　dahlin $(C_6H_{10}O_5)_n$

ダルムスタチウム　*darumusutachiumu*
　darmstadtium (former ununnilium, eka-platinum) (Ds, former Uun, Eka-Pt, element 110)

ダルシット　*darushitto*
　dulcite $HOCH_2(CHOH)_4CH_2OH$

ダルツェン縮合　*darutsenshukugou*
　Darzens condensation

ダトライト　*datoraito*
　datolite $2CaO \cdot B_2O_3 \cdot 2SiO_2 \cdot H_2O$

ダテュリン　*datyurin*
　daturine $C_{17}H_{13}NO_3$

デ de

デアミナーゼ　*deamina-ze*
　deaminase

デバイ　*debai*
　debye

デバイヒュッケルの理論　*debaihyukkerunoriron*
　Debye-Hückel theory

データ解析　*de-dakaiseki*
　data analysis

デフレグメータ　*defuregume-ta*
　dephlegmator

デフロキュレーション　*defurokyure-shon*
　deflocculation

デグラデーション　*degurade-shon*
　degradation, decomposition

デヒドラーゼ　*dehidora-ze*
　dehydrase

デヒドロ酢酸　*dehidorosakusan*
　dehydroacetic acid $C_8H_8O_4$

デヒドロ酢酸ナトリウム
　dehidorosakusannatoriumu
　sodium dehydroacetate $C_8H_7NaO_4 \cdot H_2O$

デジタル機器　*dejitarukiki*
　digital instrument

デカフルオロブタン　*dekafuruorobutan*
　decafluorobutane C_4F_{10}

デカヒドロナフタリン
　dekahidoronafutarin
　decalin, decahydronaphthalene $C_{10}H_{18}$

デカン　*dekan*
　decane $C_{10}H_{22}$

デカン酸　*dekansan*
　decanoic acid $C_9H_{19}COOH$

デカンテーション　*dekante-shon*
　decantation

デカップリング　*dekappuringu*
　decoupling

デカリン　*dekarin*
　decalin, decahydronaphthalene $C_{10}H_{18}$

デカロール　*dekaro-ru*
　decalol $C_{10}H_{17}OH$

デキストラン　*dekisutoran*
　dextran

デキストリン　*dekisutorin*
　dextrin $(C_6H_{10}O_5)_n$

デキストロース　*dekisutoro-su*
　dextrose, grape sugar $C_6H_{12}O_6$

デッキグラス　*dekkigurasu*
　cover glass

デーコン法　*de-konhou*
　Deacon process

デノボ合成　*denobogousei*
　de novo synthesis

デンシメーター　*denshime-ta-*
　densimeter

デオキシリボ核酸　*deokishiribokakusan*
　deoxyribonucleic acid, DNA

デオキシリボヌクレアーゼ
　deokishiribonukurea-ze
　deoxyribonuclease

デオキシリボース　*deokishiribo-su*
　deoxyribose $C_5H_{10}O_4$

デプロトネーション　*depurotone-shon*
　deprotonation

デプシド　*depushido*
　depside

デルビー石　*derubi-seki*
　derbylite $FeO \cdot Sb_2O_5 \cdot 5FeO \cdot TiO_2$

デルフィニジン　*derufinijin*
　delphinidine $C_{15}H_{11}O_7$

デルフィニン　*derufinin*
　delphinine $C_{33}H_{45}NO_9$

デルマトール　*derumato-ru*
　dermatol $C_6H_2(OH)_3COOBi(OH)_2$

デルモール　*derumo-ru*
　dermol $Bi(C_{15}H_9O_4)_3 \cdot Bi_2O_3$

デシベル　*deshiberu*
　decibel

デシケーター　*deshike-ta-*
　desiccator

デシン　*deshin*
　decyne $CH_3(CH_2)_7CCH$

デシル　*deshiru*
　decyl $C_{10}H_{20}-$; desyl $C_6H_5COCH(C_6H_5)-$

デシル酸　*deshirusan*
　decyl acid $C_9H_{19}COOH$

デスマーチン酸化　*desum-chinsanka*
　Dess-Martin oxidation

デスモラーゼ　*desumora-ze*
　desmolase

データ　*de-ta*
　data

データベース　*de-tabe-su*
　data base

データ構造　*de-takouzou*
　data structure

データの品質　*de-tanohinshitsu*
　data quality

データ処理　*de-tashori*
　data processing

データ収集　*de-tashuushuu*
　data collection

デトネーション　*detone-shon*
　detonation

デウインド石　*deuindoseki*
　dewinditite $3PbO \cdot 5UO_3 \cdot 2P_2O_5 \cdot 12H_2O$

デウェー石　*dewe-seki*
　deweylite $4MgO \cdot 3SiO_2 \cdot 6H_2O$

デウィー石　*dewi-seki*
　iron gymnite $4MgO \cdot 3SiO_2 \cdot 6H_2O$

ディ di

ディアフドール　*diafudo-ru*
　diaphthol $C_9H_5N(OH)(SO_3H)$

ディアフラリン　*diafurarin*
　diaphtherin $HOC_6H_4SO_3H(C_9H_6(OH)N)_2$

ディアステレオメル　*diasutereomeru*
　diastereomer

ディークマン縮合　*di-kumanshukugou*
　Dieckmann condensation

ディメンション　*dimenshon*
　dimension

ディリテュール酸　*dirityu-rusan*
　dilituric acid $C_4H_5O_3N_3$

ディールスアルダー反応　*di-rusuaruda-hannou*
　Diels-Alder reaction

ディスパージョン　*disupa-jon*
　dispersion

ディスプロシウム　*disupuroshiumu*
　dysprosium (Dy, element 66)

ディーゼル油　*di-zeruyu*
　diesel fuel

ド do

ドブニウム　*dobuniumu*
　dubnium (Db, element 105)

ドデカフルオロペンタン　*dodekafuruoropentan*
　dodecafluoropentane C_5H_{12}

ドデカン　*dodekan*
　dodecane $C_{12}H_{26}$

ドデカン酸　*dodekansan*
　dodecanoic acid $C_{11}H_{23}COOH$

ドデシル基　*dodeshiruki*
　dodecyl $C_{12}H_{25}-$

ドコサン　*dokosan*
　docosane $CH_3(CH_2)_{20}CH_3$

ドコサン酸　*dokosansan*
　docosanoic acid $C_{21}H_{43}COOH$

ドーパ　*dopa*
　dopa $C_6H_3(OH)_2CH_2CH(NH_2)COOH$

ドーパミン　*do-pamin*
　dopamine $C_8H_{11}NO_2$

ドップラー効果　*doppura-kouka*
　Doppler effect

ドライアイス　*doraiaisu*
　dry ice

ドリメタジオン　*dorimetajion*
　trimethadione $C_6H_9NO_3$

ドロマイト　*doromaito*
　dolomite $CaMg(CO_3)_2$

ドルトンの分圧の法則　*dorutonnobunatsunohousoku*
　Dalton's law of partial pressure

ドウグラスオール石　*dougurasuo-ruseki*
　douglasite $2KCl \cdot FeCl_2 \cdot 2H_2O$

ドウソン石　*dousonseki*
　dawsonite $Na_3Al(CO_3)_2 \cdot 2Al(OH)_3$

デュ dyu

デュボイシン　*dyuboishin*
　duboisine $C_{17}H_{23}NO_3$

デュモン石　*dyumonseki*
dumontite $2PbO \cdot 3UO_3 \cdot P_2O_5 \cdot 5H_2O$

デュンダス石　*dyundasuseki*
dundasite $PbAl_2(CO_3)_2(OH)_4 \cdot H_2O$

デュランゴ石　*dyurangoseki*
durangite $NaAlFAsO_4$

デュラルミン　*dyurarumin*
duralumin

デュレン　*dyuren*
durene $C_6H_2(CH_3)_4$

デュレノール　*dyureno-ru*
durenol $C_6H(CH_3)_4OH$

デュロキノン　*dyurokinon*
duroquinone $C_{10}H_{12}O_2$

デューロンペティの法則
dyu-ronpetinohousoku
Dulong and Petit's law

デュルデン石　*dyurudenseki*
durdenite $Fe_2(TeO_3)_2 \cdot 4H_2O$

デュテリウム　*dyuteriumu*
deuterium

デュテリウム置換　*dyuteriumuchikan*
deuteration, deuterization, deuterizing

エ e

エアドライヤ　*eadoraiya*
air dryer

エアロゾル　*earozoru*
aerosol

エバンス石　*ebansuseki*
evansite $2AlPO_4 \cdot 4Al(OH)_3 \cdot 12H_2O$

エベルン酸　*eberunsan*
evernic acid $C_{17}H_{16}O_7$

エボナイト　*ebonaito*
ebonite

エボニミット　*ebonimitto*
evonymit $C_8H_8(OH)_3$

エチン　*echin*
ethyne C_2H_2

エチオニン　*echionin*
ethionine $C_2H_5S(CH_2)_2CH(NH_2)COOH$

エチレン　*echiren*
ethylene $H_2C=CH_2$

エチレングリコール　*echirenguriko-ru*
ethylene glycol $HOCH_2CH_2OH$

エチレングリコールジエチルエーテル
echirenguriko-rujiechirue-teru
ethylene glycol diethyl ether

エチレングリコールジメチルエーテル
echirenguriko-rujimechirue-teru
ethylene glycol dimethyl ether

エチレングリコールモノエチルエーテル　*echirenguriko-rumonoechirue-teru*
ethylene glycol monoethyl ether,
2-ethoxy-ethanol $C_2H_5OCH_2CH_2OH$

エチレングリコールモノエチルエーテルアセテート
echirenguriko-rumonoechirue-teruasete-to
ethylene glycol monoethyl ether acetate,
2-ethoxyethyl acetate
$CH_3COOCH_2CH_2OC_2H_5$

エチレングリコールモノメチルエーテル　*echirenguriko-rumonomechirue-teru*
ethylene glycol monomethyl ether,
2-methoxyethanol $CH_3OCH_2CH_2OH$

エチレンイミン　*echirenimin*
ethyleneimine, aziridine C_2H_5N

エチレンジアミン　*echirenjiamin*
ethylenediamine $H_2N(CH_2)_2NH_2$

エチレンクロロヒドリン
echirenkurorohidorin
ethylene chlorohydrin C_2H_5OCl

エチレンオキサイド　*echirenokisaido*
ethylene oxide, oxirane C_2H_4O

エチレノキシド　*echirenokishido*
ethylene oxide C_2H_4O

エチル　*echiru*
ethyl C_2H_5-

エチルアミン　*echiruamin*
ethylamine $C_2H_5NH_2$

エチルアルコール　*echiruaruko-ru*
ethyl alcohol, ethanol C_2H_5OH

エチルベンゼン　*echirubenzen*
ethylbenzene $C_6H_5C_2H_5$

エチルビニルエーテル　*echirubinirue-teru*
ethyl vinyl ether $C_2H_5OCH=CH_2$

エチルエーテル　*echirue-teru*
diethyl ether $C_2H_5OC_2H_5$

エチル基　*echiruki*
ethyl group C_2H_5-

エチルメチルケトン　*echirumechiruketon*
ethyl methyl ketone $CH_3COC_2H_5$

エチルメルカプタン　*echirumerukaputan*
ethanethiol C_2H_5SH

エチルレッド　*echirureddo*
ethyl red $C_{23}H_{23}N_2I$

エチルセルロース　*echiruseruro-su*
ethyl cellulose

エチルスルフォン酸　*echirusurufonsan*
ethanesulfonic acid $C_2H_5SO_3H$

エフューション　*efyu-shon*
effusion

エーゲル石　*e-geruseki*
egeran $Ca_6(Al(OH,F))Al_2(SiO_4)_5$

エグレストン石　*eguresutonseki*
eglestonite Hg_4Cl_2O

エイコサン　*eikosan*
eicosane, icosane $C_{20}H_{42}$

エイコサン酸　*eikosansan*
eicosanoic acid $C_{19}H_{39}COOH$

エイコサペンタエン酸　*eikosapentaensan*
eicosapentaenoic acid $C_{19}H_{29}COOH$

エジングトン石　*ejingutonseki*
edingtonite $BaAl_2Si_3O_{10} \cdot 3H_2O$

エジリン，エジル石　*ejirin, ejiruseki*
aegirine $NaFeSi_2O_6$

エキノクロム　*ekinokuromu*
echinochrome $C_{10}H_6O_8$

エキソ形　*ekisogata*
exo form

エキソペプチダーゼ　*ekisopepuchida-ze*
exopeptidase

エキソシトシス　*ekisoshitoshisu*
exocytosis

エキタミン　*ekitamin*
echitamine $C_{22}H_{28}N_2O_4$

エキザルギン　*ekizarugin*
exalgin $CH_3CON(CH_3)C_6H_5$

エキザルトン　*ekizaruton*
exaltono $C_{15}H_{28}O$

エックス線　*ekkususen*
X-ray

エクアトリアル結合　*ekuatoriaruketsugou*
equatorial bond

エクゴニン　*ekugonin*
ecgonine $C_9H_{15}O_3N \cdot H_2O$

エクイレニン　*ekuirenin*
equilenine $C_{18}H_{18}O_2$

エクソン　*ekuson*
exon

エクソペプチダーゼ　*ekusopepuchida-ze*
exopeptidase

エマルジョン　*emarujon*
emulsion

エメチン　*emechin*
emetine $C_{29}H_{40}N_2O_4$

エメラルディン　*emerarudin*
emeraldine $C_{54}H_{29}N_8$

エメラルド　*emerarudo*
emerald

エメリー　*emeri-*
emery

エモジン　*emojin*
emodin $C_{15}H_{10}O_5$

エムプレクタイト　*emupurekutaito*
emplectite $Cu_2S \cdot Bi_2S_3$

エナメル　*enameru*
enamel

エナミン　*enamin*
enamine

エナンチオマー　*enanchioma-*
enantiomer

エナンタール　*enanta-ru*
oenanthal $CH_3(CH_2)_5CHO$

エナントアルデヒド　*enantoarudehido*
oenanthic aldehyde $CH_3(CH_2)_5CHO$

エナントール *enanto-ru*
  enanthic alcohol, 1-heptanol
  $CH_3(CH_2)_5CH_2OH$

エナント酸 *enantosan*
  oenanthic acid, n-heptanoic acid
  $CH_3(CH_2)_5COOH$

エンボライト *enboraito*
  embolite $Ag(Cl,Br)$

エンド形 *endogata*
  endo form

エンドヌクレアーゼ *endonukurea-ze*
  endonuclease

エンドペプチダーゼ *endopepuchida-ze*
  endopeptidase

エンドルフィン *endorufin*
  endorphin

エンドトキシン *endotokishin*
  endotoxin

エネルギー *enerugi-*
  energy

エネルギー分布 *enerugi-bunpu*
  energy distribution

エネルギー伝達 *enerugi-dentatsu*
  transfer of energy

エネルギー保存の法則 *enerugi-hozonnohousoku*
  law of conservation of energy

エネルギー準位 *enerugi-juni*
  energy level, term

エネルギー係数 *enerugi-keisuu*
  energy coefficient

エネルギー効率 *enerugi-kouritsu*
  energy efficiency

エネルギー量子 *enerugi-ryoushi*
  energy quantum

エネルギー障壁 *enerugi-shouheki*
  energy barrier

エネルギー収支 *enerugi-shuushi*
  energy balance

エネルギー帯 *enerugi-tai*
  energy band

エネルギー代謝 *enerugi-taisha*
  energy metabolism

エンゲルハルド石 *engeruharudoseki*
  engelhardite $ZrSiO_4$

エングラフラスコ *engurafurasuko*
  Engler flask

エングラー粘度 *engura-nendo*
  Engler viscosity

エンハンサー *enhansa-*
  enhancer

エンハンスメント効果 *enhansumentokouka*
  enhancement effect

エンジニア *enjinia*
  engineer, technician

エンジニアリング *enjiniaringu*
  engineering

エンジン油 *enjinyu*
  engine oil

エンケファリン *enkefarin*
  enkephalin

エノラート *enora-to*
  enolate

エノラーゼ *enora-ze*
  enolase

エノル型 *enorugata*
  enol form

エノル化, エノール化 *enoruka, eno-ruka*
  enolization

エンタルピー *entarupi-*
  enthalpy

エントロピー *entoropi-*
  entropy

エオジン *eojin*
  eosine $C_{20}H_8O_5Br_4$

エピゲナイト *epigenaito*
  epigenite $4Cu_2S \cdot 3FeS \cdot As_2S_5$

エピグアニン *epiguanin*
  epiguanine $C_6H_7N_5O$

エピグルコサミン *epigurukosamin*
  epiglucosamine
  $HOCH_2(CHOH)_3CH(NH_2)CHO$

エピヒドリン酸 *epihidorinsan*
  epihydrinic acid $C_3H_4O_3$

エピ位　*epii*
　epi-position

エピジジマイト　*epijijimaito*
　epididymite $Na_2 \cdot 2BeO \cdot 6SiO_2 \cdot 6H_2O$

エピカリン　*epikarin*
　epicarin $C_6H_3(OH)(COOH)CH_2C_{10}H_6OH$

エピカテキン　*epikatekin*
　epicatechol, epicatechin $C_{15}H_{14}O_6$

エピキトサミン　*epikitosamin*
　epichitosamine
　$HOCH_2(CHOH)_3CH(NH_2)CHO$

エピクロロヒドリン　*epikurorohidorin*
　epichlorohydrine C_3H_5ClO

エピマー　*epima-*
　epimer

エピマー化, エピメル化
　epima-ka, epimeruka
　epimerization

エピネフリン　*epinefurin*
　epinephrine, adrenaline
　$C_6H_3(OH)_2CH(OH)CH_2NHCH_3$

エポキシド　*epokishido*
　epoxide

エポキシ樹脂　*epokishijushi*
　epoxy resin

エポキシ化　*epokishika*
　epoxidation

エラグ酸　*eragusan*
　ellagic acid $C_{14}H_6O_8$

エライジン　*eraijin*
　elaidin, $C_{57}H_{104}O_6$

エライジン酸　*eraijinsan*
　elaidic acid
　$CH_3(CH_2)_7CH=CH(CH_2)_7COOH$

エラスチン, エラスティン
　erasuchin, erasutin
　elastin

エラスターゼ　*erasuta-ze*
　elastase

エラストマー　*erasutoma-*
　elastomer

エレクトロン　*erekutoron*
　electron

エレクトロニクス　*erekutoronikusu*
　electronics

エレクトロルミネセンス
　erekutororuminesensu
　electroluminescence

エレマイト　*eremaito*
　eremite $(Ce,Yt)_2PO_4 \cdot ThO_2 \cdot SiO_2$

エレメント　*eremento*
　element

エレミシン　*eremishin*
　elemicin $C_6H_2(OCH_3)_3CH_2=CHCH_2$

エレオマルガリン酸　*ereomarugarinsan*
　eleomargaric acid $CH_3(CH_2)_7$
　$CH=CH \cdot (CH_2)_2CH=CH(CH_2)_3COOH$

エレオステアリン酸　*ereosutearinsan*
　eleostearic acid $C_{17}H_{29}COOH$

エリチン　*erichin*
　ericin $C_6H_4(OH)COOCH_2OCH_3$

エリスレン　*erisuren*
　erythrene $H_2C=CHCH=CH_2$

エリスリット　*erisuritto*
　erythrol $CH_2OH(CHOH)_2CH_2OH$

エリスロマイシン　*erisuromaishin*
　erythromycin $C_{37}H_{67}NO_{13}$

エリスロシン　*erisuroshin*
　erythrosine $C_{20}H_8O_5I_4$

エリスロース　*erisuro-su*
　erythrose $HOCH_2(CHOH)_2CHO$

エリトロ形　*eritorogata*
　erythro-form

エリトローゼ, エリトロース
　eritoro-ze, eritoro-su
　erythrose $HOCH_2(CHOH)_2CHO$

エーロゾル　*e-rozoru*
　aerosol

エルビウム　*erubiumu*
　erbium (Er, element 68)

エルドマン試薬　*erudomanshiyaku*
　Erdmann's reagent

エルゴバシン　*erugobashin*
　ergobasine $C_{19}H_{23}N_3O_2$

エルゴチニン　*erugochinin*
　ergotinine $C_{35}H_{39}N_5O_5$

エルゴカルシフェロール
　erugokarushifero-ru
　ergocalciferol, vitamin D2 $C_{28}H_{44}O$

エルゴコルニン　*erugokorunin*
　ergocornine $C_{31}H_{39}N_5O_5$

エルゴクリプチン　*erugokuripuchin*
　ergocryptine $C_{32}H_{41}N_5O_5$

エルゴクリスチン　*erugokurisuchin*
　ergocristine $C_{35}H_{39}N_5O_5$

エルゴノヴィン　*erugonovin*
　ergonovine $C_{19}H_{23}N_3O_2$

エルゴシン, エルゴシニン
　erugoshin, erugoshinin
　ergosine $C_{30}H_{37}N_5O_5$

エルゴステリン　*erugosuterin*
　ergosterol $C_{28}H_{44}O$

エルゴステロール　*erugosutero-ru*
　ergosterin, ergosterol $C_{28}H_{44}O$

エルゴタミン　*erugotamin*
　ergotamine, ergotaminine $C_{33}H_{35}N_5O_5$

エルゴトキシン　*erugotokishin*
　ergotoxine $C_{35}H_{39}N_5O_5$

エルカ酸　*erukasan*
　erucic acid $C_{21}H_{41}COOH$

エルピダイト　*erupidaito*
　elpidite $Na_2ZrSi_6O_{15}$

エルレンマイヤーフラスコ
　erurenmaiya-furasuko
　Erlenmeyer flask

エルレンマイヤー反応
　erurenmaiya-hannou
　Erlenmeyer synthesis

エールリッヒ反応　*e-rurihhihannou*
　Ehrlich's reaction

エルステッド　*erusuteddo*
　oersted

エルヴィン　*eruvin*
　jervine $C_{26}H_{37}NO_3$

エサコニチン　*esakonichin*
　jesaconitine $C_{35}H_{49}NO_{16}$

エシェル石　*esheruseki*
　echellite $(Ca,Na_2)O \cdot 2Al_2O_3 \cdot 3SiO_2 \cdot 4H_2O$

エシュバイラークラーク反応
　eshubaira-kura-kuhannou
　Eschweiler-Clarke reaction

エスバッハ試薬　*esubahhashiyaku*
　Esbach reagent

エスクリン　*esukurin*
　aesculin, esculin $C_{15}H_{16}O_9$

エステラーゼ　*esutera-ze*
　esterase

エステル　*esuteru*
　ester

エステル価　*esuteruka*
　ester number

エステル化　*esuteruka*
　esterification

エステル化する　*esuterukasuru*
　to esterify

エステル交換反応　*esuterukoukanhannou*
　transesterification

エストン　*esuton*
　estone $Al_2O(C_2H_3O_2) \cdot 4H_2O$

エストラゴール　*esutorago-ru*
　estragole $C_6H_4(OCH_3)(CH_2CH=CH_2)$

エストラジオール　*esutorajio-ru*
　estradiol $C_{18}H_{24}O_2$

エストリオール　*esutorio-ru*
　estriol $C_{18}H_{24}O_3$

エストロン　*esutoron*
　estrone $C_{18}H_{22}O_2$

エタクリン酸　*etakurinsan*
　ethacrynic acid $C_{13}H_{12}Cl_2O_4$

エタン　*etan*
　ethane C_2H_6

エタナール　*etana-ru*
　ethanal, acetic aldehyde CH_3CHO

エタンチオール　*etanchio-ru*
　ethanethiol C_2H_5SH

エタノール　*etano-ru*
　ethyl alcohol, ethanol C_2H_5OH

エタノールアミン　*etano-ruamin*
　ethanolamine $HOCH_2CH_2NH_2$

エテン　eten
　ethylene $H_2C=CH_2$

エテニル　eteniru
　ethenyl, vinyl $H_2C=CH-$

エテノール　eteno-ru
　ethenol, vinyl alcohol $H_2C=CHOH$

エーテル　e-teru
　ether ROR

エーテル化　e-teruka
　etherification

エーテル化物　e-terukabutsu
　etherate

エーテル結合　e-teruketsugou
　ether linkage

エーテル溶性の　e-teruyouseino
　ether-soluble

エトキシド　etokishido
　ethoxide C_2H_5OM

エトリング石　etoringuseki
　ettringite $Al_2O_3 \cdot 3SO_3 \cdot 8H_2O$

エゼリン　ezerin
　eserine $C_{15}H_{21}N_3O_2$

ファ　fa

ファイバーガラス　faiba-garasu
　fiberglass

ファインケミカル　fainkemikaru
　fine chemicals

ファコライト　fakoraito
　phacolite $CaAl_2Si_4O_{12} \cdot 6H_2O$

ファクター　fakuta-
　factor

ファマチナ鉱　famachinakou
　famatinite $3Cu_2S \cdot Sb_2S_5$

ファン　fan
　fan

ファンデルワールス力　fanderuwa-rusuryoku
　van der Waals' forces

ファントホッフの法則　fantohoffunohousoku
　van't Hoff's principle

ファラチホ石　farachihoseki
　faratsihite $H_4(Al,Fe)_2Si_2O_9$

ファラデー　farade-
　faraday

ファラデー定数　farade-teisuu
　Faraday constant

ファルネソール, ファルネゾール
　faruneso-ru, farunezo-ru
　farnesol $C_{15}H_{26}O$

ファストエロー　fasutoero-
　fast yellow $C_6H_5N=NC_6H_4NH_2 \cdot HCl$

ファストレッド　fasutoreddo
　fast red A
　$(HSO_3)C_6H_4N=NC_6H_3(SO_3Na)(NH_2)$

ファウセル石　fauseruseki
　fauserite $(Mg,Mn)SO_4 \cdot 7H_2O$

ファヴォルスキー転位　favorusuki-teni
　Favorsky rearrangement

フェ　fe

フェナジン　fenajin
　phenazine $(C_6H_4N)_2$

フェナントラキノン　fenantorakinon
　phenanthraquinone $C_{14}H_8O_2$

フェナントレン　fenantoren
　phenanthrene $C_{14}H_{10}$

フェナントリジン　fenantorijin
　phenanthridine $C_{13}H_9N$

フェナントリルアミン　fenantoriruamin
　phenanthrylamine $C_{14}H_9NH_2$

フェナントロフェナジン　fenantorofenajin
　phenanthrophenazine $C_{20}H_{30}N_2$

フェナントロリン　fenantororin
　phenanthroline $C_{12}H_8N_2$

フェナントロール　fenantoro-ru
　phenanthrole $C_{14}H_9OH$

フェナサイト　*fenasaito*
　phenacite Be_2SiO_4

フェナセチン　*fenasechin*
　phenacetin $CH_3CONHC_6H_4OC_2H_5$

フェナセツール酸　*fenasetsu-rusan*
　phenaceturic acid
　$C_6H_5CH_2CONHCH_2COOH$

フェナシル基　*fenashiruki*
　phenacyl radical $C_6H_5COCH_2$-

フェナゾン　*fenazon*
　phenazone

フェネチジン　*fenechijin*
　phenetidine $H_2NC_6H_4OC_2H_5$

フェネチル基　*fenechiru*
　phenethyl radical $C_6H_5CH_2CH_2$-

フェネチルアルコール　*fenechiruaruko-ru*
　phenethyl alcohol $C_6H_5CH_2CH_2OH$

フェネトール　*feneto-ru*
　phenetole $C_6H_5OC_2H_5$

フェニチジン　*fenichijin*
　phenetidine $H_2NC_6H_4OC_2H_5$

フェニチジンレッド　*fenichijinreddo*
　phenetidine red
　$HOC_6H_4N=NC_6H_3(NO_2)OC_2H_5$

フェニレンブラウン　*fenirenburaun*
　phenylene brown
　$H_2NC_6H_4N=NC_6H_3(NH_2)_2$

フェニレンジアミン　*fenirenjiamin*
　phenylenediamine $C_6H_4(NH_2)_2$

フェニル　*feniru*
　phenyl C_6H_5-

フェニル亜フォスフィン酸
　feniruafosufinsan
　phenylphospinous acid $C_6H_5PO_2H_2$

フェニルアミン　*feniruamin*
　phenylamine, aniline $C_6H_5NH_2$

フェニルアラニン　*feniruaranin*
　phenylalanine $C_6H_5CH_2CH(NH_2)COOH$

フェニルエチレン　*feniruechiren*
　phenylethylene, styrene $C_6H_5CH=CH_2$

フェニルエチルアルコール
　feniruechiruaruko-ru
　phenylethyl alcohol $C_6H_5CH_2CH_2OH$

フェニルエーテル　*fenirue-teru*
　phenyl ether

フェニルフォスフィン　*fenirufosufin*
　phenylphosphine $C_6H_5PH_2$

フェニルフォスフィン酸
　fenirufosufinsan
　phenylphosphinic acid $C_6H_5PO(OH)_2$

フェニルグリチン　*fenirugurichin*
　phenylglycine $H_2NCH(C_6H_5)COOH/$
　$C_6H_5NHCH_2COOH$

フェニルヒドラジン　*feniruhidorajin*
　phenylhydrazine $C_6H_5NHNH_2$

フェニルヒドラジン酸塩
　feniruhidorajinensanen
　phenylhydrazine hydrochloride
　$C_6H_5NHNH_2 \cdot HCl$

フェニルヒドラゾン　*feniruhidorazon*
　phenylhydrazone $C_6H_5NHN=CRR'$

フェニルホスフィン　*feniruhosufin*
　phenylphosphine $C_6H_5PH_2$

フェニルカルビノール　*fenirukarubino-ru*
　phenylcarbinol $C_6H_5CH_2OH$

フェニルメチルケトン
　fenirumechiruketon
　phenyl methyl ketone $C_6H_5COCH_3$

フェニルメルカプタン
　fenirumerukaputan
　thiophenol, phenyl mercaptan C_6H_5SH

フェニルナトリウム　*fenirunatoriumu*
　phenylsodium C_6H_5Na

フェニルニトラミン　*fenirunitoramin*
　phenylnitramine $C_6H_5NHNO_2$

フェニルニトリル　*fenirunitoriru*
　benzonitrile C_6H_5CN

フェニルニトロサミン　*fenirunitorosamin*
　phenylnitrosamine C_6H_5NHNO

フェニル乳酸　*fenirunyuusan*
　phenyllactic acid $C_6H_5CH_2CH(OH)COOH$

フェニルオサゾン　*feniruosazon*
　phenylosazone $C_{17}H_{20}N_4O_3$

フェニルピリジン　*fenirupirijin*
　phenylpyridine $C_{11}H_9N$

フェニルリチウム *fenirurichiumu*
phenyl lithium C_6H_5Li

フェニル硫酸 *feniruryuusan*
phenylsulfuric acid $C_6H_5SO_3H$

フェニル酢酸 *fenirusakusan*
phenylacetic acid $C_6H_5CH_2COOH$

フェニルスチビン酸 *fenirusuchibinsan*
phenylstibinic acid $C_6H_5SbO_3H_2$

フェニルスルフィン酸 *fenirusurufinsan*
phenylsulfinic acid $C_6H_5SO_2H$

フェニサト *fenisato*
phoenicite $3PbO \cdot 2Cr_2O_3$

フェンコン *fenkon*
fenchone $C_{10}H_{16}O$

フェノバービタール *fenoba-bita-ru*
phenobarbital $C_{12}H_{12}N_2O_3$

フェノチアジン *fenochiajin*
phenothiazine $C_{12}H_9NS$

フェノキノン *fenokinon*
phenoquinone $C_6H_4O_2 \cdot 2C_6H_5OH$

フェノキサジン *fenokisajin*
phenoxazine $C_{12}H_9NO$

フェノキシ基 *fenokishiki*
phenoxy group

フェノコール *fenoko-ru*
phenocoll $C_2H_5OC_6H_4NHCOCH_2NH_2$

フェノラート *fenora-to*
phenolate C_6H_5OM

フェノール *feno-ru*
phenol C_6H_5OH

フェノールブルー *feno-ruburu-*
phenol blue $C_{14}H_{14}N_2O$

フェノールエステル *feno-ruesuteru*
phenolic ester $RCOOC_6H_5$

フェノールエーテル *feno-rue-teru*
phenolic ether C_6H_5OR

フェノールフタレイン *feno-rufutarein*
phenolphthalein $C_{20}H_{14}O_4$

フェノールグロコシド *feno-rugurokoshido*
phenolglucoside $C_6H_5OC_6H_{11}O_4$

フェノール樹脂 *feno-rujushi*
phenolic resin

フェノールレッド *feno-rureddo*
phenol red $C_{19}H_{14}O_5S$

フェノサフラニン *fenosafuranin*
phenosafranine $C_{18}H_{14}N_4Cl$

フェノトリアジン *fenotoriajin*
phentriazine $C_7H_6N_3$

フェノザール *fenoza-ru*
phenosal
$C_2H_5OC_6H_4NHOCC_6H_4OH \cdot H_2O$

フェライト *feraito*
ferrite α-Fe

フェランドレン *ferandoren*
phellandrene $C_{10}H_{16}$

フェラツ石 *feratsuseki*
ferrazite $3(Pb,Ba)O \cdot 2P_2O_5 \cdot 8H_2O$

フェリチン *ferichin*
ferritin

フェリ磁性 *ferijisei*
ferrimagnetism

フェーリング液 *fe-ringueki*
Fehling's solution

フェーリング試験法 *fe-ringushikenhou*
Fehling test, Fehling's reaction

フェリン酸 *ferinsan*
fellic acid $C_{23}H_{40}O_4$

フェリピリン *feripirin*
ferripyrine $2FeCl_3 \cdot 3C_{11}H_{12}ON_2$

フェリシアン化銀 *ferishiankagin*
silver ferricyanide $Ag_3Fe(CN)_6$

フェリシアン酸 *ferishiansan*
ferricyanic acid $H_3Fe(CN)_6$

フェロモン *feromon*
pheromone

フェロン *feron*
ferron $C_9H_6INO_4S$

フェロピリン *feropirin*
ferropyrine $3C_{11}H_{12}N_2O \cdot 2FeCl_3$

フェロセン *ferosen*
ferrocene $Fe(C_5H_5)_2$

フェロシアン化銀　*feroshiankagin*
silver ferrocyanide $Ag_4Fe(CN)_6 \cdot H_2O$

フェロシアン酸　*feroshiansan*
ferrocyanic acid $H_4Fe(CN)_6$

フェロスティプティン　*ferosutiputin*
ferrostyptine $(CH_2)_6N_4 \cdot HCl \cdot FeCl_3$

フェルブソナイト　*ferubusonaito*
tyrite, fergusonite $Y(Nb,Ta)O_4$

フェルミウム　*ferumiumu*
fermium (Fm, element 100)

フェルラアルデヒド　*feruraarudehido*
coniferylaldehyde, ferulaldehyde
$C_6H_3(OCH_3)(OH)CH=CHCHO$

フェルラ酸　*ferurasan*
ferulic acid
$CH_3OC_6H_3(OH)CH=CHCOOH$

フェウェル石　*feweruseki*
whewellite $CaC_2O_4 \cdot H_2O$

フェヤーフィルド石　*feya-firudoseki*
fairfieldite $Ca_2Mn(PO_4)_2 \cdot 2H_2O$

フィ fi

フィボロイン　*fiboroin*
fibroin

フィブリン　*fiburin*
fibrin

フィブリノーゲン　*fiburino-gen*
fibrinogen

フィブロイン　*fiburoin*
fibroin

フィブロネクチン　*fiburonekuchin*
fibronectin

フィチン　*fichin*
phytin e.g. $Ca_5Mg(C_6H_{12}O_{24}P_6 \cdot {}_3H_2O)_2$

フィチルアルコール　*fichiruaruko-ru*
phytol, phytyl alcohol $C_{20}H_{39}OH$

フィード　*fi-do*
feed

フィードバック　*fi-dobakku*
feedback

フィプロニル　*fipuroniru*
fipronil $C_{12}H_4Cl_2F_6N_4OS$

フィラメント　*firamento*
filament

フィリンチン酸　*firinchinsan*
filicinic acid $C_8H_{10}O_3$

フィリシン酸　*firishinsan*
filicic acid $C_8H_{10}O_3$

フィロエリトリン　*firoeritorin*
phylloerythrin $C_{33}H_{34}N_4O_8$

フィロフィリン　*firofirin*
phyllophyllin $C_{32}H_{34}N_4O_2Mg$

フィルム　*firumu*
film

フィルム蒸発器　*firumujouhatsuki*
film evaporator

フィルタ, フィルター　*firuta, firuta-*
filter

フィルタープレス　*firuta-puresu*
filter press

フィサライン　*fisarain*
physalite $Al_2O_3 \cdot SiO_2$

フィセチン　*fisechin*
fisetin $C_{15}H_{10}O_6$

フィセトール酸　*fiseto-rusan*
physetoleic acid $C_{16}H_{30}O_2$

フィッシャー合成　*fissha-gousei*
Fischer synthesis

フィッシャー式　*fissha-shiki*
Fischer formula

フィトアレキシン　*fitoarekishin*
phytoalexin

フィトール　*fito-ru*
phytol, phytyl alcohol $C_{20}H_{39}OH$

フィゾスチグミン　*fizosuchigumin*
physostigmine $C_{15}H_{21}N_3O_2$

フォ fo

フォエフェライト　*foeferaito*
hoeferite $2Fe_2O_3 \cdot 4SiO_2 \cdot 7H_2O$

フォーグト石 *fo-gutoseki*
vogtite, ferrobustamite
$Ca(Fe_2+,Ca,Mn_2+)[Si_2O_6]$

フォノン *fonon*
phonon

フォロン *foron*
phorone $(CH_3)_2C=CHCOCH=C(CH_3)_2$

フォルフベルグ鉱 *forufuberugukou*
wolfsbergite $Cu_2S \cdot Sb_2S_3$

フォルゲル石 *forugeruseki*
folgerite $(Fe,Ni)S$

フォルマミント *forumaminto*
formamint $C_{17}H_{32}O_{16}$

フォルマン *foruman*
forman $C_{10}H_{19}OCH_2Cl$

フォルマリン *forumarin*
formalin

フォルマール *foruma-ru*
formal, formaldehyde acetal $H_2C(OCH_3)_2$

フォルミチン *forumichin*
formicin $CH_3CONHCH_2OH$

フォルミン *forumin*
formine, hexamethylenetetramine $C_6H_{12}N_4$

フォルムアミド *forumuamido*
formamide $HCONH_2$

フォルムアルデヒド *forumuarudehido*
formaldehyde $HCHO$

フォルムアルデヒド溶液 *forumuarudehidoyoueki*
formaldehyde solution

フォルムアルドキシム *forumuarudokishimu*
formaldoxime $HCHNOH$

フォスフェニル酸 *fosufenirusan*
phosphenylic acid $C_6H_5PO_3H_2$

フォスフィン *fosufin*
phosphine PH_3

フォスフィン酸 *fosufinsan*
phosphinic acid H_3PO_2 (= $H_2PO(OH)$)

フォスフォン酸 *fosufonsan*
phosphonic acid H_3PO_3 (= $HPO(OH)_2$)

フォスフォリパーゼ *fosuforipa-ze*
phospholipase

フォスフォリラーゼ *fosuforira-ze*
phosphorylase

フォトダイオード *fotodaio-do*
photodiode

フ、フユ fu, fyu

フチオコール *fuchioko-ru*
phthiocol $C_{11}H_8O_3$

フガシティー *fugashiti-*
fugacity

フッ化亜鉛 *fukkaaen*
zinc fluoride ZnF_2

フッ化アンモニウム *fukkaanmoniumu*
ammonium fluoride $(NH_4)F$

フッ化物 *fukkabutsu*
fluoride

フッ化銀 *fukkagin*
silver fluoride AgF

フッ化カドミウム *fukkakadomiumu*
cadmium fluoride CdF_2

フッ化化合物 *fukkakagoubutsu*
fluorine compounds

フッ化カリウム *fukkakariumu*
potassium fluoride KF

フッ化カルシウム *fukkakarushiumu*
calcium fluoride CaF_2

フッ化マグネシウム *fukkamaguneshiumu*
magnesium fluoride MgF_2

フッ化メチル *fukkamechiru*
methyl fluoride CH_3F

フッ化ナトリウム *fukkanatoriumu*
sodium fluoride NaF

フッ化セシウム *fukkaseshiumu*
cesium fluoride CsF

フッ化水素酸 *fukkasuisosan*
hydrofluoric acid

フックの法則　*fukkunohousoku*
　Hooke's law

フーコー振子　*fu-ko-furiko*
　Foucault's pendulum

フコキサンチン　*fukokisanchin*
　fucoxanthin $C_{40}H_{56}O_6$

フコース，フコーゼ　*fuko-su, fuko-ze*
　fucose $CH_3(CHOH)_4CHO$

フクシン　*fukushin*
　fuchsine $C_{20}H_{19}N_3$

フクソン　*fukuson*
　fuchsone $C_{19}H_{14}O$

フマラーゼ　*fumara-ze*
　fumarase

フマル酸　*fumarusan*
　fumaric acid $HOOCCH=CHCOOH$

フミン　*fumin*
　humin

フミン酸　*fuminsan*
　humic acid

フムレン　*fumuren*
　humulen C_5H_8

フムリン酸　*fumurinsan*
　humulinic acid $C_{15}H_{22}O_4$

フムロン　*fumuron*
　humulon $C_{21}H_{30}O_5$

フンボルト石　*funborutoseki*
　humboldtine $FeC_2O_4 \cdot 2H_2O$

フンギステリン　*fungisuterin*
　fungisterol $C_{28}H_{48}O$

フラビン　*furabin*
　flavine $C_{12}H_{10}N_4O_2$

フラボン　*furabon*
　flavone $C_{15}H_{10}O_2$

フラボノイド　*furabonoido*
　flavonoid

フライエスレーベン石　*furaiesure-benseki*
　freieslebenite $5(Pb,Ag)S \cdot 2Sb_2S_3$

フラキシン　*furakishin*
　fraxin $C_{16}H_{18}O_{10}$

フラックス　*furakkusu*
　flux

フラクトース　*furakuto-su*
　fructose, fruit sugar
　$HOCH_2(CHOH)_3COCH_2OH$

フラン　*furan*
　furan C_4H_4O

フランカルボン酸　*furankarubonsan*
　furancarboxylic acid $C_5H_4O_3$

フランケ鉱　*furankekou*
　franckeite $5PbS \cdot 2SnS_2 \cdot Sb_2S_3$

フランクコンドン原理　*furankukondongenri*
　Franck-Condon principle

フラノース，フラノーゼ　*furano-su, furano-ze*
　furanose

フランシウム　*furanshiumu*
　francium (Fr, element 87)

フラッシング　*furasshingu*
　flushing

フラッシュ　*furasshu*
　flash

フラッシュエバポレーター　*furasshuebapore-ta-*
　flash evaporator

フラッシュ法　*furasshuhou*
　Frasch process

フラッシュ蒸発器　*furasshujouhatsuki*
　flash evaporator

フラッシュ蒸留　*furasshujouryuu*
　flash distillation

フラスコ　*furasuko*
　flask

フラウンホーフェル線　*furaunho-ferusen*
　Fraunhofer line

フラヴァニリン　*furavanirin*
　flavaniline $H_2NC_6H_4C_9H_5NCH_3$

フラヴァノン　*furavanon*
　flavanone $C_{15}H_{12}O_2$

フラヴァンスレン　*furavansuren*
　flavanthrene $C_{28}H_{12}N_2O_2$

フラヴェアン酸　*furaveansan*
　flaveanic acid $CNCCSNH_2$

フラヴォキサンチン　*furavokisanchin*
flavoxanthine $C_{40}H_{56}O_3$

フラヴォン　*furavon*
flavone $C_{15}H_{10}O_2$

フラヴォノン　*furavonon*
flavonone $C_{15}H_{12}O_2$

フラヴォノール　*furavono-ru*
flavonol $C_{15}H_{10}O_2$

フラヴォプルプリン　*furavopurupurin*
flavopurpurine $C_{14}H_6O_2(OH)_3$

フラヴォール　*furavo-ru*
flavol $C_{14}H_8(OH)_2$

フラウィアン酸　*furawiansan*
flavianic acid $C_{10}H_4(OH)(NO_2)_2(SO_3H)$

フラザン　*furazan*
furazane $C_2H_2N_2O$

フレキシブル　*furekishiburu*
flexible

フレミングの法則　*furemingunohousoku*
Fleming's rule

フレーム分析　*fure-mubunseki*
flame analysis

フレノシン　*furenoshin*
phrenosin $C_{48}H_{93}NO_9$

フレノシン酸　*furenoshinsan*
phrenosinic acid
$CH_3(CH_2)_{21}CH(OH)COOH$

フレオンガス　*fureongasu*
freon gas

フリーデルクラフツ反応　*furi-derukurafutsuhannou*
Friedel-Crafts reaction

フーリエ変換　*fu-riehenkan*
Fourier transformation

フリント　*furinto*
flint SiO_2

フリントガラス　*furintogarasu*
flint glass

フリル酸　*furirusan*
furilic acid $C_{10}H_8O_5$

フリス転位　*furisuteni*
Fries rearrangement

フリット　*furitto*
frit

フリット炉　*furittoro*
frit kiln

フローダイアグラム　*furo-daiaguramu*
flow diagram

フロイン　*furoin*
furoin $C_{10}H_8O_3$

フロキュレーション　*furokyure-shon*
flocculation

フロンティア軌道　*furontiakidou*
frontier orbital

フローパターン　*furo-pata-n*
flow pattern

フロレチン酸　*furorechinsan*
phloretinic acid $HOC_6H_4CH_2CH_2COOH$

フログルシト　*furorogurushito*
phloroglucite $C_6H_3(OH)_3$

フロロン　*furoron*
phlorone $C_6H_2O_2(CH_3)_2$

フロロル　*furororu*
phlorol $C_2H_5C_6H_4OH$

フルアルデヒド　*furuarudehido*
furfural, furfuryl aldehyde $C_5H_4O_2$

フルアヴィル　*furuaviru*
fluavil $C_{20}H_{32}O_2$

フルベン　*furuben*
fulvene C_6H_6

フルフラール　*furufura-ru*
furfural $C_5H_4O_2$

フルフリルアルコール　*furufuriruaruko-ru*
furfuryl alcohol $C_5H_6O_2$

フルフロール　*furufuro-ru*
furfural $C_5H_4O_2$

ふるい　*furui*
sieve

フルクトサミン　*furukutosamin*
fructosamine
$HOCH_2(CHOH)_3COCH_2NH_2$

フルクトース, フルクトーゼ　*furukuto-su, furukuto-ze*
fructose, fruit sugar
$HOCH_2(CHOH)_3COCH_2OH$

フルオラン　*furuoran*
　fluoran $C_{20}H_{12}O_3$

フルオレン　*furuoren*
　fluorene $C_{13}H_{10}$

フルオレノン　*furuorenon*
　fluorenone $C_{13}H_8O$

フルオレノール　*furuoreno-ru*
　fluorenol $C_{13}H_{10}O$

フルオレセイン, フルオレシイン
　furuoresein, furuoreshiin
　fluorescein $C_{20}H_{12}O_5$

フルオロベンゼン　*furuorobenzen*
　fluorobenzene C_6H_5F

フルオロフォルム　*furuoroforumu*
　fluoroform CHF_3

フルオロホウ酸, フルオロ硼酸
　furuorohousan
　tetrafluoroboric acid HBF_4

フルオロメタン　*furuorometan*
　methyl fluoride CH_3F

フルオロ酢酸　*furuorosakusan*
　fluoroacetic acid FCH_2COOH

フルオルベンゼン　*furuorubenzen*
　fluorobenzene C_6H_5F

フルオルベンゾール　*furuorubenzo-ru*
　fluorobenzene C_6H_5F

フルオルケイ酸, フルオル珪酸
　furuorukeisan
　fluorosilicic acid H_2SiF_6

フルオルスルフォン酸
　furuorusurufonsan
　fluorosulfuric acid HSO_3F

フッ素, 弗素　*fusso*
　fluorine (F, element 9)

フッ素ゴム　*fussogomu*
　fluororubber

フッ素樹脂　*fussojushi*
　fluorine resins

フスチン　*fusuchin*
　fustine $C_{36}H_{26}O_{14}$

フタラジン　*futarajin*
　phthalazine $C_8H_6N_2$

フタリド　*futarido*
　phthalide $C_8H_6O_2$

フタリル　*futariru*
　phthaloyl $C_6H_4(CO-)_2$

フタロイル　*futaroiru*
　phthaloyl $C_6H_4(CO-)_2$

フタロン酸　*futaronsan*
　phthalonic acid $HOOCC_6H_4COCOOH$

フタロシアニン　*futaroshianin*
　phthalocyanine $C_{32}H_{18}N_8$

フタルアミド　*futaruamido*
　phthalamide $C_6H_4(CONH_2)_2$

フタルアミン酸　*futaruaminsan*
　phthalamic acid $C_6H_4(CONH_2)COOH$

フタルアルデヒド　*futaruarudehido*
　phthalaldehyde $C_6H_4(CHO)_2$

フタルイミド　*futaruimido*
　phthalimide $C_8H_5NO_2$

フタルイミジン　*futaruimijin*
　phthalimidine C_8H_7NO

フタル酸, フタール酸
　futarusan, futa-rusan
　phthalic acid $C_6H_4(COOH)_2$

フタル酸ジエチル　*futarusanjiechiru*
　diethyl phthalate $C_6H_4(COOC_2H_5)_2$

フタル酸ジメチル　*futarusanjimechiru*
　dimethyl phthalate $C_6H_4(COOCH_3)_2$

フタル酸樹脂　*futarusanjushi*
　phthalic acid resin

フザリン酸　*fuzarinsan*
　fusaric acid $C_4H_9C_5H_3NCOOH$

フーゼル油　*fu-zeruyu*
　fusel oil

フューゲル石　*fyu-geruseki*
　huegelite $Pb_2(UO_2)_3(AsO_4)_2(OH)_4 \cdot 3H_2O$

ガ ga

ガドリニウム　*gadoriniumu*
　gadolinium (Gd, element 64)

ガイガーミュラー計数管
gaiga-myura-keisuukan
Geiger-Müller counter

ガム　*gamu*
gum

ガーネット　*ga-netto*
garnet

ガングリオシド　*gangurioshido*
ganglioside

ガンマ線　*ganmasen*
gamma-ray

ガノマアル石　*ganomaaruseki*
ganomalite $4CaO \cdot 6PbO \cdot 6SiO_2 \cdot 6H_2O$

ガラックス石　*garakkususeki*
galaxite, manganese spinel $MnAl_2O_4$

ガラクチン　*garakuchin*
galactin $C_{54}H_{78}O_{45}N_4$

ガラクタン　*garakutan*
galactan $(C_6H_{10}O_5)_x$

ガラクターゼ　*garakuta-ze*
galactase

ガラクトン酸　*garakutonsan*
galactonic acid $HOCH_2(CHOH)_4COOH$

ガラクトサミン　*garakutosamin*
galactosamine $C_6H_{13}NO_5$

ガラクトシダーゼ　*garakutoshida-ze*
galactosidase

ガラクトース, ガラクトーゼ
garakuto-su, garakuto-ze
galactose $C_6H_{12}O_6$

ガラクツロン酸　*garakutsuronsan*
galacturonic acid $C_6H_{10}O_7$

ガラス　*garasu*
glass

ガラス瓶　*garasubin*
glass bottle

ガラス電極　*garasudenkyoku*
glass electrode

ガラスファイバー　*garasufaiba-*
glass fiber

ガラスフラスコ　*garasufurasuko*
glass flask

ガラス化　*garasuka*
vitrification

ガラス管　*garasukan*
glass tube

ガラス濾過器　*garasurokaki*
glass filter

ガラス繊維　*garasuseni*
glass fiber

ガラス転移温度　*garasuteniondo*
glass transition temperature

ガラスウール　*garasuu-ru*
glass wool

ガレイン　*garein*
gallein $C_{20}H_{12}O_7$

ガリチン　*garichin*
gallicin, methyl gallate $C_6H_2(OH)_3COOCH_3$

ガリン　*garin*
gallin $C_{20}H_{14}O_7$

ガリピン　*garipin*
galipine $C_{20}H_{21}NO_2$

ガリウム　*gariumu*
gallium (Ga, element 31)

ガロチアニン　*garochianin*
gallocyanine $C_{15}H_{13}N_2O_5$

ガロゲン　*garogen*
gallogen $C_{14}H_6O_8 \cdot 2H_2O$

ガロタンニン酸　*garotanninsan*
gallotannic acid $C_{14}H_{10}O_9$

ガルバニ電池　*garubanidenchi*
galvanic cell

ガルバノメータ　*garubanome-ta*
galvanometer

ガルメイ鉱　*garumeikou*
calamine $ZnCO_3 + Zn_2SiO_4$

ガルヴァーニ電池　*garuva-nidenchi*
galvanic cell

ガソリン　*gasorin*
gasoline

ガス　*gasu*
gas

ガスバーナー　*gasuba-na-*
gas-burner

ガス分析　*gasubunseki*
gas analysis

ガスビュレット　*gasubyuretto*
gas buret

ガス発生　*gasuhassei*
gas formation

ガス化　*gasuka*
gasification

ガス計　*gasukei*
gas meter

ガスクロマトグラフィー
gasukuromatogurafi-
gas chromatography

ガス吸収　*gasukyuushuu*
gas absorption

ガス吸収装置　*gasukyuushuusouchi*
gas absorber

ガスマントル　*gasumantoru*
incandescent mantle

ガスマスク　*gasumasuku*
gas mask

ガス洗浄瓶　*gasusenjoubin*
gas-washing bottle

ガスタービン　*gasuta-bin*
gas turbine

ガス状の　*gasutaino*
gaseous

ガス定量　*gasuteiryou*
gasometry

ガス定数　*gasuteisuu*
gas constant

ガス灯　*gasutou*
gas-burner

ガス容量分析　*gasuyouryoubunseki*
gasometric analysis

ガッターマン反応　*gatta-manhannou*
formic acid

ガッターマンコッホの反応
gatta-mankohhonohannou
Gattermann-Koch reaction

ガウス　*gausu*
Gauss

ガウスの法則　*gausunohousoku*
Gauss law

ゲ ge

ゲイリュ石　*geiryuseki*
gaylussite $Na_2Ca(CO_3)_2 \cdot 5H_2O$

ゲージ石　*ge-jiseki*
gageite $8(Mg,Zn,Mn)O \cdot 3SiO_2 \cdot 3H_2O$

ゲーキー石　*ge-ki-seki*
geikielite $(Mg,Fe)TiO_3$

ゲンチアニン　*genchianin*
gentisin, gentianin $C_{14}H_{10}O_5$

ゲンチアノーゼ　*genchiano-ze*
gentianose $C_{18}H_{32}O_{16}$

ゲンチイン　*genchiin*
gentiin $C_{25}H_{28}O_{14}$

ゲンチオビオース, ゲンチオビオーゼ
genchiobio-su, genchiobio-ze
gentiobiose $C_{12}H_{22}O_{11}$

ゲンチオゲニン　*genchiogenin*
gentiogenin $C_{14}H_{16}O_5$

ゲンチシン　*genchishin*
gentisin, gentianin $C_{14}H_{10}O_5$

ゲンチシン酸　*genchishinsan*
gentisic acid $C_7H_6O_4$

ゲノム　*genomu*
genome

ゲンス石　*gensuseki*
genthite $2NiO \cdot 2MgO \cdot 3SiO_2 \cdot 6H_2O$

ゲラニアール　*gerania-ru*
geranial
$(CH_3)_2C=CHCH_2CH_2C(CH_3)=CHCHO$

ゲラニオール　*geranio-ru*
geraniol
$(CH_3)_2C=CHCH_2CH_2C(CH_3)=CHCH_2OH$

ゲラニウム酸　*geraniumusan*
geranic acid
$(CH_3)_2C=CH(CH_2)_2C(CH_3)=CHCOOH$

ゲーレン石　*ge-renseki*
gehlenite $CaO \cdot MgO \cdot Al_2O_3 \cdot SiO_2$

ゲレスピー石 *geresupi-seki*
gillespite FeO・BaO・4SiO$_2$

ゲル *geru*
gel, jelly

ゲルハルド石 *geruharudoseki*
cerhardite Cu(NO$_3$)$_2$・3Cu(OH)$_2$

ゲル化 *geruka*
gelation

ゲル化点 *gerukaten*
gel point

ゲル化剤 *gerukazai*
gelling agent

ゲルクロマトグラフィー
gerukuromatogurafi-
gel chromatography

ゲルマン *geruman*
germane GeH$_4$/Ge$_2$H$_6$

ゲルマニウム *gerumaniumu*
germanium (Ge, element 32)

ゲルマニウム酸 *gerumaniumusan*
germanic acid H$_4$GeO$_4$

ゲルマニウム石 *gerumaniumuseki*
germanite 5Cu$_2$S・12(Cu,Fe)S・As$_2$S$_3$・2GeS$_2$

ゲールサックの法則
ge-rusakkunohousoku
Gay-Lussac's law

ゲルセミン *gerusemin*
gelsemine C$_{20}$H$_{22}$N$_2$O$_2$

ゲルセミン酸 *geruseminsan*
gelsemic aicd C$_{10}$H$_8$O$_4$

ゲル浸透クロマトグラフィー
gerushintoukuromatogurafi-
gel permeation chromatography

ゲーリュサックの法則
ge-ryusakkunohousoku
Gay-Lussac's law

ギ gi

ギッブスの相律 *gibbusu no souritsu*
Gibbs phase rule

ギッブスの自由エネルギー
gibbusunojiyuuenerugi-
Gibbs free energy

ギベレリン *gibererin*
gibberellin

ギノヴァール *ginova-ru*
gynoval (CH$_3$)$_2$CHCH$_2$COOC$_{10}$H$_{17}$

ギ酸 *gisan*
formic acid HCOOH

ギ酸エチル *gisanechiru*
ethyl formate HCOOC$_2$H$_5$

ギタリン *gitarin*
gitaline C$_{35}$H$_{56}$O$_{12}$

ギトキシゲニン *gitokishigenin*
gitoxigenin C$_{23}$H$_{34}$O$_5$

ギトキシン *gitokishin*
gitoxin C$_{41}$H$_{64}$O$_{14}$

ゴ go

ゴム *gomu*
rubber, gum

ゴム管 *gomukan*
rubber tube

ゴムラテックス *gomuratekkusu*
rubber latex

ゴム石 *gomuseki*
gummite (Pb,Ca,Ba)SiU$_3$O$_{12}$・5H$_2$O

ゴム栓 *gomusen*
rubber stopper

ゴム絶縁 *gomuzetsuen*
rubber insulation

ゴンナルド石 *gonnarudoseki*
gonnardite (Ca,Na$_2$)$_2$Al$_2$Si$_5$O$_{15}$・5,5H$_2$O

ゴルダ石 *gorudaseki*
gordaite 3Na$_2$SO$_4$・Fe$_2$(SO$_4$)$_3$・6H$_2$O

ゴルジ体 *gorujitai*
Golgi apparatus

ゴルリン酸 *gorurinsan*
gorlic acid C$_6$H$_{11}$(CH$_2$)$_6$CH=CH(CH$_2$)$_4$COOH

ゴシポーゼ *goshipo-ze*
gossypose $C_{18}H_{32}O_{16}$

ゴーシュ形 *go-shugata*
gauche form

ゴッシペチン *gosshipechin*
gossypetin $C_{15}H_{10}O_8$

ゴッシピトリン *gosshipitorin*
gossypitrin $C_{21}H_{20}O_{13}$

ゴスラー石 *gosura-seki*
goslarite $ZnSO_4 \cdot 7H_2O$

ゴヤ石 *goyaseki*
goyazite $Ca_3Al_{10}P_2O_{23}$

グ gu

グアエトール *guaeto-ru*
guaethol $C_6H_4(OH)OC_2H_5$

グアカンフォール *guakanfo-ru*
guacamphol $C_8H_{14}COO(C_6H_4OCH_3)_2$

グアナミン *guanamin*
guanamine $C_3H_5N_5$

グアニジン *guanijin*
guanidine $HN=C(NH_2)_2$

グアニカイン *guanikain*
guanicain $C_{21}H_{26}O_3N_3Cl$

グアニン *guanin*
guanine $C_5H_5N_5O$

グアニル尿素 *guanirunyouso*
guanylurea $H_2NC(=NH)NHCONH_2$

グアニル酸 *guanirusan*
guanylic acid $C_{10}H_{14}N_5O_8P$

グアノ *guano*
guano

グアノシン *guanoshin*
guanosine $C_{10}H_{13}N_5O_5$

グアセチン *guasechin*
guacetin, guaiacetin
$C_6H_4(OH)OCH_2COONa$

グアヴァチン *guavachin*
guavacine $C_6H_9O_2N$

グアヤック酸 *guayakkusan*
guaiacic acid $C_6H_8O_3$

グアヤコン酸 *guayakonsan*
guaiaconic acid $C_{20}H_{24}O_5$

グアヤコール *guayako-ru*
guaiacol $C_6H_4(OH)OCH_3$

グアヤシル *guayashiru*
guaiacyl $(C_6H_3(OCH_3)(OH)SO_3)_2Ca$

グムマイト *gumumaito*
gummite $(Pb,Ca,Ba)SiU_3O_{12} \cdot 5H_2O$

グノスコピン *gunosukopin*
gnoscopine $C_{22}H_{23}NO_9$

グラディエント *guradiento*
gradient

グラファイト *gurafaito*
graphite

グラファイトグリース *gurafaitoguri-su*
graphite lubricant

グラフ *gurafu*
graph

ぐらい *gurai*
approximately, about, roughly

グラム *guramu*
gram

グラム当量 *guramutouryou*
gram equivalent

グラウシン *guraushin*
glaucine $C_{21}H_{25}NO_4$

グレアムの拡散の法則
gureamunokakusannohousoku
Graham's law of diffusion

グレン *guren*
grain

グリコーゲン *guriko-gen*
glycogen $(C_{10}H_{10}O_5)_n$

グリコゲナーゼ *gurikogena-ze*
glycogenase

グリコーゲン分解 *guriko-genbunkai*
glycogenolysis

グリココレイン酸 *gurikokoreinsan*
glycocholeic acid $C_{26}H_{43}NO_5$

グリココール *gurikoko-ru*
glycocoll H_2NCH_2COOH

グリココール酸 *gurikoko-rusan*
glycocholic acid $C_{26}H_{43}NO_6$

グリコン酸 *gurikonsan*
gluconic acid $HOCH_2(CHOH)_4COOH$

グリコリッド *gurikoriddo*
glycolide $C_4H_4O_4$

グリコリピド *gurikoripido*
glycolipid

グリコール *guriko-ru*
glycol $HOCH_2CH_2OH$

グリコールアルデヒド
guriko-ruarudehido
glycoaldehyde $HOCH_2CHO$

グリコルル酸 *gurikorurusan*
glycoluric acid $H_2NCONHCH_2COOH$

グリコール酸 *guriko-rusan*
glycolic acid $HOCH_2COOH$

グリコシアミン *gurikoshiamin*
glycocyamin $H_2NC(=NH)NHCH_2COOH$

グリコシダーゼ *gurikoshida-ze*
glycosidase

グリコシド結合 *gurikoshidoketsugou*
glycosidic bond, glycosidic linkage

グリコシド交換反応
gurikoshidokoukanhannou
transglycosylation

グリコシルトランスフェラーゼ
gurikoshirutoransufera-ze
glycosyltransferase

グリコザル *gurikozaru*
glycosal $C_6H_4(OH)COOC_3H_5(OH)_2$

グリニャール反応 *gurinya-ruhannou*
Grignard reaction

グリニャール試薬 *gurinya-rushiyaku*
Grignard reagent

グリオキサール *guriokisa-ru*
glyoxal $OHCCHO$

グリオキシム *guriokishimu*
glyoxime $HONCCNOH$

グリオキシル酸 *guriokishirusan*
glyoxylic acid $OHC-COOH$

グリオキザリン *guriokizarin*
glyoxaline, imidazole $C_3H_4N_2$

グリオキザル *guriokizaru*
glyoxal $OHCCHO$

グリオキザル酸 *guriokizarusan*
glyoxylic acid $OHC-COOH$

グリセリド *guriserido*
glyceride

グリセリン *guriserin*
glycerol $HOCH_2CH(OH)CH_2OH$

グリセリンアルデヒド
guriserinarudehido
glyceric aldehyde $HOCH_2CH(OH)CHO$

グリセリン酸 *guriserinsan*
glyceric acid $HOCH_2CH(OH)COOH$

グリセルアルデヒド *guriseruarudehido*
glyceraldehyde $OHCH_2CH(OH)CHO$

グリシン *gurishin*
glycine H_2NCH_2COOH

グリシルアラニン *gurishiruaranin*
glycylalanine
$H_2NCH_2CONHCH(CH_3)COOH$

グリシルグリシン *gurishirugurishin*
glycylglycine $H_2NCH_2CONHCH_2COOH$

グリース *guri-su*
grease, lubricant

グリザル *gurizaru*
glysal $C_6H_4(OH)COOCH_2CH_2OH$

グロブリン *guroburin*
globulin

グロー放電 *guro-houden*
glow discharge

グロノイン *guronoin*
glyceryl trinitrate $C_3H_5(ONO_2)_3$

グロン酸 *guronsan*
gulonic acid $HOCH_2(CHOH)_5COOH$

グローランプ *guro-ranpu*
glow lamp

グルチン *guruchin*
glutin $C_{192}H_{294}N_{60}O_{70}S$

グルカゴン *gurukagon*
glucagon

グルカン　*gurukan*
　glucan

グルカール　*guruka-ru*
　glucal $C_6H_{10}O_4$

グルカル酸　*gurukarusan*
　glucaric acid $C_6H_{10}O_8$

グルカーゼ　*guruka-ze*
　glucase

グルコアミラーゼ　*gurukoamira-ze*
　glucoamylase

グルコフラノーゼ　*gurukofurano-ze*
　glucofuranose $C_6H_{12}O_6$

グルコン酸　*gurukonsan*
　gluconic acid $HOCH_2(CHOH)_4COOH$

グルコン酸カルシウム
　gurukonsankarushiumu
　calcium gluconate $C_{12}H_{22}CaO_{14}$

グルコサミン　*gurukosamin*
　glucosamine $C_6H_{13}NO_5$

グルコサン　*gurukosan*
　glucosane $C_6H_{10}O_5$

グルコシダーゼ　*gurukoshida-ze*
　glucosidase

グルコシド　*gurukoshido*
　glucoside

グルコソン　*gurukoson*
　glucosone $HOCH_2(CHOH)_3COCHO$

グルコース　*guruko-su*
　glucose $HOCH_2(CHOH)_4CHO$

グルコースイソメラーゼ
　guruko-suisomera-ze
　glucose isomerase

グルコースオキシダーゼ
　guruko-suokishida-ze
　glucose oxidase

グルコース燐酸　*guruko-surinsan*
　glucose phosphate, Cori ester $C_6H_{13}O_9P$

グルコーゼ　*guruko-ze*
　glucose $HOCH_2(CHOH)_4CHO$

グルクロン　*gurukuron*
　glucurone $C_6H_8O_6$

グルクロン酸　*gurukuronsan*
　glucuronic acid $C_6H_{10}O_7$

グルタチオン　*gurutachion*
　glutathione $C_{10}H_{17}N_3O_6S$

グルタコン酸　*gurutakonsan*
　glutaconic acid $HOOCCH_2CH=CHCOOH$

グルタミン　*gurutamin*
　glutamine
　$H_2NOCCH_2CH_2CH(NH_2)COOH$

グルタミナーゼ　*gurutamina-ze*
　glutaminase

グルタミン酸　*gurutaminsan*
　glutamic acid
　$HOOCCH_2CH_2CH(NH_2)COOH$

グルタルアルデヒド　*gurutaruarudehido*
　glutaraldehyde $OHC(CH_2)_3CHO$

グルタル酸　*gurutarusan*
　glutaric acid $HOOC(CH_2)_3COOH$

グルテン　*guruten*
　gluten

グルテニン　*gurutenin*
　glutenin

グルテリン　*guruterin*
　glutelin

グルトーゼ　*guruto-ze*
　glutose
　$HOCH_2CH(OH)CO(CHOH)_2CH_2OH$

グリュンリング石　*guryunringuseki*
　grünlingite Bi_4TeS_3

グヴァチン　*guvachin*
　guvacine $C_6H_9NO_2$

ハ ha

ハーバーボッシュ法　*ha-ba-bosshuhou*
　Haber-Bosch process

ハード塩基　*ha-doenki*
　hard base

ハード酸　*ha-dosan*
　hard acid

ハフニウム　*hafuniumu*
　hafnium (Hf, element 72)

ハーゲンポアズイユの法則
ha-genpoazuiyunohousoku
Hagen-Poiseuille's law

ハイブリダイゼーション
haiburidaize-shon
hybridization

ハイブリッド　*haiburiddo*
hybrid

ハイブリッド分子　*haiburiddobunshi*
hybrid molecule

ハイブリッド化　*haiburiddoka*
hybridization

ハイディンゲル石　*haidingeruseki*
haidingerite $HCaAsO_4 \cdot H_2O$

ハイドロフォーミング　*haidorofo-mingu*
hydroforming

ハイドロキノン　*haidorokinon*
hydroquinone $C_6H_4(OH)_2$

はかり瓶　*hakaribin*
weighing bottle

はかりビュレット　*hakaribyuretto*
weighing burette

はかり管　*hakarikan*
weighing tube

はっか油　*hakkayu*
peppermint oil

ハックマン石　*hakkumanseki*
hackmannite $3NaAl \cdot SiO_4 \cdot Na_2S$

ハミック反応　*hamikkuhannou*
Hammick reaction

ハミルトン関数　*hamirutonkansuu*
Hamilton function

ハムリン石　*hamurinseki*
hamlinite $3Al_2O_3 \cdot 2SrO \cdot 2P_2O_5 \cdot 7H_2O$

ハンベルグ石　*hanberuguseki*
hambergite $Be_2(OH)BO_3$

ハンチのピリジン合成法
hanchinopirijingouseihou
Hantzsch's pyridine synthesis

ハネー石　*hane-seki*
hannayite $H_4(NH_4)_2Mg_3(PO_4)_4 \cdot 8H_2O$

ハンクス石　*hankususeki*
hanksite $9Na_2SO_4 \cdot 2Na_2CO_3 \cdot KCl$

ハンスディーカー反応
hansudi-ka-hannou
Hunsdiecker reaction

はん点反応　*hantenhannou*
spot reaction

ハプテン　*haputen*
hapten

ハロアルキルチオ基　*haroarukiruchioki*
haloalkylthio group

ハロアルキル基　*haroarukiruki*
haloalkyl group

ハロアルコキシ基　*haroarukokishiki*
haloalkoxy group

ハロゲン　*harogen*
halogen

ハロゲン置換　*harogenchikan*
halogenation

ハルゲンホルム　*harogenhorumu*
haloform, trihalogenomethane CHX_3

ハロゲン化　*harogenka*
halogenation

ハロゲン化亜鉛アルキル
harogenkaaenarukiru
zinc alkyl halogenide $RZnX$

ハロゲン化アルキル　*harogenkaarukiru*
alkyl halide

ハロゲン化物　*harogenkabutsu*
halide, halogenide

ハロゲン化銀　*harogenkagin*
silver halide AgX

ハロゲン化マグネシウムアルキル
harogenkamaguneshiumuarukiru
alkyl magnesium halogenide $RMgX$

ハロゲン化水素　*harogenkasuiso*
hydrogen halide HX

ハロゲン化水素酸　*harogenkasuisosan*
halogen acid

ハルマリン　*harumarin*
harmaline $C_{13}H_{14}N_2O$

ハルマロール　*harumaro-ru*
harmalol $C_{12}H_{12}ON_2$

ハルミン　*harumin*
harmine $C_{13}H_{12}N_2O$

ハルス石　*harususeki*
　hulsite (FeII,Mg)$_2$(FeIII,Sn)BO$_5$

ハルティット　*harutitto*
　hartite C$_{12}$H$_{20}$

ハルトマン石　*harutomanseki*
　hartmannite NiSb

ハルト石　*harutoseki*
　harttite (Sr,Ca)O・2Al$_2$O$_3$・P$_2$O$_5$・SO$_3$・5H$_2$O

ハルヴァックス効果　*haruvakkusukouka*
　Hallwachs effect

ハッシウム　*hasshiumu*
　hassium (Hs, element 108)

ハースティッヒ石　*ha-sutihhiseki*
　harstigite H$_7$(Ca,Mn)$_{12}$Al$_2$Si$_{10}$O$_{40}$

ハウエル石　*haueruseki*
　hauerite MnS$_2$

ハウ石　*hauseki*
　howlite Ca$_2$[SiB$_5$O$_9$(OH)$_5$]

ハザード制御　*haza-doseigyo*
　hazard control

ヘ　he

ヘデラゲニン　*hederagenin*
　hederagenin C$_{30}$H$_{48}$O$_4$

へき開　*hekikai*
　cleavage

ヘキサブロモフルオロプロパン
　hekisaburomofuruoropuropan
　hexabromofluoropropane C$_3$HFBr$_6$

ヘキサデカン　*hekisadekan*
　hexadecane C$_{16}$H$_{34}$

ヘキサデカン酸　*hekisadekansan*
　hexadecanoic acid C$_{15}$H$_{31}$COOH

ヘキサデシルアルコール
　hekisadeshiruaruko-ru
　hexadecyl alcohol, cetyl alcohol
　CH$_3$(CH$_2$)$_{15}$OH

ヘキサフルオロアセトン
　hekisafuruoroaseton
　hexafluoroacetone CO(CF$_3$)$_2$

ヘキサフルオロエタン
　hekisafuruoroetan
　hexafluoroethane C$_2$F$_6$

ヘキサクロロベンゼン
　hekisakurorobenzen
　hexachlorobenzene C$_6$Cl$_6$

ヘキサクロロエタン　*hekisakuroroetan*
　hexachloroethane C$_2$Cl$_6$

ヘキサクロロシクロヘキサン
　hekisakuroroshikurohekisan
　hexachlorocyclohexane C$_6$H$_6$Cl$_6$

ヘキサメチレンジアミン
　hekisamechirenjiamin
　hexamethylenediamine H$_2$N(CH$_2$)$_6$NH$_2$

ヘキサメチレンジイソシアネート
　hekisamechirenjiisoshiane-to
　hexamethylene diisocyanate
　OCN(CH$_2$)$_6$NCO

ヘキサメチレンテトラミン
　hekisamechirentetoramin
　hexamethylenetetramine C$_6$H$_{12}$N$_4$

ヘキサミン　*hekisamin*
　hexamine, hexamethylenetetramine
　C$_6$H$_{12}$N$_4$

ヘキサン　*hekisan*
　hexane C$_6$H$_{14}$

ヘキサンジニトリル　*hekisanjinitoriru*
　adiponitrile NC(CH$_2$)$_4$CN

ヘキサノール　*hekisano-ru*
　hexyl alcohol, hexanol C$_6$H$_{13}$OH

ヘキサン酸　*hekisansan*
　hexanoic acid C$_6$H$_{11}$COOH

ヘキセン　*hekisen*
　hexene C$_6$H$_{12}$

ヘキセストロール　*hekisesutoro-ru*
　hexestrol
　C$_6$H$_4$(OH)CH(C$_2$H$_5$)CH(C$_2$H$_5$)C$_6$H$_4$(OH)

ヘキセトン　*hekiseton*
　hexetone C$_{10}$H$_{16}$O

ヘキシン　*hekishin*
　hexine C$_6$H$_{10}$

ヘキシレン　*hekishiren*
　hexylene C$_6$H$_{12}$

ヘキシルアルコール　*hekishiruaruko-ru*
hexyl alcohol $C_6H_{13}OH$

ヘキシル酸　*hekishirusan*
hexylic acid $CH_3(CH_2)_4COOH$

ヘキソバルビトン　*hekisobarubiton*
hexobarbitone $C_{11}H_{15}O_3NNa$

ヘキソーゲン　*hekiso-gen*
hexogen $C_3H_6O_4N_6$

ヘキソン　*hekison*
hexone $(CH_3)_2CHCH_2COCH_3$

ヘキソース　*hekiso-su*
hexose $C_6H_{12}O_6$

ヘキソザン　*hekisozan*
hexosane $(C_6H_{10}O_5)_n$

ヘキソーゼ　*hekiso-ze*
hexose $C_6H_{12}O_6$

ヘック反応　*hekkuhannou*
Heck reaction

ヘクサチアン　*hekusachian*
hexacyan $C_3N_3(CN)_3$

ヘクサデカン　*hekusadekan*
hexadecane $C_{16}H_{34}$

ヘクサデカン酸　*hekusadekansan*
hexadecanoic acid $C_{15}H_{31}COOH$

ヘクサデシル　*hekusadeshiru*
hexadecyl, cetyl $C_{16}H_{33}-$

ヘクサフェニルエタン　*hekusafeniruetan*
hexaphenylethane $C_{20}H_{30}$

ヘクサヒドロベンゾール
hekusahidorobenzo-ru
hexahydrobenzene C_6H_{12}

ヘクサヒドロオキシ第二白金酸塩
hekusahidorookishidainihakkinsanen
hexahydroxyplatinate $M^I{}_2[Pt(OH)_6]$

ヘクサクロルベンゾール
hekusakurorubenzo-ru
hexachlorobenzene C_6Cl_6

ヘクサクロルエタン　*hekusakuroruetan*
hexachloroethane C_2Cl_6

ヘクサメチレン　*hekusamechiren*
hexamethylene C_6H_{12}

ヘクサメチレンジアミン
hekusamechirenjiamin
hexamethylenediamine $H_2N(CH_2)_6NH_2$

ヘクサメチレンテトラミン
hekusamechirentetoramin
urotropine, hexamethylenetetramine $C_6H_{12}N_4$

ヘクサメチルベンゾール
hekusamechirubenzo-ru
hexamethylbenzene $C_6(CH_3)_6$

ヘクサン　*hekusan*
hexane C_6H_{14}

ヘクサノール　*hekusano-ru*
hexyl alcohol, hexanol $C_6H_{13}OH$

ヘクサリン　*hekusarin*
hexalin $C_6C_{11}OH$

ヘクサール　*hekusa-ru*
hexal $C_6H_3(OH)(COOH)(SO_3H)(CH_2)_6N_4$

ヘクサトリオーゼ　*hekusatorio-ze*
hexatriose $C_{18}H_{32}O_{16}$

ヘクトリットル　*hekutorittoru*
hectoliter

ヘマチン　*hemachin*
hematin $C_{34}H_{32}N_4O_4FeOH$

ヘマチン酸　*hemachinsan*
hematinic acid $C_8H_9NO_4$

ヘマクロイン　*hemakuroin*
hemachroin, hematin $C_{34}H_{32}N_4O_4Fe(OH)$

ヘマタイト　*hemataito*
hematite, red iron ore, bloodstone $\alpha\text{-}Fe_2O_3$

ヘマテイン　*hematein*
hematein $C_{16}H_{12}O_6$

ヘマトキシリン　*hematokishirin*
hematoxylin $C_{16}H_{14}O_6 \cdot 3H_2O$

ヘメリト酸　*hemeritosan*
hemellitic acid $(CH_3)_2C_6H_3COOH$

ヘミアセタール　*hemiaseta-ru*
hemiacetal $RCH(OH)OR'$

ヘミメリテン　*hemimeriten*
hemimellitene $C_6H_3(CH_3)_3$

ヘミメリット酸　*hemimerittosan*
hemimellitic acid $C_6H_3(COOH)_3$

ヘミン　*hemin*
hemin $C_{34}H_{32}N_4O_4FeCl$

ヘミピン酸　*hemipinsan*
hemipic acid $(CH_3O)_2C_6H_2(COOH)_2$

ヘミセルロース, ヘミセルローゼ
hemiseruro-su, hemiseruro-ze
hemicellulose

ヘミテルペン　*hemiterupen*
hemiterpene C_5H_8

ヘモグロビン　*hemogurobin*
hemoglobin

ヘモピロール　*hemopiro-ru*
hemopyrrole $C_4H_2N(CH_3)_2(C_2H_5)$

ヘモポルフィリン　*hemoporufirin*
hemoporphyrin $C_{34}H_{38}N_4O_6$

ヘモルテイン　*hemorutein*
hemocutein $C_{34}H_{56}O_2$

ヘンデカン　*hendekan*
hendecane $C_{11}H_{24}$

ヘンエイコサン　*heneikosan*
henicosane $CH_3(CH_2)_{19}CH_3$

ヘニコサン　*henikosan*
henicosane $CH_3(CH_2)_{19}CH_3$

ヘンリー　*henri-*
Henry

ヘンリー反応　*henri-hannou*
Henry reaction

ヘンリーの法則　*henri-nohousoku*
Henry's law

ヘパチン　*hepachin*
hepatine $(C_6H_{10}O_5)n$

ヘパリン　*heparin*
heparin

ヘプチン　*hepuchin*
heptine C_7H_{12}

ヘプチルアルコール　*hepuchiruaruko-ru*
heptyl alcohol $C_7H_{15}OH$

ヘプチル酸　*hepuchirusan*
oenanthic acid, heptoic aicd
$CH_3(CH_2)_5COOH$

ヘプタデカン　*heputadekan*
heptadecane $C_{17}H_{36}$

ヘプタデカン　*heputadekan*
heptadecane $C_{17}H_{36}$

ヘプタデシリ酸　*heputadeshirisan*
heptadecylic acid $CH_3(CH_2)_{15}COOH$

ヘプタコンタン　*heputakontan*
heptacontane $C_{70}H_{142}$

ヘプタコサン　*heputakosan*
heptacosane $C_{27}H_{56}$

ヘプタクロロフルオロプロパン
heputakurorofuruoropuropan
heptachlorofluoropropane C_3HF_7

ヘプタメチレン　*heputamechiren*
heptamethylene C_7H_{14}

ヘプタモリブデン酸塩
heputamoribudensaen
heptamolybdate $Mo_8O_{26}{}^{4-}$

ヘプタン　*heputan*
heptane C_7H_{16}

ヘプタナフテン　*heputanafuten*
heptanaphthene C_7H_{14}

ヘプタナール　*heputana-ru*
heptanal $CH_3(CH_2)_5CHO$

ヘプタン酸　*heputansan*
heptanoic acid $C_6H_{13}COOH$

ヘプテン　*heputen*
heptene C_7H_{14}

ヘプトン酸　*heputonsan*
heptonic acid $OHCH_2(CHOH)_5COOH$

ヘプトーゼ　*heputo-ze*
heptose $C_7H_{14}O_7$

へら　*hera*
spatula

ヘレボレチン　*hereborechin*
helleboretine $C_{19}H_{30}O_5$

ヘレボレスチン　*hereboresuchin*
helleborescine $C_{30}H_{38}O_4$

ヘレニエン　*herenien*
helenien $C_{72}H_{16}O_4$

ヘレニン　*herenin*
helenin C_6H_8O

ヘリアンチン　*herianchin*
helianthine $(CH_3)_2NC_6H_4N=NC_6H_4SO_3H$

ヘリチン *herichin*
helicin $C_{13}H_{16}O_7$

ヘリトニン *heritonin*
chelidonine $C_{20}H_{19}NO_5$

ヘリウム *heriumu*
helium He (element 2)

ヘリウム核 *heriumukaku*
helium nucleus

ヘロイン *heroin*
heroin, diacetylmorphine $C_{17}H_{17}NO(C_2H_3O_2)_2$

ヘルコゾール *herukozo-ru*
helcusol $C_6H_3(OH)_2OBiOH$

ヘルミトール *herumito-ru*
helmitol $C_6H_5O_7(CH_2)_6N_4$

ヘルムホルツエネルギー *herumuhorutsuenerugi-*
Helmholtz energy, free energy

ヘルツ *herutsu*
hertz

ヘスペレチン *hesuperechin*
hesperetin $C_{16}H_{14}O_6$

ヘスペレチン酸 *hesuperechinsan*
hesperetic acid $C_6H_3(OH)(OCH_3)CH=CHCOOH$

ヘスペリデン *hesuperiden*
hesperidene $C_{10}H_{16}$

ヘスペリジン *hesuperijin*
hesperidin $C_{28}H_{34}O_{15}$

ヘテロ原子 *heterogenshi*
heteroatom

ヘテロ重合体 *heterojuugoutai*
heteropolymer

ヘテロオーキシン *heteroo-kishin*
heteroauxin $C_{10}H_{10}NO_2$

ヘテロリシス *heterorishisu*
heterolysis

ヘトフォルム *hetoforumu*
hetoform $(C_9H_7O_2)_3Bi \cdot Bi_2O_3$

ヘトラリン *hetorarin*
hetralin $C_6H_4(OH)_2 \cdot (CH_2)_6N_4$

ヒ hi

ヒアルロン酸 *hiaruronsan*
hyaluronic acid

ヒアルウロニダーゼ *hiaruuronida-ze*
hyaluronidase

ヒダントイン *hidantoin*
hydantoin $C_3H_4N_2O_2$

ヒダントイン酸 *hidantoinsan*
hydantoic acid $HOOCCH_2NHCONH_2$

ヒデナイト *hidenaito*
hiddenite $LiAl[Si_2O_6]$

ヒドノカルプス酸 *hidonokarupususan*
hydnocarpic acid $C_{16}H_{28}O_2$

ヒドラジド *hidorajido*
hydrazide $RCONHNH_2$

ヒドラジン *hidorajin*
hydrazine H_2NNH_2

ヒドラジン一水和物 *hidorajinichisuiwabutsu*
hydrazine monohydrate $N_2H_4 \cdot H_2O$

ヒドロクリル酸 *hidorakurirusan*
hydracrylic acid $HOCH_2CH_2COOH$

ヒドラセチン *hidorasechin*
hydracetin $CH_3CONHNHC_6H_5$

ヒドラスチン *hidorasuchin*
hydrastine $C_{21}H_{21}NO_6$

ヒドラスチニン *hidorasuchinin*
hydrastinine $C_{11}H_{13}NO_3$

ヒドラスチニン酸 *hidorasuchininsan*
hydrastininic acid $C_{11}H_9NO_6$

ヒドラターゼ *hidorata-ze*
hydratase

ヒドラゾベンゾール *hidorazobenzo-ru*
hydrazobenzene $C_6H_5NHNHC_6H_5$

ヒドラゾン, ヒドラゾーン *hidorazon, hidorazo-n*
hydrazone

ヒドラゾ酸 *hidorazosan*
hydrazoic acid, hydrogen azide HN_3

ヒドリド還元 *hidoridokangen*
hydride reduction

ヒドリンデン　*hidorinden*
hydrindene C_9H_{10}

ヒドロベンゾイン　*hidorobenzoin*
hydrobenzoin $C_6H_5CH(OH)CH(OH)C_6H_5$

ヒドロビリルビン　*hidorobirirubin*
hydrobilirubin $C_{32}H_{40}N_4O_7$

ヒドロゲナーゼ　*hidorogena-ze*
hydrogenase

ヒドロヒドラスチン　*hidorohidorasuchin*
hydrohydrastine $C_{11}H_{13}NO_2$

ヒドロヒノン　*hidorohinon*
hydroquinone $C_6H_4(OH)_2$

ヒドロキニン　*hidorokinin*
hydroquinine $C_{20}H_{26}N_2O_2 \cdot 2H_2O$

ヒドロキノン　*hidorokinon*
hydroquinone $C_6H_4(OH)_2$

ヒドロキサム酸　*hidorokisamusan*
hydroxamic acid RCONHOH

ヒドロキシプロピルメチルセルロース
hidorokishipuropirumehciruseruso-su
hydroxypropylmethylcellulose

ヒドロキシプロピルセルロース
hidorokishipuropiruseruro-su
hydroxypropylcellulose

ヒドロキシルアミン　*hidorokishiruamin*
hydroxylamine NH_2OH

ヒドロキシル化　*hidorokishiruka*
hydroxylation

ヒドロキシ酸　*hidorokishisan*
hydroxy acid

ヒドロクマロン　*hidorokumaron*
hydrocumarone C_8H_8O

ヒドロン　*hidoron*
hydrone

ヒドロオクソニウムイオン
hidorookusoniumuion
hydronium ion H_3O^+

ヒドロペルオキシド　*hidoroperuokishido*
hydroperoxide MOOH

ヒドロラーゼ　*hidorora-ze*
hydrolase

ヒドロール　*hidoro-ru*
hydrol H_2O_4

ヒドロゾル　*hidorozoru*
hydrosol

ヒドルアトロパ酸　*hidoruatoropasan*
hydratropic acid $C_6H_5CH(CH_3)COOH$

ヒエナ酸　*hienasan*
hyenic acid $C_{24}H_{49}COOH$

ヒエラ石　*hieraseki*
hieratite K_2SiF_6

ヒッギンス石　*higginsuseki*
higginsite $CaCu(OH)AsO_4$

ヒグリン　*higurin*
hygrine $C_8H_{15}NO$

ヒグリン酸　*higurinsan*
hygric acid $C_6H_{11}NO_2$

ヒ化亜鉛　*hikaaen*
zinc arsenide Zn_3As_2

ヒ化銅　*hikadou*
copper arsenide Cu_3As_2

ヒ化水素　*hikasuiso*
arsane AsH_3

ヒマシ油　*himashiyu*
castor oil

ひまわり油　*himawariyu*
sunflower oil

ヒンスベルグ反応　*hinsuberuguhannou*
Hinsberg reaction

ヒパコニチン　*hipakonichin*
hypaconitine $C_{33}H_{45}NO_{10}$

ヒポゲア酸　*hipogeasan*
hypogaeic acid
$CH_3(CH_2)_7CH=CH(CH_2)_5COOH$

ヒポキサンチン　*hipokisanchin*
hypoxanthine $C_5H_4N_4O$

ヒプナール　*hipuna-ru*
hypnal $C_{13}H_{15}N_2O_3Cl_3$

ヒプノン　*hipunon*
hypnone $CH_3COC_6H_5$

ヒレブランド石　*hireburandoseki*
hillebrandite Ca_2SiO_4

ヒルガルダイト　*hirugarudaito*
hilgardite $Ca_8(B_6O_{11})_3Cl_4 \cdot 4H_2O$

ヒルゲンストック石　*hirugensutokkuseki*
　hilgenstockite $4CaO \cdot P_2O_5$

ヒルスチジン　*hirusuchijin*
　hirsutidine $C_{18}H_{16}O_{17} \cdot HCl$

ヒルスチン　*hirusuchin*
　hirsutine $C_{30}H_{37}O_{17}Cl$

ビルスマイヤーハック反応　*hirusumaiya-hakkuhannou*
　Vilsmeier-Haack reaction

ヒ酸　*hisan*
　arsenic acid H_3AsO_4

ヒ酸塩　*hisanen*
　arsenate $M^I_3AsO_4$

ヒ酸ナトリウム　*hisannatoriumu*
　sodium arsenate Na_3AsO_4

ヒ素, 砒素　*hiso*
　arsenic (As, element 33)

ヒスチダーゼ　*hisuchida-ze*
　histidase

ヒスチジン　*hisuchijin*
　histidine $C_6H_9N_3O_2$

ヒスダザリン　*hisudazarin*
　hystazarin $C_{14}H_8O_4$

ひすい　*hisui*
　jade

ヒスタミン　*hisutamin*
　histamine $C_5H_9N_3$

ヒステリシス　*hisuterishisu*
　hysteresis

ヒストチム　*hisutochimu*
　histozyme

ヒストグラム　*hisutoguramu*
　histogram

ヒストン　*hisuton*
　histone

ヒートポンプ　*hi-toponpu*
　heat pump

ヒットルフの輸率　*hittorufunoyuritsu*
　Hittorf number

ヒヨスチアミン　*hiyosuchiamin*
　hyoscyamine $C_{17}H_{23}NO_3$

ヒヨスチン　*hiyosuchin*
　hyoscine $C_{17}H_{21}NO_4$

ホ ho

ホフマン脱離　*hofumandatsuri*
　Hofmann elimination

ホフマン減成　*hofumangensei*
　Hofmann degradation

ホフマン転位　*hofumanteni*
　Hofmann rearrangement

ホイヘンスの原理　*hoihensunogenri*
　Huygens principle

ホイットネー石　*hoittone-seki*
　whitneyite Cu_9As

ホーマン石　*ho-manseki*
　hohmannite $Fe_2O_3 \cdot 2SO_3 \cdot 7H_2O$

ホマリン　*homarin*
　homarine $C_7H_7NO_2$

ホミライト　*homiraito*
　homilite $FeCa_2B_2Si_2O_{10}$

ホモブレンツカテキン　*homoburentsukatekin*
　homopyrocatechol $C_6H_3(CH_3)(OH)_2$

ホモフタル酸　*homofutarusan*
　homophthalic acid $HOOCCH_2C_6H_4COOH$

ホモゲンチシン酸　*homogenchishinsan*
　homogentisic acid $C_8H_8O_4$

ホモ芳香族　*homohoukouzoku*
　homoaromatic

ホモジナイザー　*homojinaiza-*
　homogenizer

ホモジネート　*homojine-to*
　homogenate

ホモ重合　*homojuugou*
　homopolymerization

ホモポリマー　*homoporima-*
　homopolymer

ホモリシス　*homorishisu*
　homolysis

ホモログ　*homorogu*
　homologue, homologous compound

ホモサリチル酸　*homosarichirusan*
　homosalicylic acid $C_6H_3(OH)(CH_3)COOH$

ホモセリン　*homoserin*
　homoserine $HO(CH_2)_2CH(NH_2)COOH$

ホモスルファミン　*homosurufamin*
　homosulfamine
　$(H_2NCH_2)C_6H_4SO_2NH_2 \cdot HCl$

ホムトロピン　*homutoropin*
　homatropine $C_{16}H_{21}NO_3$

ホーナーエモンズ反応
　ho-na-emonzuhannou
　Horner-Emmons reaction

ホペリジン　*hoperijin*
　operidine $C_{14}H_{19}NO_2 \cdot HCl$

ホッパー　*hoppa-*
　hopper

ホップ　*hoppu*
　hop

ホープ石　*ho-puseki*
　hopeite $Zn_3(PO_4)_2 \cdot 4H_2O$

ホローファイバー　*horo-faiba-*
　hollow fiber

ホロカイン　*horokain*
　holocaine
　$CH_3C(=NC_6H_4OC_2H_5)NHC_6H_4C_2H_5$

ホルデニン　*horudenin*
　hordenine $C_6H_4(OH)CH_2CH_2N(CH_3)_2$

ホルデン石　*horudenseki*
　holdenite $8MnO \cdot 4ZnO \cdot As_2O_5 \cdot 5H_2O$

ホール効果　*ho-rukouka*
　Hall effect

ホルマリン　*horumarin*
　formalin

ホルミル基　*horumiruki*
　formyl group -CHO

ホルミウム　*horumiumu*
　holmium (Ho, element 67)

ホルモン　*horumon*
　hormone

ホルムアミド　*horumuamido*
　formamide $HCONH_2$

ホルムアルデヒド　*horumuarudehido*
　formaldehyde HCHO

ホルムアルデヒド液
　horumuarudehidoeki
　formaldehyde solution

ホルム石　*horumuseki*
　holmite NiSe

ホスファーゲン　*hosufa-gen*
　phosphagen
　$HOOCCH_2N(CH_3)C(=NH)NHPO(OH)_2$

ホスファターゼ　*hosufata-ze*
　phosphatase

ホスフィン　*hosufin*
　phosphine PH_3

ホスフィンオキシド　*hosufinokishido*
　phosphinoxide H_3PO (= H_3PO)

ホスフィン酸　*hosufinsan*
　phosphinic acid H_3PO_2 (= $H_2PO(OH)$)

ホスフォリパーゼ　*hosuforipa-ze*
　phospholipase

ホスゲン　*hosugen*
　phosgene $COCl_2$

ホスホン酸　*hosuhonsan*
　phosphonic acid H_3PO_3 (= $HPO(OH)_2$)

ホスホリパーゼ　*hosuhoripa-ze*
　phospholipase

ホットスポット　*hottosupotto*
　hot spot

ホウ酸, 硼酸　*housan*
　boric acid H_3BO_3

ホウ酸リチウム　*housanrichiumu*
　lithium borate $Li_2B_4O_7 \cdot 5H_2O$

ホウ酸トリエチル　*housantoriechiru*
　triethyl borate

ホウ素, 硼素　*houso*
　boron (B, element 5)

ヒュ hyu

ヒューム石　*hyu-museki*
　humite $Mg(Fe,OH)_2 \cdot Mg_5(SiO_4)_3$

ヒュウェット石　*hyuwettoseki*
hewettite CaO・3V$_2$O$_3$・9H$_2$O

ヒューズ　*hyu-zu*
electric fuse, fuse

イ i

イディオタイプ　*idiotaipu*
idiotype

イドーゼ　*ido-ze*
idose C$_6$H$_{12}$O$_6$

イェナガラス　*ienagarasu*
Jena glass

イヒチオール　*ihichio-ru*
ichthyol C$_{28}$H$_{42}$O$_6$N$_2$S$_3$・2H$_2$O

イヒタルガン　*ihitarugan*
ichthargan

イヒトホルム　*ihitohorumu*
ichthoform

イジングス石　*ijingususeki*
iddingsite MgO・Fe$_2$O$_3$・3SiO$_2$・4H$_2$O

イコサン　*ikosan*
eicosane, icosane C$_{20}$H$_{42}$

イクストルーダー　*ikusutoru-da-*
extruder

イミダゾール　*imidazo-ru*
imidazole C$_3$H$_4$N$_2$

イミド　*imido*
imide

イミドエステル　*imidoesuteru*
imido ester RC(=NH)OR′

イミド尿素　*imidonyouso*
iminourea, guanidin HN=C(NH$_2$)$_2$

イミド酸　*imidosan*
imidic acid RC(=NH)OH

イミン　*imin*
imine RR′C=NR″

イミノエステル　*iminoesuteru*
imido ester RC(=NH)OR′

イミノ基　*iminoki*
imino group

イミノ二酢酸　*iminonisakusan*
iminodiacetic acid
HOOCCH$_2$NHCH$_2$COOH

イナガラス　*inagarasu*
Jena glass

インベルターゼ　*inberuta-ze*
invertase

インビボ　*inbibo*
in vivo

インビトロ　*inbitoro*
in vitro

インダクタンス　*indakutansu*
inductance

インダミン　*indamin*
indamine C$_{12}$H$_{11}$N$_3$

インダン　*indan*
indane C$_9$H$_{10}$

インダンスレン　*indansuren*
indanthrene C$_{28}$H$_{14}$N$_2$O$_4$

インダンスレンブルー　*indansurenburu-*
indanthrene blue, vat Blue C$_{28}$H$_{14}$N$_2$O$_4$/
C$_{28}$H$_{12}$N$_2$O$_4$Cl$_2$

インダントロン　*indantoron*
indanthrone C$_{28}$H$_{14}$N$_2$O$_4$

インダゾール　*indazo-ru*
indazole C$_7$H$_6$N$_2$

インデン　*inden*
indene C$_9$H$_8$

インデライト，インデル石
inderaito, inderuseki
inderite Mg$_2$B$_6$O$_{11}$・15H$_2$O

インドフェニン　*indofenin*
indophenine C$_{24}$H$_{14}$O$_2$N$_2$S$_2$

インドフェノール　*indofeno-ru*
indophenol C$_{12}$H$_9$NO$_2$

インドキシル　*indokishiru*
indoxyl C$_8$H$_7$NO

インドキシル酸　*indokishirusan*
indoxylic acid C$_9$H$_7$NO$_3$

インドール　*indo-ru*
indole C$_8$H$_7$N

インドール酢酸　*indo-rusakusan*
indoleacetic acid C$_{10}$H$_9$NO$_2$

イニシエーション　*inishie-shon*
　initiation

イニシエーター　*inishie-ta-*
　initiator, starter

インジアナ石　*injianaseki*
　indianaite $Al_2O_3 \cdot 2SiO_2$

インジゴ, インジゴチン　*injigo, injigochin*
　indigo, indigo blue $C_{16}H_{10}N_2O_2$

インジゴ染色法　*injigosenshokuhou*
　indigo dyeing

インジルビン　*injirubin*
　indigo red, indirubin $C_{16}H_{10}N_2O_2$

インジウム　*injiumu*
　indium (In, element 49)

インキ　*inki*
　ink

インキュベーション　*inkyube-shon*
　incubation

インキュベーター　*inkyube-ta-*
　incubator

イノシン酸　*inoshinsan*
　inosinic acid $C_{10}H_{13}N_4O_8P$

イノシトール, イノシット, イノシットル　*inoshito-ru, inoshitto, inoshittoru*
　inositol $C_6H_6(OH)_6$

イノーゼ　*ino-ze*
　inose $C_6H_6(OH)_6$

インパルス電流　*inparusudenryuu*
　impulse current

インパルス発生器　*inparusuhasseiki*
　impulse generator

インピーダンス　*inpi-dansu*
　impedance

インシュリン　*inshurin, insurin*
　insulin $C_{257}H_{387}N_{65}O_{66}S_6$

インターフェロン　*inta-feron*
　interferon

イントロン　*intoron*
　intron

イヌリン　*inurin*
　inulin, alant starch $(C_6H_{10}O_5)_x$

インゼクター　*inzekuta-*
　injector

イオン　*ion*
　ion

イオンチャンネル　*ionchanneru*
　ion channel

イオン伝導　*iondendou*
　ionic conduction

イオン電荷　*iondenka*
　ionic charge

イオン雰囲気　*ionfuniki*
　ion atmosphere

イオン半径　*ionhankei*
　ionic radius

イオン反応　*ionhannou*
　ionic reaction

イオン平衡　*ionheikou*
　ionic equilibrium

イオン移動　*ionidou*
　ionic migration

イオニウム　*ioniumu*
　ionium

イオン重合　*ionjuugou*
　ionic polymerization

イオン化　*ionka*
　ionization

イオン化電位　*ionkadeni*
　ionization potential

イオン化エネルギー　*ionkaenerugi-*
　ionization energy

イオン化傾向　*ionkakeikou*
　ionization tendency

イオン化ポテンシャル　*ionkapotensharu*
　ionization potential

イオン結晶　*ionkesshou*
　ionic crystal

イオン結合　*ionketsugou*
　ionic bond

イオン交換　*ionkoukan*
　ion exchange

イオン交換樹脂　*ionkoukanjushi*
　ion-exchange resin

イオン交換クロマトグラフィー
ionkoukankuromatogurafi-
ion-exchange chromatography

イオン交換体　*ionkoukantai*
ion exchanger

イオン交換容量　*ionkoukanyouryou*
ion-exchange capacity

イオン格子　*ionkoushi*
ionic lattice

イオンクロマトグラフィー
ionkuromatogurafi-
ion chromatography

イオン強度　*ionkyoudo*
ionic strength

イオン濃度　*ionnoudo*
ion concentration

イオノホア　*ionohoa*
ionophore

イオノマー　*ionoma-*
ionomer

イオノン　*ionon*
ionone $C_{13}H_{20}O$

イオン積　*ionseki*
ionic product

イオン線　*ionsen*
ion beam

イオン対　*iontsui*
ion pair

イレン　*iren*
irene $C_{13}H_{18}$

イリド　*irido*
ylide

イリゲーション　*irige-shon*
irrigation

イリジン　*irijin*
iridin $C_{24}H_{26}O_{13}$

イリジン酸　*irijinsan*
iridinic acid $C_{10}H_{12}O_{16}$

イリジウム　*irijiumu*
iridium (Ir, element 77)

イリニウム　*iriniumu*
illinium, promethium (Pm, element 61)

イロン　*iron*
irone $C_{13}H_{20}O$

イルメナイト　*irumenaito*
ilmenite $FeTiO_3$

イサチン　*isachin*
isatin $C_8H_5NO_2$

イサトロパ酸　*isatoropasan*
isatropic acid $C_{18}H_{16}O_4$

イサツェン　*isatsen*
isacene $C_{24}H_{19}NO_5$

イセチオン酸　*isechionsan*
isethionic acid $HOCH_2CH_2SO_3H$

イシンダゾール　*ishindazo-ru*
isindazol $C_7H_6N_2$

イソ　*iso*
iso

イソアミレン　*isoamiren*
isoamylene C_5H_{10}

イソアミルアルコール　*isoamiruaruko-ru*
isoamyl alcohol $(CH_3)_2CHCH_2CH_2OH$

イソアミルエーテル　*isoamirue-teru*
isoamyl ether $C_5H_{11}OC_5H_{11}$

イソアミラーゼ　*isoammira-ze*
isoamylase

イソアピオール　*isoapio-ru*
isoapiol $C_{12}H_{14}O_4$

イソアスコルビン酸　*isoasukorubinsan*
isoascorbic acid $C_6H_8O_6$

イソバレルアルデヒド
isobareruarudehido
iso valeraldehyde $(CH_3)_2CHCH_2CHO$

イソバリン　*isobarin*
isovaline $H_2NC(CH_3)(C_2H_5)COOH$

イソボルネオール　*isoboruneo-ru*
isoborneol $C_{10}H_{17}OH$

イソブチレン　*isobuchiren*
isobutylene $H_2C=C(CH_3)_2$

イソブチル　*isobuchiru*
isobutyl, 2-methylpropyl $(CH_3)_2CHCH_2$-

イソブチルアミン　*isobuchiruamin*
isobutylamine $(CH_3)_2CHCH_2NH_2$

イソブチルアルデヒド *isobuchiruarudehido*
isobutyraldehyde $(CH_3)_2CHCHO$

イソブチルアルコール *isobuchiruaruko-ru*
isobutanol $(CH_3)_2CHCH_2OH$

イソブチロニトリル *isobuchirunitoriru*
isobutyronitrile $(CH_3)_2CHCN$

イソブタン *isobutan*
isobutane $HC(CH_3)_3$

イソブタノール *isobutano-ru*
isobutanol $(CH_3)_2CHCH_2OH$

イソチオ尿素 *isochionyouso*
isothiourea $H_2NC(=NH)SH$

イソチオシアン酸 *isochioshiansan*
isothiocyanic acid $HNCS$

イソチオシアン酸アリル *isochioshiansanariru*
allyl isothiocyanate $H_2C=CHCH_2N=C=S$

イソチオシアン酸塩 *isochioshiansanen*
isothiocyanate $MNCS$

イソチオシアン酸メチル *isochioshiansanmechiru*
methyl isothiocyanate CH_3NCS

イソデュロール *isodyuro-ru*
isodurene $C_6H_2(CH_3)_4$

イソエモジン *isoemojin*
isoemodin $C_{15}H_{10}O_5$

イソエルカ酸 *isoerukasan*
isoerucic acid
$CH_3(CH_2)_7CH=CH(CH_2)_{11}COOH$

イソフェルラ酸 *isoferurasan*
isoferulic acid
$6H_3(OH)(OCH_3)CH=CHCOOH$

イソフラボン *isofurabon*
isoflavone $C_{15}H_{10}O_2$

イソフタル酸 *isofutarusan*
isophthalic acid $C_6H_4(COOH)_2$

イソヘキサン *isohekisan*
isohexane $CH_3(CH_2)_2CH(CH_3)_2$

イソホロン *isohoron*
isophorone $C_9H_{14}O$

イソジアルル酸 *isojiarurusan*
isodialuric acid $C_4H_4N_2O_4$

イソ次燐酸, イソ次リン酸 *isojirinsan*
isohypophosphoric acid $H_4P_2O_6$

イソカンファン *isokanfan*
isocamphane $C_{10}H_{18}$

イソカプリル酸 *isokapurirusan*
isocaprylic acid $(CH_3)_2CH(CH_2)_4COOH$

イソカプロン酸 *isokapuronsan*
isocaproic acid $(CH_3)_2CH(CH_2)_2COOH$

イソケチン酸 *isokechinsan*
isocetinic acid $CH_3(CH_2)_{13}COOH$

イソキノリン *isokinorin*
isoquinoline C_9H_7N

イソキサチオン *isokisachion*
isoxathion $C_{13}H_{16}NO_4PS$

イソキサゾール *isokisazo-ru*
isoxazole C_3H_3NO

イソキシロール *isokishiro-ru*
isoxylol, isoxylene C_3H_3NO

イソ吉草酸 *isokissousan*
isovaleric acid $(CH_3)_2CHCH_2COOH$

イソ吉草酸エチル *isokissousanechiru*
ethyl isovalerate $(CH_3)_2CHCH_2COOC_2H_5$

イソ吉草酸メチル *isokissousanmechiru*
methyl isovalerate $(CH_3)_2CHCH_2COOCH_3$

イソコトイン *isokotoin*
isocotoin $C_6H_5COC_6H_2(OH)_2(OCH_3)$

イソ酵素 *isokouso*
isoenzyme, isozyme

イソくえん酸 *isokuensan*
isocitric acid
$HOOCCH(OH)CH(COOH)CH_2COOH$

イソクマリン *isokumarin*
isocoumarine $C_9H_6O_2$

イソクモール *isokumo-ru*
isocumene $C_6H_4CH_2CH_2CH_3$

イソクラサイト *isokurasaito*
isoclasite $Ca_2P_2O_7 \cdot Ca(OH)_2 \cdot 3H_2O$

イソクロトン酸 *isokurotonsan*
isocrotonic acid $CH_3CH=CHCOOH$

イソマルトース　*isomaruto-su*
　isomaltose

イソメラーゼ　*isomera-ze*
　isomerase

イソナフトール　*isonafuto-ru*
　isonaphthol $C_{10}H_7OH$

イソニコチン酸　*isonikochinsan*
　isonicotinic acid $C_5H_4N(COOH)$

イソニトリル　*isonitoriru*
　isonitrile, isocyanide RNC

イソニトロ　*isonitoro*
　isonitro, aci-nitro $=N(O)OH$

イソニトロソ　*isonitoroso*
　isonitroso, oxime $=NOH$

イソノナン　*isononan*
　isononane C_9H_{20}

イソ尿素　*isonyouso*
　isourea $H_2NC(=NH)OH$

イソオイゲノール　*isooigeno-ru*
　isoeugenol $C_{10}H_{12}O_2$

イソオキサゾール　*isookisazo-ru*
　isoxazole C_3H_3NO

イソオクタン　*isookutan*
　isooctane $CH_3CH(CH_3)CH_2C(CH_3)_3$

イソオクタン酸　*isookutansan*
　isooctanoic acid $C_7H_{15}COOH$

イソペンチルアルコール　*isopenchiruaruko-ru*
　isoamyl alcohol $(CH_3)_2CHCH_2CH_2OH$

イソペンタン　*isopentan*
　isopentane $CH_3CH_2CH(CH_3)_2$

イソプラール　*isopura-ru*
　isopral $Cl_3CCH(CH_3)OH$

イソプレゴン　*isopuregon*
　isopulegon $C_{10}H_{16}O$

イソプレゴール　*isopurego-ru*
　isopulegol $C_{10}H_{18}O$

イソプレン　*isopuren*
　isoprene $H_2C=C(CH_3)CH=CH_2$

イソプレンゴム　*isopurengomu*
　isoprene rubber

イソプレノイド　*isopurenoido*
　isoprenoid

イソプロチオラン　*isopurochioran*
　isoprothiolane $C_{12}H_{18}O_4S_2$

イソプロパノール　*isopuropano-ru*
　isopropyl alcohol $(CH_3)_2CHOH$

イソプロピル　*isopuropiru*
　isopropyl $(CH_3)_2CH-$

イソプロピルアルコール　*isopuropiruaruko-ru*
　isopropyl alcohol $(CH_3)_2CHOH$

イソプロピルエーテル　*isopuropirue-teru*
　isopropyl ether $(CH_3)_2CHOCH(CH_3)_2$

イソ酪酸　*isorakusan*
　isobutyric acid $(CH_3)_2CHCOOH$

イソラムネチン　*isoramunechin*
　isorhamnetin $C_{16}H_{12}O_7$

イソラムノーゼ　*isoramuno-ze*
　isorhamnose $C_6H_{12}O_5$

イソロイシン　*isoroishin*
　isoleucine
　$CH_3CH_2CH(CH_3)CH(NH_2)COOH$

イソサフロール　*isosafuro-ru*
　isosafrole $C_{10}H_{10}O_2$

イソセリン　*isoserin*
　isoserine $HOOCCH(OH)CH_2NH_2$

イソシアニド　*isoshianido*
　isocyanide, isonitrile RNC

イソシアン化物　*isoshiankabutsu*
　isocyanide, isonitrile RNC

イソシアン化エチル　*isoshiankaechiru*
　ethyl isocyanide C_2H_5NC

イソシアン酸　*isoshiansan*
　isoyanic acid $HN=C=O$

イソシアン酸塩　*isoshiansanen*
　isocyanate MNCO

イソシアン酸フェニル　*isoshiansanfeniru*
　phenyl isocyanate $C_6H_5N=C=O$

イソシアン酸メチル　*isoshiansanmechiru*
　methyl isocyanate $CH_3N=CO$

イソシアン酸シクロヘキシル　*isoshiansanshikurohekishiru*
　cyclohexyl isocyanate $C_6H_{11}NCO$

イソシアヌール酸　*isoshianu-rusan*
　isocyanuric acid, fulminuric acid
　$NCCH(NO_2)CONH_2$

イソスチルベン　*isosuchiruben*
　isostilbene $C_6H_5CH=CHC_6H_5$

イソステアリン酸　*isosutearinsan*
　isostearic acid $(CH_3)_2CH(CH_2)_{14}COOH$

イソ体　*isotai*
　isomer

イソタクチック　*isotakuchikku*
　isotactic

イソツジョン　*isotsujon*
　isothujone $C_{10}H_{16}O$

イソヴァニリン　*isovanirin*
　isovanillin,
　3-hydroxy-4-methoxybenzaldehyde
　$C_6H_3(OCH_3)(OH)CHO$

イソヴァレリアン酸　*isovareriansan*
　isovaleric acid $(CH_3)_2CHCH_2COOH$

イソヴァレロン　*isovareron*
　isovalerone, 2,6-dimethyl-4-heptanone
　$(CH_3)_2CHCH_2COCH_2CH(CH_3)_2$

イソヴァリン　*isovarin*
　isovaline, 2-amino-2-methylbutanoic acid
　$H_2NC(CH_3)(C_2H_5)COOH$

イスチジン　*isuchijin*
　istizine $C_{14}H_8O_4$

いす形　*isugata*
　chair-form

イタコン酸　*itakonsan*
　itaconic acid $HOOCC(=CH_2)CH_2COOH$

イッテルビウム　*itterubiumu*
　ytterbium (Yb, element 70)

イットリウム　*ittoriumu*
　yttrium (Y, element 39)

イットロクラサイト　*ittorokurasaito*
　yttrocrasite $Y_2O_3 \cdot TiO_2 \cdot ThO_2 \cdot H_2O$

ジャ ja

ジャボリン　*jaborin*
　jaborine $C_{22}H_{32}N_4O_{11}$

ジャーゴン　*ja-gon*
　jargon $ZrSiO_4$

ジャイプル鉱　*jaipurukou*
　jaipurite, gray cobalt ore CoS

じゃこう　*jakou*
　musk

ジャンシナイト　*janshinaito*
　janthinite $2UO_2 \cdot 7H_2O$

ジャスマール　*jasuma-ru*
　jasmal $C_9H_{10}O_2$

ジャスミナール　*jasumina-ru*
　jasmine aldehyde, jasminal
　$C_6H_5CH=C(CHO)C_5H_{11}$

ジャスミン油　*jasuminyu*
　jasmine oil

ジャスモン　*jasumon*
　jasmone $C_{11}H_{16}O$

ジ ji

ジアベチン　*jiabechin*
　diabetin, levulose $C_6H_{12}O_6$

ジアドカイト　*jiadokaito*
　diadochite $H_{24}Fe_4P_2S_2O_{29}$

ジアミド　*jiamido*
　hydrazine H_2NNH_2

ジアニスジン　*jianisujin*
　dianisidine $C_{14}H_{16}O_2N_2$

ジアリル　*jiariru*
　diallyl $H_2C=CH(CH_2)_2CH=CH_2$

ジアルデヒド　*jiarudehido*
　dialdehyde

ジアルキルアミン　*jiarukiruamin*
　dialkylamine HNR_2

ジアルキルジシラノール　*jiarukirujishirano-ru*
　dialkyldisilanol $R_2Si(OH)_2$

ジアルル酸　*jiarurusan*
　dialuric acid $C_4H_4N_2O_4$

ジアセチン　*jiasechin*
　diacetin $C_3H_5(OH)(OOCCH_3)_2$

ジアセチル　*jiasechiru*
　diacetyl $CH_3COCOCH_3$

ジアセチルジオキシム　*jiasechirujiokishimu*
　diacetyldioxime $CH_3C(=NOH)C(=NOH)CH_3$

ジアセトンアルコール　*jiasetonaruko-ru*
　diacetone alcohol $(CH_3)_2C(OH)CH_2COCH_3$

ジアスポル　*jiasuporu*
　diaspore $Al_2O_3 \cdot H_2O$

ジアスターゼ　*jiasuta-ze*
　diastase

ジアステレオ異性体　*jiasutereoiseitai*
　diastereomer

ジアステレオマー　*jiasutereoma-*
　diastereomer

ジアゾ化　*jiazoka*
　diazotization

ジアゾ化合物　*jiazokagoubutsu*
　diazo-compound

ジアゾカップリング　*jiazokappuringu*
　diazo-coupling

ジアゾメタン　*jiazometan*
　diazomethane CH_2N_2

ジアゾニウム塩　*jiazoniumuen*
　diazonium salt

ジアゾール　*jiazo-ru*
　diazole $C_3H_4N_2$

ジアゾ酢酸　*jiazosakusan*
　diazoacetic acid $N_2CHCOOH$

ジアゾ染料　*jiazosenryou*
　diazo dye

ジバナジン酸　*jibanajinsan*
　divanadic acid $H_4V_2O_7$

ジベンジル　*jibenjiru*
　dibenzyle $C_6H_5CH_2CH_2C_6H_5$

ジベンゾフラン　*jibenzofuran*
　dibenzofuran $C_{12}H_8O$

ジベンゾイル　*jibenzoiru*
　benzil, dibenzoyl $C_6H_5COCOC_6H_5$

ジベレリン　*jibererin*
　gibberellin $C_{19}H_{24}O_6$

ジビニルエーテル　*jibinirue-teru*
　divinyl ether $H_2C=CHOCH=CH_2$

ジーボーギウム　*ji-bo-giumu*
　seaborgium (former eka-tungsten) (Sg, element 106)

ジボラン　*jiboran*
　diborane B_2H_6

ジブロミン　*jiburomin*
　dibromin $C_4H_2O_3N_2Br_2$

ジブロモフルオロメタン　*jiburomofuruorometan*
　dibromofluoromethane $CHFBr_2$

ジブロモジフルオロメタン　*jiburomojifuruorometan*
　difluorodibromomethan CBr_2F_2

ジブロモクロロメタン　*jiburomokurorometan*
　chlorodibromomethane $CHBr_2Cl$

ジブロモペンタフルオロプロパン　*jiburomopentafuruoropuropan*
　dibromopentafluoropropane $C_3HBr_2F_5$

ジブロモテトラフルオロプロパン　*jiburomotetorafuruoropuropan*
　dibromotetrafluoropropane $C_3H_2F_4Br_2$

ジブロムベンゾール　*jiburomubenzo-ru*
　dibromobenzene $C_6H_4Br_2$

ジチアジン　*jichiajin*
　dithiazine $C_3H_3NS_2$

ジチオン酸　*jichionsan*
　dithionic acid $H_2S_2O_6$

ジチオール　*jichio-ru*
　dithiol $C_6H_3(CH_3)(SH)_2$

ジチゾン　*jichizon*
　dithizone $C_6H_5NHNHCSN=NC_6H_5$

ジエチレングリコール　*jiechirenguriko-ru*
　diethylene glycol $HOCH_2CH_2OCH_2CH_2OH$

ジエチレングリコールモノエチルエーテル　*jiechirenguriko-rumonoechirue-teru*
　diethylene glycol monoethyl ether $HO(CH_2CH_2O)_2C_2H_5$

ジエチレンジアミン　*jiechirenjiamin*
　diethylene diamin, piperazine $C_4H_{10}N_2$

ジエチレントリアミン　*jiechirentoriamin*
　diethylenetriamine $H_2N(CH_2CH_2NH)_2H$

ジエチル亜鉛　*jiechiruaen*
　zinc diethyl $Zn(C_2H_5)_2$

ジエチルアミン　*jiechiruamin*
　diethylamine $HN(C_2H_5)_2$

ジエチルエーテル　*jiechirue-teru*
　diethyl ether $C_2H_5OC_2H_5$

ジエチルフタレート　*jiechirufutare-to*
　diethyl phthalate $C_6H_4(COOC_2H_5)_2$

ジエチルケトン　*jiechiruketon*
　diethyl ketone $(C_2H_5)_2CO$

ジエタノールアミン　*jietano-ruamin*
　diethanolamine $HN(CH_2CH_2OH)_2$

ジフェニリン　*jifenirin*
　diphenyline $H_2NC_6H_4C_6H_4NH_2$

ジフェニル　*jifeniru*
　biphenyl $(C_6H_5)_2$

ジフェニルアミン　*jifeniruamin*
　diphenylamine $NH(C_6H_5)_2$

ジフェニルエーテル　*jifenirue-teru*
　diphenyl ether $C_6H_5OC_6H_5$

ジフェニルメタン　*jifenirumetan*
　diphenylmethane $(C_6H_5)_2CH_2$

ジフェニル尿素　*jifenirunyouso*
　diphenylurea $CH_2ON_2(C_6H_5)_2$

ジフェノール　*jifeno-ru*
　biphenol $(C_6H_4OH)_2$

ジフェン酸　*jifensan*
　diphenic acid $HOOC-C_6H_4-C_6H_4-COOH$

ジフルオロジブロモメタ　*jifuruorojiburomometan*
　difluorodibromomethan CBr_2F_2

ジフルオロメタン　*jifuruorometan*
　difluoromethane CH_2F_2

ジフルオロ酢酸　*jifuruorosakusan*
　difluoroacetic acid $F_2CHCOOH$

ジギタリン　*jigitarin*
　digitalin $C_{35}H_{56}O_{14}$

ジギタローゼ　*jigitaro-ze*
　digitalose $C_9H_{14}O_5$

ジギトゲニン　*jigitogenin*
　digitogenin $C_{27}H_{44}O_5$

ジギトキシゲニン　*jigitokishigenin*
　digitoxigenin $C_{23}H_{34}O_4$

ジギトキシン　*jigitokishin*
　digitoxin $C_{41}H_{64}O_{13}$

ジギトニン　*jigitonin*
　digitonin $C_{56}H_{92}O_{29}$

ジゴキシゲニン　*jigokishigenin*
　digoxigenin $C_{23}H_{34}O_5$

ジゴキシン　*jigokishin*
　digoxin $C_{41}H_{64}O_{14}$

ジグリコール酸　*jiguriko-rusan*
　diglycolic acid $HOOCCH_2OCH_2COOH$

ジグリセリン　*jiguriserin*
　diglycerol $C_6H_{12}O_3$

ジヒドロフラン　*jihidorofuran*
　dihydrofuran C_4H_6O

ジヒドロ葉酸還元酵素
　jihidoroyousankangenkouso
　dihydrofolate reductase

ジホスホン酸　*jihosuhonsan*
　diphosphonic acid $H_4P_2O_5$

ジイソブチレン　*jiisobuchiren*
　diisobutylene $(CH_3)_3CCH=C(CH_3)_2$

ジイソブチルアミン　*jiisobuchiruamin*
　diisobutylamine $HN(CH_2CH(CH_3)_2)_2$

ジイソブチルケトン　*jiisobuchiruketon*
　diisobutyl ketone $((CH_3)_2CHCH_2)_2CO$

ジイソプロピルアミン　*jiisopuropiruamin*
　diisopropylamine $HN(CH(CH_3)_2)_2$

ジイソプロピルエーテル
　jiisopuropirue-teru
　diisopropyl ether $(CH_3)_2CHOCH(CH_3)_2$

ジジモライト　*jijimoraito*
　didymolithe $2CaO \cdot 3Al_2O_3 \cdot 9SiO_2$

ジカルボン酸　*jikarubonsan*
　dicarboxylic acid

ジケイ酸, ジ珪酸　*jikeisan*
　disilicic acid $H_2Si_2O_5$

ジケトン　*jiketon*
　diketone

ジケトピペラジン　*jiketopiperajin*
　diketopiperazine $C_4H_7N_2O_2$

ジキセナイト *jikisenaito*
dixenite $MnSiO_3 \cdot 2Mn_2(OH)(AsO_3)_2$

ジクロム酸 *jikuromusan*
dichromic acid $H_2Cr_2O_7$

ジクロム酸塩 *jikuromusanen*
dichromate $H_2Cr_2O_7$

ジクロロアセチレン *jikuroroasechiren*
dichloroacetylene C_2Cl_2

ジクロロベンゼン *jikurorobenzen*
dichlorobenzene $C_6H_4Cl_2$

ジクロロブロモメタン *jikuroroburomometan*
dichlorobromomethane $CHBrCl_2$

ジクロロフルオロプロパン *jikurorofuruoropuropan*
dichlorofluoropropane $C_3H_5Cl_2F$

ジクロロヘキサフルオロプロパン *jikurorohekisafuruoropuropan*
dichlorohexafluoropropane $C_3Cl_2F_6$

ジクロロジフルオロエタン *jikurorojifuruoroetan*
dichlorodifluoroethane $C_2H_2Cl_2F_2$

ジクロロジフルオロメタン *jikurorojifuruorometan*
dichlorodifluoromethane CCl_2F_2

ジクロロメタン *jikurorometan*
dichloromethane CH_2Cl_2

ジクロロ酢酸 *jikurorosakusan*
dichloroacetic acid $Cl_2CHCOOH$

ジクロロテトラフルオロエタン *jikurorotetorafuruoroetan*
dichlorotetrafluoroethane $C_2Cl_2F_4$

ジクロルアミンT *jikuroruamin T*
dichloramine-T $CH_3PheSO_2NCl_2$

ジクロルベンゾール *jikurorubenzo-ru*
dichlorobenzene $C_6H_4Cl_2$

ジクロル酢酸 *jikurorusakusan*
dichloroacetic acid $Cl_2CHCOOH$

ジメチロール尿素 *jimechiro-runyouso*
dimethylurea $H_2NCONMe_2$

ジメチル亜鉛 *jimechiruaen*
dimethylzinc $Zn(CH_3)_2$

ジメチルアミン *jimechiruamin*
dimethylamine $NH(CH_3)_2$

ジメチルアニリン *jimechiruanirin*
dimethylaniline $C_6H_5N(CH_3)_2$

ジメチルアセタール *jimechiruaseta-ru*
dimethylacetal $RCH(OCH_3)_2$

ジメチルエーテル *jimechirue-teru*
dimethyl ether CH_3OCH_3

ジメチルエトキシシラン *jimechiruetokishishiran*
dimethylethoxysilane $C_4H_{12}OSi$

ジメチルフォルムアミド *jimechiruforumuamido*
dimethylformamide, DMF $HCON(CH_3)_2$

ジメチル硫酸 *jimechiruryuusan*
dimethyl sulphate $SO_2(OCH_3)_2$

ジメチル水銀 *jimechirusuigin*
dimethyl mercury $Hg(CH_3)_2$

ジメチルスルホキシド *jimechirusuruhokishido*
dimethyl sulfoxide $SO(CH_3)_2$

ジメチルフタレート *jimehcirufutare-to*
dimethyl phthalate $C_6H_4(COOCH_3)_2$

ジーメンス *ji-mensu*
Siemens

ジメルカブロール *jimerukaburo-ru*
dimercaprol $HOCH_2CH(SH)CH_2SH$

ジーン *ji-n*
gene

ジーンエンジニアリング *ji-nenjiniaringu*
genetic engineering

ジンゲロン *jingeron*
zingerone
$C_6H_3(OH)(OCH_3)CH_2CH_2COCH_3$

ジンギベレン *jingiberen*
zingiberene $C_{15}H_{24}$

ジニトロベンゼン *jinitorobenzen*
dinitrobenzene $C_6H_4(NO_2)_2$

ジニトロベンゾール *jinitorobenzo-ru*
dinitrobenzene $C_6H_4(NO_2)_2$

ジニトロフェノール *jinitorofeno-ru*
dinitrophenol $C_4H_3(OH)(NO_2)_2$

ジンコーン　*jinko-n*
　zincon $C_{20}H_{16}N_4O_6S$

ジンテーゼ　*jinte-ze*
　synthesis

ジオゲナール　*jiogena-ru*
　diogenal $C_{11}H_{16}N_2O_3Br_2$

ジオキサン　*jiokisan*
　dioxan $C_4H_8O_2$

ジオキシアセトン　*jiokishiaseton*
　dihydroxyacetone $HOCH_2COCH_2OH$

ジオクザン　*jiokuzan*
　dioxan $C_4H_8O_2$

ジオニン　*jionin*
　dionin $C_{19}H_{18}NO_3 \cdot HCl \cdot 2H_2O$

ジオスメチン　*jiosumechin*
　diosmetine $C_{16}H_{12}O_6$

ジペンテン　*jipenten*
　dipentene $C_{10}H_{16}$

ジペプチダーゼ　*jipepuchida-ze*
　dipeptidase

ジペプチド　*jipepuchido*
　dipeptide

ジプロペジン　*jipuropejin*
　dipropaesin $CO(NHC_6H_4COOC_3H_7)_2$

ジプロサール　*jipurosa-ru*
　diplosal $C_6H_4(OH)COOC_6H_4COOH$

ジラード試薬　*jira-doshiyaku*
　Girard's reagent
　$H_2NNHCOCH_2N(CH_3)_3Cl$

ジルコン　*jirukon*
　zirconium (Zr, element 40)

ジルコニア　*jirukonia*
　zirconia ZrO_2

ジルコニウム　*jirukoniumu*
　zirconium (Zr, element 40)

ジルコニウムアミド　*jirukoniumuamido*
　zirconium amide $Zr(NH_2)_4$

ジサイクロヘキシルカルボジイミド　*jisaikurohekishirukarubojiimido*
　dicyclohexylcarbodiimide
　$C_6H_{11}N=C=CC_6H_{11}$

ジシアン　*jishian*
　cyanogen $(CN)_2$

ジシアンジアミド　*jishianjiamido*
　dicyandiamide, cyanoguanidine
　$H_2NC(=NH)NHCN$

ジシクロヘキシルアミン　*jishikurohekishiruamin*
　dicyclohexylamine $HN(C_6H_{11})_2$

ジシクロヘキシルカルボジイミド　*jishikurohekishirukarubojiimido*
　dicyclohexylcarbodiimide
　$C_6H_{11}N=C=CC_6H_{11}$

ジシクロペンタジエン　*jishikuropentajien*
　dicyclopentadiene $C_{10}H_{12}$

ジシクロペンタジエニル鉄　*jishikuropentajienirutetsu*
　ferrocene, dicyclopentadienyl iron

ジシラン　*jishiran*
　disilane Si_2H_6

ジシロクサン　*jishirokusan*
　disiloxane Si_2H_6O

ジスプロポーショネーション　*jisupuropo-shone-shon*
　disproportionation

ジスプロシウム　*jisupuroshiumu*
　dysprosium (Dy, element 66)

ジスルフィド　*jisurufido*
　disulfide

ジスルフィド結合　*jisurufidoketsugou*
　disulfide bridge, disulfide linkage

ジタイン　*jitain*
　ditaine $C_{22}H_{28}N_2O_4$

ジテルペン　*jiterupen*
　diterpene

ジウレチン　*jiurechin*
　diuretin $C_7H_7N_4O_2Na \cdot C_6H_4(OH)COONa$

ジヴァナジン酸　*jivanajinsan*
　divanadic acid $H_4V_2O_7$

ジヴィニル　*jiviniru*
　divinyl $H_2C=CHCH=CH_2$

ジヴィニルエーテル　*jivinirue-teru*
　divinyl ether $H_2C=CHOCH=CH_2$

ジヨードメタン　*jiyo-dometan*
　diiodomethane CH_2I_2

ジユレア　*jiyurea*
　diurea $H_2NCONHNHCONH_2$

ジョ jo

ジョンソン塩　*jonsonen*
　Johnson's salt KI_3

ジョセ石　*joseseki*
　joseite $Bi_8Te_3S_2$

ジュ ju

ジュラルミン　*jurarumin*
　duralumin

ジュレン　*juren*
　durene $C_6H_2(CH_3)_4$

ジュロキノン　*jurokinon*
　duroquinone $C_{10}H_{12}O_2$

ジュール　*ju-ru*
　joule

ジュール熱　*ju-runetsu*
　Joule's heat

ジュールの法則　*ju-runohousoku*
　Joule's law

ジュールトムソン効果　*ju-rutomusonkouka*
　Joule-Thomson effect

ジュウテリオクロロホルム　*juuteriokurorohorumu*
　deuterochloroform $CDCl_3$

ジュウテリウム　*juuteriumu*
　deuterium

ジュワー瓶　*juwa-bin*
　Dewar vessel, Dewar flask

カ ka

カバーガラス　*kaba-garasu*
　cover glass

カーバイド　*ka-baido*
　carbide, carbonide $M^I{}_2C_2$, CaC_2

カーベン　*ka-ben*
　carbene

カビ　*kabi*
　mold

カビ毒　*kabidoku*
　mycotoxin

カービノール　*ka-bino-ru*
　carbinol CH_3OH

カーボン　*ka-bon*
　carbon (C, element 6)

カーボンブラック　*ka-bonburakku*
　carbon black

カーボンファイバー　*ka-bonfaiba-*
　carbon fiber

カーボングラファイト　*ka-bongurafaito*
　graphite

カブレラ石　*kabureraseki*
　cabrerite $(Ni,Mg)_3(AsO_4)_2 \cdot 8H_2O$

カチオン　*kachion*
　cation

カチオン重合　*kachionjuugou*
　cationic polymerization

カチオンラジカル　*kachionrajikaru*
　cation radical

カダベリン　*kadaberin*
　cadaverine $H_2N(CH_2)_5NH_2$

カドミウム　*kadomiumu*
　cadmium (Cd, element 48)

カドミウム中毒　*kadomiumuchuudoku*
　cadmium poisoning

カドミウム標準電池　*kadomiumuhyoujundenchi*
　cadmium standard cell

カフェイン　*kafein*
　caffeine, coffeine $C_8H_{10}N_4O_2$

カフェ酸　*kafesan*
　caffeic acid $(HO)_2C_6H_3CH=CHCOOH$

カハバニライド　*kahabaniraido*
　carbanilide $C_6H_5NHCONHC_6H_5$

カイナイト　*kainaito*
　kainite $MgSO_4 \cdot KCl \cdot 3H_2O$

カイネチン　*kainechin*
　kinetin $C_{10}H_9N_5O$

カインカ酸　*kainkasan*
　cahincic acid $C_{40}H_{64}O_{18}$

カイロリン　*kairorin*
　kairoline $C_{10}N_{13}N$

カイザー石　*kaiza-seki*
　kayserite $Al_2O_3 \cdot H_2O$

カジネン　*kajinen*
　cadinene $C_{15}H_{24}$

カカイン　*kakain*
　cacaine $C_7H_8N_4O_2$

カカオ脂　*kakaoshi*
　cacao butter

カコジル酸　*kakojirusan*
　cacodylic acid $(CH_3)_2AsOOH$

カコジル水素化　*kakojirusuisoka*
　cacodyl hydride $(CH_3)_2AsH$

カモサイト　*kamosaito*
　chamosite
　$15(Fe,Mg)O \cdot 5Al_2O_3 \cdot 11SO_2 \cdot 16H_2O$

カナダバルサム　*kanadabarusamu*
　Canada balsam

カナディン　*kanadin*
　canadine $C_{20}H_{21}NO_4$

カナマイシン　*kanamaishin*
　kanamycin $C_{18}H_{36}N_4O_{11}$

カーナル石　*ka-naruseki*
　carnallite $MgCl_2 \cdot KCl \cdot 6H_2O$

カーネギー石　*ka-negi-seki*
　carnegieite $Na_2O \cdot Al_2O_3 \cdot 2SiO_2$

カンファン酸　*kanfansan*
　camphanic acid $C_{10}H_{14}O_4$

カンフェン　*kanfen*
　camphene $C_{10}H_{16}$

カンフォロン酸　*kanforonsan*
　camphoronic acid $(CH_3)_2C(COOH)$
　$C \cdot (CH_3)(COOH)CH_2COOH$

カンフアン　*kanfuan*
　camphane $C_{10}H_{18}$

カンフレン　*kanfuren*
　camphrene $C_9H_{14}O$

カニッツァーロ反応, カニツァロ反応, カニッツァロ反応　*kanissha-rohannou, kanisharohannou, kanissharohannou*
　Cannizzaro reaction

カニザロー反応　*kanizaro-hannou*
　Cannizzaro reaction

カンナベン　*kannaben*
　cannabene $C_{18}H_{20}$

カンナビノール　*kannabino-ru*
　cannabinol $C_{19}H_{24}O$

カーン石　*ka-nseki*
　cahnite $4CaO \cdot B_2O_3 \cdot As_2O_5 \cdot 4H_2O$

カンセリン　*kanserin*
　cancerine $C_8H_5NO_3$

カンタレーン　*kantare-n*
　cantharene C_8H_{12}

カンタリジン　*kantarijin*
　cantharidine $C_{10}H_{12}O_4$

カンタリジン酸　*kantarijinsan*
　canthardic acid $C_{10}H_{14}O_5$

カップリング　*kappuringu*
　coupling

カップリング値　*kappuringuchi*
　coupling value

カップリング法　*kappuringuhou*
　coupling process

カップリング剤　*kappuringuzai*
　coupling agent

カプリンアルデヒド　*kapurinarudehido*
　capric aldehyde $CH_3(CH_2)_8CHO$

カプリニトリル　*kapurinitoriru*
　caprinitrile $CH_3(CH_2)_8CN$

カプリン酸　*kapurinsan*
　capric acid $CH_3(CH_2)_8COOH$

カプリルアルコール　*kapuriruaruko-ru*
　capryl alcohol $CH_3(CH_2)_6CH_2OH$

カプリル酸　*kapurirusan*
　caprylic acid $CH_3(CH_2)_6COOH$

カプロノン　*kapuronon*
　capronone $(C_5H_{11})_2CO$

カプロン酸 *kapuronsan*
caproic acid $CH_3(CH_2)_4COOH$

カプロラクタム *kapurorakutamu*
caprolactam $C_6H_{11}NO$

カプロラクトン *kapurorakuton*
caprolactone $C_6H_{10}O_2$

カプサイチン *kapusaichin*
capsaicin $C_{18}H_{27}NO_3$

カプサンチン *kapusanchin*
capsanthin $C_{40}H_{56}O_3$

カプセル *kapuseru*
capsule

カプシド *kapushido*
capsid

カラベアス石 *karaberasuseki*
calaverite $AuTe_2$

カラベリン *karaberin*
calabrine $C_{15}H_{21}N_3O_2$

カラコレス石 *karakoresuseki*
caracolite $Pb(OH)Cl \cdot Na_2SO_4$

カラメル *karameru*
caramel

カラム *karamu*
column

カラムクロマトグラフィー *karamukuromatogurafi-*
column chromatography

カラムリアクター *karamuriakuta-*
column reactor

カラン *karan*
carane $C_{10}H_{18}$

カラット *karatto*
carat

カレドニア石 *karedoniaseki*
caledonite $2(Pb,Cu)O \cdot SO_3 \cdot H_2O$

カレン *karen*
carene $C_{10}H_{16}$

カリ *kari*
potash K_2CO_3

カリフォルニウム *kariforuniumu*
californium (Cf, element 98)

カリガラス *karigarasu*
potash glass

カリ肥料 *karihiryou*
potash fertilizer

カリホルニウム *karihoruniumu*
californium (Cf, element 98)

カリ明礬 *karimyouban*
potash alum $KAl(SO_4)_2 \cdot 12H_2O$

カリオフィレン *kariofiren*
caryophyllene $C_{20}H_{32}O_2$

カリオフィリン *kariofirin*
caryophyllin $C_{30}H_{48}O_3$

カリ石 *kariseki*
kalinite $KAl(SO_4)_2 \cdot 12H_2O$

カリセッケン *karisekken*
potash soap

カリ硝石 *karishouseki*
salpeter KNO_3

カリウム *kariumu*
potassium (K, element 19)

カリウムアミド *kariumuamido*
potassium amide KNH_2

カロチン *karochin*
carotene $C_{40}H_{56}$

カロチノイド *karochinoido*
carotinoid

カロメル *karomeru*
calomel Hg_2Cl_2

カロメル電極 *karomerudenkyoku*
calomel electrode

カロン *karon*
carone $C_{10}H_{16}O$

カロリー *karori-*
calorie

カロ酸 *karosan*
Caro's acid, peroxysulfuric acid H_2SO_5

カロテン *karoten*
carotene $C_{40}H_{56}$

カルバクロール *karubakuro-ru*
carvacrol $C_6H_3(CH_3)(C_3H_7)OH$

カルバミド *karubamido*
carbamide, urea $CO(NH_2)_2$

カルバミン酸 *karubaminsan*
carbamic acid H_2NCOOH

カルバミン酸エチル *karubaminsanechiru*
ethyl carbamate $H_2NCOOC_2H_5$

カルバミン酸塩 *karubaminsanen*
carbamate $RR'N$-$COOR''$

カルバミル *karubamiru*
carbamyl $H_2NC(O)$-

カルバニル *karubaniru*
carbanil, phenyl isocyanate C_6H_5NCO

カルバニル酸 *karubanirusan*
carbanilic acid $C_6H_5NHCOOH$

カルバゾール *karubazo-ru*
carbazole $C_{12}H_9N$

カルベン *karuben*
carbene

カルベーン *karube-n*
carvene, hesperidene $C_{10}H_{16}$

カルベニウムイオン *karubeniumuion*
carbenium ion

カルビミド *karubimido*
carbimide $HNCO$

カルビトール *karubito-ru*
carbitol $HO(CH_2)_2O(CH_2)_2OC_2H_5$

カルボアニオン *karuboanion*
carbanion

カルボヒドラーゼ *karubohidora-ze*
carbohydrase

カルボジフェニルイミド
karubojifeniruimido
carbodiphenylimide $C_6H_5N=C=NC_6H_5$

カルボジイミド *karubojiimido*
carbodiimide $HN=C=NH$

カルボカチオン *karubokachion*
carbocation

カルボキシ *karubokishi*
carboxy -$COOH$

カルボキシメチルセルロース
karubokishimechiruseruro-su
carboxymethylcellulose

カルボキシペプチダーゼ
karubokishipepuchida-ze
carboxypeptidase

カルボキシラーゼ *karubokishira-ze*
carboxylase

カルボキシル置換 *karubokishiruchikan*
carboxylation

カルボキシル基 *karubokishiruki*
carboxylic group

カルボニル化 *karuboniruka*
carbonylation

カルボニル化合物 *karubonirukagoubutsu*
carbonyl compound

カルボニル基 *karuboniruki*
carbonyl group

カルボニル基の酸素 *karubonirukinosanso*
carbonyl oxygen

カルボニウムイオン *karuboniumuion*
carbonium ion

カルボン酸 *karubonsan*
carboxylic acid

カルボン酸アミド *karubonsanamido*
carboxamide $RCONH_2$

カルボン酸塩 *karubonsanen*
carboxylate

カルボン酸エステル *karubonsanesuteru*
carboxylic acid ester R-$COOR'$

カルボオキシ桂皮酸 *karubookishikeihisan*
carboxycinnamic acid
$C_6H_4(COOH)CH=CHCOOH$

カルデル石 *karuderuseki*
calderite $MnII_3FeIII_2[SiO_4]_3$

カルフィッシャー法 *karufissha-hou*
Karl-Fischer technique

カルジノフィリン *karujinofirin*
carzinophilin $C_{31}H_{33}N_3O_{11}$

カルジオリピン *karujioripin*
cardiolipin

カルコン *karukon*
chalcone $C_6H_5CH=CHC(=O)C_6H_5$

カルコライト *karukoraito*
chalcolite $Cu(UO_2)P_2O_8 \cdot 8H_2O$

カルミン酸 *karuminsan*
carminic acid $C_{22}H_{20}O_{13}$

カルモジュリン　*karumojurin*
　calmodulin

カルナウバ酸　*karunaubasan*
　carnaubic acid, tetracosanoic acid
　$C_{23}H_{47}COOH$

カルナウビルアルコール
　karunaubiruaruko-ru
　carnaubyl aclohol $C_{24}H_{49}OH$

カルニチン　*karunichin*
　carnitine $C_7H_{15}NO_3$

カルニン　*karunin*
　carnin $C_7H_8N_4O_3$

カルノーサイクル　*karuno-saikuru*
　Carnot cycle

カルノシン　*karunoshin*
　carnosine $C_9H_{14}N_4O_3$

カルパイン　*karupain*
　carpaine $C_{14}H_{25}NO_2$

カルシフェロール　*karushifero-ru*
　calciferol $C_{28}H_{44}O$

カルシトニン　*karushitonin*
　calcitonine $C_{151}H_{226}N_{40}O_{45}S_3$

カルシウム　*karushiumu*
　calcium (Ca, element 20)

カルシウムチャンネル
　karushiumuchanneru
　calcium channel

カルシウムイオン　*karushiumuion*
　calcium ion

カルシウムカーバイド
　karushiumuka-baido
　calcium carbide CaC_2

カルシウム管　*karushiumukan*
　calcium tube

カルヴォン　*karuvon*
　carvone $C_{10}H_{14}O$

カルヴォール　*karuvo-ru*
　carvol $C_{10}H_{14}O$

カルヴォール，カルヴォン
　karuvo-ru, karuvon
　carvol, carvone $C_{10}H_{14}O$

カルウェオール　*karuweo-ru*
　carveol $C_{10}H_{16}O$

カセイアルカリ　*kaseiarukari*
　caustic alkali

カセイカリ　*kaseikari*
　caustic potash KOH

カセイソーダ　*kaseiso-da*
　caustic soda NaOH

カセロール　*kasero-ru*
　casserole

か焼マグネシア　*kashoumaguneshia*
　calcined magnesia MgO

カソード液　*kaso-doeki*
　catholyte

カソード還元　*kaso-dokangen*
　cathodic reduction

カソロ石　*kasoroseki*
　casolit $PbO \cdot UO_3 \cdot SiO_2 \cdot H_2O$

カッシア油　*kasshiayu*
　cassia oil

カスケード法　*kasuke-dohou*
　cascade process

カスナー法　*kasuna-hou*
　Castner-Keller mercury process

カストール石　*kasuto-ruseki*
　castorite, castor $LiAlSi_2O_5$

カタラーゼ　*katara-ze*
　catalase

カタルチン酸　*kataruchinsan*
　cathartic acid $C_{30}H_{36}NO_5$

カテキン　*katekin*
　catechin $C_{15}H_{16}O_6$

カテコール　*kateko-ru*
　catechol $C_6H_4(OH)_2$

カテコールアミン　*kateko-ruamin*
　catecholamine

カテネーション　*katene-shon*
　catenation

カテプシン　*katepushin*
　cathepsin

カートリッジ　*ka-torijji*
　cartridge

カットガラス　*kattogarasu*
　cut glass

カヤプテン　*kayaputen*
　cajaputene $C_{10}H_{16}$

カヤプトール　*kayaputo-ru*
　cajeputole $C_{10}H_{18}O$

カゼイン　*kazein*
　casein

ケ ke

ケーブル　*ke-buru*
　cable

ケエルセタゲチン　*keerusetagechin*
　quercetagetin $C_{15}H_{10}O_8$

ケイ皮酸　*keihisan*
　cinnamic acid $C_6H_5CH=CHCOOH$

ケイ皮油　*keihiyu*
　cinnamon oil

けい石　*keiishi*
　Silica stone

ケイ酸, 珪酸　*keisan*
　silicic acid H_4SiO_4

ケイ酸塩, 珪酸塩　*keisanen*
　silicate $M^I_4SiO_4/M^I_2SiO_3/M^I_3SiO_5$

けい砂　*keisha*
　quartz sand, silica sand

ケイ素, 珪素　*keiso*
　silicon (Si, element 14)

ケーキング　*ke-kingu*
　caking

ケーク　*ke-ku*
　cake

ケクレの式　*kekurenoshiki*
　Kekulé formula

ケミカル　*kemikaru*
　chemical

ケンフェロール　*kenfero-ru*
　kaempferol $C_{15}H_{10}O_6$

ケントロル石　*kentororuseki*
　kentrolite $3PbO \cdot 2Mn_2O_3 \cdot 3SiO_2$

ケラチン　*kerachin*
　keratin $C_{41}H_{71}N_{12}O_{14}S$

ケラシン　*kerashin*
　kerasin $C_{48}H_{93}NO_8$

ケルビン度　*kerubindo*
　Kelvin degree

ケールダルフラスコ　*ke-rudarufurasuko*
　Kjeldahl apparatus, Kjeldahl flask

ケルメサイト　*kerumesaito*
　antimony blende, kermesite
　$2Sb_2S_3 \cdot Sb_2O_3$

ケルン石　*kerunseki*
　kernite $Na_2B_4O_7 \cdot 4H_2O$

ケルセチン　*kerusechin*
　quercetin $C_{15}H_{10}O_7$

ケルセメトリン　*kerusemetorin*
　quercimetrine $C_{21}H_{20}O_{12}$

ケルシット　*kerushitto*
　quercitol $C_6H_7(OH)_5$

ケタジン　*ketajin*
　ketazine $R_2C=N-N=CR_2$

ケテン　*keten*
　ketene $RR'C=C=O$

ケトエノル互変異性　*ketoenorugoheinisei*
　keto-enol-tautomerism

ケト形　*ketogata*
　keto form

ケトヘクソーゼ　*ketohekuso-ze*
　ketohexose $HOCH_2CO(CHOH)_3CH_2OH$

ケト化合物　*ketokagoubutsu*
　keto compound

ケト開裂　*ketokairetsu*
　ketone cleavage

ケトキシム　*ketokishimu*
　ketoxime

ケトグリタール酸　*ketokurita-rusan*
　ketoglutaric acid
　$HOOC(CO)(CH_2)_2COOH$

ケトン　*keton*
　ketone $RCOR'$

ケトン分解　*ketonbunkai*
　ketone cleavage

ケトン基　*ketonki*
　keto group, carbonyl group, oxo group -CO-

ケトペントーゼ　*ketopento-ze*
　ketopentose $HOCH_2(CHOH)_2COCH_2OH$

ケトプロピオン酸　*ketopuropionsan*
　pyroracemic acid $CH_3COCOOH$

ケト酸　*ketosan*
　keto acid

ケトース　*keto-su*
　ketose

ケトテトロース　*ketotetoro-su*
　ketotetrose $HOCH_2CH(OH)COCH_2OH$

ケトトリオース　*ketotorio-su*
　ketotriose $HOCH_2COCH_2OH$

キ　ki

キアニジン　*kianijin*
　kyanidine $C_{15}H_{10}O_6 \cdot HCl$

キチン　*kichin*
　chitin $(C_8H_{13}NO_5)_x$

キチナーゼ　*kichina-ze*
　chitinase

キミルアルコール　*kimiruaruko-ru*
　chimyl alcohol
　$CH_3(CH_2)_{15}OCH_2CH(OH)CH_2OH$

キモトリプシン　*kimotoripushin*
　chymotrypsin

キナルジン　*kinarujin*
　quinaldine $C_{10}H_9N$

キナ酸　*kinasan*
　quinic acid $C_6H_7(OH)_4COOH$

キナーゼ　*kina-ze*
　kinase

キナゾリン　*kinazorin*
　quinazoline $C_8H_6N_2$

キンドリン　*kindorin*
　quindoline $C_{15}H_{10}N_2$

キネチン　*kinechin*
　kinetin $C_{10}H_9N_5O$

キネン　*kinen*
　quinene $C_{20}H_{22}N_2O$

キンヒドロン　*kinhidoron*
　quinhydrone $C_6H_4O_2 \cdot C_6H_4(OH)_2$

キニチン　*kinichin*
　quinicine $C_{20}H_{24}N_2O_2$

キニジン　*kinijin*
　quinidine $C_{20}H_{24}N_2O_2$

キニン　*kinin*
　quinine $C_{20}H_{24}N_2O_2$

キニノン　*kininon*
　quininone $C_{20}H_{22}N_2O_2$

キニン酸　*kininsan*
　quininic acid $CH_3OC_9H_5NCOOH$

キニザリン　*kinizarin*
　quinizarin $C_{14}H_8O_4$

キノフェン　*kinofen*
　quinophen $C_{15}H_{10}NCOOH$

キノフォルム　*kinoforumu*
　quinoform C_9H_5OICIN

キノキサリン, キノクサリン　*kinokisarin, kinokusarin*
　quinoxaline $C_8H_6N_2$

キノン　*kinon*
　quinone $C_6H_4O_2$

キノン式　*kinonshiki*
　quinoid

キノリジン　*kinorijin*
　quinolizine C_9H_9N

キノリン　*kinorin*
　quinoline C_9H_7N

キノリンエロー　*kinorinero-*
　quinoline yellow

キノリン酸　*kinorinsan*
　quinolinic acid $C_5H_3N(COOH)_2$

キノール　*kino-ru*
　quinol $C_6H_4(OH)_2$

キノソール　*kinoso-ru*
　quinosol $(C_9H_7NO)_2 \cdot H_2SO_4$

キノトキシン　*kinotokishin*
　quinotoxine $C_{20}H_{24}N_2O_2$

キノヴィン　*kinovin*
　quinovin $C_{30}H_{48}O_8$

キノヴォーゼ　*kinovo-ze*
quinovose $C_6H_{12}O_5$

キヌレニン　*kinurenin*
kynurenine $C_6H_4(NH_2)C(COOH)=CHCH \cdot (NH_2)COOH$

キヌレン酸　*kinurensan*
kynurenic acid $C_9H_5N(OH)COOH$

キヌリン　*kinurin*
kynurine C_9H_7NO

キップの装置　*kippunosouchi*
Kipp's apparatus

キラリティー　*kirariti-*
chirality

キラル　*kiraru*
chiral

キラル中心　*kiraruchuushin*
chiral center

キレート　*kire-to*
chelate

キレート化　*kire-toka*
chelation

キレート化合物　*kire-tokagoubutsu*
chelate compound

キレート結合　*kire-toketsugou*
chelate bond

キレート効果　*kire-tokouka*
chelation effect

キレート試薬　*kire-toshiyaku*
chelating reagent

キレート滴定　*kire-totekitei*
chelatometric titration

キリアニ合成法　*kirianigouseihou*
Kiliani synthesis

キロボルト　*kiroboruto*
kilovolt

キロワット　*kirowatto*
kilowatt

キルヒホッフの法則　*kiruhihoffunohousoku*
Kirchhoff's law

キサンチン　*kisanchin*
xanthine $C_5H_4N_4O_2$

キサンタリン　*kisantarin*
xanthaline $C_{20}H_{19}NO_5$

キサンテン　*kisanten*
xanthene $C_{13}H_{10}O$

キサントアミド　*kisantoamido*
xanthamide $H_2NCSOC_2H_5$

キサントフィル　*kisantofiru*
xanthophyll $C_{40}H_{54}(OH)_2$

キサントゲン酸　*kisantogensan*
xanthogenic acid ROCSSH

キサントゲン酸塩　*kisantogensanen*
xanthate, xanthogenate ROC(S)SR

キサントゲン酸ナトリウム　*kisantogensannatoriumu*
sodium xanthogenate $SC(OC_2H_5)SNa$

キサントヒドロール　*kisantohidoro-ru*
xanthydrol $C_{13}H_{10}O_2$

キサントケラチン　*kisantokerachin*
xanthokeratin $C_5H_{19}N_4O_4$

キサントキシレン　*kisantokishiren*
xanthoxylene $C_{10}H_{16}$

キサントン　*kisanton*
xanthone $C_{13}H_8O_2$

キサントプクシン　*kisantopukushin*
xanthopuccine $C_{20}H_{21}NO_4$

キサントプルプリン　*kisantopurupurin*
xanthopurpurine $C_{14}H_6O_2(OH)_2$

キサントプテリン　*kisantoputerin*
xanthopterine $C_6H_5N_5O_2 \cdot H_2O$

キサントシン　*kisantoshin*
xanthosine $C_{10}H_{12}N_4O_6$

キサントトキシン　*kisantotokishin*
xanthotoxin $C_{12}H_8O_4$

キセノン　*kisenon*
xenon (Xe, element 54)

キセロフォルム　*kiseroforumu*
xeroform $Bi_2O_3 \cdot C_6H_2Br_3(OH)$

キシラン　*kishiran*
xylan $(C_5H_8O_4)_n$

キシラナーゼ　*kishirana-ze*
xylanase

キシレノール *kishireno-ru*
　xylenol $C_6H_3(CH_3)_2OH$

キシリジン, キシリデン
　kishirijin, kishiriden
　xylidine $C_6H_3(CH_3)_2NH_2$

キシリジン酸 *kishirijinsan*
　xylidinic acid $C_6H_3(CH_3)(COOH)_2$

キシリレングリコール
　kishirirenguriko-ru
　xylylene glycol $C_6H_4(CH_2OH)_2$

キシリルアミン *kishiriruamin*
　xylylamine $C_6H_4(CH_3)CH_2NH_2$

キシリルアルコール *kishiriruaruko-ru*
　xylyl alcohol $C_6H_4(CH_3)CH_2OH$

キシリル酸 *kishirirusan*
　xylylic acid $C_6H_3(CH_3)_2COOH$

キシロキノン *kishirokinon*
　xyloquinone $C_6H_2O_2(CH_3)_2$

キシロン酸 *kishironsan*
　xylonic acid $HOCH_2(CHOH)_3COOH$

キシロール *kishiro-ru*
　xylene $C_6H_4(CH_3)_2$

キシロールシン *kishiro-rushin*
　xylorcinol $C_6H_2(CH_3)_2(OH)_2$

キシロース, キシローゼ
　kishiro-su, kishiro-ze
　xylose $C_5H_{10}O_5$

ガス温度計 *kitaiondokei*
　gas thermometer

キテニル *kiteniru*
　quitenine $C_{19}H_{19}N_2O_4$

キトール *kito-ru*
　kitol $C_{40}H_{58}(OH)_2$

キトサミン *kitosamin*
　chitosamine $C_6H_{11}O_5NH_2$

キトサン *kitosan*
　chitosan

キトーゼ *kito-ze*
　chitose $C_6H_{10}O_5$

キーゼル石 *ki-zeruseki*
　kieserite $MgSO_4 \cdot H_2O$

コ ko

こはく酸イミド *kobakusanimido*
　succinimide $C_4H_5NO_2$

コバラミン *kobaramin*
　cobalamine

コバルト *kobaruto*
　cobalt (Co, element 27)

コバルトガラス *kobarutogarasu*
　cobalt glass

コーブル石 *ko-buruseki*
　kobellite $2PbS \cdot (Bi,Sb)_2S_3$

コチニン *kochinin*
　cotinine $C_{10}H_{12}N_2O$

コダミン *kodamin*
　codamine $C_{20}H_{25}NO_4$

コデイン *kodein*
　codeine $C_{18}H_{21}NO_3$

コドン *kodon*
　codon

こはく *kohaku*
　succinite, amber $C_{10}H_{16}O$

コハク酸 *kohakusan*
　succinic acid $HOOCCH_2CH_2COOH$

コハク酸一ナトリウム
　kohakusanichinatoriumu
　monosodium succinate

コハク酸二ナトリウム
　kohakusanninatoriumu
　disodium succinate

コーヘン石 *ko-henseki*
　cohenite $(Fe,Ni,Co)_3C$

コッホ酸 *kohhosan*
　Koch's acid $C_{10}H_4(NH_2)(SO_3H)_3$

コヒーレンス *kohi-rensu*
　coherence

コヒーレント光 *kohi-rentokou*
　coherent light

コープ転移 *kohpu ten-i*
　cope rearrangement

コイル *koiru*
　coil

コカイン　*kokain*
　cocaine $C_{17}H_{21}NO_4$

コキンボ石　*kokinboseki*
　coquimbite $Fe_2(SO_4)_3$

コック　*kokku*
　cock

コクラウリン　*kokuraurin*
　cocculaurine $C_{17}H_{19}NO_3$

コークス　*ko-kusu*
　coke

コメン酸　*komensan*
　comenic acid $C_6H_4O_5$

コンバーター　*konba-ta-*
　converter

コンビナトリー化学　*konbinatori-kagaku*
　combinatorial chemistry

コンダクタンス　*kondakutansu*
　conductance

コンデンサ, コンデンサー
　kondensa, kondensa-
　condenser

コンディショニング　*kondishoningu*
　conditioning

コンドロイチン　*kondoroichin*
　chondroitin

コンドロシン　*kondoroshin*
　chondrosin $C_{12}H_{21}NO_{11}$

コーネル石　*ko-neruseki*
　connellite $CuSO_4 \cdot CuCl_2 \cdot 19Cu(OH)_2 \cdot H_2O$

コンヒドリン　*konhidorin*
　conhydrine $C_8H_{17}N$

コンヒニン　*konhinin*
　conchinine $C_{20}H_{24}N_2O_2$

コニフェリン　*koniferin*
　coniferin, abietene $C_{16}H_{22}O_8 \cdot 2H_2O$

コニフェリルアルコール
　koniferiruaruko-ru
　coniferyl alcohol $C_6H_3(OCH_3)(OH)CH=CHCH_2OH$

コニイン　*koniin*
　coniine $C_8H_{17}N$

コニリン　*konirin*
　conyrine $C_8H_{11}N$

コニセイン　*konisein*
　coniceine $C_8H_{15}N$

コンキオリン　*konkiorin*
　conchiolin $C_{30}H_{48}N_9O_{11}$

コンクリート　*konkuri-to*
　concrete

コンプトン効果　*konputonkouka*
　Compton effect

コンプトン石　*konputonseki*
　comptonite $(Ca,Na_2)Al_2Si_2O_8 \cdot 2,5H_2O$

コンタクト接着剤　*kontakutosecchakuzai*
　contact adhesive

コントラスト　*kontorasuto*
　contrast

コンヴァラリン　*konvararin*
　convallarin $C_{34}H_{62}O_{11}$

コンヴォルブリン　*konvoruburin*
　convolvulin $C_{31}H_{50}O_{16}$

コーンウォール石　*ko-nwo-ruseki*
　cornwallite $Cu_3As_2O_8 \cdot 2Cu(OH)_2 \cdot H_2O$

コペリジン　*koperijin*
　coppellidin $C_8H_{17}N$

コポリマー　*koporima-*
　copolymer

コプロポルフィリン　*kopuroporufirin*
　coproporphyrin $C_{36}H_{36}N_4O_8$

コプロスタン　*kopurosutan*
　coprostane $C_{27}H_{48}$

コラーゲン　*kora-gen*
　collagen

コラン　*koran*
　cholane $C_{24}H_{22}$

コランダム　*korandamu*
　corundum Al_2O_3

コラン酸　*koransan*
　cholanic acid $C_{24}H_{40}O_2$

コラリン　*korarin*
　corallin $C_{20}H_{16}O_3$

コラル酸　*korarusan*
　cholalic acid $C_{23}H_{36}(OH)_3COOH$

コレイン酸 *koreinsan*
　choleic acid

コレスチンブラウ　*koresuchinburau*
　coelestine blue $C_{17}H_{18}N_3O_3$

コレスタン　*koresutan*
　cholestane $C_{27}H_{48}$

コレスタノール　*koresutano-ru*
　cholestanol $C_{27}H_{47}OH$

コレステリン　*koresuterin*
　cholesterol, cholesterin $C_{27}H_{45}OH$

コレステロール　*koresutero-ru*
　cholesterol, cholesterin $C_{27}H_{45}OH$

コレテリン　*koreterin*
　choletelin $C_{16}H_{18}N_2O_6$

コリブルビン　*koriburubin*
　corybulbine $C_{18}H_{15}N(OH)(OCH_3)_3$

コリダリン　*koridarin*
　corydaline $C_{18}H_{15}N(OCH_3)_4$

コリフィン　*korifin*
　coryfin $C_{14}H_{16}O_3$

コリジン　*korijin*
　collidine $C_8H_{11}N$

コリン　*korin*
　choline $HOCH_2CH_2N(CH_3)_3OH$; corrin $C_{19}H_{22}N_4$

コリンエステラーゼ　*korinesutera-ze*
　choline esterase

コリオリ力　*koriorika*
　Coriolis force

コリオリの力　*koriorinochikara*
　coriolis' force

コーリタイト　*ko-ritaito*
　bismuth blende $Bi_4(SiO_4)_3$

コロファナイト　*korofanaito*
　collophanite $Ca_3(PO_4)_2 \cdot H_2O$

コロフェン　*korofen*
　colophene $C_{20}H_{32}$

コロフォニン　*korofonin*
　colophonin $C_{16}H_{22}O_3 \cdot H_2O$

コロフォン酸　*korofonsan*
　colophonic acid $C_{20}H_{30}O_2$

コロイド　*koroido*
　colloid

コロイド化学　*koroidokagaku*
　colloid chemistry

コロイド金　*koroidokin*
　colloidal gold

コロイド金属　*koroidokinzoku*
　colloidal metal

コロイド粒子　*koroidoryuushi*
　colloidal particle

コロイド溶液　*koroidoyoueki*
　colloidal solution

コロジオン　*korojion*
　collodion

コロジウム　*korojiumu*
　collodion

コロンビン　*koronbin*
　colombin $C_{21}H_{22}O_7$

コロネン　*koronen*
　coronene $C_{24}H_{12}$

コーロンエネルギー　*ko-ronenerugi-*
　Coulomb energy

コーロン力　*ko-ronryoku*
　Coulomb's force

コロラド石　*kororadoseki*
　coloradoite HgTe

コロシンチン　*koroshinchin*
　colocynthin $C_{56}H_{84}O_{23}$

コロシンテイン　*koroshintein*
　colocynthein $C_{44}H_{64}O_{13}$

コルベック石　*korubekkuseki*
　kolbeckite $ScPO_4 \cdot 2H_2O$

コルベの反応　*korubenohannou*
　Kolbe synthesis

コルベシュミット反応
　korubeshumittohannou
　Kolbe-Schmitt reaction

コルチコステロン　*koruchikosuteron*
　corticosterone $C_{21}H_{30}O_4$

コルチゾン　*koruchizon*
　cortisone $C_{21}H_{28}O_5$

コルデイン　*korudein*
　cordeine $C_6H_3(OH)(CH_3)COOC_6H_2Br_3$

コルドール　*korudo-ru*
　cordol $C_6H_4(OH)COOC_6H_2Br_3$

コルヒチン　*koruhichin*
　colchicine $C_{22}H_{25}NO_6$

コルコ栓　*korukosen*
　cork stopper

コルク酸　*korukusan*
　suberic acid $HOOC(CH_2)_6COOH$

コルンビン　*korunbin*
　columbin $C_{21}H_{22}O_7$

コルネライト　*koruneraito*
　kornelite $Fe_2(SO_4)_3 \cdot 7,5H_2O$

コルネ石　*koruneseki*
　cornetite $Cu_3(PO_4)_2 \cdot 3Cu(OH)_2$

コール酸　*ko-rusan*
　cholic acid $C_{23}H_{36}(OH)_3COOH$

コールタール　*ko-ruta-ru*
　coal tar

コッシン　*kosshin*
　cossin $C_{31}H_{38}O_{10}$

コスマット石　*kosumattoseki*
　kossmatite $H_{18}Mg_3Ca_7Al_6Si_7O_{42}F$

コタルニン　*kotarunin*
　cotarnine $C_{12}H_{15}NO_4$

コーティング　*ko-tingu*
　coating

コトイン　*kotoin*
　cotoin $C_6H_2(OCH_3)(OH)_2COC_6H_5$

コットン効果　*kottonkouka*
　Cotton effect

コットレル集じん器　*kottorerushuujinki*
　Cottrell precipitator

コウジ酸　*koujisan*
　kojic acid $C_6H_6O_4$

ク ku

クアッシン　*kuasshin*
　quassine $C_{10}H_{12}O_3$

クベビン　*kubebin*
　cubebin $C_{10}H_{10}O_3$

クエン酸　*kuensan*
　citric acid $HOOCCH_2C(OH)(COOH)CH_2COOH$

クエン酸第一鉄ナトリウム
　kuensandaiichitetsunatoriumu
　sodium ferrous citrate $C_6H_5O_7Fe(II)Na$

クエン酸第二鉄　*kuensandainitetsu*
　ferric citrate $C_6H_5O_7Fe(III)$

クエン酸塩　*kuensanen*
　citrate $C_3H_4(OH)(COOM^I)_3$

クエン酸一カリウム　*kuensanichikariumu*
　monopotassium citrate $C_6H_7O_7K$

クエン酸カリウム　*kuensankariumu*
　potassium citrate

クエン酸カルシウム　*kuensankarushiumu*
　calcium citrate

クエン酸三カリウム　*kuensansankariumu*
　tripotassium citrate $C_6H_5O_7K_3$

クエン酸トリエチル　*kuensantoriechiru*
　triethyl citrate $C_3H_4(OH)(COOC_2H_5)_3$

クエテナ石　*kuetenaseki*
　quetenite $MgO \cdot Fe_2O_3 \cdot 3SO_3 \cdot 13H_2O$

クッフェロン　*kufferon*
　cupferron $C_6H_5N(O)NO \cdot NH_4$

クマラン　*kumaran*
　coumaran C_8H_8O

クマリン　*kumarin*
　coumarin $C_9H_6O_2$

クマリン酸, クマル酸　*kumarinsan, kumarusan*
　coumarinic acid $C_6H_4(OH)CH=CHCOOH$

クマロン　*kumaron*
　cumarone C_8H_6O

クメン　*kumen*
　cumene $C_6H_5CH(CH_3)_2$

クミン酸　*kuminsan*
　cuminic acid $(CH_3)_2CHC_6H_4COOH$

クモール　*kumo-ru*
　cumene $C_6H_5CH(CH_3)_2$

クムレン　*kumuren*
　cumulene

クネーフェナーゲルの反応
kune-fena-gerunohannou
Knoevenagel condensation

クニチン　*kunichin*
cnicin $C_{42}H_{56}O_{15}$

クノルの反応　*kunorunohannou*
Knorr synthesis

クペロン　*kuperon*
cupferron $C_6H_5N(O)NO \cdot NH_4$

クプレイン　*kupurein*
cupreine $C_{19}H_{22}N_2O_2$

クプレン　*kupuren*
cuprene $C_{14}H_{12}$

クプロン　*kupuron*
cupron $C_6H_5CH(OH)C(=NOH)C_6H_5$

クラッチ　*kuracchi*
clutch

くらい　*kurai*
approximately, about, roughly

クライゼン縮合　*kuraizenshukugou*
Claisen condensation

クライゼン転位　*kuraizenteni*
Claisen rearrangement

クラジウスクラペイロンの式
kurajiusukurapeironnoshiki
Clausius-Clapeyron's equation

クラッキング　*kurakkingu*
cracking

クラメル石　*kurameruseki*
kramerite $Na_2O \cdot 2CaO \cdot 5BO_3 \cdot 10H_2O$

クランプ　*kuranpu*
clamp

クラプロト石　*kurapurotoseki*
klaprotholite $3Cu_2S \cdot 2Bi_2S_3$

クラスター　*kurasuta-*
cluster

クラスター分析　*kurasuta-bunseki*
cluster analysis

クラスター化合物　*kurasuta-kagoubutsu*
cluster compound

クラウンエーテル　*kuraune-teru*
crown ether

クレアチン　*kureachin*
creatine $H_2NC(=NH)N(CH_3)CH_2COOH$

クレアチニン　*kureachinin*
creatinine $C_4H_7ON_3$

クレーブ酸　*kure-busan*
Cleve's acid $XC_{10}H_6NSO_3H$ [X=OH, NH$_2$]

クレドネル石　*kuredoneruseki*
crednerite $CuO \cdot Mn_2O_3$

クレメンゼン還元　*kuremenzenkangen*
Clemmensen reduction

クレメルス石　*kuremerususeki*
kremersite $KCl \cdot NH_4Cl \cdot FeCl_2 \cdot H_2O$

クレーム　*kure-mu*
claim

クレネル鉱　*kurenerukou*
krennerite $(Au,Ag)Te_2$

クレンメ　*kurenme*
clamp

クレオゾール　*kureozo-ru*
creosol $C_6H_3(CH_3)(OH)(OCH_3)$

クレシルアルコール　*kureshiruaruko-ru*
cresol $C_6H_4(OH)(CH_3)$

クレソチン酸　*kuresochinsan*
cresotic acid $CH_3C_6H_3(OH)COOH$

クレザチン　*kurezachin*
cresatin $CH_3COOC_6H_4CH_3$

クレザロール　*kurezaro-ru*
cresalol $HOC_6H_4COOC_6H_4CH_3$

クレゾラミン　*kurezoramin*
cresamine $C_6H_4(OH)COOC_6H_4CH_3$

クレゾール　*kurezo-ru*
cresol $C_6H_4(OH)(CH_3)$

クレゾルシン　*kurezorushin*
cresorcinol $CH_3C_6H_3(OH)_2$

クリオフィン　*kuriofin*
kryofin $C_6H_4(OC_2H_5)(NHCOCH_2OCH_3)$

クリップ　*kurippu*
clip

クリプトン　*kuriputon*
krypton (Kr, element 36)

クリプトンランプ　*kuriputonranpu*
krypton lamp

クリプトピン　*kuriputopin*
cryptopine $C_{21}H_{23}NO_5$

クリサジン　*kurisajin*
chrysazin $C_{14}H_8O_4$

クリサミン酸　*kurisaminsan*
chrysammic acid $C_{14}H_2(OH)(NO_2)_4O_2$

クリサニリン　*kurisanirin*
chrysaniline $C_{15}H_{16}N_3$

クリサンテミン　*kurisantemin*
chrysanthemine $C_{21}H_{21}O_{11}Cl$

クリセン　*kurisen*
chrysene $C_{18}H_{12}$

クリシン　*kurishin*
chrysin $C_{15}H_{10}O_4$

クリソファン酸　*kurisofansan*
chrysophanic acid $C_{15}H_{10}O_4$

クリソイジン　*kurisoijin*
chrysoidine $C_6H_5N{=}NC_6H_3(NH_2)_2 \cdot HCl$

クリソタイル　*kurisotairu*
chrysotile $3MgO \cdot 2SiO_2 \cdot 2H_2O$

クリスタライザー　*kurisutaraiza-*
crystallizer

クリスタリン　*kurisutarin*
crystalline

クリスタリット　*kurisutaritto*
crystallite

クリスタル　*kurisutaru*
crystal

クリスタルガラス　*kurisutarugarasu*
crystal glass

クリスタルヴァイオレット
kurisutaruvaioretto
crystal violet $C_{25}H_{30}N_3Cl$

クリストバル石，クリストバライト
kurisutobaruseki, kurisutobaraito
cristobalite SiO_2

クリヤラッカー　*kuriyarakka-*
clear lacquer

クロコン酸　*kurokonsan*
croconic acid $C_5H_2O_5$

クロマチン　*kuromachin*
chromatin

クロマイト　*kuromaito*
chromite, chromate(III) $M^I{}_3[Cr(OH)_6]$

クロマン　*kuroman*
chroman $C_9H_{10}O$

クロマトフォアー　*kuromatofoa-*
chromatophore

クロマトグラフィ，クロマトグラフィー
kuromatogurafi, kuromatogurafi-
chromatography

クロマトグラフ分離
kuromatogurafubunri
chromatographic separation

クロマトグラフ用カラム
kuromatogurafuyoukaramu
chromatographic column

クロマトグラム　*kuromatoguramu*
chromatogram

クロメン　*kuromen*
chromene C_9H_8O

クーロメーター　*ku-rome-ta-*
coulometer

クーロメトリー　*ku-rometori-*
coulometry

クロミタイト　*kuromitaito*
chromitite $Fe_2O_3 \cdot Cr_2O_3$

クロモン　*kuromon*
chromone $C_9H_6O_2$

クロモトロプ酸　*kuromotoropusan*
chromotropic acid $C_{10}H_4(OH)_2(SO_3H)_2$

クロム　*kuromu*
chromium (Cr, element 24)

クロム媒染　*kuromubaisen*
chrome mordanting

クロムブラック　*kuromuburakku*
chrome black $C_{27}H_{18}N_4O_7S$

クロム明礬　*kuromumyouban*
chrome alum $KCr(SO_4)_2 \cdot 12H_2O$

クロム酸　*kuromusan*
chromic acid H_2CrO_4

クロム酸亜鉛 *kuromusanaen*
zinc chromate $ZnCrO_4$

クロム酸アンモニウム
kuromusananmoniumu
ammonium chromate $(NH_4)_2CrO_4$

クロム酸バリウム *kuromusanbariumu*
barium chromate $BaCrO_4$

クロム酸第二水銀 *kuromusandainisuigin*
mercury(II) chromate $HgCrO_4$

クロム酸第二錫 *kuromusandainisuzu*
stannic chromate $Sn(CrO_4)_2$

クロム酸塩 *kuromusanen*
chromate $M^I_2CrO_4$

クロム酸銀 *kuromusangin*
silver chromate Ag_2CrO_4

クロム酸カリウム *kuromusankariumu*
potassium chromate K_2CrO_4

クロム酸ナトリウム
kuromusannatoriumu
sodium chromate Na_2CrO_4

クロム染料 *kuromusenryou*
chrome dye

クロム処理 *kuromushori*
chroming

クロム鉄鉱 *kuromutekkou*
chromite

クロム浴 *kuromuyoku*
chrome bath

クーロン *ku-ron*
coulomb

クーロン引力 *ku-roninryoku*
Coulomb attraction

クーロン力 *ku-ronka*
Coulomb force

クーロンの法則 *ku-ronnohousoku*
Coulomb's law

クロラミン *kuroramin*
chloramine $ClNH_2$

クロラムフェニコール
kuroramufeniko-ru
chloramphenicol $C_{11}H_{12}Cl_2N_2O_5$

クロラル, クロラール *kuroraru, kurora-ru*
chloral, trichloroacetaldehyde CCl_3CHO

クロロアセチルクロライド
kuroroasechirukuroraido
chloroacetyl chloride $ClCH_2COCl$

クロロアセトアミド *kuroroasetoamido*
chloroacetamide $ClCH_2CONH_2$

クロロアセトン *kuroroaseton*
chloroacetone CH_3COCH_2Cl

クロロアセトニトリル
kuroroasetonitoriru
chloroacetonitrile $ClCH_2CN$

クロロベンゾール, クロロベンゼン
kurorobenzo-ru, kurorobenzen
chlorobenzene C_6H_5Cl

クロロエタン *kuroroetan*
ethyl chloride C_2H_5Cl

クロロフィル *kurorofiru*
chlorophyll

クロロフォルム *kuroroforumu*
chloroform $CHCl_3$

クロロホルム *kurorohorumu*
chloroform $CHCl_3$

クロロジフルオロメタン
kurorojifuruorometan
chlorodifluoromethane $CHClF_2$

クロロメタン *kurorometan*
methyl chloride CH_3Cl

クロロニトロベンゼン
kuroronitorobenzen
chloronitorobenzene $C_6H_4NO_2Cl$

クロロプレン *kuroropuren*
chloroprene $H_2C=CClCH=CH_2$

クロロプレンゴム *kuroropurengomu*
chloroprene rubber

クロロ硫酸 *kurororyuusan*
chlorosulfuric acid $SO_2(OH)Cl$

クロロ酢酸 *kurorosakusan*
chloroacetic acid $ClCH_2COOH$

クロロ酢酸エチル *kurorosakusanechiru*
ethyl chloroacetate $ClCH_2COOC_2H_5$

クロロスルホン酸　*kurorosuruhonsan*
chlorosulfuric acid $SO_2(OH)Cl$

クロロトリフルオロメタン
kurorotorifuruorometan
chlorotrifluoromethan $CClF_3$

クロルアミン　*kuroruamin*
chloramine $ClNH_2$

クロルアニル　*kuroruaniru*
chloranil $C_6Cl_4O_2$

クロルアゼーン T　*kuroruaze-n*
chloramine T $C_7H_7SO_2NClNa$

クロルベンゾール, クロールベンゾール
kurorubenzo-ru, kuro-rubenzo-ru
chlorobenzene C_6H_5Cl

クロル蟻酸　*kurorugisan*
chloroformic acid $ClCOOH$

クロル蟻酸エステル　*kurorugisanesuteru*
chloroformic ester $ClCOOR$

クロルクロム酸　*kurorukuromusan*
chlorochromic acid $HCrO_3Cl$

クロルメタン　*kurorumetan*
methyl chloride CH_3Cl

クロル酢酸　*kurorusakusan*
chloroacetic acid $ClCH_2COOH$

クロルスルフォン酸　*kurorusurufonsan*
chlorosulfuric acid $SO_2(OH)Cl$

クロルヨードフォルム
kuroruyo-doforumu
chloriodform $CHCl_2I$

クロセチン　*kurosechin*
crocetin $C_{20}H_{24}O_4$

クロセイン酸　*kuroseinsan*
croceic acid $C_{10}H_6(OH)SO_3H$

クロシン　*kuroshin*
crocin $C_{20}H_{24}O_4$

クロティルアルコール　*kurotiruaruko-ru*
crotyl alcohol $CH_3CH=CHCH_2OH$

クロート法　*kuro-tohou*
Claude process

クロトンアルデヒド　*kurotonarudehido*
crotonaldehyde $CH_3CH=CHCHO$

クロトノール酸　*kurotono-rusan*
crotonolic acid $CH_3CH=C(CH_3)COOH$

クロトン酸　*kurotonsan*
crotonic acid $CH_3CH=CHCOOH$

クロトン酸メチル　*kurotonsanmechiru*
methyl crotonate $CH_3CH=CHCOOCH_3$

クルチウス転位　*kuruchiusuteni*
Curtius rearrangement

クルックス石　*kurukkususeki*
crookesite $(Cu,Tl,Ag)_2Se$

クルクマ　*kurukuma*
curcuma, Indian saffron Cu_3SnS_4

クルクミン　*kurukumin*
curcumin
$(C_6H_3(OH)(OCH_3)CH=CHCO)_2CH_2$

クルティウス転位　*kurutiusuteni*
Curtius rearrangement

クスコニン　*kusukonin*
cuskonine $C_{23}H_{26}O_4$

クスパリン　*kusuparin*
cusparine $C_{19}H_{17}NO_3$

クウェブラカミン　*kuweburakamin*
quebrachamine $C_{21}H_{26}N_2O_3$

クウェブラキン　*kuweburakin*
quebrachine $C_{21}H_{26}N_2O_3$

キャ kya

キャビテーション　*kyabite-shon*
cavitation

キャパシタンス　*kyapashitansu*
capacitance

キャプタン　*kyaputan*
captan $C_9H_8Cl_3NO_2S$

キャリア　*kyaria*
carrier

キャリヤー　*kyariya-*
carrier

キャリヤーガス　*kyariya-gasu*
carrier gas

キュ kyu

キューバ鉱　*kyu-bakou*
  cubanite $CuS \cdot Fe_4S_5$

キュリー　*kyuri-*
  curie

キュリーの法則　*kyuri-nohousoku*
  Curie's law

キュリー石　*kyuri-seki*
  curite $2PbO \cdot 5UO_3 \cdot 4H_2O$

キュリウム　*kyuriumu*
  curium (Cm, element 96)

マ ma

マッフル炉　*maffururo*
  muffle furnace

マグマ　*maguma*
  magma

マグネサイト　*magunesaito*
  magnesite $MgCO_3$

マグネシア　*maguneshia*
  magnesia MgO

マグネシアアルバ　*maguneshiaaruba*
  basic magnesium carbonate (magnesia alba) $4MgCO_3 \cdot Mg(OH)_2 \cdot 4H_2O$

マグネシウム　*maguneshiumu*
  magnesium (Mg, element 12)

マグネシウム合金　*maguneshiumugoukin*
  magnesium alloy

マイエルホッフェル石　*maieruhofferuseki*
  meyerhofferite $Ca_2B_6O_6(OH)_{10} \cdot 2H_2O$

マイカ　*maika*
  mica

マイケル付加　*maikerufuka*
  Michael addition

マイコバクテリウム　*maikobakuteriumu*
  mycobacterium

マイコトキシン　*maikotokishin*
  mycotoxin

マイクロフィルム　*maikurofirumu*
  microfilm

マイクロ波　*maikuroha*
  microwave

マイクロ波分光　*maikurohabunkou*
  microwave spectroscopy

マイクロ波加熱装置　*maikurohakanetsusouchi*
  microwave heating device

マイクロカプセル　*maikurokapuseru*
  microcapsule

マイクロピペット　*maikuropipetto*
  micropipette

マイトマイシン　*maitomaishin*
  mitomycin

マイトネリウム　*maitoneriumu*
  meitnerium (former eka-iridium) (Mt, former Eka-Ir, element 109)

マイヤース石　*maiya-suseki*
  meyersite $AlPO_4 \cdot 2H_2O$

マーカライト　*ma-karaito*
  mercallite $KHSO_4$

マックス石　*makkususeki*
  maxite $PbSO_4 \cdot 2PbCO_3 \cdot Pb(OH)_2$

マクロ分析　*makurobunseki*
  macroanalysis

マクロファージ　*makurofa-ji*
  macrophage

マクログロブリン　*makuroguroburin*
  macroglobulin

マクロライド　*makuroraido*
  macrolide

マクルリン　*makururin*
  maclurin $C_6H_3(OH)_2COC_6H_2(OH)_3$

マクスウェルボルツマン分布　*makusuweruborutsumanbunpu*
  Maxwell-Boltzmann distribution

マクスウェル分布法則　*makusuwerubunpuhousoku*
  Maxwell distribution law

マナカン鉱　*manakankou*
　menaccanite $FeTiO_3$

マナンドナ石　*manandonaseki*
　manadonite $H_{24}Li_4Al_{14}b_4Si_6O_{23}$

マンデル酸　*manderusan*
　mandelic acid $C_6H_5CH(OH)COOH$

マンデル酸ニトリル　*manderusannitoriru*
　mandelonitrile $C_6H_5CH(OH)CN$

マンドラゴリン　*mandoragorin*
　mandragorine $C_{17}H_{23}NO_3$

マンガン　*mangan*
　manganese (Mn, element 25)

マンガン重石　*manganjuuseki*
　huebnerite $MnWO_4$

マンガン乾電池　*mangankandenchi*
　manganese dry cell

マンガン鉄重石　*mangantetsujuuseki*
　wolframite $(Mn,Fe)WO_4$

マンニッヒ反応　*mannihhihannou*
　Mannich reaction

マンニトール　*mannito-ru*
　mannitol $HOCH_2(CHOH)_4CH_2OH$

マンノヘプチット　*mannohepuchitto*
　mannoheptitol $C_7H_9(OH)_7$

マンノヘプトーゼ　*mannoheputo-ze*
　mannoheptose $HOCH_2(CHOH)_5CHO$

マンノン酸　*mannonsan*
　mannonic acid $HOCH_2(CHOH)_4COOH$

マンノース　*manno-su*
　mannose $C_6H_{12}O_6$

マンノテトローゼ　*mannotetoro-ze*
　mannotetrose $C_{24}H_{42}O_{21}$

マンノトリオーゼ　*mannotorio-ze*
　mannotriose $C_{18}H_{32}O_{16}$

マンノーゼ　*manno-ze*
　mannose $C_6H_{12}O_6$

マノメーター　*manome-ta-*
　manometer

マニュアルコントロール
　manyarukontoro-ru
　manual control

マラドラ石　*maradoraseki*
　mallardite $MnSO_4 \cdot 7H_2O$

マラカイト　*marakaito*
　malachite $Cu(OH)_2 \cdot CuCO_3$

マラカイトグリーン　*marakaitoguri-n*
　malachite green $C_{23}H_{25}N_2Cl$

マラリン　*mararin*
　malarin $C_{16}H_{17}NO$

マレイン酸　*mareinsan*
　maleic acid $HOOCCH=CHCOOH$

マレイン酸塩　*mareinsanen*
　maleate (salt of maleic acid)
　$MOOCCH=CHCOOM$

マレイン酸エステル　*mareinsanesuteru*
　maleate (ester of maleic acid)
　$ROOCCH=CHCOOR$

マレイン酸ジエチル　*mareinsanjiechiru*
　diethyl maleate
　$C_2H_5OOCCH=CHCOOC_2H_5$

マリオットの法則　*mariottonohousoku*
　Mariotte's law

マロフェン　*marofen*
　mallophene $C_{11}H_{12}N_5Cl$

マロナミド　*maronamido*
　malonamide $CH_2(CONH_2)_2$

マロンニトリル　*maronnitoriru*
　malononitrile $CH_2(CN)_2$

マロン酸　*maronsan*
　malonic acid $CH_2(COOH)_2$

マロン酸塩　*maronsanen*
　malonate (salt of malonic acid)
　$CH_2(COOR)_2$

マロン酸エステル　*maronsanesuteru*
　malonic ester $CH_2(COOR)_2$

マロン酸ジエチル　*maronsanjiechiru*
　diethyl malonate $CH_2(COOC_2H_5)_2$

マロン酸ジメチルエステル
　maronsanjimechiruesuteru
　dimethyl malonate $CH_2(COOCH_3)_2$

マロン酸ジニトリル　*maronsanjinitoriru*
　malononitrile $CH_2(CN)_2$

マルチン石　*maruchinseki*
　martinite $H_2Ca_5(PO_4)_4$

マルドナイト　*marudonaito*
　maldonite Au_2Bi

マルファニル　*marufaniru*
　marfanil $C_6H_4(CH_2NH_2)(SO_2NH_2)$

マルガリン　*marugarin*
　margarin $C_3H_5(OOC_{17}H_{32})_3$

マルガリン酸　*marugarinsan*
　margaric acid $CH_3(CH_2)_{15}COOH$

マルガリン酸ニトリル
　marugarinsannitoriru
　margaronitrile $CH_3(CH_2)_{15}CN$

マルガロサン石　*marugarosanseki*
　margarosanite $Pb(Ca,Mn)_2[Si_3O_9]$

マルコーヴニコフの法則
　maruko-vunikofunohousoku
　Markovnikov's rule

マルターゼ　*maruta-ze*
　maltase

マルテンサイト　*marutensaito*
　martensite α-Fe

マルトデキストリン　*marutodekisutorin*
　maltodextrin, amyloin

マルトン酸　*marutonsan*
　maltonic acid $HOCH_2(CHOH)_4COOH$

マルトース, マルトーゼ
　maruto-su, maruto-ze
　maltose $C_{12}H_{22}O_{11}$

マルヴィジン　*maruvijin*
　malvidin $C_{17}H_{24}O_9$

マルヴィン　*maruvin*
　malvine $C_{29}H_{35}O_{17}Cl$

マーシャルの酸　*ma-sharunosan*
　peroxodisulfuric acid, Marshall's acid $H_2S_2O_8$

マシン油　*mashinyu*
　machine oil

マーシュ試験　*ma-shushiken*
　Marsh's test

マソン石　*masonseki*
　masonite $FeAl_2[(OH)_2,O,SiO_4]$

マスチック　*masuchikku*
　mastic

マスカニー石　*masukani-seki*
　mascagnite $(NH_4)_2SO_4$

マスキング　*masukingu*
　masking

マスキング剤　*masukinguzai*
　masking reagent

マスタードガス　*masuta-dogasu*
　mustard gas, dichlorodiethyl sulfide $S(CH_2CH_2Cl)_2$

マトリックス　*matorikkusu*
　matrix

マゼンタ　*mazenta*
　magenta $C_{20}H_{20}N_3Cl$

メ me

メバロン酸　*mebaronsan*
　mevalonic acid
　$HOCH_2CH_2C(OH)(CH_3)CH_2COOH$

メチオニン　*mechionin*
　methionine $CH_3SCH_2CH_2CH(NH_2)COOH$

メチラール　*mechira-ru*
　methylal $CH_2(OCH_3)_2$

メチラート　*mechira-to*
　methylate CH_3OM

メチレンブルー　*mechirenburu-*
　methylene blue $C_{16}H_{18}N_3SCl$

メチロール　*mechiro-ru*
　methylol $HOCH_2$-

メチル　*mechiru*
　methyl CH_3-

メチルアミン　*mechiruamin*
　methylamine CH_3NH_2

メチルアルブチン　*mechiruarubuchin*
　methyl arbutin $C_{13}H_{19}O_7$

メチルアルデヒド　*mechiruarudehido*
　formaldehyde $HCHO$

メチルアルコール　*mechiruaruko-ru*
　methanol CH_3OH

メチルアセチレン　*mechiruasechiren*
　propyne CH_3CCH

メチルビニルエーテル *mechirubinirue-teru*
methyl vinyl ether $H_2C=CHOCH_3$

メチルブルー *mechiruburu-*
methyl blue $C_{37}H_{27}O_9N_3S_3Na_2$

メチルエチルケトン *mechiruechiruketon*
ethyl methyl ketone $CH_3COC_2H_5$

メチルエーテル *mechirue-teru*
dimethyl ether, methyl ether CH_3OCH_3

メチルフェニルケトン *mechirufeniruketon*
acetophenone, phenylmethyl ketone $C_6H_5COCH_3$

メチルグリコシド *mechirugurikoshido*
methyl glucoside

メチルグリオクザル *mechiruguriokuzaru*
methyl glyoxal CH_3COCHO

メチルヒドラジン *mechiruhidorajin*
methylhydrazine CH_3NHNH_2

メチルヒドロキシルアミン *mechiruhidorokishiruamin*
methoxyamine CH_3ONH_2

メチルイソブチルケトン *mechiruisobuchiruketon*
methyl isobutyl ketone $CH_3COCH_2CH(CH_3)_2$

メチルイソチオシアナート *mechiruisochioshiana-to*
methyl isothiocyanate CH_3NCS

メチルイソシアネート *mechiruisoshiane-to*
methyl isocyanate CH_3NCS

メチル化 *mechiruka*
methylation

メチル基 *mechiruki*
methyl group CH_3-

メチルオレンジ *mechiruorenji*
methyl orange $(CH_3)_2NC_6H_4N=NC_6H_4SO_3Na$

メチルレッド *mechirureddo*
methyl red $C_{15}H_{15}N_3O_2$

メチルセルロース *mechiruseruro-su*
methylcellulose

メチルシクロヘキサン *mechirushikurohekisan*
methyl cyclohexane C_7H_{14}

メチルシクロヘキサノン *mechirushikurohekisanon*
methylcyclohexanone $C_6H_9O(CH_3)$

メチルスルフォン酸 *mechirusurufonsan*
methyl sulfonic acid CH_3SO_3H

メチルトリクロルシラン *mechirutorikurorushiran*
methyltrichlorosilane Cl_3SiCH_3

メガサイクル *megasaikuru*
megahertz

メイラード反応 *meira-dohannou*
Maillard reaction

メカニズム *mekanizumu*
mechanism

メコニジン *mekonijin*
meconidine $C_{21}H_{23}NO_4$

メコニン *mekonin*
meconin $C_{10}H_{10}O_4$

メコン酸 *mekonsan*
meconic acid $C_7H_4O_7 \cdot 3H_2O$

メコシアニン *mekoshianin*
mecocyanin $C_{27}H_{30}O_{16}$

メーマック石 *me-makkuseki*
meymacite $WO_3 \cdot H_2O$

メナジオン *menajion*
menadione $C_{11}H_8O_2$

メナキノン *menakinon*
menaquinone $C_{46}H_{64}O_2$

メンチルアルコール *menchiruaruko-ru*
menthyl alcohol $C_{10}H_{19}OH$

メンデレビウム *menderebiumu*
mendelevium (Md, element 101)

メンデレエフ石 *mendereefuseki*
mendelejevite $CaO \cdot U_3O_8 \cdot TiO_2 \cdot Nb_2O_5$

メネギニ石 *meneginiseki*
meneghinite $4PbS \cdot Sb_2S_3$

メネティサイト *menetisaito*
minetisite $PbCl_2 \cdot 3Pb_3(AsO_4)_2$

メニアンチン menianchin
menyantin $C_{33}H_{50}O_{14}$

メンジップ石 menjippuseki
mendipite $2PbO \cdot PbCl_2$

メノリシン menorishin
menolysine $C_{22}H_{30}N_2O_4$

メノザール menoza-ru
menosal $C_{18}H_{26}O_3$

メンタン mentan
menthane $C_{10}H_{20}$

メンタノン mentanon
menthanone $C_{10}H_{18}O$

メンタノール mentano-ru
menthanol $C_{10}H_{19}OH$

メンテン menten
menthene $C_{10}H_{18}$

メントン menton
menthone $C_{10}H_{18}O$

メントール mento-ru
menthol $C_{10}H_{19}OH$

メラミン meramin
melamine $C_3N_6H_6$

メラミン樹脂 meraminjushi
melamine resin

メラムピリット, メラムピリン meramupiritto, meramupirin
melampyrine, melampyrite $HOCH_2(CHOH)_4CH_2OH$

メラニン meranin
melanine, diphenylguanidine $HN=C(NHPhe)_2$

メラニリン meranirin
melaniline $C_6H_5NHC(NH_2)=NC_6H_5$

メレン meren
melene $C_{20}H_{60}$

メリビオーゼ meribio-ze
melibiose $C_{12}H_{22}O_{11}$

メリファナイト merifanaito
meliphanite $PbMoO_4$

メリロト酸 merirotosan
melilotic acid $HOC_6H_4CH_2CH_2COOH$

メリル石 meriruseki
merrillite $3CaO \cdot Na_2O \cdot P_2O_5$

メリー石 meri-seki
melite $2(Al,Fe)_2O_3 \cdot SiO_2 \cdot 8H_2O$

メリシルアルコール merishiruaruko-ru
melissic alcohol $CH_3(CH_2)_{28}CH_2OH$

メリシトーゼ merishito-ze
melecitose $C_{18}H_{32}O_{16} \cdot 2H_2O$

メリテン meriten
mellitene $C_6(CH_3)_6$

メリトリオーゼ meritorio-ze
melitriose $C_{18}H_{22}O_{16} \cdot 5H_2O$

メリトーゼ merito-ze
melitose $C_{12}H_{22}O_{11}$

メリット酸 merittosan
mellitic acid $C_6(COOH)_6$

メロファン酸 merofansan
mellophanic acid $C_6H_2(COOH)_4$

メロン meron
mellone $C_{19}H_{13}$

メロネス鉱 meronesukou
melonite $NiTe_2$

メルカプタン merukaputan
thiol, mercaptan RSH

メルカプタール merukaputa-ru
mercaptal, thioacetal $RCH(SR')_2$

メルカプト基 merukaputoki
mercapto group $-SH$

メルカプト酢酸 merukaputosakusan
mercaptoacetic acid $HSCH_2COOH$

メルクプリン merukupurin
mercupurin $C_{14}H_{25}O_5NHg$

メルペリジン meruperijin
merperidine $C_{14}H_{19}NO_2$

メールワインポンドルフバーレー還元 me-ruwainpondorufuba-re-kangen
Meerwein-Ponndorf-Verley reduction

メサコニチン mesakonichin
mesaconitine $C_{34}H_{47}NO_{11}$

メサコン酸 mesakonsan
mesaconic acid $HOOCC(CH_3)=CHCOOH$

メシチレン　*meshichiren*
mesitylene $C_6H_3(CH_3)_3$

メシチレン酸　*meshichirensan*
mesitylenic acid $(CH_3)_2C_6H_3COOH$

メシチル基　*meshichiruki*
mesityl radical $(CH_3)_3C_6H_2-$

メシチルオキシド　*meshichiruokishido*
mesityl oxide $(CH_3)_2C=CHCOCH_3$

メシジン　*meshijin*
mesidine $H_2NC_6H_2(CH_3)_3$

メシル基　*meshiruki*
mesyl group $-SO_2CH_3$

メシトール　*meshito-ru*
mesitol $(CH_3)_3C_6H_2OH$

メソフェーズ　*mesofe-zu*
mesophase

メソ形　*mesogata*
meso form

メソメリー効果　*mesomeri-kouka*
resonance effect, mesomeric effect

メソン　*meson*
methone $C_8H_{12}O_2$

メソ酒石酸　*mesoshusekisan*
mesotartaric acid $HOOC(CHOH)_2COOH$

メソシュウ酸, メソ蓚酸　*mesoshuusan*
mesoxalic acid $CO(COOH)_2$

メッシュ　*messhu*
mesh

メスフラスコ　*mesufurasuko*
measuring flask

メスピペット　*mesupipetto*
measuring pipette

メスシリンダー　*mesushirinda-*
measuring cylinder

メタ　*meta*
meta

メタアンチモン酸　*metaanchimonsan*
metaantimonic acid $HSbO_3$

メタアルデヒド　*metaarudehido*
metaldehyde $(CH_3CHO)_n$

メタバナジン酸　*metabanajinsan*
metavanadic acid $(HVO_3)_n$

メタボライト　*metaboraito*
metabolite

メタブルッシュ石　*metaburusshuseki*
metabrushite $2HCaPO_4 \cdot 3H_2O$

メタチアゾール　*metachiazo-ru*
metathiazole C_3H_3NS

メタチタン酸　*metachitansan*
metatitanic acid H_2TiO_3

メタチタン酸マグネシウム
metachitansanmaguneshiumu
magnesium metatitanate Mg_2TiO_4

メタ鉛酸　*metaensan*
metaplumbic acid H_2PbO_3

メタフェン　*metafen*
metaphen $C_6H_2(CH_3)(Ona)(HgOH)(NO_2)$

メタゲルマニウム酸
metagerumaniumusan
metagermanic acid H_2GeO_3

メタゲルマニウム酸塩
metagerumaniumusanen
metagermanate M_2GeO_3

メタヘウイット石　*metaheuittoseki*
metahewettite $CaO \cdot 3V_2O_5 \cdot 9H_2O$

メタホウ酸, メタ硼酸　*metahousan*
metaboric acid HBO_2

メタ位　*metai*
meta-position

メタジゲルマニウム酸塩
metajigerumaniumusanen
metadigermanate $M^I_2Ge_2O_5$

メタジルコン酸　*metajirukonsan*
metazirconic acid H_2ZrO_3

メタジルコン酸塩　*metajirukonsanen*
metazirconate $M^I_2ZrO_3$

メタジルコン酸ナトリウム
metajirukonsannatoriumu
sodium metazirconate Na_2ZrO_3

メタキシロール　*metakishiro-ru*
metaxylene $C_6H_4(CH_3)_2$

メタクレゾール　*metakurezo-ru*
metacresol $C_6H_4(CH_3)(OH)$

メタクリロニトリル　*metakurironitoriru*
methacrylonitrile $H_2C=C(CH_3)CN$

メタクリルアルデヒド
metakuriruarudehido
methacrolein, methacrylaldehyde
$H_2C=C(CH_3)CHO$

メタクリル樹脂 *metakurirujushi*
methacrylic resin

メタクリルニトリル *metakurirunitoriru*
methacrylonitrile $H_2C=C(CH_3)CN$

メタクリル酸 *metakurirusan*
methacrylic acid $H_2C=C(CH_3)COOH$

メタクリル酸エチル *metakurirusanechiru*
ethyl methacrylate $H_2C=C(CH_3)COOC_2H_5$

メタクリル酸メチル
metakurirusanmechiru
methyl methacrylate $H_2C=C(CH_3)COOCH_3$

メタクロレイン *metakurorein*
methacrylaldehyde $H_2C=C(CH_3)CHO$

メタン *metan*
methane CH_4

メタナール *metana-ru*
methanal $HCHO$

メタンフェタミン *metanfetamin*
methamphetamine
$C_6H_5CH_2CH(CH_3)NHCH_3$

メタンガス *metangasu*
methane gas, marsh gas CH_4

メタニルエロー *metaniruero-*
metanil yellow, acid yellow
$C_6H_5NHC_6H_4SO_3Na$

メタニル酸 *metanirusan*
metanilic acid $H_2NC_6H_4SO_3H$

メタノリシス *metanorishisu*
methanolysis

メタノール *metano-ru*
methanol CH_3OH

メタンスルホン酸 *metansuruhonsan*
methanesulfonic acid $(CH_3)SO_2(OH)$

メタンスルホン酸メチル
metansuruhonsanmechiru
methyl methanesulfonate
$(CH_3)SO_2(OCH_3)$

メタリン酸, メタ燐酸 *metarinsan*
metaphosphoric acid $(HPO_3)_n$

メタ燐酸ナトリウム
metarinsannatoriumu
calgon, sodium hexametaphosphate
$(NaPO_3)_{15-20}$

メタロセン *metarosen*
metallocene $M^I(C_5H_5)_n$

メタ三硼酸ナトリウム
metasanhousannatoriumu
sodium metatriborate $Na_3B_3O_6$

メタセチン *metasechin*
methacetin $CH_3CONHC_6H_4OCH_3$

メタセシス反応 *metaseshisuhannou*
metathesis reaction

メタセトン *metaseton*
metacetone $C_2H_5COC_2H_5$

メタスズ酸, メタ錫酸 *metasuzusan*
metastannic acid H_2SnO_3

メタ錫酸ナトリウム
metasuzusannatoriumu
sodium metastannate Na_2SnO_3

メタ錫酸リチウム *metasuzusanrichiumu*
lithium metastannate Li_2SnO_3

メタタンタル酸塩 *metatantarusanen*
metatantalate M^ITaO_3

メタテトラケイ酸, メタテトラ珪酸
metatetorakeisan
metatetrasilicic acid $H_6Si_4O_{11}$

メタヴァナジン酸 *metavanajinsan*
metavanadic acid $(HVO_3)_n$

メタゾイネル石 *metazoineruseki*
metazeunerite $Cu[UO_2,AsO_4]_2 \cdot 8H_2O$

メトキシド *metokishido*
methoxide CH_3OM

メトキシ基 *metokishiki*
methoxy group CH_3O-

メートル *me-toru*
meter

メートル法 *me-toruhou*
metric system

メートル制 *me-torusei*
metric system

メツカリン *metsukarin*
mescaline $C_{11}H_{17}NO_3$

メゾエリトリット　*mezoeritoritto*
　mesoerythrite $HOCH_2(CHOH)_2CH_2OH$

メゾイノシット　*mezoinoshitto*
　mesoinosite $C_6H_6(OH)_6$

メゾメリー　*mezomeri-*
　mesomerism, resonance

メゾルチン　*mezoruchin*
　mesorcinol $C_6H(OH)_2(CH_3)_3$

メゾタン　*mezotan*
　mesotan $HOC_6H_4COOCH_2OCH_3$

ミ mi

ミエモ石　*miemoseki*
　miemite $CaCO_3 \cdot MgCO_3$

ミエローマ　*miero-ma*
　myeloma

ミジン　*mijin*
　mydine $C_9H_{11}O_2N$

ミカエリスアルブーゾフ反応
　mikaerisuarubu-zofuhannou
　Michaelis-Arbusov reaction

ミカエリスメンテンの式
　mikaerisumentennoshiki
　Michaelis-Menten equation

ミキサー　*mikisa-*
　mixer

ミコバクテリア　*mikobakuteria*
　mycobacterium

ミクラーのヒドロール
　mikura-nohidoro-ru
　Michler's hydrol $(C_6H_4N(CH_3)_2)_2CH(OH)$

ミクラーのケトン　*mikura-noketon*
　Michler's ketone $(C_6H_4N(CH_3)_2)_2CO$

ミクロバール　*mikuroba-ru*
　microbar

ミクロビュレット　*mikurobyuretto*
　microburette

ミクログラム　*mikuroguramu*
　microgram

ミクロ構造　*mikurokouzou*
　microstructure

ミクロリットル　*mikurorittoru*
　microliter

ミナサイト　*minasaito*
　minasite $2Al_2O_3 \cdot 3H_2O$

ミナスラグラ石　*minasuraguraseki*
　minasragrite $V_2O_4 \cdot 3SO_3 \cdot 16H_2O$

ミネラル　*mineraru*
　mineral

ミネサイト　*minesaito*
　minesite $KHSO_4$

ミニッツ　*minittsu*
　minute

ミオゲン　*miogen*
　myogen

ミオグロビン　*miogurobin*
　myoglobin

ミオシン　*mioshin*
　myosin

ミラー石　*mira-seki*
　milarite
　$K_2O \cdot 4CaO \cdot 4BeO \cdot Al_2O_3 \cdot 24SiO_2 \cdot H_2O$

ミリボルト　*miriboruto*
　millivolt

ミリグラム　*miriguramu*
　milligram

ミリメートル　*mirime-toru*
　millimeter

ミリモル　*mirimoru*
　millimole

ミリポアフィルター　*miripoafiruta-*
　millipore filter

ミリリットル　*miririttoru*
　milliliter

ミリセチン　*mirisechin*
　myricetin $C_{15}H_{10}O_8$

ミリシン　*mirishin*
　myricin $C_{13}H_{31}COOC_{30}H_{61}$

ミリシルアルコール　*mirishiruaruko-ru*
　myricyl alcohol $C_{31}H_{63}OH$

ミリスチンアルコール *mirisuchinaruko-ru*
myristic alcohol $CH_3(CH_2)_{12}CH_2OH$

ミリスチン酸 *mirisuchinsan*
myristic acid $CH_3(CH_2)_{12}COOH$

ミリスティコール *mirisutiko-ru*
myristicol $C_{10}H_{16}O$

ミリストン *mirisuton*
myriston $(C_{13}H_{27})_2CO$

ミロン塩基 *mironenki*
Millon's base Hg_2NOH

ミロン酸 *mironsan*
myronic acid $C_{10}H_{19}NS_2O_{10}$

ミル *miru*
mill

ミルセン *mirusen*
myrcene $(CH_3)_2C=CH(CH_2)_2C(=CH_2)CH=CH_2$

ミセル *miseru*
micelle

ミセル形成 *miserukeisei*
micelle formation

ミトコンドリア *mitokondoria*
mitochondria

モ mo

モーア塩 *mo-aen*
Mohr's salt $(NH_4)_2Fe(SO_4)_2 \cdot 6H_2O$

モデル *moderu*
model

モデル化 *moderuka*
modelling

モジュール *moju-ru*
module

モナルディン *monarudin*
monardain $C_{44}H_{45}O_{23}Cl$

モナ石 *monaseki*
monite $Ca_3(PO_4)_2 \cdot H_2O$

モンド法 *mondohou*
Mond process

モネタ石 *monetaseki*
monetite $CaH[PO_4]$

モニモス鉱 *monimosukou*
monimolite $Pb_3Sb_2O_8$

モノアジン *monoajin*
monazine C_5H_5N

モノアセチン *monoasechin*
monoacetin $CH_3COOCH_2CH(OH)CH_2OH$

モノアゾール *monoazo-ru*
monazol C_4H_4NH

モノチオグリコール *monochioguriko-ru*
monothioglycol $HOCH_2CH_2SH$

モノチオ炭酸 *monochiotansan*
monothiocarbonic acid H_2CO_2S

モノエタノールアミン *monoetano-ruamin*
ethanolamine $HOCH_2CH_2NH_2$

モノゲルマン *monogeruman*
monogermane GeH_4

モノグリセリド *monoguriserido*
monoglyceride

モノカルボン酸 *monokarubonsan*
monocarboxylic acid

モノクロム酸 *monokuromusan*
monochromic acid, chromic acid H_2CrO_4

モノマー *monoma-*
monomer

モノヌクレオチド *mononukureochido*
mononucleotide

モノオキザリン *monookizarin*
monoxalin $HOOCCOOCH_2CH(OH)CH_2OH$

モノオレイン *monoorein*
monoolein $C_{17}H_{31}COOCH_2CH(OH)CH_2OH$

モノシラン *monoshiran*
monosilane SiH_4

モノステアリン *monosutearin*
monostearin $C_{17}H_{35}COOCH_2CH(OH)CH_2OH$

モノテルペン *monoterupen*
monoterpene

モントロイ石　*montoroiseki*
　montroydite HgO

モンゼ塩　*monzeen*
　Monsel's salt $Fe_4O(SO_4)_5$

モーラル　*mo-raru*
　molar

モレキュラーシーブ　*morekyura-shi-bu*
　molecular sieve

モリブデン　*moribuden*
　molybdenum (Mo, element 42)

モリブデン鉛鉱　*moribudenenkou*
　wulfenite $PbMoO_4$

モリブデン酸　*moribudensan*
　molybdic acid H_2MoO_4

モリブデン酸アンモニウム
　moribudensananmoniumu
　ammonium molybdate $(NH_4)_2MoO_4$

モリブデン酸塩　*moribudensanen*
　molybdate M_2MoO_4

モリブデン酸ナトリウム
　moribudensannatoriumu
　sodium molybdate Na_2MoO_4

モーリン　*mo-rin*
　morin $C_{15}H_{10}O_7$

モリンガタンニン酸　*moringatanninsan*
　moringatannic acid
　$C_6H_3(OH)_2COC_6H_2(OH)_3$

モリサイト　*morisaito*
　molysite $FeCl_3$

モロミ　*moromi*
　mash

モル　*moru*
　mole

モル分率　*morubunritsu*
　mole fraction

モル伝導率　*morudendouritsu*
　molar conductivity

モル液　*morueki*
　molar solution

モルフィン　*morufin*
　morphine $C_{17}H_{19}NO_3$

モルフォリン　*moruforin*
　morpholine C_4H_9NO

モルフォール　*morufo-ru*
　morphol $C_{14}H_8(OH)_2$

モルヒネ　*moruhine*
　morphine $C_{17}H_{19}NO_3$

モルホリン　*moruhorin*
　morpholine C_4H_9NO

モル百分率　*moruhyakubunritsu*
　molar percentage

モル吸収係数　*morukyuushuukeisuu*
　molar absorption coefficient

モル濃度　*morunoudo*
　molarity

モル質量　*morushitsuryou*
　molar weight, relative molar mass

モル体積　*morutaiseki*
　molar volume

モルタル　*morutaru*
　mortar

モルウェン石　*moruwenseki*
　morvenite $(K_2,Ba)Al_2Si_5O_{14}\cdot 5H_2O$

モス石　*mosuseki*
　mossite $(Fe,Mn)(Nb,Ta)_2O_6$

モーター油　*mo-ta-yu*
　motor oil

モーヴェイン　*mo-vein*
　mauve $C_{26}H_{23}N_4Cl$

ム mu

ムチン酸　*muchinsan*
　mucic acid $HOOC(CHOH)_4COOH$

ムコイド　*mukoido*
　mucoid

ムコン酸　*mukonsan*
　muconic acid
　HOOCCH=CHCH=CHCOOH

ムコ蛋白　*mukotanpaku*
　mucoprotein

ムコ多糖　*mukotatou*
　mucopolysaccharide, glycosaminoglycan

ムヌマン石 *munumanseki*
muthmannite (Ag,Au)Te

ムラミン酸 *muraminsan*
muramic acid $C_9H_{17}NO_7$

ムレキシド *murekishido*
murexide $C_8H_4N_5O_6NH_4 \cdot H_2O$

ムル石 *muruseki*
mullite

ムシン *mushin*
mucin

ムスカリン *musukarin*
muscarine $C_8H_{19}NO_3$

ムスコン *musukon*
muscone $C_{16}H_{30}O$

ナ na

ナチュラルサイエンス *nachurarusaiensu*
natural science

ナフチジン *nafuchijin*
naphthidine $C_{20}H_{12}(NH_2)_2$

ナフチオン酸 *nafuchionsan*
naphthionic acid $H_2NC_{10}H_6SO_3H$

ナフチル *nafuchiru*
naphthyl $C_{10}H_7$-

ナフチルアミン *nafuchiruamin*
naphthylamine $C_{10}H_7NH_2$

ナフサ *nafusa*
petroleum naphtha

ナフスリタム *nafusuritamu*
naphsultam $C_{10}H_7NSO_2$

ナフタジン *nafutajin*
naphthazine $C_{20}H_{12}N_2$

ナフタミド *nafutamido*
naphthamide $C_{10}H_7CONH_2$

ナフタリン, ナフタレン
nafutarin, nafutaren
naphthalene $C_{10}H_8$

ナフタリンエロー *nafutarinero-*
naphthalene yellow $C_{10}H_6O(NO_2)_2$

ナフタリンレッド *nafutarinreddo*
naphthalene red $C_{30}H_{21}N_4Cl$

ナフタリン酸 *nafutarinsan*
naphthalenic acid $C_{10}H_6O_3$

ナフタリンスルフォン酸
nafutarinsurufonsan
naphthalenesulfonic acid $C_{10}H_7SO_3H$

ナフタルデヒド *nafutarudehido*
naphthaldehyde $C_{10}H_7CHO$

ナフタール酸 *nafuta-rusan*
naphthalic acid $C_{10}H_6(COOH)_2$

ナフタセン *nafutasen*
naphthacene $C_{18}H_{12}$

ナフタザリン *nafutazarin*
naphthazarine $C_{10}H_4O_2(OH)_2$

ナフテン *nafuten*
naphthene CnH_{2n}

ナフテン酸 *nafutensan*
naphthenic acid

ナフテン酸亜鉛 *nafutensanaen*
zinc naphthenate

ナフテン酸マグネシウム
nafutensanmaguneshiumu
magnesium naphthenate

ナフトアミド *nafutoamido*
naphthoamide $C_{10}H_7CONH_2$

ナフト工酸 *nafutoesan*
naphthoic acid $C_{10}H_7COOH$

ナフトヒドロキノン *nafutohidorokinon*
naphthohydroquinone $C_{10}H_8O_2$

ナフトイル *nafutoiru*
naphthoyl $C_{10}H_7CO$-

ナフトキノン *nafutokinon*
naphthoquinone $C_{10}H_6O_2$

ナフトン *nafuton*
naphtone $C_{10}H_8O$

ナフトオキシ *nafutookishi*
naphthoxy $C_{10}H_7O$-

ナフトール *nafuto-ru*
naphthol $C_{10}H_7OH$

ナフトールブルー *nafuto-ruburu-*
naphthol blue $C_{18}H_{16}N_3O$

ナフトールエロー　*nafuto-ruero-*
naphthol yellow, acid yellow $C_{10}H_4N_2O_8SNa_2$

ナフトールグリーン　*nafuto-ruguri-n*
naphthol green, acid green $C_{10}H_5(NO)(OH)(SO_3H)$

ナフトールジスルフォン酸　*nafuto-rujisurufonsan*
naphtholdisulfonic acid $C_{10}H_5(OH)(SO_3H)_2$

ナフトールオレンジ　*nafuto-ruorenji*
naphthol orange $HOC_{10}H_6N=NC_6H_4SO_3H$

ナフトール染料　*nafuto-rusenryou*
naphthol dye

ナフトールスルフォン酸　*nafuto-rusurufonsan*
naphtholsulfonic acid $C_{10}H_6(OH)SO_3H$

ナフトールザロール　*nafuto-ruzaro-ru*
naphthosalol $C_6H_4(OH)COOC_{10}H_7$

ナイロン　*nairon*
nylon

ナイルブルー　*nairuburu-*
nile blue $C_{20}H_{19}N_3O$

ナイトロジェンマスタード　*naitorojenmasuta-do*
nitrogen mustard $CH_3N(CH_2CH_2Cl)_2 \cdot HCl$

ナイトロ・マグネサイト　*naitorosmagunesaito*
nitromagnesite $Mg(NO_3)_2 \cdot nH_2O$

ナメシ剤　*nameshizai*
tanning agent

ナンジニン　*nanjinin*
nandinine $C_{19}H_{19}NO_4$

ナントコ石　*nantokoseki*
nantokite $CuCl$

ナパーム　*napa-mu*
napalm

ナペリン　*naperin*
napeline $C_{32}H_{43}NO_{10}$

ナポリ石　*naporiseki*
naplite $3NaAlSiO_4 \cdot CaSO_4$

ナリンゲニン　*naringenin*
naringenin $C_{15}H_{12}O_5$

ナリンギン　*naringin*
naringin $C_{21}H_{26}O_{11}$

ナルコチン　*narukochin*
narcotine $C_{22}H_{23}NO_7$

ナルセイン　*narusein*
narceine $C_{23}H_{27}NO_8 \cdot 3H_2O$

ナソン石　*nasonseki*
nasonite $Pb_4(PbCl)_2 \cdot Ca(Si_2O_7)_3$

ナタロイン　*nataroin*
nataloin $C_{34}H_{38}O_{15}$

ナトリウム　*natoriumu*
sodium (Na, element 11)

ナトリウムアルコラート　*natoriumuarukora-to*
sodium alcoholate C_2H_5ONa

ナトリウムエトキシド　*natoriumuetokishido*
sodium ethylate C_2H_5ONa

ナトリウムフェノラート　*natoriumufenora-to*
sodium phenolate C_6H_5ONa

ナトリウムメチラート　*natoriumumechira-to*
sodium methylate $NaOCH_3$

ナトリウム明礬　*natoriumumyouban*
sodium alum $NaAl(SO_4)_2 \cdot 12H_2O$

ネ ne

ねじれ　*nejire*
torsion

ネオヘキサン　*neohekusan*
neohexane $(CH_3)_3CCH_2CH_3$

ネオイソメントール　*neoisomento-ru*
neoisomenthol $C_{10}H_{19}OH$

ネオジム　*neojimu*
neodymium (Nd, element 60)

ネオマイシン　*neomaishin*
neomycin neomycin A: $C_{12}H_{26}N_4O_6$
neomycin B, neomycin C: $C_{23}H_{46}N_6O_{13}$

ネオメントール　*neomento-ru*
neomenthol $C_{10}H_{20}O$

ネオン *neon*
neon (Ne, element 10)

ネオンランプ *neonranpu*
neon lamp

ネオペンチル *neopenchiru*
neopentyl $-CH_2C(CH_3)_3$

ネオペンチルアルコール
neopenchiruaruko-ru
neopentyl alcohol $(CH_3)_3CH_2OH$

ネオペンタン *neopentan*
neopentane $CH_3C(CH_3)_2CH_3$

ネオプレン *neopuren*
neoprene $[-CH_2CCl=CHCH_2-]_n$

ネオシアニン *neoshianin*
neocyanine $C_{25}H_{33}N_2O_2I$

ネプシニウム *nepushiniumu*
neptunium (Np, element 93)

ネロリン *nerorin*
neroline $C_{12}H_{12}O$

ネロール *nero-ru*
nerol
$(CH_3)_2C=CHCH_2CH_2C(CH_3)=CHCH_2OH$

ネルンスト式 *nerunsutoshiki*
Nernst equation

ネルヴォン *neruvon*
nervone $C_{48}H_{91}NO_8$

ネルヴォン酸 *neruvonsan*
nervonic acid
$CH_3(CH_2)_7CH=CH(CH_2)_{13}COOH$

ニ ni

ニグラニリン *niguranirin*
nigraniline $C_{48}H_{36}N_8$

ニグロシン *niguroshin*
nigrosine $C_{38}H_{27}N_3$

にじみ *nijimi*
bleeding

ニッケル *nikkeru*
nickel (Ni, element 28)

ニッケルカルボニル *nikkerukaruboniru*
nickel carbonyl $Ni(CO)_4$

ニッケルテトラカルボニル
nikkerutetorakaruboniru
nickel tetracarbonyl $Ni(CO)_4$

ニコチン *nikochin*
nicotine $C_{10}H_{14}N_2$

ニコチンアミド *nikochinamido*
nicotinamide $C_5H_4NCONH_2$

ニコチン硫酸塩 *nikochinryuusanen*
nicotine sulfate $(C_{10}H_{14}N_2)_2 \cdot H_2SO_4$

ニコチン酸 *nikochinsan*
nicotinic acid C_5H_4NCOOH

ニコチン酸アミド *nikochinsanamido*
nicotinamide $C_5H_4NCONH_2$

ニコチリン *nikochirin*
nicotyrine $C_{10}H_{10}N_2$

にくずく油 *nikuzukuyu*
nutmeg oil

ニンヒドリン *ninhidorin*
ninhydrine $C_9H_6O_4$

ニンヒドリン反応 *ninhidorinhannou*
ninhydrine reaction

ニオベ油 *niobeyu*
niobe oil $C_6H_5COOCH_3$

ニオブ *niobu*
niobium (Nb, element 41)

ニオブ酸塩 *niobusanen*
niobate $M^I_8Nb_6O_{19}$

ニオフォルム *nioforumu*
nioform C_9H_7NOICl

ニルヴァニン *niruvanin*
nirvanin $C_{14}H_{22}N_3O_4 \cdot HCl$

ニルヴァノール *niruvano-ru*
nirvanol $C_{11}H_{12}O_2N_2$

ニトラミド *nitoramido*
nitramide H_2NNO_2

ニトラミン *nitoramin*
nitramine $RNHNO_2$

ニトラニリド *nitoranirido*
nitranilide $C_6H_5NHNO_2$

ニトラーゼ *nitora-ze*
nitrase

ニトリラーゼ *nitorira-ze*
nitrilase

ニトリロスルフォン酸
nitorirosurufonsan
nitrilosulfonic acid $N(SO_3H)_3$

ニトリル *nitoriru*
nitrile RCN; nitryl NO_2^+

ニトリル三酢酸 *nitorirusansakusan*
nitriloacetic acid $N(CH_2COOH)_3$

ニトロ *nitoro*
nitro NO_2-

ニトロアミン *nitoroamin*
nitramine $RNHNO_2$

ニトロベンゾール, ニトロベンゼン
nitorobenzo-ru, nitorobenzen
nitrobenzene $C_6H_5NO_2$

ニトロブロモフォルム
nitoroburomoforumu
nitrobromoform Br_3CNO_2

ニトロエタン *nitoroetan*
nitroethane $C_2H_5NO_2$

ニトロフェノール *nitorofeno-ru*
nitrophenol $C_6H_4(OH)NO_2$

ニトロフォスカ *nitorofosuka*
nitrophoska

ニトログアニジン *nitoroguanijin*
nitroguanidine $HN=C(NH_2)NHNO_2$

ニトログリコール *nitoroguriko-ru*
nitroglycol $C_2H_4(ONO_2)_2$

ニトログリセリン *nitoroguriserin*
nitroglycerin $C_3H_5(ONO_2)_3$

ニトロ化 *nitoroka*
nitration

ニトロ化合物 *nitorokagoubutsu*
nitro compound

ニトロ化剤 *nitorokazai*
nitrating agent

ニトロ基 *nitoroki*
nitro group $-NO_2$

ニトロキノン *nitorokinon*
nitroquinone $C_6H_3NO_4$

ニトロメーター *nitorome-ta-*
nitrometer

ニトロメタン *nitorometan*
nitromethane CH_3NO_2

ニトロン *nitoron*
nitron $C_{20}H_{16}N_4$

ニトロン酸 *nitoronsan*
nitronic acid $R=NO(OH)$

ニトロパラフィン *nitoroparafin*
nitroparaffin

ニトロル酸 *nitororusan*
nitrolic acid $RC(=NOH)NO_2$

ニトロサミン *nitorosamin*
nitrosamine $RR'NNO$

ニトロセルロース, ニトロセルローゼ
nitoroseruro-su, nitoroseruro-ze
nitrocellulose, cellulose nitrate

ニトロシル硫酸 *nitoroshiruryuusan*
nitrosylsulfuric acid $NOHSO_4$

ニトロソアミン *nitorosoamin*
nitrosamine $RR'NNO$

ニトロソベンゾール, ニトロソベンゼン *nitorosobenzo-ru, nitorosobenzen*
nitrosobenzene C_6H_5NO

ニトロソ化 *nitorosoka*
nitrosation

ニトロソ基 *nitorosoki*
nitroso group $-NO$

ニトロソメタン *nitorosometan*
nitrosomethane CH_3NO

ニトロトルエン *nitorotoruen*
nitrotoluene $C_6H_4(CH_3)NO_2$

ノ no

ノーベリウム *no-beriumu*
nobelium (No, element 102)

ノイベルグのエステル
noiberugunoesuteru
Neuberg ester $C_6H_{13}O_9P$

ノイラミン酸　*noiraminsan*
neuraminic acid $C_9H_{17}NO_8$

ノイリジン　*noirijin*
neuridine $C_5H_{14}N_2$

ノイリン　*noirin*
neurine $H_2C=CHN(CH_3)_3OH$

ノイロジン　*noirojin*
neurodine $CH_3COOC_6H_4NHCOOC_2H_5$

ノナデカン　*nonadekan*
nonadecane $C_{19}H_{40}$

ノナン　*nonan*
nonane C_9H_{20}

ノナン酸　*nonansan*
nonanoic acid $C_8H_{17}COOH$

ノンデシルアルコール　*nondeshiruaruko-ru*
nonadecyl alcohol $CH_3(CH_2)_{17}CH_2OH$

ノンデシル酸　*nondeshirusan*
nonadecanoic acid $CH_3(CH_2)_{17}COOH$

ノニルアルコール　*noniruaruko-ru*
nonyl alcohol $CH_3(CH_2)_7CH_2OH$

ノニル酸　*nonirusan*
nonylic acid $CH_3(CH_2)_7COOH$

ノントロナイト　*nontoronaito*
nontronite $H_4Fe_2Si_2O_9$

ノピン酸　*nopinsan*
nopinic acid $C_{10}H_{16}O_3$

ノルアドレナリン　*noruadorenarin*
norepinephrine
$(HO)_2C_6H_3CH(OH)CH_2NH_2$

ノルバリン　*norubarin*
norvaline, 2-aminopentanoic acid
$CH_3(CH_2)_2CH(NH_2)COOH$

ノルビキシン　*norubikishin*
norbixin $C_{22}H_{26}(COOH)_2$

ノルデンショルド石　*norudenshorudoseki*
nordenskioldine $CaSn(BO_3)_2$

ノルエピネフリン　*noruepinefurin*
norepinephrine
$(HO)_2C_6H_3CH(OH)CH_2NH_2$

ノルカンファン　*norukanfan*
norcamphane $C_{10}H_{18}$

ノルカラン　*norukaran*
norcarane C_7H_{12}

ノルム　*norumu*
norm, standard, specification

ノルロイシン　*noruroishin*
norleucine $CH_3(CH_2)_3CH(NH_2)COOH$

ノルヴァリン　*noruvarin*
norvaline $CH_3(CH_2)_2CH(NH_2)COOH$

ノソフェン　*nosofen*
nosophen $C_{20}H_{10}O_4I_2$

ノヴァスピリン　*novasupirin*
novaspirin $C_{23}H_{16}O_9$

ノヴァトファン　*novatofan*
novatophan $C_{18}H_{15}NO_2$

ノズル　*nozuru*
nozzle

ヌ nu

ヌクレアーゼ　*nukurea-ze*
nuclease

ヌクレイン　*nukurein*
nuclein

ヌクレイン酸　*nukureinsan*
nucleic acid

ヌクレオチド　*nukureochido*
nucleotide

ヌクレオシダーゼ　*nukureoshida-ze*
nucleosidase

ヌクレオシド　*nukureoshido*
nucleoside

ぬれ　*nure*
wetting

ヌシン　*nushin*
nucin $C_{10}H_6O_3$

ニュ nyu

ニューランダー試薬　*nyu-randa-shiyaku*
Nylanders reagent

ニューロン   *nyu-ron*
neurone

ニュートン   *nyu-ton*
newton

ニュートン環   *nyu-tonkan*
Newton rings

ニュートンの流体   *nyu-tonnoryuutai*
Newtonian fluid, Newtonian liquid

ニュートンの運動法則   *nyu-tonnoundouhousoku*
Newton's laws of motion

ニュートン力学   *nyu-tonrikigaku*
Newtonian mechanics

ニュートラルレッド   *nyu-torarureddo*
neutral red, toluylene red $C_{15}H_{16}N_4 \cdot HCl$

ニュートリノ   *nyu-torino*
neutrino

ニュートロン   *nyu-toron*
neutron

オ o

オーバーフロー   *o-ba-furo-*
overflow

オボアルブミン   *oboarubumin*
ovalbumin

オボムコイド   *obomukoido*
ovomucoid

オボムシン   *obomushin*
ovomucin

オイボルニル   *oiboruniru*
eubornyl $CH_3CH(CH_3)CHBrCOO(C_{10}H_{17}O)$

オイデルモール   *oiderumo-ru*
eudermol $C_{10}H_{14}N_2C_6H_4(OH)COOH$

オイフォリン   *oiforin*
euphorine $C_2H_5OCONHC_6H_5$

オイフタルミン   *oifutarumin*
euphthalmine $C_{17}H_{25}NO_3$

オイガロール   *oigaro-ru*
eugallol $CH_3COOC_6H_3(OH)_2$

オイゲノール   *oigeno-ru*
eugenol $HOC_6H_3(OCH_3)CH_2CH=CH_2$

オイグフォルム   *oiguforumu*
euguforme $C_6H_3(OH)(OCH_3)COCH_3$

オイヒニン   *oihinin*
euchinine $C_{23}H_{28}N_2O_4$

オイカリプテン   *oikariputen*
eucalyptene $C_{10}H_{16}$

オイカリプトール   *oikariputo-ru*
eucalyptole $C_{10}H_{18}O$

オイカトロピン   *oikatoropin*
eucatropine $C_{17}H_{25}NO_3$

オイキサンチン酸   *oikisanchinsan*
euxanthic acid $C_{19}H_{16}O_{10} \cdot 3H_2O$

オイキサントン   *oikisanton*
euxanthone $C_{12}H_7O_4$

オイコディン   *oikodin*
eucodine $C_{19}H_{24}NO_3Br$

オイコール   *oiko-ru*
eucol $CH_3COOC_6H_4OCH_3$

オイミドリン   *oimidorin*
eumydrin $C_{19}H_{29}N_2O_6$

オイナトロール   *oinatoro-ru*
eunatrol $C_{17}H_{33}COONa$

オイオニミト   *oionimito*
euonymit $C_6H_8(OH)_6$

オイピトン   *oipiton*
eupittone $C_{25}H_{26}O_9$

オイポルフィン   *oiporufin*
euporphine $C_{18}H_{20}NO_2Br$

オイレゾール   *oirezo-ru*
euresol $H_2CCOOC_6H_4OH$

オイロフィン   *oirofin*
europhen $C_{22}H_{29}O_2I$

オイル   *oiru*
oil

オイルペイント   *oirupeinto*
oil paint

オイルシェール   *oirushe-ru*
oil shale

オイートストーン・ブリッジ
oi-tosuto-nsburijji
Wheatstone bridge

オージェ電子 *o-jedenshi*
Auger electron

オキサイド *okisaido*
oxide

オキサジン *okisajin*
oxazine C_4H_5NO

オキサミド *okisamido*
oxamide, oxalic acid diamide $H_2NOCCONH_2$

オキサミド酸 *okisamidosan*
oxamic acid, oxalic acid monoamide $H_2NOCCOOH$

オキサン *okisan*
oxane $C_5H_{10}O$

オキサリル *okisariru*
oxalyl (-OC-CO-)

オキサロ酢酸 *okisarosakusan*
oxalacetic acid $HOOCCOCH_2COOH$

オキサゾリジン *okisazorijin*
oxazolidine C_3H_7NO

オキサゾリン *okisazorin*
oxazoline C_3H_5NO

オキサゾール *okisazo-ru*
oxazole C_3H_3NO

オキセタン *okisetan*
oxetane C_3H_6O

オキシダント *okishidanto*
oxidizing agent

オキシダーゼ *okishida-ze*
oxidase

オキシ塩化硫黄 *okishienkaiou*
sulfur oxychloride $S_2O_5Cl_2/S_2OCl_4/S_2O_3Cl_4$

オキシ塩化珪素 *okishienkakeiso*
silicon oxychloride $Si_2OCl_6/Si_4O_4Cl_8$

オキシ塩化燐 *okishienkarin*
phosphorus oxychloride $POCl_3$

オキシ塩化セレン *okishienkaseren*
selenium oxychloride $SeOCl_2$

オキシ塩素化 *okishiensoka*
oxychlorination

オキシゲナーゼ *okishigena-ze*
oxygenase

オキシヒドラスチニン *okishihidorasuchinin*
oxyhydrastinine $C_{11}H_{11}NO_3$

オキシヒドロキノン *okishihidorokinon*
hydroxyhydroquinone $C_6H_3(OH)_3$

オキシム *okishimu*
oxime $RC(=NOH)R'$

オキシン *okishin*
oxine C_9H_7NO

オーキシン *o-kishin*
auxin $C_{10}H_9NO_2$

オキシナルコチン *okishinarukochin*
oxynacotine $C_{22}H_{23}NO_8$

オキシプロリン *okishipurorin*
hydroxyproline $C_6H_7N(OH)COOH$

オキシ水銀化 *okishisuiginka*
oxymercuration

オキシテトラサイクリン *okishitetorasaikurin*
oxytetracycline

オキシトシン *okishitoshin*
oxytocin $C_{43}H_{66}N_{12}O_{12}S_2$

オキソニウム塩 *okisoniumuen*
oxonium salt

オキソニウムイオン *okisoniumuion*
oxonium ion H_3O^+

オキソ酸 *okisosan*
oxygen acid

オクチン *okuchin*
octine C_7H_{14}

オクチレン *okuchiren*
octylene $C_8H_{16}O_8$

オクチルアルコール *okuchiruaruko-ru*
octyl alcohol $C_8H_{17}OH$

オクダデシレン *okudadeshiren*
octadecylene $C_{18}H_{36}$

オクサントラノール *okusantorano-ru*
oxanthranol, anthrahydroquinone $C_{14}H_{10}O_2$

オクソニウムイオン　*okusoniumuion*
oxonium ion H_3O^+

オクタデカン　*okutadekan*
octadecane $C_{18}H_{38}$

オクタモリブデン酸塩
okutamoribudensanen
octamolybdate $Mo_7O_{24}^{6-}$

オクタン　*okutan*
octane C_8H_{18}

オクタン価　*okutanka*
octane number

オクタノール　*okutano-ru*
octyl alcohol $C_8H_{17}OH$

オクタン酸　*okutansan*
octanoic acid $C_7H_{15}COOH$

オクテット　*okutetto*
octet

オクザニリド　*okuzanirido*
oxanilide $C_6H_5NHOCCONHC_6H_5$

オクザリル尿素　*okuzarirunyouso*
oxalylurea $C_3H_2N_2O_3$

オクザルル酸　*okuzarurusan*
oxaluric acid $H_2NCONHOCCOOH$

オマール　*oma-ru*
omal $C_6H_2(OH)Cl_3$

オーム　*o-mu*
ohm

オームの法則　*o-munohousoku*
Ohm's law

オナント酸　*onantosan*
oenanthic acid, heptoic acid
$CH_3(CH_2)_5COOH$

オングストローム　*ongusutoro-mu*
angström

オンラインテスト　*onraintesuto*
online testing

オパール　*opa-ru*
opal $SiO_2 \cdot nH_2O$

オペロン　*operon*
operon

オピアン酸　*opiansan*
opianic acid $(CH_3O)_2C_6H_2(CHO)(COOH)$

オピエート　*opie-to*
opiate

オーラミン　*o-ramin*
auramine $((CH_3)_2NC_6H_4)_2C=NH$

オレアノール酸　*oreano-rusan*
oleanoic acid $C_{30}H_{48}O_3$

オレフィン　*orefin*
olefin C_nH_{2n}

オレイン　*orein*
olein

オレイン酸　*oreinsan*
oleic acid
$CH_3(CH_2)_7CH=CH(CH_2)_7COOH$

オレイン酸デシル　*oreinsandeshiru*
decyl oleate
$CH_3(CH_2)_7CH=CH(CH_2)_7COOC_{10}H_{21}$

オレイン酸ナトリウム
oreinsannatoriumu
sodium oleate
$CH_3(CH_2)_7CH=CH(CH_2)_7COONa$

オレイルアルコール　*oreiruaruko-ru*
oleyl alcohol
$CH_3(CH_2)_7CH=CH(CH_2)_7CH_2OH$

オレンジレッド　*orenjireddo*
orange red, minium, lead(II,IV) oxide Pb_3O_4

オーレオマイシン　*o-reomaishin*
aureomycin $C_{22}H_{23}ClN_2O_8$

オリベイラ石　*oribeiraseki*
oliveiraite $3ZrO_2 \cdot _2ThO_2 \cdot 2H_2O$

オリベトール　*oribeto-ru*
olivetol $C_{11}H_{16}O_2$

オリブ油　*oribuyu*
olive oil

オリエント石　*orientoseki*
orientite $Ca_4Mn_4(SiO_4)_5 \cdot 4H_2O$

オリフィス　*orifisu*
orifice

オリガーゼ　*origa-ze*
oligase

オリゴマー　*origoma-*
oligomer

オリゴマー化　*origoma-ka*
oligomerization

オリゴヌクレオチド　*origonukureochido*
oligonucleotide

オリゴ糖　*origotou*
oligosaccharide

オーリン　*o-rin*
aurine $C_{19}H_{14}O_3$

オロト酸　*orotosan*
orotic acid $C_5H_4N_2O_4$

オルチン　*oruchin*
orthin $HOC_6H_3(NHNH_2)COOH$

オルガノシロキサン　*oruganoshirokisan*
organosiloxane

オルニチン　*orunichin*
ornithine $H_2N(CH_2)_2CH(NH_2)COOH$

オルセイン　*orusein*
orcein $C_{28}H_{24}N_2O_7$

オルセーン　*oruse-n*
orthene $C_6H_4Cl_2$

オルセリン酸　*oruserinsan*
orsellinic acid $CH_3C_6H_2(OH)_2(COOH)$

オルシン　*orushin*
orcine $CH_3C_6H_3(OH)_2$

オルシンアルデヒド　*orushinarudehido*
orcine aldehyde $C_6H_2(CH_3)(OH)_2CHO$

オルシノール　*orushino-ru*
orcinol $C_6H_3(CH_3)(OH)_2$

オルト　*oruto*
ortho

オルトアミノベンゾールスルフォン酸
orutoaminobenzo-rusurufonsan
orthanilic acid $H_2NC_6H_4SO_3H$

オルトバナジン酸　*orutobanajinsan*
orthovanadic acid, vanadic acid H_3VO_4

オルトバナジン酸塩　*orutobanajinsanen*
orthovanadate, vanadate $M^I{}_3VO_4$

オルトチタン酸　*orutochitansan*
orthotitanic acid H_4TiO_4

オルトチタン酸マグネシウム
orutochitansanmaguneshiumu
magnesium orthotitanate $MgTiO_3$

オルト鉛酸塩　*orutoensanen*
orthoplumbate $M^I{}_4PbO_4$

オルトエステル　*orutoesuteru*
orhocarboxylic ester $RC(OR)_3$

オルトフェニルフェノール
orutofenirufeno-ru
rtho-phenylphenol $C_6H_4(C_6H_5)OH$

オルトフォルム　*orutoforumu*
orthoform $C_6H_3(NH_2)(OH)COOCH_3$

オルト蟻酸　*orutogisan*
orthoformic acid H_4CO_3

オルトギ酸トリエチル
orutogisantoriechiru
triethyl orthoformate $HC(OC_2H_5)_3$

オルトホウ酸, オルト硼酸　*orutohousan*
orthoboric acid H_3BO_3

オルト位　*orutoi*
ortho-position

オルト異性体　*orutoiseitai*
ortho isomer

オルトジルコン酸　*orutojirukonsan*
orthozirconic acid H_4ZrO_4

オルトジルコン酸塩　*orutojirukonsanen*
orthozirconate $M^I{}_4ZrO_4$

オルトジルコン酸ナトリウム
orutojirukonsannatoriumu
sodium orthozirconate Na_4ZrO_4

オルト過ヨウ素酸, オルト過沃素酸
orutokayousosan
orthoperiodic acid H_5IO_6

オルトケイ酸, オルト珪酸　*orutokeisan*
orthosilicic acid, silicic acid H_4SiO_4

オルトキノン　*orutokinon*
orthoquinone $C_6H_4O_2$

オルトクレゾール　*orutokurezo-ru*
orthocresol $C_6H_4(OH)CH_3$

オルトリン酸, オルト燐酸　*orutorinsan*
orthophosphoric acid, phosphoric acid
H_3PO_4

オルト酢酸　*orutosakusan*
orthoacetic acid $CH_3C(OH)_3$

オルト酸　*orutosan*
ortho acid

オルト硝酸　*orutoshousan*
orthonitric acid H_3NO_4

オルト水素　*orutosuiso*
ortho-hydrogen

オルトスズ酸, オルト錫酸　*orutosuzusan*
orthostannic acid H_4SnO_4

オルトタンタル酸塩　*orutotantarusanen*
orthotantalate $M^I_4Ta_2O_7$

オルトテルル酸　*orutoterurusan*
orthotelluric acid H_6TeO_6

オルトバナジン酸　*orutovanajinsan*
orthovanadic acid, vanadic acid H_3VO_4

オルトヴァナジン酸塩
orutovanajinsanen
orthovanadate, vanadate $M^I_3VO_4$

オサゾン　*osazon*
osazone $C_6H_5NHN=CHC(R)=NNHC_6H_5$

オシメン　*oshimen*
ocimene $C_{10}H_{16}$

オシリトリン　*oshiritorin*
osyritrine $C_{27}H_{30}O_{16}$

オシログラフ　*oshirogurafu*
oscillograph

オソトリアゾール　*osotoriazo-ru*
osotriazole $C_2N_3H_3$

オスミウム　*osumiumu*
osmium (Os, element 76)

オスミウム酸　*osumiumusan*
osmic acid H_2OsO_4

オスミウム酸塩　*osumiumusanen*
osmate $M^I_2[OsO_4(OH)_2]$

オスミウム酸カリウム
osumiumusankariumu
potassium osmate(VI) $K_2[OsO_2(OH)_4]$

オストラジオール　*osutorajio-ru*
oestradiol $C_{18}H_{24}O_2$

オストリオール　*osutorio-ru*
oestriol $C_{18}H_{24}O_3$

オストロン　*osutoron*
oestrone $C_{18}H_{22}O_2$

オートクレーブ　*o-tokure-bu*
autoclave

オートメーション　*o-tome-shon*
automation

オウステナイト　*ousutenaito*
austenite

オヴォフラビン　*ovofurabin*
ovoflavin $C_{17}H_{20}N_4O_6$

オヴォムコイド　*ovomukoido*
ovomucoid

オヴォムーシン　*ovomu-shin*
ovomucin

オゾケライト　*ozokeraito*
ozocerite

オゾン　*ozon*
ozone O_3

オゾン発生器　*ozonhasseiki*
ozonizer

オゾン化　*ozonka*
ozonization

パ pa

パーフルオロブタン　*pa-furuorobutan*
decafluorobutane C_4F_4

パーフルオロエチレン　*pa-furuoroechiren*
tetrafluoroethylene $F_2C=CF_2$

パーフルオロエタン　*pa-furuoroetan*
hexafluoroethane C_2F_6

パーフルオロヘキサン
pa-furuorohekisan
tetradecafluorohexane C_6F_{14}

パーフルオロメタン　*pa-furuorometan*
tetrafluoromethane CF_4

パーフルオロペンタン　*pa-furuoropentan*
dodecafluoropentane C_5F_{12}

パイ電子　*paidenshi*
pi-electron

パイロ電気　*pairodenki*
pyroelectricity

パイロファナイト　*pairofanaito*
pyrophanite $MnTiO_3$

パイロクロイト　*pairokuroito*
pyrochroite $Mn(OH)_2$

パイロスマライト　*pairosumaraito*
pyrosmalite $(Fe,Mn)_{10}Si_8O_{25}Cl_2 \cdot 7H_2O$

パイロットプラント　*pairottopuranto*
pilot plant

パーキン反応　*pa-kinhannou*
Perkin reaction

パッキング　*pakkingu*
packing

パーク法　*pa-kuhou*
Parkes process

パークロロエチレン　*pa-kuroroechiren*
tetrachloroethylene C_2Cl_4

パークロロエタン　*pa-kuroroetan*
hexachloroethane C_2Cl_6

パーマネントレッド　*pa-manentoreddo*
permanent red
$HOC_{10}H_6N=NC_6H_3(CH_3)NO_2$

パーム油　*pa-muyu*
palm oil

パンクレアチン　*pankureachin*
pancreatin

パントテン酸　*pantotensan*
panthothenic acid $C_9H_{17}NO_5$

パパイン　*papain*
papain

パパヴェリン　*papaverin*
papaverine $C_{20}H_{21}NO_4$

パラ　*para*
para

パラアミノ安息香酸
paraaminoansokukousan
p-amino benzoic acid $C_6H_4(NH_2)COOH$

パラアミノフェノール
paraaminofeno-ru
rodinal, p-aminophenol $H_2NC_6H_4OH$

パラアルデヒド　*paraarudehido*
paraldehyde $C_6H_{12}O_3$

パラアルドール　*paraarudo-ru*
paraldol $(CH_3CH(OH)CH_2CHO)_2$

パラアセトアルデヒド
paraasetoarudehido
paraaldehyde, paraacetaldehyde
$(CH_3CHO)_3$

パラバン酸　*parabansan*
parabanic acid $C_3H_2N_2O_3$

パラブチルアルデヒド
parabuchiruarudehido
parabutyraldehyde $(cC_3H_7CHO)_3$

パラチオン　*parachion*
parathion $S=P((OC_2H_5)_2)OC_2H_4NO_2$

パラフィン　*parafin*
paraffin

パラフィン炭化水素　*parafintankasuiso*
paraffin hydrocarbon

パラフィンワックス　*parafinwakkusu*
parafin wax

パラフィン油　*parafinyu*
paraffin oil

パラフィン族炭化水素
parafinzokutankasuiso
paraffin hydrocarbon

パラフォルムアルデヒド
paraforumuarudehido
paraformaldehyde $(HCHO)_n$

パラフクシン　*parafukushin*
parafuchsine $HOC(C_6H_4NH_2)_3$

パラヘリウム　*paraheriumu*
parahelium

パラヘ石　*paraheseki*
palacheite $2MgO \cdot Fe_2O_3 \cdot 4SO_3 \cdot 15H_2O$

パラホープ石　*paraho-puseki*
parahopeite $Zn_3(PO_4)_3 \cdot 4H_2O$

パラホルムアルデヒド
parahorumuarudehido
paraformaldehyde $(HCHO)n$

パラ位　*parai*
para-position

パラジウム　*parajiumu*
palladium (Pd, element 46)

パラキノン　*parakinon*
paraquinone $C_6H_4O_2$

パラキサンチン　*parakisanchin*
paraxanthine $C_7H_8N_4O_2$

パラキシロール　*parakishiro-ru*
paraxylene $C_6H_4(CH_3)_2$

パラキシル酸　*parakishirusan*
paraxylic acid $C_6H_3(CH_3)_2COOH$

パラコン酸　*parakonsan*
paraconic acid $C_5H_6O_4$

パラコトイン　*parakotoin*
paracotoin $C_{12}H_8O_4$

パラクレゾール　*parakurezo-ru*
paracresol $C_6H_4(CH_3)OH$

パラメーター　*parame-ta-*
parameter

パラミン　*paramin*
paramin $C_6H_4(NH_2)_2$

パラモルファン　*paramorufan*
paramorphane $C_{17}H_{21}NO_2 \cdot HCl$

パラモルフィン　*paramorufin*
paramorphine $C_{17}H_{15}(OCH_3)_2NO$

パラヌクレイン　*paranukurein*
paranuclein $C_{30}H_{52}O_{17}N_9P_3$

パラ乳酸　*paranyuusan*
paralactic acid $C_3H_6O_3$

パラオキシ安息香酸エチル
paraokishiansokukousanechiru
ethyl p-hydroxybenzoate
$C_6H_4(OH)COOC_2H_5$

パラレッド　*parareddo*
para red $C_{10}H_6(OH)NNC_6H_4NO_2$

パラサイト　*parasaito*
parasite

パラ石　*paraseki*
palaite $5Mno \cdot 2P_2O_5 \cdot 4H_2O$

パラシメン　*parashimen*
paracymene $C_6H_4(CH_3)CH(CH_3)_2$

パラシモール　*parashimo-ru*
paracymene $C_6H_4(CH_3)CH(CH_3)_2$

パラ酒石酸　*parashusekisan*
paratartaric acid $C_4H_6O_4$

パラソルビン酸　*parasorubinsan*
parasorbic acid $C_6H_8O_2$

パルス減衰時間　*parasugensuijikan*
pulse decay time

パラ水素　*parasuiso*
parahydrogen

パリチン　*parichin*
paricine $C_{16}H_{18}N_2O$

パリティ　*pariti*
parity

パールクノル反応　*pa-rukunoruhannou*
Paal-Knorr reaction

パルミチン　*parumichin*
palmitin, tripalmitin, glycerol tripalmitate
$C_3H_5(OOCC_{15}H_{31})_3$

パルミチン酸　*parumichinsan*
palmitic acid, n-hexadecanoic acid
$CH_3(CH_2)_{14}COOH$

パルミチン酸亜鉛　*parumichinsanaen*
zinc palmitate $(CH_3(CH_2)_{14}COO)_2Zn$

パルミチン酸アルミニウム
parumichinsanaruminiumu
aluminium palmitate
$(CH_3(CH_2)_{14}COO)_3Al$

パルミチン酸塩　*parumichinsanen*
palmitate $CH_3(CH_2)_{14}COOH$

パルミチルアルコール
parumichiruaruko-ru
palmityl alcohol $CH_3(CH_2)_{15}OH$

パルミエリ石　*parumieriseki*
palmierite $(K,Na)_2Pb[SO_4]_2$

パルミトン　*parumiton*
palmitone $C_{31}H_{62}O$

パルミトール酸　*parumito-rusan*
palmitolic acid $C_{15}H_{27}COOH$

パルプ　*parupu*
pulp

パルス　*parusu*
pulse

パルス長　*parusuchou*
pulse length

パルス波　*parusuha*
pulse wave

パルス放射線　*parusuhoushasen*
pulsed radiation

パルス特性　*parusutokusei*
pulse characteristics

パルヴォリン　*paruvorin*
parvoline $C_9H_{13}N$

パーセント　*pa-sento*
　percent

パーソンス石　*pa-sonsuseki*
　parsonite $2PbO \cdot UO_3 \cdot P_2O_5 \cdot H_2O$

パッセエン系列　*passeenkeiretsu*
　Paschen series

パスカル　*pasukaru*
　pascal

パスタ　*pasuta*
　paste

パストリゼーション　*pasutorize-shon*
　pasteurization

パターノ石　*pata-noseki*
　paternoite $MgB_8O_{13} \cdot 4H_2O$

パテント　*patento*
　patent

パウリの原理　*paurinogenri*
　Pauli principle

パワーソース　*pawa-so-su*
　power source

ペ pe

ペガナイト　*peganaito*
　peganite $AlPO_4 \cdot Al(OH)_3$

ペーハー　*pe-ha-*
　pH value

ペイント　*peinto*
　paint

ペーキンの反応　*pe-kinnohannou*
　Perkin's reaction

ペクチン　*pekuchin*
　pectin

ペクチナーゼ　*pekuchina-ze*
　pectinase

ペクチノーゼ　*pekuchino-ze*
　arabinose, pectinose $C_5H_{10}O_5$

ペクチン酸　*pekuchinsan*
　pectic acid

ペクトーゼ　*pekuto-ze*
　pectose

ペンチン　*penchin*
　pentyne C_5H_8

ペンチレン　*penchiren*
　pentylene C_5H_{10}

ペンチル　*penchiru*
　pentyl, amyl C_5H_{11}-

ペンチルアルコール　*penchiruaruko-ru*
　pentanol $C_5H_{11}OH$

ペンチット　*penchitto*
　pentite $HOCH_2(CHOH)_3CH_2OH$

ペニシリン　*penishirin*
　penicillin

ペニシリン酸　*penishirinsan*
　penicillic acid $C_8H_{10}O_4$

ペニシリウム属　*penishiriumuzoku*
　penicillium

ペンキ　*penki*
　paint

ペンタブロモジフルオロプロパン　*pentaburomojifuruoropuropan*
　pentabromodifluoropropane $C_3HF_2Br_5$

ペンタチオン酸　*pentachionsan*
　pentathionic acid $H_2S_5O_6$

ペンタデカン　*pentadekan*
　pentadecane $C_{15}H_{32}$

ペンタデカン酸　*pentadekansan*
　pentadecanoic acid $C_{14}H_{29}COOH$

ペンタデシル酸　*pentadeshirusan*
　pentadecylic acid $CH_3(CH_2)_{13}COOH$

ペンタエリスリトール　*pentaerisurito-ru*
　pentaerythritol $C(CH_2OH)_4$

ペンタエリトリット, ペンタエリトリトール　*pentaeritoritto, pentaeritorito-ru*
　pentaerythritol, pentaerythrite $C(CH_2OH)_4$

ペンタフルオロエタン　*pentafuruoroetan*
　pentafluoroethane C_2HF_5

ペンタジエン　*pentajien*
　pentadiene C_5H_8

ペンタジイン　*pentajiin*
　pentadiine C_5H_4

ペンタクロロフェノール　*pentakkurorofeno-ru*
　pentachlorophenol $C_6Cl_5(OH)$

ペンタコサン　*pentakosan*
pentacosane $CH_3(CH_2)_{23}CH_3$

ペンタクロロベンゼン
pentakurorobenzen
pentachlorobenzene C_6HCl_5

ペンタクロロナフタレン
pentakuroronafutaren
pentachloronaphthalen $C_{10}H_3Cl_5$

ペンタクロルエタン　*pentakuroruetan*
pentachlorethane C_2HCl_5

ペンタクロルフェノール
pentakurorufeno-ru
pentachlorphenol C_6Cl_5OH

ペンタメチレングリコール
pentamechirenguriko-ru
pentamethylene glycol $HO(CH_2)_5OH$

ペンタメチレンジアミン
pentamechirenjiamin
pentamethylendiamine $H_2N(CH_2)_5NH_2$

ペンタン　*pentan*
pentane C_5H_{12}

ペンタナール　*pentana-ru*
pentanal C_4H_9CHO

ペンタンチオール　*pentanchio-ru*
pentanthiol $C_5H_{11}SH$

ペンタンジカルボン酸
pentanjikarubonsan
pimelic acid, pentanedioic acid
$HOOC(CH_2)_5COOH$

ペンタノン　*pentanon*
pentanone $C_5H_{10}O$

ペンタノール　*pentano-ru*
pentanol $C_5H_{11}OH$

ペンタール　*penta-ru*
pental $CH_3CH=C(CH_3)_2$

ペンタシラン　*pentashiran*
pentasilane Si_5H_{12}

ペンターゼ　*penta-ze*
pentase

ペンテン　*penten*
pentene C_5H_{10}

ペントバルビタール　*pentobarubita-ru*
pentobarbital $C_{11}H_{18}N_2O_3$

ペントン酸　*pentonsan*
pentonic acid $HOCH_2(CHOH)_3COOH$

ペントース　*pento-su*
pentose $C_5H_{10}O_5$

ペントーザン　*pento-zan*
pentosan

ペントーゼ　*pento-ze*
pentose $C_5H_{10}O_5$

ペオニジン　*peonijin*
peonidin $C_{16}H_{13}O_6Cl$

ペオニン　*peonin*
paeonin $C_{28}H_{33}O_{16}Cl$

ペオノール　*peono-ru*
paeonol $C_9H_{10}O_3$

ペーパークロマトグラフィー
pe-pa-kuromatogurafi-
paper chromatography

ペプチダーゼ　*pepuchida-ze*
peptidase

ペプチド　*pepuchido*
peptide

ペプチド合成　*pepuchidogousei*
peptide synthesis

ペプチド合成機　*pepuchidogouseiki*
peptide synthesizer

ペプチド結合　*pepuchidoketsugou*
peptide bond

ペプシン　*pepushin*
pepsin

ペプシノゲン，ペプシノーゲン
pepushinogen, pepushino-gen
pepsinogen

ペプトン　*peputon*
peptone

ペラルゴン　*perarugon*
pelargone $CH_3(CH_2)_7CO(CH_2)_7CH_3$

ペラルゴンアルデヒド
perarugonarudehido
pelargonaldehyde $CH_3(CH_2)_7CHO$

ペラルゴニジン　*perarugonijin*
pelargonidin $C_{15}H_{10}O_5 \cdot HCl$

ペラルゴニン　*perarugonin*
pelargonin $C_{27}H_{30}O_{15} \cdot HCl$

ペラルゴン酸 *perarugonsan*
pelargonic acid $CH_3(CH_2)_7COOH$

ペレイリン *pereirin*
pereirine $C_{19}H_{24}N_2O$

ペレット *peretto*
pellet

ペレットミル *perettomiru*
pellet mill

ペリ環状反応 *perikanjouhannou*
pericyclic reaction

ペリクラース *perikura-su*
periclasite MgO

ペリックリン *perikurin*
pericline $Na_2Al_2Si_{16}O_6$

ペリミジン *perimijin*
perimidine $C_{11}H_8N_2$

ペリラアルデヒド *periraarudehido*
perillaaldehyde $C_{10}H_{14}O$

ペリレン *periren*
perylene $H_6C_{10}=C_{10}H_6$

ペリサイト *perisaito*
perseite $HOCH_2(CHOH)_5CH_2OH$

ペリ酸 *perisan*
peri acid $H_2NC_{10}H_6SO_3H$

ペロチン *perochin*
pellotine $C_{13}H_{16}O_2N$

ペロニン *peronin*
peronine $C_{24}H_{25}NO_3 \cdot HCl$

ペルオキシダーゼ *peruokishida-ze*
peroxidase

ペルオキシクロム酸塩 *peruokishikuromusanen*
peroxychromate $M^I_3CrO_8, M^I_2CrO_6$

ペルオキシ酸 *peruokishisan*
peroxy acid

ペルオキシ硝酸 *peruokishishousan*
peroxynitric acid HNO_4

ペルオキソ一硫酸 *peruokisoichiryuusan*
peroxysulfuric acid H_2SO_5

ペルオキソ二硫酸 *peruokisoniryuusan*
peroxodisulfuric acid, Marshall's acid $H_2S_2O_8$

ペルオキソリン酸 *peruokisorinsan*
peroxophosphoric acid H_3PO_5

ペースト *pe-suto*
paste

ペトリ皿、ペトリー皿 *petorizara, petori-zara*
Petri dish

ペツニジン *petsunijin*
petunidine $C_{16}H_{12}O_7 \cdot HCl$

ピ pi

ピバロイル基 *pibaroiruki*
pivaloyl radical $(CH_3)_3CCO-$

ピバル酸 *pibarusan*
pivalic acid $(CH_3)_3CCOOH$

ピッチ *picchi*
pitch

ピッチブレンド *picchiburendo*
pitchblende UO_2

ピッチサイト *picchisaito*
pitticite $Fe_2O_3 \cdot SO_3 \cdot As_2O_5 \cdot H_2O$

ピグメント *pigumento*
pigment

ピコリン *pikorin*
picoline $C_5H_4NCH_3$

ピコリン酸 *pikorinsan*
picolinic acid C_5H_4NCOOH

ピーク *pi-ku*
peak

ピーク値 *pi-kuchi*
peak value

ピクラコニチン *pikurakonichin*
picraconitine $C_{32}H_{43}NO_{10}$

ピクラミド *pikuramido*
picramide $H_2NC_6H_2(NO_2)_2$

ピクラミン酸 *pikuraminsan*
picramic acid $H_2NC_6H_2(NO_2)_2OH$

ピクラート *pikura-to*
picrate $MOC_6H_2(NO_2)_3$

ピクラトール　*pikurato-ru*
　picratol

ピクリン酸　*pikurinsan*
　picric acid $HOC_6H_2(NO_2)_3$

ピクリン酸塩　*pikurinsanen*
　picrate $MOC_6H_2(NO_2)_3$

ピクリン酸ナトリウム
　pikurinsannatoriumu
　soium picrate $C_6H_2(NO_2)_3ONa$

ピクリル基　*pikuriruki*
　picryl radical $-C_6H_2(NO_3)_3$

ピクロファルマコライオ
　pikurofarumakoraito
　picropharmacolite $(Ca,Mg)_3As_2O_8 \cdot 6H_2O$

ピクロクロチン　*pikurokurochin*
　picrocrocin $C_{16}H_{26}O_7$

ピクロロン酸　*pikuroronsan*
　picolonic acid $C_{10}H_8N_4O_5$

ピクロール　*pikuro-ru*
　picrol $C_6H(I)_2(OH)_2(SO_3K)$

ピクロスミン　*pikurosumin*
　picrosmine $H_4Mg_3Si_2O_9 \cdot 2H_2O$

ピクロトキシン　*pikurotokishin*
　picrotoxine $C_{30}H_{34}O_{13}$

ピマール酸　*pima-rusan*
　pimaric acid $C_{20}H_{30}O_2$

ピメント油　*pimentoyu*
　pimento oil

ピメライト　*pimeraito*
　pimelite
　$Al_2O_3 \cdot Fe_2O_3 \cdot NiO \cdot MgO \cdot CaO \cdot SiO_2 \cdot H_2O$

ピメリンケトン　*pimerinketon*
　pimelic ketone $C_6H_{10}O$

ピメリン酸　*pimerinsan*
　pimelic acid $HOOC(CH_2)_5COOH$

ピナキオライト　*pinakioraito*
　pinakiolite $3MgO \cdot B_2O \cdot MnO_3 \cdot Mn_2O_3$

ピナコリン　*pinakorin*
　pinacoline $(CH_3)_3CCOCH_3$

ピナコリルアルコール
　pinakoriruaruko-ru
　pinacolyl alcohol $(CH_3)_3CCH(OH)CH_3$

ピナコロン　*pinakoron*
　pinacolone $CH_3COC(CH_3)_3$

ピナコール　*pinako-ru*
　pinacol $(CH_3)_2C(OH)C(OH)(CH_3)_2$

ピナコール転位　*pinako-ruteni*
　pinacol rearrangement

ピナクロム　*pinakuromu*
　pinachrome $C_{24}H_{25}N_2I$

ピナン　*pinan*
　pinane $C_{10}H_{18}$

ピナヴェルドール　*pinaverudo-ru*
　pinaverdol $C_{22}H_{21}N_2I$

ピネン　*pinen*
　pinene $C_{10}H_{16}$

ピニトール　*pinito-ru*
　pinitol $C_6H_6(OH)_5OCH_3$

ピノカルヴェオール　*pinokaruveo-ru*
　pinocarveol $C_{11}H_{16}O$

ピノカルヴォン　*pinokaruvon*
　pinocarvon $C_{10}H_{13}O$

ピノン酸　*pinonsan*
　pinonic acid $C_{10}H_{16}O_3$

ピノール　*pino-ru*
　pinole $C_{10}H_{16}O$

ピンピネリン　*pinpinerin*
　pimpinelllin $C_{13}H_{12}O_4$

ピン酸　*pinsan*
　pinic acid $C_8H_{14}O_3$

ピンセット　*pinsetto*
　pincette

ピンタド石　*pintadoseki*
　pintadoite $2CaO \cdot V_2O_5 \cdot 9H_2O$

ピオチアニン　*piochianin*
　pyocyanine $C_{14}H_{14}NO_3$

ピオクタニン　*piokutanin*
　pyoctanin $C_{24}H_{29}NO_3$

ピペラジン　*piperajin*
　piperazine $C_4H_{10}N_2$

ピペリジン　*piperijin*
　piperidine $C_5H_{10}NH$

ピペリジル基　*piperijiruki*
　piperidyl radical $-C_4H_3N_2$

ピペリン　*piperin*
　piperine $C_{17}H_{19}NO_3$

ピペリン酸　*piperinsan*
　piperic acid $C_{11}H_9O_2COOH$

ピペリレン　*piperiren*
　piperylene $H_2C=CHCH=CHCH_3$

ピペリトン　*piperiton*
　piperitone $C_{10}H_{15}O$

ピペリトール　*piperito-ru*
　piperitol $C_{10}H_{18}O$

ピペロナール　*piperona-ru*
　piperonal $C_8H_6O_3$

ピペロニルアルコール　*piperoniruaruko-ru*
　piperonyl alcohol $C_8H_8O_3$

ピペロニル酸　*piperonirusan*
　piperonylic acid $C_7H_5O_2COOH$

ピペロヴァチン　*piperovachin*
　piperovatine $C_{16}H_{21}NO_2$

ピペット　*pipetto*
　pipette

ピラジン　*pirajin*
　pyrazine $C_4H_4N_2$

ピラジンアミド　*pirajinamido*
　pyranzinamide $C_5H_5N_3O$

ピラミドン　*piramidon*
　pyramidone $C_{13}H_{17}N_3O$

ピラン　*piran*
　pyran C_5H_6O

ピランチン　*piranchin*
　pyrantin $C_{12}H_{13}NO_3$

ピランジン　*piranjin*
　peyranzin $C_{11}H_{12}NO$

ピラノース　*pirano-su*
　pyranose

ピラノーゼ　*pirano-ze*
　pyranose

ピラントロン　*pirantoron*
　pyranthrene $C_{30}H_{16}$

ピラトリドン　*piratoridon*
　pyranthridone $C_{29}H_{33}NO_2$

ピラゾリン　*pirazorin*
　pyrazoline $C_3H_6N_2$

ピラゾロン　*pirazoron*
　pyrazolone $C_3H_4N_2O$

ピラゾール　*pirazo-ru*
　pyrazole $C_3H_4N_2$

ピラゾールブルー　*pirazo-ruburu-*
　pyrazole blue $C_{20}H_{16}N_4O_2$

ピレン　*piren*
　pyrene $C_{16}H_{10}$

ピレスリン　*piresurin*
　pyrethrin $C_{10}H_{17}CONHCH_2CH(CH_3)_2$

ピレトリン　*piretorin*
　pyrethrin

ピリダジン　*piridajin*
　pyridazine $C_4H_4N_2$

ピリドキサール　*piridokisa-ru*
　pyridoxal $C_8H_9NO_3$

ピリドキシン　*piridokishin*
　pyridoxal $C_8H_9NO_3$

ピリドン　*piridon*
　pyridone C_5H_5NO

ピリドール　*pirido-ru*
　pyridol $C_5H_4N(OH)$

ピリジン　*pirijin*
　pyridine C_5H_5N

ピリジン塩基　*pirijinenki*
　pyridine base

ピリジンカルボン酸　*pirijinkarubonsan*
　pyridinecarboxylic acid C_5H_4NCOOH

ピリジルアミン　*pirijiruamin*
　pyridylamine $C_5H_4NNH_2$

ピリジル基　*pirijiruki*
　pyridyl radical C_5H_4N-

ピリミジン　*pirimijin*
　pyrimidine $C_4H_4N_2$

ピリリウム塩基　*piririumuenki*
　pyrylim base $C_5H_6O_2$

ピロアンチモン酸　*piroanchimonsan*
　pyroantimonic acid $H_4Sb_2O_7$

ピロ亜硫酸　*piroaryuusan*
　disulfurous acid $H_2S_2O_5$

ピロ電気　*pirodenki*
pyroelectricity

ピロガロール　*pirogaro-ru*
pyrogallol $C_6H_3(OH)_3$

ピロガロールフタレイン　*pirogaro-rufutarein*
pyrogallophthalein $C_{20}H_{12}O_7$

ピログルタミン酸　*pirogurutaminsan*
pyroglutamic acid $C_5H_7NO_2$

ピロ硼酸, ピロホウ酸　*pirohousan*
tetraboric acid, pyroboric acid $B_4O_5(OH)_2$

ピロジン　*pirojin*
pyrodin $C_8H_{10}N_2O$

ピロカルビン　*pirokarubin*
pilocarpine $C_{11}H_{16}N_2O_2$

ピロカルピジン　*pirokarupijin*
pilocarpidine $C_8H_{13}N_2COOH$

ピロカテコール　*pirokateko-ru*
pyrocatechol $C_6H_4(OH)_2$

ピロコール　*piroko-ru*
pyrocoll $C_{10}H_6N_2O_2$

ピロメコン酸　*piromekonsan*
pyromeconic acid $C_5H_4O_3$

ピロメリト酸　*piromeritosan*
pyromellitic acid $C_6H_2(COOH)_4$

ピロン　*piron*
pyrone $C_5H_4O_2$

ピロリジン　*pirorijin*
pyrrolidine C_4H_9N; pyrrolizine C_7H_7N

ピロリン　*pirorin*
pyrroline C_4H_7N

ピロリン酸, ピロ燐酸　*pirorinsan*
pyrophosphoric acid, diphosphoric acid $H_4P_2O_7$

ピロリン酸第一鉄　*pirorinsandaiichitetsu*
ferrous pyrophosphate $Fe_2P_2O_7$

ピロリン酸第二鉄　*pirorinsandainitetsu*
ferric pyrophosphate $Fe_4(P_2O_7)_3 \cdot xH_2O$

ピロ燐酸塩　*pirorinsanen*
pyrophosphate(V) $M^I_4P_2O_7$

ピロ燐酸ナトリウム　*pirorinsannatoriumu*
sodium pyrophosphate $Na_4P_2O_7$

ピロリン酸四カリウム　*pirorinsanshikariumu*
potassium pyrophosphate $K_4P_2O_7 \cdot 3H_2O$

ピロ燐酸テトラエチル　*pirorinsantetoraechiru*
tetraethyl pyrophosphate $(C_2H_5)_4P_2O_7$

ピロリレン　*piroriren*
pyrrolylene $H_2C=CHCH=CH_2$

ピロール　*piro-ru*
pyrrole C_4H_5N

ピロヴァナジン酸　*pirovanajinsan*
pyrovanadic acid $H_4V_2O_7$

ピロヴィンアルデヒド　*pirovinarudehido*
pyruvaldehyde CH_3COCHO

ピロザール　*piroza-ru*
pyrosal $C_{20}H_{20}O_5$

ピルヴィン酸　*piruvinsan*
pyruvic acid $CH_3COCOOH$

ピルヴォニトリル　*piruvonitoriru*
pyrovonitrile CH_3COCN

ピセン　*pisen*
picene $C_{22}H_{14}$

ピスダサイト　*pisudasaito*
pistacite $3Al_2O_3 \cdot 4CaO \cdot 6SiO_2 \cdot H_2O$

ピタイン　*pitain*
pitayn $C_{20}H_{24}N_2O_2 \cdot 2,5H_2O$

ピーターソン反応　*pi-ta-sonhannou*
Peterson reaction

ピトー管　*pito-kan*
pitot tube

ポ po

ポアソン比, ポアッソン比　*poasonhi, poassonhi*
Poisson ratio

ポアズイユの法則　*poazuiyunohousoku*
Poiseuille's law

ポドリア石　*podoriaseki*
podolite $3Ca_3(PO_4)_2 \cdot CaCO_3$

ポイセダニン　*poisedanin*
peucedanin $C_{16}H_{16}O_4$

ポイゾン　*poizon*
poison

ポンド　*pondo*
pound

ポンプ　*ponpu*
pump

ポプリン　*popurin*
popullin $C_{20}H_{22}O_8 \cdot 2H_2O$

ポーラログラフィー　*po-rarogurafi-*
polarography

ポリ　*pori*
poly

ポリアージライト　*poria-jiraito*
polyargyrite $Ag_{24}Sb_2S_{15}$

ポリアクリルアミド　*poriakuriruamido*
polyacrylamide $[-CH_2CH(CONH_2)-]_n$

ポリアクリルニトリル　*poriakurirunitoriru*
polyacrylnitrile $[-CH_2CH(CN)-]_n$

ポリアクリル酸　*poriakurirusan*
polyacrylic acid $[-CH_2CH(COOH)-]$

ポリアクリル酸エステル　*poriakurirusanesuteru*
polyacrylic ester

ポリアクリル酸樹脂　*poriakurirusanjushi*
polyacrylic resin

ポリアミド　*poriamido*
polyamide

ポリアミド樹脂　*poriamidojushi*
polyamide resin

ポリアミン　*poriamin*
polyamine

ポリビニルアルコール　*poribiniruaruko-ru*
polyvinyl alcohol $[-CH_2CH(OH)-]_n$

ポリブタジエン　*poributajien*
polybutadiene $[-CH_2CH=CHCH_2-]_n/$
$[-CH_2C(CH=CH_2)-]_n$

ポリブテン　*poributen*
polybutene $[-CH_2CH(C_2H_5)-]_n$

ポリチオン酸　*porichionsan*
polythionic acid $H_2S_nO_6$

ポリチオン酸　*porichionsan*
polythionic acid

ポリエチレン　*poriechiren*
polyethylene

ポリエチレングリコール　*poriechirenguriko-ru*
polyethylene glycol

ポリエチレン樹脂　*poriechirenjushi*
polyethylene resins

ポリエン　*porien*
polyene

ポリ塩化ビフェニル　*porienkabifeniru*
polychlorinated biphenyls

ポリ塩化ビニリデン　*porienkabiniriden*
polyvinylidene chloride $[-CH_2CCl_2-]_n$

ポリ塩化ビニル, ポリ塩化ビニール　*porienkabiniru, porienkabini-ru*
polyvinyl chloride $[-CH_2CHCl-]_n$

ポリエステル　*poriesuteru*
polyester

ポリエステル繊維　*poriesuteruseni*
polyester fiber

ポリエーテル　*porie-teru*
polyether

ポリフェノール　*porifeno-ru*
polyphenol

ポリゲルマン　*porigeruman*
polygermane $(GeH_2)_n$

ポリゴン　*porigon*
polygon

ポリヘクソーゼ　*porihekuso-ze*
polyhexose $(C_6H_{10}O_5)_n$

ポリイソブチレン　*poriisobuchiren*
polyisobutene $[-CH_2C(CH_3)_2-]_n$

ポリイソプレン　*poriisopuren*
polyisoprene $[-CH_2CH(CH_3)=CHCH_2-]_n$

ポリジマイト　*porijimaito*
polydimite $(Ni,Co)_4S_5$

ポリカーボネート　*porika-bone-to*
polycarbonate

ポリ珪酸 *porikeisan*
polysilicic acid

ポリクロム酸塩 *porikuromusanen*
polychromate $M^I_2[Cr_nO_{3n+1}]$

ポリマー *porima-*
polymer

ポリマーディスパージョン *porima-disupa-jon*
polymer dispersion

ポリメラーゼ *porimera-ze*
polymerase

ポリメタクリレート *porimetakurire-to*
polymethacrylates $[-CH_2C(CH_3)(COOR)-]_n$

ポリメタクリル酸 *porimetakurirusan*
polymethacrylic acid
$[-CH_2C(CH_3)(COOH)-]_n$

ポリモリブデン酸塩 *porimoribudensanen*
polymolybdate e.g. $Mo_7O_{24}^{6-}$, $Mo_8O_{26}^{4-}$

ポリモリフィズム *porimorifizumu*
polymorphism

ポリヌクレオチド *porinukureochido*
polynucleotide

ポリオキシエチレン *poriokishiechiren*
polyoxyethylene

ポリオキシエチレンアルキルエーテル
poriokishiechirenarukirue-teru
polyoxyethylene alkyl ethers

ポリオレフィン *poriorefin*
polyolefin

ポリペプチド *poripepuchido*
polypeptide

ポリポル酸 *poriporusan*
polyporic acid $C_{16}H_{12}O_4$

ポリプレン *poripuren*
polyprene, polyisoprene
$[-CH_2CH(CH_3)=CHCH_2-]_n$

ポリプロピレン *poripuropiren*
polypropylene, PP $[-CH_2CH(CH_3)-]_n$

ポリリン酸, ポリ燐酸 *poririnsan*
polyphosphoric acid $H_{n+2}P_nO_{3n+1}$

ポリ硫酸 *poriryuusan*
polysulfuric acid $H_2S_nO_{3n+1}$

ポリ酢酸ビニル, ポリ酢酸ビニール
porisakusanbiniru, porisakusanbini-ru
polyvinyl acetate $[-CH_2CH(OOCCH_3)-]_n$

ポリ酸 *porisan*
polyacid

ポリシロキサン *porishirokisan*
polysiloxane $[-R_2SiO-]_n$

ポリスチレン *porisuchiren*
polystyrene, PS $[-CH_2-CH(C_6H_5)-]_n$

ポリスチレン樹脂 *porisuchirenjushi*
polystyrene resins

ポリスチロール *porisuchiro-ru*
polystyrene, PS $[-CH_2-CH(C_6H_5)-]_n$

ポリスルフィド *porisurufido*
polysulfide $M^I_2S_n$

ポリテルペン *poriterupen*
polyterpene $[C_{10}H_{16}]_n$

ポリウレタン *poriuretan*
polyurethane

ポリウレタン樹脂 *poriuretanjushi*
polyurethane resin

ポロニウム *poroniumu*
polonium (Po, element 84)

ポルフィン *porufin*
porphine $C_{20}H_{14}N_4$

ポルフィラシン *porufirashin*
porphyrazine $C_{32}H_{18}N_8$

ポルフィリン *porufirin*
porphine $C_{20}H_{14}N_4$

ポルックス石 *porukkususeki*
peollucit $H_2Cs_4Al_4Si_9O_{27}$

ポテンシャル *potensharu*
potential, voltage

ポテンシャル面 *potensharumen*
potential surface

ポテンシャル障壁 *potensharushouheki*
potential barrier

ポリビニルピロリドン *probinirupiroridon*
polyvinyl pyrrolidone $[-CH_2CH(C_4H_6ON)-]_n$

プ pu

プッチャー石　puccha-seki
pucherite $BiVO_4$

プメラー転位　pumera-teni
Pummerer rearrangement

プニチン　punichin
punicine $C_8H_{15}NO$

プラチナイト　purachinaito
platynite $PbS \cdot Bi_2Se_3$

プラグ　puragu
plug

プライマー　puraima-
primer, undercoat

プランバーン　puranba-n
lead tetrahydride, plumbane PbH_4

プランボフェライト　puranboferaito
plumboferrite $PbO \cdot 2Fe_2O_3$

プランクの作用量子
purankunosayouryoushi
Planck's quantum

プランク定数　purankuteisuu
Planck's constant

プラセボ　purasebo
placebo

プラセオジム　puraseojimu
praseodymium (Pr, element 59)

プラスチック　purasuchikku
plastic

プラスチック製造工業
purasuchikkuseizoukougyou
plastics manufacturing industry

プラスモヒン　purasumohin
plasmoquine $C_{19}H_{29}N_3O$

プラスター　purasuta-
plaster

プラスティック　purasutikku
plastic

プラストメル　purasutomeru
plastomer, thermoplastic

プラットナー石　purattona-seki
plattnerite PbO_2

プラズマ　purazuma
plasma

プレゴン　puregon
pulegone $C_{10}H_{16}O$

プレグナン　puregunan
pregnane $C_{21}H_{36}$

プレグナンジオール　puregunanjio-ru
pregnanediol $C_{21}H_{34}(OH)_2$

プレグナノロン　puregunanoron
pregnanolone $C_{21}H_{32}O_2$

プレグネノロン　puregunenoron
pregnenolone $C_{21}H_{32}O_2$

プレニチル酸　purenichirusan
prehnitylic acid $C_6H_2(CH_3)_3COOH$

プレニトール　purenito-ru
prehnitene $C_6H_2(CH_3)_4$

プレニト酸　purenitosan
prehnitic acid $C_6H_2(COOH)_4$

プレポリマー　pureporima-
prepolymer

プレサイト　puresaito
plessite $NiAs_2 \cdot NiS_2$

プレス機　puresuki
pressing machine

プリメチン　purimechin
primetin $C_{15}H_{10}O_4$

プリン　purin
purine $C_5H_4N_4$

プリン塩基　purinenki
purine base

プリンス反応　purinsuhannou
Prins reaction

プロビタミン　purobitamin
provitamin

プロチウム　purochiumu
protium

プロダクト　purodakuto
product

プロゲステロン　purogesuteron
progesterone $C_{21}H_{30}O_2$

プロイドナイト　puroidonaito
proidonite SiF_4

プロジェクト文書　*purojekutobunsho*
project documentation

プロジェクト工学　*purojekutokougaku*
project engineering

プロカイン　*purokain*
procaine $H_2NC_6H_4COOCH_2CH_2N(C_2H_5)_2$

プロ酵素　*purokouso*
proenzyme

プロメチウム　*puromechiumu*
promethium (Pm, element 61)

プロミン　*puromin*
promin $C_{24}H_{34}N_2O_{18}S_2Na_2$

プロモーター　*puromo-ta-*
promoter

プロンビエール石　*puronbie-ruseki*
plombierite $Ca_5H_2[Si_3O_9]_2 \cdot 6H_2O$

プロパジエン　*puropajien*
propadiene, allene $H_2C=C=CH_2$

プロパミン　*puropamin*
propamin $C_{18}H_{26}N_2 \cdot H_2SO_4$

プロパン　*puropan*
propane $CH_3CH_2CH_3$

プロパナール　*puropana-ru*
propanal CH_3CH_2CHO

プロパノン　*puropanon*
acetone CH_3COCH_3

プロパルギルアルデヒド
puroparugiruarudehido
propargyl aldehyde $HCCCHO$

プロパルギルアルコール
puroparugiruaruko-ru
propargyl alcohol $HCCCH_2OH$

プロパルギル基　*puroparugiruki*
propargyl radical $-CH_2CCH$

プロパルギル酸　*puroparugirusan*
propargylic acid $HCCCOOH$

プロペン　*puropen*
propene $CH_3CH=CH_2$

プロペナール　*puropena-ru*
propenal $H_2C=CHCHO$

プロペニル　*puropeniru*
propenyl C_3H_5-

プロペノール　*puropeno-ru*
propenol C_3H_5OH

プロペン酸　*puropensan*
propenoic acid, acrylic acid $H_2C=CHCOOH$

プロペシン　*puropeshin*
propaesin $(NH_2)C_6H_4COOC_3H_7$

プロピン　*puropin*
propyne CH_3CCH

プロピニル　*puropiniru*
propinyl $CHCCH_2-$

プロピン酸　*puropinsan*
propiolic acid $HCCHCOOH$

プロピオフェノン　*puropiofenon*
propiophenone $C_2H_5COC_6H_5$

プロピオン　*puropion*
propione $C_2H_5COC_2H_5$

プロピオンアミド　*puropionamido*
propionamide $CH_3CH_2CONH_2$

プロピオンアルデヒド
puropionarudehido
propanal CH_3CH_2CHO

プロピオニトリル　*puropionitoriru*
propionitrile CH_3CH_2CN

プロピオン酸　*puropionsan*
propionic acid CH_3CH_2COOH

プロピオン酸エチル　*puropionsanechiru*
ethyl propionate $C_2H_5COOC_2H_5$

プロピオン酸無水物
puropionsanmusuibutsu
propionic anhydride $(CH_3CH_2CO)_2O$

プロピオン酸ナトリウム
puropionsannatoriumu
sodium propionate CH_3CH_2COONa

プロピレン　*puropiren*
propene, propylene $CH_3CH=CH_2$

プロピレングリコール
puropirenguriko-ru
propylene glycol $C_3H_6(OH)_2$

プロピレングリコールモノメチルエーテル　*puropirenguriko-rumonomechirue-teru*
propylene glycol monomethyl ether
$C_3H_6(OH)(OCH_3)$

プロピレンイミン　*puropirenimin*
propylenimine C_7H_7N

プロピレンオキサイド, プロピレンオキシド
puropirenokisaido, puropirenokishido
propylene oxide C_3H_6O

プロピリジン　*puropirijin*
propylidene $CH_3CH_2CH=$

プロピリジンイミン　*puropirijinimin*
propylidene imine $CH_3CH_2CH=NH$

プロピル　*puropiru*
propyl C_3H_7-

プロピルアミン　*puropiruamin*
propylamine $C_3H_7NH_2$

プロピルアルコール　*puropiruaruko-ru*
propyl alcohol C_3H_7OH

プロポキシ基　*puropokishiki*
propoxy radical C_3H_7O-

プロポナール　*puropona-ru*
proponal $C_{10}H_{16}N_2O_3$

プロプロキサンチン　*puropurokisanchin*
purpuroxanthin $C_{14}H_8O_4$

プロラクチン　*purorakuchin*
prolactin

プロレイン　*purorein*
prolein $C_{17}H_{33}COOC_3H_6OH$

プロリン　*purorin*
proline $C_5H_9NO_2$

プロリナーゼ　*purorina-ze*
prolinase

プロリシン　*purorishin*
prolysine $C_8H_{13}N_3O_4$

プロセシング　*puroseshingu*
processing

プロセッシング　*purosesshingu*
processing

プロセス　*purosesu*
process

プロセスコントロール
purosesukontoro-ru
process control

プロセス制御　*purosesuseigyo*
process control

プロシロクサン　*puroshirokusan*
prosiloxan H_2SiO

プロソパイト　*purosopaito*
prosopite $CaF_2 \cdot Al(F,OH)_3$

プロスタグランジン　*purosutaguranjin*
prostaglandin

プロステアリン　*purosutearin*
prostearin $C_{17}H_{35}COOCH_2CH(OH)CH_3$

プロテアーゼ　*purotea-ze*
protease

プロテアーゼ阻害剤　*purotea-zesogaizai*
protease inhibitor

プロトアクチニウム　*purotoakuchiniumu*
protactinium (Pa, element 91)

プロトカテク酸, プロトカテキュ酸
purotokatekusan, purotokatekyusan
protocatechuic acid $C_6H_3(OH)_2COOH$

プロトカテキュアルデヒド
purotokatekyuarudehido
protocatechualdehyde $C_6H_3(OH)_2CHO$

プロトコイトン　*purotokoiton*
protocotoin $C_{16}H_{14}O_6$

プロトン　*puroton*
proton

プロトン付加　*purotonfuka*
protonation

プロトン移動　*purotonidou*
proton shift

プロトン磁気共鳴　*purotonjikikyoumei*
proton magnetic resonance, PMR

プロトン性溶媒　*purotonseiyoubai*
protic solvent

プロトピン　*purotopin*
protopine $C_{20}H_{19}NO_5$

プロトリシス　*purotorishisu*
protolysis

プロトロンビン　*purotoronbin*
prothrombin

プロトセトラル酸　*purotosetorarusan*
protocetraric acid $C_{16}H_{14}O_9$

プロトトロピー　*purototoropi-*
prototropy

プルガチン　*purugachin*
　purgatin $C_{18}H_{12}O_7$

プルゲン　*purugen*
　purgen $C_{20}H_{13}O_3$

プルナシン　*purunashin*
　prunasin $C_{14}H_{17}NO_6$

プルネチン　*purunechin*
　prunetin $C_{16}H_{12}O_5$

プルニトリン　*purunitorin*
　prunitrin $C_{22}H_{22}O_{10}$

プルプリン　*purupurin*
　purpurin $C_{14}H_8O_5$

プルプロガリン　*purupurogarin*
　purpurogallin $C_{11}H_8O_5$

プルプルカルミン　*purupurukarumin*
　purple carmine $C_8H_4N_5O_6 \cdot NH_4 \cdot H_2O$

プルプル酸　*purupurusan*
　purpuric acid $C_8H_5N_5O_6$

プルトニウム　*purutoniumu*
　plutonium (Pu, element 94)

プシカイン　*pushikain*
　psicaine $C_{21}H_{25}NO_{10}$

プシコーゼ　*pushiko-ze*
　psicose $C_6H_{12}O_6$

プソイドアコニチン　*pusoidoakonichin*
　pseudaconitine $C_{36}H_{49}NO_{12}$

プソイドアトロピン　*pusoidoatoropin*
　pseudoatropin $C_{17}H_{23}NO_3$

プソイドエフェドリン　*pusoidoefedorin*
　pseudoephedrine
　$C_6H_5CH(OH)CH(CH_3)NHCH_3$

プソイドイオノン　*pusoidoionon*
　pseudoionone $C_{13}H_{20}O$

プソイドキサンチン　*pusoidokisanchin*
　pseudoxanthine $C_4H_5N_5O$

プソイドコデイン　*pusoidokodein*
　pseudo-codeine $C_{18}H_{21}NO_3$

プソイドクメン　*pusoidokumen*
　pseudocumene $C_6H_3(CH_3)_3$

プソイドクメノール　*pusoidokumeno-ru*
　pseudo-cumenol $C_6H_2(CH_3)_3OH$

プソイドクミジン　*pusoidokumijin*
　pseudocumidine $H_2NC_6H_2(CH_3)_3$

プソイドクモール　*pusoidokumo-ru*
　pseudo-cumen $C_6H_3(CH_3)_3$

プソイドニトリル　*pusoidonitoriru*
　pseudonitrile RNC

プソイドニトロール　*pusoidonitoro-ru*
　pseudonitrole $R_2C(NO)NO_2$

プソイドペリチエリン
　pusoidoperichierin
　pseudopelletierine $C_9H_{15}NO \cdot H_2O$

プソイドトロピン　*pusoidotoropin*
　pseudotropine $C_8H_{15}NO$

プテリジン　*puterijin*
　pteridine $C_6H_4N_4$

プトレッシン　*putoresshin*
　putrescine $H_2N(CH_2)_4NH_2$

ラ　ra

ラベンダー油　*rabenda-yu*
　lavender oil

ラベル　*raberu*
　label

ラボラトリー　*raboratori-*
　laboratory

ラブ、ラボ　*rabu, rabo*
　laboratory (abbr.)

ラード　*ra-do*
　lard

ラドン　*radon*
　radon (Rn, element 86)

ラフィネート　*rafine-to*
　raffinate

ラフィノース, ラフィノーゼ
　rafino-su, rafino-ze
　raffinose $C_{18}H_{32}O_{16} \cdot 5H_2O$

ラグ相　*ragusou*
　lag phase

ライフサイエンス　*raifusaiensu*
　life sciences

ライフサイクル　*raifusaikuru*
life cycle

ライマーチーマン反応
raima-chi-manhannou
Reimer-Tiemann reaction

ライマン系列　*raimankeiretsu*
Lyman series

ライン鉱　*rainkou*
reinite $FeWO_4$

ラジアン　*rajian*
rad

ラジカル　*rajikaru*
radical

ラジカル反応　*rajikaruhannou*
radical reaction

ラジカルイオン　*rajikaruion*
radical ion

ラジカル重合　*rajikarujuugou*
radical polymerization

ラジカルスカベンジャー
rajikarusukabenja-
radical scavenger

ラジオグラフ　*rajiogurafu*
radiograph

ラジオグラム　*rajioguramu*
radioigram

ラジオイムノアッセイ　*rajioimunoassei*
radioimmunoassay

ラジオカーボンテスト　*rajioka-bontesuto*
radiocarbon test

ラジオメーター　*rajiome-ta-*
radiometer

ラジオオートグラフィー　*rajioo-togurafi-*
radioautography

ラジウム　*rajiumu*
radium (Ra, element 88)

ラジウム放射線　*rajiumuhoushasen*
radium radiation

ラッカー　*rakka-*
lacquer

ラッカーエナメル　*rakka-enameru*
lacquer enamel

ラッカーゼ　*rakka-ze*
laccase

ラッコール　*rakko-ru*
laccol $C_{23}H_{36}O_2$

ラック　*rakku*
rack

ラック酸　*rakkusan*
laccinic acid $C_{16}H_{12}O_8$

ラクチド　*rakuchido*
lactide

ラクチム　*rakuchimu*
lactim

ラクモイド　*rakumoido*
lacmoid $(HO)_2C_6H_3N(C_6H_2(OH)_3)_2$

ラクタミド　*rakutamido*
lactamide $CH_3CH(OH)CONH_2$

ラクタム　*rakutamu*
lactam

ラクターゼ　*rakuta-ze*
lactase

ラクトアルブミン　*rakutoarubumin*
lactalbumin

ラクトビオーゼ　*rakutobio-ze*
lactobiose $C_{12}H_{22}O_{11} \cdot H_2O$

ラクトフラビン, ラクトフラヴィン
rakutofurabin, rakutofuravin
riboflavine, lactoflavine $C_{17}H_{20}N_4O_6$

ラクトグロブリン　*rakutoguroburin*
lactoglobulin

ラクトグルコーゼ　*rakutoguruko-ze*
lactoglucose $C_6H_{12}O_6$

ラクトン　*rakuton*
lactone

ラクトニトリル　*rakutonitoriru*
lactic nitrile $CH_3CH(OH)CN$

ラクトン酸　*rakutonsan*
lactone acid

ラクトール　*rakuto-ru*
lactol $CH_3CH(OH)COOC_{10}H_7$

ラクトース　*rakuto-su*
lactose $C_{12}H_{22}O_{11}$

ラマン効果　*ramankouka*
　Raman effect

ラマンスペクトル　*ramansupekutoru*
　Raman spectrum

ラメラ　*ramera*
　lamella

ラミネーション　*ramine-shon*
　lamination, layering

ラムナジン　*ramunajin*
　rhamnazin $C_{17}H_{14}O_7$

ラムネチン　*ramunechin*
　rhamnetin $C_{16}H_{12}O_7$

ラムニノーゼ　*ramunino-ze*
　rhamninose $C_{18}H_{32}O_{14}$

ラムニット　*ramunitto*
　rhamnitol, rhamnite $CH_3(CHOH)_4CH_2OH$

ラムノーゼ　*ramuno-ze*
　rhamnose $C_6H_{12}O_5$

ラムゼー石　*ramuze-seki*
　ramsayite $Na_2O \cdot 2SiO_2 \cdot 2TiO_2$

ラナーク石　*rana-kuseki*
　lanarkit Pb_2OSO_4

ランベルトベールの法則　*ranberutobe-runohousoku*
　Lambert-Beer's law

ランダム誤差　*randamugosa*
　random error

ランダムコイル　*randamukoiru*
　random coil

ランドール反応　*rando-ruhannou*
　Landolt reaction

ラネーニッケル　*rane-nikkeru*
　Raney nickel

ラングバイン石　*rangubainseki*
　langbeinite $K_2Mg_2[SO_4]_3$

ラングミュアの吸着等温式，ラングミュラー吸着等温式
　rangumyuanokyuuchakutouonshiki, rangumyura-kyuuchakutouonshiki
　Langmuir adsorption isotherm

ランメルスベルグ石
　ranmerusuberuguseki
　rammelsbergite $NiAs_2$

ランニング　*ranningu*
　run, running

ラノリン　*ranorin*
　lanoline, wool fat

ラノステリン　*ranosuterin*
　lanosterol $C_{30}H_{50}O$

ランスフォルド石　*ransuforudoseki*
　lansfordite $MgCO_3 \cdot 5H_2O$

ランタン　*rantan*
　lanthanum (La, element 57)

ランタニド　*rantanido*
　lanthanide

ランタニド族　*rantanidozoku*
　lanthanide group

ランタノイド　*rantanoido*
　lanthanoide

ラパコール　*rapako-ru*
　lapachol $C_{15}H_{14}O_3$

ラルン石　*rarunseki*
　larnite Ca_2SiO_4

ラサーフォード　*rasa-fo-do*
　rutherford

ラセマーゼ　*rasema-ze*
　racemase

ラセミ型　*rasemigata*
　racemic form

ラセミ化　*rasemika*
　racemization

ラセミ化合物　*rasemikagoubutsu*
　racemic compound

ラセミ混合物　*rasemikongoubutsu*
　racemate

ラセミの　*rasemino*
　racemic

ラセミ体　*rasemitai*
　racemic modification

ラセモ　*rasemo*
　racemo

らせん軸　*rasenjiku*
　screw axis

らせん構造　*rasenkouzou*
　helical structure

ラーセン石　*ra-senseki*
　larsenite $PbZnSiO_4$

ラシヒリング, ラシッヒリング
　rashihiringu, rashihhiringu
　Raschig ring

ラソライト　*rasoraito*
　rasorite $Na_2B_4O_7 \cdot 4H_2O$

ラッセルソーンダース結合
　rasseruso-nda-suketsugou
　Russell-Saunders coupling

ラス鉱　*rasukou*
　rathite $3PbS \cdot 2As_2S_3$

ラテックス　*ratekkusu*
　latex

ラテックス塗料　*ratekkusutoryou*
　latex paint

ラテックス増粘剤　*ratekkusuzounenzai*
　latex thickener

ラテライト　*rateraito*
　laterite

ラーテレロ石　*ra-tereroseki*
　larderellite $(NH_4)_2B_{10}O_{16} \cdot 5H_2O$

ラウダニン, ラウダニジン
　raudanin, raudanijin
　laudanidine, laudanine $C_{20}H_{25}NO_4$

ラウダノシン　*raudanoshin*
　laudanosine $C_{21}H_{27}NO_4$

ラウリン　*raurin*
　laurin $C_3H_5(CH_3(CH_2)_{10}COO)_3$

ラウリン酸　*raurinsan*
　lauric acid $CH_3(CH_2)_{10}COOH$

ラウリン酸塩　*raurinsanen*
　laurate $CH_3(CH_2)_{10}COOR$

ラウリン酸ヘキシル　*raurinsanhekishiru*
　hexyl laurate $CH_3(CH_2)_{10}COOC_6H_{13}$

ラウリルアルコール　*rauriruaruko-ru*
　lauryl alcohol $CH_3(CH_2)_{10}CH_2OH$

ラウロテタニン　*raurotetanin*
　laurotestanine $C_{17}H_{22}NO_3$

ラウルの法則, ラウールの法則
　raurunohousoku, rau-runohousoku
　Raoult's law

ラウタロ石　*rautaroseki*
　lautarite $Ca(IO_3)_2$

ラヴェンダー油　*ravenda-yu*
　lavender oil

ラザホージウム　*razaho-jiumu*
　rutherfordium (Rf, element 104)

レ re

レアメタル　*reametaru*
　rare metal

レベリング　*reberingu*
　leveling

レベル検出器　*reberukenshutsuki*
　level sensor

レベル図　*reberuzu*
　level diagram

レービ石　*re-biseki*
　mesoline $CaAl_2Si_3O_{10} \cdot 5H_2O$

レビー石　*rebi-seki*
　levynite $CaAl_2Si_3O_{10} \cdot 5H_2O$

レブリン酸　*reburinsan*
　levulinic acid $HOOCCH_2CH_2COCH_3$

レチン　*rechin*
　retine $C_{18}H_{28}$

レチナール　*rechina-ru*
　retinal, vitamin A aldehyde $C_{20}H_{28}O$

レチノイド　*rechinoido*
　retinoid

レチノール　*rechino-ru*
　retinol, vitamin A $C_{20}H_{29}OH$

レダクターゼ　*redakuta-ze*
　reductase

レッドシフト　*reddoshifuto*
　red shift

レドックス電位　*redokkusudeni*
　redox potential

レドックス反応　*redokkusuhannou*
　redox reaction

レドックス重合　*redokkusujuugou*
　redox polymerization

レドックス系　*redokkusukei*
oxidation-reduction system

レドックス指示薬　*redokkusushijiyaku*
redox indicator

レドックス触媒　*redokkusushokubai*
redox catalyst

レエメル石　*reemeruseki*
roemerite $FeSO_4 \cdot Fe_2(SO_4)_2 \cdot 14H_2O$

レフォルマツキー反応
reforumatsuki-hannou
Reformatsky's reaction

レグランダイト　*regurandaito*
legrandite $Zn_2(OH)(AsO_4) \cdot H_2O$

レイアウト　*reiauto*
layout, design, draft, plan

レイモンジ石　*reimonjiseki*
raimondite $2FeO_3 \cdot 3SO_2 \cdot 7H_2O$

レイン　*rein*
rhein $C_{15}H_8O_6$

レイノルズ数　*reinoruzusuu*
Reynolds number

レイン酸　*reinsan*
rheinic acid $C_{15}H_{10}O_4$

レジスター　*rejisuta-*
register

レカノール酸　*rekano-rusan*
lecanoric acid $C_{16}H_{14}O_7$

レコント石　*rekontoseki*
lecontite $(NH_4)NaSO_4 \cdot 2H_2O$

レクチン　*rekuchin*
lectin

レーマー石　*re-ma-seki*
ferromerite $FeSO_4 \cdot Fe_2(SO_4)_3 \cdot 14H_2O$

レモン油　*remonyu*
lemon oil

レナーダイト　*rena-daito*
renardite $PbO \cdot 4UO_3 \cdot P_2O_5 \cdot 9H_2O$

レンベルグ石　*renberuguseki*
lembergite $4Na_2Al_2Si_2O_8 \cdot 5H_2O$

レンドゲン線放射, レントゲン線光
rendogensenhousha
Roentgen radiation

レニン　*renin*
renin

レニウム　*reniumu*
rhenium (Re, element 75)

レントゲン　*rentogen*
roentgen

レントゲニウム　*rentogeniumu*
roentgenium (former unununium, eka-gold) (Rg, former Uuu, Eka-Au, element 111)

レントゲン線　*rentogensen*
X-ray

レンツの法則　*rentsunohousoku*
Lenz's law

レオメーター　*reome-ta-*
rheometer

レオロジー　*reoroji-*
rheology

レパルギル酸　*reparugirusan*
lepargylic acid, heptane-1,7-dicarboxylic acid $HOOC(CH_2)_7COOH$

レピドクロサイト　*repidokurosaito*
lipidocrocite γ-FeOOH

レピジン　*repijin*
lipidine $C_{10}H_9N$

レッペ反応　*reppehannou*
Reppe process

レプレッサー　*repuressa-*
repressor

レプリカ　*repurika*
replica

レプタゾール　*reputazo-ru*
leptazol $C_6H_{10}N_4$

レプトン　*reputon*
lepton

レセプター　*reseputa-*
receptor

レセルピン　*reserupin*
reserpine $C_{33}H_{40}N_2O_9$

レシーバー　*reshi-ba-*
receiver, collector

レシチン　*reshichin*
lecithin

レシノイド　*reshinoido*
　resinoid

レソルシノール　*resorushino-ru*
　resorcinol $C_6H_4(OH)_2$

レターダー　*reta-da-*
　retarder

レターデーション　*reta-de-shon*
　retardation

レテン　*reten*
　retene $C_{18}H_{18}$

レトロウィルス　*retorowirusu*
　retrovirus

レトルト　*retoruto*
　retort

レトルトカーボン　*retorutoka-bon*
　retort carbon

レーウ石　*re-useki*
　loeweite $Na_2SO_4 \cdot MgSO_4 \cdot 2,55H_2O$

レヴォグルコサン　*revogurukosan*
　levoglucosan $C_6H_9O_5$

レヴリン　*revurin*
　levuline $(C_6H_{10}O_5)_n$

レヴリン酸　*revurinsan*
　levulinic acid $HOOCCH_2CH_2COCH_3$

レヴロサン　*revurosan*
　levulosan $C_6H_{10}O_6$

レーヤー石　*re-ya-seki*
　reyerite $Ca_2[Si_4O_{10}] \cdot H_2O$

レーヨン　*re-yon*
　rayon

レーザー　*re-za-*
　laser

レーザービーム　*re-za-bi-mu*
　laser beam

レーザー磁気共鳴法　*re-za-jikikyoumeihou*
　laser magnetic resonance, LMR

レーザーラマン分光法
　re-za-ramanbunkouhou
　laser Raman spectroscopy

レーザー誘起蛍光　*re-za-yuukikeikou*
　laser-induced fluorescence, LIF

レゾール　*rezo-ru*
　resol

レゾルシンブルー　*rezorushinburu-*
　resorcine blue $C_{12}H_9NO_4$

レゾルシンエロー　*rezorushinero-*
　resorcine yellow
　$C_6H_3(OH)_2N=NC_6H_4SO_3Na$

レゾルシンフタレイン　*rezorushinfutarein*
　resorcinolphthalein $C_{20}H_{12}O_5$

レゾルシル酸　*rezorushirusan*
　resorcylic acid $C_6H_3(OH)_2COOH$

リ ri

リアクション　*riakushon*
　reaction

リアクター　*riakuta-*
　reactor

リアクタンス　*riakutansu*
　reactance

リアーゼ　*ria-ze*
　lyase

リバーサイド石　*riba-saidoseki*
　riversideite $2CaSiO_3 \cdot H_2O$

リーベルマン反応　*ri-berumanhannou*
　Liebermann's reaction

リービッヒ冷却器　*ri-bihhireikyakuki*
　Liebig condenser

リビングポリマー　*ribinguporima-*
　living polymer

リボフラビン, リボフラヴィン
　ribofurabin, ribofuravin
　riboflavine, vitamin B2 $C_{17}H_{20}N_4O_6$

リボ核酸　*ribokakusan*
　ribonucleic acid, RNA

リボン酸　*ribonsan*
　ribonic acid $HOCH_2(CHOH)_3COOH$

リボヌクレアーゼ　*ribonukurea-ze*
　ribonuclease

リボヌクレオチド　*ribonukureochido*
　ribonucleotide

リボヌクレオシド　*ribonukureoshido*
ribonucleoside

リボソーム　*riboso-mu*
ribosome

リボース　*ribo-su*
ribose $HOCH_2(CHOH)_3CHO$

リボーゼ　*ribo-ze*
ribose $HOCH_2(CHOH)_3CHO$

リボゾームリボ核酸
ribozo-muribokakusan
ribosomal ribonucleic acid, rRNA

リチンエライジン酸　*richineraijinsan*
ricinelaidic acid $C_{18}H_{34}O_3$

リチニン　*richinin*
lichenin $(C_6H_{10}O_5)_n$; ricinine Cu_4Te_3

リチノレイン　*richinorein*
ricinolein $C_{57}H_{104}O_9$

リチノール酸　*richino-rusan*
ricinoleic acid $CH_3(CH_2)_5CH(OH)$
$CH_2CH=CH(CH_2)_7COOH$

リチウム　*richiumu*
lithium Li (element 3)

リフォーミング　*rifo-mingu*
reforming

リガンド　*rigando*
ligand

リガンド非結合　*rigandohiketsugo*
unliganded

リガーゼ　*riga-ze*
ligase

リグニン　*rigunin*
lignin

リグノセリン酸　*rigunoserinsan*
lignoceric acid $CH_3(CH_2)_{22}COOH$

リグノセルロース, リグノセルローズ
rigunoseruro-su, rigunoseruro-zu
lignocellulose

リーグラー試薬　*ri-gura-shiyaku*
Riegler's reagent

リグロイン　*riguroin*
ligroine

リヒテル石　*rihiteruseki*
richterite $(Mg,Mn,Ca)SiO_3$

リホーミング　*riho-mingu*
reforming

リジン　*rijin*
lysin $H_2N(CH_2)_4CH(NH_2)COOH$

リキッド　*rikiddo*
liquid, fluid

リキッドグリース　*rikiddoguri-su*
liquid grease

リキソン酸　*rikisonsan*
lyxononic acid $HOCH_2(CHOH)_3COOH$

リキソーゼ　*rikiso-ze*
lyxose $HOCH_2(CHOH)_3CHO$

リコポジン　*rikopojin*
lycopodin $C_{32}H_{52}N_2O_3$

リコリン　*rikorin*
lycorine $C_{32}H_{32}N_2O_8$

リミティングファクター
rimitingufakuta-
limiting factor

リモネン　*rimonen*
limonene $C_{10}H_{16}$

リモートセンサー　*rimo-tosensa-*
remote sensor

リモートスイッチ　*rimo-tosuicchi*
remote switch

リムーバー　*rimu-ba-*
remover

リン　*rin*
phosphorus (P, element 15)

リナロオール, リナロール
rinaroo-ru, rinaro-ru
linalool $C_{10}H_{18}O$

リンデ法　*rindehou*
Linde process, Linde cycle

リネン　*rinen*
linen

リンゴ酸　*ringosan*
malic acid $HOOCCH(OH)CH_2COOH$

リンゴ酸アミド　*ringosanamido*
malic amide $H_2NCOCH_2CH(OH)CONH_2$

リング　*ringu*
ring, nucleus, cycle

リンホカイン　*rinhokain*
lymphokine

リンホトキシン　*rinhotokishin*
lymphotoxin

リンカー　*rinka-*
linker

リン化亜鉛　*rinkaaen*
zinc phosphide Zn_3P_2

リン化アルミニウム　*rinkaaruminiumu*
aluminum phosphide AlP

りん灰石　*rinkaiseki*
apatite $Ca_5(PO_4)_3(OH,Cl,F)$

リンコマイシン　*rinkomaishin*
lincomycin $C_{18}H_{34}N_2O_6S$

りん光　*rinkou*
phosphorescence

リンク　*rinku*
link

リノレン酸　*rinorensan*
linolenic acid $C_{17}H_{29}COOH$

リノリウム　*rinoriumu*
linoleum

リノル酸, リノール酸　*rinorusan, rino-rusan*
linoleic acid $C_{17}H_{31}COOH$

リンパ　*rinpa*
lymph

リンパ系　*rinpakei*
lymphatic system

リンパ球　*rinpakyuu*
lymphocyte

リン酸　*rinsan*
phosphoric acid H_3PO_4

リン酸ホウ素　*rinsanhouso*
boron phosphate BPO_4

リン酸一水素カルシウム　*rinsanichisuisokarishiumu*
calcium monohydrogen phosphate $CaHPO_4$

リン酸カルシウム　*rinsankarushiumu*
calcium phosphate $Ca_3(PO_4)_2$

リン酸二水素アンモニウム　*rinsannisuisoanmoniumu*
ammonium dihydrogen phosphate $(NH_4)_2HPO_4$

リン酸二水素カリウム　*rinsannisuisokariumu*
potassium dihydrogen phosphate KH_2PO_4

リン酸二水素ナトリウム　*rinsannisuisonatoriumu*
sodium dihydrogen phosphate NaH_2PO_4

リン酸三カリウム　*rinsansankariumu*
tripotassium phospahte K_3PO_4

リン酸三カルシウム　*rinsansankarushiumu*
tricalcium phosphate $Ca_3(PO_4)_2$

リン酸三ナトリウム　*rinsansannatoriumu*
trisodium phosphate Na_3PO_4

リン酸水素カルシウム　*rinsansuisokarushiumu*
dibasic calcium phosphate $CaHPO_4$

リン酸水素二アンモニウム　*rinsansuisonianmoniumu*
diammonium hydrogen phosphate $(NH_4)_2HPO_4$

リン酸水素二ナトリウム　*rinsansuisoninatoriumu*
dipotassium hydrogen phosphate Na_2HPO_4

リン酸トリフェニル　*rinsantorifeniru*
triphenyl phosphate $PO(OC_6H_5)_3$

リン酸トリメチル　*rinsantorimechiru*
trimethyl phosphate $PO(OCH_3)_3$

リパーゼ　*ripa-ze*
lipase

リポイド　*ripoido*
lipoid

リポキシゲナーゼ　*ripokishigena-ze*
lipoxygenase

リポコルチン　*ripokoruchin*
lipocortin

リポオキシゲナーゼ　*ripookishigena-ze*
lipoxygenase

リポプロテイド　*ripopuroteido*
lipoprotein, lipoproteid

リポ酸　*riposan*
　lipoic acid

リポ蛋白室　*ripotanpakushitsu*
　lipoprotein

リポ多糖　*ripotatou*
　lipopolysaccharide

リプレッサー　*ripuressa-*
　repressor

リレー　*rire-*
　relay

リサーチ　*risa-chi*
　research

リサイクルする　*risaikurusuru*
　to recycle, to reuse

リサージ　*risa-ji*
　litharge PbO

リセルギン酸　*riseruginsan*
　lysergic acid $C_{16}H_{16}N_2O_6$

リセタール　*riseta-ru*
　lycetol $C_{10}H_{20}N_2O_6$

リセトール　*riseto-ru*
　lycetol $C_{10}H_{20}N_2O_6$

リシノール酸　*rishino-rusan*
　ricinoleic acid $CH_3(CH_2)_5CH(OH) \cdot CH_2CH=CH(CH_2)_7COOH$

リソゾーム　*risozo-mu*
　lysosome

リスキールド石　*risuki-rudoseki*
　liskeardite $(Al,Fe)AsO_4 \cdot 2(Al,Fe)(OH)_3 \cdot 5H_2O$

リトコール酸　*ritoko-rusan*
　lithocholic acid $C_{24}H_{40}O_3$

リトマス　*ritomasu*
　litmus

リトマス紙　*ritomasushi*
　litmus paper

リトポン　*ritopon*
　lithophone ZnS-BaSO$_4$-mixture

リットル　*rittoru*
　liter

リットルフラスコ　*rittorufurasuko*
　liter flask

リウネブルグ石　*riuneburuguseki*
　lüneburgite $3MgO \cdot B_2O_3 \cdot P_2O_5 \cdot 8H_2O$

リヴァノール　*rivano-ru*
　rivanol $C_{15}H_{15}N_3O$

リゾチーム　*rizochi-mu*
　lysozyme

ロ ro

ロベラニジン　*roberanijin*
　lobelanidine $C_{22}H_{29}NO_2$

ロベラニン　*roberanin*
　lobelanine $C_{22}H_{25}NO_2$

ロベリン　*roberin*
　lobeline $C_{22}H_{27}NO_2$

ロビンソンのエステル　*robinsonnoesuteru*
　Robinson ester $C_6H_{10}O_5(OPO_3H_2)$

ローダミン　*ro-damin*
　rhodamine $C_{17}H_{18}N_2OCl$

ロダナンモン　*rodananmon*
　ammonium rhodanate NH_4NCS

ロダン酸アンモニウム　*rodansananmoniumu*
　ammonium rhodanate NH_4NCS

ロドキサンチン　*rodokisanchin*
　rhodoxanthin $C_{40}H_{50}O_2$

ロドポルフィリン　*rodoporufirin*
　rhodoporphyrine $C_{31}H_{34}N_4O_4$

ロドプシン　*rodopushin*
　rhodopsin

ロドリン酸　*rodorinsan*
　rhoduline acid $C_{20}H_{11}N(OH)_2(SO_3H)_2$

ロエアジン　*roeajin*
　rhoeadine $C_{21}H_{21}NO_6$

ロフィン　*rofin*
　lophine $C_{21}H_{16}N_2$

ロイカート反応　*roika-tohannou*
　Leuckart reaction

ロイコアニリン　*roikoanirin*
　leucoaniline $CH(C_6H_4NH_2)_3$

ロイコ塩基　*roikoenki*
　leuco-base

ロイコマイシン　*roikomaishin*
　leucomycin

ロイコプテリン　*roikoputerin*
　leucopterin $C_6H_5N_5O_3$

ロイコリン　*roikorin*
　leucoline C_9H_7N

ロイマチン　*roimachin*
　rheumatin $C_{34}H_{34}N_2O_7$

ロイナガス　*roinagasu*
　leuna gas

ロイシン　*roishin*
　leucine $(CH_3)_2CHCH_2CH(NH_2)COOH$

ロイシン酸　*roishinsan*
　leucinic acid
　$(CH_3)_2CHCH_2CH(OH)COOH$

ロジン　*rojin*
　rosin

ロジナール　*rojina-ru*
　rodinal $H_2NC_6H_4OH$

ロジノール　*rojino-ru*
　rhodinol
　$(CH_3)_2C=CH(CH_2)_2CH(CH_3)(CH_2)_2OH$

ロジウム　*rojiumu*
　rhodium (Rh, element 45)

ロジウム酸塩　*rojiumusanen*
　rhodate $M^I_2RhO_3$

ロジゾン酸　*rojizonsan*
　rhodizonic acid $C_6H_2O_6$

ロージョン　*ro-jon*
　lotion

ロクセリン　*rokuserin*
　roccelline $C_{10}H_6(OH)N=NC_{10}H_6(SO_3Na)$

ローマー石　*ro-ma-seki*
　roemerite $Fe_3[SO_4]_4 \cdot 14H_2O$

ロメ石　*romeseki*
　romeite $5CaO \cdot 3Sb_2O_5$

ロンガリット　*rongaritto*
　rongalite $HOCH_2SO_2Na \cdot nH_2O$

ロオジザイド　*roojizaido*
　rhodizite $(Li,Na,K,H)_2O \cdot 2Al_2O_3 \cdot 3B_2O_3$

ロビンソン環化反応　*ropinsonkankahannou*
　Robinson annulation

ロープ　*ro-pu*
　rope

ローラー, ローラ　*ro-ra-, ro-ra*
　roller

ローラーミル　*ro-ra-miru*
　roller mill

ロレチン　*rorechin*
　loretin $C_9H_6INO_4S$

ローレンシウム　*ro-renshiumu*
　lawrencium (Lr, element 103)

ロレンツェ鉱　*rorentsekou*
　lorenzenite $Na_2Ti_2Si_2O_9$

ローレンツ変換　*ro-rentsuhenkan*
　Lorentz transformation

ローリングミル　*ro-ringumiru*
　rolling mill

ロール　*ro-ru*
　roll

ロールクラッシャー　*ro-rukurasha-*
　roll crusher

ロールンゼン石　*ro-runzenseki*
　lorenzenite $Na_2(TiO)_2 \cdot Si_2O_7$

ローサルファ油　*ro-sarufayu*
　low-sulfphur oil

ロセル酸　*roserusan*
　rocellic acid $C_{20}H_{20}O_7$

ローシュミッド数　*ro-shumiddosuu*
　Loschmidt number

ローソン石　*ro-sonseki*
　lawsonite $CaO \cdot Al_2O_3 \cdot 2SiO_2 \cdot H_2O$

ロスマリン油　*rosumarinyu*
　rosemary oil

ロタメーター　*ro-tame-ta-*
　rotameter

ロータリーキルン　*ro-tari-kirun*
　rotary kiln

ロテノン　*rotenon*
　rotenone $C_{23}H_{22}O_6$

ロテノン　*rotenon*
　rotenone $C_{23}H_{22}O_6$

ロート　*ro-to*
　funnel

ロート台　*ro-todai*
　funnel stand

ロートフラビン　*ro-tofurabin*
　lotoflavin $C_{15}H_{10}O_6$

ロートフラスコ　*ro-tofurasuko*
　funnel flask

ロート状の　*ro-tojouno*
　funnel-shaped

ロツシン　*rotsushin*
　lotusin $C_{28}H_{31}NO_6$

ロットナンバー　*rottonanba-*
　batch number

ローザニリン　*ro-zanirin*
　rosaniline $C_{20}H_{21}N_3O$

ローゼンムンドの還元法
　ro-zenmundonokangenhou
　Rosenmund reduction

ロゾファン　*rozofan*
　losophan $C_6H(CH_3)(I)_3(OH)$

ロゾール酸　*rozo-rusan*
　rosolic acid $C_{20}H_{16}O_3$

ローズアミン　*ro-zuamin*
　rosamine $C_{23}H_{23}N_2OCl$

ローズアニリン　*ro-zuanirin*
　rosaniline $C_{20}H_{21}N_3O$

ローズマリン油　*ro-zumarinyu*
　rosemary oil

ローズ油　*ro-zuyu*
　rose oil

ル　ru

ルベアン酸　*rubeansan*
　rubeanic acid $H_2NSCCSNH_2$

ルベリトリン酸　*ruberitorinsan*
　ruberythric acid $C_{26}H_{28}O_{14}$

ルビー　*rubi-*
　ruby Al_2O_3 + chromium

ルビアジン　*rubiajin*
　rubiadine $C_{17}H_{14}O_2$

ルビジウム　*rubijiumu*
　rubidium (Rb, element 37)

ルブレン　*ruburen*
　rubrene $C_{42}H_{28}$

ルブリン石　*ruburinseki*
　lublinite $CaCO_3$

ルチジン　*ruchijin*
　lutidine $C_5H_3N(CH_3)_2$

ルチジン酸　*ruchijinsan*
　lutidinic acid $C_7H_5NO_4$

ルチン　*ruchin*
　rutin $C_{27}H_{30}O_{16}$

ルーチン分析　*ru-chinbunseki*
　routine analysis

ルチル　*ruchiru*
　rutile TiO_2

ルフィガルス酸　*rufigarususan*
　rufigallic acid $C_{14}H_8O_8$

ルフィオピン　*rufiopin*
　rufiopin $C_{14}H_8O_6$

ルフォル　*ruforu*
　rufol $C_{14}H_{10}O_2$

ルゴール液　*rugo-rueki*
　Lugol's solution

ルイス塩基　*ruisuenki*
　Lewis base

ルイス酸　*ruisusan*
　Lewis acid

ルイス石　*ruisuseki*
　lewisite $5CaO \cdot 2TiO_2 \cdot 3Sb_2O_5$

ルクランシェ電池　*rukuranshedenchi*
　Leclanché cell

ルクス　*rukusu*
　lux

ルーメン　*ru-men*
　lumen

ルミクロム　*rumikuromu*
　lumichrome $C_{12}H_{10}N_4O_2$

ルミネセンス　*ruminesensu*
　luminescence

ルミノール　*rumino-ru*
　luminol $C_8H_7N_3O_2$

ルミステリン　*rumisuterin*
　lumisterol $C_{28}H_{44}O$

ルペタジン　*rupetajin*
　lupetazin $C_6H_{14}N_2$

ルピゲニン　*rupigenin*
　lupigenine $C_{14}H_{12}O_6$

ルピイン　*rupiin*
　lupiin $C_{29}H_{32}O_{16} \cdot 7H_2O$

ルピニジン　*rupinijin*
　lupinidine $C_{15}H_{26}N_2$

ルピニン　*rupinin*
　lupinine $C_{10}H_{19}NO$

ループ　*ru-pu*
　loop

ルシャトリエの原理, ルシャトウリエの原理　*rushatorienogenri, rushatourienogenri*
　Le Chatelier's principle

ルシン石　*rushinseki*
　lucinite $AlPO_4 \cdot 2H_2O$

ルテチウム　*rutechiumu*
　lutetium (Lu, element 71)

ルテイン　*rutein*
　lutein $C_{40}H_{56}O_2$

ルテカルピン　*rutekarupin*
　ruteacarpin $C_{18}H_{13}N_3O$

ルテニウム　*ruteniumu*
　ruthenium (Ru, element 44)

ルテオリン　*ruteorin*
　luteolin $C_{15}H_{10}O_6$

リョ ryo

りょう面体の　*ryoumentaino*
　rhombohedral

リュ ryu

リュードベリ定数　*ryu-doberiteisuu*
　Rydberg constant

リュイスの塩基　*ryuisunoenki*
　Lewis base

リュイスの酸　*ryuisunosan*
　Lewis acid

リュコアニリン　*ryukoanirin*
　leucoaniline $CH(C_6H_4NH_2)_3$

リュコオーリン　*ryukoo-rin*
　leucoaurine $CH(C_6H_4OH)_3$

サ sa

サバジン　*sabajin*
　sabadine $C_{29}H_{51}NO_8$

サバジニン　*sabajinin*
　sabardinine $C_{27}H_{45}NO_8$

さび止め　*sabidome*
　rust proofing

さび止め塗料　*sabidometoryou*
　anticorrosive paint

さび止め剤　*sabidomezai*
　rust preventive, rust inhibitor, corrosion inhibitor, antirust agent

サビナン　*sabinan*
　sabinane, thujane $C_{10}H_{18}$

サビネン　*sabinen*
　sabinene, thujene $C_{10}H_{16}$

サビネン酸　*sabinensan*
　sabinenic acid $C_{10}H_{16}O_3$

サブユニット　*sabuyunitto*
　subunit

サファイヤ　*safaiya*
　sapphire Al_2O_3

サファーリン　*safa-rin*
　saphirine $Mg_5Al_{12}Si_2O_{17}$

サフラン　*safuran*
　saffron

サフラナール　*safurana-ru*
　safranal $C_{10}H_{14}O$

サフラニン　*safuranin*
　safranine

サフレン　*safuren*
　safren $C_{10}H_{16}$

サフロール　*safuro-ru*
　safrole $C_{10}H_{10}O_2$

サイフォン　*saifon*
　siphon

サイクリックの　*saikurikkuno*
　cyclic

サイクロスポリンA　*saikurosuporin A*
　cyclosporine A $C_{62}H_{111}N_{11}O_{12}$

サイクロトロン　*saikurotoron*
　cyclotron

サイクル　*saikuru*
　cycle

サイクルタイム　*saikurutaimu*
　cycle time

サイトカイニン　*saitokainin*
　cytokinin

サイズ　*saizu*
　size

サッカラート　*sakkara-to*
　saccharate $C_{12}H_{22}O_{11} \cdot CaO \cdot H_2O$

サッカラーゼ　*sakkara-ze*
　saccharase

サッカリド　*sakkarido*
　saccharide

サッカリン　*sakkarin*
　saccharine $C_7H_5NO_3S$

サッカリンナトリウム
　sakkarinnatoriumu
　saccharin sodium $C_7H_4NNaO_3S$

サッカリンナトリウム二水和物
　sakkarinnatoriumunisuiwabutsu
　saccharin sodium dihydrate
　$C_7H_4NNaO_3S \cdot 2H_2O$

サッカロ燐酸　*sakkarorinsan*
　saccharophosphoric acid
　$C_{12}H_{21}O_{10} \cdot H_2PO_4$

サマリウム　*samariumu*
　samarium (Sm, element 62)

サモール　*samo-ru*
　samol $C_{17}H_{24}O_3$

サーモスタット　*sa-mosutatto*
　thermostat

サムソン石　*samusonseki*
　samsonite $2Ag_2S \cdot MnS \cdot Sb_2S_3$

サンブシン　*sanbushin*
　sambucine $C_{27}H_{31}O_{15}Cl$

サンダラック　*sandarakku*
　sandarac

サンダラック樹脂　*sandarakkujushi*
　sandarac resin

サンドイッチ構造　*sandoicchikouzou*
　sandwich structure

サンドマイヤー反応　*sandomaiya-hannou*
　Sandmeyer reaction

サンギ銀鉱　*sangiginkou*
　sanguinite $_3Ag_2S \cdot As_2S_3$

サンギナリン　*sanginarin*
　sanguinarine $C_{20}H_{14}NO_4$

サノクリジン　*sanokurijin*
　sanocrysin $Na_2S_2O_3 \cdot NaAuS_2O_3$

サンプリング　*sanpuringu*
　sampling

サンプル　*sanpuru*
　sample

サンタレン　*santaren*
　santalene $C_{15}H_{24}$

サンタリン　*santarin*
　santalin $C_{15}H_{14}O_4$

サンタロール　*santaro-ru*
　santalol $C_{15}H_{24}O$

サンタル酸　*santarusan*
　santalic acid $C_{15}H_{14}O_5$

サントニン　*santonin*
　santonin $C_{15}H_{18}O_3$

サントニン酸　*santoninsan*
　santoninic acid $C_{15}H_{20}O_4$

サントニン酸ラクトン　santoninsanrakuton
santonic lactone $C_{15}H_{18}O_3$

サポゲニン　sapogenin
sapogenin $C_{14}H_{22}O_2$

サポニン　saponin
saponin

サポタリン　sapotarin
sapotalene $C_{13}H_{14}$

サラセトール, サラントール　saraseto-ru, saranto-ru
salacetol, salantol $HOC_6H_4COOCH_2COCH_3$

サラゾロン　sarazoron
salazolon $C_{18}H_{18}N_2O_3$

サリチルアルデヒド　sarichiruarudehido
salicylaldehyde $C_6H_4(OH)COOH$

サリチルアルコール　sarichiruaruko-ru
salicyl alcohol $C_6H_4(OH)CH_2OH$

サリチルル酸　sarichirurusan
salicyluric acid $C_6H_4(OH)CONHCH_2COOH$

サリチル酸　sarichirusan
salicylic acid $C_6H_4(OH)COOH$

サリチル酸メチルエステル　sarichirusanmechiruesuteru
methyl salicylate $C_6H_4(OH)COOCH_3$

サリゲニン　sarigenin
salicyl alcohol $C_6H_4(OH)CH_2OH$

サリメントール　sarimento-ru
salimenthol $C_6H_4(OH)COOC_{10}H_{19}$

サリン　sarin
sarin $(CH_3)_2CHOP(O)(CH_3)F$

サリシン　sarishin
salicin $C_{13}H_{18}O_7$

サルファダイアジン　sarufadaiajin
sulfadiazine $C_6H_4(NH_2)SO_2NHC_4H_3N_2$

サルフェーション　sarufe-shon
sulfation

サルフォボーライト　sarufobo-raito
sulfoborite $2MgSO_4 \cdot 4MgHBO_3 \cdot 7H_2O$

サルヒプノン　saruhipunon
salhypnone $C_{15}H_{12}O_4$

サルキン　sarukin
sarcine $C_5H_4N_4O$

サルコミン　sarukomin
salcomine $C_{16}H_{14}O_2N_2Co$

サルコプサイド　sarukopusaido
sarcopside $(Fe,Mn,Ca)_3[PO_4]_2$

サルコライト　sarukoraito
sacolite $(Ca,Na_2)_3Al_2(SiO_4)_3$

サルコシン　sarukoshin
sarcosine CH_3NHCH_2COOH

サルミン　sarumin
salmine $C_{30}H_{57}N_{14}O_6$

サルモンス石　sarumonsuseki
salmonsite $Fe_2O_3 \cdot 9MnO \cdot _4P_2O_5 \cdot 14H_2O$

サルササポニン　sarusasaponin
sarsasaponin $C_{45}H_{74}O_{17}$

サルシン　sarushin
sarcine $C_5H_4N_4O$

サルソリン　sarusorin
salsoline $C_{11}H_{15}NO_2$

サルトリウス石　sarutoriususeki
sartorite $PbS \cdot As_2S_3$

サルヴァルサン　saruvarusan
salvarsan $C_{12}H_{12}N_2O_2As_2 \cdot 2HCl$

セ se

セバジン　sebajin
cevadine $C_{32}H_{49}NO_9$

セバジリン　sebajirin
cevadilline $C_{34}H_{58}O_8N$

セバシン酸　sebashinsan
sebacic acid $HOOC(CH_2)_8COOH$

セチルアルコール　sechiruaruko-ru
cetyl alcohol $CH_3(CH_2)_{15}OH$

セダノリド　sedanorido
sedanolid $C_{12}H_{18}O_2$

セドヘプチット　sedohepuchitto
sedoheptitol $HOCH_2(CHOH)_5CH_2OH$

セドヘプトーゼ　sedoheputo-ze
sedoheptose $HOCH_2CO(CHOH)_4CH_2OH$

セファイリン　*sefairin*
　cephaeline $C_{28}H_{38}N_2O_4$

セファロスポリン　*sefarosuporin*
　cephalosporin

セグメント　*segumento*
　segment

セコンド　*sekondo*
　second

セクレチン　*sekurechin*
　secretin

セメンタイト　*sementaito*
　cementite Fe_3C

セメント　*semento*
　cement

セーメット　*se-metto*
　cermet, ceramic metal

セミアセタール　*semiaseta-ru*
　hemiacetal $RCH(OH)OR'$

セミディン　*semidin*
　semidine $C_6H_5NHC_6H_4NH_2$

セミカルバチド, セミカルバジド
　semikarubachido, semikarubajido
　semicarbazide $H_2NCONHNH_2$

セミカルバゾーン　*semikarubazo-n*
　semicarbazone $R_2C=N-NHCONH_2$

セミキノン　*semikinon*
　semiquinone

セミノーゼ　*semino-ze*
　seminose, D-mannose $C_6H_{12}O_6$

セミテルペン　*semiterupen*
　semiterpene

セムセー石　*semuse-seki*
　semseyite $Pb_9Sb_8S_{21}$

センチポアズ　*senchipoazu*
　centipoise

せん断　*sendan*
　shear

せん断安定度　*sendananteido*
　shear stability

せん断応力　*sendanouryoku*
　shear stress

センサー　*sensa-*
　sensor

センシトメトリー　*senshitometori-*
　sensitometry

セパレーター　*separe-ta-*
　separator

セポール石　*sepo-ruseki*
　syepoorite CoS_2

セライト　*seraito*
　celite

セラックニス　*serakkunisu*
　shellac varnish

セラコレイン酸　*serakoreinsan*
　selacholeic acid
　$CH_3(CH_2)_7CH=CH(CH_2)_{13}COOH$

セラミックス　*seramikkusu*
　ceramics, pottery

セラ石　*seraseki*
　sellaite MgF_2

セレブロシド　*sereburoshido*
　cerebroside

セレン　*seren*
　selenium (Se, element 34)

セレン銅銀鉱　*serendoukinkou*
　eucairite $CuAgSe$

セレン鉛鉱　*serenenkou*
　clausthalite $PbSe$

セレニウム　*sereniumu*
　selenium (Se, element 34)

セレン化物　*serenkabutsu*
　selenide M^I_2Se

セレン化銀　*serenkagin*
　silver selenide Ag_2Se

セレン化インジウム　*serenkainjiumu*
　indium selenide In_2Se_3

セレン化カドミウム　*serenkakadomiumu*
　cadmium selenide $CdSe$

セレン化水銀（ＩＩ）　*serenkasuigin(II)*
　mercuric selenide $HgSe$

セレン化水素　*serenkasuiso*
　hydrogen selenide H_2Se

セレノフェン　*serenofen*
　selenophene C_4H_4Se

セレン酸　*serensan*
　selenic acid H_2SeO_4

セレン酸ジルコニウム　*serensanjirukoniumu*
　zirconium selenate $Zr(SeO_4)_2$

セレン酸ナトリウム　*serensannatoriumu*
　sodium selenate Na_2SeO_4

セレシン　*sereshin*
　ceresin

セリグマン石　*serigumanseki*
　seligmannite $PbCuAsS_3$

セリン　*serin*
　serine $HOCH_2CH(NH_2)COOH$

セリルアルコール　*seriruaruko-ru*
　ceryl alcohol $CH_3(CH_2)_{24}CH_2OH$

セリシン　*serishin*
　sericin

セリウム　*seriumu*
　cerium (Ce, element 58)

セロビアーゼ　*serobia-ze*
　cellobiase

セロビオース, セロビオーゼ　*serobio-su, serobio-ze*
　cellobiose

セロチン酸　*serochinsan*
　cerotic acid $CH_3(CH_2)_{24}COOH$

セロファン　*serofan*
　cellophane

セロハン　*serohan*
　cellophane

セロテイン酸　*seroteinsan*
　cerotic acid $CH_3(CH_2)_{24}COOH$

セロテン　*seroten*
　cerotene $C_{27}H_{54}$

セロトニン　*serotonin*
　serotonin $C_{10}H_{12}N_2O$

セロトリオーゼ　*serotorio-ze*
　cellotriose $C_{18}H_{32}O_{16}$

セローゼ　*sero-ze*
　cellose $C_{12}H_{22}O_{11}$

セルベレチン　*seruberechin*
　cerberetin $C_{19}H_{26}O_4$

セルラーラバー　*serura-raba-*
　cellular rubber

セルラーゼ　*serura-ze*
　cellulase

セルロイド　*seruroido*
　celluloid

セルロース, セルローゼ　*seruro-su, seruro-ze*
　cellulose $(C_6H_{10}O_5)_n$

セルローズプラスチック　*seruro-zupurasuchikku*
　cellulose plastic

セル石　*seruseki*
　cerite $2CaO \cdot 3M_2O_3 \cdot 6SiO_2 \cdot 3H_2O$

セサミン　*sesamin*
　sesamin $C_{18}H_{16}O_5$

セシウム　*seshiumu*
　cesium (Cs, element 55)

セスキ　*sesuki*
　sesqui

セスキテルペン　*sesukiterupen*
　sesquiterpene $C_{15}H_{24}$

セタン　*setan*
　cetane, hexadecane $CH_3(CH_2)_{14}CH_3$

セタン価　*setanka*
　cetane number

セタノール　*setano-ru*
　cetyl alcohol, hexadecyl alcohol $CH_3(CH_2)_{15}OH$

セテイン酸　*seteinsan*
　cetylic acid $C_{15}H_{31}COOH$

セテン　*seten*
　cetene $CH_3(CH_2)_{13}CH=CH_2$

セテン価　*setenka*
　cetene number

セッティング　*settingu*
　setting

セット時間　*settojikan*
　setting time

シャ sha

シャモット　*shamotto*
　chamotte

シャープレス酸化　*sha-puresusanka*
　Sharpless oxidation

シャラ石　*sharaseki*
　schallerite
　$6MnO \cdot Mn_2(OH)_4As_2O_3 \cdot 6SiO_3 \cdot 3H_2O$

シャルルの法則　*sharurunohousoku*
　Charles' law

シャッター　*shatta-*
　shutter

シャッタク石　*shattakuseki*
　shattuckite $2CuSiO_3 \cdot H_2O$

シェ she

シェーフェル塩　*she-feruen*
　Schaeffer's salt $C_{10}H_6(OH)(SO_3Na)$

シェーフェル酸　*she-ferusan*
　Schaeffer's acid $C_{10}H_6(OH)(SO_3H)$

シェーカー　*she-ka-*
　shaker

シェーニット　*she-nitto*
　schoenite $K_2Mg[SO_4]_2 \cdot 6H_2O$

シェルロース　*sheruro-su*
　cellulose $(C_6H_{10}O_5)_n$

シ shi

シアメリド　*shiamerido*
　cyamelide $C_3H_3N_3O_3$

シアナミド　*shianamido*
　cyanamide H_2NCN

シアニン　*shianin*
　cyanine $C_{27}H_{31}O_{16}Cl$

シアン化亜鉛　*shiankaaen*
　zinc cyanide $Zn(CN)_2$

シアン化バリウム　*shiankabariumu*
　barium cyanide $Ba(CN)_2$

シアン化物　*shiankabutsu*
　cyanide MCN

シアン化フェニル　*shiankafeniru*
　benzonitrile C_6H_5CN

シアン化銀　*shiankagin*
　silver cyanide AgCN

シアン化カリウム　*shiankakariumu*
　potassium cyanide KCN

シアン化カルシウム　*shiankakarushiumu*
　calcium cyanide $Ca(CN)_2$

シアン化ナトリウム　*shiankanatoriumu*
　sodium cyanide NaCN

シアン化水素　*shiankasuiso*
　hydrogen cyanide HCN

シアン化水素酸　*shiankasuisosan*
　hydrocyanic acid HCN

シアン化ヴィニル　*shiankaviniru*
　vinylcyanide $H_2C=CHCN$

シアノフォルム　*shianoforumu*
　cyanoform $CH(CN)_3$

シアノライト　*shianoraito*
　cyanolite $4CaO \cdot 7SiO_2 \cdot 5H_2O$

シアン酢酸　*shiansakusan*
　cyanoacetic acid $NCCH_2COOH$

シアン酸　*shiansan*
　cyanic acid NCOH

シアン酸塩　*shiansanen*
　cyanate MOCN

シアヌル酸　*shianurusan*
　tricyanic acid $C_3H_3N_3O_3 \cdot 2H_2O$

シアヌール酸　*shianu-rusan*
　cyanuric acid $C_3H_3N_3O_3$

シアヌル酸塩化物　*shianurusanenkabutsu*
　cyanuric chlorid $C_3N_3Cl_3$

シアルル酸　*shiarurusan*
　dialuric acid $C_4H_4N_2O_4$

シアル酸　*shiarusan*
　sialic acid $C_{11}H_{19}NO_9$

シーボーギウム *shi-bo-giumu*
    seaborgium (former eka-tungsten) (Sg, element 106)

シチジン *shichijin*
    cytidine $C_9H_{13}N_3O_5$

シチジル酸 *shichijirusan*
    cytidylic acid $C_9H_{14}N_3O_8P$

シチシン *shichishin*
    cytisine $C_{11}H_{14}N_2O$

シドナール *shidona-ru*
    sidonal $(C_6H_7(OH)_4COOH)_2 \cdot C_4H_{10}N_2$

シッフの塩基 *shiffunoenki*
    Schiff's base

シフト試薬 *shifutoshiyaku*
    shift reagent

シグマトロピー転位 *shigumatoropi-teni*
    sigmatropic rearrangement

シカゴブルー *shikagoburu-*
    chicago blue $C_{34}H_{28}N_6O_{16}S_4$

しきい値 *shikiichi*
    threshold value

シキミ酸 *shikimisan*
    shikimic acid $C_6H_2(OH)_3COOH$

シックナー *shikkuna-*
    thickener

シコニン *shikonin*
    shikonin $C_{16}H_{17}O_5$

シクノジマイト *shikunojimaito*
    sychnodymite $(Co,Cu)_4S_3$

シクラン *shikuran*
    cyclic hydrocarbon, cyclane

シクロアルカン *shikuroarukan*
    cycloalkane CnH_{2n}

シクロアルケニル基 *shikuroarukeniruki*
    cycloalkenyl group

シクロアルキル *shikuroarukiru*
    cycloalkyl

シクロブタン *shikurobutan*
    cyclobutane C_4H_8

シクロブテン *shikurobuten*
    cyclobutene C_4H_6

シクロデキストリン *shikurodekisutorin*
    cyclodextrin

シクロヘキサン *shikurohekisan*
    cyclohexane C_6H_{12}

シクロヘキサノン *shikurohekisanon*
    cyclohexanone $C_6H_{10}O$

シクロヘキサノンオキシム *shikurohekisanonokishimu*
    cyclohexanonoxime

シクロヘキサノール *shikurohekisano-ru*
    cyclohexanol $C_6H_{12}O$

シクロヘキセン *shikurohekisen*
    cyclohexene C_6H_{10}

シクロヘキシルアミン *shikurohekishiruamin*
    cyclohexylamine $C_6H_{11}NH_2$

シクロヘクサジエン *shikurohekusajien*
    cyclohexadiene C_6H_8

シクロヘクサン *shikurohekusan*
    cyclohexane C_6H_{12}

シクロヘクサノン *shikurohekusanon*
    cyclohexanone $C_6H_{10}O$

シクロヘクサノール *shikurohekusano-ru*
    cyclohexanol $C_6H_{12}O$

シクロヘプタン *shikuroheputan*
    suberane, cycloheptane C_7H_{14}

シクロヘプタノール *shikuroheputano-ru*
    suberol, cycloheptanol $C_7H_{13}OH$

シクロヘプテン *shikuroheputen*
    suberene, cycloheptene C_7H_{12}

シクロオクタン *shikurookutan*
    cyclooctane C_8H_{16}

シクロオクタテトラエン *shikurookutatetoraen*
    cyclooctatetraene C_8H_8

シクロオレフィン *shikuroorefin*
    cycloolefine CnH_{2n-2}

シクロパラフィン *shikuroparafin*
    cycloalkane CnH_{2n}

シクロペンタジエン *shikuropentajien*
    cyclopentadiene C_5H_6

シクロペンタン *shikuropentan*
    cyclopentane C_5H_{10}

シクロペンタノン　*shikuropentanon*
　cyclopentanone C_5H_8O

シクロペンタノール　*shikuropentano-ru*
　cyclopentanol C_5H_9OH

シクロペンテン　*shikuropenten*
　cyclopentene C_5H_8

シクロペタジエン　*shikuropetajien*
　cyclopentadiene C_5H_6

シクロプロパン　*shikuropuropan*
　cyclopropane C_3H_6

シクロプロパノン　*shikuropuropanon*
　cyclopropanone C_3H_4O

シクロトリシロクサン
　shikurotorishirokusan
　cyclotrisiloxane $Si_3O_3R_3$

シーマン反応　*shi-manhannou*
　Schiemann reaction

シマリン　*shimarin*
　cymarin $C_{30}H_{40}O_9$

シメン　*shimen*
　cymene, isopropyltoluene
　$C_6H_4(CH_3)CH(CH_3)_2$

シーメンスマルチン法
　shi-mensumaruchinhou
　Siemens-Martin process

シモンズスミス反応
　shimonzusumisuhannou
　Simmons-Smith reaction

シムノール　*shimuno-ru*
　scymnol $C_{27}H_{46}O_5$

シナミン　*shinamin*
　sinamine $C_4H_6N_2$

シナピン　*shinapin*
　sinapine $C_{16}H_{24}NO_5$

シナピン酸　*shinapinsan*
　sinapic acid
　$HO(CH_3O)_2C_6H_2CH=CHCOOH$

シンチレーション　*shinchire-shon*
　scintillation

シネオール　*shineo-ru*
　cineole $C_{10}H_{18}O$

シンゲナイト　*shingenaito*
　syngenite $CaSO_4 \cdot K_2SO_4 \cdot H_2O$

シニグリン　*shinigurin*
　sinigrine $C_{10}H_{18}NO_{10}S_2K$

シンジアゾ酸　*shinjiazosan*
　syn-diazo acid

シンジオタクチック　*shinjiotakuchikku*
　syndiotactic

シンカリン　*shinkarin*
　sincaline $HOCH_2CH_2N(CH_3)_3OH$

シンキサイト　*shinkisaito*
　synchisite $CeCa[F,(CO_3)_2]$

シンコフェン　*shinkofen*
　cinchophen $C_{16}H_{11}NO_2$

シンコメロン酸　*shinkomeronsan*
　cinchomeronic aicd $C_5H_3N(COOH)_2$

シンコニジン　*shinkonijin*
　cinchonidine $C_{19}H_{22}N_2O$

シンコニン　*shinkonin*
　cinchonine $C_{19}H_{22}N_2O$

シンコニン酸　*shinkoninsan*
　cinchonic acid C_9H_6NCOOH

シンコニシン　*shinkonishin*
　cinchonicine $C_{19}H_{24}N_2O$

シンコティン　*shinkotin*
　cinchotine $C_{20}H_{24}N_2O_2$

シンコトキシン　*shinkotokishin*
　cinchotoxine $C_{19}H_{24}N_2O$

シンナメイン　*shinnamein*
　cinnamein $C_6H_5CH=CHCOOCH_2C_6H_5$

シンナミルアルコール
　shinnamiruaruko-ru
　cinnamyl alcohol $C_6H_5CH=CHCH_2OH$

シンナミル基　*shinnamiruki*
　cinnamyl group $-CH_2CH=CHC_6H_5$

シンナムアルデヒド　*shinnamuarudehido*
　cinnamaldehyde $C_6H_5CH=CHCHO$

シンナピリン　*shinnapirin*
　cinnapyrine
　$C_6H_5CH=CHCOOC_6H_4NHCONH_2$

シンノリン　*shinnorin*
　cinnoline $C_8H_6N_2$

シノメニン　*shinomenin*
　sinomenine $C_{19}H_{23}NO_4$

シノリン　shinorin
　cinnoline $C_8H_6N_2$

シンプレサイト　shinpuresaito
　symplesite $Fe_3As_2O_8 \cdot 8H_2O$

シンタリン　shintarin
　decamethylenediguanidine dihydrochloride
　$H_2NC(=NH)NH(CH_2)_{10}NHC$
　$(=NH) \cdot NH_2 \cdot 2HCl$

シペルメトリン　shiperumetorin
　cypermethrin $C_{22}H_{19}NO_3Cl_2$

シピルス石　shipirususeki
　sipylite $ErNbO_4$

シラビオーゼ　shirabio-ze
　scillabiose $C_{12}H_{22}O_{11}$

シーライト　shi-raito
　celite

シラン　shiran
　silane $SinH_{2n+2}$

シラノール　shirano-ru
　silanol R_3SiOH

シーラント　shi-ranto
　sealant

シラレン　shiraren
　scillaren $C_{36}H_{52}O_{13}$

シラリジン　shirarijin
　scillaridin $C_{24}H_{30}O_3$

シラザン　shirazan
　silazane $H_3Si(NHSiH_2)nNHSiH_3$

シリカ　shirika
　silica SiO_2

シリカガラス　shirikagarasu
　silica glass

シリカゲル　shirikageru
　silica gel

シリコン　shirikon
　silicon

シリコーン　shiriko-n
　silicone

シリコーンゴム　shiriko-ngomu
　silicone rubber

シリコーングリース　shiriko-nguri-su
　silicone grease

シリコーン樹脂　shiriko-njushi
　silicone resin

シリコンカーバイド　shirikonka-baido
　silicon carbide SiC

シリコーン油　shiriko-nyu
　silicone oil

シリコール　shiriko-ru
　silicol R_3SiOH

シリコ蓚酸　shirikoshuusan
　silicooxalic acid $H_2Si_2O_4$

シリンダー　shirinda-
　cylinder

シリンダーオイル, シリンダー油
　shirinda-oiru, shirinda-yu
　cylinder oil

シリンガ酸　shiringasan
　syringic acid $(CH_3O)_2C_6H_2(OH)COOH$

シリンゲニン　shiringenin
　syringenin
　$C_6H_2(OH)(OCH_3)_2CH=CHCH_2OH$

シリンギジン　shiringijin
　syringidin $C_{17}H_{14}O_7 \cdot HCl$

シリンギン　shiringin
　syringin $C_{17}H_{24}O_9 \cdot H_2O$

シーリング　shi-ringu
　sealing

シリレン　shiriren
　silylene $H_2Si=$

シリユクロロフォルム　shiriyukuroroforumu
　silicochloroform $SiHCl_3$

シロキサン　shirokisan
　siloxane $(-SiRR-O-)_n$

シロクセン　shirokusen
　siloxene $[H_3Si(OH)_3]_n$

シロップ　shiroppu
　syrup

シルメル鉱　shirumerukou
　schirmerite $(Ag_2,Pb)S \cdot _2Bi_2S_3$

シルト岩　shirutogan
　siltstone

シルヴァン　shiruvan
　silvan $C_4H_3OCH_3$

シルヴェストリ石　*shiruvesutoriseki*
silvestrite Fe_5N_2

シルヴィン酸　*shiruvinsan*
sylvic acid $C_{20}H_{30}O_2$

シサル麻　*shisaruasa*
sisal

シスチン　*shisuchin*
cystine $(SCH_2CH(NH_2)COOH)_2$

シス型, シス形　*shisugata*
cis form

シス位　*shisui*
cis-position

シスモンダ石　*shisumondaseki*
sismondine $H_2(Fe,Mg)Al_2SiO_7$

シス立体配置　*shisurittaihaichi*
cis-configuration

システイン　*shisutein*
cysteine $HSCH_2CH(NH_2)COOH$

システム　*shisutemu*
system

シストランス異性　*shisutoransuisei*
cis-trans isomerism

シトキニン　*shitokinin*
cytokinin

シトラコン酸　*shitorakonsan*
citraconic acid $HOOCC(CH_3)=CHCOOH$

シトラール　*shitora-ru*
citral $C_9H_{15}CHO$

シトレン　*shitoren*
citrene $C_{10}H_{16}$

シトロフェン　*shitorofen*
citrophene $((C_2H_5O)C_6H_4NHCO)_3C_3H_4OH$

シトロネラール　*shitoronera-ru*
citronellal $C_{10}H_{18}O$

シトロネラ油　*shitoronerayu*
citronella oil

シトロネロール　*shitoronero-ru*
citronellol $C_{10}H_{20}O$

シトロネル酸　*shitoronerusan*
citronellic acid
$CH_2C(CH_3)(CH_2)_3CH(CH_3)CH_2COOH$

シトルリン　*shitorurin*
citrulline $H_2NCONH(CH_2)_3CH(NH_2)COOH$

シトシン　*shitoshin*
cytosine $C_4H_5N_3O$

シトステリン　*shitosuterin*
sitosterol $C_{29}H_{50}O$

シトステロール　*shitosutero-ru*
sitosterol $C_{29}H_{50}O$

ショ sho

ショア硬さ, ショア硬度　*shoakatasa, shoakatado*
Shore hardness

ショ糖　*shotou*
sucrose, saccharose $C_{12}H_{22}O_{11}$

ショッテンバウマン反応　*shottenbaumanhannou*
Schotten-Baumann reaction

しょう液　*shoueki*
serum

ショウガオール　*shougao-ru*
shogaol $C_{17}H_{24}O_3$

しょうのう油　*shounouyu*
camphor oil

シュ shu

シュガー　*shuga-*
sugar

シュークロース　*shu-kuro-su*
sucrose, table sugar $C_{12}H_{22}O_{11}$

シュミット反応　*shumittohannou*
Schmidt reaction

シュネーベルグ石　*shune-beruguseki*
schneebergite $(Ca,Fe)_2Sb_2O_3$

シュライベルス石　*shuraiberususeki*
schreibersite $(Fe,Ni,Co)_3P$

シュレッテル石　*shuretteruseki*
schrötterite $8Al_2O_3 \cdot 3SiO_2 \cdot 3H_2O$

シュリップ塩　*shurippuen*
Schlippe's salt $Na_3SbS_4 \cdot 9H_2O$

シュタウディンガーの式
shutaudinga-noshiki
Staudinger equation

シュテファンボルツマンの法則
shutefanborutsumannohousoku
Stefan-Boltzmann law

シュッタフェル石　*shuttaferuseki*
staffelite $Ca_5F(PO_4)_3 \cdot nCaCO_3 \cdot H_2O$

シュウ酸　*shuusan*
oxalic acid HOOCCOOH

シュウ酸塩　*shuusanen*
oxalate $M^I_2C_2O_4$

シュウ酸カリウム　*shuusankariumu*
potassium oxalate $K_2C_2O_4$

シュウ酸カルシウム　*shuusankarushiumu*
calcium oxolate CaC_2O_4

シュウ酸ナトリウム　*shuusannatoriumu*
sodium oxalate $Na_2C_2O_4$

シュヴァイツアー試薬
shuvaitsua-shiyaku
Schweizer's reagent, cuprammonium hydroxide solution $[CU(NH_3)_4](OH)_2$

シュワルチエンベルグ石
shuwaruchienberuguseki
schwartzenberegite $2PbO \cdot 3PbCl_2 \cdot PbI_2O_6$

ソ so

ソブレノン　*soburenon*
sobrerone $C_{10}H_{16}O$

ソブレロール　*soburero-ru*
sobrerol $C_{10}H_{17}O_2$

ソダ水　*sodamizu*
soda water

ソーダ明礬　*so-damyouban*
sodium alum, soda alum $NaAl(SO_4)_2 \cdot 12H_2O$

ソーダパルプ　*so-daparupu*
soda pulp

ソーダ石灰　*so-dasekkai*
soda lime $NaOH + Ca(OH)_2$ mixture

ソーダ硝酸　*so-dashousan*
Chile salpeter $NaNO_3$

ソフォリン　*soforin*
sophorine $C_{11}H_{14}N_2O$

ソフト塩基　*sofutoenki*
soft base

ソフト酸　*sofutosan*
soft acid

ソッジー石　*sojji-seki*
soddite $12UO_3 \cdot 5SiO_2 \cdot 14H_2O$

ソーキング　*so-kingu*
soaking, steeping

ソックスレー抽出器
sokkusure-chuushutsuki
Soxhlet apparatus, Soxhlet extractor

ソマトスタチン　*somatosutachin*
somatostatin $C_{76}H_{104}N_{18}O_{19}S_2$

ソーピング　*so-pingu*
soaping

ソープ反応　*so-puhannou*
Thorpe reaction

ソラニジン　*soranijin*
solanidine $C_{27}H_{43}NO$

ソラニン　*soranin*
solanine $C_{45}H_{73}NO_{15}$

ソラリゼーション　*sorarize-shon*
solarization

ソリア　*soria*
thoria ThO_2

ソルバイト　*sorubaito*
sorbitol $HOCH_2(CHOH)_3CH_2OH$

ソルベー法　*sorube-hou*
Solvay process

ソルビン　*sorubin*
sorbin $C_6H_{12}O_6$

ソルビン酸　*sorubinsan*
sorbic acid $CH_3CH=CHCH=CHCOOH$

ソルビン酸カリウム　*sorubinsankariumu*
　potassium sorbate C_5H_7COOK

ソルビトール　*sorubito-ru*
　sorbitol $HOCH_2(CHOH)_4CH_2OH$

ソルビット　*sorubitto*
　sorbitol $HOCH_2(CHOH)_4CH_2OH$

ソルボース, ソルボーゼ
　sorubo-su, sorubo-ze
　sorbose $C_6H_{12}O_6$

ソルロール　*soruro-ru*
　solurol $C_{30}H_{26}N_4O_{15} \cdot 2P_2O_5$

ス su

スベラン　*suberan*
　suberane, cycloheptane C_7H_{14}

スベレン　*suberen*
　suberene, cycloheptene C_7H_{12}

スベリン酸　*suberinsan*
　suberic acid $HOOC(CH_2)_6COOH$

スベリルアミン　*suberiruamin*
　suberyl amine $C_7H_{13}NH_2$

スベロン　*suberon*
　suberone, cycloheptanone $C_7H_{12}O$

スベロール　*subero-ru*
　suberol, cycloheptanol $C_7H_{13}OH$

スチビン　*suchibin*
　stibine SbH_3

スチフニン酸　*suchifuninsan*
　styphnic acid $C_6H(OH)_2(NO_2)_3$

スチグマステロール　*suchigumasutero-ru*
　stigmasterol $C_{29}H_{48}O$

スチヒト石　*suchihitoseki*
　stichtite $Mg_6Cr_2(OH)_{16}(CO_3) \cdot 4H_2O$

スチクマステリン　*suchikumasuterin*
　stigmasterol $C_{29}H_{48}O$

スチピタチン酸　*suchipitachinsan*
　stipitatic acid $C_8H_6O_5$

スチレン　*suchiren*
　styrene $C_6H_5CH=CH_2$

スチレンブタジエンゴム
　suchirenbutajiengomu
　styrene-butadiene rubber, SBR

スチレン樹脂　*suchirenjushi*
　styrene resin

スチリルアルコール　*suchiriruaruko-ru*
　styrone $C_6H_5CH=CHCH_2OH$

スチリルケトン　*suchiriruketon*
　styryl ketone $(C_6H_5CH=CH)_2CO$

スチリル基　*suchiriruki*
　styryl radical $C_6H_5CH=CH-$

スチロン　*suchiron*
　styrone $C_6H_5CH=CHCH_2OH$

スチロレン　*suchiroren*
　vinylbenzene $C_6H_5CH=CH_2$

スチロール　*suchiro-ru*
　styrene $C_6H_5CH=CH_2$

スチルベン　*suchiruben*
　stilbene $C_6H_5CH=CHC_6H_5$

スフェエライト　*sufeeraito*
　sphaerite $4AlPO_4 \cdot 6Al(OH)_3$

スフェノクラース　*sufenokura-su*
　sphenoclase $(Mg,Ca,Fe,Mn)_6Al_2Si_6O_{11}$

スフェロフロル　*suferofuroru*
　sphaerophorol $C_{13}H_{20}O_2$

スフィンゴミエリン　*sufingomierin*
　sphingomyelin

スフィンゴシン　*sufingoshin*
　sphingosine $CH_3(CH_2)_{12}CH=CHCH$
　$(OH) \cdot CH(NH_2)CH_2OH$

スフィンゴ糖脂質　*sufingotoushishitsu*
　glycosphingolipid

スガリン　*sugarin*
　sugarine $C_6H_5SO_2NHCH_3$

スイッチ　*suicchi*
　switch, controller, button

スカベンシャー　*sukabensha-*
　scavenger

スカッチー石　*sukacchi-seki*
　scacchite $MnCl_2$

スカンジウム　*sukanjiumu*
　scandium (Sc, element 21)

スカトール *sukato-ru*
scatole C_9H_9N

スケール *suke-ru*
scale

スケールアップ *suke-ruappu*
scale up

スキメチン *sukimechin*
skimetine $C_9H_6O_3$

スキミン *sukimin*
skimmine $C_{15}H_{16}O_8$

スコパリン *sukoparin*
scoparin $C_{22}H_{22}O_{11}$

スコピン *sukopin*
scopine $C_8H_{13}NO_2$

スコポラミン *sukoporamin*
scopolamine $C_{17}H_{21}NO_4$

スコポリン *sukoporin*
scopoline $C_8H_{13}NO_2$

スクアラン *sukuaran*
squalane $C_{30}H_{62}$

スクアレン *sukuaren*
squalene $C_{30}H_{50}$

スクラッバー *sukurabba-*
scrubber

スクラップ *sukurappu*
scrap, waste

スクラウプのキノリン合成法 *sukuraupunokinoringouseihou*
Skraup quinoline synthesis

スクレロプロテイド, スクレロプロテイン *sukureropuroteido, sukureropurotein*
scleroprotein

スクリーン *sukuri-n*
screen

スクリーニング *sukuri-ningu*
screening

スクロドウスカ石 *sukurodousukaseki*
sklodowskite $MgO \cdot 2UO_3 \cdot 2SiO_2 \cdot 7H_2O$

スクロール *sukuro-ru*
sucrol $C_2H_5OC_6H_4NHCONH_2$

スクローゼ *sukuro-ze*
sucrose, table sugar $C_{12}H_{22}O_{11}$

スクリュークランプ *sukuryu-kuranpu*
screw clamp

スクシナミド *sukushinamido*
succinamide $H_2NOC(CH_2)_2CONH_2$

スクシナルデヒド *sukushinarudehido*
succinaldehyde $OHC(CH_2)_2CHO$

スクシニミド *sukushinimido*
succinimide $C_4H_5NO_2$

スクシニル基 *sukushiniruki*
succinyl radical $-(O)C(CH_2)_2C(O)-$

スクシノニトリル *sukushinonitoriru*
succinonitrile $NC(CH_2)_2CN$

スクテラレイン *sukuterarein*
scutellarein $C_{15}H_{10}O_6$

スクテラリン *sukuterarin*
scutellarin $C_{21}H_{18}O_{12}$

スミラチン *sumirachin*
smilacin, salseparin, parillin $C_{18}H_{30}O_6$

スミス石 *sumisuseki*
smithite $AgAsS_2$

スモッグ *sumoggu*
smog canopy

スパンゴ石 *supangoseki*
spangolite $Cu_6AlClSO_4 \cdot 9H_2O$

スパルテイン *suparutein*
sparteine $C_{15}H_{26}N_2$

スパテル, スパーテル *supateru, supa-teru*
spatula

スペクトル *supekutoru*
spectrum

スペクトル分析 *supekutorubunseki*
spectrum analysis

スペクトル配列 *supekutoruhairetsu*
spectral distribution

スペクトル感度 *supekutorukando*
spectral sensitivity

スペクトル密度 *supekutorumitsudo*
spectral density

スペクトル線 *supekutorusen*
spectral line

スペクトル図表　*supekutoruzuhyou*
  spectrum atlas

スペンサー石　*supensa-seki*
  spencerite $Zn_3(PO_4)_2 \cdot Zn(OH)_2 \cdot 2H_2O$

スペルミジン　*superumijin*
  spermidine $H_2N(CH_2)_3NH(CH_2)_4NH_2$

スペルミン　*superumin*
  spermine $C_{10}H_{26}N_4$

スペーサー　*supe-sa-*
  spacer

スピン　*supin*
  spin

スピネル　*supineru*
  spinel, magnesium aluminate $MgAl_2O_4$

スピン編極　*supinhenkyoku*
  spin polarization

スピン軌道結合　*supinkidouketsugou*
  spin-orbit coupling

スピン軌道相互作用
supinkidousougosayou
  spin-orbit interaction

スピン量子数　*supinryoushisuu*
  spin quantum number

スピンスピン結合　*supinsupinketsugou*
  spin-spin coupling

スピンスピン相互作用
supinsupinsougosayou
  spin-spin interaction, spin-over interaction

スピラン　*supiran*
  spirane, spiro compound

スピロフォルム　*supiroforumu*
  spiroform $CH_3COOC_6H_4COOC_6H_5$

スピロ原子　*supirogenshi*
  spiro atom

スピロ化合物　*supirokagoubutsu*
  spirane, spiro compound

スピロザール　*supiroza-ru*
  glycol monosalicylate, spirosal
  $C_6H_4(OH)COOCH_2CH_2OH$

スポディオス石　*supodiosuseki*
  spodiosite $Ca_2[F,PO_4]$

スポンジゴム　*suponjigomu*
  foamed rubber

スプライシング　*supuraishingu*
  splicing

スプラレナリン　*supurarenarin*
  adrenaline, suprarenalin
  $C_6H_3(OH)_2CH(OH)CH_2NHCH_3$

スプレー塔　*supure-tou*
  spray tower

スラリー　*surari-*
  slurry

スレオニン　*sureonin*
  threonine $CH_3CH(OH)CH(NH_2)COOH$

スレート　*sure-to*
  slate

スリナミン　*surinamin*
  surinamine
  $C_6H_4(OH)CH_2CH(NHCH_3)COOH$

スルファチアゾール　*surufachiazo-ru*
  sulfathiazole $C_6H_4(NH_2)SO_2NHC_3H_2NS$

スルファジアジン　*surufajiajin*
  sulfadiazine $C_6H_4(NH_2)SO_2NHC_4H_3N_2$

スルファジアゾール　*surufajiazo-ru*
  sulfadiazole $C_6H_4(NH_2)SO_2NHC_3H_4N_2$

スルファメラジン　*surufamerajin*
  sulfamerazine $C_6H_4(NH_2)SO_2NHC_5H_5N_2$

スルファミド　*surufamido*
  sulfamide $SO_2(NH_2)_2$

スルファミノール　*surufamino-ru*
  sulfaminol $C_{12}H_9SO_3$

スルファミン酸　*surufaminsan*
  sulfamic acid HSO_3NH_2

スルファニル酸　*surufanirusan*
  sulfanilic acid $H_2NC_6H_4SO_3H$

スルファニル酸アミド
surufanirusanamido
  sulfanilamide $C_6H_4(NH_2)SO_2NH_2$

スルファピリジン　*surufapirijin*
  sulfapyridine $H_2NC_6H_4SO_2NHC_5H_4N$

スルファルセノール　*surufaruseno-ru*
  sulfarsenol $C_{14}H_{14}N_2O_8S_2Na_2As_2$

スルファセトアミド　*surufasetoamido*
  sulfacetoamide $S(CH_2CONH_2)_2$

スルフェン酸　*surufensan*
  sulfenic acid $RSOH$

スルフィド　*surufido*
　sulfide M^I_2S

スルフィン　*surufin*
　sulfine, sulfonium compound R_3SX

スルフィニル　*surufiniru*
　sulfinyl $SO=$

スルフィン酸　*surufinsan*
　sulfinic acid RSO_2H

スルフォ　*surufo*
　sulfo- $-SO_3H$

スルフォン化　*surufonka*
　sulfonation

スルフォン酸　*surufonsan*
　sulfonic acid, sulfo acid RSO_3H

スルフアルデヒド　*surufuarudehido*
　sulfaldehyde $RCHS$

スルホ基　*suruhoki*
　sulfo group $-SO_2OH$

スルホキシド　*suruhokishido*
　sulfoxide R_2SO

スルホキシル酸　*suruhokishirusan*
　sulfoxylic acid, hyposulfurous acid H_2SO_2

スルホン　*suruhon*
　sulfone R_2SO_2

スルホン化　*suruhonka*
　sulfonation

スルホン酸　*suruhonsan*
　sulfonic acid RSO_3H

スルホン酸塩　*suruhonsanen*
　sulfonate $RSO_2(OR)$

スルホラン　*suruhoran*
　sulfolane $C_4H_8SO_2$

スルホ酸化　*suruhosanka*
　sulfoxidation

スタビライザー　*sutabiraiza-*
　stabilizer

スタキドリン　*sutakidorin*
　stachydrine $C_7H_{13}NO_2$

スタキオーゼ　*sutakio-ze*
　stachyose $C_{24}H_{42}O_{21}$

スタンピング　*sutanpingu*
　stamping

スターリング石　*suta-ringuseki*
　sterlingite $(Zn,Mn^{2+},Fe^{2+})O$

スターリン石　*suta-rinseki*
　stirlingite Mn_2SiO_4

スターター　*suta-ta-*
　initiator, starter

ステアリン　*sutearin*
　stearin

ステアリンアルデヒド　*sutearinarudehido*
　stearaldehyde $CH_3(CH_2)_{16}CHO$

ステアリン酸　*sutearinsan*
　stearic acid $CH_3(CH_2)_{16}COOH$

ステアリン酸亜鉛　*sutearinsanaen*
　zinc stearate $(CH_3(CH_2)_{16}COO)_2Zn$

ステアリン酸アルミニウム　*sutearinsanaruminiumu*
　aluminum stearate

ステアリン酸ブチル　*sutearinsanbuchiru*
　butyl stearate $CH_3(CH_2)_{16}COOC_4H_9$

ステアリン酸塩　*sutearinsanen*
　stearate $CH_3(CH_2)_{16}COOM$

ステアリン酸カルシウム　*sutearinsankarishiumu*
　calcium stearate $(CH_3(CH_2)_{16}COO)_2Ca$

ステアリン酸マグネシウム　*sutearinsanmaguneshiumu*
　magnesium stearate $(CH_3(CH_2)_{16}COO)_2Mg$

ステアリン酸ナトリウム　*sutearinsannatoriumu*
　sodium stearate $CH_3(CH_2)_{16}COONa$

ステアリルアルコール　*suteariruaruko-ru*
　stearyl alcohol $CH_3(CH_2)_{17}OH$

ステアロキシル酸　*sutearokishirusan*
　stearoxylic acid $CH_3(CH_2)_7COCO(CH_2)_7COOH$

ステアロン　*sutearon*
　stearone $C_{35}H_{70}O$

ステアロニトリル　*sutearonitoriru*
　stearonitrile $CH_3(CH_2)_{16}CN$

ステアロル酸　*sutearorusan*
　stearolic acid $CH_3(CH_2)_7CC(CH_2)_7COOH$

ステイン　sutein
stain

ステインレススティール　suteinresusuti-ru
stainless steel

ステンドガラス　sutendogarasu
stained glass

ステンレススティール　sutenresusuti-ru
stainless steel

ステラー石　sutera-seki
stellerite $CaAl_2Si_7O_{18} \cdot 7H_2O$

ステリン　suterin
sterol

ステリルアルコール　suterirusaruko-ru
stearyl alcohol $CH_3(CH_2)_{16}CH_2OH$

ステロイド　suteroido
steroid

ステロイドホルモン　suteroidohorumon
steroid hormone

ステロイド核　suteroidokaku
steroid skeleton

ステロール　sutero-ru
sterol

ステルコル石　suterukoruseki
stercorite $NaNH_4HPO_4 \cdot 4H_2O$

ステルンベルグ鉱　suterunberugukou
sternbergite, iron silver glance $AgFe_2S_3$

スチビン　sutibin
stibane, antimony hydride SbH_3

スチビオタンタル石　sutibiotantaruseki
stibiotantalite $(SbO_2)_2(Ta,Nb)_2O_6$

ステフニン酸　sutifuninsan
styphnic acid $C_6H(OH)_2(NO_2)_3$

スチグマステロール　sutigumasutero-ru
stigmasterol $C_{29}H_{48}O$

スチプチシン　sutiputishin
stypticine $C_{12}H_{15}NO_4 \cdot HCl$

スチプトール　sutiputo-ru
styptol $C_6H_4(COOH)_2 \cdot (C_{12}H_{15}NO_4)_2$

スチラコール　sutirako-ru
styracol $C_{16}H_{14}O_3$

スチラシン　sutirashin
styracine $C_6H_5CH=CHCOOCH_2CH=CHC_6H_5$

ストークスの法則　suto-kusunohousoku
Stoke law

ストレッカー反応　sutorekka-hannou
Strecker reaction

ストレング石　sutorenguseki
strengite $FePO_4 \cdot H_2O$

ストレプチジン　sutorepuchijin
streptidine $C_8H_{18}N_6O_4$

ストレプトマイシン　sutoreputomaishin
streptomycin $C_{21}H_{39}N_7O_{12}$

ストレプトスライシン　sutoreputosuraishin
streptothricin $C_{19}H_{34}N_4O_8$

ストレプトーゼ　sutoreputo-ze
streptose $C_6H_{10}O_5$

ストリエゴ石　sutoriegoseki
strigovite $H_4(Fe,Mn)_2Si_2O_{11}$

ストリキニン　sutorikinin
strychnine $C_{21}H_{22}N_2O_2$

ストリキニーネ　sutorikini-ne
strychnine $C_{21}H_{22}N_2O_2$

ストリッパー　sutorippa-
stripper

ストリッピング　sutorippingu
stripping

ストリップ　sutorippu
strip

ストロファンチジン　sutorofanchijin
strophanthidin $C_{23}H_{32}O_6$

ストロファンチン　sutorofanchin
strophanthin $C_{30}H_{47}O_{12}$

ストロファントシド　sutorofantoshido
strophantohoside $C_{42}H_{64}O_{19}$

ストロファントトリオーゼ　sutorofantotorio-ze
strophanthotriose $C_{19}H_{34}O_{14}$

ストロンチウム　sutoronchiumu
strontium (Sr, element 38)

ストロンチウム鉱　sutoronchiumukou
strontianite $SrCO_3$

スツリン　*sutsurin*
　sturine $C_{36}H_{89}N_{10}O_7$

スワーン酸化　*suwa-nsanka*
　Swern oxidation

スズ, 錫　*suzu*
　tin (Sn, element 50)

スズ酸　*suzusan*
　stannic acid $H_2SnO_3/H_2Sn(OH)_6$

スズ酸塩　*suzusanen*
　stannate $M^I_2Sn(OH)_6/M^I_2SnO_3$

スズ酸カリウム　*suzusankariumu*
　pottassium stannate $K_2[Sn(OH)_6]$

スズ石　*suzuseki*
　cassiterite SnO_2

タ ta

たばこ　*tabako*
　tabacco

タブレット　*taburetto*
　tablet, pill

タフスルトン　*tafusuruton*
　naphsultone $C_{10}H_6O_3S$

タガトーゼ　*tagato-ze*
　tagatose $C_6H_{12}O_6$

タキアファルタイド　*takiafarutaido*
　tachyaphaltite $(Zr,Th)_2SiO_{10}$

タキハイドライト　*takihaidoraito*
　tachydrite $CaCl_2 \cdot 2MgCl_2 \cdot 12H_2O$

タキオール　*takio-ru*
　tachiol AgF

タキシン　*takishin*
　taxine $C_{37}H_{51}NO_{10}$

タマン石　*tamanseki*
　tamanite $3(Ca,Fe)O \cdot P_2O_5 \cdot 4H_2O$

タマルガル石　*tamarugaruseki*
　tamarugite $NaAl[SO_4] \cdot 6H_2O$

ターモナトライト　*ta-monatoraito*
　thermonatrite $Na_2CO_3 \cdot H_2O$

ターナー石　*ta-na-seki*
　turnerite $CdPO_4$

タングステン　*tangusuten*
　tungsten, wolfram (W, element 74)

タングステン酸　*tangusutensan*
　tungstic acid H_2WO_4

タングステン酸銀　*tangusutensangin*
　silver tungstate Ag_2WO_4

タングステン酸ナトリウム　*tangusutensannatoriumu*
　sodium tungstate Na_2WO_4

タンナーゼ　*tanna-ze*
　tannase

タンニゲン　*tannigen*
　tannogen $C_{18}H_{14}O_{11}$

タンニン　*tannin*
　tannin

タンニン媒染　*tanninbaisen*
　tannin mordanting

タンニン酸　*tanninsan*
　tannic acid

タンノフォルム　*tannoforumu*
　tannoform $C_{29}H_{20}O_8$

タンノン　*tannon*
　tannon $C_{48}H_{42}N_4O_{27}$

タンノゼン　*tannozen*
　tannogen $C_{18}H_{14}O_{11}$

ターンオーバー　*ta-no-ba-*
　turnover

たんぱく石　*tanpakuseki*
　opal

タンパク質　*tanpakushitsu*
　protein

タンタル　*tantaru*
　tantalum (Ta, element 73)

タンタル酸　*tantarusan*
　tantalic acid $HTaO_3$

タンタル酸塩　*tantarusanen*
　tantalate $M^ITaO_3/M^I_4Ta_2O_7/M^I_8Ta_6O_{19}$

タンタル石　*tantaruseki*
　tantalite $(Fe,Mn)((Ta,Nb)O_3)_2$

タピオ石　*tapioseki*
　tapiolite $(Fe,Mn)(Nb,Ta)_2O_6$

タラクサンチン　*tarakusanchin*
　taraxanthin $C_{40}H_{56}O_4$

タラメリ石　*tarameriseki*
　taramellite $Ba_4Fe_5Si_{10}O_{31}$

タリン　*tarin*
　thalline $C_{10}H_{13}NO$

タリング石　*taringuseki*
　tallingite $CuCl_2 \cdot 4Cu(OH)_2 \cdot H_2O$

タリウム　*tariumu*
　thallium (Tl, element 81)

タロ粘液酸　*taronenekisan*
　talomucic acid $HOOC(CHOH)_4COOH$

タロン酸　*taronsan*
　talonic acid $HOCH_2(CHOH)_4COOH$

タローゼ　*taro-ze*
　talose $HOCH_2(CHOH)_4CHO$

タール　*ta-ru*
　tar

タルク　*taruku*
　talc $Mg_3[(OH)_2/Si_4O_{10}]/$
　$3MgO \cdot 4SiO_2 \cdot H_2O$

タール酸　*ta-rusan*
　tar acid

タルトロニル尿素　*tarutoronirunyouso*
　tartronoylruea, dialuric acid $C_4H_4N_2O_4$

タルトロン酸　*tarutoronsan*
　tartronic acid $HOOCCH(OH)COOH$

タウリン　*taurin*
　taurine $H_2NCH_2CH_2SO_3H$

タヴィストック石　*tavisutokkuseki*
　tavistockite $Ca_3(PO_4)_2 \cdot 2Al(OH)_3$

たわみ性　*tawamisei*
　flexibility

テ　te

テバイン　*tebain*
　thebaine $C_{19}H_{21}NO_3$

テベニン　*tebenin*
　thebenin $C_{19}H_{21}NO_3$

テフロン　*tefuron*
　teflon $[-CF_2-CF_2-]_n$

テイン　*tein*
　theine, caffeine $C_8H_{10}O_2N_4$

テクネチウム　*tekunechiumu*
　technetium (Tc, element 43)

テクトリゲニン　*tekutorigenin*
　tectorigenin $C_{16}H_{12}O_6$

テンゲル石　*tengeruseki*
　tengerite $Y_2CO_3 \cdot nH_2O$

テオブロミン　*teoburomin*
　theobromine $C_7H_8N_4O_2$

テオブロモーゼ　*teoburomo-ze*
　theobromose $LiC_7H_7N_4O_2$

テオチン　*teochin*
　theocine $C_7H_8N_4O_2 \cdot H_2O$

テオフィリン　*teofirin*
　theophylline $C_7H_8N_4O_2$

テオナトリウム　*teonatoriumu*
　theophylline-sodium $C_7H_7N_4O_2Na$

テラコン酸　*terakonsan*
　teraconic acid
　$(CH_3)_2C=C(COOH)CH_2COOH$

テレビン油　*terebinyu*
　turpentine oil

テレフタルデヒド　*terefutarudehido*
　terephthalaldehyde $C_6H_4(CHO)_2$

テレフタル酸　*terefutarusan*
　terephthalic acid $C_6H_4(COOH)_2$

テレフタル酸ジメチル
　terefutarusanjimechiru
　dimethyl terephthalate $C_6H_4(COOCH_3)_2$

テロマー　*teroma-*
　telomer

テロメリゼーション　*teromerize-shon*
　telomerization

テルビン　*terubin*
　terpine, terpinol $C_{10}H_{20}O_2$

テルビウム　*terubiumu*
　terbium (Tb, element 65)

テルフェニル　*terufeniru*
　terphenyl C_8H_{14}

テルミエール石 *terumie-ruseki*
termierite $Al_2O_3 \cdot 6SiO_2 \cdot 18H_2O$

テルミット *terumitto*
thermite

テルミット法 *terumittohou*
thermite process

テルモジン *terumojin*
thermodin
$C_6H_4(OC_2H_5)N(COCH_3)COOC_2H_5$

テルパン *terupan*
terpane $C_{10}H_{20}$

テルペン *terupen*
terpene

テルペンアルコール *terupenaruko-ru*
terpene alkohol $C_{10}H_{17}OH$

テルペンチン *terupenchin*
turpentine

テルペニル酸 *terupenirusan*
terpenylic acid $C_8H_{12}O_4$

テルペン化学 *terupenkagaku*
terpen chemistsry

テルペノイド化合物 *terupenoidokagoubutsu*
terpenoids

テルピネン *terupinen*
terpinene $C_{10}H_{16}$

テルピネオール *terupineo-ru*
terpineol $C_{10}H_{17}OH$

テルピノレン *terupinoren*
terpinolene $C_{10}H_{16}$

テルル *teruru*
tellurium (Te, element 52)

テルルビズマス鉱 *terurubizumasukou*
tetradymite Bi_2Te_3

テルル化物 *terurukabutsu*
telluride M^I_2Te

テルル化第二水銀 *terurukadainisuigin*
mercury(II) telluride HgTe

テルル化鉛 *terurukaen*
lead telluride PbTe

テルル化銀 *terurukagin*
silver telluride Ag_2Te

テルル化カドミウム *terurukakadomiumu*
cadmium telluride CdTe

テルル酸 *terurusan*
telluric acid H_6TeO_6

テルル酸塩 *terurusanen*
tellurate $M^I_6TeO_6$

テルル酸銀 *terurusangin*
silver tellurate Ag_6TeO_6

テスト *tesuto*
test

テストステロン *tesutosuteron*
testosterone $C_{19}H_{28}O_2$

テタニン *tetanin*
tetanine $C_{13}H_{30}N_2O_4$

テトラ *tetora*
tetra

テトラアルキルアンモニウム塩 *tetoraarukiruanmoniumuen*
tetraalkyl ammonium salt R_4NX

テトラアルキルシラン *tetoraarukirushiran*
tetraalkylsilane R_4Si

テトラブロモメタン *tetoraburomometan*
carbon tetrabromide CBr_4

テトラブロムエチレン *tetoraburomuechiren*
tetrabromoethylene $Br_2C=CBr_2$

テトラチオ砒酸塩 *tetorachiohisanen*
tetrathioarsenate $M^I_3AsS_4$

テトラデカフルオロヘキサン *tetoradekafuruorohekisan*
tetradecafluorohexane C_6H_{14}

テトラデカン *tetoradekan*
tetradecane $CH_3(CH_2)_{12}CH_3$

テトラデカン酸 *tetoradekansan*
tetradecanoic acid, myristic acid
$CH_3(CH_2)_{12}COOH$

テトラデシルアルコール *tetoradeshiruaruko-ru*
tetradecyl alcohol $CH_3(CH_2)_{12}CH_2OH$

テトラドイテロ酢酸 *tetoradoiterosakusan*
tetradeuteroacetic acid CD_3COOD

テトラエチル鉛　*tetoraechirunamari*
　tetraethyllead Pb$(C_2H_5)_4$

テトラエチル錫　*tetoraechirusuzu*
　tetraethyl tin $(C_2H_5)_4$Sn

テトラエトキシシラン
　tetoraetokishishiran
　tetraethyl silicate Si$(OC_2H_5)_4$

テトラフェニルヒドラジン
　tetorafeniruhidorajin
　tetraphenylhydrazine $(C_6H_5)_2$N=N$(C_6H_5)_2$

テトラフェニルメタン
　tetorafenirumetan
　tetraphenylmethane C$(C_6H_5)_4$

テトラフルオロエチレン
　tetorafuruoroechiren
　tetrafluoroethylene F_2C=CF_2

テトラフルオロメタン
　tetorafuruorometan
　tetrafluoromethane CF_4

テトラフルオルエチレン
　tetorafuruoruechiren
　tetrafluoroethylene F_2C=CF_2

テトラフルオルエタン
　tetorafuruoruetan
　tetrafluoroethane $C_2H_2F_4$

テトラヒドロベンゾール
　tetorahidorobenzo-ru
　tetrahydrobenzene, cyclohexene C_6H_{10}

テトラヒドロフラン　*tetorahidorofuran*
　tetrahydrofuran C_4H_8O

テトラヒドロキノリン
　tetorahidorokinorin
　tetrahydroquinoline $C_9H_{11}N$

テトラヒドロナフタリン
　tetorahidoronafutarin
　tetrahydronaphthalene, tetraline $C_{10}H_{12}$

テトラジン　*tetorajin*
　tetrazine $C_2H_2N_4$

テトラカルボン酸　*tetorakarubonsan*
　tetracarboxylic acid

テトラケイ酸, テトラ珪酸　*tetorakeisan*
　tetrasilicic acid $H_4Si_4O_{13}$

テトラコンタン　*tetorakontan*
　tetracontane $C_{40}H_{82}$

テトラコサン　*tetorakosan*
　tetracosane $C_{24}H_{50}$

テトラコサン酸　*tetorakosansan*
　tetracosanoic acid $C_{23}H_{49}$COOH

テトラクロム酸塩　*tetorakuromusanen*
　tetrachromate $M^I_2Cr_4O_{13}$

テトラクロロ銅酸塩
　tetorakurorodousanen
　tetrachlorocuprate $M^I_2CuCl_4$

テトラクロロエチレン
　tetorakuroroechiren
　tetrachloroethylene C_2Cl_4

テトラクロロフルオロプロパン
　tetorakurorofuruorupuropan
　tetrachlorofluoropropane $C_3H_3Cl_4F$

テトラクロロジフルオロエタン
　tetorakurorojifuruoroetan
　tetrachlorodifluoroethane $C_2Cl_4F_2$

テトラクロロナフタレン
　tetorakuroronafutaren
　tetrachloronaphthalen $C_{10}H_4Cl_4$

テトラクロロシラン　*tetorakuroroshiran*
　silicon tetrachloride $SiCl_4$

テトラクロルベンゾール
　tetorakurorubenzo-ru
　tetrachlorbenzene $C_6H_2Cl_4$

テトラクロルエチレン
　tetorakuroruechiren
　tetrachloroethylene Cl_2C=CCl_2

テトラクロルエタン　*tetorakuroruetan*
　tetrachlorethane $C_2H_2Cl_4$

テトラクロルキノン　*tetorakurorukinon*
　tetrachloroquinone $C_6Cl_4O_2$

テトラクロル金酸　*tetorakurorukinsan*
　tetrachloroauric acid HAuCl_4・4H_2O

テトラクロルメタン　*tetorakurorumetan*
　tetrachloromethane, carbon tetrachloride CCl_4

テトラマンノシド　*tetoramannoshido*
　tetramonnoside $C_{24}H_{42}O_{21}$

テトラメチレングリコール
　tetoramechirenguriko-ru
　tetramethylene glycol HO$(CH_2)_4$OH

テトラメチレンジアミン
tetoramechirenjiamin
tetramethylenediamine $H_2N(CH_2)_4NH_2$

テトラメチルアンモニウム化合物
tetoramechiruanmoniumukagoubutsu
tetramethylammonium compound $N(CH_3)_4X$

テトラメチルベンゾール
tetoramechirubenzo-ru
tetramethylbenzene $C_6H_2(CH_3)_4$

テトラメチルビアルシン
tetoramechirubiarushin
tetramethyl diarsyl, cacodyl $(CH_3)_2AsAs(CH_3)_2$

テトラメチル鉛   *tetoramechiruen*
tetramethyllead $Pb(CH_3)_4$

テトラメチルエチレンジアミン
tetoramechirusechirenjiamin
tetramethylethylene diamine $(CH_3)_2NCH_2CH_2N(CH_3)_2$

テトラメチルシラン   *tetoramechirushiran*
tetramethylsilane $Si(CH_3)_4$

テトラメチル錫   *tetoramechirusuzu*
tetramethylstannane, tetramethyl tin $Sn(CH_3)_4$

テトラメタリン酸, テトラメタ燐酸
tetorametarinsan
tetrametaphosphoric acid $H_4P_4O_{12}$

テトラメトキシシラン
tetorametokishishiran
tetramethyl silicate $Si(OCH_3)_4$

テトラニトロメタン   *tetoranitorometan*
tetranitromethane $C(NO_2)_4$

テトラニトロトルオール
tetoranitorotoruo-ru
tetranitrotoluene, TNT $C_6H(CH_3)(NO_2)_4$

テトラブロモジフルオロプロパン
tetorapuromojifuruoropuropan
tetrabromodifluoropropane $C_3H_2F_2Br_4$

テトラリン   *tetorarin*
tetrahydronaphthalene, tetraline $C_{10}H_{12}$

テトラリン   *tetorarin*
tetralin

テトラリン酸, テトラ燐酸   *tetorarinsan*
tetraphosphoric acid $H_6P_4O_{13}$

テトラ硫酸   *tetoraryuusan*
tetrasulfuric acid $H_2S_4O_{13}$

テトラサイクリン   *tetorasaikurin*
tetracycline $C_{22}H_{24}N_2O_8$

テトラセン   *tetorasen*
tetrazene, naphthacene $C_{18}H_{12}$

テトラソン   *tetorason*
tetrazone $R_2NN=NNR_2$

テトラヨードエチレン
tetorayo-doechiren
tetraiodethylene $CI_2=CI_2$

テトラヨードピロル   *tetorayo-dopiroru*
tetraiodpyrrole C_4HNI_4

テトラヨード水銀酸塩
tetorayo-dosuiginsanen
tetraiodomercurate $M^I_2HgI_4$

テトラヨージン   *tetorayo-jin*
tetraiodine $C_{20}H_9I_4Na_2 \cdot 3H_2O$

テトラザン   *tetorazan*
tetrazane, bihydrazine $H_2N\text{-}NHNH\text{-}NH_2$

テトラゾール   *tetorazo-ru*
tetrazole CH_2N_4

テトリル   *tetoriru*
tetryl, tetralite $C_6H_2(NO_2)_3(N(CH_3)NO_2)$

テトリット   *tetoritto*
tetrite $HOCH_2(CHOH)_2CH_2OH$

テトロドトキシン   *tetorodotokishin*
tetrodotoxin $C_{11}H_{17}N_3O_3$

テトロン   *tetoron*
tetron $C_{10}H_{13}N$

テトロナール   *tetorona-ru*
tetronal $CH_3CH_2C(C_2H_5)(SO_2C_2H_5)_2$

テトロン酸   *tetoronsan*
tetronic acid $C_4H_4O_3$

テトロール   *tetoro-ru*
tetrol C_4H_4

テトロール酸   *tetoro-rusan*
tetrolic acid $CH_3CCCOOH$

テトロース, テトローゼ
tetoro-su, tetoro-ze
tetrose $C_4H_8O_4$

テイ ti

ティグリンアルデヒド　*tigurinarudehido*
　tiglaldehyde $CH_3CH=C(CH_3)CHO$

ティグリン酸　*tigurinsan*
　tiglic acid $CH_3CH=C(CH_3)COOH$

ティンカール　*tinka-ru*
　tincal $Na_2B_4O_7 \cdot 10H_2O$

ト to

トービアス酸　*to-biasusan*
　Tobias' acid $C_{10}H_6(NH_2)SO_3H$

トコフェロール　*tokofero-ru*
　tocopherol, vitamin E

トコーナル鉱　*toko-narukou*
　tocornalite $(Ag,Hg)I$

トクサフェン　*tokusafen*
　toxaphene $C_{10}H_{10}Cl_8$

トーマス法　*to-masuhou*
　Thomas process

トムスン効果　*tomusunkouka*
　Thomson effect, Kelvin effect

トムゼン石　*tomuzenseki*
　tomsenolite $NaF \cdot CaF_2 \cdot AlF_3 \cdot H_2O$

トンキノール　*tonkino-ru*
　tonquinol $C_6H(CH_3)(NO_2)_3C_4H_9$

トンネルダイオード　*tonnerudaio-do*
　tunnel diode

トンネル効果　*tonnerukouka*
　tunnelling effect

トパズ　*topazu*
　topaz $Al_2SiO_4(F,OH)_2$

トポ化学　*topokagaku*
　topochemistry

トッピング　*toppingu*
　topping

トランジスター　*toranjisuta-*
　transistor

トランキライザー　*torankiraiza-*
　tranquilizer

トランス　*toransu*
　trans

トランスアミナーゼ　*toransuamina-ze*
　transaminase

トランスフェラーゼ　*toransufera-ze*
　transferase

トランスフェリン　*toransuferin*
　transferrin

トランスグルコシダーゼ
　toransugurukoshida-ze
　transglucosidase

トランスグルタミナーゼ
　toransugurutamina-ze
　transglutaminase

トランス異性体　*toransuiseitai*
　trans isomer

トラス　*torasu*
　trass

トレエーゲルの塩基　*toree-gerunoenki*
　Troeger's base $C_{17}H_{18}N_2$

トレゲル石　*toregeruseki*
　trögerite $(UO_2)_3(AsO_4)_2 \cdot 12H_2O$

トレンス試薬　*torensushiyaku*
　Tollen's reagent

トレオ形　*toreogata*
　threo-form

トレオニン　*toreonin*
　threonine $CH_3CH(OH)CH(NH_2)COOH$

トレオース，トレオーゼ
　toreo-su, toreo-ze
　threose $OHC(CH(OH))_2CH_2OH$

トレーサー　*tore-sa-*
　tracer

トリアジン　*toriajin*
　triazine $C_3H_3N_3$

トリアコンタン　*toriakontan*
　triacontane $CH_3(CH_2)_{28}CH_3$

トリアミノベンゾール
　toriaminobenzo-ru
　triaminobenzene $C_6H_3(NH_2)_3$

トリアミノ燐酸　*toriaminorinsan*
triamidophosphoric acid $O=P(NH_2)_3$

トリアミルアミン　*toriamiruamin*
triamylamine $N(C_5H_{11})_3$

トリアルキルアルミニウム　*toriarukiruaruminiumu*
trialkylaluminium AlR_3

トリアルキル硼素　*toriarukiruhouso*
trialkylboron R_3B

トリアルキルクロルシラン　*toriarukirukurorushiran*
trialkylchlorosilane R_3SiCl

トリアルキルスルフォニウム塩　*toriarukirusurufoniumuen*
trialkylsulfonium salt R_3SX

トリアセチン　*toriasechin*
triacetin, glycerol triacetate
$(CH_3COO)_3C_3H_5$

トリアセテート繊維　*toriasete-toseni*
triacetate fiber

トリアセトアミド　*toriasetoamido*
triacetamide $(CH_3CO)_3N$

トリアセト酢酸エチルエステル
toriasetosakusanechiruesuteru
triacetoacetic ester $(CH_3CO)_3CCOOC_2H_5$

トリアザン　*toriazan*
triazane H_2NNHNH_2

トリアゼン　*toriazen*
triazene $H_2NN=NH$

トリアゾベンゾール　*toriazobenzo-ru*
triazobenzene $C_6H_5N_3$

トリアゾール　*toriazo-ru*
triazole $C_2H_3N_3$

トリベンジルアミン　*toribenjiruamin*
tribenzylamine $N(CH_2C_6H_5)_3$

トリベンジルフォスフィン
toribenjirufosufin
tribenzylphosphine $P(CH_2C_6H_5)_3$

トリブチルアミン　*toribuchiruamin*
tributylamine $N(C_4H_9)_3$

トリブチルスズフルオリド
toribuchirusuzufuruorido
tributyltin fluoride $SnF(C_4H_9)_3$

トリブチルスズラウレート
toribuchirusuzuraure-to
tributyltin laurate
$CH_3(CH_2)_{10}COOSn(C_4H_9)_3$

トリブロモジフルオロエタン
toriburomojifuruoroetan
tribromodifluoroethane $C_2HF_2Br_3$

トリブロモメタン　*toriburomometan*
tribromomethane; bromoform $CHBr_3$

トリブロモテトラフルオロプロパン
toriburomotetorafuruoropuropan
tribromotetrafluoropr opane $C_3HF_4Br_3$

トリブロムベンゾール
toriburomubenzo-ru
tribromobenzene $C_6H_3Br_3$

トリブロムエタノール
toriburomuetano-ru
tribromoethanol CBr_3CH_2OH

トリブロムヒドリン
toriburomuhidorin
tribromohydrine $BrCH_2CHBrCH_2Br$

トリブロムザロール
toriburomuzaro-ru
tribromsalol $C_6H_4(OH)COOC_6H_2Br_3$

トリチアン　*torichian*
trithiane $C_3H_6S_3$

トリチル基　*torichiruki*
trityl radical $(C_6H_5)_3C-$

トリチウム　*torichiumu*
tritium

トリデカン　*toridekan*
tridecane $C_{13}H_{28}$

トリドデカン　*toridekan*
tridecane $C_{13}H_{28}$

トリデカノン　*toridekanon*
tridecanone $C_{13}H_{26}O$

トリデカノール　*toridekano-ru*
tridecyl alcohol, tridecanol
$CH_3(CH_2)_{11}CH_2OH$

トリデカン酸　*toridekansan*
tridecanoic acid $C_{12}H_{25}COOH$

トリデシレン　*torideshiren*
tridecylene $C_{13}H_{20}$

トリデシルアルコール　*torideshiruaruko-ru*
　tridecyl alcohol, tridecanol
　$CH_3(CH_2)_{11}CH_2OH$

トリデシル酸　*torideshirusan*
　tridecanoic acid, tridecylic acid
　$CH_3(CH_2)_{11}COOH$

トリドイテロベンゾール
　toridoiterobenzo-ru
　trideuterobenzene $C_6H_3D_3$

トリドイテロ酢酸　*toridoiterosakusan*
　trideuteroacetic acid CD_3COOH

トリエチレングリコール
　toriechirenguriko-ru
　triethylene glycol $HO(C_2H_4O)_3H$

トリエチレンテトラミン
　toriechirentetoramin
　triethylene tetramine $H_2N(C_2H_4NH)_3H$

トリエチルアミン　*toriechiruamin*
　triethylamine $N(C_2H_5)_3$

トリエチルシラノール
　toriechirushirano-ru
　triethylsilanol $(C_2H_5)_3SiOH$

トリエタノールアミン　*torietano-ruamin*
　triethanolamine $N(CH_2CH_2OH)_3$

トリフェニレン　*torifeniren*
　triphenylene $C_{18}H_{12}$

トリフェニルアミン　*torifeniruamin*
　triphenylamine $N(C_6H_5)_3$

トリフェニルアルシン　*torifeniruarushin*
　triphenylarsine $As(C_6H_5)_3$

トリフェニルフォスフィン
　torifenirufosufin
　triphenylphosphine $P(C_6H_5)_3$

トリフェニルフォスフィンオキシド
　torifenirufosufinokishido
　triphenylphosphin oxide $PO(C_6H_5)_3$

トリフェニルホスフィン　*torifeniruhosufin*
　triphenylphosphine $P(C_6H_5)_3$

トリフェニルホスフィンオキシド
　torifeniruhosufinokishido
　triphenylphosphin oxide $PO(C_6H_5)_3$

トリフェニルメタン　*torifenirumetan*
　triphenylmethane $CH(C_6H_5)_3$

トリフルオロメタン　*torifuruorometan*
　trifluoromethane CHF_3

トリゲミン　*torigemin*
　trigemin $C_{17}H_{22}N_3O_2Cl_3 \cdot H_2O$

トリゲルマン　*torigeruman*
　trigermane Ge_3H_8

トリゴネリン　*torigonerin*
　trigonelline $C_7H_{12}NO_2$

トリグリセリド　*toriguriserido*
　triglyceride, glycerol triester $(RCOO)_3C_3H_5$

トリグリセリン　*toriguriserin*
　triglycerine
　$C_3H_5(OH)_2OC_3H_5(OH)OC_3H_5(OH)_2$

トリイソアミルアミン　*toriisoamiruamin*
　triisoamylamine $N(CH_2CH_2CH(CH_3)_2)_3$

トリイソブチルアミン
　toriisobuchiruamin
　triisobutylamine $N(CH_2CH(CH_3)_2)_3$

トリジン　*torijin*
　tolidine $C_6H_3(CH_3)(NH_2)$-$C_6H_3(NH_2)CH_3$

トリカルビイミド　*torikarubiimido*
　tricarbimide $C_3H_3N_3O_3$

トリカルボン酸回路　*torikarubonsankairo*
　citrate cycle, tricarboxylic acid cycle

トリケイ酸, トリ珪酸　*torikeisan*
　trisilicic acid $H_4Si_3O_8/H_8Si_3O_{10}$

トリキノイル　*torikinoiru*
　triquinoyl C_6O_6

トイコサン　*torikosan*
　tricosane $CH_3(CH_2)_{21}CH_3$

トリコサン　*torikosan*
　tricosane $CH_3(CH_2)_{11}CH_3$

トリコサノン　*torikosanon*
　tricosanone $C_{23}H_{46}O$

トリクロム酸塩　*torikuromusanen*
　trichromate $M^I_2Cr_3O_{10}$

トリクロロアセトアルデヒド
　torikuroroasetoarudehido
　trichloroacetaldehyde, chloral Cl_3CCHO

トリクロロエチレン　*torikuroroechiren*
　trichloroethylene $CHCl{=}CCl_2$

トリクロロエタン　*torikuroroetan*
　trichloroethane $C_2H_3Cl_3$

トリクロロフルオロメタン
torikurorofuruorometan
trichlorofluoromethane

トリクロロフルオロプロパン
torikurorofuruoropuropan
trichlorofluoropropane $C_3H_4Cl_3F$

トリクロロニトロメタン
torikuroronitorometan
trichloronitromethane CCl_3NO_2

トリクロロ酢酸　*torikurorosakusan*
trichloroacetic acid Cl_3CCOOH

トリクロロシラン　*torikuroroshiran*
trichlorosilane $SiHCl_3$

トリクロロテトラフルオロプロパン
torikurorotetorafuruoropuropan
trichlorotetrafluoropropane $C_3HCl_3F_4$

トリクロルアミン　*torikuroruamin*
nitrogen trichloride NCl_3

トリクロルアセトアルデヒド
torikuroruasetoarudehido
chloral, trichloroacetaldehyde CCl_3CHO

トリクロルベンゾール　*torikurorubenzo-ru*
trichlorbenzene $C_6H_3Cl_3$

トリクロルエチレン　*torikuroruechiren*
trichloroethylene $CHCl=CCl_2$

トリクロルエタン　*torikuroruetan*
trichloroethane $C_2H_3Cl_3$

トリクロルメタン　*torikurorumetan*
chloroform, trichlormethane $CHCl_3$

トリクロルニトロメタン
torikurorunitorometan
trichloronitromethane Cl_3CNO_2

トリクロル酢酸　*torikurorusakusan*
trichloroacetic acid Cl_3CCOOH

トリクロルシラン　*torikurorushiran*
trichlorosilane, silicochloroform $SiHCl_3$

トリメチレン　*torimechiren*
cyclopropane, trimethylene C_3H_6

トリメチレンブロミド
torimechirenburomido
trimethylene bromide $BrCH_2CH_2CH_2Br$

トリメチレンチアニド
torimechirenchianido
trimethylene cyanide $NC(CH_2)_3CN$

トリメチレングリコール
torimechirenguriko-ru
trimethylene glycol $HOCH_2CH_2CH_2OH$

トリメチレンイミン　*torimechirenimin*
trimethyleneimine C_3H_7N

トリメチレンジアミン
torimechirenjiamin
trimethylenediamine $H_2N(CH_2)_3NH_2$

トリメチレンクロリド
torimechirenkurorido
trimethylene chloride $ClCH_2CH_2CH_2Cl$

トリメチレンクロルヒドリン
torimechirenkuroruhidorin
trimethylene chlorohydrin
$ClCH_2CH_2CH_2OH$

トリメチレンオキシド
torimechirenokishiddo
trimethylene oxide C_3H_6O

トリメチレントリアミン
torimechirentoriamin
trimethylentriamine $(CH_2NH)_3$

トリメチルアミン　*torimechiruamin*
trimethylamine $N(CH_3)_3$

トリメチルアルミニウム
torimechiruaruminiumu
trimethylaluminium $Al(CH_3)_3$

トリメチルアルシン　*torimechiruarushin*
trimethylarsine $As(CH_3)_3$

トリメチルアセトアルデヒド
torimechiruasetoarudehido
pivalaldehyde, trimethylacetaldehyde
$(CH_3)_3CCHO$

トリメチルベンゼン　*torimechirubenzen*
trimethylbenzene $CH_3(CH_3)_3$

トリメチルフォスフィン
torimechirufosufin
trimethylphosphine $P(CH_3)_3$

トリメチルグロコーゼ
torimechiruguroko-ze
trimethylglucose $C_6H_9O_3(OCH_3)_3$

トリメチルホスファイト
torimechiruhosufaito
trimethyl phosphite $P(OCH_3)_3$

トリメチル硼素, トリメチルホウ素 *torimechiruhouso*
trimethylborane $B(CH_3)_3$

トリメチルスチビン *torimechirusutibin*
trimethylstibine $Sb(CH_3)_3$

トリメリット酸 *torimerittosan*
trimellitic acid $C_6H_3(COOH)_3$

トリメシン酸 *torimeshinsan*
trimesic acid $C_6H_3(COOH)_3$

トリニトリン *torinitorin*
trinitrin $C_2H_5(ONO_2)_3$

トリニトロベンゾール *torinitorobenzo-ru*
trinitrobenzene $C_6H_3(NO_2)_3$

トリニトロフェノール *torinitorofeno-ru*
picric acid $C_6H_2(OH)(NO_2)_3$

トリニトログリセリン *torinitoroguriserin*
trinitroglycerine $C_5H_3(ONO_2)_3$

トリニトロキシロール *torinitorokishiro-ru*
trinitroxylene $C_6H(CH_3)_2(NO_2)_3$

トリニトロクレゾール *torinitorokurezo-ru*
trinitrocresol $C_6H(OH)(CH_3)(NO_2)_3$

トリニトロメタン *torinitorometan*
trinitromethane $CH(NO_2)_3$

トリニトロセルローゼ *torinitoroseruro-ze*
trinitrocellulose $[C_6H_7O_5(NO_2)_3]x$

トリニトロトルエン *torinitorotoruen*
trinitrotoluene $C_6H_2(CH_3)(NO_2)_3$

トリオキサン *toriokisan*
trioxane $C_3H_6O_3$

トリオナール *toriona-ru*
trional $H_5C_2(CH_3)C(SO_2C_2H_5)_2$

トリオレイン *toriorein*
triolein
$(CH_3(CH_2)_7CH=CH(CH_2)_7COO)_3C_3H_5$

トリオース, トリオーゼ *torio-su, torio-ze*
triose, glycerose $HOCC(OH)HCH_2OH$

トリパンロート *toripanro-to*
trypan red $C_{32}H_{24}N_6O_{15}S_5$

トリパルミチン *toriparumichin*
tripalmitin $C_3H_5(OOCC_{15}H_{31})_3$

トリプイー石 *toripui-seki*
tripuhyite $2FeO \cdot Sb_2O_5$

トリプレット *toripuretto*
triplet

トリプロイダイド *toripuroidaido*
triploidite $(Mn,Fe)_2[OH,PO_4]$

トリプロムフェノール *toripuromufeno-ru*
tribromophenol $C_6H_2Br_3OH$

トリプロムメタン *toripuromumetan*
bromoform, tribromomethane $CHBr_3$

トリプロム酢酸 *toripuromusakusan*
tribromoacetic acid Br_3CCOOH

トリプシン *toripushin*
trypsin

トリプシノーゲン *toripushino-gen*
trypsinogen

トリプターゼ *toriputa-ze*
tryptase

トリプトファン *toriputofan*
tryptophane $C_8H_6NCH_2CH(NH_2)COOH$

トリリン酸 *toririnsan*
triphosphoric acid $H_5P_3O_{10}$

トリルアルデヒド *toriruarudehido*
tolylaldehyde $C_6H_4(CH_3)CHO$

トリル基 *toriruki*
tolyl radical

トリ硫酸 *toriryuusan*
trisulfuric acid $H_2S_3O_{10}$

トリサチン *torisachin*
trisatin $C_{26}H_{20}NO_6$

トリシアン酸 *torishiansan*
tricyanic acid $C_3H_3N_3O_3 \cdot 2H_2O$

トリシラン *torishiran*
trisilane Si_3H_8

トリステアリン *torisutearin*
tristearine $(C_{17}H_{35}COO)_3C_3H_5$

トリテルペン *toriterupen*
triterpene

トリトン *toriton*
triton

トリトール *torito-ru*
trinitrotoluene $C_6H(CH_3)_2(NO_2)_3$

トリウム　*toriumu*
　thorium (Th, element 90)

トリヨードメタン　*toriyo-dometan*
　triiodomethan, iodoform CHI_3

トロンビン　*toronbin*
　thrombine

トロンボキナーゼ　*toronbokina-ze*
　thrombokinase

トロパコカイン　*toropakokain*
　tropacocaine $C_{15}H_{19}NO_2$

トロパン　*toropan*
　tropane $C_8H_{15}N$

トロパ酸　*toropasan*
　tropic acid $C_6H_5CH(CH_2OH)COOH$

トロピジン　*toropijin*
　tropidine $C_8H_{13}N$

トロピン　*toropin*
　tropine $C_8H_{15}NO$

トロピノン　*toropinon*
　tropinone $C_8H_{13}NO$

トロピン酸　*toropinsan*
　tropinic acid $C_6H_{11}N(COOH)_2$

トロピリジン　*toropirijin*
　tropilidene C_7H_8

トロピリウムイオン　*toropiriumuion*
　tropylium ion C_7H_7

トロポン　*toropon*
　tropone C_7H_6O

トロポリン　*toroporin*
　tropoline $C_7H_{13}NO$

トロポロン　*toroporon*
　tropolone $C_7H_6O_2$

トル　*toru*
　torr

トルアミド　*toruamido*
　toluamide $C_6H_4(CH_3)CONH_2$

トルエン　*toruen*
　toluene $C_6H_5CH_3$

トルエンスルホン酸　*toruensuruhonsan*
　toluenesulfonic acid $C_6H_4(CH_3)SO_3H$

トルイジン　*toruijin*
　toluidine $C_6H_4(NH_2)CH_3$

トルイジンレッド　*toruijinsreddo*
　toluidine red $C_{17}H_{13}O_3N_3$

トルイレンブル　*toruirenburu*
　toluylene blue $C_{15}H_{19}N_2Cl$

トルイレンレッド　*toruirenreddo*
　toluylene red, neutral red $C_{15}H_{16}N_4 \cdot HCl$

トルイル酸　*toruirusan*
　toluic acid $C_6H_4(CH_3)COOH$

トルキナルジン　*torukinarujin*
　toluquinaldine $C_{11}H_{11}N$

トルキノン　*torukinon*
　toluquione $CH_3C_6H_3O_2$

トルキシン酸　*torukishinsan*
　truxinic acid $C_{16}H_{14}(COOH)_4$

トルキシル酸　*torukishirusan*
　truxillic acid $C_{16}H_{14}(COOH)_4$

トルオール　*toruo-ru*
　toluene $C_6H_5CH_3$

トルオルカルボン酸　*toruorukarubonsan*
　toluic acid $C_6H_4(CH_3)COOH$

トルオルスルフォクロリド　*toruorusurufokurorido*
　toluenesulfonyl chloride $C_6H_4(CH_3)SO_2Cl$

トルオルスルフォンアミド　*toruorusurufonamido*
　toluenesulfonamide $C_6H_4(CH_3)SO_2NH_2$

トルオルスルフォン酸　*toruorusurufonsan*
　toluenesulfonic acid $C_6H_4(CH_3)SO_3H$

トルル酸　*torurusan*
　methylhippuric acid $C_6H_4(CH_3)CONHCH_2COOH$

トルサフラニン　*torusafuranin*
　tolusafranine $C_{20}H_{19}N_4Cl$

トール油　*to-ruyu*
　tall oil

トシル化　*toshiruka*
　tosylation

トシル基　*toshiruki*
　tosyl radical $C_6H_4(CH_3)SO_2$-

トウモロコシ油　*toumorokoshiyu*
  corn oil

トウィッチェル分解　*towiccherubunkai*
  Twitchell splitting

ツヨン　*tsuyon*
  thujone $C_{10}H_{16}O$

ツゼン　*tsuzen*
  thujene $C_{10}H_{16}$

ツ　tsu

ツアイゼ塩　*tsuaizeen*
  Zeise's salt $K[PtCl_3C_2H_4]$

ツエン　*tsuen*
  thujene $C_{10}H_{16}$

ツイルアルコール　*tsuiruaruko-ru*
  thujylalcohol $C_{10}H_{17}OH$

ツジャン　*tsujan*
  thujane $C_{10}H_{18}$

ツパ酸　*tsupasan*
  tubaic acid $C_{11}H_{11}O_2COOH$

ツリシン　*tsurishin*
  turicine $C_7H_{13}NO_3 \cdot H_2O$

ツリウム　*tsuriumu*
  thulium (Tm, element 69)

ツルヴァン　*tsuruvan*
  sylvan C_5H_6O

ツソール　*tsuso-ru*
  tussol $C_{19}H_{20}NO_4$

ツトカイン　*tsutokain*
  tutocaine $C_{14}H_{22}O_2N_2 \cdot HCl$

ツウィーゼル石　*tsuwi-zeruseki*
  zwieselite $(Fe^{2+},Mn^{2+})_2[(F,OH)/PO_4]$

つや消し　*tsuyakeshi*
  delustering

つや消し面　*tsuyakeshimen*
  dull surface

つや消しペンキ　*tsuyakeshipenki*
  mat paint

つや消し剤　*tsuyakeshizai*
  delustering agent

ツヤン　*tsuyan*
  thujane $C_{10}H_{18}$

ウ　u

ウアバゲニン　*uabagenin*
  ouabagenin $C_{23}H_{34}O_8$

ウアバイン　*uabain*
  ouabain $C_{29}H_{44}O_{12} \cdot 8H_2O$

ウエーバー　*ue-ba-*
  weber

ウギ反応　*ugihannou*
  Ugi reaction

ウイルス　*uirusu*
  virus

ウマンゴ鉱　*umangokou*
  umangite, native copper selenide Cu_3Se

ウンベリフェロン　*unberiferon*
  umbelliferone $C_9H_6O_3$

ウンベル酸　*unberusan*
  umbellic acid $C_6H_3(OH)_2CH=CHCOOH$

ウンデカン　*undekan*
  undecane $C_{11}H_{24}$

ウンデカノール　*undekano-ru*
  undecanol $C_{11}H_{23}OH$

ウンデカン酸　*undekansan*
  undecanoic acid $C_{10}H_{21}COOH$

ウンデセン　*undesen*
  undecene $C_{11}H_{22}$

ウンデシル酸　*undeshirusan*
  undecylic acid $CH_3(CH_2)_9COOH$

ウンウンビウム　*ununbiumu*
  ununbium, eka-mercury (Uub, Eka-Hg, element 112)

ウンウンヘキシウム　*ununhekishiumu*
  ununhexium, eka-polonium (Uuh, Eka-Po, element 116, unconfirmed)

ウンウンヘクシウム *ununhekushiumu*
ununhexium, eka-polonium (Uuh, Eka-Po, element 116, unconfirmed)

ウンウンクアジウム *ununkuajiumu*
ununquadium, eka-lead (Uuq, Eka-Pb, element 114)

ウンウンニリウム *ununniriumu*
darmstadtium (former ununnilium, eka-platinum) (Ds, former Uun, Eka-Pt, element 110)

ウンウンペンチウム *ununpenchiumu*
ununpentium, eka-bismuth (Uup, Eka-Bi, element 115, unconfirmed)

ウンウンセプチウム *ununsepuchiumu*
ununseptium, eka-astat (Uus, Eka-At, element 117, undiscovered)

ウンウントリウム *ununtoriumu*
ununtrium, eka-thallium (Uut, Eka-Tl, element 113, undiscovered)

ウンウンウニウム *unununiumu*
roentgenium (former unununium, eka-gold) (Rg, former Uuu, Eka-Au, element 111)

ウラン *uran*
uranium (U, element 92)

ウラニウム *uraniumu*
uranium (U, element 92)

ウラン酸 *uransan*
uranic acid $UO_2(OH)_2$

ウラン酸塩 *uransanen*
uranate $M^I_2UO_4$, $M^I_2U_2O_7$

ウラシル *urashiru*
uracil $C_4H_4N_2O_2$

ウレアーゼ *urea-ze*
urease

ウレイド *ureido*
ureide $RCONHCONH_2$

ウレタン *uretan*
urethane, carbamate H_2NCOOR

ウレタンゴム *uretangomu*
urethane rubber

ウレタン樹脂 *uretanjushi*
urethane resin

ウレタンポリマー *uretanporima-*
urethane polymer

ウリジン *urijin*
uridine $C_9H_{12}N_2O_6$

ウリジル酸 *urijirusan*
uridylic acid $C_9H_{13}N_2O_9P$

ウロビリン *urobirin*
urobilin $C_{33}H_{40}N_4O_6$

ウロキナーゼ *urokina-ze*
urokinase

ウロキサンチン *urokisanchin*
uroxanthin $C_8H_6O_4NSK$

ウロン酸 *uronsan*
uronic acid $OHC(CHOH)_nCOOH$

ウロプロゴール *uropurogo-ru*
uropurgol $C_7H_6O_7(CH_2)_6N_4$

ウロシン *uroshin*
urosin $C_6H_7(OH)_4COOLi$

ウロトロピン *urotoropin*
urotropine $C_6H_{12}N_4$

ウルフェナイト *urufenaito*
wulfenite $PbMoO_4$

ウルマン反応 *urumanhannou*
Ullmann reaction

ウルマン鉱 *urumankou*
ullmannite $NiSbS$

ウルミン酸 *uruminsan*
ulmic acid $C_{20}H_{14}O_6$

ウルリッヒ石 *ururihhiseki*
ulrichite, uranite UO_2

ウルツァイト *urutsaito*
wurtzite ZnS

ウルツフィッティッヒ反応 *urutsufittihhihannou*
Wurtz-Fittig synthesis

ウルツ鉱 *urutsukou*
wurtzite ZnS

ウシン石 *ushinseki*
ussingite $Na_2[OH,AlSi_3O_8]$

ウスニン酸 *usuninsan*
usninic acid $C_{18}H_{16}O_7$

ウヴァナイト　*uvanaito*
　uvanite $2UO_3 \cdot 3V_2O_5 \cdot 15H_2O$

ウヴィチン酸　*uvichinsan*
　uvitic acid $C_6H_3(CH_3)(COOH)_2$

ウヴィトン酸　*uvitonsan*
　uvitonic acid $C_5H_2N(CH_3)(COOH)_2$

うわぐすり　*uwagusuri*
　glaze

ヴァ va

ヴァナディナイト　*vanadinaito*
　vanadinite $Pb_5[Cl/(VO_4)_3]$

ヴァナジン酸　*vanajinsan*
　vanadic acid H_3VO_4

ヴァナジン酸塩　*vanajinsanen*
　vanadate $M^I_3VO_4$

ヴァナジン酸ナトリウム　*vanajinsannatoriumu*
　sodium vanadate $Na_3VO_4 \cdot 10H_2O$

ヴァナジウム　*vanajiumu*
　vanadium (V, element 23)

ヴァニリン　*vanirin*
　vanillin $C_6H_3(OH)(OCH_3)CHO$

ヴァニリンアルデヒド　*vanirinarudehido*
　vanillin $C_6H_3(OH)(OCH_3)CHO$

ヴァニリンアルコール　*vanirinaruko-ru*
　vanillyl alcohol $C_6H_3(OH)(OCH_3)CH_2OH$

ヴァニリン酸　*vanirinsan*
　vanillic acid $C_6H_3(OH)(OCH_3)COOH$

ヴァラッハの反応　*varahhanohannou*
　Wallach reaction

ヴァレン　*vareren*
　valeren C_5H_{10}

ヴァレリアン酸　*vareriansan*
　valeric acid $CH_3(CH_2)_3COOH$

ヴァレリアン酸塩　*vareriansanen*
　valerate $CH_3(CH_2)_3COOM$

ヴァレリジン　*varerijin*
　valeridine $CH_3(CH_2)_3CONHCH_4OC_2H_5$

ヴァレリレン　*vareriren*
　valerylene $CH_3CCCH_2CH_3$

ヴァレロフェノン　*varerofenon*
　valerophenone $C_6H_5CO(CH_2)_3CH_3$

ヴァレロン　*vareron*
　valerone $C_4H_9COC_4H_9$

ヴァレロニトリル　*vareronitoriru*
　valeronitrile $CH_3(CH_2)_3CN$

ヴァレロラクタム　*varerorakutamu*
　valerolactam C_5H_9NO

ヴァレロラクトン　*varerorakuton*
　valerolactone $C_5H_8O_2$

ヴァレルアルデヒド　*vareruarudehido*
　valeraldehyde $CH_3(CH_2)_3CHO$

ヴァリドール　*varido-ru*
　validol $(CH_3)_2CHCH_2COOC_{10}H_{19}$

ヴァリン　*varin*
　valine $(CH_3)_2CHCH(NH_2)COOH$

ヴァリサン　*varisan*
　valisan $(CH_3)_2CHCHBrCOOC_{10}H_{17}$

ヴァシチン　*vashichin*
　vasicine $C_{11}H_{12}N_2O$

ヴェ ve

ヴェラトリジン　*veratorijin*
　veratridine $C_{32}H_{49}NO_9$

ヴェラトリン酸　*veratorinsan*
　veratric acid $C_6H_3(OCH_3)_2COOH$

ヴェラトロール　*veratoro-ru*
　veratrole $C_6H_4(OCH_3)_2$

ヴェラトルムアルデヒド　*veratorumuarudehido*
　veratraldehyde $C_6H_3(OCH_3)_2CHO$

ヴェラトルム酸　*veratorumusan*
　veratric acid $C_6H_3(OCH_3)_2COOH$

ヴェロナール　*verona-ru*
　veronal $C_8H_{12}N_2O_3$

ヴィ vi

ヴィクトリアブルー　*vikutoriaburu-*
　Victora blue $C_{33}H_{31}N_3 \cdot HCl$

ヴィニル　*viniru*
　vinyl $H_2C=CH-$

ヴィニルアミン　*viniruamin*
　vinylamine $H_2C=CHNH_2$

ヴィニルアルコール　*viniruaruko-ru*
　vinyl alcohol $H_2C=CHOH$

ヴィニルベンゾール　*vinirubenzo-ru*
　vinylbenzene $C_6H_5CH=CH_2$

ヴィニルエステル　*viniruesuteru*
　vinylester $RCOOCH=CH_2$

ヴィニルエーテル　*vinirue-teru*
　vinyl ether $H_2C=CHOCH=CH_2$

ヴィニル樹脂　*vinirujushi*
　vinyl resin

ヴィニル酢酸　*vinirusakusan*
　vinylacetic acid $H_2C=CHCH_2COOH$

ヴィオラキサンチン　*viorakisanchin*
　violaxanthin $C_{40}H_{56}O_4$

ヴィオルール酸　*vioru-rusan*
　violuric acid $C_4H_3N_3O_4$,

ヴィリアウム石　*viriaumuseki*
　villiaumite NaF

ヴィールス　*vi-rusu*
　virus

ヴィシン　*vishin*
　vicine $C_{10}H_{16}N_4O_7 \cdot H_2O$

ヴィスコーゼ　*visuko-ze*
　viscose

ヴィタカンファー　*vitakanfa-*
　vitacamphor $C_{10}H_{14}O_2$

ヴィタミン　*vitamin*
　vitamin

ヴィヨーム石　*viyo-museki*
　villiaumite NaF

ヴォ vo

ヴォーレル石　*vo-reruseki*
　woehlerite $Na_5Ca_{10}Nb_2Zr_3F_3Si_{10}O_{42}$

ヴォルタメーター　*vorutame-ta-*
　voltameter

ヴォルタ石　*vorutaseki*
　voltaite $5(K_2,Fe)O \cdot 2(Al,Fe)_2O_3 \cdot 10SO_3 \cdot 15H_2O$

ヴォルト　*voruto*
　volt

ヴ、ヴュ vu, vyu

ヴルピン酸　*vurupinsan*
　vulpinic acid, chrysopicrin $C_{19}H_{14}O_5$

ヴュルツフィッティッヒの合成　*vyurutsufittihhinogousei*
　Wurtz-Fittig-synthesis

ワ wa

ワード石　*wa-doseki*
　wardite $NaAl_3[(OH)_4,(PO_4)_2] \cdot 2H_2O$

ワーグナーメーヤワイン転位　*wa-guna-me-yawainteni*
　Wagner-Meerwein rearrangement

ワグネルメールバイン転位　*wagunerume-rubainteni*
　Wagner-Meerwein rearrangement

ワグネル石　*waguneruseki*
　wagnerite $Mg_3(PO_4)_2 \cdot MgF_2$

ワイセルベルグ岩　*waiseruberugugan*
　weissbergite $TlSbS_2$

ワイス石　*waisuseki*
　weissite Cu_xTe

ワッカー酸化　*wakka-sanka*
　Wacker oxidation

ワックス　*wakkusu*
　wax

ワクチン　*wakuchin*
　vaccine

ワニス　*wanisu*
　varnish

ワルデンの反転　*warudennohanten*
　Walden inversion

ワセリン　*waserin*
　paraffin jelly, vaseline

ワット　*watto*
　watt

ウエ we

ウェーバー　*we-ba-*
　weber

ウェルネライト, ウェルネル石
　weruneruito, weruneruseki
　wernerite

ウィ wi

ウィチヘン鉱　*wichihenkou*
　wittichenite $Cu_3BiS_3/3Cu_2S \cdot Bi_2S_3$

ウィレマイト　*wiremaito*
　willemite Zn_2SiO_4

ウィリアムソン合成　*wiriamusongousei*
　Williamson ether synthesis

ウィリヤマ鉱　*wiriyamakou*
　willyamite $(Co,Ni)SbS$

ウィルゲロード反応　*wirugero-dohannou*
　Willgerodt reaction

ウィルト石　*wirutoseki*
　wiluit $Ca_6(Al(OH,F)) \cdot Al_2(SiO_4)_5$

ウィッティヒ反応　*wittihihannou*
　Wittig reaction

ウィッティヒ転位　*wittihiteni*
　Wittig rearrangement

ウィット石　*wittoseki*
　wittite $5PbS \cdot 3Bi_2(S,Se)_3$

ウォ wo

ウォールチーグラー反応
　wo-ruchi-gura-hannou
　Wohl-Ziegler reaction

ウォルフキッシュナー還元
　worufukisshuna-kangen
　Wolff-Kishner reduction

ウォルフの転位　*worufunoteni*
　Wolff rearrangement

ウォルフラム　*worufuramu*
　tungsten, wolfram (W, element 74)

ヤ ya

ヤコベゼン反応　*yakobezenhannou*
　Jacobsen reaction

ヤンゴニン　*yangonin*
　yangonin $C_{15}H_{14}O_4$

ヤーングリース　*ya-nguri-su*
　yarn grease

ヤング率　*yanguritsu*
　Young's modulus

ヤーンテラー効果　*ya-ntera-kouka*
　Jahn-Teller effect

ヤパコニチン　*yapakonichin*
　japaconitine $C_{34}H_{47}NO_{11}$

ヤパコニン　*yapakonin*
　japaconine $C_{26}H_{41}NO_{10}$

ヤラピン酸　*yarapinsan*
　jalapic acid $C_{17}H_{30}O_9$

ヤラヤラ　*yarayara*
　yara-yara $C_{11}H_{10}O$

ヤトレン　*yatoren*
　yatren, ferron $C_9H_6INO_4S$

ヨ yo

ヨーチオン　*yo-chion*
　iothion $CH_2ICH(OH)CH_2I$

ヨーダル　*yo-daru*
　iodal Cl_3CHO

ヨード　*yo-do*
　iodine I (element 53)

ヨードアミル　*yo-doamiru*
　iodamyl $C_5H_{11}I$

ヨードアンチフェブリン　*yo-doanchifeburin*
　jodantifebrin C_8H_8NOI

ヨードアニソール　*yo-doaniso-ru*
　jodanisol $C_6H_4(CH_3)I$

ヨードベンゾール, ヨードベンゼン　*yo-dobenzo-ru, yo-dobenzen*
　iodobenzene C_6H_5I

ヨードチンキ　*yo-dochinki*
　iodine tincture

ヨードエオシン　*yo-doeoshin*
　iodeosine $C_{20}H_6O_5I_4Na_2$

ヨードエタン　*yo-doetan*
　ethyl iodide C_2H_5I

ヨードフェン　*yo-dofen*
　iodophen $C_{20}H_{10}O_4I_4$

ヨードフォルム　*yo-doforumu*
　iodoform CHI_3

ヨードゴルゴ酸　*yo-dogorugosan*
　iodogorgoic acid
　$C_6H_2I_2(OH)CH_2CH(NH_2)COOH$

ヨードホルマール　*yo-dohoruma-ru*
　iodoformal $C_6H_{12}N_4 \cdot C_2H_5I \cdot CHI_3$

ヨードホルミン　*yo-dohorumin*
　iodoformin $C_6H_{12}N_4 \cdot CHI_3$

ヨードホルム　*yo-dohorumu*
　iodoform CHI_3

ヨードホルム反応　*yo-dohorumuhannou*
　iodoform reaction

ヨード化　*yo-doka*
　iodination, iodization

ヨードカリ, ヨードカリウム　*yo-dokari, yo-dokariumu*
　potassium iodide KI

ヨードメタン　*yo-dometan*
　methyl iodide CH_3I

ヨードニウム塩基　*yo-doniumuenki*
　iodonium base

ヨードピリン　*yo-dopirin*
　iodopyrin $C_{11}H_{11}N_2OI$

ヨードリン　*yo-dorin*
　iodoline $C_9H_7N \cdot CH_3Cl \cdot ICl$

ヨドール　*yodo-ru*
　iodol C_4HNI_4

ヨード酢酸　*yo-dosakusan*
　iodoacetic acid ICH_2COOH

ヨードゾール　*yo-dozo-ru*
　iodozol $C_6H_2I_2(OH)(SO_3H)$

ヨハンゼン石　*yohanzenseki*
　johannsenite $MnCa(SiO_3)_2$

ヨヒンビン　*yohinbin*
　yohimbine $C_{21}H_{26}N_2O_3$

ヨージヴァル　*yo-jivaru*
　jodival $(CH_3)_2CHICONHCONH_2$

ヨーネン　*yo-nen*
　ionene $C_{13}H_{18}$

ヨオロピウム　*yooropiumu*
　europium Eu (element 63)

ヨウ化エチル　*youkaechiru*
　ethyl iodide C_2H_5I

ヨウ化銀　*youkagin*
　silver iodide AgI

ヨウ化カリウム　*youkakariumu*
　potassium iodide KI

ヨウ化カルシウム　*youkakarushiumu*
　calcium iodide CaI_2

ヨウ化メチル　*youkamechiru*
　methyl iodide CH_3I

ヨウ化水素酸　*youkasuisosan*
　hydriodic acid HI

ヨウ酸　*yousan*
　hydroiodic acid HI

ヨウ素, 沃素　*youso*
　iodine (I, element 53)

ヨウ素酸, 沃素酸　*yousosan*
　iodic acid HIO_3

ヨウ素酸銀　*yousosangin*
　silver iodate $AgIO_3$

ヨウ素酸カリウム　*yousosankariumu*
　potassium iodate KIO_3

ヨウ素酸ナトリウム　*yousosannatoriumu*
　sodium iodate $NaIO_3$

ヨーゾール　*yo-zo-ru*
　iosol $(C_6H_2(CH_3)(OI)(C_3H_7))_2$

ユ　yu

ユビキノン　*yubikinon*
　ubiquinone $C_{59}H_{90}O_4$ (coenzyme Q)

ユグロン　*yuguron*
　juglone $C_{10}H_6O_3$

ユージアライト　*yu-jiaraito*
　eudyalite $(Na,K,H)_{13}(Ca,Fe)_6(Si,Zr)_{20}O_{52}Cl$

ユーコライト　*yu-koraito*
　eucolite $(Na,K,H)_{13}(Ca,Fe)_6(Si,Zr)_{20}O_{52}Cl$

ユークラス　*yu-kurasu*
　euclase $Be(AlOH)SiO_4$

ユークリプタイト　*yu-kuriputaito*
　eukryptite $LiAlSiO_4$

ユークロイト　*yu-kuroito*
　euchroite $Cu_3(AsO_4)_2 \cdot Cu(OH)_2 \cdot 6H_2O$

ユークセイメイト　*yu-kuseimeito*
　euxenite $(Y,Ca,Ce)(Nb,Ta,Ti)_2O_6$

ゆらぎ　*yuragi*
　fluctuation

ゆらぎ電流　*yuragidenryuu*
　fluctuating current

ユウロピウム　*yu-ropiumu*
　europium Eu (element 63)

ユーロピウム　*yu-ropiumu*
　europium Eu (element 63)

ザ　za

ザビノール　*zabino-ru*
　sabinol $C_{10}H_{16}O$

ザブロミン　*zaburomin*
　sabromine $C_{44}H_{82}O_4Br_4Ca$

ザンドマイヤー反応　*zandomaiya-hannou*
　Sandmeyer reaction

ザンソフィライト　*zansofiraito*
　xanthophyllite $H_8(Mg,Ca)_{14}Al_{16}Si_5O_{52}$

ザンソコン鉱　*zansokonkou*
　xanthocon $3Ag_2S \cdot As_2S_3$

ザリブロミン　*zariburomin*
　salibromin $C_6H_2Br_2(OH)COOCH_3$

ザロフェン　*zarofen*
　salophene
　$C_6H_4(OH)COOC_6H_4NH(COCH_3)$

ザロール　*zaro-ru*
　salol $C_6H_4(OH)COOC_6H_5$

ザルコシン　*zarukoshin*
　sarcosine CH_3NHCH_2COOH

ゼ　ze

ゼアキサンチン　*zeakisanchin*
　zeaxanthin $C_{40}H_{56}O_2$

ゼーベック効果　*ze-bekkukouka*
　Seebeck effect, thermoelectric effect

ゼファロウィッチ石　*zefarowicchiseki*
　zepharovichite $AlPO_4 \cdot 3H_2O$

ゼーマン効果　*ze-mankouka*
　Zeeman effect

ゼノン　*zenon*
　xenon (Xe, element 54)

ゼオライト　*zeoraito*
　zeolite

ゼラチン　*zerachin*
　gelatin

ゼラチン化　*zerachinka*
　gelation, gelling

ゼリー　*zeri-*
　jelly

ゼロ点　*zeroten*
　zero point

ゾ zo

ゾイネル石　*zoineruseki*
　zeunerite Cu[UO$_2$,AsO$_4$]$_2$·12H$_2$O

ゾーン電気泳動　*zo-ndenkieidou*
　zone electrophoresis

ゾノトラ石　*zonotoraseki*
　xonotlite 5CaSiO$_3$·H$_2$O

ゾル　*zoru*
　sol

ズ zu

ズブクチン　*zubukuchin*
　subcutin C$_{15}$H$_{18}$NO$_6$S

ずり速度　*zurisokudo*
　shear rate

ズルチン　*zuruchin*
　dulcin C$_2$H$_5$OC$_6$H$_4$NHCONH$_2$

ズルフォキシド　*zurufokishido*
　sulfoxide R$_2$SO

ズルフォキシル酸　*zurufokishirusan*
　sulfoxylic acid S(OH)$_2$

ズルフォン　*zurufon*
　sulfone R$_2$SO$_2$

ズルフォニル基　*zurufoniruki*
　sulfonyl -SO$_2$-

ズルフォニウム塩　*zurufoniumuen*
　sulfonium salt

ズルフォニウム塩基　*zurufoniumuenki*
　sulfonium base

ズルフォラン　*zuruforan*
　sulfolane C$_4$H$_8$SO$_2$

ズルフォヴィン酸　*zurufovinsan*
　sulfovinic acidsàure C$_2$H$_5$(HSO$_4$)

8
Dictionary Part II: Scientific Terms Beginning with Basic *kanji*

For explanations see Chapter 6.

8.1
Scientific Terms Beginning with *kanji* for Figures and Quantities

---
一 one

一原子分子　*ichgenshibunshi*
monoatomic molecule

一分子反応　*ichibunshihannou*
monomolecular reaction

一分子層　*ichibunshisou*
monomolecular layer

一塩化アセトン　*ichienkaaseton*
chloroacetone CH_3COCH_2Cl

一塩化物　*ichienkabutsu*
monochloride

一塩化硫黄　*ichienkaiou*
sulfur monochloride S_2Cl_2

一塩化ジルコニウム　*ichienkajirukoniumu*
zirconium(I) chloride $ZrCl$

一塩化沃素　*ichienkayouso*
iodine chloride ICl

一塩基酸　*ichienkisan*
monobasic acid

一方向弁　*ichihoukouben*
one-way valve

一方向中継器　*ichihoukouchuukeiki*
outgoing unit

一方向中継線　*ichihoukouchuukeisen*
one-way trunk

一次電池　*ichijidenchi*
primary cell

一次反応　*ichijihannou*
first-order reaction

一時硬度　*ichijikoudo*
temporary hardness

一次構造　*ichijikouzou*
primary structure

一重項酸素　*ichijuukousanso*
singlet oxygen

一重線　*ichijuusen*
singlet

一燐酸　*ichirinsan*
monophosphoric acid, phosphoric acid H_3PO_4

一硫化炭素　*ichiryuukatanso*
carbon monosulfide CS

一酸化イッテルビウム　*ichisankaitterubiumu*
ytterbium monoxide YbO

Japanese-English Chemical Dictionary. Edited by Markus Gewehr
Copyright © 2008 WILEY-VCH Verlag GmbH & Co. KGaA, Weinheim
ISBN: 978-3-527-31293-1

一水化物 *ichisuikabutsu*
monohydrate

一価アルコール *ikkaaruko-ru*
monohydric alcohol

一回量 *ikkairyou*
single dose

一価の *ikkano*
monovalent, monohydric

一半 *ippan*
sesqui

一酸塩基 *issanenki*
monoacid base

一酸化物 *issankabutsu*
monoxide

一酸化窒素 *issankachisso*
nitrogen monoxide NO

一酸化チタン *issankachitan*
titanium monoxide TiO

一酸化鉛 *issankaen*
lead(II) oxide PbO

一酸化白金 *issankahakkin*
platinum monoxide PtO

一酸化珪素 *issankakeiso*
silicon monoxide SiO

一酸化二窒素 *issankanichisso*
nitrous oxide N_2O

一酸化ニッケル *issankanikkeru*
nickel(II) oxide NiO

一酸化ルビジウム *issankarubijiumu*
rubidium(I) oxide Rb_2O_2

一酸化錫 *issankasuzu*
stannous oxide SnO

一酸化炭素 *issankatanso*
carbon monoxide CO

一酸化炭素中毒 *issankatansochuudoku*
carbon monoxide poisoning

一酸化鉄 *issankatetsu*
iron(II) oxide FeO

一定温度 *itteiondo*
constant temperature

二 two

二亜ホスホン酸 *niahosuhonsan*
diphosphonous acid $H_4P_2O_3$

二亜硫酸 *niaryuusan*
disulfurous acid $H_2S_2O_5$

二亜硫酸塩 *niaryuusanen*
disulfite $M^I_2S_2O_5$

二亜硫酸カリウム *niaryuusankariumu*
potassium disulfite $K_2S_2O_5$

二亜硫酸ナトリウム *niaryuusannatoriumu*
sodium disulfite $Na_2S_2O_5$

二分子反応 *nibunshihannou*
bimolecular reaction

二分子過程 *nibunshikatei*
bimolecular process

二分子系 *nibunshikei*
bimolecular system

二分子求核置換 *nibunshikyuukakuchikan*
bimolecular nucleophilic substitution

二分子層 *nibunshisou*
bilayer

二チオン酸バリウム *nichionsanbariumu*
barium dithionate $BaS_2O_6 \cdot 2H_2O$

二チオン酸塩 *nichionsanen*
dithionate $M^I_2S_2O_6$

二チオン酸ナトリウム *nichionsannatoriumu*
sodium dithionate,
sodium disulfate(V) $Na_2S_2O_6$

二チオ炭酸 *nichiotansan*
dithiocarbonic acid,
xanthogenic acid HOC(S)SH

二窒素配位子 *nichissohaiishi*
dinitrogen ligand

二窒素錯体 *nichissosakutai*
dinitrogen complex

二段階 *nidankai*
two-step

二段階変換 *nidankaihenkan*
two-step conversion

二段階加水分解　nidankaikasuibunkai
two-step hydrolysis

二段階過程　nidankaikatei
two-step process

二電子反応　nidenshihannou
two-electron reaction

二電子還元　nidenshikangen
two-electron reduction

二塩化チタン　nienkachitan
titanium dichloride $TiCl_2$

二塩化白金　nienkahakkin
platinous chloride $PtCl_2$

二塩化硫黄　nienkaiou
sulfur dichloride SCl_2

二塩化ジルコニウム　nienkajirukoniumu
zirconium(II) chloride $ZrCl_2$

二塩化プロピレン　nienkapuropiren
propylene dichloride $CH_3CHClCH_2Cl$

二塩化錫　nienkasuzu
stannous chloride $SnCl_2$

二塩基酸　nienkisan
dibasic acid

二塩基性リン酸カリウム
nienkiseirinsankariumu
dibasic potassium phosphate K_2HPO_4

二塩基性リン酸ナトリウム
nienkiseirinsannatoriumu
disodium hydrogen phosphate Na_2HPO_4

二フッ化二酸素, 二弗化二酸素
nifukkanisanso
dioxygen difluoride O_2F_2

二元液体　nigenekitai
binary liquid

二元合金　nigengoukin
binary alloy

二元化合物　nigenkagoubutsu
binary compound

二元系　nigenkei
binary system

二元機能触媒　nigenkinoushoukubai
bifunctional catalyst

二元金属クラスタ　nigenkinzokukurasuta
bimetallic cluster

二元金属硫化物　nigenkinzokuryuukabutsu
binary metal sulfide

二元錯体　nigensakutai
binary complex

二原子　nigenshi
diatomic

二原子分子　nigenshibunshi
diatomic molecule

二原子酸素　nigenshisanso
dioxygen

二元溶媒　nigenyoubai
binary solvent

二ホウ化物, 二硼化物　nihoukabutsu
diboride

二ホウ化チタン, 二硼化チタン
nihoukachitan
titanium diboride TiB_2

二ホウ化クロム, 二硼化クロム
nihoukakuromu
chromium diboride CrB_2

二次元ゲル分析　nijigengerubunseki
two-dimensional gel analysis

二次元磁化　nijigenjika
two-dimensional magnetization

二次元重合　nijigenjuugou
two-dimensional polymerization

二次元結晶構造　nijigenkesshoukouzou
two-dimensional crystal structure

二次元クロマトグラフィー
nijigenkuromatogurafi-
two-dimensional chromatography

二次反応　nijihannou
second order reaction

二次効果　nijikouka
secondary effect

二次構造　nijikouzou
secondary structure

二次生成物　nijiseiseibutsu
secondary product

二次転移　nijiteni
second order transition

二重アルキル化　nijuuarukiruka
double alkylation

二重置換 *nijuuchikan*
double substitution

二重反転 *nijuuhanten*
twofold inversion

二重イオン化 *nijuuionka*
double ionization

二重イオン交換 *nijuuionkoukan*
double-ion-exchange

二重異性化 *nijuuiseika*
twofold isomerization

二重環化 *nijuukanka*
twofold cyclization

二重結合 *nijuuketsugou*
double bond

二重結合異性化 *nijuuketsugouiseika*
double-bond isomerization

二重結合系 *nijuuketsugoukei*
double bond system

二重結合化合物 *nijuuketsukagoubutsu*
double-bond compound

二重極 *nijuukyoku*
dipole

二重プロトン化 *nijuupurotonka*
double protonation

二重らせん *nijuurasen*
double helix

二重らせん構造 *nijuurasenkouzou*
double helical structure

二重露光 *nijuurokou*
double exposure

二重鎖 *nijuusa*
double chain

二重水酸化物 *nijuusankabutsu*
double hydroxide

二重線 *nijuusen*
doublet

二重層 *nijuusou*
double layer

二重相 *nijuusou*
double phase

二重水素 *nijuusuiso*
dihydrogen

二価アルコール *nikaaruko-ru*
dihydric alcohol, diol

二価塩基 *nikaenki*
divalent base

二価フェノール *nikafeno-ru*
dihydric phenol

二価イオン *nikaion*
bivalent ion

二価金属 *nikakinzoku*
divalent metal

二価金属錯体 *nikakinzokusakutai*
divalent metal complex

二核 *nikaku*
binuclear

二核銅錯体 *nikakudousakutai*
binuclear copper complex

二核化合物 *nikakukagoubutsu*
dinuclear compound

二核金属錯体 *nikakukinzokusakutai*
binuclear metal complex

二核ニッケル錯体 *nikakunikkerusakutai*
binuclear nickel complex

二核パラジウム錯体 *nikakuparajiumusakutai*
binuclear palladium complex

二核ロジウム錯体 *nikakurojiumusakutai*
dinuclear rhodium complex

二核ルテニウム錯体 *nikakuruteniumusakutai*
binuclear ruthenium complex

二核錯体 *nikakusakutai*
binuclear complex

二官能性分子 *nikannouseibunshi*
bifunctional molecule

二官能性配位子 *nikannouseihaiishi*
bifunctional ligand

二官能性化合物 *nikannouseikagoubutsu*
bifunctional compound

二官能性モノマ *nikannouseimonoma*
bifunctional monomer

二官能性触媒 *nikannouseishokubai*
bifunctional catalyst

二価の　*nikano*
  bivalent, dihydric

二環ラクタム　*nikanrakuktam*
  bicyclic lactam

二環ラクトン　*nikanrakuton*
  bicyclic lactone

二環性複素環化合物
  nikanseifukusokankagoubutsu
  biheterocycle

二環式化合物　*nikanshikikagoubutsu*
  bicyclic compound

二価パラジウム　*nikaparjiumu*
  divalent palladium

二価酸　*nikasan*
  divalent acid

二価遷移金属錯体　*nikasenikinzokusakutai*
  divalent transition metal complex

二ケイ酸リチウムガラス
  nikeisanrichiumugarasu
  lithium disilicate glass

二金属錯体　*nikinzokusakutai*
  bimetallic complex

二コバルトオクタカルボニル
  nikobarutookutakaruboniru
  dicobalt octacarbonyl

二コバルト錯体　*nikobarutosakutai*
  dicobalt complex

二クロム錯体　*nikuromusakutai*
  dichromium complex

二クロム酸アンモニウム
  nikuromusananmoniumu
  ammonium dichromate $(NH_4)_2Cr_2O_7$

二クロム酸　*nikuromusanen*
  dichromic acid $H_2Cr_2O_7$

二クロム酸塩　*nikuromusanen*
  dichromate $M^I_2Cr_2O_7$

二クロム酸カリウム
  nikuromusankariumu
  potassium dichromate $K_2Cr_2O_7$

二クロム酸ナトリウム
  nikuromusannatoriumu
  sodium dichromate $Na_2Cr_2O_7 \cdot 2H_2O$

二クロム酸ピリジニウム
  nikuromusanpirijiniumu
  pyridinium dichromate $(C_5H_5N)_2H_2Cr_2O_7$

二クロム酸リチウム
  nikuromusanrichiumu
  lithium dichromate $Li_2Cr_2O_7$

二極性の　*nikyokuseino*
  dipolar

二級アミド　*nikyuuamido*
  secondary amide

二燐酸, 二リン酸　*nirinsan*
  pyrophosphoric acid,
  diphosphoric acid $H_4P_2O_7$

二燐酸（Ⅰ）　*nirinsan(I)*
  diphosphonous acid $H_4P_2O_3$

二燐酸（Ⅱ）　*nirinsan(II)*
  hypodiphosphorous acid $H_4P_2O_4$

二燐酸（Ⅱ,Ⅳ）　*nirinsan(II,IV)*
  diphosphoric(II,IV) acid $H_4P_2O_5$

二燐酸（Ⅲ,Ⅳ）　*nirinsan(III,IV)*
  diphosphoric(III,IV) acid $H_4P_2O_6$

二燐酸（Ⅲ,Ⅴ）　*nirinsan(III,V)*
  isohypophosphoric acid,
  diphosphoric(III,V) acid $H_4P_2O_6$

二燐酸（Ⅳ）　*nirinsan(IV)*
  hypodiphosphoric acid $H_4P_2O_6$

二リン酸亜鉛　*nirinsanaen*
  zinc phosphate $Zn_3(PO_4)_2$

二燐酸カリウム　*nirinsankariumu*
  potassium diphosphate $K_4P_2O_7$

二燐酸ナトリウム　*nirinsannatoriumu*
  sodium pyrophosphate $Na_4P_2O_7$

二量化　*niryouka*
  dimerization

二量化生成物　*niryoukaseiseibutsu*
  dimerization product

二量体　*niryoutai*
  dimer

二量体化　*niryoutaika*
  dimerization

二量体化合物　*niryoutaikagoubutsu*
  dimer compound

二量体解離 *niryoutaikairi*
dimer dissociation

二量体形成 *niryoutaikeisei*
dimer formation

二量体結合 *niryoutaiketsugou*
dimer bond

二量体錯体 *niryoutaisakutai*
dimeric complex

二硫化物 *niryuukabutsu*
disulfide $M^I_2S_2$

二硫化白金 *niryuukahakkin*
platinum disulfide PtS_2

二硫化三ニッケル *niryuukasannikkeru*
nickel sulfide Ni_3S_2

二硫化セレン *niryuukaseren*
selenium sulfide SeS_2

二硫化炭素 *niryuukatanso*
carbon disulfide CS_2

二硫酸 *niryuusan*
disulfuric acid $H_2S_2O_7$

二硫酸塩 *niryuusanen*
disulfate $M^I_2S_2O_7$

二硫酸カリウム *niryuusankariumu*
potassium disulfate $K_2S_2O_7$

二酸塩基 *nisanenki*
diacid base

二酸化物 *nisankabutsu*
dioxide

二酸化窒素 *nisankachisso*
nitrogen dioxide NO_2

二酸化チタン *nisankachitan*
titanium dioxide TiO_2

二酸化チタン顔料 *nisankachitanganryou*
titanium dioxide pigment

二酸化塩素 *nisankaenso*
chlorine dioxide ClO_2

二酸化塩素漂白 *nisankaensohyouhaku*
chlorine dioxide bleaching

二酸化硫黄 *nisankaiou*
sulfur dioxide SO_2

二酸化硫黄吸収 *nisankaioukyuushuu*
sulfur dioxide absorption

二酸化ジルコニウム *nisankajirukoniumu*
zirconium(IV) oxide ZrO_2

二酸化ケイ素, 二酸化珪素 *nisankakeiso*
silicon dioxide SiO_2

二酸化マンガン *nisankamangan*
manganese(IV) oxide MnO_2

二酸化鉛 *nisankanamari*
lead dioxide PbO_2

二酸化ニッケル *nisankanikkeru*
nickel dioxide NiO_2

二酸化セレン *nisankaseren*
selenium(IV) oxide SeO_2

二酸化セリウム *nisankaseriumu*
cerium dioxide CeO_2

二酸化炭素 *nisankatanso*
carbon dioxide CO_2

二酸化炭素固定 *nisankatansokotei*
carbon dioxide fixation

二酸化炭素吸着 *nisankatansokyuuchaku*
carbon dioxide adsorption

二酸化炭素濃度 *nisankatansonoudo*
carbon dioxide concentration

二酸化炭素錯体 *nisankatansosakutai*
carbon dioxide complex

二酸化テルル *nisankateruru*
tellurium dioxide TeO_2

二酸化ウラン *nisankauran*
uranous oxide UO_2

二酸素錯体 *nisansosakutai*
dioxygen complex

二成分化合物 *niseibunkagoubutsu*
binary compound

二成分系 *niseibunkei*
two-phase-system, binary system

二成分混合系 *niseibunkongoukei*
two component mixed system

二成分触媒 *niseibunshokubai*
binary catalyst

二色性 *nishokusei*
dichroism

二臭化物 *nishuukabutsu*
dibromide

二臭素化 *nishuusoka*
dibromination

二相分離装置 *nisoubunrisouchi*
two-phase separator

二相法 *nisouhou*
two-phase process

二相系触媒 *nisoukeishokubai*
two-phase catalyst

二相混合 *nisoukongou*
two-phase mixture

二層構造 *nisoukouzou*
bilayer structure

二相溶液 *nisouyoueki*
biphasic solution

二水化物 *nisuikabutsu*
dihydrate

二水素化物 *nisuisokabutsu*
dihydride M^IH_2, $M^{II}H_2$

二糖 *nitou*
disaccharide

二糖類誘導体 *nitouruidoutai*
disaccharide derivative

二糖類フラグメント
nitouruifuragumento
disaccharide fragment

二糖類単位 *nitouruitani*
disaccharide unit

二ヨウ化物, 二沃素化物 *niyoukabutsu*
diiodide

二座配位子 *nizahaiishi*
bidentate ligand

二座ホスフィン配位子
nizahosufinhaiishi
bidentate phosphine ligand

三 three

三分子反応 *sanbunshihannou*
trimolecular reaction

三弗化塩素 *sandaikaenso*
chlorine trifluoride ClF_3

三弗化燐, 三フッ化リン *sandaikarin*
phosphorous trifluoride PF_3

三塩化アンチモン *sanenkaanchimon*
antimony trichloride $SbCl_3$

三塩化ビスマス *sanenkabisumasu*
bismuth trichloride $BiCl_3$

三塩化窒素 *sanenkachisso*
nitrogen trichloride NCl_3

三塩化チタン *sanenkachitan*
titanium trichloride $TiCl_3$

三塩化ヒ素, 三塩化砒素 *sanenkahiso*
arsenic trichloride $AsCl_3$

三塩化ホウ素, 三塩化硼素
sanenkahouso
boron trichloride BCl_3

三塩化ジルコニウム *sanenkajirukoniumu*
zirconium(III) chloride $ZrCl_3$

三塩化クロム *sanenkakukromu*
chromium trichloride $CrCl_3$

三塩化リン *sanenkarin*
phosphorous trichloride PCl_3

三塩化燐 *sanenkarin*
phosphorous trichloride PCl_3

三塩化ヴィニル *sanenkaviniru*
vinyl trichloride $CHCl_2CH_2Cl$

三塩化沃素 *sanenkayouso*
iodine trichloride ICl_3

三塩基酸 *sanenkisan*
tribasic acid

三フッ化アンチモン *sanfukkaanchimon*
antimony trifluoride SbF_3

三弗化窒素, 三フッ化窒素
sanfukkachisso
nitrogen trifluoride NF_3

三フッ化ヒ素, 三フッ化砒素
sanfukkahiso
arsenic trifluoride AsF_3

三フッ化ホウ素, 三弗化硼素
sanfukkahouso
boron trifluoride BF_3

三フッ化ホウ素エーテル錯塩
sanfukkahousoe-terusakuen
boron trifluoride ethyl ether complex

三元分塩 *sangenbunen*
ternary salt

三元系 *sangenkei*
ternary system

三原子の *sangenshino*
triatomic

三ホウ化クロム, 三硼化クロム
sanhoukakuromu
chromium triboride Cr_3B_2

三員環 *saninkan*
three-membered ring

三次元構造 *sanjigenkouzou*
three-dimensional structure

三次反応 *sanjihannou*
third order reaction

三次構造 *sanjikouzou*
tertiary structure

三重結合 *sanjuuketsugou*
triple bond

三重線 *sanjuusen*
triplet

三重水素 *sanjuusuiso*
tritium

三価アルコール *sankaaruko-ru*
trivalent alcohol

三角フラスコ *sankakufurasuko*
Erlenmeyer flask

三価の *sankano*
trivalent, trihydric

三環式化合物 *sankanshikikagoubutsu*
tricyclic compound

三脚 *sankyaku*
tripod

三極管 *sankyokukan*
triode

三メタ燐酸, 三メタリン酸
sanmetarinsan
trimetaphosphoric acid $H_3P_3O_9$

三二硫化チタン *sanniryuukachitan*
titanium sesquisulfide Ti_2S_3

三二酸化物 *sannisankabutsu*
sesquioxide

三二酸化チタン *sannisankachitan*
titanium sesquioxide Ti_2O_3

三二酸化白金 *sannisankahakkin*
platinum sesquioxide Pt_2O_3

三二酸化ニッケル *sannisankanikkeru*
nickel sesquioxide Ni_2O_3

三二酸化オスミウム
sannisankaosumiumu
osmium sesquioxide Os_2O_3

三二酸化硫酸 *sannisankaryuusan*
sulfur sesquioxide S_2O_3

三二酸化鉄 *sannisankatetsu*
iron(III) oxide Fe_2O_3

三方晶系 *sanpoushoukei*
trigonal system

三燐酸 *sanrinsan*
triphosphoric acid $H_5P_3O_{10}$

三量体 *sanryoutai*
trimer

三硫化アンチモン *sanryuukaanchimon*
antimony trisulfide Sb_2S_3

三硫化ビスマス *sanryuukabisumasu*
bismuth trisulfide Bi_2S_3

三硫化物 *sanryuukabutsu*
trisulfide, tersulfide

三硫化ヒ素, 三硫化砒素 *sanryuukahiso*
arsenic trisulfide As_2S_3

三硫化二リン, 三硫化二燐
sanryuukanirin
phosphorous trisulfide P_2S_3

三硫化リン, 三硫化燐 *sanryuukarin*
phosphorous trisulfide P_2S_3

三酢酸グリセリン *sansakusanguriserin*
glyceryl triacetate $(CH_3COO)_3C_3H_5$

三酸塩基 *sansanenki*
triacid base

三酸化アンチモン *sansankaanchimon*
antimonous oxide Sb_2O_3

三酸化ビスマス *sansankabisumasu*
bismuth trioxide Bi_2O_3

三酸化物 *sansankabutsu*
trioxide

三酸化窒素 *sansankachisso*
nitrous acid anhydride, nitrogen trioxide N_2O_3

三酸化チタン *sansankachitan*
titanium peroxide TiO_3

三酸化白金 *sansankahakkin*
platinum trioxide PtO_3

三酸化ヒ素, 三酸化砒素 *sansankahiso*
arsenic trioxide As_2O_3

三酸化硫黄 *sansankaiou*
sulfur trioxide, sulfuric acid anhydride SO_3

三酸化クロム *sansankakuromu*
chromium trioxide CrO_3

三酸化モリブデン *sansankamoribuden*
molybdenum trioxide MoO_3

三酸化二窒素 *sansankanichisso*
nitrous acid anhydride, dinitrogen trioxide N_2O_3

三酸化二ホウ素, 三酸化二硼素 *sansankanihouso*
boron trioxide B_2O_3

三酸化リン, 三酸化燐 *sansankarin*
phosphorous(III) oxide P_2O_3/P_4O_6

三酸化セレン *sansankaseren*
selenium trioxide SeO_3

三酸化テルル *sansankateruru*
tellurium trioxide TeO_3

三斜晶系 *sanshashoukei*
triclinic system

三色性 *sanshokusei*
trichroism

三硝酸グリセリン *sanshousanguriserin*
glycerol trinitrate $C_3H_5(ONO_2)_3$

三臭化ビスマス *sanshuukabisumasu*
bismuth tribromide $BiBr_3$

三臭化砒素 *sanshuukahiso*
arsenic tribromide $AsBr_3$

三臭化ホウ素, 三臭化硼素 *sanshuukahouso*
boron tribromide BBr_3

三臭化リン, 三臭化燐 *sanshuukarin*
phosphorous tribromide PBr_3

三水化物 *sansuikabutsu*
trihydrate

三炭糖 *santantou*
triose $C_3H_6O_3$

三糖 *santou*
trisaccharide $C_{18}H_{32}O_{16}$

四 four

四分法 *shibunhou*
quartering

四塩化アセチレン *shienkaasechiren*
1,1,2,2-tetrachloroethane $CHCl_2CHCl_2$

四塩化物 *shienkabutsu*
tetrachloride

四塩化チタン *shienkachitan*
titanium tetrachloride $TiCl_4$

四塩化エタン *shienkaetan*
tetrachloroethane $C_2H_2Cl_4$

四塩化白金 *shienkahakkin*
platinic chloride $PtCl_4$

四塩化硫黄 *shienkaiou*
sulfur tetrachloride SCl_4

四塩化ジルコニウム *shienkajirukoniumu*
zirconium(IV) chloride $ZrCl_4$

四塩化珪素, 四塩化ケイ素 *shienkakeiso*
silicon tetrachloride $SiCl_4$

四塩化モリブデン *shienkamoribuden*
molybdenum tetrachloride $MoCl_4$

四塩化セレン *shienkaseren*
selenium tetrachloride $SeCl_4$

四塩化シラン *shienkashiran*
silicon tetrachloride $SiCl_4$

四塩化錫 *shienkasuzu*
stannic chloride $SnCl_4$

四塩化炭素 *shienkatanso*
carbon tetrachloride CCl_4

四塩化ウラン *shienkauran*
uranium tetrachloride UCl_4

四塩基酸 *shienkisan*
tetrabasic acid

四フッ化ホウ酸カルシウム *shifukkahousankarushiumu*
　calcium tetrafluoroborate Ca(BF$_4$)$_2$

四弗化硫黄　*shifukkaiou*
　sulfur tetrafluoride SF$_4$

四フッ化ケイ素, 四弗化珪素　*shifukkakeiso*
　silicon tetrafluoride SiF$_4$

四弗化炭素, 四フッ化炭素　*shifukkatanso*
　carbon tetrafluoride CF$_4$

四弗化ウラン　*shifukkauran*
　uranium trafluoride UF$_4$

四面体　*shimentai*
　tetrahedron

四酸塩基　*shisanenki*
　tetracid base

四酸化物　*shisankabutsu*
　tetroxide

四酸化硫黄　*shisankaiou*
　sulfur tetroxide,
　peroxosulfur(VI) oxide SO$_4$

四酸化二窒素　*shisankanichisso*
　dinitrogen tetroxide N$_2$O$_4$

四酸化オスミウム　*shisankaosumiumu*
　osmium tetroxide OsO$_4$

四酸化燐　*shisankarin*
　phosphorus tetroxide P$_2$O$_4$

四酸化三鉛　*shisankasannamari*
　lead(II,IV) oxide Pb$_3$O$_4$

四酸化三鉄　*shisankasantetsu*
　triiron tetraoxide Fe$_3$O$_4$

四臭化炭素　*shishuukatanso*
　carbon tetrabromide CBr$_4$

四原子の　*yongenshino*
　tetraatomic

四員環　*yoninkan*
　four-membered ring

四次構造　*yonjikouzou*
　quaternary structure

四重硼酸　*yonjuuhousan*
　tetraboric acid B$_4$O$_5$(OH)$_2$

四重極　*yonjuukyoku*
　quadrupole

四重極モーメント　*yonjuukyokumo-mento*
　quadruple moment

四重線　*yonjuusen*
　quadruplet

四価アルコール　*yonkaaruko-ru*
　tetravalent alcohol

四価の　*yonkano*
　quadrivalent, tetravalent, tetrahydric

四極放射　*yonkyokuhousha*
　quadrupole radiation

四極結合定数　*yonkyokuketsugouteisuu*
　quadrupole coupling constant

四極子共鳴　*yonkyokukyoumei*
　quadrupole resonance

四極子　*yonkyokushi*
　quadrupole

四極子モーメント　*yonkyokushimo-mento*
　quadroupole moment

四量体　*yonryoutai*
　tetramer

四糖　*yontou*
　tetrasaccharide C$_{24}$H$_{42}$O$_{21}$

五　five

五塩化アンチモン　*goenkaanchimon*
　antimony(V) chloride SbCl$_5$

五塩化ヒ素　*goenkahiso*
　pentachloroarsorane AsCl$_5$

五塩化ニオブ　*goenkaniobu*
　niobium pentachloride NbCl$_5$

五塩化リン, 五塩化燐　*goenkarin*
　phosphorus pentachloride PCl$_5$

五フッ化ヒ素, 五弗化ヒ素　*gofukkahiso*
　arsenic pentafluoride AsF$_5$

五フッ化リン, 五弗化燐　*gofukkarin*
　phosphorus pentafluoride PF$_5$

五フッ化ホウ素, 五弗化沃素 *gofukkayouso*
iodine pentafluoride IF_5

五ホウ酸, 五硼酸 *gohousan*
pentaboric acid $B_5O_6(OH)_3$

五員環 *goinkan*
five-membered ring

五重線 *gojuusen*
quintet

五価アルコール *gokaaruko-ru*
pentavalent alcohol

五価の *gokano*
pentavalent, pentahydric

五硫化アンチモン *goryuukaanchimon*
antimony pentasulfide Sb_2S_5

五硫化窒素 *goryuukachisso*
nitrogen pentasulfide N_2S_5

五硫化砒素 *goryuukahiso*
arsenic(V)sulfide As_2S_5

五硫化二リン, 五硫化二燐
goryuukanirin
phosphorus pentasulfide P_4S_{10}

五酸化アンチモン *gosankaanchimon*
antimony pentoxide Sb_2O_5

五酸化バナジウム *gosankabanajiumu*
vanadium pentoxide V_2O_5

五酸化窒素 *gosankachisso*
nitric anhydride, nitrogen pentoxide N_2O_5

五酸化ニヒ素, 五酸化二砒素
gosankanihiso
arsenic pentoxide As_2O_5

五酸化リン, 五酸化燐 *gosankarin*
phosphorus pentaoxide P_2O_5

五酸化タンタル *gosankatantaru*
tantalum pentoxide Ta_2O_5

五酸化ヴァナジウム *gosankavanajiumu*
vanadium pentoxide V_2O_5

五臭化燐 *goshuukarin*
phosphorus pentabromide PBr_5

五水灰硼石 *gosuihaihouseki*
pentahydroborite $CaB_2O(OH)_6 \cdot 2H_2O$

五炭糖 *gotantou*
pentose $C_5H_{10}O_5$

六 six

六塩化ブタジエン *rokuenkabutajien*
hexachlorobutadiene C_4Cl_6

六塩化エタン *rokuenkaetan*
hexachloroethane C_2Cl_6

六塩化白金（IV）カリウム
rokuenkahakkin(IV)kariumu
potassium hexachloroplatinate(IV) K_2PtCl_6

六塩化珪素, 六塩化ケイ素
rokuenkakeiso
silicon hexachloride Si_2Cl_6

六塩化ロジウム（III）酸アンモニウム *rokuenkarojiumu*
(III)-sananmoniiumu
triammoniumhexachlororhodate(III)

六塩化ウラン *rokuenkauran*
uranium hexachloride UCl_6

六塩化ウォルフラム
rokuenkaworufuramu
tungsten hexachloride WCl_6

六弗化砒酸銀 *rokufukkahisangin*
silver hexafluoroarsenate $AgAsF_6$

六弗化砒酸ナトリウム
rokufukkahisannatoriumu
sodium hexafluoroarsenate $NaAsF_6$

六弗化砒酸リチウム
rokufukkahisanrichiumu
lithium hexafluoroarsenate $LiAsF_6$

六フッ化硫黄, 六弗化硫黄 *rokufukkaiou*
sulfur hexafluoride SF_6

六フッ化ケイ酸亜鉛, 六弗化珪酸亜鉛
rokufukkakeisanaen
zinc hexafluorosilicate $ZnSiF_6$

六フッ化ケイ酸マグネシウム, 六弗化珪酸マグネシウム
rokufukkakeisanmaguneshiumu
magnesium hexafluorosilicate $MgSiF_6$

六フッ化セレン, 六弗化セレン
rokufukkaseren
selenium hexafluoride SeF_6

六弗化ウラン *rokufukkauran*
uranium hexafluoride UF_6

六原子の　*rokugenshino*
hexatomic

六配位錯体　*rokuhaiisakutai*
six-coordinated complex

六方格子　*rokuhoukoushi*
hexagonal lattice

六員環　*rokuinkan*
six-membered ring

六重線　*rokujuusen*
sextet

六価の　*rokukano*
hexavalent, hexahydric

六面体　*rokumentai*
hexahedron

六量体　*rokuryoutai*
hexamer

六水化物　*rokusuikabutsu*
hexahydrate

六水石　*rokusuiseki*
hexahydrite $MgSO_4 \cdot 6H_2O$

六水和物　*rokusuiwabutsu*
hexahydrate

六炭糖　*rokutantou*
hexose $C_6H_{12}O_6$

六座配位子　*rokuzahaiishi*
hexadentate ligand

六方晶系　*roppoushoukei*
hexagonal system

七 seven

七重線　*shichijuusen*
septet

七価の　*shichikano*
heptavalent, heptahydric

七面体　*shichimentai*
heptahedron

七硫化物　*shichiryuukabutsu*
heptasulfide

七酸化硫黄　*shichisankaiou*
disulfur heptoixde S_2O_7

七酸化二マンガン　*shichisankanimangan*
manganese(VII) oxide Mn_2O_7

八 eight

八員環　*hachiinkan*
eight-membered ring

八重項　*hachijuukou*
octet

八面沸石　*hachimenfusseki*
faujasite
$(Na_2,Ca,Mg)_{3.5}[Al_7Si_{17}O_{48}] \cdot 32(H_2O)$

八面体　*hachimentai*
octahedron

八価の　*hakkano*
octavalent, octahydric

九 nine

九価の　*kyuukano*
nonavalent, nonahydric

十 ten

十字石　*juujiseki*
staurolite
$(Fe^{2+},Mg)_2Al_9(Si,Al)_4O_{20}(OH,O)_4$

半 half, semi

半電池　*handenchi*
half cell, half element, single-electrode system

半導体　*handoutai*
semiconductor

半減期　*hangenki*
　half life

半自動の　*hanjidouno*
　semiautomatic

半径　*hankei*
　radius

半結晶性の　*hankesshouseino*
　semicrystalline

半金属元素　*hankinzokugenso*
　semimetal element

半極性結合　*hankyokuseiketsugou*
　semipolar bond

半流動体　*hanryuudoutai*
　semiliquid

半水化物　*hansuikabutsu*
　hemihydrate

半透膜　*hantoumaku*
　semipermeable membrane

半透性　*hantousei*
　semipermeability

単　one, single

単分散　*tanbunsan*
　monodispersion

単分子反応　*tanbunshihannou*
　unimolecular reaction

単分子膜　*tanbunshimaku*
　monolayer

単分子層　*tanbunshisou*
　monomolecular layer

単分子層表面　*tanbunshisouhyoumen*
　monolayer surface

単独重合　*tandokujuugou*
　homopolymerization

単独重合速度　*tandokujuugousokudo*
　homopolymerization rate

単形　*tangata*
　monomorphism

単原子分子　*tangenshibunshi*
　monoatomic molecule

単位　*tani*
　unit

単一層　*tanissou*
　monolayer

単一　*tanitsu*
　single, simple, individual, sole

単一ビーズ　*tanitsubi-zu*
　single bead

単一単分子層　*tanitsubunshisou*
　single monolayer

単一分子　*tanitsubushi*
　single molecule

単一電池　*tanitsudenchi*
　single cell

単一電極　*tanitsudenkyoku*
　single electrode

単一電子系　*tanitsudenshikei*
　one-electron system

単一エナンチオマ　*tanitsuenanchioma*
　single enantiomer

単一元素　*tanitsugenso*
　single element

単一波長　*tanitsuhachou*
　single frequency

単一反応　*tanitsuhannou*
　single reaction

単一異性体　*tanitsuiseitai*
　single isomer

単一ジアステレオマ　*tanitsujiasutereoma*
　single diastereomer

単一化　*tanitsuka*
　simplification, unification

単一化合物　*tanitsukagoubutsu*
　single compound

単一カラム　*tanitsukaramu*
　single column

単一結晶　*tanitsukesshou*
　single crystal

単一結合　*tanitsuketsugou*
　single bond, single binding

単一高分子鎖　*tanitsukoubunshisa*
　single polymer chain

単一命令操作 *tanitsumeireisousa*
single-step operation, step-by-step operation

単一モード *tanitsumo-do*
single mode

単一ヌクレオチド *tanitsunukureochido*
single nucleotide

単一パラメータ *tanitsuparame-ta*
single parameter

単一パルス *tanitsuparusu*
single pulse

単一レベル *tanitsureberu*
single level

単一酸化物 *tanitsusankabutsu*
single oxide

単一性 *tanitsusei*
unitary

単一成分 *tanitsuseibun*
single component

単一セル *tanitsuseru*
one cell

単一試験 *tanitsushiken*
single test

単一種 *tanitsushu*
single species

単一溶媒 *tanitsuyoubai*
single solvent

単条ポリマー *tanjouporima-*
single-strand polymer

単純分子 *tanjunbunshi*
simple molecule

単純液体 *tanjunekitai*
simple liquid

単純反応 *tanjunhannou*
simple reaction

単純化 *tanjunka*
simplification

単純気体 *tanjunkitai*
simple gas

単純混合物 *tanjunkongoubutsu*
simple mixture

単純模型 *tanjunmokei*
simple model

単純理論 *tanjunriron*
simple theory

単環 *tankan*
monocycle

単環式化合物 *tankanshikikagoubutsu*
monocyclic compound

単結晶 *tankesshou*
single crystal

単結晶合成 *tankesshougousei*
single crystal synthesis

単結晶面 *tankesshoumen*
single crystal face

単結晶粒子 *tankesshouryuushi*
monocrystalline particle

単結晶性 *tankesshousei*
monocrystallinity

単結晶製造 *tankesshouseisou*
monocrystallization

単結合 *tanketsugou*
single bond

単極電位 *tankyokudeni*
single electrode potential

単離 *tanri*
isolation

単離物 *tanributsu*
isolated material

単離法 *tanrihou*
isolation method

単離過程 *tanrikatei*
isolation process

単量体 *tanryoutai*
monomer

単量体合成 *tanryoutaigousei*
monomer synthesis

単量体反応性 *tanryoutaihannousei*
monomer reactivity

単量体混合物 *tanryoutaikongoubutsu*
monomer mixture

単量体濃度 *tanryoutainoudo*
monomer concentration

単量体錯体　*tanryoutaisakutai*
　monomer complex

単量体組成　*tanryoutaisosei*
　monomer composition

単量体単位　*tanryoutaitani*
　monomeric unit

単量体添加　*tanryoutaitenka*
　monomer addition

単量体前駆体　*tanryoutaizenkutai*
　monomer precursor

単繊維　*tanseni*
　single fiber

単斜晶系　*tanshashoukei*
　monoclinic system

単斜晶結晶　*tanshashoukesshou*
　monoclinic crystal

単色　*tanshoku*
　monochromatic

単色光　*tanshokukou*
　monochromatic light

単色の　*tanshokuno*
　monochrome

単色性　*tanshokusei*
　monochromaticity

単層　*tansou*
　monolayer

単相　*tansou*
　single phase

単体　*tantai*
　simple substance

単糖　*tantou*
　monosaccharide

単座配位子　*tanzahaiishi*
　monodentate ligand

原 mono

原型　*genkei*
　prototype, model

原形　*genkei*
　original form

原形質分離　*genkeishitsubunri*
　plasmolysis

原鉱　*genkou*
　crude ore, raw ore

原理　*genri*
　theory, principle

原料　*genryou*
　raw material

原石　*genseki*
　crude ore, raw ore

原繊維　*genseni*
　raw fiber

原子　*genshi*
　atom

原子爆弾　*genshibakudan*
　atomic bomb

原子番号　*genshibangou*
　atomic number

原子エネルギー　*genshienerugi-*
　atomic energy

原子半径　*genshihankei*
　atomic radius

原子順位　*genshijuni*
　atomic number

原子価　*genshika*
　valence, valency

原子価結合法　*genshikaketsugouhou*
　valence-bond method

原子核　*genshikaku*
　atomic nucleus

原子殻　*genshikaku*
　atomic shell

原子核分裂　*genshikakubunretsu*
　nuclear fission, atom splitting

原子核エネルギー準位　*genshikakuenerugi-juni*
　nuclear energy level

原子核反応　*genshikakuhannou*
　nuclear reaction

原子核化学　*genshikakukagaku*
　nuclear chemistry

原子間距離 *genshikankyori*
    interatomic distance

原子価説 *genshikasetsu*
    valence theory

原子価式 *genshikashiki*
    valence formula

原子結合 *genshiketsugou*
    atomic bond, atomic linkage

原子軌道 *genshikidou*
    atomic orbital

原子格子 *genshikoushi*
    atomic lattice

原子吸光分析 *genshikyuukoubunseki*
    atomic absorption spectrometry

原子模型 *genshimokei*
    atomic model

原子力学 *genshirikigaku*
    atomic mechanics

原子炉 *genshiro*
    nuclear reactor

原子炉安全性 *genshiroanzensei*
    reactor safety

原子炉事故 *genshirojiko*
    reactor accident

原子論 *genshiron*
    atomic theory

原子力 *genshiryoku*
    atomic power

原子力発電所 *genshiryokuhatsudensho*
    nuclear power plant

原子量 *genshiryou*
    atomic weight

原子散乱因子 *genshisanraninshi*
    atomic scattering factor

原子線 *genshisen*
    atomic line

原子説 *genshisetsu*
    atomic theory

原子質量 *genshishitsuryou*
    atomic mass

原子質量単位 *genshishitsuryoutani*
    atomic mass unit

原子スペクトル *genshisupekutoru*
    atomic spectrum

原子単位 *genshitani*
    atomic unit

原子容 *genshiyou*
    atomic volume

原色 *genshoku*
    primary color

原炭 *gentan*
    raw coal

原糖 *gentou*
    raw sugar, unrefined sugar

原油 *genyu*
    crude oil

原材料 *genzairyou*
    raw material

多 poly

多分子層 *tabunshisou*
    multilayer

多塩基 *taenki*
    polybasic acid, polyacid

多塩基酸 *taenkisan*
    polybasic acid

多原子 *tagenshi*
    polyatomic

多原子分子 *tagenshibunshi*
    polyatomic molecule

多辺形 *tahenkei*
    polygon

多陰イオン *tainion*
    polyanion

多員環 *tainkan*
    many-membered ring

多重結合 *tajuuketsugou*
    multiple bond

多重項 *tajuukou*
    multiplet

多重線 *tajuusen*
    multiplet

多価アルコール　*takaaruko-ru*
  polyhydric alcohol, polyvalent alcohol, polyol

多価フェノール　*takafeno-ru*
  polyhydric phenol, polyphenol

多角形　*takakkei*
  polygon

多角体　*takakutai*
  polyhedron

多官能化合物　*takannoukagoubutsu*
  multifunctional compound

多価の　*takano*
  polyvalent, polyhydric

多環式化合物　*takanshikikagoubutsu*
  polycyclic compound

多形現象　*takeigenshou*
  polymorphism

多形の　*takeino*
  polymorphic

多珪酸　*takeisan*
  polysilicic acid

多結晶体、多結晶体　*takesshou, takesshoutai*
  polycrystal

多孔性　*takousei*
  porosity

多孔性材料　*takouseizairyou*
  porous materials

多面体　*tamentai*
  polyhedron

多硫化アンモニウム　*taryuukaanmoniumu*
  ammonium polysulfide

多硫化物　*taryuukabutsu*
  polysulfide $M^I_2 Sn$

多硫化カリウム　*taryuukakariumu*
  potassium polysulfide

多硫化ナトリウム　*taryuukanatoriumu*
  sodium polysulfide

多酸塩基　*tasanenki*
  polyacid base

多酸化物　*tasankabutsu*
  polyoxide

多色性　*tashokusei*
  polychromatism

多層　*tasou*
  multilayer

多水化物　*tasuikabutsu*
  polyhydrat

多点吸着　*tatenkyuuchaku*
  multipoint adsorption

多糖類　*tatourui*
  polysaccharide

多陽イオン　*tayouion*
  polycation

多様性　*tayousei*
  variability, diversity

8.2
Scientific Terms Beginning with *kanji* for Chemical Elements

硼 boron (^{5}B)

硼化水素　*houkasuiso*
  boron hydride

硼珪酸　*houkeisanen*
  borosilicate

硼珪酸ガラス　*houkeisangarasu*
  borosilicate glass

硼酸　*housan*
  boric acid H_3BO_3

硼酸塩　*housanen*
  borate $M^I_3 BO_3$

硼酸マグネシウム　*housanmaguneshiumu*
  magnesium borate $Mg_3(BO_3)_2$

硼酸リチウム　*housanrichiumu*
  lithium borate Li_3BO_3

硼酸石　*housanseki*
　　sassolite H_3BO_3

硼酸トリエチル　*housantoriechiru*
　　triethyl borate $B(OC_2H_5)_3$

硼酸トリメチル　*housantorimechiru*
　　trimethyl borate $B(OCH_3)_3$

硼砂　*housha*
　　borax $Na_2B_4O_7 \cdot 10H_2O$

硼素　*houso*
　　boron (B, element 5)

炭 carbon (^{6}C) – 炭化 carbonization

炭化　*tanka*
　　carbonization

炭化物　*tankabutsu*
　　carbide

炭化チタン　*tankachitan*
　　titanium carbide TiC

炭化ホウ素, 炭化硼素　*tankahouso*
　　boron carbide B_4C

炭化カルシウム　*tankakarushiumu*
　　calcium carbide CaC_2

炭化ケイ素　*tankakeiso*
　　silicon carbide SiC

炭化温度　*tankaondo*
　　carbonization temperature

炭化生成物　*tankaseiseibutsu*
　　carbonization product

炭化水素　*tankasuiso*
　　hydrocarbon

炭化水素分析　*tankasuisobunseki*
　　hydrocarbon analysis

炭化水素フラグメント
　　tankasuisofuragumento
　　hydrocarbon fragment

炭化水素ガス　*tankasuisogasu*
　　hydrocarbon gas

炭化水素源　*tankasuisogen*
　　hydrocarbon source

炭化水素合成　*tankasuisogousei*
　　hydrocarbon synthesis

炭化水素異性体　*tankasuisoiseitai*
　　hydrocarbon isomer

炭化水素蒸気　*tankasuisojouki*
　　hydrocarbon vapour

炭化水素活性化　*tankasuisokasseika*
　　hydrocarbon activation

炭化水素基　*tankasuisoki*
　　hydrocarbon radical

炭化水素気体　*tankasuisokitai*
　　hydrocarbon gas

炭化水素混合物　*tankasuisokongoubutsu*
　　hydrocarbon mixture

炭化水素工業　*tankasuisokougyou*
　　hydrocarbon industry

炭化水素構造　*tankasuisokouzou*
　　hydrocarbon structure

炭化水素燃料　*tankasuisonenryou*
　　hydrocarbon fuel

炭化水素燃焼　*tankasuisonenshou*
　　hydrocarbon combustion

炭化水素熱分解　*tankasuisonetsubunkai*
　　hydrocarbon pyrolysis

炭化水素濃度　*tankasuisonoudo*
　　hydrocarbon concentration

炭化水素ポリマ　*tankasuisoporima*
　　hydrocarbon polymer

炭化水素ラジカル　*tankasuisorajikaru*
　　hydrocarbon radical

炭化水素流体　*tankasuisoryuutai*
　　hydrocarbon fluid

炭化水素鎖　*tankasuisosa*
　　hydrocarbon chain

炭化水素酸化　*tankasuisosanka*
　　hydrocarbon oxidation

炭化水素相互作用　*tankasuisosougosayou*
　　hydrocarbon interaction

炭化水素油　*tankasuisoyu*
　　hydrocarbon oil

炭化水素誘導体　*tankasuisoyuudoutai*
　hydrocarbon derivative

炭化する　*tankasuru*
　to carbonate

炭 carbon (^{6}C) – 炭酸 carbonic acid

炭酸　*tansan*
　carbonic acid H_2CO_3

炭酸亜鉛　*tansanaen*
　zinc carbonate $ZnCO_3$

炭酸アンモニウム　*tansananmoniumu*
　ammonium carbonate $(NH_4)_2CO_3$

炭酸バリウム　*tansanbariumu*
　barium carbonate $BaCO_3$

炭酸第一鉄　*tansandaiichitetsu*
　iron(II) carbonate $FeCO_3$

炭酸銅　*tansandou*
　copper carbonate $CuCO_3/Cu_2CO_3$

炭酸同化作用　*tansandoukasayou*
　carbon dioxide assimilation

炭酸塩　*tansanen*
　carbonate $M^I_2CO_3$

炭酸塩化　*tansanenka*
　carbonization

炭酸塩水溶液　*tansanensuiyoueki*
　aqueous carbonate solution

炭酸塩溶液　*tansanenyoueki*
　carbonate solution

炭酸塩融解　*tansanenyuukai*
　carbonate fusion

炭酸エステル　*tansanesuteru*
　carbonic acid ester, carbonate

炭酸銀　*tansangin*
　silver carbonate Ag_2CO_3

炭酸イオン　*tansanion*
　carbonate ion

炭酸イットリウム　*tansanittoriumu*
　yttrium carbonate $Y_2(CO_3)_3$

炭酸ジアルキル　*tansanjiarukiru*
　dialkyl carbonate $CO(OR)_2$

炭酸ジエチル　*tansanjiechiru*
　diethyl carbonate $CO(OC_2H_5)_2$

炭酸ジメチル　*tansanjimechiru*
　dimethyl carbonate

炭酸ジルコニル　*tansanjirukoniru*
　zirconyl carbonate $3ZrO_2 \cdot H_2CO_3/2ZrO_2 \cdot H_2CO_3 \cdot 7H_2O$

炭酸化　*tansanka*
　carbonation

炭酸カリウム　*tansankariumu*
　potassium carbonate, potash K_2CO_3

炭酸カリウム水溶液
　tansankariumusuiyoueki
　aqueous potassium carbonate solution

炭酸カルシウム　*tansankarushiumu*
　calcium carbonate $CaCO_3$

炭酸マグネシウム　*tansanmaguneshiumu*
　magnesium carbonate $MgCO_3$

炭酸マンガン　*tansanmangan*
　manganese carbonate $MnCO_3$

炭酸ナトリウム　*tansannatoriumu*
　soda, sodium carbonate Na_2CO_3

炭酸ナトリウム水溶液
　tansannatoriumusuiyoueki
　aqueous sodium carbonate solution

炭酸リチウム　*tansanrichiumu*
　lithium carbonate Li_2CO_3

炭酸水　*tansansui*
　carbonated water

炭酸水素アンモニウム
　tansansuisoanmoniumu
　ammonium hydrogen carbonate NH_4HCO_3

炭酸水素塩　*tansansuisoen*
　hydrogen carbonate, bicarbonate M^IHCO_3

炭酸水素ナトリウム
　tansansuisonatoriumu
　sodium hydrogen carbonate $NaHCO_3$

炭酸ストロンチウム
　tansansutoronchiumu
　strontium carbonate $SrCO_3$

炭酸鉄　*tansantetsu*
　iron carbonate $FeCO_3/Fe_2(CO_3)_3$

炭 carbon (⁶C) – 炭素

炭素　*tanso*
   carbon (C, element 6)

炭素アーク　*tansoa-ku*
   carbon arc

炭素ブラシ, 炭素ブラッシュ
   tansoburashi, tansoburasshu
   carbon brush

炭素置換基　*tansochikanki*
   carbon substituent

炭素中心　*tansochuushin*
   carbon center

炭素電極　*tansodenkyoku*
   carbon electrode

炭素同位体　*tansodouitai*
   carbon isotope

炭素同位体分離　*tansodouitaibunri*
   carbon isotope separation

炭素同位体比　*tansodouitaihi*
   carbon isotope ratio

炭素含量　*tansoganryou*
   carbon content

炭素源　*tansogen*
   carbon source

炭素原子　*tansogenshi*
   carbon atom

炭素配位子　*tansohaiishi*
   carbon ligand

炭素反応　*tansohannou*
   carbon reaction

炭素被覆　*tansohifuku*
   carbon coating

炭素イオン　*tansoion*
   carbon ion

炭素化合物　*tansokagoubutsu*
   carbon compound

炭素核　*tansokaku*
   carbon nucleus

炭素環　*tansokan*
   carbon ring

炭素還元　*tansokangen*
   carbon reduction

炭素環系　*tansokankei*
   cyclic carbon system

炭素環ヌクレオシド　*tansokannukureoshido*
   carbocyclic nucleoside

炭素環生成反応　*tansokanseiseihannou*
   carbocyclization

炭素環式化合物　*tansokanshikikagoubutsu*
   carbocyclic compound

炭素研究　*tansokenkyuu*
   carbon research

炭素結晶　*tansokesshou*
   carbon crystal

炭素結合　*tansoketsugou*
   carbon bond, carbon linkage

炭素固体　*tansokotai*
   carbon solid

炭素固定　*tansokotei*
   carbon fixation

炭素鋼　*tansokou*
   carbon steel

炭素交換　*tansokoukan*
   carbon exchange

炭素クラスタ　*tansokurasuta*
   carbon cluster

炭素吸着　*tansokyuuchaku*
   carbon adsorption

炭素求核試薬　*tansokyuukakushiyaku*
   carbon nucleophile

炭素吸収　*tansokyuushuu*
   carbon absorption

炭素マトリックス　*tansomatorikkusu*
   carbon matrix

炭素モレキュラシーブ　*tansomorekyurashi-bu*
   carbon molecular sieve

炭素ナノ繊維　*tansonanoseni*
   carbon nanofiber

炭素燃焼　*tansonenshou*
   carbon combustion

炭素二重結合　*tansonijuuketsugou*
　carbon double bond

炭素の循環　*tansonojunkan*
　carbon cycle

炭素濃度　*tansonoudo*
　carbon concentration

炭素ラジカル　*tansorajikaru*
　carbon radical

炭素鎖　*tansosa*
　carbon chain

炭素鎖長　*tansosachou*
　carbon chain length

炭素三重結合　*tansosanjuuketsugou*
　carbon triple bond

炭素酸化　*tansosanka*
　carbon oxidation

炭素製品　*tansoseihin*
　carbon product

炭素繊維　*tansoseni*
　carbon fiber

炭素シグマ結合　*tansoshigumaketsugou*
　carbon sigma bond

炭素質相　*tansoshitsusou*
　carbonaceous phase

炭素相　*tansosou*
　carbon phase

炭素相互作用　*tansosougosayou*
　carbon interaction

炭素水素結合　*tansosuisoketsugou*
　carbon-hydrogen bond

炭素数　*tansosuu*
　carbon number

炭素多重結合　*tansotajuuketsugou*
　carbon multiple bond

炭素炭素結合　*tansotansoketsugou*
　carbon-carbon linkage

炭素炭素二重結合
　tansotansonijuuketsugou
　carbon-carbon double bond

炭素炭素三重結合
　tansotansosanjuuketsugou
　carbon-carbon triple bond

炭素炭素単結合　*tansotansotanketsugou*
　carbon-carbon single bond

炭素添加　*tansotenka*
　carbon addition

炭素層　*tansousou*
　carbon layer

炭　carbon (^{6}C) – 炭水

炭水化物　*tansuikabutsu*
　carbohydrate

炭水化物分析　*tansuikabutsubunseki*
　carbohydrate analysis

炭水化物分子　*tansuikabutsubunshi*
　carbohydrate molecule

炭水化物デンドリマ
　tansuikabutsudendorima
　carbohydrate dendrimer

炭水化物合成　*tansuikabutsugousei*
　carbohydrate synthesis

炭水化物配位子　*tansuikabutsuhaiishi*
　carbohydrate ligand

炭水化物重合体　*tansuikabutsujuugoutai*
　carbohydrate polymer

炭水化物混合物
　tansuikabutsukongoubutsu
　carbohydrate mixture

炭水化物鎖　*tansuikabutsusa*
　carbohydrate chain

炭水化物側鎖　*tansuikabutsusokusa*
　carbohydrate side chain

珪　silicon (^{14}Si)

珪灰石　*keikaiseki*
　wollastonite $CaSiO_3$

珪酸亜鉛鉱　*keisanaenkou*
　willemite Zn_2SiO_4

珪酸アルミニウム　*keisanaruminiumu*
　aluminium silicate $Al_2(SiO_3)_3$

珪酸ガラス　*keisangarasu*
silicate glass $M^I_4SiO_4$

珪酸ゲル　*keisangeru*
silica gel

珪酸ジルコニウム　*keisanjirukoniumu*
zirconium silicate $ZrSiO_4$

珪酸カルシウム　*keisankarushiumu*
calcium silicate $CaSiO_3$

珪砂　*keisha*
quartz sand, silica sand

珪素, ケイ素　*keiso*
silicon (Si, element 14)

珪素樹脂　*keisojushi*
silicone resin

鉛白　*enpaku*
white lead $2PbCO_3 \cdot Pb(OH)_2$

鉛酸塩　*ensanen*
plumbate

鉛室　*enshitsu*
lead chamber

鉛室式　*enshitsushiki*
chamber process

鉛丹　*entan*
minium, lead(II,IV)oxide Pb_3O_4

鉛蓄電池　*namarichikudenchi*
lead storage battery

鉛ガラス　*namarigarasu*
lead glass

錫 tin (^{50}Sn)

錫, スズ　*suzu*
tin (Sn, element 50)

錫箔　*suzuhaku*
tin foil

錫酸　*suzusan*
stannic acid $H_2SnO_3/H_2Sn(OH)_6$

錫酸塩　*suzusanen*
stannate $M^I_2Sn(OH)_6/M^I_2SnO_3$

錫酸ナトリウム　*suzusannatoriumu*
sodium hexahydroxostannate(IV) $Na_2Sn(OH)_6$

錫浴　*suzuyokuk*
tin bath

鉛 lead (^{46}Pb)

鉛　*namari*
lead (Pb, element 82)

鉛中毒　*enchuudoku*
lead poisoning

鉛鉱　*enkou*
lead deposit

窒 nitrogen (^{7}N)

窒化　*chikka*
nitriding

窒化物　*chikkabutsu*
nitride M^I_3N

窒化銀　*chikkagin*
silver nitride Ag_3N

窒化ホウ素, 窒化硼素　*chikkahouso*
boron nitride BN

窒化硫黄　*chikkaiou*
sulfur nitride

窒化ケイ素, 窒化桂素　*chikkakeiso*
nitrogen silicide, silicon nitride Si_3N_4

窒化機構　*chikkakikou*
nitridation mechanism

窒化マグネシウム　*chikkamaguneshiumu*
magnesium nitride Mg_3N_2

窒化ナトリウム　*chikkanatoriumu*
sodium nitride Na_3N

窒化リチウム　*chikkarichiumu*
lithium nitride Li_3N

窒化水素酸　*chikkasuisosan*
hydrazoic acid, hydrogen azide HN_3

窒化炭素 *chikkatanso*
carbon nitride C_3N_4

窒素 *chisso*
nitrogen (N, element 7)

窒素分圧 *chissobunatsu*
nitrogen partial pressure

窒素分析 *chissobunseki*
nitrogen analysis

窒素置換基 *chissochikanki*
nitrogen substituent

窒素同位体 *chissodouitai*
nitrogen isotope

窒素同化 *chissodouka*
nitrogen assimilation, nitrogen fixation

窒素導入 *chissodounyuu*
nitrogen introduction

窒素塩基 *chissoenki*
nitrogen base

窒素複素環 *chissofukusokan*
nitrogen heterocycle

窒素含量 *chissoganryou*
nitrogen content

窒素含有化合物 *chissoganyuukagoubutsu*
nitrogen containing compound

窒素含有炭化水素 *chissoganyuutankasuiso*
nitrogen-containing hydrocarbon

窒素ガス *chissogasu*
nitrogen gas N_2

窒素源 *chissogen*
nitrogen source

窒素原子 *chissogenshi*
nitrogen atom

窒素配位子 *chissohaiishi*
nitrogen ligand

窒素反転 *chissohanten*
nitrogen inversion

窒素平衡 *chissoheikou*
nitrogen balance

窒素肥料 *chissohiryou*
nitrogenous fertilizer

窒素保護基 *chissohogoki*
nitrogen protecting group

窒素移動 *chissoidou*
nitrogen transfer

窒素イオン *chissoion*
nitrogen ion

窒素イリド *chissoirido*
nitrogen ylide

窒素化合物 *chissokagoubutsu*
nitrogenous compound

窒素環 *chissokan*
nitrogen ring

窒素官能基 *chissokannouki*
nitrogen functionality

窒素結合開裂 *chissoketsugoukairetsu*
nitrogen bond cleavage

窒素キレート配位子 *chissokire-tohaiishi*
nitrogen chelating ligand

窒素固定 *chissokotei*
nitrogen assimilation, nitrogen fixation

窒素供与体 *chissokyouyotai*
nitrogen donor

窒素吸着 *chissokyuuchaku*
nitrogen adsorption

窒素吸収 *chissokyuushuu*
nitrogen absorption

窒素マトリックス *chissomatorikkusu*
nitrogen matrix

窒素の循環 *chissonojunkan*
nitrogen cycle

窒素ラジカル *chissorajikaru*
nitrogen radical

窒素サイクル *chissosaikuru*
nitrogen cycle

窒素三重結合 *chissosanjuuketsugou*
nitrogen triple-bond

窒素酸化物 *chissosankabutsu*
nitrogen oxide NOx

窒素相互作用 *chissosougosayou*
nitrogen interaction

窒素定量 *chissoteiryou*
 nitrogen determination, nitrogen analysis

窒素添加 *chissotenka*
 nitrogen addition

窒素誘導体 *chissoyuudoutai*
 nitrogen derivative

窒素族 *chissozoku*
 nitrogen group

燐 phosphorus (^{15}P)

燐, リン *rin*
 phosphorus (P, element 15)

燐化亜鉛 *rinkaaen*
 zink phosphide Zn_3P_2

燐化物 *rinkabutsu*
 phosphide M^I_3P

燐化インジウム *rinkainjiumu*
 indium phosphide InP

燐化ジルコニウム *rinkajirukoniumu*
 zirconium phosphid ZrP_2

燐化カルシウム *rinkakarushiumu*
 calcium phosphide Ca_3P

燐光 *rinkou*
 phosphorescence

燐酸, リン酸 *rinsan*
 phosphoric acid H_3PO_4

燐酸亜鉛 *rinsanaen*
 zinc phosphate $Zn_3(PO_4)_2$

燐酸アンモニウム *rinsananmoniumu*
 ammonium phosphat $(NH_4)_3PO_4$

燐酸アルミニウム *rinsanaruminiumu*
 aluminium phosphate $AlPO_4$

燐酸塩 *rinsanen*
 phosphate N_3PO_4

燐酸塩化 *rinsanenka*
 phosphatizing

燐酸エステル *rinsanesuteru*
 phosphoric ester $PO(OR)_3$

燐酸銀 *rinsangin*
 silver phosphate Ag_3PO_4

燐酸肥料 *rinsanhiryou*
 phosphate fertilizer

燐酸ジルコニウム *rinsanjirukoniumu*
 zirconium phosphate $Zr_3(PO_4)_4$

燐酸カリウム *rinsankariumu*
 potassium phosphate K_3PO_4

燐酸カルシウム *rinsankarushiumu*
 calcium phosphate $Ca_3(PO_4)_2$

燐酸クレアチン *rinsankureachin*
 phosphagen, phosphocreatine
 $HOOCCH_2N(CH_3)C(=NH)NHPO(OH)_2$

燐酸ナトリウム *rinsannatoriumu*
 sodium phosphate Na_3PO_4

燐酸二水素アンモニウム
 rinsannisuisoanmoniumu
 ammonium dihydrogenphosphate
 $NH_4H_2PO_4$

燐酸二水素カリウム
 rinsannisuisokariumu
 potassium dihydrogenphosphate KH_2PO_4

燐酸二水素カルシウム
 rinsannisuisokarushiumu
 calcium dihydrogenphosphate
 $Ca(H_2PO_4)_2$

燐酸二水素ナトリウム
 rinsannisuisonatoriumu
 sodium dihydrogenphosphate NaH_2PO_4

燐酸水素カルシウム
 rinsansuisokarushiumu
 calcium hydrogenphosphate $CaHPO_4$

燐酸水素二ナトリウム
 rinsansuisoninatoriumu
 disodium hydrogenphosphate Na_2HPO_4

燐酸ストロンチウム
 rinsansutoronchiumu
 strontium phosphate $Sr_3(PO_4)_2$

燐酸トリアルキル *rinsantoriarukiru*
 trialkyl phosphate $PO(OR)_3$

燐酸トリエチル *rinsantoriechiru*
 triethyl phosphate $PO(OC_2H_5)_3$

燐酸トリフェニル *rinsantorifeniru*
 triphenyl phosphate $PO(OC_6H_5)_3$

燐酸トリメチル *rinsantorimechiru*
 trimethyl phosphate $PO(OCH_3)_3$

燐脂質 *rinshishitsu*
 phospholipide, phosphatide

燐蛋白 *rintanpaku*
 phosphoprotein

燐タンパク質 *rintanpakushitsu*
 phosphoprotein

砒 arsenic (^{33}As)

砒化水素 *hikasuiso*
 arsane AsH_3

砒化鉄 *hikatetsu*
 iron arsenide $FeAs$

砒酸 *hisan*
 arsenic acid H_3AsO_4

砒酸塩 *hisanen*
 arsenate $M^I_3AsO_4$

砒酸ナトリウム *hisannatoriumu*
 sodium arsenate Na_3AsO_4

砒酸トリエチル *hisantoriechiru*
 triethyl arsenate $(C_2H_5)_3AsO_4$

砒素, ヒ素 *hiso*
 arsenic (As, element 33)

砒素中毒 *hisochuudoku*
 arsenic poisoning

酸 oxygen (^{8}O)
[for 酸化 see below]

酸 *san*
 acid

酸アチド *sanachido*
 acid azide $RCON_3$

酸アミド *sanamido*
 acid amide $RCONH_2$

酸アニリド *sananirido*
 acid anilide $RCONHC_6H_5$

酸洗い *sanarai*
 pickling

酸洗い剤 *sanaraizai*
 pickling agent

酸分 *sanbun*
 acid content

酸分解 *sanbunkai*
 acid cleavage, cleavage by acids

酸中毒 *sanchuudoku*
 acid poisoning

酸度 *sando*
 acidity

酸塩化物 *sanenkabutsu*
 acid chloride $RCOCl$

酸塩基平衡 *sanenkiheikou*
 acid-base balance

酸塩基指示薬 *sanenkishijiyaku*
 acid-base indicator

酸塩基触媒 *sanenkishokubai*
 acid-base catalyst

酸塩基滴定 *sanenkitekitei*
 acid-base titration

酸塩基対 *sanenkitsui*
 acid-base pair

酸弗化物, 酸フッ化物 *sanfukkabutsu*
 acid fluoride $RCOF$

酸ハロゲン化物 *sanharogenkabutsu*
 acid halide $RCOX$

酸ヒドラジド *sanhidorajido*
 acid hydrazide $RCONHNH_2$

酸イミド *sanimido*
 acid imide

酸価 *sanka*
 acid number

酸基 *sanki*
 acid radical

酸強度 *sankyoudo*
 acid strength

酸味 *sanmi*
 acidity, sourness

酸無水物 *sanmusuibutsu*
 acid anhydride

酸のイオン化平衡 *sannoionkaheikou*
 ionic equilibrium of acid

酸敗 *sanpai*
acidification

酸率 *sanritsu*
acidity, acid number

酸硫化燐 *sanryuukarin*
phosphorus sulfoxide $P_4S_4O_6$

酸性, 酸性度 *sansei, sanseido*
acidity

酸性亜硫酸ナトリウム
sanseiaryuusannatoriumu
sodium bisulfite $NaHSO_3$

酸性中心 *sanseichuushin*
acid center

酸性土壌 *sanseidojou*
acid soil

酸性塩 *sanseien*
acid salt

酸性岩 *sanseigan*
acid rock

酸性白土 *sanseihakudo*
acid clay

酸性反応 *sanseihannou*
acid reaction

酸性肥料 *sanseihiryou*
acid fertilizer

酸性化 *sanseika*
acidification

酸性化剤 *sanseikazai*
acidifying agent

酸性係数 *sanseikeisuu*
acidity coefficient

酸性の *sanseino*
acidic

酸性硫酸カリウム *sanseiryuusankariumu*
potassium hydrogensulfate $KHSO_4$

酸性酸化物 *sanseisankabutsu*
acidic oxide

酸生成 *sanseisei*
acid formation

酸性染料 *sanseisenryou*
acid dye

酸性色 *sanseishoku*
acid color

酸性炭酸塩 *sanseitansanen*
bicarbonate, hydrogencarbonate M^IHCO_3

酸性炭酸カリウム *sanseitansankariumu*
acid potassium carbonate $KHCO_3$

酸性炭酸ナトリウム
sanseitansannatoriumu
acid sodium carbonate $NaHCO_3$

酸性雨 *sanseiu*
acid rain

酸触媒 *sanshokubai*
acid catalyst

酸処理 *sanshori*
acid treatment

酸臭化物 *sanshuukabutsu*
acid bromide RCOBr

酸素 *sanso*
oxygen (O, element 8)

酸素瓶 *sansobin*
oxygen cylinder

酸素電極 *sansodenkyoku*
oxygen electrode

酸素含量 *sansoganryou*
oxygen content

酸素飽和 *sansohouwa*
oxygen saturation

酸素価 *sansoka*
oxygen value

酸素化 *sansoka*
oxygenation

酸素化合物 *sansokagoubutsu*
oxygen compound

酸素欠乏 *sansoketsubou*
oxygen deficiency

酸素呼吸 *sansokokyuu*
aerobic respiration

酸素供給 *sansokyoukyuu*
oxygen supply

酸素吸収 *sansokyuushuu*
oxygen absorption

酸素濃度　*sansonoudo*
　oxygen concentration

酸素酸　*sansosan*
　oxygen acid

酸素消費　*sansoshouhi*
　oxygen consumption

酸素担体　*sansotantai*
　oxygen carrier

酸素要求　*sansoyoukyuu*
　oxygen demand

酸素族　*sansozoku*
　oxygen group

酸水素炎　*sansuisoen*
　oxyhydrogen flame

酸滴定　*santekitei*
　acidimetry

酸浴　*sanyoku*
　acid bath

酸っぱい　*suppai*
　acidic

酸 oxygen (^{8}O) – 酸化 oxidation

酸化　*sanka*
　oxidation

酸化亜鉛　*sankaaen*
　zinc oxide ZnO

酸化安定性　*sankaanteisei*
　oxidative stability, oxidation resistance

酸化アルミニウム　*sankaaruminiumu*
　aluminium oxide Al_2O_3

酸化バリウム　*sankabariumu*
　barium oxide BaO

酸化ベリリウム　*sankabeririumu*
　beryllia, beryllium oxide BeO

酸化防止剤　*sankaboushizai*
　antioxidant, oxidation inhibitor

酸化物　*sankabutsu*
　oxide

酸化物イオン　*sankabutsuion*
　oxide ion

酸化物触媒　*sankabutsushokubai*
　oxide catalyst

酸化物層　*sankabutsusou*
　oxide layer

酸化チメル　*sankachimeru*
　dimethyl ether, methyl oxide CH_3OCH_3

酸化窒素　*sankachisso*
　nitrogen oxide NO_x

酸化窒素（II）　*issankachisso(II)*
　nitrogen monoxide NO

酸化チタン（IV）　*sankachitan(IV)*
　titanium dioxide TiO_2

酸化第一銅　*sankadaiichidou*
　cuprous oxide Cu_2O

酸化第一ニッケル　*sankadaiichinikkeru*
　nickel(II) oxide NiO

酸化第一水銀　*sankadaiichisuigin*
　mercurous oxide Hg_2O

酸化第一スズ, 酸化第一錫　*sankadaiichisuzu*
　stannous oxide SnO

酸化第一鉄　*sankadaiichitetsu*
　iron(II) oxide FeO

酸化第二銅　*sankadainidou*
　cupric oxide CuO

酸化第二ニッケル　*sankadaininikkeru*
　nickel sesquioxide Ni_2O_3

酸化第二水銀　*sankadainisuigin*
　mercury(II) oxide HgO

酸化第二スズ, 酸化第二錫　*sankadainisuzu*
　stannic oxide SnO_2

酸化第二タリウム　*sankadainitariumu*
　thallium(III) oxide Tl_2O_3

酸化第二鉄　*sankadainitetsu*
　iron(III) oxide Fe_2O_3

酸化電位　*sankadeni*
　oxidation potential

酸化銅（I）　*sankadou(I)*
　cuprous oxide Cu_2O

酸化銅（II）　*sankadou(II)*
　cupric oxide CuO

酸化エチレン　*sankaechiren*
oxirane, ethylene oxide C_2H_4O

酸化フォスフィン　*sankafosufin*
phosphinoxide R_3PO

酸化形　*sankagata*
oxidized form

酸化銀　*sankagin*
silver oxide Ag_2O

酸化標白　*sankahyouhaku*
oxidation bleaching

酸化硫黄　*sankaiou*
sulfur oxide

酸化イッテルビウム　*sankaitterubiumu*
ytterbium oxide Yb_2O_3

酸化イットリウム　*sankaittoriumu*
yttrium oxide Y_2O_3

酸化還元電池　*sankakangendenchi*
oxidation-reduction cell

酸化還元電位　*sankakangendeni*
oxidation-reduction potential

酸化還元反応　*sankakangenhannou*
oxidation-reduction reaction

酸化還元平衡　*sankakangenheikou*
redox equilibrium

酸化還元系　*sankakangenkei*
oxidation-reduction system

酸化還元酵素　*sankakangenkouso*
oxidoreductase, redox enzyme

酸化還元指示薬　*sankakangenshijiyaku*
redox indicator

酸化還元触媒　*sankakangenshokubai*
redox catalyst

酸化還元滴定　*sankakangentekitei*
oxidation-reduction titration

酸化カルシウム　*sankakarushiumu*
calcium oxide CaO

酸化酵素　*sankakouso*
oxidizing enzyme

酸化クロム（Ⅲ）　*sankakuromu(III)*
chromium(III) oxide Cr_2O_3

酸化マグネシウム　*sankamaguneshiumu*
magnesium oxide MgO

酸化メシチル　*sankameshichiru*
mesityl oxide $(CH_3)_2C=CHCOCH_3$

酸化ナトリウム　*sankanatoriumu*
sodium oxide Na_2O

酸化熱　*sankanetsu*
oxidation heat

酸化ニッケル（ⅠⅠ）　*sankanikkeru(II)*
nickelous oxide NiO

酸化ポリメチレン　*sankaporimechiren*
polymethylene oxide $(CH_2)_nO$

酸化プラセオジム　*sankapuraseojimu*
praseodymium oxide Pr_2O_3

酸化プロペン　*sankapuropen*
propene oxide C_3H_6O

酸化ランタン　*sankarantan*
lanthanum oxide La_2O_3

酸化燐　*sankarin*
phosphorus oxide

酸化サマリウム　*sankasamariumu*
samarium oxide Sm_2O_3

酸化作用　*sankasayou*
oxidation process

酸化セルロース　*sankaseruro-su*
oxycellulose

酸化試験　*sankashiken*
oxidation test

酸化水銀（ⅠⅠ）　*sankasuigin(II)*
mercury(II) oxide HgO

酸化する　*sankasuru*
to oxidize

酸化ストロンチウム　*sankasutoronchiumu*
strontium oxide SrO

酸化数　*sankasuu*
oxidation number

酸化錫（ⅠⅠ），酸化スズ（ⅠⅠ）　*sankasuzu(II)*
stannous oxide SnO

酸化錫（ⅠⅤ），酸化スズ（ⅠⅤ）　*sankasuzu(IV)*
stannic oxide SnO_2

酸過多　*sankata*
hyperacidity

酸化テルビウム　*sankaterubiumu*
terbium oxide Tb_2O_3

酸化テトラメチレン
sankatetoramechiren
tetramethylene oxide C_4H_8O

酸化鉄（ⅠⅠ）　*sankatetsu(II)*
iron(II) oxide

酸化鉄（ⅠⅠⅠ）　*sankatetsu(III)*
iron(III) oxide Fe_2O_3

酸化トリアルキルフォスフィン
sankatoriarukirufosufin
trialkylphosphinoxide R_3PO

酸化トリウム　*sankatoriumu*
thoria ThO_2

酸化ツリウム　*sankatsuriumu*
thulium oxide Tm_2O_3

酸化沃素, 酸化ヨウ素　*sankayouso*
iodine oxide

酸化剤　*sankazai*
oxidant

硫　sulfur (^{16}S)

硫黄　*iou*
sulfur (S, element 16)

硫黄置換基　*iouchikanki*
sulfur substituent

硫黄ドナー　*ioudona-*
sulfur donor

硫黄同位体　*ioudouitai*
sulfur isotope

硫黄複素環化合物
ioufukusokankagoubutsu
sulfur heterocycles

硫黄含有量　*iouganyuuryou*
sulfur content

硫黄原子　*iougenshi*
sulfur atom

硫黄配位子　*iouhaiishi*
sulfur ligand

硫黄化学発光　*ioukagakuhakkou*
sulfur chemiluminescence

硫黄化合物　*ioukagoubutsu*
sulfur compound

硫黄加硫　*ioukaryuu*
sulfur vulcanization

硫黄結合　*iouketsugou*
sulfur bond

硫黄結合開裂　*iouketsugoukairetsu*
sulfur bond cleavage

硫黄結合形成　*iouketsugoukeisei*
sulfur bond formation

硫黄混合物　*ioukongoubutsu*
sulfur mixture

硫黄吸着　*ioukyuuchaku*
sulfur adsorption

硫黄二重結合　*iounijuuketsugou*
sulfur double bond

硫黄濃度　*iounoudo*
sulfur concentration

硫黄細菌　*iousaikin*
sulfur bacterium

硫黄錯体　*iousakutai*
sulfur complex

硫黄酸化物　*iousankabutsu*
sulfur oxide

硫黄添加　*ioutenka*
sulfur addition

硫酸エチル　*ryusanechiru*
diethyl sulfate $(C_2H_5)_2SO_4$

硫安ニッケル鉱　*ryuuannikkerukou*
ullmannite NiSbS

硫銅鉱　*ryuudoukou*
domeykite Cu_3As

硫化　*ryuuka*
sulfurization

硫化亜鉛　*ryuukaaen*
zinc sulfide ZnS

硫化アンチモン　*ryuukaanchimon*
antimony trisulfide Sb_2S_3

硫化アンモニウム　*ryuukaanmoniumu*
ammonium sulfide $(NH_4)_2S$

硫化アルミニウム　*ryuukaaruminiumu*
aluminium sulfide Al_2S_3

硫化バリウム　*ryuukabariumu*
barium sulfide BaS

硫化ビスマス　*ryuukabisumasu*
bismuth sulfide Bi_2S_3

硫化物　*ryuukabutsu*
sulfide M^I_2S

硫化物沈澱　*ryuukabutsuchinden*
sulfide precipitation

硫化物イオン　*ryuukabutsuion*
sulfide ion

硫化物錯体　*ryuukabutsusakutai*
sulfide complex

硫化窒素　*ryuukachisso*
nitrogen sulfide

硫化第一銅　*ryuukadaiichidou*
copper(I) sulfide Cu_2S

硫化第一白金　*ryuukadaiichihakkin*
platinous sulfide PtS

硫化第一錫　*ryuukadaiichisuzu*
stannous sulfide SnS

硫化第一鉄　*ryuukadaiichitetsu*
ferrous sulfide FeS

硫化第二銅　*ryuukadainidou*
copper(II) sulfide CuS

硫化第二白金　*ryuukadainihakkin*
platinum disulfide PtS_2

硫化第二錫　*ryuukadainisuzu*
stannic sulfide SnS_2

硫カドミウム鉱　*ryuukadomiumukou*
greenockite CdS

硫化銅　*ryuukadou*
copper sulfide Cu_2S/CuS

硫化鉛　*ryuukaen*
lead sulfide PbS

硫化フェニル　*ryuukafeniru*
phenyl sulfide $(C_6H_5)_2S$

硫化銀　*ryuukagin*
argentite, silver sulfide Ag_2S

硫化イットリウム　*ryuukaittoriumu*
yttrium sulfide Y_2S_3

硫化ジアルキル　*ryuukajiarukiru*
dialkyl sulfide, thioether R_2S

硫化ジルコニル　*ryuukajirukoniru*
zirconyl sulfide $ZrOS$

硫化カドミウム　*ryuukakadomiumu*
cadmium sulfide CdS

硫化カリウム　*ryuukakariumu*
potassium sulfide K_2S

硫化カルシウム　*ryuukakarushiumu*
calcium sulfide CaS

硫化メチル　*ryuukamechiru*
methyl sulfide $S(CH_3)_2$

硫化ナトリウム　*ryuukanatoriumu*
sodium sulfide Na_2S

硫化水素　*ryuukasuiso*
hydrogen sulfide H_2S

硫化水素塩　*ryuukasuisoen*
hydrogen sulfide M^IHS

硫化水素酸　*ryuukasuisosan*
hydrosulfuric acid $H_2S \cdot aq$

硫化ストロンチウム　*ryuukasutoronchiumu*
strontium sulfide SrS

硫化タングステン鉱　*ryuukatangusutenkou*
tungstenite WS_2

硫化テルル　*ryuukateruru*
tellurium sulfide TeS_2

硫化鉄　*ryuukatetsu*
iron sulfide

硫化油　*ryuukayu*
sulfurized oil

硫酸　*ryuusan*
sulfuric acid H_2SO_4

硫酸亜鉛　*ryuusanaen*
zinc sulfate $ZnSO_4$

硫酸アンモニウム　*ryuusananmoniumu*
ammonium sulfate $(NH_4)_2SO_4$

硫酸アンモニウム肥料　*ryuusananmoniumuhiryou*
ammonium sulfate fertilizer

硫酸アルミニウム　*ryuusanaruminiumu*
aluminium sulfate $Al_2(SO_4)_3$

硫酸バリウム　*ryuusanbariumu*
barium sulfate $BaSO_4$

硫酸ビスマス　*ryuusanbisumasu*
bismuth sulfate $Bi_2(SO_4)_3$

硫酸第一水銀　*ryuusandaiichisuigin*
mercury(I) sulfate Hg_2SO_4

硫酸第一錫　*ryuusandaiichisuzu*
stannous sulfate $SnSO_4$

硫酸第一鉄　*ryuusandaiichitetsu*
iron(II) sulfate $FeSO_4$

硫酸第二水銀　*ryuusandainisuigin*
mercury(II) sulfate $HgSO_4$

硫酸第二錫　*ryuusandainisuzu*
stannic sulfate $Sn(SO_4)_2$

硫酸第二鉄　*ryuusandainitetsu*
iron(III) sulfate $Fe_2(SO_4)_3$

硫酸銅　*ryuusandou*
copper sulfate $CuSO_4$

硫酸塩　*ryuusanen*
sulfate $M^I_2SO_4$

硫酸塩一水和物　*ryuusanenissuiwabutsu*
sulfate monohydrate

硫酸塩混合物　*ryuusanenkongoubutsu*
sulfate mixture

硫酸塩パルプ　*ryuusanenparupu*
sulfate pulp

硫酸塩錯体　*ryuusanensakutai*
sulfate complex

硫酸塩添加　*ryuusanentenka*
sulfate addition

硫酸塩溶液　*ryuusanenyoueki*
sulfate solution

硫酸エステル　*ryuusanesuteru*
sulfuric acid ester

硫酸銀　*ryuusangin*
silver sulfate Ag_2SO_4

硫酸イオン　*ryuusanion*
sulfate ion

硫酸イットリウム　*ryuusanittoriumu*
yttrium sulfate $Y_2(SO_4)_3$

硫酸ジアルキル　*ryuusanjiarukiru*
dialkyl sulfate R_2SO_4

硫酸ジエチル　*ryuusanjiechiru*
diethyl sulfate $(C_2H_5)_2SO_4$

硫酸ジメチル　*ryuusanjimechiru*
dimethyl sulfate $(CH_3)_2SO_4$

硫酸ジルコニウム　*ryuusanjirukoniumu*
zirconium sulfate $Zr(SO_4)_2 \cdot 4H_2O$

硫酸化　*ryuusanka*
sulfation

硫酸カドミウム　*ryuusankadomiumu*
cadmium sulfate $CdSO_4$

硫酸化反応　*ryuusankahannou*
sulfation reaction

硫酸化オリゴ糖　*ryuusankaorigotou*
sulfated oligosaccharide

硫酸カリウム　*ryuusankariumu*
potassium sulfate K_2SO_4

硫酸カルシウム　*ryuusankarushiumu*
calcium sulfate $CaSO_4$

硫酸化ステロイド　*ryuusankasuteroido*
sulfated steroid

硫酸化炭水化物　*ryuusankatansuikabutsu*
sulfated carbohydrate

硫酸化油　*ryuusankayu*
sulfated oil

硫酸化誘導体　*ryuusankayuudoutai*
sulfated derivative

硫酸基　*ryuusanki*
sulfate group

硫酸コンドロイチン
ryuusankondoroichin
chondroitin sulfate

硫酸混合物　*ryuusankongoubutsu*
sulfuric acid mixture

硫酸マグネシウム
ryuusanmaguneshiumu
magnesium sulfate $MgSO_4$

硫酸メチル　*ryuusanmechiru*
methyl sulfate $(CH_3)_2SO_4$

硫酸鉛（ⅠⅠ）　*ryuusannamari(II)*
lead(II) sulfate $PbSO_4$

硫酸ナトリウム　*ryuusannatoriumu*
　sodium sulfate Na_2SO_4

硫酸ニッケル（ⅠⅠ）　*ryuusannikkeru(II)*
　nickel(II) sulfate $NiSO_4$

硫酸濃度　*ryuusannoudo*
　sulfuric acid concentration

硫酸リチウム　*ryuusanrichiumu*
　lithium sulfate Li_2SO_4

硫酸セシウム　*ryuusanseshiumu*
　cesium sulfate Cs_2SO_4

硫酸水素カリウム　*ryuusansuisokariumu*
　potassium hydrogensulfate $KHSO_4$

硫酸水素ナトリウム
　ryuusansuisonatoriumu
　sodium hydrogensulfate $NaHSO_4$

硫酸水溶液　*ryuusansuiyoueki*
　sulfuric acid aqueous solution

硫酸ストロンチウム
　ryuusansutoronchiumu
　strontium sulfate $SrSO_4$

硫酸鉄（ⅠⅠ）　*ryuusantetsu(II)*
　ferrous sulfate $FeSO_4$

硫酸鉄（ⅠⅠⅠ）　*ryuusantetsu(III)*
　ferric sulfate $Fe_2(SO_4)_3$

硫酸ウラン鉱物　*ryuusanurankoubutsu*
　zippeite $K_3(H_2O)_3[(UO_2)_4/(SO_4)_2/O_3(OH)]$

硫炭酸塩　*ryuutansanen*
　thiocarbonate $M^I_2CSO_2$

弗　fluorine (^{9}F)

弗化アンモニウム　*fukkaanmoniumu*
　ammonium fluoride NH_4F

弗化窒素　*fukkachisso*
　nitrogen fluoride NF_3

弗化第一水銀　*fukkadaiichisuigin*
　dimercury difluoride Hg_2F_2

弗化第二銅　*fukkadainidou*
　copper(II) fluoride CuF_2

弗化銀　*fukkagin*
　silver fluoride AgF

弗化硼素, フッ化ホウ素　*fukkahouso*
　boron fluoride BF_3

弗化イッテルビウム　*fukkaitterubiumu*
　ytterbium fluoride YbF_3

弗化カリウム　*fukkakariumu*
　potassium fluoride KF

弗化カルシウム　*fukkakarushiumu*
　calcium fluoride CaF_2

弗化珪素, フッ化ケイ素　*fukkakeiso*
　silicon fluoride SiF_4/Si_2F_6

弗化ナトリウム　*fukkanatoriumu*
　sodium fluoride NaF

弗化沃素　*fukkayouso*
　iodine fluoride IF_5, IF_7

弗酸　*fussan*
　hydrofluoric acid HF

弗素, フッ素　*fusso*
　fluorine (F, element 9)

弗素置換　*fussochikan*
　fluorination, substitution by fluorine

塩　chlorine (^{17}Cl)
[for 塩化 see below]

塩　*en*
　salt

塩分　*enbun*
　salt content, salinity

塩基　*enki*
　base

塩基度　*enkido*
　basicity

塩基配列　*enkihairetsu*
　base sequence

塩基価　*enkika*
　base number

塩基性度　*enkiseido*
　basicity, basic strength

塩基性塩　*enkiseien*
　basic salt

塩基性岩　*enkiseigan*
　basic rock

塩基性顔料　*enkiseiganryou*
　basic pigment

塩基性の　*enkiseino*
　basic

塩基性酢酸銅　*enkiseisakusandou*
　basic copper acetate $Cu(C_2H_3O_2)_2 \cdot CuO/$
　$2Cu(C_2H_3O_2)_2 \cdot CuO$

塩基性酸化物　*enkiseisankabutsu*
　basic oxide

塩基性染料　*enkiseisenryou*
　basic dye

塩基性色　*enkiseishoku*
　basic color

塩基性触媒　*enkiseishokubai*
　basic catalyst

塩基性炭酸銅　*enkiseitansandou*
　basic copper carbonate $CuCO_3 \cdot Cu(OH)_2/$
　$2CuCO_3 \cdot Cu(OH)_2$

塩基性炭酸鉛　*enkiseitansanen*
　basic lead carbonate $2PbCO_3 \cdot Pb(OH)_2$

塩基触媒　*enkishokubai*
　base catalyst

塩基対　*enkitsui*
　base pair

塩効果　*enkouka*
　salt effect

塩橋　*enkyou*
　salt bridge

塩素酸アルミニウム　*enosanaruminiumu*
　aluminium chlorate $Al(ClO_3)_3$

塩類効果　*enruikouka*
　salt effect

塩酸　*ensan*
　hydrochloric acid HCl

塩酸塩　*ensanen*
　hydrochloride

塩酸ヒドロキシルアミン
　ensanhidorokishiruamin
　hydroxylamine hydrochloride $NH_2OH \cdot HCl$

塩酸メパクリン　*ensanmepakurin*
　mepacrin hydrochloride
　$C_{22}H_{30}N_3OCl \cdot 2HCl \cdot 2H_2O$

塩酸メペリジン　*ensanmeperijin*
　meperidin hydrochloride $C_{15}H_{21}NO \cdot HCl$

塩酸モルヒネ, 塩酸モルフィン
　ensanmoruhine, ensanmorufin
　morphine hydrochloride $C_{17}H_{19}NO_3 \cdot HCl$

塩素　*enso*
　chlorine (Cl, element 17)

塩素置換　*ensochikan*
　chlorination, substitution by chlorine

塩素ガス　*ensogasu*
　chlorine gas

塩素ガス中毒　*ensogasuchuudoku*
　chloride gas poisoning

塩素漂白　*ensohyouhaku*
　chlorine bleaching

塩素化　*ensoka*
　chlorination

塩素化ゴム　*ensokagomu*
　chlorinated rubber

塩素酸　*ensosan*
　chloric acid $HClO_3$

塩素酸アンモニウム　*ensosananmoniumu*
　ammonium chlorate NH_4ClO_3

塩素酸塩　*ensosanen*
　chlorate M^IClO_3

塩素酸銀　*ensosangin*
　silver chlorate $AgClO_3$

塩素酸カリウム　*ensosankariumu*
　potassium chlorate $KClO_3$

塩素酸ナトリウム　*ensosannatoriumu*
　sodium chlorate $NaClO_3$

塩素消剤　*ensoshouzai*
　antichlor

塩素水　*ensosui*
　chlorine water

塩素滴定　*ensotekitei*
　chlorometry

塩水　*ensui*
　brine, salt water

塩 chlorine (^{17}Cl) – 塩化

塩化亜鉛　*enkaaen*
　zinc chloride $ZnCl_2$

塩化アンチモン　*enkaanchimon*
　antimony chloride $SbCl_3/SbCl_5$

塩化アンモニウム　*enkaanmoniumu*
　ammonium chloride NH_4Cl

塩化アリル　*enkaariru*
　allyl chloride $H_2C=CHCH_2Cl$

塩化アルミニウム　*enkaaruminiumu*
　aluminium chloride $AlCl_3$

塩化アセチル　*enkaasechiru*
　acetyl chloride CH_3COCl

塩化アセチルコリン　*enkaasechirukorin*
　acetylcholinchloride
　$CH_3COOCH_2CH_2N(CH_3)_3Cl$

塩化アシル　*enkaashiru*
　acyl chloride $RCOCl$

塩化バリウム　*enkabariumu*
　barium chloride $BaCl_2$

塩化ベンジル　*enkabenjiru*
　benzyl chloride $C_6H_5CH_2Cl$

塩化ベンザル　*enkabenzaru*
　benzal chloride $C_6H_5CHCl_2$

塩化ベンゾイル　*enkabenzoiru*
　benzoyl chloride C_6H_5COCl

塩化ベリリウム　*enkabeririumu*
　beryllium chloride $BeCl_2$

塩化ベルベリン　*enkaberuberin*
　berberine hydrochloride $C_{20}H_{18}NO_4Cl$

塩化ビニリデン　*enkabiniriden*
　vinylidene choride $H_2C=CCl_2$

塩化ビニル　*enkabiniru*
　vinyl chloride $H_2C=CHCl$

塩化ボルニル　*enkaboruniru*
　bornyl chloride $C_{10}H_{16}HCl$

塩化物　*enkabutsu*
　chloride

塩化チアミン　*enkachiamin*
　thiamine hydrochloride $C_{12}H_{17}N_4OS·HCl$

塩化チオカルボニル　*enkachiokaruboniru*
　thiophosgene $CSCl_2$

塩化チオニル　*enkachioniru*
　thionyl chloride $SOCl_2$

塩化第一銅　*enkadaiichidou*
　copper(I) chloride $CuCl$

塩化第一白金　*enkadaiichihakkin*
　platinous chloride $PtCl_2$

塩化第一金水素酸　*enkadaiichikinsuisosan*
　chloroaurous acid $HAuCl_2$

塩化第一水銀　*enkadaiichisuigin*
　mercury(I) chloride Hg_2Cl_2

塩化第一スズ　*enkadaiichisuzu*
　tin dichloride $SnCl_2$

塩化第一錫　*enkadaiichisuzu*
　stannous chloride $SnCl_2$

塩化第一鉄　*enkadaiichitetsu*
　ferrous chloride $FeCl_2$

塩化第二銅　*enkadainidou*
　copper(II) chloride $CuCl_2$

塩化第二白金　*enkadainihakkin*
　platinic chloride $PtCl_4$

塩化第二金　*enkadainikin*
　auric chloride $AuCl_3$

塩化第二水銀　*enkadainisuigin*
　mercury(II) chloride $HgCl_2$

塩化第二スズ　*enkadainisuzu*
　tin tetrachloride $SnCl_4$

塩化第二錫　*enkadainisuzu*
　stannic chloride $SnCl_4$

塩化第二鉄　*enkadainitetsu*
　iron trichloride $FeCl_3$

塩化銅（Ⅰ）　*enkadou(I)*
　cuprous chloride $CuCl$

塩化銅（ⅠⅠ）　*enkadou(II)*
　cupric chloride $CuCl_2$

塩化エチレン　*enkaechiren*
　ethylene chloride,
　1,2-dichloroethane $ClCH_2CH_2Cl$

塩化エチル　*enkaechiru*
　ethyl chloride C_2H_5Cl

塩化鉛　*enkaen*
　lead chloride $PbCl_2/PbCl_4$

塩化フェニル　*enkafeniru*
　phenyl chloride C_6H_5Cl

塩化フォルミル　*enkaforumiru*
　formyl chloride $HCOCl$

塩化フォスフェニル　*enkafosufeniru*
　phosphenyl chloride $C_6H_5PCl_2$

塩化フタリル　*enkafutariru*
　phthaloyl chloride $C_6H_4(COCl)_2$

塩化銀　*enkagin*
　silver chloride $AgCl$

塩化白金　*enkahakkin*
　platinum chloride

塩化ホスホリル　*enkahosuhoriru*
　phosphoryl chloride, phosphorus trichloride oxide $POCl_3$

塩化硫黄　*enkaiou*
　sulfur chloride S_2Cl_2

塩化イソブチル　*enkaisobuchiru*
　isobutyl chloride $(CH_3)_2CHCH_2Cl$

塩化イソプロピル　*enkaisopuropiru*
　isopropyl chloride $(CH_3)_2CHCl$

塩化イッテルビウム　*enkaitterubiumu*
　ytterbium chloride $YbCl_3$

塩化イットリウム　*enkaittoriumu*
　yttrium chloride YCl_3

塩化ジルコニル　*enkajirukoniru*
　zirconyl chloride $ZrOCl_2 \cdot 8H_2O$

塩化カドミウム　*enkakadomiumu*
　cadmium chloride $CdCl_2$

塩化カコジル　*enkakakojiru*
　cacodyl chloride, chlorodimethylarsine $(CH_3)_2AsCl$

塩化カプロイル　*enkakapuroiru*
　capryl chloride $CH_3(CH_2)_6CH_2Cl$

塩化カリウム　*enkakariumu*
　potassium chloride KCl

塩化カルボニル　*enkakaruboniru*
　phosgene $COCl_2$

塩化カルシウム　*enkakarushiumu*
　calcium chloride $CaCl_2$

塩化カルシウム管　*enkakarushiumukan*
　calcium chloride tube

塩化金（ＩＩＩ）　*enkakin(III)*
　gold(III) chloride $AuCl_3$

塩化金酸　*enkakinsan*
　chlorauric acid $HAuCl_4 \cdot 4H_2O$

塩化金酸カリウム　*enkakinsankariumu*
　potassium aurate $KAuCl_4 \cdot 0,5H_2O$

塩化キシリレン　*enkakishiriren*
　xylylene chloride $(CH_3)_2C_6H_2Cl_2/(CH_3)(CH_2Cl)C_6H_3Cl/C_6H_4(CH_2Cl)_2$

塩化キシリル　*enkakishiriru*
　xylyl chloride $(CH_3)_2C_6H_3Cl$

塩化クロム（ＩＩ）　*enkakuromu(II)*
　chromium(II)chloride $CrCl_2$

塩化クロム（ＩＩＩ）　*enkakuromu(III)*
　chromium(III)chloride $CrCl_3$

塩化クロルアセチル　*enkakuroruasechiru*
　chloroacetyl chloride $ClCH_2COCl$

塩化マグネシウム　*enkamaguneshiumu*
　magnesium chloride $MgCl_2$

塩化マンガン（ＩＩ）　*enkamangan(II)*
　manganese(II) chloride $MnCl_2$

塩化メチレン　*enkamechiren*
　methylene chloride CH_2Cl_2

塩化メチル　*enkamechiru*
　methyl chloride CH_3Cl

塩化ナトリウム　*enkanatoriumu*
　sodium chloride $NaCl$

塩化ニトリル　*enkanitoriru*
　nitryl chloride NO_2Cl

塩化ニトロシル　*enkanitoroshiru*
　nitrosyl chloride $NOCl$

塩化オクソニウム　*enkaokusoniumu*
　oxonium chloride H_3OCl

塩化オクザリル　*enkaokuzariru*
　oxalyl chloride $(COCl)_2$

塩化パラジウム　*enkaparajiumu*
　palladium chloride

塩化パラルゴニル　*enkapararugoniru*
　pelargonyl chloride $CH_3(CH_2)_7COCl$

塩化パルミチル　*enkaparumichiru*
　palmityl chloride $CH_3(CH_2)_{14}COCl$

塩化ピバロイル　*enkapibaroiru*
　pivaloyl chloride $C(CH_3)_3COCl$

塩化ピクリン　*enkapikurin*
　picryl chloride $ClC_6H_2(NO_2)_3$

塩化プロピオニル　*enkapuropioniru*
　propionyl chloride CH_3CH_2COCl

塩化プロピリジン　*enkapuropirijin*
　propylidene chloride $CH_3CH_2CHCl_2$

塩化プロピル　*enkapuropiru*
　propyl chloride C_3H_7Cl

塩化リチウム　*enkarichiumu*
　lithium chloride $LiCl$

塩化燐　*enkarin*
　phosphorus chloride PCl_3/PCl_5

塩化ロジウム　*enkarojiumu*
　rhodium chloride $RhCl_3$

塩化ルビジウム　*enkarubijiumu*
　rubidium chloride $RbCl$

塩化酢酸　*enkasakusan*
　chloroacetic acid $ClCH_2COOH$

塩化セミカルバチド
　enkasemikarubachido
　semicarbazide hydrochloride
　$H_2NCONHNH_2 \cdot HCl$

塩化セリウム　*enkaseriumu*
　cerium chloride $CeCl_3$

塩化シアン　*enkashian*
　cyanogen chloride $ClCN$

塩化シアヌリル　*enkashianuriru*
　cyanuric chloride $C_3N_3Cl_3$

塩化臭素　*enkashuuso*
　bromine chloride $BrCl$

塩化水銀（Ⅰ）　*enkasuigin(I)*
　mercury(I)chloride Hg_2Cl_2

塩化水銀（Ⅱ）　*enkasuigin(II)*
　mercury(II) chloride $HgCl_2$

塩化水素　*enkasuiso*
　hydrogen chloride HCl

塩化水素酸　*enkasuisosan*
　hydrochloric acid HCl

塩化スルフリル　*enkasurufuriru*
　sulfuryl chloride SO_2Cl_2

塩化ステアリル　*enkasuteariru*
　stearoyl chloride $CH_3(CH_2)_{16}COCl$

塩化ストロンチウム　*enkasutoronchiumu*
　strontium chloride $SrCl_2$

塩化スズ（Ⅱ）,塩化錫（Ⅱ）
　enkasuzu(II)
　stannous chloride $SnCl_2$

塩化スズ（Ⅳ）,塩化錫（Ⅳ）
　enkasuzu(IV)
　stannic chloride $SnCl_4$

塩化炭化水素　*enkatankasuiso*
　chlorinated hydrocarbon

塩化テルル　*enkateruru*
　tellurium chloride $TeCl_2/TeCl_4$

塩化鉄（Ⅲ）　*enkatetsu(III)*
　ferric chloride $FeCl_3$

塩化トリエチルアミン
　enkatoriechiruamin
　triethylamine hydrochloride $N(C_2H_5)_3 \cdot HCl$

塩化トリフェニルメチル
　enkatorifenirumechiru
　triphenylmethyl chloride $(C_6H_5)_3CCl$

塩化トリメチルアンモニウム
　enkatorimechiruanmoniumu
　trimethylammonium chloride $N(CH_3)_3 \cdot HCl$

塩化トリル　*enkatoriru*
　tolyl chloride $CH_3C_6H_4Cl$

塩化トルイル　*enkatoruiru*
　toluoyl chloride $CH_3C_6H_4COCl$

塩化ツボクラリン　*enkatsubokurarin*
　tubocurarine chloride $C_{37}H_{42}N_2O_6Cl_2I$

塩化ウラニル　*enkauraniru*
　uranyl chloride UO_2Cl_2

塩化ヴァレリル　*enkavareriru*
　valeryl chloride $CH_3(CH_2)_3COCl$

塩化ヴィニリデン　*enkaviniriden*
　vinylidene choride $H_2C=CCl_2$

塩化ヴィニル　*enkaviniru*
　vinyl chloride $H_2C=CHCl$

塩化ヨヒンビン　*enkayohinbin*
yohimbine hydrochloride
$C_{21}H_{26}N_2O_3 \cdot HCl$

塩化沃素　*enkayouso*
iodine chloride ICl/ICl_3

塩化ザルコシン　*enkazarukoshin*
sarcosine hydrochloride
$CH_3NHCH_2COOH \cdot HCl$

臭　bromine (^{35}Br)

臭銀鉱　*shuuginkou*
bromyrite AgBr

臭化亜鉛　*shuukaaen*
zinc bromide $ZnBr_2$

臭化アンモニウム　*shuukaanmoniumu*
ammonium bromide NH_4Br

臭化アリル　*shuukaariru*
allyl bromide $H_2C=CHCH_2Br$

臭化アルミニウム　*shuukaaruminiumu*
aluminum bromide $AlBr_3$

臭化アセチル　*shuukaasechiru*
acetyl bromide CH_3COBr

臭化ベンジル　*shuukabenjiru*
benzyl bromide $C_6H_5CH_2Br$

臭化物　*shuukabutsu*
bromide $M^IBr/M^{II}Br_2/M^{III}Br_3$

臭化第一銅　*shuukadaiichidou*
cuprous bromide CuBr

臭化第一水銀　*shuukadaiichisuigin*
mercury(I) bromide Hg_2Br_2

臭化第一錫　*shuukadaiichisuzu*
tin(II) bromide $SnBr_2$

臭化第一鉄　*shuukadaiichitetsu*
ferrous bromide $FeBr_2$

臭化第二銅　*shuukadainidou*
cupric bromide $CuBr_2$

臭化第二水銀　*shuukadainisuigin*
mercury(II) bromide $HgBr_2$

臭化第二錫　*shuukadainisuzu*
tin(IV) bromide $SnBr_4$

臭化第二鉄　*shuukadainitetsu*
ferric bromide $FeBr_3$

臭化エチル　*shuukaechiru*
ethyl bromide C_2H_5Br

臭化フェニルマグネシウム
shuukafenirumaguneshiumu
phenyl magnesium bromide C_6H_5MgBr

臭化銀　*shuukagin*
silver bromide AgBr

臭化硫黄　*shuukaiou*
sulfur bromide S_2Br_2

臭化イッテルビウム　*shuukaitterubiumu*
ytterbium bromide $YbBr_3$

臭化カドミウム　*shuukakadomiumu*
cadmium bromide $CdBr_2$

臭化珪素, 臭化ケイ素　*shuukakeiso*
silicon bromide $SiBr_4/Si_2Br_6$

臭化マグネシウム　*shuukamaguneshiumu*
magnesium bromide $MgBr_2$

臭化メチレン　*shuukamechiren*
methylene bromide CH_2Br_2

臭化メチル　*shuukamechiru*
methyl bromide CH_3Br

臭化ナトリウム　*shuukanatoriumu*
sodium bromide NaBr

臭化リチウム　*shuukarichiumu*
lithium bromide LiBr

臭化燐　*shuukarin*
phosphorus bromide PBr_3, PBr_5

臭化酸基　*shuukasanki*
acid bromide RCOBr

臭化水素　*shuukasuiso*
hydrogen bromide HBr

臭化水素酸　*shuukasuisosan*
hydrobromic acid HBr

臭化水素酸塩　*shuukasuisosanen*
hydrobromide

臭気　*shuuki*
bad smell, odor

臭酸　*shuusan*
hydrobromic acid HBr

臭素 *shuuso*
  bromine (Br, element 35)

臭素置換 *shuusochikan*
  bromination

臭素価 *shuusoka*
  bromine number

臭素化 *shuusoka*
  bromination

臭素酸 *shuusosan*
  bromic acid $HBrO_3$

臭素酸塩 *shuusosanen*
  bromate M^IBrO_3

臭素酸ナトリウム *shuusosannatoriumu*
  sodium bromate $NaBrO_3$

臭素水 *shuusosui*
  bromine water

臭素定量 *shuusoteiryou*
  bromometry

臭素滴定 *shuusotekitei*
  bromometry

臭化カリウム *syuukakariumu*
  potassium bromide KBr

臭素酸カリウム *syuusosankariumu*
  potassium bromate(V) $KBrO_3$

沃 iodine (^{53}I)

沃化ナトリウム *yokukanatoriumu*
  sodium iodide NaI

沃化, ヨウ化 *youka*
  iodination

沃化亜鉛 *youkaaen*
  zinc iodide ZnI_2

沃化物 *youkabutsu*
  iodide

沃化物イオン *youkabutsuion*
  iodide ion

沃化第一水銀 *youkadaiichisuigin*
  mercurous iodide Hg_2I_2

沃化第一錫 *youkadaiichisuzu*
  stannous iodide SnI_2

沃化第二水銀 *youkadainisuigin*
  mercuric iodide HgI_2

沃化第二錫 *youkadainisuzu*
  stannic iodide SnI_4

沃化エチル *youkaechiru*
  ethyl iodide C_2H_5I

沃化鉛 *youkaen*
  lead iodide PbI_2

沃化銀 *youkagin*
  silver iodide AgI

沃化イッテルビウム *youkaitterubiumu*
  ytterbium iodide YbI_3

沃化カリウム *youkakariumu*
  potassium iodide KI

沃化金 *youkakin*
  gold iodide AuI/AuI_3

沃化マグネシウム *youkamaguneshiumu*
  magnesium iodide Mg_2I

沃化メチル *youkamechiru*
  methyl iodide CH_3I

沃化リチウム *youkarichiumu*
  lithium iodide LiI

沃化シアン *youkashian*
  cyanogen iodide ICN

沃化水素 *youkasuiso*
  hydrogen iodide HI

沃化水素酸 *youkasuisosan*
  hydriodic acid HI

沃酸, ヨウ酸 *yousan*
  hydroiodic acid HI

沃素, ヨウ素 *youso*
  iodine (I, element 53)

沃素反応 *yousohannou*
  iodine reaction

沃素イオン *yousoion*
  iodine ion I^-

沃素価 *yousoka*
  iodine number

沃素化 *yousoka*
  iodination

沃素化する *yousokasuru*
  to iodize

沃素酸, ヨウ素酸　*yousosan*
  iodic acid HIO_3

沃素酸塩　*yousosanen*
  iodate M^IIO_3

沃素酸カリウム　*yousosankariumu*
  potassium iodate KIO_3

沃素酸化滴定　*yousosankatekitei*
  iodimetry

沃素酸ナトリウム　*yousosannatoriumu*
  sodium iodate $NaIO_3$

沃素水　*yousosui*
  iodine water

沃素滴定　*yousotekitei*
  iodometry

鉄　iron (^{26}Fe)

鉄器　*tekki*
  ironware

鉄筋コンクリート　*tekkinkonkuri-to*
  reinforced concrete

鉄鉱, 鉄鉱石　*tekkou, tekkouseki*
  iron ore

鉄工所　*tekkoujo*
  ironworks

鉄板　*teppan*
  iron plate

鉄鎖　*tessa*
  iron chain

鉄錯体　*tessakutai*
  iron complex

鉄石　*tesseki*
  iron stone

鉄心　*tesshin*
  iron core

鉄　*tetsu*
  iron (Fe, element 26)

鉄（ＩＩ）塩　*tetsu(II)en*
  iron(II) salt, ferrous salt

鉄（ＩＩＩ）塩　*tetsu(III)en*
  iron (III) salt, ferric salt

鉄媒染　*tetsubaisen*
  iron mordanting

鉄バクテリア　*tetsubakuteria*
  iron bacterium

鉄電極　*tetsudenkyoku*
  iron electrode

鉄不純物　*tetsufujunbutsu*
  iron impurity

鉄腐食　*tetsufushoku*
  iron corrosion

鉄含有酸化物　*tetsuganyuusankabutsu*
  iron containing oxide

鉄原子　*tetsugenshi*
  iron atom

鉄合金　*tetsugoukin*
  iron alloy

鉄配位子　*tetsuhaiishi*
  iron ligand

鉄表面層　*tetsuhyoumensou*
  iron surface layer

鉄磁性　*tetsujisei*
  ferromagnetism

鉄磁性の　*tetsujiseino*
  ferromagnetic

鉄重石　*tetsujuuseki*
  ferberite $FeWO_4$

鉄カーボニル　*tetsuka-boniru*
  iron carbonyl

鉄化合物　*tetsukagoubutsu*
  iron compound

鉄還元　*tetsukangen*
  iron reduction

鉄かんらん石　*tetsukanranseki*
  fayalite Fe_2SiO_4

鉄カルベン錯体　*tetsukarubensakutai*
  iron carbene complex

鉄カルボニル　*tetsukaruboniru*
  iron carbonyl

鉄結合蛋白　*tetsuketsugoutanpaku*
  iron-binding protein

鉄金属間化合物 **tetsukinzokukankagoubutsu**
iron intermetallic compound

鉄キレート **tetsukire-to**
iron chelate

鉄クラスタ **tetsukurasuta**
iron cluster

鉄マンガン重石 **tetsumanganjuuseki**
wolframite (Mn,Fe)WO$_4$

鉄ノナカーボニル **tetsunonaka-boniru**
diiron nonacarbonyl Fe$_2$(CO)$_9$

鉄濃度 **tetsunoudo**
iron concentration

鉄ペンタカーボニル，鉄ペンタカルボニル **tetsupentaka-boniru, tetsupentakaruboniru**
iron pentacarbonyl Fe(CO)$_5$

鉄ポルフィリン **tetsuporufirin**
iron porphyrin

鉄リュイス酸 **tetsuryuisusan**
iron Lewis acid

鉄硫化物 **tetsuryuukabutsu**
iron sulfide FeS

鉄錆 **tetsusabi**
iron rust

鉄細菌 **tetsusaikin**
iron bacterium

鉄酸塩 **tetsusanen**
ferrate

鉄酸化物 **tetsusankabutsu**
iron oxide

鉄酸素複合体 **tetsusansofukugoutai**
iron-oxygen complex

鉄触媒 **tetsushokubai**
iron catalyst

鉄スラグ **tetsusuragu**
iron slag

鉄タングスタイト **tetsutangusutaito**
ferritungstite e.g. Fe$_2$O$_3$・WO$_3$・6H$_2$O

鉄トリカルボニル **tetsutorikaruboniru**
iron tricarbonyl Fe(CO)$_3$

鉄材 **tetsuzai**
iron material

鉄族 **tetsuzoku**
iron group

銅 copper (^{29}Cu)

銅 **dou**
copper (Cu, element 29)

銅（I）塩 **dou(I)en**
copper(I) salt, cuprous salt

銅（II）塩 **dou(II)en**
copper (II) salt, cupric salt

銅安法 **douanhou**
cuprous ammoniacal process

銅アンモニア法 **douanmoniahou**
cuprous ammoniacal process

銅分離 **doubunri**
copper separation

銅置換 **douchikan**
copper substitution

銅同位体 **doudouitai**
copper isotope

銅塩 **douen**
copper salt

銅複合体 **doufukugoutai**
copper complex

銅複合材 **doufukugouzai**
copper composite

銅腐食 **doufushoku**
copper corrosion

銅含有化合物 **douganyuukagoubutsu**
copper containing compound

銅原子 **dougenshi**
copper atom

銅箔 **douhaku**
copper foil

銅イオン交換 **douionkoukan**
copper ion exchange

銅イオン錯体 **douionsakutai**
copper ion complex

銅化 **douka**
cuprization

銅カチオン　*doukachion*
　　copper cation

銅化合物　*doukagoubutsu*
　　copper compound

銅カルコゲン化物　*doukarukogenkabutsu*
　　copper chalcogenide

銅鉱　*doukou*
　　copper ore

銅吸着　*doukyuuchaku*
　　copper adsorption

銅二量体　*douniryoutai*
　　copper dimer

銅ポルフィリン　*douporifirin*
　　copper porphyrin

銅錯体　*dousakutai*
　　copper complex

銅錯体触媒　*dousakutaishokubai*
　　copper complex catalyst

銅酸塩　*dousanen*
　　cuprate(II), cuprite $M^I_2[Cu(OH)_4]$

銅精錬　*douseiren*
　　copper refining

銅触媒　*doushokubai*
　　copper catalyst

銅スラグ　*dousuragu*
　　copper slag

銀　silver (^{47}Ag)

銀　*gin*
　　silver (Ag, element 47)

銀アマルガム　*ginamarugamu*
　　silver amalgam

銀アミド　*ginamido*
　　silver amide $AgNH_2$

銀アセチリド　*ginasechirido*
　　silver acetylide Ag_2C_2

銀塩　*ginen*
　　silver salt

銀箔　*ginhaku*
　　silver leaf, silver foil

銀鉱　*ginkou*
　　silver ore

銀鏡反応　*ginkyouhannou*
　　silver mirror reaction

銀箔　*ginpaku*
　　silver foil

銀シアン化カリウム　*ginshiankakariumu*
　　silver potassium cyanide $KAg(CN)_2$

銀滴定　*gintekitei*
　　argentometry

銀黝銅鉱　*ginyuudoukou*
　　freibergite $(Ag,Cu,Fe)_{12}(Sb,As)_4S_{13}$

金　gold (^{79}Au)

金型　*kanagata*
　　metal pattern, metal mold

金気　*kanake*
　　metallic taste, taste of iron

金錆　*kanasabi*
　　rust

金　*kin*
　　gold (Au, element 79)

金原子　*kingenshi*
　　metallic atom

金銀　*kingin*
　　gold and silver

金箔　*kinhaku*
　　gold leaf

金鉱　*kinkou*
　　gold ore

金粉　*kinpun*
　　gold dust

金酸　*kinsan*
　　auric acid, gold hydroxide $Au(OH)_3$

金酸塩　*kinsanen*
　　aurate $M^I[Au(OH)_4]$

金石　*kinseki*
　　minerals and rocks

金浴　*kinyoku*
　　metallic bath

金属 *kinzoku*
　metal

金属アミド　*kinzokuamido*
　metallic amide M^INH_2

金属アルキル　*kinzokuarukiru*
　metallic alkyl

金属アルコラート　*kinzokuarukora-to*
　metallic alcoholate

金属アザイド　*kinzokuazaido*
　metallic azide M^IN_3

金属導体　*kinzokudoutai*
　metallic conductor

金属塩　*kinzokuen*
　metallic salt

金属元素　*kinzokugenso*
　metallic element

金属箔　*kinzokuhaku*
　metallic foil

金属被覆　*kinzokuhifuku*
　metallic coating

金属疲労　*kinzokuhirou*
　metal fatigue

金属化　*kinzokuka*
　metallization

金属カーボニル　*kinzokuka-boniru*
　metal carbonyl

金属間　*kinzokukan*
　intermetallic

金属間化合物　*kinzokukankagoubutsu*
　intermetallic compound

金属カルボニル　*kinzokukaruboniru*
　metal carbonyl

金属結合　*kinzokuketsugou*
　metallic bond

金属コンダクタンス　*kinzokukondakutansu*
　metallic conduction

金属酵素　*kinzokukouso*
　metalloenzyme

金属光沢　*kinzokukoutaku*
　metallic luster

金属ナトリウム　*kinzokunatoriumu*
　metallic sodium

金属熱気　*kinzokunekki*
　metallic vapor

金属燐　*kinzokurin*
　metallic phosphorus

金属錯塩　*kinzokusakuen*
　metallic complex salt

金属錯体　*kinzokusakutai*
　metal complex

金属酸化物　*kinzokusankabutsu*
　metallic oxide

金属製　*kinzokusei*
　made of metal

金属組織学　*kinzokusoshikigaku*
　metallography

金属タンパク質　*kinzokutanpakushitsu*
　metalloprotein

金属浴　*kinzokuyoku*
　metallic bath

金剛石　*kongouseki*
　diamond

金剛砂　*kongousha*
　emery

8.3
Scientific Terms Beginning with Characters Frequently Appearing in the Initial Position of Chemical Terms

化　change, -ization

化学　*kagaku*
　chemistry

化学分解　*kagakubunkai*
　chemical decomposition, chemolysis

化学分析　*kagakubunseki*
　chemical analysis

化学物質　*kagakubusshitsu*
　chemical

化学物理学　*kagakubutsurigaku*
　physical chemistry

化学エネルギー　*kagakuenerugi-*
　chemical energy

化学元素　*kagakugenso*
　chemical element

化学現像　*kagakugenzou*
　chemical development

化学合成　*kagakugousei*
　chemical synthesis

化学はかり　*kagakuhakari*
　chemical balance

化学発光　*kagakuhakkou*
　chemiluminescence

化学反応　*kagakuhannou*
　chemical reaction

化学反応動力学　*kagakuhannoudourikigaku*
　chemical reaction kinetics

化学反応工学　*kagakuhannoukougaku*
　chemical reaction engineering

化学反応力学　*kagakuhannourikigaku*
　chemical reaction kinetics

化学反応式　*kagakuhannoushiki*
　chemical reaction formula

化学平衡　*kagakuheikou*
　chemical equilibrium

化学変化　*kagakuhenka*
　chemical change

化学品　*kagakuhin*
　chemical

化学肥料　*kagakuhiryou*
　chemical fertilizer

化学方程式　*kagakuhouteishiki*
　chemical equation

化学院　*kagakuin*
　chemistry institute

化学イオン化　*kagakuionka*
　chemical ionization

化学純の　*kagakujunno*
　chemically pure

化学受容体　*kagakujuyoutai*
　chemoreceptor

化学化合物　*kagakukagoubutsu*
　chemical compound

化学研究　*kagakukenkyuu*
　chemical research

化学結合　*kagakuketsugou*
　chemical bond

化学記号　*kagakukigou*
　chemical symbol

化学機械　*kagakukikai*
　chemical machinery

化学工学　*kagakukougaku*
　chemical engineering, chemical technology

化学工業　*kagakukougyou*
　chemical industry

化学工場　*kagakukoujou*
　chemical factory

化学構造　*kagakukouzou*
　chemical structure

化学吸着　*kagakukyuuchaku*
　chemisorption

化学熱力学　*kagakunetsurikigaku*
　chemical thermodynamics

化学汚染　*kagakuosen*
　chemical pollution

化学パルポ　*kagakuparupo*
　chemical pulp

化学ポテンシャル　*kagakupotensharu*
　chemical potential

化学プロセス　*kagakupurosesu*
　chemical process

化学レセプター　*kagakureseputa-*
　chemoreceptor

化学レーザー　*kagakure-za-*
　chemical laser

化学力学　*kagakurikigaku*
　chemical dynamics

化学ルミネセンス　*kagakuruminesensu*
  chemiluminescence
化学療法　*kagakuryouhou*
  chemotherapy
化学量論　*kagakuryouron*
  stoichiometry
化学量数　*kagakuryousuu*
  stoichiometric number
化学作用　*kagakusayou*
  chemical action
化学製品　*kagakuseihin*
  chemical
化学製法　*kagakuseihou*
  chemical process, chemical preparation
化学生産　*kagakuseisan*
  chemical production
化学繊維　*kagakuseni*
  chemical fiber
化学者　*kagakusha*
  chemist
化学シフト　*kagakushifuto*
  chemical shift
化学試験　*kagakushiken*
  chemical experiment
化学式　*kagakushiki*
  chemical formula
化学式量　*kagakushikiryou*
  chemical formular weight
化学進化　*kagakushinka*
  chemical evolution
化学親和力　*kagakushinwaryoku*
  chemical affinity
化学組成　*kagakusosei*
  chemical composition
化学定数　*kagakuteisuu*
  chemical constant
化学的安定性　*kagakutekianteisei*
  chemical stability
化学的純度　*kagakutekijundo*
  chemical purity
化学的な　*kagakutekina*
  chemical

化学的酸素要求量
  kagakutekisansoyoukyuuryou
  chemical oxygen demand
化学天秤　*kagakutenbin*
  chemical balance
化学当量　*kagakutouryou*
  chemical equivalent
化学薬品　*kagakuyakuhin*
  chemical
化合　*kagou*
  combination
化合物　*kagoubutsu*
  compound
化合熱　*kagounetsu*
  heat of combination
化工　*kakou*
  chemical engineering (abbr.)
化成　*kasei*
  change, transformation, formation
化成物　*kaseibutsu*
  industrial chemical
化成肥料　*kaseihiryou*
  chemical fertilizer
化成工業　*kaseikougyou*
  chemical industry
化石燃料　*kasekinenryou*
  fossil fuel
化粧品　*keshouhin*
  cosmetic

水　water

水面　*minamo, suimen*
  water surface
水　*mizu*
  water H_2O
水時計　*mizudokei*
  water clock
水グラス　*mizugurasu*
  water glass $M^I_2O \cdot (SiO_2)_x$

水ジャケット　*mizujaketto*
　water jacket

水の硬度　*mizunokoudo*
　water harndess

水の軟化　*mizunonanka*
　water softening

水の軟化剤　*mizunonankazai*
　water softener

水の面　*mizunoomo*
　water surface

水圧　*suiatsu*
　water pressure

水圧計　*suiatsukei*
　water pressure gauge

水瓶　*suibin*
　water bottle

水分　*suibun*
　water content

水分析　*suibunseki*
　water analysis

水鉛　*suien*
　molybdenum (Mo, element 42)

水鉛鉛鉱　*suienenkou*
　wulfenite $PbMoO_4$

水銀　*suigin*
　mercury (Hg, element 80)

水銀圧力計　*suiginatsuryokukei*
　mercury manometer

水銀中毒　*suiginchuudoku*
　mercury poisoning

水銀電池　*suigindenchi*
　mercury cell

水銀合金　*suigingoukin*
　mercury alloy

水銀蒸気　*suiginjouki*
　mercury vapor

水銀化　*suiginka*
　mercuration

水銀温度計　*suiginondokei*
　mercurial thermometer

水銀灯　*suigintou*
　mercury lamp

水銀浴　*suiginyoku*
　mercury bath

水砒亜鉛鉱　*suihiaenkou*
　adamine $4ZnO \cdot As_2O_5 \cdot H_2O$

水位　*suii*
　water level

水位計　*suiikei*
　water gauge

水蒸気　*suijouki*
　steam, water vapor

水蒸気分解　*suijoukibunkai*
　steam cracking

水蒸気蒸留　*suijoukijouryuu*
　steam distillation

水蒸気浴　*suijoukiyoku*
　steam bath

水化物　*suikabutsu*
　hydrate

水解物　*suikaibutsu*
　hydrolysate

水密　*suimitsu*
　waterproof

水温　*suion*
　water temperature

水冷　*suirei*
　water-cooling

水力学　*suirikigaku*
　hydraulics

水量計　*suiryoukei*
　water gauge

水流　*suiryuu*
　water current

水流ポンプ　*suiryuuponpu*
　water jet pump

水酸イオン　*suisanion*
　hydroxide ion OH^-

水酸化　*suisanka*
　hydroxylation

水酸化亜鉛　*suisankaaen*
　zinc hydroxide $Zn(OH)_2$

水酸化アンモニウム　*suisankaanmoniumu*
ammonium hydroxide NH₄OH

水酸化アルミニウム　*suisankaaruminiumu*
aluminium hydroxide Al(OH)₃

水酸化バリウム　*suisankabariumu*
barium hydroxide Ba(OH)₂

水酸化ベリリウム　*suisankabeririumu*
beryllium hydroxide Be(OH)₂

水酸化ビスマス　*suisankabisumasu*
bismuth hydroxyde Bi(OH)₃

水酸化物　*suisankabutsu*
hydroxide

水酸化物イオン　*suisankabutsuion*
hydroxide ion, hydroxyl ion OH⁻

水酸化チタン　*suisankachitan*
titanium hydroxide Ti(OH)₄

水酸化第一銅　*suisankadaiichidou*
copper(I) hydroxide CuOH

水酸化第一鉄　*suisankadaiichitetsu*
ferrous hydroxide Fe(OH)₂

水酸化第二銅　*suisankadainidou*
copper(II) hydroxide Cu(OH)₂

水酸化第二鉄　*suisankadainitetsu*
ferric hydroxide Fe(OH)₃

水酸化ガリウム　*suisankagariumu*
gallium hydroxide Ga(OH)₃

水酸化ジルコニウム　*suisankajirukoniumu*
zirconium hydroxide Zr(OH)₄

水酸化カドミウム　*suisankakadomiumu*
cadmium hydroxide Cd(OH)₂

水酸化カリウム　*suisankakariumu*
potassium hydroxide KOH

水酸化カルシウム　*suisankakarushiumu*
calcium hydroxide Ca(OH)₂

水酸化クロム　*suisankakuromu*
chromium hydroxide Cr(OH)₃

水酸化マグネシウム　*suisankamaguneshiumu*
magnesium hydroxide, magnesium hydrate, brucite Mg(OH)₂

水酸化ナトリウム　*suisankanatoriumu*
sodium hydroxide NaOH

水酸化ニッケル（Ⅱ）　*suisankanikkeru (II)*
nickel hydroxide Ni(OH)₂

水酸化リチウム　*suisankarichiumu*
lithium hydroxide LiOH

水酸化セシウム　*suisankaseshiumu*
cesium hydroxide CsOH

水酸化ストロンチウム　*suisankasutoronchiumu*
strontium hydroxide Sr(OH)₂

水酸化テトラアルキルアンモニウム　*suisankatetoraarukiruanmoniumu*
tetraalkylammonium hydroxide R₄NOH

水酸基　*suisanki*
hydroxyl group

水性　*suisei*
aqueous, hydrous, watery

水性ガス　*suiseigasu*
water gas

水性乳剤　*suiseinyuuzai*
aqueous emulsion

水性塗料　*suiseitoryou*
water paint

水質　*suishitsu*
water quality

水質汚濁　*suishitsuodaku*
water polution

水晶　*suishou*
crystal, quartz

水素　*suiso*
hydrogen (H, element 1)

水素電極　*suisodenkyoku*
hydrogen electrode

水素ガス　*suisogasu*
hydrogen gas

水素原子　*suisogenshi*
hydrogen atom

水素原子核　*suisogenshikaku*
hydrogen nucleus

水素原子供与体　*suisogenshikyouyotai*
hydrogen donor

水素イオン　*suisoion*
　hydrogen ion, proton H^+

水素イオン濃度　*suisoionnoudo*
　hydrogen ion concentration

水素受容体　*suisojuyoutai*
　hydrogen acceptor

水素価　*suisoka*
　hydrogen value

水素化　*suisoka*
　hydrogenation

水素化アルミニウムリチウム
　suisokaaruminiumurichiumu
　lithium aluminium hydride $LiAlH_4$

水素化分解　*suisokabunkai*
　hydrogenolysis

水素化物　*suisokabutsu*
　hydride

水素化ホウ素ナトリウム
　suisokahousonatoriumu
　sodium borohydride $NaBH_4$

水素化ホウ素リチウム
　suisokahousorichiumu
　lithium borohydride $LiBH_4$

水素化ジルコニウム　*suisokajirukoniumu*
　zirconium hydride ZrH_2

水素化カルシウム　*suisokakarushiumu*
　calcium hydride CaH_2

水素化ナトリウム　*suisokanatoriumu*
　sodium hydride NaH

水素化パラジウム　*suisokaparajiumu*
　palladium hydride Pd_2H

水素化リチウム　*suisokarichiumu*
　lithium hydride LiH

水素化ストロンチウム
　suisokasutoronchiumu
　strontium hydride SrH_2

水素化油　*suisokayu*
　hydrogenated oil

水素結合　*suisoketsugou*
　hydrogen bond

水素供与体　*suisokyouyotai*
　hydrogen donor

水素担体　*suisotantai*
　hydrogen carrier

水素添加　*suisotenka*
　hydrogenation

水和物　*suiwabutsu*
　hydrate

水和エネルギー　*suiwaenerugi-*
　hydration energy

水和熱　*suiwanetsu*
　heat of hydration

水和作用　*suiwasayou*
　hydration

水浴　*suiyoku*
　water bath

水溶液　*suiyoueki*
　aqueous solution

水溶性ビタミン　*suiyouseibitamin*
　water-soluble vitamin

水溶性の　*suiyouseino*
　water-soluble

水溶性溶剤　*suiyouseiyouzai*
　water-soluble solvent

中　neutral, center, inside

中毒　*chuudoku*
　poisoning

中毒反応　*chuudokuhannou*
　toxic reaction

中毒化学　*chuudokukagaku*
　toxicological chemistry

中毒量　*chuudokuryou*
　toxic dose

中位環　*chuuikan*
　medium ring

中員環ラクトン　*chuuinkanrakuton*
　medium-sized lactone

中員環サイズ　*chuuinkansaizu*
　medium ring size

中間物　*chuukanbutsu*
　intermediate

中間電極 *chuukandenkyoku*
bipolar electrode

中間反応 *chuukanhannou*
intermediate reaction

中間膜 *chuukanmaku*
interlayer

中間生成物 *chuukanseiseibutsu*
intermediate

中間相 *chuukansou*
mesophase

中間体 *chuukantai*
intermediate

中間体合成 *chuukantaigousei*
intermediate synthesis

中間体濃度 *chuukantainoudo*
intermediate concentration

中空ガラス *chuukuugarasu*
hollow glass

中空繊維 *chuukuuseni*
hollow fiber

中空糸 *chuukuushi*
hollow fiber

中空炭素繊維 *chuukuutansoseni*
hollow carbon fiber

中温度 *chuuondo*
moderate temperature

中性 *chuusei*
neutrality

中性アミノ酸 *chuuseiaminosan*
neural amino acid

中性分子 *chuuseibunshi*
neutral molecule

中性窒素 *chuuseichisso*
neutral nitrogen

中性塩 *chuuseien*
neutral salt

中性塩化物 *chuuseienkabutsu*
neutral chloride

中性フィルター *chuuseifiruta-*
neutral filter

中性ガス *chuuseigasu*
neutral gas

中性形 *chuuseigata*
neutral form

中性原子 *chuuseigenshi*
neutral atom

中性ゲスト分子 *chuuseigesutobunshi*
neutral guest molecule

中性配位子 *chuuseihaiishi*
neutral ligand

中性反応 *chuuseihannou*
neutral reaction

中性表面 *chuuseihyoumen*
neutral surface

中性条件 *chuuseijouken*
neutral condition

中性化 *chuuseika*
neutralization

中性化合物 *chuuseikagoubutsu*
neutral compound

中性クラスタ *chuuseikurasuta*
neutral cluster

中性ミセル *chuuseimiseru*
neutral micelle

中性の *chuuseino*
neutral, indifferent, uncharged

中性パラジウム錯体
chuuseiparajiumusakutai
neutral palladium complex

中性ｐＨ *chuuseiph*
neutral pH

中性ポリマ *chuuseiporima*
neural polymer

中性ラジカル *chuuseirajikaru*
neutral radical

中性錯体 *chuuseisakutai*
neutral complex

中性酸化物 *chuuseisankabutsu*
neutral oxide

中性成分 *chuuseiseibun*
neutral component

中性生成物 *chuuseiseiseibutsu*
neutral product

中性子 *chuuseishi*
  neutron

中性脂肪 *chuuseishibou*
  neutral fat

中性子エネルギー *chuuseishienerugi-*
  neutron energy

中性子回折 *chuuseishikaisetsu*
  neutron diffraction

中性水 *chuuseisui*
  neutral water

中性水溶液 *chuuseisuiyoueki*
  neutral aqueous solution

中性担体 *chuuseitantai*
  neutral carrier

中性多糖類 *chuuseitatourui*
  neutral polysaccharide

中性点 *chuuseiten*
  neutral point

中性溶液 *chuuseiyoueki*
  neutral solution

中性有機分子 *chuuseiyuukibunshi*
  neutral organic molecule

中心 *chuushin*
  center

中心原子 *chuushingenshi*
  central atom

中心位 *chuushini*
  centric position

中心軸 *chuushinjiku*
  medial axis

中心窩洞 *chuushinkadou*
  central cavity

中心金属原子 *chuushinkinzokugenshi*
  central metal atom

中心金属イオン *chuushinkinzokuion*
  central metal ion

中心錯体 *chuushinsakutai*
  center complex

中心性キラリティー *chuushinseikirariti-*
  central chirality

中心線 *chuushinsen*
  center line

中心炭素 *chuushintanso*
  central carbon

中和 *chuuwa*
  neutralization

中和度 *chuuwado*
  neutralization degree

中和反応 *chuuwahannou*
  neutralization reaction

中和法 *chuuwahou*
  neutralization process

中和曲線 *chuuwakyokusen*
  neutralization curve

中和熱 *chuuwanetsu*
  heat of neutralization

中和プロセス *chuuwapurosesu*
  neutralization process

中和性 *chuuwasei*
  neutralizing property

中和剤 *chuuwazai*
  neutralizer

分 part, degree, minute, rate, change

分 *bun*
  minute

分圧 *bunatsu*
  partial pressure

分別 *bunbetsu*
  fractionation

分別沈澱 *bunbetsuchinden*
  fractional precipitation

分別抽出 *bunbetsuchuushutsu*
  fractional extraction

分別フラスコ *bunbetsufurasuko*
  fractionating flask

分別蒸留 *bunbetsujouryuu*
  fractional distillation

分別結晶 *bunbetsukesshou*
  fractional crystallization

分別濾過 *bunbetsuroka*
  fractional filtration

分別昇華 *bunbetsushouka*
fractional sublimation

分別晶出 *bunbetsushoushutsu*
fractional crystallization

分液フラスコ *bunekifurasuko*
separating vessel

分液ロート *bunekiro-to*
separating funnel

分液漏斗 *bunekirouto*
separating funnel

分化 *bunka*
differentiation

分解 *bunkai*
decomposition, breakdown, cracking, splitting

分解圧 *bunkaiatsu*
decomposition pressure

分解ガス *bunkaigasu*
cracked gas

分解法 *bunkaihou*
cracking process

分解能 *bunkainou*
resolving power

分解炉 *bunkairo*
cracking furnace

分解産物 *bunkaisanbutsu*
decomposition product

分解油 *bunkaiyu*
cracked oil

分解残油 *bunkaizanyu*
cracked residue

分画 *bunkaku*
fraction, fractionation

分岐 *bunki*
divergence

分岐鎖 *bunkisa*
branched chain

分光分布 *bunkoubunpu*
spectral distribution

分光分析 *bunkoubunseki*
spectroscopic analysis, spectrochemical analysis

分光学 *bunkougaku*
spectroscopy

分光法 *bunkouhou*
spectroscopy

分光化学系列 *bunkoukagakukeiretsu*
spectrochemical series

分光計 *bunkoukei*
spectrometer

分光器 *bunkouki*
spectroscope

分光光度分析 *bunkoukoudobunseki*
spectrophotometric analysis

分光光度計 *bunkoukoudokei*
spectrophotometer

分光写真 *bunkoushashin*
spectrogram

分光写真器 *bunkoushashinki*
spectrograph

分光測光 *bunkousokkou*
spectrophotometry

分光測定 *bunkousokutei*
spectrometry

分極 *bunkyoku*
polarization

分極率 *bunkyokuritsu*
polarizability

分配関数 *bunpaikansuu*
partition function, distribution function

分配係数 *bunpaikeisuu*
partition coefficient

分配クロマトグラフィー *bunpaikuromatogurafi-*
partition chromatography

分泌 *bunpitsu*
secretion

分泌腺 *bunpitsusen*
secretory gland

分布 *bunpu*
distribution

分布係数 *bunpukeisuu*
distribution coefficient, partition coefficient

分布曲線　*bunpukyokusen*
distribution curve

分布則　*bunpusoku*
distribution law, partition law

分裂　*bunretsu*
splitting, fission, cleavage

分裂エネルギー　*bunretsuenerugi-*
fission energy

分裂法　*bunretsuhou*
cracking process

分裂過程　*bunretsukatei*
breakage process

分裂生成物　*bunretsuseiseibutsu*
cleavage product, fission product

分離　*bunri*
separation, isolation, segregation

分離媒体　*bunribaitai*
separation medium

分離度　*bunrido*
degree of isolation

分離反応器　*bunrihannouki*
separation reactor

分離法　*bunrihou*
separation method

分離方法　*bunrihouhou*
separation method, isolation method

分離器　*bunriki*
separator

分離効果　*bunrikouka*
separation effect

分離膜　*bunrimaku*
separation membrane

分離プロセス　*bunripurosesu*
separation process

分離力　*bunriryoku*
separating power

分離精製　*bunriseisei*
purification by separation

分離速度　*bunrisokudo*
separation speed

分離促進　*bunrisokushin*
induced separation

分離相　*bunrisou*
separated phase

分離する　*bunrisuru*
to isolate, to separate

分類　*bunrui*
classification

分類学　*bunruigaku*
systematics

分溜管　*bunryuukan*
dephlegmator

分留カラム, 分留塔, 分留管
bunryuukaramu, bunryuutou, bunryuukan
fractionating column, separation column

分散　*bunsan*
dispersion

分散媒　*bunsanbai*
dispersion medium

分散度　*bunsando*
degree of dispersion

分散系　*bunsankei*
dispersed system

分散力　*bunsanryoku*
dispersion force

分散性　*bunsansei*
dispersibility

分散染料　*bunsansenryou*
dispersed dye

分散相　*bunsansou*
dispersed phase

分散剤　*bunsanzai*
dispersant, dispersing agent

分析　*bunseki*
analysis

分析学　*bunsekigaku*
analytics

分析誤差　*bunsekigosa*
analytical error

分析法　*bunsekihou*
analytical method

分析表　*bunsekihyou*
analysis table

分析化学　bunsekikagaku
analytical chemistry

分析結果　bunsekikekka
analytical result

分析する　bunsekisuru
to analyze

分析天秤　bunsekitenbin
analytical balance

分子　bunshi
molecule

分子分光学　bunshibunkougaku
molecular spectroscopy

分子伝導度　bunshidendoudo
molecular conductivity

分子動力学　bunshidourikigaku
molecular dynamics

分子ふるい　bunshifurui
molecular sieve

分子配置　bunshihaichi
molecular arrangement, molecular configuration

分子配向　bunshihaikou
molecular orientation

分子比　bunshihi
molar ratio, mole ratio

分子イオン　bunshiion
molecular ion

分子蒸留　bunshijouryuu
molecular distillation

分子化合物　bunshikagoubutsu
molecular compound

分子拡散　bunshikakusan
molecular diffusion

分子間反応　bunshikanhannou
intermolecular reaction

分子間の　bunshikanno
intermolecular

分子間力　bunshikanryoku
intermolecular force

分子間縮合　bunshikanshukugou
intermolecular condensation

分子間水素結合　bunshikansuisoketsugou
intermelecular hydrogen bridge

分子軌道　bunshikidou
molecular orbital

分子コロイド　bunshikoroido
molecular colloid

分子構造　bunshikouzou
molecular structure

分子クラスター　bunshikurasuta-
molecular cluster

分子吸光係数　bunshikyuukoukeisuu
molar extinction coefficient

分子模型　bunshimokei
molecular model

分子内塩　bunshinaien
inner salt

分子内反応　bunshinaihannou
intramolecular reaction

分子内無水物　bunshinaimusuibutsu
intramolecular anhydride

分子内の　bunshinaino
intramolecular

分子内錯塩　bunshinaisakuen
inner complex salt

分子内縮合　bunshinaishukugou
intramolecular condensation

分子内転位　bunshinaiteni
intramolecular rearrangement

分子熱　bunshinetsu
molar heat

分子の大きさ　bunshinoookisa
molecular size

分子量　bunshiryou
molecular weight

分子量分布　bunshiryoubunpu
molecular-weight distribution

分子流　bunshiryuu
molecular flow

分子生物学　bunshiseibutsugaku
molecular biology

分子性結晶　bunshiseikesshou
molecular crystal

分子線　*bunshisen*
　molecular beam

分子説　*bunshisetsu*
　molecular theory

分子式　*bunshishiki*
　molecular formula

分子スペクトル　*bunshisupekutoru*
　molecular spectrum

分子容　*bunshiyou*
　molar volume

分銅　*fundou*
　weight

石 stone

石　*ishi*
　stone, brick, rock

石墨　*sekiboku*
　graphite

石英　*sekiei*
　quartz SiO_2

石英ガラス　*sekieigarasu*
　silica glass, quartz glass

石英結晶　*sekieikesshou*
　quartz crystal

石英水銀アーク灯　*sekieisuigina-kutou*
　quartz mercury arc lamp

石版印刷　*sekihaninsatsu*
　lithography

石綿　*sekimen*
　asbestos

石蝋　*sekirou*
　paraffin, paraffin wax

石炭　*sekitan*
　coal, mineral coal

石炭分析　*sekitanbunseki*
　coal analysis

石炭ガス　*sekitangasu*
　coal gas

石炭化　*sekitanka*
　coalification

石炭酸　*sekitansan*
　phenol C_6H_5OH

石炭酸塩　*sekitansanen*
　phenolate C_6H_5OM

石炭水素添加　*sekitansuisotenka*
　coal hydrogenation

石油　*sekiyu*
　petroleum, mineral oil

石油ベンジン　*sekiyubenjin*
　petroleum benzine

石油エーテル　*sekiyue-teru*
　petroleum ether

石油化学　*sekiyukagaku*
　petrochemistry

石油化学工業　*sekiyukagakukougyou*
　petrochemical industry

石油化学製品　*sekiyukagakuseihin*
　petrochemicals

石油コークス　*sekiyuko-kusu*
　petroleum coke

石油留分　*sekiyuryuubun*
　petroleum fraction

石油酸　*sekiyusan*
　petroleum acid

石油精製　*sekiyuseisei*
　oil refining

石灰　*sekkai*
　lime

石灰フェライト　*sekkaiferaito*
　calcioferrite $Ca_3Fe_2(PO_4)_4 \cdot Fe(OH)_3 \cdot 8H_2O$

石灰岩　*sekkaigan*
　limestone $CaCO_3$

石灰肥料　*sekkaihiryou*
　lime fertilizer

石灰石　*sekkaiseki*
　limestone $CaCO_3$

石灰質の水　*sekkaishitsunomizu*
　calcerous water, hard water

石灰水　*sekkaisui*
　lime water

石灰トリウム石　*sekkaitoriumuseki*
　calciothorite $SiO_2 \cdot ThO_2 \cdot (Ce,Y)_2O_3 \cdot Al_2O_3 \cdot Mn_2O_3 \cdot CaO \cdot Na_2O \cdot H_2O$

石鹸　*sekken*
　soap

石器　*sekki*
　stoneware

生　bio-, life

生糸　*kiito*
　raw silk

生ゴム　*namagomu*
　crude rubber, raw rubber

生分解　*seibunkai*
　biodegradation

生分解性の　*seibunkaiseino*
　biodegradable

生物分解　*seibutsubunkai*
　biodegradation

生物物理学　*seibutsubutsurigaku*
　biophysics

生物学的平衡　*seibutsugakutekiheikou*
　biological equilibrium

生物学的活性度　*seibutsugakutekikasseido*
　biological activity

生物発光　*seibutsuhakkou*
　bioluminescence

生物発生　*seibutsuhassei*
　biogenesis

生物変換　*seibutsuhenkan*
　bioconversion

生物価　*seibutsuka*
　biological value

生物活性物質　*seibutsukasseibusshitsu*
　biologically active compound

生物検定　*seibutsukentei*
　bioassay

生物工学　*seibutsukougaku*
　bioengineering, biotechnology

生物濃縮　*seibutsunoushuku*
　bioaccumulation

生物量　*seibutsuryou*
　biomass

生物測定　*seibutsusokutei*
　biometry, biometrics

生物体　*seibutsutai*
　organism

生物有機化学　*seibutsuyuukikagaku*
　bioorganic chemistry

生エネルギー　*seienerugi-*
　bioenergetics

生元素　*seigenso*
　bioelement

生合成　*seigousei*
　biosynthesis

生化学　*seikagaku*
　biochemistry

生化学的分解　*seikagakutekibunkai*
　biochemical decomposition

生化学的変換　*seikagakutekihenkan*
　biochemical transformation

生化学的プロセス　*seikagakutekipurosesu*
　biochemical process

生活反応　*seikatsuhannou*
　vital reaction

生活環　*seikatsukan*
　life cycle

生活形　*seikatsukei*
　lifeform

生命科学　*seimeikagaku*
　life sciences

生理学　*seirigaku*
　physiology

生理学的平衡　*seirigakutekiheikou*
　physiological equilibrium

生理化学　*seirikagaku*
　physiological chemistry

生理的食塩水　*seiritekishokuensui*
　physiological saline solution

生産 *seisan*
production, manufacturing

生産分析 *seisanbunseki*
production analysis

生産物 *seisanbutsu*
product

生産物阻害 *seisanbutsusogai*
product inhibition

生産技術 *seisangijutsu*
production technology

生産費 *seisanhi*
production costs

生産方法 *seisanhouhou*
production process, manufacturing process

生産過剰 *seisankajou*
overproduction

生産効率 *seisankouritsu*
production effciency

生産プロセス *seisanpurosesu*
manufacturing process

生産サイクル *seisansaikuru*
production cycle

生産性 *seisansei*
productivity

生産線 *seisansen*
production line

生産設備 *seisansetsubi*
production facility

生産材料 *seisanzairyou*
production material

生成物 *seiseibutsu*
product

生成エンタルピー *seiseientarupi-*
enthalpy of formation

生成熱 *seiseinetsu*
heat of formation

生成定数 *seiseiteisuu*
formation constant

生体分子 *seitaibunshi*
biomolecule

生態系 *seitaikei*
ecosystem

生体コロイド *seitaikoroido*
biocolloid

生体高分子 *seitaikoubunshi*
biopolymer

生体内の *seitainaino*
in vivo

生体力学 *seitairikigaku*
biomechanics

生体制御 *seitaiseigyo*
biological control

生体触媒 *seitaishokubai*
biocatalyst

生存率 *seizonritsu*
survival rate, survival ratio

生薬 *shouyaku*
crude drug

平 stereo, three dimensional

平調 *heichou*
flatness, plane

平方根 *heihoukon*
square root

平方メートル *heihoume-toru*
square meter

平均分子量 *heikinbunshiryou*
average molecular weight

平均値 *heikinchi*
mean value

平均電荷 *heikindenka*
averaged charge

平均沸点 *heikinfutten*
average boiling point

平均偏差 *heikinhensa*
mean deviation

平均寿命 *heikinjumyou*
average lifetime

平均重合度 *heikinjuugoudo*
  mean degree of polymerization
平均化 *heikinka*
  averaging
平均活量 *heikinkatsuryou*
  mean activity
平均勾配 *heikinkoubai*
  average gradient
平均力 *heikinryoku*
  mean force
平均粒度 *heikinryuudo*
  average particle size
平均振幅 *heikinshinpuku*
  mean amplitude
平均速度 *heikinsokudo*
  mean velocity
平衡 *heikou*
  equilibrium
平衡分析 *heikoubunseki*
  equilibrium analysis
平衡電位 *heikoudeni*
  equilibrium potential
平衡反応 *heikouhannou*
  equilibrium reaction
平衡蒸留 *heikoujouryuu*
  equilibrium distillation
平衡状態 *heikoujoutai*
  steady state, equilibrium state
平衡状態図 *heikoujoutaizu*
  equilibrium diagram
平衡化学反応 *heikoukagakuhannou*
  equilibrium reaction
平衡系 *heikoukei*
  equilibrium system
平衡混合物 *heikoukongoubutsu*
  equilibrium mixture
平衡構造 *heikoukouzou*
  equilibrium structure
平衡吸着 *heikoukyuuchaku*
  equilibrium adsorption
平衡濃度 *heikounoudo*
  equilibrium concentration

平衡パラメータ *heikouparame-ta*
  equilibrium parameter
平衡酸性度 *heikousanseidou*
  equilibrium acidity
平行線 *heikousen*
  parallel lines
平衡シフト *heikoushifuto*
  equilibrium shift
平衡シミュレーション *heikoushimyure-shon*
  equilibrium simulation
平衡組成 *heikousosei*
  equilibrium composition
平衡定数 *heikouteisuu*
  equilibrium constant
平衡点 *heikouten*
  equilibrium point
平衡溶解性 *heikouyoukaisei*
  equilibrium solubility
平面 *heimen*
  plane surface
平面分子 *heimenbunshi*
  planar molecule
平面偏光 *heimenhenkou*
  plane-polarized light
平面状表面 *heimenjouhyoumen*
  planar surface
平面角 *heimenkaku*
  plane angle
平面キラリティー *heimenkirariti-*
  planar chirality
平面構造 *heimenkouzou*
  planar structure
平面対称 *heimentaishou*
  plane symmetry
平温 *heion*
  normal temperature
平織 *hiraori*
  plain weave
平底フラスコ *hirazokofurasuko*
  flat bottom flask

立 stand up

立方結晶　*rippoukesshou*
　cubic crystal

立方根　*rippoukon*
　cubic root

立方格子　*rippoukoushi*
　cubic lattice, cubic structure

立方メートル　*rippoume-toru*
　cubic meter

立方センチメートル　*rippousenchime-toru*
　cubic centimeter

立方体　*rippoutai*
　cube

立方体構造　*rippoutaikouzou*
　cubic structure

立体　*rittai*
　three-dimensional

立体電子的因子　*rittaidenshitekiinshi*
　stereoelectronic factor

立体電子的効果　*rittaidenshitekikouka*
　stereoelectronic effect

立体電子的制御　*rittaidenshitekiseigyo*
　stereoelectronic control

立体配置反転　*rittaihaichihanten*
　configuration inversion

立体配置解析　*rittaihaichikaiseki*
　configuration analysis

立体配置制御　*rittaihaichiseigyo*
　configuration control

立体配座　*rittaihaiza*
　conformation

立体配座安定化　*rittaihaizaanteika*
　conformer stabilization

立体配座秩序　*rittaihaizachitsujo*
　conformational order

立体配座平衡　*rittaihaizaheikou*
　conformational equilibrium

立体配座変化　*rittaihaizahenka*
　conformational change

立体配座異性　*rittaihaizaisei*
　conformational isomerism

立体配座緩和　*rittaihaizakanwa*
　conformational relaxation

立体配座制御　*rittaihaizaseigyo*
　conformation control

立体因子　*rittaiinshi*
　steric factor

立体異性　*rittaiisei*
　stereoisomerism

立体異性中心　*rittaiiseichuushin*
　stereoisomeric center

立体異性体　*rittaiiseitai*
　stereoisomer

立体化学　*rittaikagaku*
　stereochemistry

立体化学分析　*rittaikagakubunseki*
　stereochemical analysis

立体化学過程　*rittaikagakukatei*
　stereochemical process

立体化学構造　*rittaikagakukouzou*
　stereochemical structure

立体化学モデル　*rittaikagakumoderu*
　stereochemical model

立体化学的過程　*rittaikagakutekikatei*
　stereochemical course

立体化学的効果　*rittaikagakutekikouka*
　stereochemical effect

立体化学的制御　*rittaikagakutekiseigyo*
　stereochemical control

立体化学的性質　*rittaikagakutekiseishitsu*
　stereochemical property

立体規則性　*rittaikisokusei*
　stereoregular

立体効果　*rittaikouka*
　steric effect

立体構造　*rittaikouzou*
　space structure

立体構造解析　*rittaikouzoukaiseki*
　stereoanalysis

立体制御付加　*rittaiseigyofuka*
　stereocontrolled addition

立体制御環化　*rittaiseigyokanka*
stereocontrolled cyclization

立体制御的合成　*rittaiseigyotekigousei*
stereocontrolled synthesis

立体制御全合成　*rittaiseigyozengousei*
stereocontrolled total synthesis

立体選択性　*rittaisentakusei*
stereoselectivity

立体選択的アルドール反応
rittaisentakutekiarudo-ruhannou
stereoselective aldol reaction

立体選択的アルキル化
rittaisentakutekiarukiruka
stereoselective alkylation

立体選択的置換　*rittaisentakutekichikan*
stereoselective substitution

立体選択的エポキシ化
rittaisentakutekiepokishika
stereoselective epoxidation

立体選択的エステル化
rittaisentakutekiesuteruka
stereoselective esterification

立体選択的付加　*rittaisentakutekifuka*
stereoselective addition

立体選択的付加反応
rittaisentakutekifukahannou
stereoselective addition reaction

立体選択的合成　*rittaisentakutekigousei*
stereoselective synthesis

立体選択的グリコシル化
rittaisentakutekigurikoshiruka
stereoselective glycosylation

立体選択的反応　*rittaisentakutekihannou*
stereocontrolled reaction

立体選択的変換　*rittaisentakutekihenkan*
stereoselective transformation

立体選択的ヒドロキシル化
rittaisentakutekihidorokishiruka
stereoselective hydroxylation

立体選択的異性化　*rittaisentakutekiiseika*
stereoselective isomerization

立体選択的重合　*rittaisentakutekijuugou*
stereoselective polymerization

立体選択的開環　*rittaisentakutekikaikan*
stereoselective ring opening

立体選択的開裂　*rittaisentakutekikairetsu*
stereoselective cleavage

立体選択的還元　*rittaisentakutekikangen*
stereoselective reduction

立体選択的還元剤
rittaisentakutekikangenzai
stereoselective reducing agent

立体選択的環状付加
rittaisentakutekikanjoufuka
stereoselective cycloaddition

立体選択的官能化
rittaisentakutekikannouka
stereoselective functionalization

立体選択的カップリング
rittaisentakutekikappuringu
stereoselective coupling

立体選択的ルート　*rittaisentakutekiru-to*
stereoselective route

立体選択的水素化　*rittaisentakutekisuisoka*
stereoselective hydrogenation

立体選択的転位　*rittaisentakutekiteni*
stereoselective rearrangement

立体式　*rittaishiki*
space formula

立体障害　*rittaishougai*
steric hindrance

立体相互作用　*rittaisougosayou*
steric interaction

立体的安定化　*rittaitekianteika*
steric stabilization

立体的かさ高さ　*rittaitekikasatakasa*
steric bulkiness

立体的構造　*rittaitekikouzou*
stereostructure

立体的性質　*rittaitekiseishitsu*
steric property

立体特異性　*rittaitokuisei*
stereospecificity

立体特異性重合　*rittaitokuiseijuugou*
stereospecific polymerization

立体特異的付加　*rittaitokuitekifuka*
　stereospecific addition

立体特異的合成　*rittaitokuitekigousei*
　stereospecific synthesis

立体特異的グリコシル化
　rittaitokuitekigurikoshiruka
　stereospecific glycosylation

立体特異的還元　*rittaitokuitekikangen*
　stereospecific reduction

立体特異的な　*rittaitokuitekina*
　stereospecific

立体特異的転位　*rittaitokuitekiteni*
　stereospecific rearrangement

立体特異的全合成　*rittaitokuitekizengousei*
　stereospecific total synthesis

立体誘導　*rittaiyuudou*
　stereoinduction

有　exist, occur

有限要素　*yuugenyouso*
　finite element

有機亜鉛化合物　*yuukiaenkagoubutsu*
　organozinc compound

有機アニオン　*yuukianion*
　organic anion

有機アルミニウム化合物
　yuukiaruminiumukagoubutsu
　organoaluminum compound

有機媒質　*yuukibaishitsu*
　organic medium

有機分析　*yuukibunseki*
　organic analysis

有機分子　*yuukibunshi*
　organic molecule

有機分子結晶　*yuukibunshikesshou*
　organic molecular crystal

有機分子構造　*yuukibunshikouzou*
　organic molecular structure

有機分子錯体　*yuukibunshisakutai*
　organic molecular complex

有機物, 有機物質　*yuukibutsu, yuukibusshitsu*
　organic substance

有機物液体　*yuukibutsuekitai*
　organic liquid

有機物層　*yuukibutsusou*
　organic layer

有機物水溶液　*yuukibutsusuiyoueki*
　aqueous organic mixture

有機着色物質　*yuukichakushokubusshitsu*
　organic coloring matter

有機置換　*yuukichikan*
　organic substitution

有機窒素化合物　*yuukichissokagoubutsu*
　organic nitrogen compound

有機電解液　*yuukidenkaieki*
　organic electrolyte solution

有機電解質　*yuukidenkaishitsu*
　organic electrolyte

有機銅　*yuukidou*
　metalorganic copper

有機銅化合物　*yuukidoukagoubutsu*
　organocopper compound

有機銅錯体　*yuukidousakutai*
　organocopper complex

有機銅酸塩　*yuukidousanen*
　organocuprate

有機導体　*yuukidoutai*
　organic conductor

有機塩　*yuukien*
　organic salt

有機鉛　*yuukien*
　organolead

有機鉛化合物　*yuukienkagoubutsu*
　organolead compound

有機塩基　*yuukienki*
　organic base

有機塩素　*yuukienso*
　organic chlorine

有機塩素化合物　*yuukiensokagoubutsu*
　chlorinated organic compound

有機塩素汚染物質 yuukiensoosenbusshitsu
chlorinated organic pollutant

有機塩溶媒 yuukienyoubai
organochloride solvent

有機フッ素化合物, 有機弗素化合物
yuukifussokagoubutsu
organofluorine compound

有機顔料 yuukiganryou
organic pigment

有機ゲルマニウム化合物
yuukigerumaniumukagoubutsu
organogermanium compound

有機ゴミ yuukigomi
organic waste

有機合成 yuukigousei
organic synthesis

有機配位子 yuukihaiishi
organic ligand

有機白金化合物 yuukihakkinkagoubutsu
organoplatinum compound

有機反応物 yuukihannoubutsu
organic reactants

有機ハロゲン化物 yuukiharogenkabutsu
organic halide

有機発色団 yuukihasshokudan
organic chromophore

有機変換 yuukihenkan
organic transformation

有機肥料 yuukihiryou
organic fertilizer

有機ヒ素化合物, 有機砒素化合物
yuukihisokagoubutsu
organoarsenic compound

有機ホウ素化合物, 有機硼素化合物
yuukihousokagoubutsu
organoboron compound

有機ホウ素酸, 有機硼素酸
yuukihousosan
organoboronic acid

有機イオン yuukiion
organic ion

有機イオン交換体 yuukiionkoukantai
organic ion exchanger

有機硫黄 yuukiiou
organic sulfur

有機硫黄化合物 yuukiioukagoubutsu
organic sulfur compound

有機異性体 yuukiiseitai
organic isomer

有機樹脂 yuukijushi
organic resin

有機重合体 yuukijuugoutai
organic polymer

有機受容体 yuukijuyoutai
organic acceptor

有機化学 yuukikagaku
organic chemistry

有機化学反応 yuukikagakuhannou
organic chemical reaction

有機化学者 yuukikagakusha
organic chemist

有機化学薬品 yuukikagakuyakuhin
organic chemicals

有機化合物 yuukikagoubutsu
organic compound

有機化合物分析 yuukikagoubutsubunseki
analysis of organic compound

有機化合物分子 yuukikagoubutsubunshi
organic molecule

有機化合物構造異性体
yuukikagoubutsukouzouiseitai
structural isomer of organic compound

有機化合物吸着
yuukikagoubutsukyuuchaku
organic compound adsorption

有機化合物定量 yuukikagoubutsuteiryou
organic compound determination

有機官能化 yuukikannouka
organic functionalization

有機官能基 yuukikannouki
organic functional group

有機過酸化物 yuukikasankabutsu
organic peroxide

有機基　yuukiki
organic group

有機金　yuukikin
organogold

有機金属アニオン　yuukikinzokuanion
organometallic anion

有機金属分子　yuukikinzokubunshi
organometallic molecule

有機金属中間体　yuukikinzokuchuukantai
organometallic intermediate

有機金属デンドリマ
yuukikinzokudendorima
organometallic dendrimer

有機金属液晶　yuukikinzokuekishou
metallorganic liquid crystal

有機金属付加　yuukikinzokufuka
organometal addition

有機金属フッ化物, 有機金属弗化物
yuukikinzokufukkabutsu
organometallic fluoride

有機金属フラグメント
yuukikinzokufuragumento
organometallic fragment

有機金属合成　yuukikinzokugousei
metallorganic synthesis

有機金属配位子　yuukikinzokuhaiishi
metallorganic ligand

有機金属反応　yuukikinzokuhannou
metallorganic reaction

有機金属反応機構
yuukikinzokuhannoukikou
metallorganic reaction mechanism

有機金属ハロゲン化物
yuukikinzokuharogenkabutsu
metallorganic halide

有機金属化合物　yuukikinzokukagoubutsu
organometallic compound

有機金属高分子　yuukikinzokukoubunshi
organometallic polymer

有機金属錯体　yuukikinzokusakutai
organometallic complex

有機金属酸化物　yuukikinzokusankabutsu
metallorganic oxide

有機金属試薬　yuukikinzokushiyaku
organometallic reagent

有機金属誘導体　yuukikinzokuyuudoutai
organometallic derivative

有機コバルト錯体　yuukikobarutosakutai
organocobalt complex

有機混和剤　yuukikonwazai
organic admixture

有機固体　yuukikotai
organic solid

有機高分子　yuukikoubunshi
organic macromolecule

有機光伝導体　yuukikoudendoutai
organic photoconductor

有機共重合体　yuukikyoujuugoutai
organic copolymer

有機供与体　yuukikyouyotai
organic donor

有機吸着物　yuukikyuuchakubutsu
organic adsorbate

有機マグネシウム化合物
yuukimaguneshiumukagoubutsu
organomagnesium compound

有機マンガン化合物
yuukimangankagoubutsu
organomanganese compound

有機燃料　yuukinenryou
organic fuel

有機ニッケル化合物
yuukinikkerukagoubutsu
organonickel compound

有機の　yuukino
organic

有機汚染物質　yuukiosenbusshitsu
organic pollutant

有機パラジウム錯体
yuukiparajiumusakutai
organopalladium complex

有機ポリマー　yuukiporima-
organic polymer

有機ポリシラン　*yuukiporishiran*
organic polysilane

有機ラジカル　*yuukirajikaru*
organic radical

有機リチウム化合物
yuukirichiumukagoubutsu
organolithium compound

有機燐酸配位子　*yuukirinhaiishi*
organophosphorus ligand

有機リン化合物, 有機燐化合物
yuukirinkagoubutsu
organophosphorus compound

有機リン酸エステル, 有機燐酸エステル
yuukirinsanesuteru
organophosphate

有機リン配位子, 有機燐配位子
yuukirinsanhaiishi
organophosphorus ligand

有機リン試薬, 有機燐試薬
yuukirinshiyaku
organophosphorus reagent

有機燐酸試薬　*yuukirinshiyaku*
organophosphorus reagent

有機ロジウム錯体　*yuukirojiumusakutai*
organo-rhodium complex

有機類似体　*yuukiruijitai*
organic analogue

有機ルテニウム錯体
yuukiruteniumusakutai
organoruthenium complex

有機粒子　*yuukiryuushi*
organic particle

有機流体　*yuukiryuutai*
organic fluid

有機錯体　*yuukisakutai*
organic complex

有機酸　*yuukisan*
organic acid

有機酸塩　*yuukisanen*
organic acid salt

有機酸化体　*yuukisankatai*
organic oxidant

有機酸混合物　*yuukisankongoubutsu*
organic acid mixture

有機酸無水物　*yuukisanmusuibutsu*
organic acid anhydride

有機酸素　*yuukisanso*
organic oxygen

有機成分　*yuukiseibun*
organic component

有機繊維　*yuukiseni*
organic fiber

有機遷移金属錯体
yuukisenikinzokusakutai
organic transition metal complex

有機セレン化合物　*yuukiserenkagoubutsu*
organic selenium compound

有機色素　*yuukishikiso*
organic dye

有機シリコン化合物
yuukishirikonkagoubutsu
organosilicon compound

有機試料　*yuukishiryou*
organic sample

有機質　*yuukishitsu*
organic substance

有機質量分析　*yuukishitsuryoubunseki*
organic mass spectrometry

有機試薬　*yuukishiyaku*
organic reagent

有機臭素　*yuukishuuso*
organic bromine

有機臭素化合物　*yuukishuusokagoubutsu*
organo bromide compound

有機水銀化合物　*yuukisuiginkagoubutsu*
organic mercury compound

有機スズ化合物　*yuukisuzukagoubutsu*
organotin compound

有機体　*yuukitai*
organism

有機単分子層　*yuukitanbunshisou*
organic monolayer

有機単量体　*yuukitanryoutai*
organic monomer

有機炭酸エステル　*yuukitansanesuteru*
organic carbonate

有機炭素　*yuukitanso*
organic carbon

有機的な　*yuukitekina*
organic

有機添加物　*yuukitenkazai*
organic additive

有機テルル化合物　*yuukiterurukagoubutsu*
organotellurium compound

有機鉄化合物　*yuukitetsukagoubutsu*
organoiron compound

有機鉄錯体　*yuukitetsusakutai*
organic iron complex

有機薬品合成　*yuukiyakuhinhannou*
synthesis of organic chemicals

有機溶媒　*yuukiyoubai*
organic solvent

有機溶媒抽出　*yuukiyoubaichuushutsu*
organic solvent extraction

有機溶媒蒸気　*yuukiyoubaijouki*
organic solvent vapor

有機溶媒可溶性　*yuukiyoubaikayousei*
organic solvent solubility

有機溶媒系　*yuukiyoubaikei*
organic solvent system

有機溶媒混合物　*yuukiyoubaikongoubutsu*
organic solvent mixture

有機溶媒水溶液　*yuukiyoubaisuiyoueki*
organic solvent aqueous solution

有機溶液　*yuukiyoueki*
organic solution

有機陽イオン　*yuukiyouion*
organic cation

有機ヨウ素化合物, 有機沃素化合物　*yuukiyousokagoubutsu*
iodinated organic compound

有機溶剤　*yuukiyouzai*
organic solvent

有機材料　*yuukizairyou*
organic material

有機前駆体　*yuukizenkutai*
organic precursor

有機絶縁体　*yuukizetsuentai*
organic insulator

有孔フィルム　*yuukoufirumu*
perforated film

有効イオン電荷　*yuukouiondenka*
effective ionic charge

有効イオン半径　*yuukouionhankei*
effective ionic radius

有効量　*yuukouryou*
effective dose

有効性　*yuukousei*
efficiency

有効質量　*yuukoushitsuryou*
effective mass

有効速度　*yuukousokudo*
effective rate

有効特性　*yuukoutokusei*
effective property

有極結合　*yuukyokuketsugou*
polar bond, polar linkage

有極性分子　*yuukyokuseibunshi*
polar molecule

有理関数　*yuurikansuu*
rational function

有糸分裂　*yuushibunretsu*
mitosis

気　spirit

気圧　*kiatsu*
atmospheric pressure

気圧変化　*kiatsuhenka*
pressure change

気圧計　*kiatsukei*
barometer

気液平衡　*kiekiheikou*
gas-liquid equilibrium

気液熱交換器 *kiekinetsukoukanki*
　vapor heat exchanger

気泡 *kihou*
　air bubble

気泡粘度計 *kihounendokei*
　bubble viscometer

気泡塔 *kihoutou*
　bubble column

気化器 *kikaki*
　carbureter

気孔率 *kikouritsu*
　porosity

気硬性 *kikousei*
　air hardening

気温 *kion*
　air temperature

気温調節装置 *kionchousetsusouchi*
　air conditioner

気相 *kisou*
　vapor phase

気相分解 *kisoubunkai*
　vapor-phase cracking

気相酸化 *kisousanka*
　gas-phase oxidation

気態 *kitai*
　gaseous state

気体 *kitai*
　gas

気体放電 *kitaihouden*
　gaseous discharge

気体計 *kitaikei*
　aerometer

気体密度 *kitaimitsudo*
　vapor density

気体燃料 *kitainenryou*
　gaseous fuel

気体の *kitaino*
　gaseous

気体の圧力 *kitainoatsuryoku*
　gas pressure

気体の法則 *kitainohousoku*
　gas law

気体の溶解度 *kitainoyoukaido*
　solubility of gas

気体温度計 *kitaiondokei*
　gas thermometer

気体定数 *kitaiteisuu*
　gas constant

気体透過性 *kitaitoukasei*
　gas permeability

光 light, ray

光分解 *hikaribunkai*
　photodecomposition, photolysis

光エネルギー *hikarienerugi-*
　light energy

光ファイバ *hikarifaiba*
　optical fiber

光ファイバーケーブル *hikarifaiba-ke-buru*
　fiber-optic cable

光イオン化 *hikariionka*
　photoionization

光重合 *hikarijuugou*
　photopolymerization

光吸収 *hikarikyuushuu*
　light absorption

光の反射 *hikarinohansha*
　light reflection, luminous reflectance

光の回折 *hikarinokaisetsu*
　light diffraction

光の散乱 *hikarinosanran*
　light scattering

光の強さ *hikarinotsuyosa*
　light intensity

光触媒反応 *hikarishokubaihannou*
　photocatalysis

光増感 *hikarizoukan*
　photosensitization

光増感剤　*hikarizoukanzai*
　photosensitizer

光電分光光度計　*koudenbunkoukoudokei*
　photoelectric spectrophotometer

光電池　*koudenchi*
　photocell, photoelectric cell

光伝導性　*koudendousei*
　photoconductivity

光伝導セル　*koudendouseru*
　photoconductive cell

光電管　*koudenkan*
　phototube

光電気現象　*koudenkigenshou*
　photoelectricity, photoelectric effect

光電気化学電池　*koudenkikagakudenchi*
　photoelectrochemical cell

光電コンダクタンス　*koudenkondakutansu*
　photoelectric conductance

光電光度計　*koudenkoudokei*
　photoelectric photometer

光電効果　*koudenkouka*
　photoelectric effect

光電離　*koudenri*
　photoionization

光電セル　*koudenseru*
　photocell

光電子　*koudenshi*
　photoelectron

光電子分光学, 光電子分光法　*koudenshibunkougaku, koudenshibunkouhou*
　photoelectron spectroscopy

光電子放出　*koudenshihoushutsu*
　photoelectron emission

光電測光法　*koudensokkouhou*
　photoelectric photometry

光度　*koudo*
　luminous intensity

光度計　*koudokei*
　photometer

光導電セル　*koudoudenseru*
　photoconductive cell

光延反応　*kouenhannou*
　Mitsunobu reaction

光学　*kougaku*
　optics

光学分割　*kougakubunkatsu*
　optical resolution

光学ガラス　*kougakugarasu*
　optical glass

光学異性　*kougakuisei*
　optical isomerism

光学異性体　*kougakuiseitai*
　optical isomer

光学軸　*kougakujiku*
　optical axis

光学純度　*kougakujundo*
　optical purity

光学活性　*kougakukassei*
　optical activity

光学活性体　*kougakukasseitai*
　optically acitve substance

光学機械　*kougakukikai*
　optical instrument

光学密度　*kougakumitsudo*
　optical density

光学繊維　*kougakuseni*
　optical fiber

光学的分析　*kougakutekibunseki*
　optical analysis

光学的活性　*kougakutekikassei*
　optical activity

光学的性質　*kougakutekiseishitsu*
　optical property

光源　*kougen*
　light source

光合成　*kougousei*
　photosynthesis

光反応　*kouhannou*
　photoreaction

光軸　*koujiku*
　optical axis

光化学　*koukagaku*
　photochemistry

光化学反応 koukagakuhannou
  photochemical reaction

光化学平衡 koukagakuheikou
  photochemical equilibrium,
  photostationary state

光化学効果 koukagakukouka
  photochemical effect

光化学的開始反応
  koukagakutekikaishihannou
  photochemical initiation

光化学誘導 koukagakuyuudou
  photochemical induction

光起電力 koukidenryoku
  photoelectromotive forcs

光起電力効果 koukidenryokukouka
  photovoltaic effect

光鹵石 kouroseki
  carnallite $KMgCl_3 \cdot 6(H_2O)$

光量子 kouryoushi
  photon

光酸化 kousanka
  photooxidation

光散乱 kousanran
  light scattering

光線 kousen
  light ray

光線反応 kousenhannou
  light reaction

光線石 kousenseki
  actinolite $Ca_2(Mg,Fe^{2+})_5Si_8O_{22}(OH)_2$

光線速度 kousensokudo
  ray velocity

光子 koushi
  photon

光子エネルギー koushienerugi-
  photon energy

光子吸収 koushikyuushuu
  photon absorption

光心 koushin
  optical center

光速 kousoku
  light velocity

芳 aromatic

芳香 houkou
  aroma

芳香化学物質 houkoukagakubusshitsu
  aromatic chemicals

芳香核 houkoukaku
  aromatic nucleus

芳香環 houkoukan
  aromatic ring

芳香性 houkousei
  aromaticity

芳香族 houkouzoku
  aromatic

芳香族アミド houkouzokuamido
  aromatic amide

芳香族アミン houkouzokuamin
  aromatic amine

芳香族アミノ酸 houkouzokuaminosan
  aromatic amino acid

芳香族アルデヒド houkouzokuarudehido
  aromatic aldehyde

芳香族アルカロイド
  houkouzokuarukaroido
  aromatic alkaloid

芳香族アルケン houkouzokuaruken
  aromatic alkene

芳香族アルコール houkouzokuaruko-ru
  aromatic alcohol

芳香族分子 houkouzokubunshi
  aromatic molecule

芳香族置換 houkouzokuchikan
  aromatic substitution

芳香族置換反応
  houkouzokuchikanhannou
  aromatic substitution reaction

芳香族置換基 houkouzokuchikanki
  aromatic substituent

芳香族チオン houkouzokuchion
  aromatic thione

芳香族チオール houkouzokuchio-ru
  aromatic thiol

芳香族中間体　*houkouzokuchuukantai*
aromatic intermediate

芳香族塩基　*houkouzokuenki*
aromatic base

芳香族エステル　*houkouzokuesuteru*
aromatic ester

芳香族エーテル　*houkouzokue-teru*
aromatic ether

芳香族複素環　*houkouzokufukusokan*
aromatic heterocyclic ring

芳香族フッ素化合物
houkouzokufussokagoubutsu
aromatic fluorine compound

芳香族ゲスト分子
houkouzokugesutobunshi
aromatic guest molecule

芳香族配位子　*houkouzokuhaiishi*
aromatic ligand

芳香族ハロゲン化物
houkouzokuharogenkabutsu
aromatic halide

芳香族イソシアニド
houkouzokuisoshianido
aromatic isocyanide

芳香族ジアミン　*houkouzokujiamin*
aromatic diamine

芳香族ジアルデヒド
houkouzokujiarudehido
aromatic dialdehyde

芳香族ジアゾニウム塩
houkouzokujiazoniumuen
aromatic diazonium salt

芳香族ジカルボン酸
houkouzokujikarubonsan
aromatic dicarboxylic acid

芳香族ジクロリド　*houkouzokujikurorido*
aromatic dichloride

芳香族ジスルフィド
houkouzokujisurufido
aromatic disulfide

芳香族化　*houkouzokuka*
aromatization

芳香族化合物　*houkouzokukagoubutsu*
aromatic compound

芳香族化反応　*houkouzokukahannou*
aromatization reaction

芳香族環　*houkouzokukan*
aromatic ring

芳香族カルボニル化合物
houkouzokukarubonirukagoubutsu
aromatic carbonyl compound

芳香族カルボン酸
houkouzokukarubonsan
aromatic carboxylic acid

芳香族系　*houkouzokukei*
aromatic system

芳香族ケトン　*houkouzokuketon*
aromatic ketone

芳香族基　*houkouzokuki*
aromatic group

芳香族コポリエステル
houkouzokukoporiesuteru
aromatic copolyester

芳香族構造　*houkouzokukouzou*
aromatic structure

芳香族ニトリル　*houkouzokunitoriru*
aromatic nitrile

芳香族ニトロソ化合物
houkouzokunitorosokagoubutsu
aromatic nitroso compound

芳香族オリゴマ　*houkouzokuorigoma*
aromatic oligomer

芳香族ポリアミド　*houkouzokuporiamido*
aromatic polyamide

芳香族ポリアミジン
houkouzokuporiamijin
aromatic polyamidine

芳香族ポリベンゾオキサゾール
houkouzokuporibenzookisazo-ru
aromatic polybenzoxazole

芳香族ポリエステル
houkouzokuporiesuteru
aromatic polyester

芳香族ポリイミド　*houkouzokuporiimido*
aromatic polyimide

芳香族ポリケチド
houkouzokuporikechido
aromatic polyketide

芳香族ポリマ *houkouzokuporima*
aromatic polymer

芳香族ポリスルホン
houkouzokuporisuruhon
aromatic polysulphone

芳香族ポリ尿素 *houkouzokuporiyouso*
aromatic polyurea

芳香族プロペレン *houkouzokupuroperen*
aromatic propellene

芳香族ラジカルアニオン
houkouzokurajikaruanion
aromatic radical anion

芳香族ラジカルカチオン
houkouzokurajikarukachion
aromatic radical cation

芳香族酸 *houkouzokusan*
aromatic acid

芳香族性 *houkouzokusei*
aromaticity

芳香族成分 *houkouzokuseibun*
aromatic component

芳香族側鎖 *houkouzokusokusa*
aromatic side chain

芳香族スルホン酸
houkouzokusuruhonsan
aromatic sulfonic acid

芳香族スルホン酸塩
houkouzokusuruhonsanen
aromatic sulfonate

芳香族臭素化 *houkouzokusuusoka*
aromatic bromination

芳香族炭化水素 *houkouzokutankasuiso*
aromatic hydrocarbon

芳香族単量体 *houkouzokutanryoutai*
aromatic monomer

芳香族炭素 *houkouzokutanso*
aromatic carbon

芳香族溶媒 *houkouzokuyoubai*
aromatic solvent

放 emit, release

放電 *houden*
electric discharge

放電反応器 *houdenhannouki*
discharge reactor

放電状態 *houdenjoutai*
discharged state

放電管 *houdenkan*
discharge tube

放電器 *houdenki*
discharger

放電曲線 *houdenkyokusen*
discharge curve

放電オゾン発生器 *houdenozonhasseiki*
discharge ozonizer

放電速度 *houdensokudo*
discharge rate

放電特性 *houdentokusei*
discharge characteristic

放熱蛇管 *hounetsudakan*
heating coil

放熱コイル *hounetsukoiru*
heating coil

放熱体 *hounetsutai*
heater

放線菌 *housenkin*
actinomyces

放射 *housha*
radiation

放射分析 *houshabunseki*
radiometric analysis

放射エネルギー *houshaenerugi-*
radiant energy, radiation energy

放射エネルギー吸収
houshaenerugi-kyuushuu
radiative energy absorption

放射平衡 *houshaheikou*
radioactive equilibrium

放射状線 *houshajousen*
radial line

放射化学　*houshakagaku*
radiochemistry

放射化学分離　*houshakagakubunri*
radiochemical separation

放射化学分析　*houshakagakubunseki*
radiochemical analysis

放射化学法　*houshakagakuhou*
radiochemical technique

放射化学実験室　*houshakagakujikkenshitsu*
radiochemical laboratory

放射化学者　*houshakagakusha*
radiochemist

放射化学的安定性
houshakagakutekianteisei
radiochemical stability

放射化学的分析　*houshakagakutekibunseki*
radiochemical analysis

放射化学的測定　*houshakagakutekisokutei*
radiochemical measurement

放射化学的定量　*houshakagakutekiteiryou*
radiochemical determination

放射吸収　*houshakyuushuu*
radiation absorption

放射免疫検定　*houshamenekikentei*
radioimmunoassay

放射の法則　*houshanohousoku*
radiation law

放射能　*houshanou*
radioactivity

放射性　*houshasei*
radioactivity

放射性物質　*houshaseibusshitsu*
radioactive substance

放射性沈積物　*houshaseichinsekibutsu*
radioactive deposit

放射性同位元素　*houshaseidouigenso*
radioactive isotope

放射性同位体　*houshaseidouitai*
radioactive isotope

放射性原子　*houshaseigenshi*
radioactive atom

放射性元素　*houshaseigenso*
radioactive element

放射性廃棄物　*houshaseihaikibutsu*
radioactive waste

放射性発光　*houshaseihakkou*
radioactive emission

放射性標識法　*houshaseihyoushikihou*
radiolabeling method

放射性標識化合物
houshaseihyoushikikagoubutsu
radiolabelled compound

放射性イオン　*houshaseiion*
radioactive ion

放射性の　*houshaseino*
radioactive

放射性汚染　*houshaseiosen*
radioactive contamination

放射性リガンド　*houshaseirigando*
radioactive ligand

放射性粒子　*houshaseiryuushi*
radioactive particle

放射性指示薬　*houshaseishijiyaku*
radioactive indicator

放射性試料　*houshaseishiryou*
radioactive sample

放射性炭素　*houshaseitanso*
radioactive carbon

放射性炭素分析　*houshaseitansobunseki*
radiocarbon analysis

放射性炭素法　*houshaseitansohou*
radiocarbon method

放射性トレーサ　*houshaseitore-sa*
radioactive indicator

放射線　*houshasen*
radiation

放射線分解　*houshasenbunkai*
radiolysis

放射線分解法　*houshasenbunkaihou*
radiological technique

放射線分解研究　*houshasenbunkaikenkyuu*
radiological investigation

放射線学 *houshasengaku*
　radiology

放射線合成 *houshasengousei*
　radiation synthesis

放射線標識 *houshasenhyoushiki*
　radiation marker

放射遷移 *houshaseni*
　radioactive transformation

放射線重合 *houshasenjuugou*
　radiation polymerization

放射線化学 *houshasenkagaku*
　radiation chemistry

放射線加硫 *houshasenkaryuu*
　radiation vulcanization

放射線硬化 *houshasenkouka*
　radiation curing

放射線写真 *houshasenshashin*
　radiograph

放射誘起 *houshayuuki*
　radiation induction

放出 *houshutsu*
　emission

放出パラメータ *houshutsuparame-ta*
　emission parameter

放出粒子 *houshutsuryuushi*
　emitted particle

放出指数 *houshutsushisuu*
　emission index

放出スペクトル *houshutsusupekutoru*
　emission spectrum

重 heavy

重亜硫酸塩 *juuaryuusanen*
　hydrogen sulfite $M^I HSO_3$

重亜硫酸ナトリウム *juuaryuusannatoriumu*
　sodium bisulfite $NaHSO_3$

重土 *juudo*
　barite $BaSO_4$

重土水 *juudosui*
　barium hydroxide solution, baryta water $Ba(OH)_2$

重液 *juueki*
　heavy liquid

重原子 *juugenshi*
　heavy atom

重合 *juugou*
　polymerization

重合度 *juugoudo*
　degree of polymerization

重合反応 *juugouhannou*
　polymerization reaction

重合平衡 *juugouheikou*
　polymerization equilibrium

重合機構 *juugoukikou*
　mechanism of polymerization

重合酸 *juugousan*
　polyporic acid $C_{18}H_{12}O_4$

重合速度 *juugousokudo*
　polymerization velocity, rate of polymerization

重合する *juugousuru*
　to polymerize

重合体 *juugoutai*
　polymer

重合体化学 *juugoutaikagaku*
　polymer chemistry

重合抑制剤 *juugouyokuseizai*
　polymerization inhibitor

重環式化合物 *juukanshikikagoubutsu*
　polycyclic compound

重金属 *juukinzoku*
　heavy metal

重クロム酸 *juukuromusan*
　dichromic acid $H_2Cr_2O_7$

重クロム酸アンモニウム *juukuromusananmoniumu*
　ammonium dichromate $(NH_4)_2Cr_2O_7$

重クロム酸塩 *juukuromusanen*
　dichromate $M^I_2Cr_2O_7$

重クロム酸銀 *juukuromusangin*
　silver dichromate $Ag_2Cr_2O_7$

重クロム酸カリウム *juukuromusankariumu*
potassium dichromate $K_2Cr_2O_7$

重クロム酸ナトリウム *juukuromusannatoriumu*
sodium dichromate $Na_2Cr_2O_7 \cdot 2H_2O$

重力 *juuryoku*
gravity

重力計 *juuryokukei*
gravimeter

重量 *juuryou*
weight

重量分析 *juuryoubunseki*
gravimetric analysis

重量平均分子量 *juuryouheikinbunshiryou*
weight-average molecular weight

重量平均重合度 *juuryouheikinjuugoudo*
weight-average degree of polymerization

重量比 *juuryouhi*
weight ratio

重量百分比 *juuryouhyakubunhi*
weight percentage

重量の増加 *juuryounozouka*
weight increase

重量パーセント *juuryoupa-sento*
weight percentage

重量滴定 *juuryoutekitei*
gravimetric titration

重硫酸塩 *juuryuusanen*
bisulfate M^IHSO_4

重硫酸ナトリウム *juuryuusannatoriumu*
sodium hydrogensulfate $NaHSO_4$

重酒石酸塩 *juusansekisanen*
bitartrate $HOOC(CHOH)_2COOM$

重酸素 *juusanso*
heavy oxygen

重石 *juuseki*
scheelite $CaWO_4$

重心 *juushin*
center of gravity, mass center

重晶石 *juushouseki*
barite $BaSO_4$

重縮合 *juushukugou*
polycondensation

重曹 *juusou*
sodium bicarbonate $NaHCO_3$

重層効果 *juusoukouka*
interlayer effect

重水 *juusui*
heavy water D_2O

重水素 *juusuiso*
heavy hydrogen, deuterium

重水素化 *juusuisoka*
deuteration

重水素化合物 *juusuisokagoubutsu*
deuterized compound

重炭酸第一鉄 *juutansandaiichitetsu*
ferrous bicarbonate $Fe(HCO_3)_2$

重炭酸塩 *juutansanen*
bicarbonate, hydrogencarbonate M^IHCO_3

重炭酸ナトリウム *juutansannatoriumu*
sodium hydrogencarbonate $NaHCO_3$

重油 *juuyu*
heavy oil

重なり *kasanari*
overlap

重さ *omosa*
weight

核 nucleus

核 *kaku*
nucleus

核爆発 *kakubakuhatsu*
nuclear explosion

核分裂 *kakubunretsu*
nuclear fission

核物理学 *kakubutsurigaku*
nuclear physics

核電荷 *kakudenka*
nuclear charge

核エネルギー *kakuenerugi-*
nuclear energy

核外電子 *kakugaidenshi*
outer electron

核廃棄物 *kakuhaikibutsu*
nuclear waste

核半径 *kakuhankei*
nuclear radius

核反応 *kakuhannou*
nuclear reaction

核反応炉 *kakuhannouro*
nuclear reactor

核磁気共鳴 *kakujikikyoumei*
nuclear magnetic resonance (NMR)

核磁気共鳴分光計
kakujikikyoumeibunkoukei
nuclear magnetic resonance spectrometer

核磁気モーメント *kakujikimo-mento*
nuclear magnetic moment

核磁子 *kakujishi*
nuclear magneton

核準位 *kakujuni*
nuclear level

核化学 *kakukagaku*
nuclear chemistry

核間距離 *kakukankyori*
internuclear distance

核膜 *kakumaku*
nuclear membrane

核モーメント *kakumo-mento*
nuclear moment

核オーバーハウザー効果
kakuo-ba-hauza-kouka
nuclear Overhauser effect, NOE

核酸 *kakusan*
nucleic acid

核酸分解酵素 *kakusanbunkaikouso*
nuclease

核酸塩基 *kakusanenki*
nucleobase, nuclein base

核生成 *kakuseisei*
nucleation, nucleus formation

核戦力 *kakusenryoku*
nuclear force

核四極子モーメント
kakushikyokushimo-mento
nuclear quadrupole moment

核スピン *kakusupin*
nuclear spin

核スピン量子数 *kakusupinryoushisuu*
nuclear spin quantum number

核タンパク質 *kakutanpakushitsu*
nucleoprotein

核融合 *kakuyuugou*
nuclear fusion

核融合反応 *kakuyuugouhannou*
nuclear fusion reaction

配 ligand, coordination

配置 *haichi*
configuration

配置安定性 *haichianteisei*
configurational stability

配置混合 *haichikongou*
configuration mixing

配置特性 *haichitokusei*
configuration characteristic

配合 *haigou*
formulation, blending

配合物 *haigoubutsu*
compound

配合技術 *haigougijutsu*
formulation technology

配合変数 *haigouhensuu*
compounding variable

配合禁忌 *haigoukinki*
incompatibility

配合成分 *haigouseibun*
compounding ingredient

配合組成 *haigousosei*
formulation

配合剤 *haigouzai*
formulation

配位　*haii*
　coordination

配位アニオン　*haiianion*
　coordinating anion

配位原子団　*haiigenshidan*
　coordinate atomic group

配位反応　*haiihannou*
　coordination reaction

配位変化　*haiihenka*
　coordination change

配位異性体　*haiiiseitai*
　coordination isomer

配位状態　*haiijoutai*
　coordination state

配位重合　*haiijuugou*
　coordination polymerization

配位化学　*haiikagaku*
　coordination chemistry

配位化合物　*haiikagoubutsu*
　coordination compound

配位結合　*haiiketsugou*
　coordinate bond

配位高分子　*haiikoubunshi*
　coordination polymer

配位サイト　*haiisaito*
　coordination site

配位錯体　*haiisakutai*
　coordinated complex

配位性溶媒　*haiiseiyoubai*
　coordinating solvent

配位子　*haiishi*
　ligand

配位子場　*haiishiba*
　ligand field

配位子場解析　*haiishibakaiseki*
　ligand field analysis

配位子場強度　*haiishibakyoudo*
　ligand field strength

配位子場パラメータ　*haiishibaparame-ta*
　ligand field parameter

配位子場理論　*haiishibariron*
　ligand field theory

配位子置換　*haiishichikan*
　ligand substitution

配位子解離　*haiishikairi*
　ligand dissociation

配位子カップリング　*haiishikappuringu*
　ligand coupling

配位子活性化　*haiishikasseika*
　ligand activation

配位子系　*haiishikei*
　ligand system

配位子結合部位　*haiishiketsugoubui*
　ligand binding site

配位子結合特性　*haiishiketsugoutokusei*
　ligand bonding property

配位式　*haiishiki*
　coordination formula

配位子効果　*haiishikouka*
　ligand effect

配位子交換　*haiishikoukan*
　ligand exchange

配位子交換反応　*haiishikoukanhannou*
　ligand exchange reaction

配位子交換クロマトグラフィー
　haiishikoukankuromatogurafi-
　ligand exchange chromatography

配位子構造　*haiishikouzou*
　ligand structure

配位子濃度　*haiishinoudo*
　ligand concentration

配位子酸化　*haiishisanka*
　ligand oxidation

配位子設計　*haiishisekkei*
　ligand design

配位子相互作用　*haiishisougosayou*
　ligand interaction

配位触媒　*haiishokubai*
　coordination catalyst

配位水　*haiisui*
　coordinated water

配位数　*haiisuu*
　coordination number

配管系 *haikankei*
  piping system

配管材料 *haikanzairyou*
  piping material

配向 *haikou*
  orientation

配向安定性 *haikouanteisei*
  orientation stability

配向分布 *haikoubunpu*
  orientation distribution

配向分析 *haikoubunseki*
  orientation analysis

配向分子 *haikoubunshi*
  oriented molecule

配向因子 *haikouinshi*
  orientation factor

配向状態 *haikoujoutai*
  oriented state

配向結晶化 *haikoukesshouka*
  orientation crystallization

配向効果 *haikoukouka*
  orientation effect

配向構造 *haikoukouzou*
  oriented structure

配列 *hairetsu*
  sequence

配位子制御 *haishiseigyo*
  ligand control

配糖体 *haitoutai*
  glucoside

配座 *haiza*
  conformation

配座安定性 *haizaanteisei*
  conformational stability

配座エネルギー *haizaenerugi-*
  conformational energy

配座異性化 *haizaiseika*
  conformational isomerization

配座異性体 *haizaiseitai*
  conformer, conformational isomer

配座可動性 *haizakadousei*
  conformational flexibility

配座解析 *haizakaiseki*
  conformational analysis

配座固定 *haizakotei*
  conformational lock

配座転移 *haizateni*
  conformational transition

流 flow, current

流れ *nagare*
  flowing, stream, current

流れ実験 *nagarejikken*
  flow experiment

流れプロセス *nagarepurosesu*
  flow process

流れ特性 *nagaretokusei*
  flowing characteristic

流れ図 *nagarezu*
  flow sheet, flow chart, flow diagram

流動電位 *ryuudoudeni*
  streaming potential

流動度 *ryuudoudo*
  fluidity

流動液体 *ryuudouekitai*
  flowing liquid

流動化 *ryuudouka*
  fluidization

流動乾燥 *ryuudoukansou*
  fluidized drying

流動化速度 *ryuudoukasokudo*
  fluidization velocity

流動化剤 *ryuudoukazai*
  flow agent

流動混合 *ryuudoukongou*
  flow mixing

流動パラフィン *ryuudouparafin*
  liquid paraffin

流動性 *ryuudousei*
  flowability

流動セル *ryuudouseru*
  flow cell

流動触媒　*ryuudoushokubai*
　　fluid catalyst

流動床　*ryuudoushou*
　　fluid bed

流動層　*ryuudousou*
　　fluid bed

流動層反応器　*ryuudousouhannouki*
　　fluid bed reactor

流動層燃焼　*ryuudousounetsushou*
　　fluidized bed combustion

流動体　*ryuudoutai*
　　liquid

流動点　*ryuudouten*
　　pour point

流動溶液　*ryuudouyoueki*
　　fluid solution

流銀ゲルマニウム鉱　*ryuugingeruaniumukou*
　　argyrodite $Ag_2S \cdot Ag_2GeS_3$

流銀鉱　*ryuuginkou*
　　acanthite Ag_2S

流マンガン鉱　*ryuumangankou*
　　alabandine MnS

流入液　*ryuunyuueki*
　　influent

流量　*ryuuryou*
　　flow rate

流量計　*ryuuryoukei*
　　flowmeter

流量測定　*ryuuryousokutei*
　　flow measurement

流線流　*ryuusenryuu*
　　streamline flow

流速　*ryuusoku*
　　current velocity

流体　*ryuutai*
　　fluid

流体媒質　*ryuutaibaishitsu*
　　fluid media

流体分散　*ryuutaibunsan*
　　fluid dispersion

流体動力学　*ryuutaidourikigaku*
　　hydrodynamics

流体磁気波　*ryuutaijikin*
　　hydromagnetic wave

流体化　*ryuutaika*
　　fluidization

流体回路　*ryuutaikairo*
　　fluid circuit

流体混合　*ryuutaikongou*
　　fluid mixing

流体境膜　*ryuutaikyoumaku*
　　fluid film

流体摩擦　*ryuutaimasatsu*
　　fluid friction

流体流れ　*ryuutainagare*
　　fluid flow

流体粘度比　*ryuutainendohi*
　　fluid viscosity ratio

流体波　*ryuutaipa*
　　fluid wave

流体力学　*ryuutairikigaku*
　　fluid mechanics

流体脂質　*ryuutaishishitsu*
　　fluid lipid

流体速度　*ryuutaisokudo*
　　fluid velocity

流体相　*ryuutaisou*
　　fluid phase

流体相平衡　*ryuutaisouheikou*
　　fluid-phase equilibrium

流通反応装置　*ryuutsuuhannousouchi*
　　flow reactor

流通法　*ryuutsuuhou*
　　flow method

粒　particle, grain, drop

粒度　*ryuudo*
　　grain size, particle size

粒度分布　*ryuudobunpu*
　　particle size distribution

粒度測定法 *ryuudousokuteihou*
particle-size measurement

粒状度 *ryuujoudo*
granularity

粒状複合材料 *ryuujoufukugouzairyou*
grained composite

粒状混合物 *ryuujoukongoubutsu*
granular mixture

粒状の *ryuujouno*
granular

粒状性 *ryuujousei*
graininess

粒状層 *ryuujousou*
granular layer

粒状炭素 *ryuujoutanso*
granular carbon

粒径 *ryuukei*
grain size, particle size

粒径拡大 *ryuukeikakudai*
size enlargement

粒子 *ryuushi*
particle, corpuscle, grain, atomic particle

粒子ビーム質量分析 *ryuushibi-mushitsuryoubunseki*
particle beam mass spectrometry

粒子分配 *ryuushibunpai*
particle partitioning

粒子分布 *ryuushibunpu*
particle distribution

粒子分離 *ryuushibunriki*
particle separation

粒子分散 *ryuushibunsan*
particle dispersion

粒子分析 *ryuushibunseki*
particle analysis

粒子電荷 *ryuushidenka*
particle charge

粒子フィルタ *ryuushifiruta*
granular filter

粒子複合材料 *ryuushifukugouzairyou*
particle composite

粒子型 *ryuushigata*
particle shape

粒子凝集 *ryuushigyoushuu*
particle agglomeration

粒子放出 *ryuushihoushutsu*
particle emission

粒子表面 *ryuushihyoumen*
particulate surface

粒子蒸発 *ryuushijouhatsu*
particle evaporation

粒子充填接着剤 *ryuushijuutensecchakuzai*
particle-filled adhesive

粒子拡散係数 *ryuushikakusankeisuu*
particle diffusivity

粒子加速器 *ryuushikasokuki*
particle accelerator

粒子径 *ryuushikei*
particle size, particle diameter

粒子径分析 *ryuushikeibunseki*
particle size analysis

粒子計数器 *ryuushikeisuuki*
particle counter

粒子懸濁物 *ryuushikendakubutsu*
particle suspension

粒子検出器 *ryuushikenshutsuki*
particle detector

粒子混合物 *ryuushikongoubutsu*
particle mixture

粒子コーティング *ryuushiko-tingu*
particle coating

粒子構造 *ryuushikouzou*
grain structure

粒子クラスタ *ryuushikurasuta*
particle cluster

粒子吸着 *ryuushikyuuchaku*
particle adsorption

粒子密度 *ryuushimitsudo*
particle density

粒子濃度 *ryuushinoudo*
particle concentration

粒子パッキング *ryuushipakkingu*
particle packing

粒子流　*ryuushiryuu*
　　particle flow

粒子粒子衝突　*ryuushiryuushishoutotsu*
　　particle-particle collision

粒子線　*ryuushisen*
　　particle beam

粒子質量　*ryuushishitsuryou*
　　particle mass

粒子速度　*ryuushisokudo*
　　particle velocity

粒子相互作用　*ryuushisougosayou*
　　particle interaction

粒子スケール　*ryuushisuke-ru*
　　particle scale

粒子数　*ryuushisuu*
　　particle count

粒子体積　*ryuushitaiseki*
　　particle volume

粒子特性評価　*ryuushitokuseihyouka*
　　particle characterization

粒子溶解性　*ryuushiyoukaisei*
　　particle solubility

液　liquid, fluid

液液抽出　*ekiekichuushutsu*
　　liquid-liquid extraction

液肥　*ekihi*
　　liquid manure

液状　*ekijou*
　　liquid state

液状肥料　*ekijouhiryou*
　　liquid manure

液状空気　*ekijoukuuki*
　　liquid air

液状の　*ekijouno*
　　liquid, fluid

液化　*ekika*
　　liquefaction

液化ガス　*ekikagasu*
　　liquefied gas

液化機　*ekikaki*
　　liquefier, condenser, plasticizer

液化力　*ekikaryoku*
　　liquefying power

液化する　*ekikasuru*
　　to liquefy, to fluidify

液面　*ekimen*
　　liquid surface

液面計　*ekimenkei*
　　level gage

液量計　*ekiryoukei*
　　liquid measurement device

液晶　*ekishou*
　　liquid crystal

液相　*ekisou*
　　liquid phase

液相法　*ekisouhou*
　　liquid phase method

液体　*ekitai*
　　liquid

液体アンモニア　*ekitaianmonia*
　　liquid ammonia

液体窒素　*ekitaichisso*
　　liquid nitrogen

液体廃棄物　*ekitaihaikibutsu*
　　liquid waste

液体金属　*ekitaikinzoku*
　　liquid metal

液体クロマトグラフィー
　　ekitaikuromatogurafi-
　　liquid chromatography

液体空気　*ekitaikuuki*
　　liquid air

液体燃料　*ekitainenryou*
　　liquid fuel

液体酸素　*ekitaisanso*
　　liquid oxygen

液体洗剤　*ekitaisenzai*
　　liquid detergent

液体水素　*ekitaisuiso*
　　liquid hydrogen

液体炭酸　*ekitaitansan*
　liquid carbon dioxide

液糖　*ekitou*
　liquid sugar

硝　nitrate

硝化　*shouka*
　nitration

硝化バクテリヤ　*shoukabakuteriya*
　nitrifying bacteria

硝化作用　*shoukasayou*
　nitrification

硝気　*shouki*
　nitrous fumes mixture $NO + N_2O_4$

硝酸　*shousan*
　nitric acid HNO_3

硝酸亜鉛　*shousanaen*
　zinc nitrate $Zn(NO_3)_2$

硝酸アンモニウム　*shousananmoniumu*
　ammonium nitrate NH_4NO_3

硝酸アルミニウム　*shousanaruminiumu*
　aluminium nitrate $Al(NO_3)_3$

硝酸バリウム　*shousanbariumu*
　barium nitrate $Ba(NO_3)_2$

硝酸ベリリウム　*shousanbeririumu*
　beryllium nitrate $Be(NO_3)_2$

硝酸ビスマス　*shousanbisumasu*
　bismuth nitrate $Bi(NO_3)_3$

硝酸第一水銀　*shousandaiichisuigin*
　mercurous nitrate $HgNO_3$

硝酸第二水銀　*shousandainisuigin*
　mercuric nitrate $Hg(NO_3)_2$

硝酸澱粉　*shousandenpun*
　starch nitrate $(C_6H_7N_3O_9)_n$

硝酸銅（ⅠⅠ）　*shousandou(II)*
　copper(II)nitrate $Cu(NO_3)_2$

硝酸エチル　*shousanechiru*
　ethyl nitrate $C_2H_5ONO_2$

硝酸塩　*shousanen*
　nitrate M^INO_3

硝酸エステル　*shousanesuteru*
　nitric acid ester RNO_3

硝酸銀　*shousangin*
　silver nitrate $AgNO_3$

硝酸イオン　*shousanion*
　nitrate ion

硝酸イットリウム　*shousanittoriumu*
　yttrium nitrate $Y(NO_3)_3 \cdot 4H_2O$

硝酸ジルコニル　*shousanjirukoniru*
　zirconyl nitrate $ZrO(NO_3)_2$

硝酸カドミウム　*shousankadomiumu*
　cadmium nitrate $Cd(NO_3)_2$

硝酸カリウム　*shousankariumu*
　potassium nitrate KNO_3

硝酸カルシウム　*shousankarushiumu*
　calcium nitrate $Ca(NO_3)_2$

硝酸菌　*shousankin*
　nitrate bacterium

硝酸クロム（ⅠⅠⅠ）　*shousankuromu(III)*
　chromium(III) nitrate $Cr(NO_3)_3$

硝酸鉛（ⅠⅠ）　*shousannamari(II)*
　lead(II) nitrate $Pb(NO_3)_2$

硝酸ナトリウム　*shousannatoriumu*
　sodium nitrate $NaNO_3$

硝酸パラジウム　*shousanparajiumu*
　palladium nitrate $Pd(NO_3)_2$

硝酸リチウム　*shousanrichiumu*
　lithium nitrate $LiNO_3$

硝酸ルビジウム　*shousanrubijiumu*
　rubidium nitrate $RbNO_3$

硝酸セリウム　*shousanseriumu*
　cerium nitrate $Ce(NO_3)_4/Ce(NO_3)_3$

硝酸セルロース　*shousanseruro-su*
　nitrocellulose

硝酸ストロンチウム　*shousansutoronchiumu*
　strontium nitrate $Sr(NO_3)_2$

硝酸ウラニル　*shousanuraniru*
　uranyl nitrate $UO_2(NO_3)_2$

硝石　*shouseki*
　salpeter KNO_3

硝酸第一鉄　*sousandaiichitetsu*
　ferrous nitrate $Fe(NO_3)_2$

硝酸第二鉄　*sousandainitetsu*
　ferric nitrate $Fe(NO_3)_3$

酢　sour, vinegar

酢化分解　*sakukabunkai*
　acetolysis

酢酸　*sakusan*
　acetic acid CH_3COOH

酢酸亜鉛　*sakusanaen*
　zinc acetate $Zn(CH_3COO)_2$

酢酸アミル　*sakusanamiru*
　amyl acetate $CH_3COOC_5H_{11}$

酢酸アンモニウム　*sakusananmoniumu*
　ammonium acetate CH_3COONH_4

酢酸アルミニウム　*sakusanaruminiumu*
　aluminium acetate $(CH_3COO)_3Al$

酢酸バリウム　*sakusanbariumu*
　barium acetate $Ba(CH_3COO)_2$

酢酸ベンジル　*sakusanbenziru*
　benzyl acetate $CH_3COOCH_2C_6H_5$

酢酸ビニル　*sakusanbiniru*
　vinyl acetate $CH_3COOCH=CH_2$

酢酸ビスマス　*sakusanbisumasu*
　bismuth acetate $Bi(CH_3COO)_3$

酢酸ボルニル　*sakusanboruniru*
　bornyl acetate $CH_3COOC_{10}H_{17}$

酢酸ブチル　*sakusanbuchiru*
　butyl acetate $CH_3COOC_4H_9$

酢酸第一鉄　*sakusandaiichitetsu*
　ferrous acetate $Fe(CH_3COO)_2$

酢酸第二鉄　*sakusandainitetsu*
　ferric acetate $Fe(CH_3COO)_3$

酢酸銅　*sakusandou*
　copper acetate $Cu(CH_3COO)_2$

酢酸エチル　*sakusanechiru*
　ethyl acetate $CH_3COOC_2H_5$

酢酸塩　*sakusanen*
　acetate

酢酸エステル　*sakusanesuteru*
　acetic ester $CH_3COOC_2H_5/CH_3COOR$

酢酸フェニル　*sakusanfeniru*
　phenyl acetate $CH_3COOC_6H_5$

酢酸銀　*sakusangin*
　silver acetate CH_3COOAg

酢酸グアヤコール　*sakusanguayako-ru*
　guaiacol acetate $CH_3COOC_6H_4OCH_3$

酢酸発酵　*sakusanhakkou*
　acetic fermentation

酢酸イソブチル　*sakusanisobuchiru*
　isobutyl acetate $CH_3COOCH_2CH(CH_3)_2$

酢酸イソプロピル　*sakusanisopuropiru*
　isopropyl acetae $CH_3COOCH(CH_3)_2$

酢酸イットリウム　*sakusanittoriumu*
　yttrium acetate $Y(C_2H_3O_2)_3 \cdot 4H_2O$

酢酸ジルコニル　*sakusanjirukoniru*
　zirconyl acetate $ZrO(CH_3COO)_2$

酢酸カリウム　*sakusankariumu*
　potassium acetate CH_3COOK

酢酸カルシウム　*sakusankarushiumu*
　calcium acetate $Ca(CH_3COO)_2$

酢酸菌　*sakusankin*
　acetobacter

酢酸マグネシウム　*sakusanmaguneshiumu*
　magnesium acetate $Mg(CH_3COO)_2$

酢酸メチル　*sakusanmechiru*
　methyl acetate CH_3COOCH_3

酢酸モルフィン　*sakusanmorufin*
　morphine acetate $CH_3COOC_{17}H_{18}NO_2$

酢酸鉛（II）　*sakusannamari(II)*
　lead(II) acetate $(CH_3COO)_2Pb$

酢酸ナトリウム　*sakusannatoriumu*
　sodium acetate CH_3COONa

酢酸ペンチル　*sakusanpenchiru*
　pentyl acetate $CH_3COOC_5H_{11}$

酢酸プロピル　*sakusanpuropiru*
　propyl acetate $CH_3COOC_3H_7$

酢酸リナリル　*sakusanrinariru*
　linalyl acetate $CH_3COOC_{10}H_{17}$

酢酸繊維素 *sakusanseniso*
cellulose acetate

酢酸セルロース *sakusanseruro-su*
cellulose acetate

酢酸水銀 *sakusansuigin*
mercurous acetate CH_3COOHg

酢酸ストロンチウム *sakusansutoronchiumu*
strontium acetate $Sr(CH_3COO)_2$

酢酸テルピニル *sakusanterupiniru*
terpinyl acetate $CH_3COOC_{10}H_{17}$

酢酸ウラニル *sakusanuraniru*
uranyl acetate $UO_2(CH_3COO)_2$

酢酸ヴァニリン *sakusanvanirin*
vanillin acetate
$CH_3COOC_6H_3(CHO)(OCH_3)$

酢酸ヴィニル *sakusanviniru*
vinyl acetate $CH_3COOCH=CH_2$

酢 *su*
vinegar

酢漬け *sutsuke*
pickling

結 bond, linkage

結晶 *kesshou*
crystal

結晶場 *kesshouba*
crystal field

結晶場理論 *kesshoubariron*
crystal field theory

結晶中心 *kesshouchuushin*
crystal center

結晶度 *kesshoudo*
crystallinity

結晶学 *kesshougaku*
crystallography

結晶ガラス *kesshougarasu*
crystal glass

結晶形 *kesshougata*
crystal form

結晶軸 *kesshoujiku*
crystal axis

結晶状態 *kesshoujoutai*
crystalline state

結晶化 *kesshouka*
crystallization

結晶化度 *kesshoukado*
crystallinity

結晶化学 *kesshoukagaku*
crystal chemistry

結晶解析 *kesshoukaiseki*
crystal analysis

結晶化水 *kesshoukasui*
crystallization water

結晶系 *kesshoukei*
crystal system

結晶器 *kesshouki*
crystallizer

結晶格子 *kesshoukoushi*
crystal lattice

結晶構造 *kesshoukouzou*
crystal structure

結晶構造分析 *kesshoukouzoubunseki*
crystal structure analysis

結晶面 *kesshoumen*
crystal face

結晶熱 *kesshounetsu*
crystallization heat

結晶性 *kesshousei*
crystallizability

結晶性物質 *kesshouseibusshitsu*
crystalline substance

結晶性重合体 *kesshouseijuugoutai*
crystalline polymer

結晶性の *kesshouseino*
crystalline

結晶生成 *kesshouseisei*
crystallization

結晶質の *kesshoushitsuno*
crystalline

結晶相 *kesshousou*
crystalline phase

結晶水　*kesshousui*
   crystal water

結晶する　*kesshousuru*
   to crystallize

結晶点　*kesshouten*
   crystallizing point

結晶皿　*kesshouzara*
   crystallizing dish

結合　*ketsugou*
   bond, linkage, combination, coupling

結合部位　*ketsugoubui*
   binding site

結合長　*ketsugouchou*
   bond length

結合エネルギー　*ketsugouenerugi-*
   bond energy

結合半径　*ketsugouhankei*
   bond radius

結合法則　*ketsugouhousoku*
   associative law

結合次数　*ketsugoujisuu*
   bond order

結合解裂　*ketsugoukairetsu*
   bond cleavage

結合解離エネルギー
   ketsugoukairienerugi-
   bond dissociation energy

結合角　*ketsugoukaku*
   bond angle

結合軌道　*ketsugoukidou*
   bonding orbital

結合交代　*ketsugoukoutai*
   bond alternation

結合距離　*ketsugoukyori*
   bond distance

結合モーメント　*ketsugoumo-mento*
   bond moment

結合力　*ketsugouryoku*
   binding strength

結合性軌道関数　*ketsugouseikidoukansuu*
   bonding orbital

結合水　*ketsugousui*
   bound water

結合定数　*ketsugouteisuu*
   coupling constant, binding constant

結合剤　*ketsugouzai*
   binder

電　electricity
[for 電気, 電子 and 電解 see below]

電圧　*denatsu*
   voltage

電圧計　*denatsukei*
   voltmeter

電場　*denba*
   electric field

電場活性化　*denbakasseika*
   electric field activation

電場勾配　*denbakoubai*
   electric field gradient

電場強化　*denbakyouka*
   electric field intensification

電場促進　*denbasokushin*
   electric field enhancement

電場強さ　*denbatsuyosa*
   electric field intensity

電着　*denchaku*
   electrodeposition

電着被覆　*denchakuhifuku*
   electrodeposited coating

電着金属　*denchakukinzoku*
   electrodeposited metal

電着特性　*denchakutokusei*
   electrodeposition property

電池　*denchi*
   battery

電池電圧　*denchidenatsu*
   cell voltage

電池モジュール　*denchimoju-ru*
   battery module

電池試験 *denchishiken*
battery testing

電鋳 *denchuu*
electroforming, electrocasting

電導度 *dendoudo*
conductivity

電導度計 *dendoudokei*
conductometer

電導度係数 *dendoudokeisuu*
conductance coefficient

電導度測定 *dendoudosokutei*
conductance measurement

電導度滴定法 *dendoudotekiteihou*
conductometric titration

電導度特性 *dendoudotokusei*
conductometric property

電導率測定 *dendouritsusokutei*
conductometric measurement

電位 *deni*
electric potential

電位壁 *deniheki*
potential barrier

電位勾配 *denikoubai*
potential gradient, voltage gradient

電位曲線 *denikyokusen*
potential curve

電位差 *denisa*
potential difference

電位差分析 *denisabunseki*
potentiometric analysis

電位差法 *denisahou*
potentiometric method

電位差計 *denisakei*
potentiometer

電位差測定 *denisasokutei*
potentiometry

電位差測定研究 *denisasokuteikenkyuu*
potentiometric study

電位差測定センサ *denisasokuteisensa*
potentiometric sensor

電位差滴定 *denisatekitei*
potentiometric titration

電位差滴定曲線 *denisatekiteikyokusen*
potentiometric titration curve

電位測定 *denisokutei*
potentiometry

電位特性 *denitokusei*
potential characteristic

電磁場 *denjiba*
electromagnetic field

電磁波 *denjiha*
electromagnetic wave

電磁波照射 *denjihashousha*
electromagnetic irradiation

電磁放射線 *denjihoushasen*
electromagnetic radiation

電磁結合 *denjikeshigou*
magnetic coupling

電磁気学 *denjikigaku*
electromagnetism

電磁気の *denjikino*
electromagnetic

電磁誘導 *denjiyuudou*
electromagnetic induction

電荷 *denka*
electric charge

電荷安定化 *denkaanteika*
charge stabilization

電界分布 *denkabunpu*
electric field profile

電荷中和 *denkachuuwa*
charge neutralization

電界エミッタ *denkaemitta*
field emitter

電荷非局在化 *denkahikyokuzaika*
charge delocalization

電界 *denkai*
electric field

電荷移動 *denkaidou*
charge tranfer

電荷移動錯体 *denkaidousakutai*
charge-transfer complex

電界放射 *denkaihousha*
field emission

電界イオン化　*denkaiionka*
　field ionization

電荷拡散　*denkakakusan*
　charge diffusion

電荷感受性　*denkakanjusei*
　charge sensitivity

電荷交換　*denkakoukan*
　charge exchange, charge transfer

電荷局在化　*denkakyokkuzaika*
　charge localization

電荷密度　*denkamitsudo*
　charge density

電荷密度分布　*denkamitsudobunpu*
　charge density distribution

電荷量　*denkaryou*
　charge

電荷生成　*denkaseisei*
　charge generation

電荷測定　*denkasokutei*
　electric charge measurement

電荷数　*denkasuu*
　charge number, ionic valency

電顕　*denken*
　electron microscope

電極　*denkyoku*
　electrode

電極分極　*denkyokubunkyoku*
　electrode polarization

電極分離　*denkyokubunri*
　electrode separation

電極電位　*denkyokudeni*
　electrode potential

電極反応　*denkyokuhannou*
　electrode reaction

電極表面　*denkyokuhyoumen*
　electrode surface

電極回転　*denkyokukaiten*
　electrode rotation

電極還元　*denkyokukangen*
　electrode reduction

電極過程　*denkyokukatei*
　electrode process

電極構成　*denkyokukousei*
　electrode configuration

電極接触　*denkyokusesshoku*
　electrode contact

電極触媒　*denkyokushokubai*
　electrocatalyst

電極特性　*denkyokutokusei*
　electrode characteristics

電極材料　*denkyokuzairyou*
　electrode material

電熱原子化　*dennetsugenshika*
　electrothermal atomization

電熱原子吸光分光法　*dennetsugenshikyuukoubunkouhou*
　electrothermal atomic absorption spectrometry

電熱原子吸収　*dennetsugenshikyuushuu*
　electrothermal atomic absorption

電熱工学　*dennetsukougaku*
　electrothermics

電熱の　*dennetsuno*
　electrothermic

電波　*denpa*
　electric wave

電離　*denri*
　electrolytic dissociation, ionization

電離度　*denrido*
　ionization degree

電離性放射線　*denridoseihoushasen*
　ionizing radiation

電離平衡　*denriheikou*
　electric dissociation equilibrium

電離放射線　*denrihoushasen*
　ionizing radiation

電離熱　*denrinetsu*
　heat of ionization

電離定数　*denriteisuu*
　electric dissociation constant

電力　*denryoku*
　electric power

電力不足　*denryokufusoku*
　power shortage

電力源　*denryokugen*
　power source

電力ヒューズ　*denryokuhyu-zu*
　power fuse

電力ケーブル　*denryokuke-buru*
　power cable

電力供給　*denryokukyoukyuu*
　power supply

電量分析　*denryoubunseki*
　coulometry

電量計　*denryoukei*
　coulometer

電量滴定　*denryoutekitei*
　coulometric titration

電流　*denryuu*
　electric current

電流電圧曲線　*denryuudenatsukyokusen*
　current-voltage curve

電流電位曲線　*denryuudenikyokusen*
　electric current-potential curve

電流計　*denryuukei*
　amperemeter

電流効率　*denryuukouritsu*
　current efficiency

電流密度　*denryuumitsudo*
　current density

電流の強さ　*denryuunotsuyosa*
　current, current strength, amperage

電流測定　*denryuusokutei*
　amperometry

電流滴定　*denryuutekitei*
　amperometric titration

電流滴定法　*denryuutekiteihou*
　amperometric assay

電流滴定指示薬　*denryuutekiteishijiyaku*
　amperometric indicator

電融　*denyuu*
　electromelting

雷銀　*raigin*
　fulminating silver, silver fulminate AgCNO

雷酸　*raisan*
　fulminic acid HCNO

雷酸塩　*raisanen*
　fulminate M^ICNO

雷酸金　*raisankin*
　fulminating gold

電　electricity – 電気

電気　*denki*
　electricity

電気安全　*denkianzen*
　electrical safety

電気分解　*denkibunkai*
　electrolysis

電気分解槽　*denkibunkaisou*
　electrolyser

電気分析　*denkibunseki*
　electroanalysis

電気分析研究　*denkibunsekikenkyuu*
　electroanalytical study

電気物理的特性　*denkibutsuritekitokusei*
　electrophysical property

電気デバイス　*denkidebaisu*
　electrical device

電気伝導　*denkidendou*
　electric conduction

電気伝導度分析　*denkidendoudobunseki*
　conductometric analysis

電気伝導度測定　*denkidendoudosokutei*
　conductimetry

電気伝導度測定セル　*denkidendoudosokuteiseru*
　conductance cell

電気伝導度滴定　*denkidendoudotekitei*
　conductometric titration

電気伝導法　*denkidendouhou*
　conductometry

電気伝導率　*denkidendouritsu*
　electric conductivity

電気伝導性重合体　*denkidendouseijuugoutai*
　electrically conducting polymer

電気電子 *denkidenshi*
electrical and electronic

電気泳動 *denkieidou*
electrophoresis

電気泳動分離 *denkieidoubunri*
electrophoretic separation

電気泳動分析 *denkieidoubunseki*
electrophoretic analysis

電気泳動移動度 *denkieidouidoudo*
electrophoretic mobility

電気泳動検出 *denkieidoukenshutsu*
electrophoretic detection

電気泳動毛管 *denkieidoumoukan*
electrophoresis capillary

電気泳動測定 *denkieidousokutei*
electrophoretic measurement

電気泳動特性 *denkieidoutokusei*
electrophoretic characteristics

電気泳動輸送 *denkieidouyusou*
electrophoretic transport

電気エネルギー *denkienerugi-*
electric energy

電気噴霧 *denkifunmu*
electro-spraying

電気原子価 *denkigenshika*
electrovalence

電気合成 *denkigousei*
electrosynthesis

電気発光 *denkihakkou*
electroluminescence

電気発生 *denkihassei*
electrogeneration

電気陰性置換基 *denkiinseichikanki*
electronegative substituent

電気陰性度 *denkiinseido*
electronegativity

電気状態 *denkijoutai*
electrical state

電気荷電量 *denkikadenryou*
amount of electric charge

電気化学 *denkikagaku*
electrochemistry

電気化学バイオセンサ *denkikagakubaiosensa*
electrochemical biosensor

電気化学分析 *denkikagakubunseki*
electrochemical analysis

電気化学デバイス *denkikagakudebaisu*
electrochemical device

電気化学電池 *denkikagakudenchi*
electrochemical cell

電気化学エネルギー貯蔵 *denkikagakuenerugi-chozou*
electrochemical energy storage

電気化学フィルタ *denkikagakufiruta*
electrochemical filter

電気化学現象 *denkikagakugenshou*
electrochemical phenomenon

電気化学合成 *denkikagakugousei*
electrochemical synthesis

電気化学反応 *denkikagakuhannou*
electrochemical reaction

電気化学反応器 *denkikagakuhannouki*
electrochemical reactor

電気化学反応性 *denkikagakuhannousei*
electrochemical reactivity

電気化学インピーダンス *denkikagakuinpi-dansu*
electrochemical impedance

電気化学実験 *denkikagakujikken*
electrochemical experiment

電気化学重合 *denkikagakujuugou*
electrochemical polymerization

電気化学還元 *denkikagakukangen*
electrochemical reduction

電気化学環境 *denkikagakukankyou*
electrochemical environment

電気化学活性 *denkikagakukassei*
electrochemical activity

電気化学系列 *denkikagakukeiretsu*
electrochemical series

電気化学ポテンシャル *denkikagakupotensharu*
electrochemical potential

電気化学プロセス *denkikagakupurosesu*
electrochemical process

電気化学センサ *denkikagakusensa*
electrochemical sensor

電気化学セル *denkikagakuseru*
electrochemical cell

電気化学式 *denkikagakushiki*
electrochemical formula

電気化学触媒作用 *denkikagakushokubaisayou*
electrochemical catalysis

電気化学処理 *denkikagakushori*
electrochemical processing

電気化学速度論 *denkikagakusokudoron*
electrochemical kinetics

電気化学装置 *denkikagakusouchi*
electrochemical apparatus

電気化学定数 *denkikagakuteisuu*
electrochemical constant

電気化学的不動態化 *denkikagakutekifudoutaika*
electrochemical passivation

電気化学特性 *denkikagakutokusei*
electrochemical characteristics

電気化学当量 *denkikagakutouryou*
electrochemical equivalent

電気化学溶解 *denkikagakuyoukai*
electrochemical dissolution

電気化学容量 *denkikagakuyouryou*
electrochemical capacity

電気回路 *denkikairo*
electric circuit

電気還元 *denkikangen*
electroreduction

電気感受率 *denkikanjuritsu*
electric susceptiblity

電気器具 *denkikigu*
electric appliance

電気機器 *denkikiki*
electronic instruments

電気工学 *denkikougaku*
electrical engineering

電気光学ポリマ *denkikougakuporima*
electrooptic polymer

電気工作物 *denkikousakubutsu*
electric facility

電気光散乱 *denkikousanran*
electric light scattering

電気クロマトグラフィ *denkikuromatogurafi*
electric chromatography

電気めっき *denkimekki*
electroplating, electrodeposition

電気毛管現象 *denkimoukangenshou*
electrocapillarity

電気パルス *denkiparusu*
electric pulse

電気炉 *denkiro*
electric furnace

電気製品 *denkiseihin*
electronic goods

電気浸透 *denkishintou*
electroosmosis

電気浸透分離 *denkishintoubunri*
electroosmotic separation

電気浸透流動 *denkishintouryuudou*
electroosmotic flow

電気触媒反応性 *denkishokubaihannousei*
electrocatalytic activity

電気触媒性能 *denkishokubaiseinou*
electrocatalytic performance

電気素量 *denkisoryou*
elementary electric charge

電気双極子 *denkisoukyokushi*
electric dipole

電気双極子モーメント *denkisoukyokushimo-mento*
electric dipole moment

電気スパーク *denkisupa-ku*
electric spark

電気抵抗 *denkiteikou*
electric resistance

電気的現象 *denkitekigenshou*
electrical phenomenon

電気的インピーダンストモグラフィー *denkitekiinpi-dansutomogurafi-*
electrical impedance tomography

電気的陰性の *denkitekiinseino*
electronegative

電気的結合 *denkitekiketsugou*
electrical coupling

電気的強度 *denkitekikyoudo*
electrical strength

電気的酸化 *denkitekisanka*
electrooxidation

電気的性質 *denkitekiseishitsu*
electrical property

電気的シミュレーション *denkitekishimyure-shon*
electrical simulation

電気的触媒作用 *denkitekishokubaisayou*
electrocatalysis

電気的促進 *denkitekisokushin*
electrically enhancement

電気滴定 *denkitekitei*
electrometric titration

電気的陽性の *denkitekiyouseino*
electropositive

電気透析 *denkitouseki*
electrodialysis

電気透析器 *denkitousekiki*
electrodialyzer

電気透析膜 *denkitousekimaku*
electrodialysis membrane

電気透析セル *denkitousekiseru*
electrodialysis cell

電気容量 *denkiyouryou*
electric capacity

電気融合法 *denkiyuugouhou*
electric fusion

電気絶縁破壊 *denkizetsuenhakai*
electrical breakdown

電気絶縁被覆 *denkizetsuenhifuku*
electrical insulation coating

電気絶縁性 *denkizetsuensei*
electrical insulation

電気絶縁材 *denkizetsuenzai*
electric insulator

電 electricity – 電子

電子 *denshi*
electron

電子ビーム *denshibi-mu*
electron beam, electron ray

電子分光 *denshibunkou*
electron spectroscopy

電子クラスタ *denshicurasuta*
electron cluster

電子伝導 *denshidendou*
electronic conduction

電子電子二重共鳴 *denshidenshinijuukyoumei*
electron-electron double resonance, ELDOR

電子伝達 *denshidentatsu*
electron transport

電子伝達系 *denshidentatsukei*
electron transport system

電子伝達体 *denshidentatsutai*
electron carrier

電子データ処理 *denshide-tashori*
electronic data processing

電子エネルギー *denshienerugi-*
electronic energy

電子エネルギー損失分光 *denshienerugi-sonshitsubunkou*
electron energy loss spectroscopy, EELS

電子付加 *denshifuka*
electron addition

電子付着反応 *denshifukahannou*
electron attachment reaction

電子不足分子 *denshifusokubunshi*
electron-deficient molecule

電子不足結合 *denshifusokuketsugou*
electron-deficient bond

電子配置 *denshihaichi*
electron configuration

電子捕獲 *denshihokaku*
electron capture

電子移動 *denshiidou*
electron transfer

電子引力 *denshiinryoku*
electron attraction, electron-withdrawing

電子イオン化 *denshiionka*
electron ionization

電子イオン化質量分析
denshiionkashitsuryoubunseki
electron ionization mass spectrometry

電子異性体 *denshiiseitai*
electronic isomer

電子常磁性共鳴 *denshijoujiseikyoumei*
electron paramagnetic resonance, EPR

電子状態 *denshijoutai*
electronic state

電子受容体 *denshijuyoutai*
electron acceptor

電子回路 *denshikairo*
electric circuit

電子回折 *denshikaisetsu*
electron diffraction

電子過剰 *denshikajou*
electron excess

電子殻 *denshikaku*
electron shell

電子核二重共鳴 *denshikakunijuukyoumei*
electron-nuclear double resonance, ENDOR

電子管 *denshikan*
electron tube

電子還元 *denshikangen*
electron reduction

電子計算機 *denshikeisanki*
computer

電子計算処理 *denshikeisanshori*
electronic data processing

電子顕微鏡 *denshikenbikyou*
electron microscope

電子軌道 *denshikidou*
electron orbit

電子機器 *denshikiki*
electronic equipment

電子孔 *denshikou*
electron hole

電子工学 *denshikougaku*
electronics

電子工学者 *denshikougakusha*
electrical engineer

電子交換 *denshikoukan*
electron exchange

電子構造 *denshikouzou*
electronic structure

電子供与体 *denshikyouyotai*
electron donor

電子吸引基 *denshikyuuinki*
electron-withdrawing group

電子求引性 *denshikyuuinsei*
electron attraction, electron-withdrawing

電子求引性置換基
denshikyuuinseichikanki
electron-withdrawing substituent

電子密度 *denshimitsudo*
electron density

電子モデル *denshimoderu*
electronic model

電子の閉殻 *denshinoheikaku*
closed shell

電子論 *denshiron*
electron theory

電子性質 *denshiseishitsu*
electronic property

電子線 *denshisen*
electron beam, electron ray

電子線回折 *denshisenkaisetsu*
electron ray diffraction

電子センサ *denshisensa*
electrosensor

電子写真 *denshishashin*
electrophotograph

電子親和力　*denshishinwaryoku*
　electron affinity

電子衝撃質量分析法
　denshishougekishitsuryoubunseki
　electron impact mass spectrometry

電子衝撃質量スペクトル
　denshishougekishitsuryousupekutoru
　electron impact mass spectrum

電子素子　*denshisoshi*
　electronic element

電子スペクトル　*denshisupekutoru*
　electronic spectrum

電子スピン共鳴　*denshisupinkyoumei*
　electron spin resonance (ESR)

電子スピン共鳴スペクトル
　denshisupinkyoumeisupekutoru
　electron spin resonance spectrum

電子数　*denshisuu*
　electron number

電子的置換基効果
　denshitekichikankikouka
　electronic substituent effect

電子的過程　*denshitekikatei*
　electronic process

電子特性　*denshitokusei*
　electronic character

電子トンネル化　*denshitonneruka*
　electron tunneling

電子透過性　*denshitoukasei*
　electron transmission

電子対　*denshitsui*
　electron pair

電子対結合　*denshitsuiketsugou*
　electron-pair bond

電子運動　*denshiundou*
　electron motion

電子溶媒和　*denshiyoubaiwa*
　electron solvation

電子要求性　*denshiyoukyuusei*
　electron demand

電子輸送性質　*denshiyusouseishitsu*
　electronic transport property

電　electricity – 電解

電解　*denkai*
　electrolysis

電解分離　*denkaibunri*
　electrolytic separation

電解分析　*denkaibunseki*
　electrolytic analysis

電解抽出　*denkaichuushutsu*
　electrolytic extraction

電解電池　*denkaidenchi*
　electrolytic battery

電解電流　*denkaidenryuu*
　electrolytic current

電解液　*denkaieki*
　electrolyte

電解反応　*denkaihannou*
　electrolytic reaction

電解法　*denkaihou*
　electrolytic method

電解重合　*denkaijuugou*
　electrochemical polymerization

電解重量分析　*denkaijuuryoubunseki*
　electrogravimetric analysis

電解還元　*denkaikangen*
　electrolytic reduction

電解カルボキシル化
　denkaikarubokishiruka
　electrochemical carboxylation

電解化成　*denkaikasei*
　electrolytic formation

電解過程　*denkaikatei*
　electrolysis process

電解酸化　*denkaisanka*
　electrolytic oxidation

電解精錬　*denkaiseiren*
　electrolytic refining

電解セル　*denkaiseru*
　electrolytic cell

電解接触酸化　*denkaisesshokusanka*
　electrocatalytic oxidation

電解質 *denkaishitsu*
   electrolyte

電解質イオン *denkaishitsuion*
   electrolyte ion

電解質混合物 *denkaishitsukongoubutsu*
   electrolyte mixture

電解質効果 *denkaishitsukouka*
   electrolyte effect

電解質濃度 *denkaishitsunoudo*
   electrolyte concentration

電解質流動 *denkaishitsuryuudou*
   electrolyte flow

電解質セル *denkaishitsuseru*
   electrolyte cell

電解質抵抗 *denkaishitsuteikou*
   electrolyte resistance

電解質溶媒 *denkaishitsuyoubai*
   electrolyte solvent

電解質溶液 *denkaishitsuyoueki*
   electrolyte solution

電解処理 *denkaishori*
   electrolytic treatment

電解槽 *denkaisou*
   electrolytic bath, electrolytic cell

電解水素 *denkaisuiso*
   electrolytic hydrogen

電解水素化 *denkaisuisoka*
   electrolytic hydrogenation

電解浴 *denkaiyoku*
   electrolytic bath, electrolytic cell

電解有機合成 *denkaiyuukigousei*
   electroorganic synthesis

溶 melt, dissolve

溶媒 *youbai*
   solvent

溶媒分子 *youbaibunshi*
   solvent molecule

溶媒抽出 *youbaichuushutsu*
   solvent extraction

溶媒付加体 *youbaifukatai*
   solvent adduct

溶媒蒸気 *youbaijouki*
   solvent vapour

溶媒化分解 *youbaikabunkai*
   solvolysis

溶媒回収 *youbaikaishuu*
   solvent recovery

溶媒系 *youbaikei*
   solvent system

溶媒ケージ *youbaike-ji*
   solvent cage

溶媒混合物 *youbaikongoubutsu*
   solvent mixture

溶媒効果 *youbaikouka*
   solvent effect

溶媒構造 *youbaikouzou*
   solvent structure

溶媒極性 *youbaikyokusei*
   solvent polarity

溶媒強度 *youbaikyoudo*
   solvent strength

溶媒濃度 *youbainoudo*
   solvent concentration

溶媒パラメータ *youbaiparame-ta*
   solvent parameter

溶媒サイズ *youbaisaizu*
   solvent size

溶媒制御 *youbaiseigyo*
   solvent control

溶媒シフト *youbaishifuto*
   solvent shift

溶媒システム *youbaishisutemu*
   solvent system

溶媒組成 *youbaisosei*
   solvent composition

溶媒トラッピング *youbaitorappingu*
   solvent trapping

溶媒和 *youbaiwa*
   solvation

溶媒和分子 *youbaiwabunshi*
   solvated molecule

溶媒和物　*youbaiwabutsu*
　solvate

溶媒和電子　*youbaiwadenshi*
　solvated electron

溶媒和エネルギー　*youbaiwaenerugi-*
　solvation energy

溶媒和エンタルピー　*youbaiwaentarupi-*
　solvation enthalpy

溶媒和イオン　*youbaiwaion*
　solvated ion

溶媒和研究　*youbaiwakenkyuu*
　solvation study

溶媒和効果　*youbaiwakouka*
　solvation effect

溶媒和模型　*youbaiwamokei*
　solvation model

溶媒和力　*youbaiwaryoku*
　solvation force

溶媒和錯体　*youbaiwasakutai*
　solvated complex

溶媒和相互作用　*youbaiwasougosayou*
　solvation interaction

溶液　*youeki*
　solution, dilution

溶液安定性　*youekianteisei*
　solution stability

溶液合成　*youekigousei*
　solution synthesis

溶液反応　*youekihannou*
　solution reaction

溶液平衡　*youekiheikou*
　solution equilibrium

溶液比粘度　*youekihiendo*
　specific viscosity of solution

溶液法　*youekihou*
　solution method

溶液状態　*youekijoutai*
　solution-state

溶液重合　*youekijuugou*
　solution polymerization

溶液結晶化　*youekikesshouka*
　solution crystallization

溶液混合　*youekikongou*
　solution mixing

溶液密度　*youekimitsudo*
　solution density

溶液粘度　*youekinendo*
　solution viscosity

溶液濃度　*youekinoudo*
　solution concentration

溶液流　*youekiryuu*
　solution flow

溶液組成　*youekisosei*
　solution composition

溶液相　*youekisou*
　solution phase

溶解　*youkai*
　dissolution

溶解窒素　*youkaichisso*
　dissolved nitrogen

溶解度　*youkaido*
　solubility

溶解度限界　*youkaidogenkai*
　solubility limit

溶解度平衡　*youkaidoheikou*
　solubility equilibrium

溶解度係数　*youkaidokeisuu*
　solubility coefficient

溶解度曲線　*youkaidokyokusen*
　solubility curve

溶解度パラメータ　*youkaidoparame-ta*
　solubility parameter

溶解度積　*youkaidoseki*
　solubility product

溶解度測定　*youkaidosokutei*
　solubility measurement

溶解度定数　*youkaidoteisuu*
　solubility constant

溶解度特性　*youkaidotokusei*
　solubility characteristic

溶解塩　*youkaien*
　dissolved salt

溶解エンタルピー　*youkaientarupi-*
　dissolution enthalpy

溶解限度 *youkaigendo*	溶接 *yousetsu*
solubility limit	welding
溶解法 *youkaihou*	溶接電極 *yousetsudenkyoku*
dissolution method	welding electrode
溶解効果 *youkaikouka*	溶接条件 *yousetsujoukei*
dissolution effect	welding conditions
溶解熱 *youkainetsu*	溶質 *youshitsu*
heat of dissolution	solute
溶解力 *youkairyoku*	溶質分子 *youshitsubunshi*
dissolving power	solute molecule
溶解酸素 *youkaisanso*	溶出 *youshutsu*
dissolved oxygen	elution
溶解性 *youkaisei*	溶出液 *youshutsueki*
solubility	eluate
溶解積 *youkaiseki*	溶出剤 *youshutsuzai*
solubility product	eluent
溶解速度 *youkaisokudo*	溶体 *youtai*
dissolution rate	solution, dilution
溶解速度論 *youkaisokudoron*	溶融 *youyuu*
dissolution kinetics	melting, fusion
溶解する *youkaisuru*	溶融アルミナ *youyuuarumina*
to dissolve	fused alumina
溶血 *youketsu*	溶融銅 *youyuudou*
hemolysis	molten copper
溶菌現象 *youkingenshou*	溶融塩 *youyuuen*
bacteriolysis	molten salt
溶鉱炉 *youkouro*	溶融鉛 *youyuuen*
blast furnace	molten lead
溶鉱炉ガス *youkourogasu*	溶融法 *youyuuhou*
blast furnace gas	fusion method
溶離 *youri*	溶融状態 *youyuujoutai*
elution	molten state
溶離液 *yourieki*	溶融重合 *youyuujuugou*
eluant	melt polymerization
溶離クロマトグラフィー *yourikuromatogurafi-*	溶融マグネシア *youyuumaguneshia*
elution chromatography	fused magnesia
溶離剤 *yourizai*	溶融粘度 *youyuunendo*
eluent	melt viscosity
溶性サッカリン *youseisakkarin*	溶融炉 *youyuuro*
soluble saccharin	melting furnace
	溶融ソーダ *youyuuso-da*
	molten soda

溶融体粘度　*youyuutainendo*
　melt viscosity

溶融炭酸塩　*youyuutansanen*
　molten carbonate

溶融鉄　*youyuutetsu*
　fused iron

溶融材料　*youyuuzairyou*
　molten material

溶剤　*youzai*
　solvent, dissolver

溶剤抽出　*youzaichuushutsu*
　solvent extraction

溶剤蒸発　*youzaihatsuki*
　solvent evaporation

溶剤平衡　*youzaiheikou*
　solvent balance

溶剤蒸気　*youzaijouki*
　solvent vapor

溶剤回収　*youzaikaishuu*
　solvent recovery

溶剤精製　*youzaiseisei*
　solvent refining

溶存　*youzon*
　dissolution

溶存イオン　*youzonion*
　dissolved ion

溶存気体　*youzonkitai*
　dissolved gas

溶存二酸化炭素　*youzonnisankatanso*
　dissolved carbon dioxide

溶存酸素　*youzonsanso*
　dissolved oxygen

溶存有機物質　*youzonyuukibusshitsu*
　dissolved organic matter

蒸　steam, heat

蒸着　*jouchaku*
　evaporation

蒸発　*jouhatsu*
　evaporation, vaporization

蒸発エンタルピー　*jouhatsuentarupi-*
　enthalpy of vaporization

蒸発フラスコ　*jouhatsufurasuko*
　evaporating flask

蒸発減　*jouhatsugen*
　evaporation loss

蒸発乾固　*jouhatsukanko*
　evaporation to dryness

蒸発器　*jouhatsuki*
　evaporator

蒸発熱　*jouhatsunetsu*
　heat of vaporization

蒸発温度　*jouhatsuondo*
　evaporation temperature

蒸発点　*jouhatsuten*
　evaporation point

蒸発残分　*jouhatsuzanbun*
　residue on evaporation

蒸発皿　*jouhatsuzara*
　evaporating dish

蒸気　*jouki*
　vapor

蒸気圧　*joukiatsu*
　vapor pressure

蒸気圧降下　*joukiatsukouka*
　depression of vapor pressure

蒸気ボイラ　*joukiboira*
　steam boiler

蒸気発生器　*joukihasseiki*
　steam generator

蒸気滅菌　*joukimekkin*
　steam sterilization

蒸気相　*joukisou*
　vapor phase

蒸気タービン　*joukita-bin*
　steam turbine

蒸気浴　*joukiyoku*
　steam bath

蒸溜, 蒸留　*jouryuu*
　distillation

蒸溜物　*jouryuubutsu*
　distillation product

蒸溜フラスコ　*jouryuufurasuko*
　distillation flask

蒸溜減　*jouryuugen*
　distillation loss

蒸溜管, 蒸留管　*jouryuukan*
　distillation tube

蒸溜器, 蒸留器　*jouryuuki*
　distillation apparatus

蒸溜水, 蒸留水　*jouryuusui*
　distilled water

蒸留する　*jouryuusuru*
　to distil

蒸溜搭, 蒸留搭　*jouryuutou*
　distillation column

燃　burn, glow

燃料　*nenryou*
　fuel, combustible

燃料アルコール　*nenryouaruko-ru*
　fuel alcohol

燃料バーナ　*nenryouba-na*
　fuel burner

燃料分析　*nenryoubunseki*
　fuels analysis

燃料電池　*nenryoudenchi*
　fuel cell

燃料電池電解質
　nenryoudenchidenkaishitsu
　fuel-cell electrolyte

燃料電池電極　*nenryoudenchidenkyoku*
　fuel cell electrode

燃料電池型　*nenryoudenchikei*
　fuel cell type

燃料電池型反応器
　nenryoudenchikeihannouki
　fuel cell reactor

燃料エタノール　*nenryouetano-ru*
　fuel ethanol

燃料比　*nenryouhi*
　fuel ratio

燃料蒸気　*nenryoujouki*
　fuel vapor

燃料混合物　*nenryoukongoubutsu*
　fuel mixture

燃料炉　*nenryouro*
　fuel furnace

燃料セル　*nenryouseru*
　fuel cell

燃料炭化水素　*nenryoutankasuiso*
　fuel hydrocarbon

燃料添加物　*nenryoutenkabutsu*
　fuel additive

燃料油　*nenryouyu*
　fuel oil

燃焼　*nenshou*
　burning, combustion

燃焼ボート　*nenshoubo-to*
　combustion boat

燃焼分析　*nenshoubunseki*
　combustion analysis

燃焼ガス　*nenshougasu*
　combustion gas

燃焼反応　*nenshouhannou*
　combustion reaction

燃焼域　*nenshouiki*
　burning zone

燃焼時間　*nenshoujikan*
　burning time

燃焼管　*nenshoukan*
　combustion tube

燃焼管理　*nenshoukanri*
　combustion control

燃焼過程　*nenshoukatei*
　burning process

燃焼型　*nenshoukei*
　ignition type

燃焼器効率　*nenshoukikouritsu*
　combustor efficiency

燃焼熱　*nenshounetsu*
　heat of combustion

燃焼温度　*nenshouondo*
　burning temperature

燃焼率　*nenshouritsu*
　burning rate

燃焼炉　*nenshouro*
　combustion furnace, burner

燃焼粒子　*nenshouryuushi*
　burning particle

燃焼サイクル　*nenshousaikukru*
　combustion cycle

燃焼生成物　*nenshouseiseibutsu*
　combustion product

燃焼試験　*nenshoushiken*
　combustion test

燃焼室　*nenshoushitsu*
　combustion chamber

燃焼触媒　*nenshoushokubai*
　combustion catalyst

燃焼速度　*nenshousokudo*
　burning velocity

燃焼帯　*nenshoutai*
　combustion zone

燃焼点　*nenshouten*
　fire point, ignition point

熱　heat, temperature

熱発生　*neppatsusei*
　calorification, heat generation

熱風乾燥　*neppuukansou*
　hot-air drying

熱水　*nessui*
　hot water

熱　*netsu*
　heat

熱安定性　*netsuanteisei*
　thermal stability

熱安定性重合体　*netsuanteiseijuugoutai*
　thermally stable polymer

熱媒染　*netsubaisen*
　hot mordanting

熱爆発　*netsubakuhatsu*
　thermal explosion

熱膨張　*netsubouchou*
　thermal expansion

熱膨張係数　*netsubouchoukeisuu*
　thermal expansion coefficient

熱分解　*netsubunkai*
　pyrolysis, thermal cracking

熱分解ガス　*netsubunkaigasu*
　pyrolysis gas

熱分解反応　*netsubunkaihannou*
　pyrolysis reaction

熱分解反応速度　*netsubunkaihannousokudo*
　pyrolysis kinetics

熱分解実験　*netsubunkaijikken*
　pyrolysis experiment

熱分解蒸留　*netsubunkaijouryuu*
　pyrogenic distillation

熱分解炉　*netsubunkairo*
　pyrolysis furnace

熱分解生成物　*netsubunkaiseiseibutsu*
　pyrolysis product

熱分解質量分析計　*netsubunkaishitsuryoubunsekikei*
　pyrolysis mass spectrometer

熱分解質量スペクトル　*netsubunkaishitsuryousupekutoru*
　pyrolysis mass spectrum

熱分解装置　*netsubunkaisochi*
　pyrolyzer

熱分解速度論　*netsubunkaisokukdoron*
　pyrolysis kinetics

熱分解炭素　*netsubunkaitanso*
　pyrolytic carbon

熱分解油　*netsubunkaiyu*
　pyrolytic oil

熱分解残留物　*netsubunkaizanryuubutsu*
　pyrolysis residue

熱分布　*netsubunpu*
　heat distribution

熱分析　*netsubunseki*
　thermal analysis

熱弾性　*netsudansei*
　thermoelasticity

熱電池　*netsudenchi*
　thermal cell

熱伝導　*netsudendou*
　heat conduction

熱伝導率　*netsudendouritsu*
　thermal conductivity

熱電気　*netsudenki*
　thermoelectricity

熱電効果　*netsudenkouka*
　thermoelectric effect

熱電列　*netsudenretsu*
　thermoelectric series

熱伝達　*netsudentatsu*
　heat transfer

熱伝達係数　*netsudentatsukeisuu*
　heat transfer coefficient

熱電対　*netsudentsui*
　thermocouple, thermoelement

熱電材料　*netsudenzairyou*
　thermoelectric material

熱導体　*netsudoutai*
　heat conductor

熱エネルギー　*netsuenerugi-*
　thermal energy

熱不安定　*netsufuantei*
　thermal instability

熱輻射　*netsufukusha*
　thermal radiation

熱フラグメンテーション
　netsufuragumente-shon
　thermal fragmentation

熱含量　*netsuganryou*
　heat content

熱源　*netsugen*
　heat sorce

熱合成　*netsugousei*
　thermal synthesis

熱破壊　*netsuhakai*
　thermal destruction

熱発火　*netsuhakka*
　thermal ignition

熱反応　*netsuhannou*
　thermal reaction

熱反応器　*netsuhannouki*
　thermal reactor

熱反応速度　*netsuhannousokkudo*
　thermokinetic

熱平衡　*netsuheikou*
　thermal equilibrium

熱変性　*netsuhensei*
　thermal denaturation

熱放射　*netsuhousha*
　thermal radiation, heat radiation

熱放出　*netsuhoushutsu*
　thermal emission

熱放出速度　*netsuhoushutsusokudo*
　heat release rate

熱イオン化　*netsuionka*
　thermal ionization

熱イオン化質量分析法
　netsuionkashitsuryoubunsekihou
　thermal ionization mass spectrometry

熱異性化　*netsuiseika*
　thermal isomerization

熱重合　*netsujuugou*
　thermal polymerization

熱重量分析　*netsujuuryoubunseki*
　thermogravimetry

熱重縮合　*netsujuushukugou*
　thermal polycondensation

熱化学　*netsukagaku*
　thermochemistry

熱化学反応　*netsukagakuhannou*
　thermochemical reaction

熱化学変換　*netsukagakuhenkan*
　thermochemical conversion

熱化学方程式　*netsukagakuhouteishiki*
　thermochemical equation

熱化学プロセス　*netsukagakupurosesu*
　thermochemical process

熱化学処理　*netsukagakushori*
　thermochemical processing

熱化学的安定性　*netsukagakutekianteisei*
　thermochemical stability

熱化学的分解　*netsukagakutekibunkai*
　thermochemical decomposition

熱化学的解析　*netsukagakutekikaiseki*
　thermochemical analysis

熱化学的特性　*netsukagakutekitokusei*
　thermochemical property

熱可逆性ゲル　*netsukagyakuseigeru*
　thermoreversible gel

熱開裂　*netsukairetsu*
　thermal cleavage

熱解析　*netsukaiseki*
　thermal analysis

熱開始　*netsukaishi*
　thermal initiation

熱核反応　*netsukakuhannou*
　thermonuclear reaction

熱拡散　*netsukakusan*
　thermal diffusion

熱架橋　*netsukakyou*
　thermal crosslinking

熱勘定　*netsukanjou*
　heat balance

熱環状付加　*netsukanjoufuka*
　thermal cycloaddition

熱加硫　*netsukaryuu*
　hot vulcanization

熱可塑　*netsukaso*
　thermoplastic

熱可塑樹脂　*netsukasojushi*
　thermoplastic resin

熱可塑性　*netsukasosei*
　thermoplasticity

熱可塑性物質　*netsukasoseibusshitsu*
　thermoplastic

熱可塑性樹脂　*netsukasoseijushi*
　thermoplastic resin

熱可塑性樹脂コンポジット
　netsukasoseijushikonpojitto
　thermoplastic composite

熱可塑性高分子　*netsukasoseikoubunshi*
　thermoplastic polymer

熱可塑性の　*netsukasoseino*
　thermoplastic, heat-deformable

熱可塑性オレフィン　*netsukasoseiorefin*
　thermoplastic olefin

熱可塑性ポリエステル
　netsukasoseiporiesuteru
　thermoplastic polyester

熱可塑性ポリオレフィン
　netsukasoseiporiorefin
　thermoplastic polyolefin

熱可塑性プラスチック材料
　netsukasoseipurasuchikkuzairyou
　thermoplastic material

熱可塑性繊維　*netsukasoseiseni*
　thermoplastic fiber

熱活性化　*netsukasseika*
　thermal activation

熱活性化エネルギー
　netsukasseikaenerugi-
　thermal activation energy

熱加水分解　*netsukasuibunkai*
　thermal hydrolysis

熱機械応力　*netsukikaiouryoku*
　thermomechanical stress

熱機械試験　*netsukikaishiken*
　thermomechanical test

熱機械的安定性　*netsukikaitekianteisei*
　thermomechanical stability

熱コイル　*netsukoiru*
　heat coil

熱効果　*netsukouka*
　heat effect

熱硬化　*netsukouka*
　thermosetting

熱硬化物　*netsukoukabutsu*
　thermosetting material

熱交換器　*netsukoukanki*
　heat exchanger

熱硬化性　*netsukoukasei*
　duroplasticity, thermosetting

熱硬化性樹脂　*netsukoukaseijushi*
　thermosetting resin

熱硬化性重合体　*netsukoukaseijuugoutai*
　thermosetting polymer

熱硬化性ポリエステル
　netsukoukaseiporiesuteru
　thermoset polyester

熱硬化性材料　*netsukoukaseizairyou*
　thermoset material

熱効率　*netsukouritsu*
　thermal efficiency

熱空気加硫　*netsukuukikaryuu*
　hot-air vulvanization, dry heat curing

熱吸着　*netsukyuuchaku*
　heat absorption

熱吸収　*netsukyuushuu*
　heat absorption

熱プラスチック　*netsupurasuchikku*
　thermoplastic

熱プロセス　*netsupurosesu*
　thermal process

熱ラセミ化　*netsurasemika*
　thermal racemization

熱力学　*netsurikigaku*
　thermodynamics

熱力学状態　*netsurikigakujoutai*
　thermodynamic state

熱力学支配　*netsurikigakushihai*
　thermodynamic control

熱力学的研究　*netsurikigakutekikenkyuu*
　thermodynamic study

熱力学的キャラクタリゼーション
　netsurikigakutekikyarakutarize-shon
　thermodynamic characterization

熱力学的な　*netsurikigakutekina*
　thermodynamic

熱量　*netsuryou*
　heat quantity

熱量計　*netsuryoukei*
　calorimeter

熱量測定　*netsuryousokutei*
　calorimetry

熱量測定曲線　*netsuryousokuteikyokusen*
　calorimetric curve

熱量滴定　*netsuryoutekitei*
　calorimetric titration

熱流束　*netsuryuusoku*
　heat flux

熱サイクル　*netsusaikuru*
　heat cycle

熱酸化　*netsusanka*
　thermal oxidation

熱酸化安定性　*netsusankaanteisei*
　thermal oxidative stability

熱酸化分解　*netsusankabunkai*
　thermo-oxidative degradation

熱成形性　*netsuseikeisei*
　thermoformability

熱処理　*netsushori*
　heat treatment

熱衝撃　*netsushougeki*
　thermal shock

熱衝撃抵抗性　*netsushougekiteikousei*
　thermal shock resistance

熱収支　*netsushuushi*
　heat balance

熱塑性　*netsusosei*
　thermoplasticity

熱相転移　*netsusouteni*
　thermal phase transition

熱水素化分解　*netsusuisokabunkai*
　thermal hydrocracking

熱水浴　*netsusuiyoku*
　hot-water bath

熱抵抗　*netsuteikou*
　thermal resistance

熱的転位　*netsuteketeni*
　thermal rearrangement

熱的エネルギー　*netsutekienerugi-*
　thermal energy

熱的方法　*netsutekihouhou*
　thermal method

熱的開環　*netsutekikaikan*
　thermal ring opening

熱的乾燥　*netsutekikansou*
　thermal drying

熱的結合　*netsutekiketsugou*
　thermal coupling

熱的刺激　*netsutekishigeki*
　thermal stress

熱的特性化　*netsutekitokuseika*
　heat characterization

熱天秤　*netsutenbin*
　thermobalance

熱転化　*netsutenka*
　thermal conversion

熱容量　*netsuyouryou*
　heat capacity

熱誘起　*netsuyuuki*
　thermal induction

熱絶縁体　*netsuzetsuentai*
　heat insulator

8.4
Scientific Terms Beginning with Characters Representing Important Prefixes for Chemical Words

不 anti-, non-

不安定, 不安定性　*fuantei, fuanteisei*
　instability

不安定分子　*fuanteibunshi*
　unstable molecule

不安定平衡　*fuanteiheikou*
　instable equilibrium

不安定状態　*fuanteijoutai*
　metastable state

不安定な　*fuanteina*
　unstable

不動態　*fudoutai*
　passive state, passivity

不動態化　*fudoutaika*
　passivation

不飽和　*fuhouwa*
　unsaturation

不飽和化合物　*fuhouwakagoubutsu*
　unsaturated compound

不飽和溶液　*fuhouwayoueki*
　unsaturated solution

不純物　*fujunbutsu*
　impurity

不可逆反応　*fukagyakuhannou*
　irreversible reaction

不可逆変化　*fukagyakuhenka*
　irreversible change

不可逆過程　*fukagyakukatei*
　irreversible process

不可逆過程熱力学
　fukagyakukateinetsukagaku
　irreversible thermodynamics

不可逆阻害　*fukagyakusogai*
　irreversible inhibition

不確定性原理　*fukakuteiseigenri*
　uncertainty prinipcle

不感時間　*fukanjikan*
　dead time, downtime, shut-down period

不完全な　*fukanzenna*
　incomplete, imperfect

不完全燃焼　*fukanzennenshou*
　incomplete combustion

不可視線　*fukashisen*
　invisible rays

不活性物質　*fukasseibusshitsu*
　inactive substance

不活性液体　*fukasseiekitai*
　inert fluid

不活性ガス　*fukasseigasu*
　inert gas

不活性化　*fukasseika*
　inactivation

不活性化剤 *fukasseikazai*
deactivator

不活性気体 *fukasseikitai*
inert gas

不活性の *fukasseino*
inactive, inert

不活性溶媒 *fukasseiyoubai*
inactive solvent

不均一系重合 *fukinitsujuugou*
heterogeneous polymerization

不均一重合 *fukinitsujuugou*
heterogeneous polymerization

不均一系 *fukinitsukei*
heterogeneous system

不均一系反応 *fukinitsukeihannou*
heterogeneous reaction

不均一流れ *fukinitsunagare*
inhomogeneous flow

不均一性 *fukinitsusei*
heterogeneity

不均一触媒作用 *fukinitsushokubaisayou*
heterogeneous catalysis

不均質, 不均質性
fukinshitsu, fukinshitsusei
heterogeneity, inhomogeneity

不均質系 *fukinshitsukei*
heterogeneous system

不規則の *fukisokuno*
irregular

不燃性 *funensei*
incombustibility

不燃性の *funenseino*
noncombustible

不燃性材料, 不燃性物質
funenseizairyou, funenseibusshitsu
incombustible material

不良解析 *furyoukaiseki*
failure analysis

不酸化性の *fusankaseino*
unoxidizable

不斉分子 *fuseibunshi*
chiral molecule

不斉原子 *fuseigenshi*
asymmetric atom

不斉合成 *fuseigousei*
asymmetric synthesis

不斉化合物 *fuseikagoubutsu*
asymmetric compound

不斉還元 *fuseikangen*
asymmetric reduction

不斉水解 *fuseisuikai*
asymmetric hydrolysis

不斉炭素原子 *fuseitansogenshi*
asymmetric carbon atom

不斉誘導 *fuseiyuudou*
asymmetric induction

不浸透性 *fushintousei*
impermeability

不自然 *fushizen*
unnatural, artificial

不対称 *futaishou*
asymmetry

不対称化合物 *futaishoukagoubutsu*
asymmetric compound

不適合 *futekigou*
incompatibility

不凍液 *futoueki*
non-freezing solution

不透過性 *futoukasei*
impermeability

不凍剤 *futouzai*
antifreezing agent

不対電子 *futsuidenshi*
unpaired electron

不和合性 *fuwagousei*
incompatibility

不溶物 *fuyoubutsu*
insoluble substance

不溶性 *fuyousei*
insolubility

不溶性の *fuyouseino*
insoluble

不溶残渣 *fuyouzansa*
insoluble residue

不確定性　*kufakuteisei*
　　uncertainty

同　iso-, equal

同位元素　*douigenso*
　　isotope
同位元素分離　*douigensobunri*
　　isotope separation
同位元素混合物　*douigensokongoubutsu*
　　isotope mixture
同位元素効果　*douigensokouka*
　　isotope effect
同位効果　*douikouka*
　　isotope effect
同位体　*douitai*
　　isotope
同位体分布　*douitaibunpu*
　　isotope distribution
同位体分離　*douitaibunri*
　　isotope separation
同位体分析　*douitaibunseki*
　　isotope analysis
同位体分子　*douitaibunshi*
　　isotopic molecule
同位体置換　*douitaichikan*
　　isotope substitution
同位体原子　*douitaigenshi*
　　isotopic atom
同位体比率　*douitaihiritsu*
　　isotope ratio
同位体標識　*douitaihyoushiki*
　　isotopic labeling
同位体標識化合物　*douitaihyoushikikagoubutsu*
　　isotopically labelled compound
同位体純度　*douitaijundo*
　　isotopic purity
同位体混合物　*douitaikongoubutsu*
　　isotope mixture

同位体効果　*douitaikouka*
　　isotope effect
同位体交換　*douitaikoukan*
　　isotopic exchange
同位体濃縮　*douitainoushuku*
　　isotope concentration
同位体シフト　*douitaishifuto*
　　isotopic shift
同位体特性化　*douitaitokuseika*
　　isotope characterization
同一構造　*douitsukouzou*
　　identical structure
同時分解　*doujibunkai*
　　simultaneous decomposition
同時分離　*doujibunri*
　　simultaneous separation
同時置換　*doujichikan*
　　simultaneous substitution
同時フローインジェクション　*doujifuro-injekushon*
　　simultaneous flow injection
同時合成　*doujigousei*
　　simultaneous synthesis
同時反応　*doujihannou*
　　simultaneous reaction
同時蒸留　*doujijouryuu*
　　simultaneous distillation
同時還元　*doujikangen*
　　simultaneous reduction
同軸　*doujiku*
　　coaxial
同時吸収　*doujikyuushuu*
　　simultaneous absorption
同時熱分解　*doujinetsubunkai*
　　simultaneous pyrolysis
同時試験　*doujishiken*
　　simultaneous test
同時操作　*doujisousa*
　　simultaneous operation
同時添加　*doujitenka*
　　simultaneous addition

同重体 *doujuutai*
　isobar

同化作用 *doukasayou*
　assimilation

同型 *doukei*
　isomorphism

同形 *doukei*
　isomorphism

同形化合物 *doukeikagoubutsu*
　isomorphic compound

同形体 *doukeitai*
　isomorphous substance

同期蛍光 *doukikeikou*
　synchronous fluorescence

同期検波 *doukikenpa*
　synchronous detection

同極化合物 *doukyokukagoubutsu*
　homopolar compound

同性元素 *douseigenso*
　isotope

同素体 *dousotai*
　allotrope

同定 *doutei*
　identification

同定方法 *douteihouhou*
　identification method

同定実験 *douteijikken*
　identification experiment

同族体化合物 *douzokukagoubutsu*
　homologue, homologous compound

同族体分布 *douzokuktaibunpu*
　homologue distribution

同族列 *douzokuretsu*
　homologous series

同族体 *douzokutai*
　homologous compound

亜 next, -ous

亜アンチモン酸 *aanchimonsan*
　antimonous acid H_3SbO_3

亜分画 *abunkaku*
　subfraction

亜鉛 *aen*
　zinc (Zn, element 30)

亜鉛アミド *aenamido*
　zinc amide $Zn(NH_2)_2$

亜鉛アルキル *aenarukiru*
　zinc alkyl ZnR_2

亜鉛中毒 *aenchuudoku*
　zinc poisoning

亜鉛エチル *aenechiru*
　zinc ethyl $Zn(C_2H_5)_2$

亜鉛合金 *aengoukin*
　zinc alloy

亜鉛華 *aenka*
　zinc oxide, zinc white ZnO

亜鉛華軟膏 *aenkanankou*
　zinc oxide ointment

亜鉛明礬 *aenmyouban*
　zinc alum $Al_2Zn(SO_4)_4$

亜・鉛酸 *aensan*
　plumbous acid, plumbic(II) acid $Pb(OH)_2$

亜鉛酸 *aensan*
　zincic acid, zinc hydroxide H_2ZnO_2/ $Zn(OH)_2$

亜鉛酸塩 *aensanen*
　zincate $M^I_2[Zn(OH)_4]$

亜・鉛酸ナトリウム *aensannatoriumu*
　sodium plumbite, sodium plumbate(II) Na_2PbO_2

亜塩素酸 *aensosan*
　chlorous acid $HClO_2$

亜塩素酸塩 *aensosanen*
　chlorite M^IClO_2

亜塩素酸塩漂白 *aensosanhyouhaku*
　chlorite bleaching

亜塩素酸ナトリウム *aensosannatoriumu*
　sodium chlorite $NaClO_2$

亜フォスフィン酸 *afosufinsan*
　phosphinous acid H_3PO (= $H_2P(OH)$)

亜灰長石 *ahainagaseki*
　bytownite $(Na,Ca)(Al,Si)_4O_8$

亜砒藍鉄鉱　*ahirantekkou*
parasymplesite $Fe_3(AsO_4)_2 \cdot 8H_2O$

亜ヒ酸, 亜砒酸　*ahisan*
arsenious acid H_3AsO_3

亜砒酸銅（II）　*ahisandou(II)*
copper(II) arsenite $Cu_3(AsO_3)_2$

亜ヒ酸塩, 亜砒酸塩　*ahisanen*
arsenite, arsenate(III) $M^IH_2AsO_3/$
$M^I_2HAsO_3/M^I_3AsO_3$

亜砒酸銀　*ahisangin*
silver arsenite Ag_3AsO_3

亜ヒ酸カリウム　*ahisankariumu*
potassium arsenite

亜ヒ酸ナトリウム, 亜砒酸ナトリウム
ahisannatoriumu
sodium arsenite

亜砒酸三ナトリウム
ahisansannatoriumu
trisodium arsenite Na_3AsO_3

亜砒酸トリエチル　*ahisantoriechiru*
triethyl arsenite $(C_2H_5)_3AsO_3$

亜ホスフィン酸　*ahosufinsan*
phosphinous acid H_3PO ($= H_2P(OH)$)

亜ホスホン酸　*ahosuhonsan*
phosphonous acid H_3PO_2 ($= HP(OH)_2$)

亜ホウ酸, 亜硼酸　*ahousan*
borous acid

亜ジチオン酸　*ajichionsan*
dithinous acid, disulfuric(III) acid $H_2S_2O_4$

亜ジチオン酸ナトリウム
ajichionsannatoriumu
sodium dithionite $Na_2S_2O_4$

亜クロム酸　*akuromusan*
chromous acid $HCrO_2$

亜クロム酸亜鉛　*akuromusanaen*
zinc chromite $ZnCr_2O_4$

亜クロム酸塩　*akuromusanen*
chromite, chromate(III) $M^I_3[Cr(OH)_6]$

亜麻に油　*amaniyu*
linseed oil

亜・鉛酸塩　*anamarisanen*
plumbite $M^I_2PbO_2$

亜二燐酸　*anirinsan*
diphosphorous acid $H_4P_2O_5$

亜リン酸, 亜燐酸　*arinsan*
phosphorous acid H_3PO_3 ($= P(OH)_3$)

亜燐酸塩　*arinsanen*
phosphite $M^I_3PO_3$

亜燐酸トリエチル　*arinsantoriechiru*
triethyl phosphite $(C_2H_5)_3PO_3$

亜燐酸トリフェニル　*arinsantorifeniru*
triphenyl phosphite $(C_6H_5)_3PO_3$

亜リン酸トリメチル, 亜燐酸トリメチル
arinsantorimechiru
trimethyl phosphite $P(OCH_3)_3$

亜硫酸　*aryuusan*
sulfurous acid H_2SO_3

亜硫酸アンモニウム
aryuusananmoniumu
ammonium sulfite $(NH_4)_2SO_3$

亜硫酸バリウム　*aryuusanbariumu*
barium sulfite $BaSO_3$

亜硫酸塩　*aryuusanen*
sulfite $M^I_2SO_3$

亜硫酸カリウム　*aryuusankariumu*
potassium sulfite K_2SO_3

亜硫酸メチル　*aryuusanmechiru*
methyl sulfite $(CH_3)_2SO_3$

亜硫酸ナトリウム　*aryuusannatoriumu*
sodium sulfite Na_2SO_3

亜硫酸水素塩　*aryuusansuisoen*
hydrogen sulfite M^IHSO_3

亜硫酸水素カリウム
aryuusansuisokariumu
potassium hydrogensulfite $KHSO_3$

亜硫酸水素カルシウム
aryuusansuisokarushiumu
calcium hydrogen sulfite $Ca(HSO_3)_2$

亜硫酸水素ナトリウム
aryuusansuisonatoriumu
sodium bisulfite $NaHSO_3$

亜酸化物　*asankabutsu*
suboxide

亜酸化窒素　*asankachisso*
　dinitrogen monoxide, nitrogen(I)oxide N_2O

亜酸化銀　*asankagin*
　silver suboxide e.g. Ag_6O_2

亜酸化鉛　*asankanamari*
　lead suboxide Pb_2O

亜酸化ルビジウム　*asankarubijiumu*
　rubidium suboxide Rb_9O_2

亜酸化セシウム　*asankaseshiumu*
　cesium suboxide $Cs_{11}O_3$

亜酸化炭素　*asankatanso*
　carbon suboxide C_3O_2

亜セレン酸　*aserensan*
　selenious acid H_2SeO_3

亜セレン酸塩　*aserensanen*
　selenite $M^I_2SeO_3$

亜セレン酸カリウム　*aserensankariumu*
　potassium selenite K_2SeO_3

亜セレン酸ナトリウム　*aserensannatoriumu*
　sodium selenite Na_2SeO_3

亜セレン酸水素ナトリウム　*aserensansuisonatoriumu*
　sodium hydrogen selenite $NaHSeO_3$

亜硝酸　*ashousan*
　nitrous acid HNO_2

亜硝酸アミル　*ashousanamiru*
　amyl nitrite $C_5H_{11}NO_2$

亜硝酸アンモニウム　*ashousananmoniumu*
　ammonium nitrite NH_4NO_2

亜硝酸銅（II）　*ashousandou(II)*
　copper(II) nitrite $Cu(NO_2)_2$

亜硝酸エチル　*ashousanechiru*
　ethyl nitrite C_2H_5ONO

亜硝酸塩　*ashousanen*
　nitrite $M^I NO_2$

亜硝酸エステル　*ashousanesuteru*
　nitrous acid ester $RONO$

亜硝酸銀　*ashousangin*
　silver nitrite $AgNO_2$

亜硝酸標白　*ashousanhyouhaku*
　nitrite bleaching

亜硝酸イソブチル　*ashousanisobuchiru*
　isobutyl nitrite $(CH_3)_2CHCH_2ONO$

亜硝酸カリウム　*ashousankariumu*
　potassium nitrite KNO_2

亜硝酸カルシウム　*ashousankarushiumu*
　calcium nitrite $Ca(NO_2)_2$

亜硝酸メチル　*ashousanmechiru*
　methyl nitrite CH_3ONO

亜硝酸ナトリウム　*ashousannatoriumu*
　sodium nitrite $NaNO_2$

亜硝酸ペンチル　*ashousanpenchiru*
　amyl nitrite $C_5H_{11}NO_2$

亜硝酸リチウム　*ashousanrichiumu*
　lithium nitrite $LiNO_2$

亜硝酸セシウム　*ashousanseshiumu*
　cesium nitrite $CsNO_2$

亜硝酸ストロンチウム　*ashousansutoronchiumu*
　strontium nitrite $Sr(NO_2)_2$

亜炭　*atan*
　lignite

亜テルル酸　*aterurusan*
　tellurous acid H_2TeO_3

亜テルル酸塩　*aterurusanen*
　tellurite $M^I_2TeO_3$

亜テルル酸カリウム　*aterurusankariumu*
　potassium tellurite K_2TeO_3

亜ヨウ素酸, 亜沃素酸　*ayousosan*
　iodous acid HIO_2

非　non-

非圧縮性　*hiasshukusei*
　incompressibility

非圧縮性気体　*hiasshukuseikitai*
　incompressible gas

非圧縮性流体　*hiasshukuseiryuutai*
　incompressible fluid

非ベンゼノイド　*hibenzenoido*　nonbenzenoid	非金属物質　*hikinzokubusshitsu*　non-metallic substance

非ベンゼノイド　*hibenzenoido*
　nonbenzenoid

非置換　*hichikan*
　unsubstituted

非直線分子　*hichokusenbunshi*
　nonlinear molecule

非直線形分子　*hichokusenkeibunshi*
　nonlinear molecule

非調和性　*hichouwasei*
　anharmonicity

非調和振動　*hichouwashindou*
　anharmonic vibration

非断熱　*hidannetsu*
　nonadiabatic

非弾性衝突　*hidanseishoutotsu*
　inelastic collision

非電解質　*hidenkaishitsu*
　nonelectrolyte

非働性　*hidousei*
　inactivation

非発酵性物質　*hihakkouseibusshitsu*
　nonfermentable substance

非ヘム鉄　*hihemutetsu*
　nonhem iron

非イオン界面活性剤
　hiionkaimenkasseizai
　nonionic surface-active agent

非還元糖　*hikangentou*
　nonreducing sugar

非環式化合物　*hikanshikikagoubutsu*
　acyclic compound

非干渉性散乱　*hikanshouseisanran*
　incoherent scattering

非結合エネルギー　*hiketsugouenerugi-*
　nonbonded energy

非結合軌道　*hiketsugoukidou*
　nonbonding orbital

非結合性軌道　*hiketsugouseikidou*
　nonbonding orbital

非金属　*hikinzoku*
　nonmetal

非金属物質　*hikinzokubusshitsu*
　non-metallic substance

非金属元素　*hikinzokugenso*
　nonmetallic element

非交互炭化水素　*hikougotankasuiso*
　nonalternant hydrocarbon

非極性分子　*hikyokuseibunshi*
　nonpolar molecule

非極性化合物　*hikyokuseikagoubutsu*
　nonpolar compound

非極性結合　*hikyokuseiketsugou*
　nonpolar bond

非極性溶媒　*hikyokuseiyoubai*
　nonpolar solvent

非局在化　*hikyokuzaika*
　delocalization

非局在化エネルギー
　hikyokuzaikaenerugi-
　delocalization energy, mesomeric energy, resonance energy

非共有電子対　*hikyouyuudenshitsui*
　unshared electron pair, lone electron pair

非共有結合　*hikyouyuuketsugou*
　noncovalent bond

非ニュートン流動　*hinyu-tonryuudou*
　non-Newtonian flow

非プロトン性溶媒　*hipurotonseiyoubai*
　aprotic solvent

非線形　*hisenkei*
　nonlinear

非線形発振器　*hisenkeihasshinki*
　nonlinear oscillator

非線形振動　*hisenkeishindou*
　nonlinear vibration

非晶系　*hishoukei*
　noncrystal

非晶質　*hishoushitsu*
　noncrystalline, amorphous

非晶質固体　*hishoushitsukotai*
　noncrystalline solid

非相同の　*hisoudouno*
　heterologous

非水系重合 *hisuikeijuugou*
nonaqueous polymerization

非水溶媒 *hisuiyoubai*
nonaqueous solvent

非対称 *hitaishou*
asymmetry

非対称炭素原子 *hitaishoutansogenshi*
asymmetric carbon atom

非鉄金属 *hitetsukinzoku*
nonferrous metal

過 per-

過安息香酸 *kaansokukousan*
perbenzoic acid C_6H_5COOOH

過圧力 *kaatsuryoku*
over pressure

過チオ炭酸 *kachiotansan*
perthiocarbonic acid H_2CS_4

過電圧 *kadenatsu*
overvoltage

過塩素酸 *kaensosan*
perchloric acid $HClO_4$

過塩素酸亜鉛 *kaensosanaen*
zinc perchlorate $Zn(ClO_4)_2$

過塩素酸アンモニウム *kaensosananmoniumu*
ammonium perchlorate NH_4ClO_4

過塩素酸塩 *kaensosanen*
perchlorate M^IClO_4

過塩素酸銀 *kaensosangin*
silver perchlorate $AgClO_4$

過塩素酸カリウム *kaensosankariumu*
potassium perchlorate $KClO_4$

過塩素酸マグネシウム *kaensosanmaguneshiumu*
magnesium perchlorate $Mg(ClO_4)_2$

過塩素酸ナトリウム *kaensosannatoriumu*
sodium perchlorate $NaClO_4$

過ギ酸, 過蟻酸 *kagisan*
performic acid $HCOOOH$

過ハロゲン化 *kaharogenka*
perhalogenation

過硼酸, 過ホウ酸 *kahousan*
perboric acid

過硼酸塩 *kahousanen*
perborate

過ホウ酸ソーダ *kahousanso-da*
sodium perborate Na_3BO_3

過飽和 *kahouwa*
supersaturation

過飽和溶液 *kahouwayoueki*
supersaturated solution

過一硫酸塩 *kaichiryuusanen*
peroxosulfate(VI) $M^I_2SO_5$

過剰 *kajou*
excess

過剰生産 *kajouseisan*
overproduction

過充電 *kajuuden*
overcharge

過クロム酸 *kakuromusan*
perchromic acid H_2CrO_6, H_3CrO_8

過クロム酸（V） *kakuromusan(V)*
perchromic(V) acid H_3CrO_8

過クロム酸（VI） *kakuromusan(VI)*
perchromic(VI) acid H_2CrO_6

過クロルベンゾール *kakurorubenzo-ru*
perchlorobenzen C_6Cl_6

過クロルエチレン *kakuroruechiren*
perchloroethylene $Cl_2C=CCl_2$

過クロルエタン *kakuroruetan*
perchlorethane C_2Cl_6

過マンガン酸カリ価 *kamangansankarika*
permanganate number

過マンガン酸カリウム *kamangansankariumu*
potassium permanganate $KMnO_4$

過マンガン酸ナトリウム *kamangansannatoriumu*
sodium permanganate $NaMnO_4$

過熱　*kanetsu*
　overheating

過熱蒸気　*kanetsujouki*
　superheated vapor

過熱器　*kanetsuki*
　superheater

過二燐酸　*kanirinsan*
　peroxodiphosphoric acid $H_4P_2O_8$

過二硫酸塩　*kaniryuusanen*
　peroxodisulfate(VI) $M^I_2S_2O_8$

過冷却　*kareikyaku*
　supercooling

過燐酸　*karinsan*
　peroxophosphoric acid H_3PO_5

過硫酸　*karyuusan*
　peroxysulfuric aicd H_2SO_5

過硫酸アンモニウム　*karyuusananmoniumu*
　ammonium persulfate $(NH_4)_2S_2O_8$

過硫酸カリウム　*karyuusankariumu*
　potassium persulfate $Na_2S_2O_8$

過酢酸　*kasakusan*
　peracetic acid CH_3COOOH

過酸　*kasan*
　peroxy acid, peracid $RCOOOH$

過酸化　*kasanka*
　peroxidation

過酸化バリウム　*kasankabariumu*
　barium peroxide BaO_2

過酸化ベンゾイル　*kasankabenzoiru*
　benzoyl peroxide $(C_6H_5CO)_2O_2$

過酸化物　*kasankabutsu*
　peroxide

過酸化物価　*kasankabutsuka*
　peroxide number, peroxide value

過酸化酵素　*kasankakouso*
　peroxidase

過酸化ナトリウム　*kasankanatoriumu*
　sodium peroxide Na_2O_2

過酸化脂質　*kasankashishitsu*
　fatty acid peroxide

過酸化水素　*kasankasuiso*
　hydrogen peroxide H_2O_2

過酸性にする　*kasanseinisuru*
　to overacidify, to superacidulate

過酸症　*kasanshou*
　hyperacidity

過炭酸　*katansan*
　percarbonic acid $H_2C_2O_6$

過渡状態　*katojoutai*
　transient state, transition state

過渡的な　*katotekina*
　transient, labil

過ヨード酸　*kayo-dosan*
　periodic acid HIO_4

過ヨード酸塩　*kayo-dosanen*
　periodate M^IIO_4

過ヨウ素酸, 過沃素酸　*kayousosan*
　periodic acid HIO_4

過沃素酸塩　*kayousosanen*
　periodate M^IIO_4

過沃素酸銀　*kayousosangin*
　silver periodate $AgIO_4$

過沃素酸カリウム　*kayousosankariumu*
　potassium periodate KIO_4

脱 de-, remove, get rid of

脱着　*dacchaku*
　desorption

脱着分光法　*dacchakubunkouhou*
　desorption spectroscopy

脱着分析　*dacchakubunseki*
　desorption analysis

脱着分子　*dacchakubunshi*
　desorbed molecule

脱着現象　*dacchakugenshou*
　desorption phenomenon

脱着イオン　*dacchakuion*
　desorbed ion

脱着過程　*dacchakukatei*
　desorption process

脱着器 *dacchakuki*
  desorption device

脱着機構 *dacchakukikou*
  desorption mechanism

脱着曲線 *dacchakukyokusen*
  desorption curve

脱着熱 *dacchakunetsu*
  desorption heat

脱着サイクル *dacchakusaikuru*
  desorption cycle

脱着生成物 *dacchakuseiseibutsu*
  desorption product

脱着装置 *dacchakusouchi*
  desorber

脱着スペクトル *dacchakusupekutoru*
  desorption spectrum

脱着剤 *dacchakuzai*
  desorbent

脱窒素 *dacchisso*
  denitrification

脱窒素活性 *dacchissokassei*
  denitrification activity

脱窒 *dacchitsu*
  denitrification

脱灰 *dakkai*
  deliming

脱気 *dakki*
  degasification, degassing

脱気水 *dakkisui*
  deaerated water

脱錯体化 *dassakutaika*
  decomplexation

脱酸 *dassan*
  deacidification, deoxidation

脱酸素 *dassanso*
  deoxidation

脱酸素化 *dassansoka*
  deoxygenation

脱酸素剤 *dassansozai*
  oxygen absorber

脱脂装置 *dasshisouchi*
  degreaser

脱湿 *dasshitsu*
  dehumidification

脱脂剤 *dasshizai*
  degreasing agent

脱色 *dasshoku*
  decoloration

脱色機構 *dasshokukikou*
  bleaching mechanism

脱色剤 *dasshokuzai*
  decolorant, decolorizer

脱硝 *dasshou*
  denitrification

脱硝剤 *dasshouzai*
  denitrating agent

脱臭 *dasshuu*
  deodorization

脱臭化水素 *dasshuukasuiso*
  dehydrobromination

脱臭素 *dasshuuso*
  debromination

脱臭剤 *dasshuuzai*
  deodorant

脱水 *dassui*
  dehydration, desiccation

脱水銀 *dassuigin*
  demercuriation

脱水反応 *dassuihannou*
  dehydration reaction

脱水環化 *dassuikanka*
  cyclodehydration

脱水素カップリング反応 *dassuikappuringuhannou*
  dehydrogenative coupling reaction

脱水器 *dassuiki*
  dehydrator

脱水機 *dassuiki*
  dehydrator

脱水プロセス *dassuipurosesu*
  dehydration process

脱水生成物 *dassuiseibutsu*
  dehydration product

脱水素触媒　*dassuishokubai*
　dehydration catalyst

脱水症　*dassuishou*
　dehydration

脱水縮合　*dassuishukugou*
　dehydrating condensation

脱水素　*dassuiso*
　dehydrogenation

脱水素重合　*dassuisojuugou*
　dehydrogenative polymerization

脱水素環化　*dassuisokanka*
　dehydrocyclization

脱水素酵素　*dassuisokouso*
　dehydrogenase

脱水素縮合　*dassuisoshukugou*
　dehydrocondensation

脱水する　*dassuisuru*
　to dehydrogenate, to dehydrate, to dry

脱水剤　*dassuizai*
　dehydrating agent

脱アミノ, 脱アミノ基
　datsuamino, datsuaminoki
　deamination

脱アミノ反応　*datsuaminohannou*
　deamination reaction

脱アミノ化　*datsuaminoka*
　deamination

脱アンモニア　*datsuanmonia*
　deammoniation

脱アルキル化　*datsuarukiruka*
　dealkylation

脱アルコキシカルボニル
　datsuarukokishikruboniru
　dealkoxycarbonylation

脱アルミニウム　*datsuaruminiumu*
　dealumination

脱アシル化, 脱アセチル化
　datsuashiruka, datsuasechiruka
　deacylation

脱バインダ　*datsubainda*
　debinding

脱ベンジル　*datsubenjiru*
　debenzylation

脱分極　*datsubunkyoku*
　depolarization

脱塩　*datsuen*
　desalination, desalting

脱塩化水素　*datsuenkasuiso*
　dehydrochlorination

脱塩プロセス　*datsuenpurosesu*
　desalination process

脱塩素　*datsuenso*
　dechlorination

脱塩装置　*datsuensouchi*
　demineralizer

脱塩素剤　*datsuensozai*
　antichlor

脱塩水　*datsuensui*
　desalted water

脱エステル　*datsuesuteru*
　deesterification

脱フッ素, 脱弗素　*datsufukka*
　defluorination

脱フッ化水素　*datsufukkasuiso*
　dehydrofluorination

脱ガス　*datsugasu*
　degassing

脱グリコシル化　*datsugurikoshiruka*
　deglycosylation

脱ハロゲン反応　*datsuharogenhannou*
　dehalogenation reaction

脱ハロゲン化　*datsuharogenka*
　dehalogenation

脱ハロゲン化水素　*datsuharogenkasuiso*
　dehydrohalogenation

脱ヒドロキシル　*datsuhidorokishiru*
　dehydroxylation

脱ヒ素　*datsuhiso*
　arsenic removal

脱保護　*datsuhogo*
　deprotection

脱芳香族化　*datsuhoukouzokuka*
　dearomatization

脱ホウ素化, 脱硼素化　*datsuhousoka*
　deboronation

脱イオン化	datsuionka	deionization
脱イオン水	datsuionsui	deionized water
脱カーボン	datsuka-bon	decarbonizing
脱緩和	datsukanwa	derelaxation
脱カルボキシル基	datsukarubokishiruki	decarboxylation
脱カルボニル	datsukaruboniru	decarbonylation
脱加硫	datsukaryuu	devulcanization
脱活性化	datsukasseika	deactivation
脱金属	datsukinzoku	demetallation
脱共役	datsukyouyaku	uncoupling
脱メタン化	datsumetanka	demethanization
脱メトキシル化	datsumetokishiruka	demethoxylation
脱ニトロ化	datsunitoroka	denitration
脱ニトロ基	datsunitoroki	denitrification, denitration
脱乳化	datsunyuuka	demulsification
脱オキシム	datsuokishimu	deoximation
脱プロトン化エネルギー	datsupurotonkaenerugi-	deprotonation energy
脱ラセミ化	datsurasemika	deracemization
脱離	datsuri	elimination, desorption, separation
脱リチウム化	datsurichiumuka	delithiation
脱リグニン	datsurigunin	delignification
脱離反応	datsurihannou	elimination reaction
脱離過程	datsurikatei	elimination process
脱離基効果	datsurikikouka	leaving group effect
脱離速度	datsurisokudo	elimination kinetics
脱ろう	datsurou	dewaxing
脱硫	datsuryuu	desulfurization, devulcanization
脱硫化物	datsuryuukabutsu	desulfidation
脱硫器	datsuryuuki	desulfurizer
脱硫機構	datsuryuukikou	desulfurization mechanism
脱硫吸着剤	datsuryuukyuuchakuzai	desulfurization adsorbent
脱硫プロセス	datsuryuupurosesu	desulfurization process
脱硫生成物	datsuryuuseiseibutsu	desulfurization product
脱硫触媒	datsuryuushokubai	desulfurization catalyst
脱硫油	datsuryuuyu	desulfurized oil
脱硫剤	datsuryuuzai	desulfurizing agent, devulcanization agent
脱シアン	datsushian	decyanation
脱シリル化	datsushiriruka	desilylation
脱スルフェニル	datsusurufeniru	desulfenylation
脱スルホニル	datsusuruhoniru	desulfonylation
脱スルホン化	datsusuruhonka	desulfonation

脱炭酸　*datsutansan*
　decarboxylation

脱炭酸反応　*datsutansanhannou*
　decarbonation reaction

脱炭酸酵素　*datsutansankouso*
　decarboxylase

脱油　*datsuyu*
　deoiling

陰 negative

陰圧　*inatsu*
　low pressure, negative pressure

陰電荷　*indenka*
　negative charge

陰イオン　*inion*
　anion

陰イオン交換, 陰イオン交換体　*inionkoukan, inionkoukantai*
　anion exchange

陰極　*inkyoku*
　cathode

陰極液　*inkyokueki*
　catholyte

陰極グロー　*inkyokuguro-*
　cathode glow

陰極反応　*inkyokuhannou*
　cathodic process

陰極還元　*inkyokukangen*
　cathodic reduction

陰極励起　*inkyokureikii*
　cathode excitation

陰極線　*inkyokusen*
　cathode ray

陰性原子　*inseigenshi*
　electronegative atom

陰性元素　*inseigenso*
　electronegative element

陰性の　*inseino*
　negative, electronegative

陽 positive

陽電荷　*youdenka*
　positive electric charge

陽電極　*youdenkyoku*
　positive electrode

陽電子　*youdenshi*
　positron

陽イオン　*youion*
　cation

陽イオン分布　*youionbunpu*
　cation distribution

陽イオン置換　*youionchikan*
　cationic substitution

陽イオン過剰　*youionkajou*
　cation excess

陽イオン交換　*youionkoukan*
　cation exchange

陽イオン交換樹脂　*youionkoukanjushi*
　cation exchange resin

陽イオン交換体　*youionkoukantai*
　cation exchanger

陽イオン交換容量　*youionkoukanyouryou*
　cation exchange capacity

陽イオン濃度　*youionnoudo*
　cation concentration

陽極　*youkyoku*
　anode

陽極アルミナ　*youkyokuarumina*
　anodic alumina

陽極分極　*youkyokubunkyoku*
　anodic polarization

陽極電位　*youkyokudeni*
　anode potential

陽極液　*youkyokueki*
　anolyte

陽極不動態化　*youkyokufudoutaika*
　anodic passivation

陽極フッ素化　*youkyokufussoka*
　anodic fluorination

陽極グロー　youkyokuguro-
　anode glow

陽極反応　youkyokuhannou
　anodic process

陽極重合　youkyokujuugou
　anodic polymerization

陽極カップリング　youkyokukappuringu
　anodic coupling

陽極効果　youkyokukouka
　anode effect

陽極メトキシル化
　youkyokumetokishiruka
　anodic methoxylation

陽極モノフッ素化　youkyokumonofussoka
　anodic monofluorination

陽極酸化　youkyokusanka
　anodic oxidation

陽極酸化処理　youkyokusankashori
　anodizing

陽極生成　youkyokuseisei
　anodic formation

陽極接触　youkyokusesshoku
　anodic contact

陽極シアン化　youkyokushianka
　anodic cyanation

陽極処理　youkyokushori
　anodizing

陽極層　youkyokusou
　anodic layer

陽極炭素　youkyokutanso
　anode carbon

陽極溶解　youkyokuyoukai
　anodic dissolution

陽性原子　youseigenshi
　electropositive atom

陽性元素　youseigenso
　electropositive element

陽性の　youseino
　positive, electropositive

陽子　youshi
　proton

陽子エネルギー　youshienerugi-
　proton energy

陽子磁気共鳴スペクトル
　youshijikikyooumeisupekutoru
　proton nuclear magnetic resonance spectrum

陽子励起　youshireiki
　proton excitation

陽子線　youshisen
　proton beam

等　iso-, homo, class

第五族元素　daigozokugenso
　group V element

第一アミド　daiichiamido
　primary amide

第一アミン　daiichiamin
　primary amine RCH_2NH_2

第一アルキルアミン　daiichiarukiruamin
　primary alkyl amine

第一アルコール　daiichiaruko-ru
　primary alcohol RCH_2OH

第一銅塩　daiichidouen
　copper(I) salt, cuprous salt

第一塩　daiichien
　primary salt

第一ホスフィン　daiichihosufin
　primary phosphine

第一クロム塩　daiichikuromuen
　chromous salt, chromium(II) salt

第一水銀塩　daiichisuiginen
　mercury(I) salt, mercurous salt

第一炭素原子　daiichitansogenshi
　primary carbon atom

第一鉄塩　daiichitetsuen
　iron(II) salt, ferrous salt

第一族元素　daiichizokugenso
　group I element

第一級アミン　daiikkyuuamin
　primary amine RCH_2NH_2

第一級アルコール　*daiikkyuuaruko-ru*
primary alcohol RCH$_2$OH

第一級炭素原子　*daiikkyuutansogenshi*
primary carbon atom

第七族元素　*dainanazokugenso*
group VII element

第二アミド　*dainiamido*
secondary amide

第二アミン　*dainiamin*
secondary amine R$_2$NH

第二アルキルアミン　*dainiarukiruamin*
secondary alkylamine

第二アルコール　*dainiaruko-ru*
secondary alcohol

第二銅塩　*dainidouen*
copper (II) salt, cupric salt

第二配位子　*dainihaiishi*
secondary ligand

第二クロム塩　*dainikuromuen*
chromic salt, chromium(III) salt

第二級アミン　*dainikyuuamin*
secondary amine R$_2$NH

第二級アルコール　*dainikyuuaruko-ru*
secondary alcohol

第二級ホスフィン　*dainikyuuhosufin*
secondary phosphine

第二級炭素原子　*dainikyuutansogenshi*
secondary carbon atom

第二水銀塩　*dainisuiginen*
mercury(II) salt, mercuric salt

第二炭素原子　*dainitansogenshi*
secondary carbon atom

第二鉄塩　*dainitetsuen*
iron(III) salt, ferric salt

第二鉄触媒　*dainitetsushokubai*
ferric catalyst

第二族元素　*dainizokugenso*
group II element

第六族元素　*dairokuzokugenso*
group VI element

第三アミン　*daisanamin*
tertiary amine NR$_3$

第三アルコール　*daisanaruko-ru*
tertiary alcohol

第三塩　*daisanen*
tertiary salt

第三級アミン　*daisankyuuamin*
tertiary amine NR$_3$

第三級アルコール　*daisankyuuaruko-ru*
tertiary alcohol

第三級ブチルエーテル
daisankyuubuchirue-teru
tert-butyl ether C$_4$H$_9$OC$_4$H$_9$

第三級ホスフィン　*daisankyuuhosufin*
tertiary phosphine

第三級ホスフィン配位子
daisankyuuhosufinhaiishi
tertiary phosphine ligand

第三級炭素原子　*daisankyuutansogenshi*
tertiary carbon atom

第三炭素原子　*daisantansogenshi*
tertiary carbon atom

第三族元素　*daisanzokugenso*
group III element

第四アンモニウム塩
daishianmoniumuen
quaternary ammonium salt

第四アンモニウム塩基
daishianmoniumuenki
quaternary ammonium base

第四級アンモニウム塩
daishikyuuanmoniumuen
quaternary ammonium salt

第四級炭素　*daishikyuutanso*
quaternary carbon

第四の　*daishino*
quaternary

第四族元素　*daishizokugenso*
group IV element

第四族遷移金属
daishizokugensosenikinzoku
group IV transition metal

第四族金属 *daishizokukinzoku*
  group IV metal

等圧変化 *touatsuhenka*
  isobaric change

等圧線 *touatsusen*
  isobar

等圧式 *touatsushiki*
  isobaric

等分子溶液 *toubunshiyoueki*
  equimolecular solution

等張液 *touchoueki*
  isotonic solution

等張の *touchouno*
  isotonic

等電子 *toudenshi*
  isoelectronic

等電子化合物 *toudenshikagoubutsu*
  isoelectronic compound

等電点 *toudenten*
  isoelectric point

等電点緩衝液 *toudentenkanshoueki*
  isoelectric buffer

等電点膜 *toudentenmaku*
  isoelectric membrane

等電点の *toudentenno*
  isoelectric

等エンタルピーの *touentarupi-no*
  isenthalpic

等エントロピー *touentoropi-*
  isentropy

等エントロピー変化 *touentoropi-henka*
  isentropic change

等方性 *touhousei*
  isotropy

等方性流体 *touhouseiryuutai*
  isotropic fluid

等方性転移 *touhouseiteni*
  isotropic transition

等方性溶液 *touhouseiyoueki*
  isotropic solution

等方性材料 *touhouseizairyou*
  isotropic material

等イオン点 *touionten*
  isoionic point

等時性 *toujisei*
  isochronism

等核 *toukaku*
  homonuclear

等核二原子分子 *toukakunigenshibunshi*
  homonuclear diatomic molecule

等極開裂 *toukyokukairetsu*
  homopolar cleavage

等極結合 *toukyokuketsugou*
  homopolar bond

等吸収点 *toukyuushuuten*
  isosbestic point

等モル *toumoru*
  equimolar

等モル混合物 *toumorukongoubutsu*
  equimolar mixture

等濃度 *tounoudo*
  isoconcentration

等濃度点 *tounoudoten*
  isosbestic point

等温反応 *touonhannou*
  isothermal reaction

等温変化 *touonhenka*
  isothermal change

等温系 *touonkei*
  isothermal system

等温硬化 *touonkouka*
  isothermal curing

等温曲線 *touonkyokusen*
  isothermal curve

等温の *touonno*
  isothermal

等温酸化 *touonsanka*
  isothermal oxidation

等温性 *touonsei*
  isothermality

等温線　*touonsen*
　isotherm

等温線測定　*touonsensokutei*
　isotherm measurement

等温式　*touonshiki*
　isotherm

等温滴定　*touontekitei*
　isothermal titration

等温等圧変化　*touontouatsuhenka*
　isobaric-isothermal change

等差級数　*tousakyuusuu*
　arithmetic series

等差数列　*tousasuuretsu*
　arithmetic progression

等質　*toushitsu*
　homogeneity

等質体　*toushitsutai*
　homogeneous substance

等色性　*toushokusei*
　isochroism

等速温度　*tousokuondo*
　isokinetic temperature

等容線　*touyousen*
　isochore

超　ultra-, super-

超微粒子触媒　*choubiryuushishokubai*
　ultrafine catalyst

超微細構造　*choubisaikouzou*
　hyperfine structure

超分岐高分子　*choubunkikoubunshi*
　hyperbranched polymer

超分子デンドリマ　*choubunshidendorima*
　supramolecular dendrimer

超分子複合体　*choubunshifukugoutai*
　supramolecular complex

超分子合成　*choubunshigousei*
　supramolecular synthesis

超分子重合体　*choubunshijuugoutai*
　supramolecular polymer

超分子化合物　*choubunshikagoubutsu*
　supramolecular compound

超分子会合　*choubunshikaigou*
　supramolecular association

超分子形成　*choubunshikeisei*
　supermolecular formation

超分子効果　*choubunshikouka*
　supramolecular effect

超分子構造　*choubunshikouzou*
　supramolecular structure

超分子モデル　*choubunshimoderu*
　supermolecular model

超分子錯体　*choubunshisakutai*
　supramolecular complexe

超分子触媒　*choubunshishokubai*
　supramolecular catalyst

超分子材料　*choubunshizairyou*
　supramolecular material

超断熱　*choudannetsu*
　superinsulation

超伝導　*choudendou*
　superconductivity

超伝導結晶　*choudendoukesshou*
　superconducting crystal

超伝導性酸化物　*choudendouseisankabutsu*
　superconducting oxide

超伝導繊維　*choudendouseni*
　superconducting fiber

超伝導セラミック　*choudendouseramikku*
　superconducting ceramic

超伝導体　*choudendoutai*
　superconductive conductor

超電気伝導　*choudenkidendou*
　supraconduction

超電流　*choudenryuu*
　supercurrent

超塩基　*chouenki*
　superbase

超塩基性触媒 *chouenkiseishokubai*
superbasic catalyst

超遠心 *chouenshin*
ultracentrifugation

超遠心機 *chouenshinki*
ultracentrifuge

超原子価化合物 *chougenshikakagoubutsu*
hypervalent compound

超原子価結合 *chougenshikaketsugou*
hypervalent bond

超薄高分子膜 *chouhakukoubunshimaku*
ultrathin polymeric film

超薄膜 *chouhakumaku*
ultra-thin film

超薄層 *chouhakusou*
ultra thin layer

超純度 *choujundo*
ultra purity

超純水 *choujunsui*
ultrapure water

超重合体 *choujuugoutai*
superpolymer

超重質油 *choujuushitsuyu*
extra heavy oil

超可塑性 *choukasosei*
superplasticity

超高分子量 *choukoubunshiryou*
ultrahigh molecular weight

超高分子量重合体 *choukoubunshiryoujuugoutai*
ultrahigh molecular weight polymer

超高分子量ポリエチレン *choukoubunshiryouporiechiren*
ultrahigh molecular weight polyethylene

超高純度化 *choukoujundoka*
ultrapurification

超高感度分析 *choukoukandobunseki*
ultrahigh sensitivity analysis

超格子 *choukoushi*
superlattice

超格子反射 *choukoushihansha*
superlattice reflection

超高真空 *choukoushinkuu*
ultrahigh vacuum

超硬質 *choukoushitsu*
superhardness

超高速分光 *choukousokubunkou*
ultrafast spectroscopy

超硬材料 *choukouzairyou*
ultrahard material

超強酸 *choukyousan*
superacid

超共役 *choukyouyaku*
hyperconjugation

超共役分子 *choukyouyakubunshi*
hyperconjugated molecule

超吸収 *choukyuushuu*
superabsorption

超吸収高分子 *choukyuushuukoubunshi*
superabsorbent polymer

超吸収剤 *choukyuushuuzai*
superabsorbent

超熱分解 *chounetsubunkai*
ultrapyrolysis

超音波 *chouonpa*
ultrasonic wave

超音波分離 *chouonpabunri*
ultrasonic separation

超音波反応 *chouonpahannou*
ultrasonic reaction

超音波機器 *chouonpakiki*
ultrasonic instrument

超音波メータ *chouonpame-ta*
ultrasonic meter

超音波濾過器 *chouonparokaki*
ultrasonic filter

超音波細胞破砕装置 *chouonpasaibouhasaisouchi*
ultrasonic disintegrator, ulstrasonic device

超音波洗浄法 *chouonpasenjouhou*
ultrasonic cleaning method

超音波処理 *chouonpashori*
ultrasonication

超音波照射法　*chouonpashoushahou*
　ultrasonic irradiation

超音波スペクトロスコピー　*chouonpasupekutorosukopi-*
　ultrasonic spectroscopy

超音波透過　*chouonpatouka*
　ultrasonics transmission

超音速反応　*chouonsokuhannou*
　supersonic reaction

超音速反応性　*chouonsokuhannousei*
　supersonic reactivity

超臨界圧力　*chourinkaiatsuryoku*
　supercritical pressure

超臨界液　*chourinkaieki*
　supercritical fluid

超臨界反応　*chourinkaihannou*
　supercritical reaction

超臨界混合物　*chourinkaikongoubutsu*
　supercritical mixture

超臨界流　*chourinkairyuu*
　supercritical flow

超臨界流体反応　*chourinkairyuutaihannou*
　supercritical fluid reaction

超臨界流体溶媒　*chourinkairyuutaiyoubai*
　supercritical fluid solvent

超臨界水　*chourinkaisui*
　supercritical water

超臨界水酸化　*chourinkaisuika*
　supercritical water oxidation

超臨界水酸化反応器　*chourinkaisuikahannouki*
　supercritical water oxidation reactor

超臨界水溶液　*chourinkaisuiyoueki*
　supercritical aqueous solution

超臨界炭酸ガス　*chourinkaitansan*
　supercritical carbon dioxide

超臨界溶媒　*chourinkaiyoubai*
　supercritical solvent

超臨界溶液　*chourinkaiyoueki*
　supercritical fluid, supercritical solution

超酸　*chousan*
　superacid

超重元素　*chousangenso*
　superheavy element

超酸化　*chousanka*
　superoxidation

超酸化物　*chousankabutsu*
　superoxide

超酸性　*chousansei*
　superacidity

超酸触媒　*chousanshokubai*
　superacid catalyst

超紫外線　*choushigaisen*
　extreme ultraviolet radiation

超ウラン元素　*chouurangenso*
　transuranic element

超増感　*chouzoukan*
　hypersensitization

超イオン伝導率　*kiiondendouritsu*
　superionic conductivity

超イオン系　*kiionkei*
　superionic system

無 non-

無電解めっき　*mudenkaimekki*
　nonelectrolytic plating

無毒性　*mudokusei*
　nontoxicity

無鉛　*muen*
　lead-free

無限希釈　*mugenkishaku*
　infinite dilution

無限鎖　*mugensa*
　infinite chain

無放射遷移　*muhoushaseni*
　radiationless transition

無硫黄加硫　*muioukaryuu*
　sulfurless cure, sulfurless vulcanization

無次元　*mujigen*
　dimensionless

無次元数 *mujigensuu*
dimensionless number

無機分析 *mukibunseki*
inorganic analysis

無機分子 *mukibunshi*
inorganic molecule

無機物, 無機物質
mukibutsu, mukibusshitsu
inorganic compound

無機塩 *mukien*
inorganic salt

無機塩基 *mukienki*
inorganic base

無機フッ化物, 無機弗化物
mukifukkabutsu
inorganic fluoride

無機粉体 *mukifuntai*
inorganic powder

無機顔料 *mukiganryou*
inorganic pigment

無機ガラス *mukigarasu*
inorganic glass

無機ガス *mukigasu*
inorganic gas

無機ゲル *mukigeru*
inorganic gel

無機合成 *mukigousei*
inorganic synthesis

無機配位子 *mukihaiishi*
inorganic ligand

無機反応 *mukihannou*
inorganic reaction

無機被覆 *mukihifuku*
inorganic coating

無機肥料 *mukihiryou*
inorganic fertilizer

無機表面 *mukihyoumen*
inorganic surface

無機イオン *mukiion*
inorganic ion

無機イオン交換体 *mukiionkoukantai*
inorganic ion exchanger

無機蒸気 *mukijouki*
inorganic vapor

無機化学 *mukikagaku*
inorganic chemistry

無機化合物 *mukikagoubutsu*
inorganic compound

無機結晶 *mukikesshou*
inorganic crystal

無機混合物 *mukikongoubutsu*
inorganic mixture

無機固体 *mukikotai*
inorganic solid

無機高分子 *mukikoubunshi*
inorganic polymer

無機吸着剤 *mukikyuuchakuzai*
inorganic adsorbent

無菌の *mukinno*
aseptic, sterile, germ free

無機の *mukino*
inorganic

無菌水 *mukinsui*
sterile water

無機ポリマ *mukiporima*
inorganic polymer

無機粒子 *mukiryuudo*
inorganic particle

無機錯体 *mukisakutai*
inorganic complex

無機酸 *mukisan*
inorganic acid

無機酸化物 *mukisankabutsu*
inorganic oxide

無機性 *mukisei*
inorganic

無機繊維 *mukiseni*
inorganic fiber

無機質 *mukishitsu*
inorganic compound, mineral

無機質化 *mukishitsuka*
mineralization

無機相 *mukisou*
inorganic phase

無機添加剤 *mukitenkazai*
inorganic additive

無機薬品 *mukiyakuhin*
inorganics

無機陽イオン *mukiyouion*
inorganic cation

無効電力 *mukoudenryoku*
reactive power

無極結合 *mukyokuketsugou*
nonpolar bond, covalent bond, homopolar bond

無極性分子 *mukyokuseibunshi*
nonpolar molecule

無極性化合物 *mukyokuseikagoubutsu*
nonpolar compound

無極性溶媒 *mukyokuseiyoubai*
nonpolar solvent

無乳化剤 *munyuukazai*
emulsifier free

無性粘液酸 *museinenekisan*
pyromucic acid C_4H_3OCOOH

無性酒石酸 *museishusekisan*
pyrovinic acid $HOOCCH(CH_3)CH_2COOH$

無線波 *musenha*
radio wave

無色 *mushoku*
achromatic, colourless

無触媒反応 *mushokubaihannou*
uncatalyzed reaction

無触媒重合 *mushokubaijuugou*
noncatalytic polymerization

無色液体 *mushokuekita*
colorless liquid

無色粉末 *mushokufunmatsu*
colorless powder

無色結晶性化合物 *mushokukesshouseikagoubutsu*
colorless crystalline compound

無色の *mushokuno*
colorless, uncolored

無臭の *mushuuno*
odorless

無水アルコール *musuiaruko-ru*
absolute alcohol

無水亜硫酸 *musuiaryuusan*
sulfurous acid anhydride, sulfur dioxide SO_2

無水亜硝酸 *musuiashousan*
nitrous acid anhydride, dinitrogen trioxide N_2O_3

無水亜テルル酸 *musuiaterurusan*
tellurous anhydride TeO_2

無水物 *musuibutsu*
anhydride

無水物生成 *musuibutsuseisei*
anhydride formation

無水塩 *musuien*
anhydrous salt

無水エタノール *musuietano-ru*
absolute ethanol

無水フタル酸 *musuifutarusan*
phthalic anhydride $C_8H_4O_3$

無水コハク酸 *musuikohakusan*
succinic anhydride $C_4H_4O_3$

無水マレイン酸 *musuimareinsan*
maleic anhydride $C_4H_2O_3$

無水の *musuino*
anhydrous

無水硫酸 *musuiryuusan*
sulfur trioxide, sulfuric acid anhydride SO_3

無水酢酸 *musuisakusan*
acetic acid anhydride $(CH_3CO)_2O$

無水酸 *musuisan*
acid anhydride

無水サリチル酸 *musuisarichirusan*
salicylic anhydride $(C_6H_4(OH)CO)_2O$

無水テルル酸 *musuiterurusan*
tellurium trioxide TeO_3

無水糖 *musuitou*
anhydro-sugar

無水ウラン酸 *musuiuransan*
uranium trioxide, uranic acid anhydride UO_3

無水ヴァレリアン酸 *musuivareriansan*
  valeric anhydride $(CH_3(CH_2)_3CO)_2O$

無水溶媒 *musuiyoubai*
  anhydrous solvent

無対称 *mutaishou*
  asymmetry

無定形硫黄 *muteikeiiou*
  amophous sulfur

無定形状態 *muteikeijoutai*
  amorphous state

無定形の *muteikeino*
  amorphous

無定形炭素 *muteikeitanso*
  amorphous carbon

無溶媒系 *muyoubaikei*
  solvent-free system

9
Dictionary Part III: Further Scientific Terms Beginning with *kanji*

For explanations see Chapter 6.

9.1
kanji without Radicals

2 strokes | 入

入口弁　*iriguchiben*
　inlet valve

入口速度　*iriguchisokudo*
　inlet velocity

入力　*nyuuryoku*
　input

入射エネルギー　*nyuushaenerugi-*
　incident energy

入射光　*nyuushakou*
　incident light

3 strokes | 工 寸 大 丸

工学　*kougaku*
　engineering

工学者　*kougakusha*
　engineer, technician

工業分析　*kougyoubunseki*
　technical analysis

工業化学　*kougyoukagaku*
　chemical engineering (abbr.),
　industrial chemistry

工業計器　*kougyoukeiki*
　industrial instrument

工業生物化学　*kougyouseibutsukagaku*
　bioindustrial biochemistry

工業的な　*kougyoutekina*
　industrial

工業用アルコール　*kougyouyouaruko-ru*
　industrial alcohol

工業用ガソリン　*kougyouyougasorin*
　Industrial gasoline

工業用水　*kougyouyousui*
　industrial water

工業有機化学製品
　kougyouyuukikagakuseihin
　industrial organic chemicals

工兵　*kouhei*
　engineer, technician

工場設備　*koujousetsubi*
　plant equipment

工務　*koumu*
　engineering

Japanese-English Chemical Dictionary. Edited by Markus Gewehr
Copyright © 2008 WILEY-VCH Verlag GmbH & Co. KGaA, Weinheim
ISBN: 978-3-527-31293-1

工程　koutei
　manufacturing process

寸法安定性　sunpouanteisei
　dimensional stability

大環状分子　daikanjoubunshi
　macrocyclic molecule

大環状重合体　daikanjoujuugoutai
　macrocyclic polymer

大環状化合物　daikanjoukagoubutsu
　macrocyclic compound

大環状オリゴマ　daikanjouorigoma
　macrocyclic oligomer

大環状ラクタム　daikanjourakutan
　macrocyclic lactam

大理石　dairiseki
　marble

大豆粉　daizuabura
　soybean oil

大豆油　daizuyu
　soybean oil

大気　taiki
　atmosphere

大気圧　taikiatsu
　atmospheric pressure

大気圧イオン化　taikiatsuionka
　atmospheric ionisation

大気中酸素　taikichuusanso
　atmospheric oxygen

大気メタン　taikimetan
　atmospheric methane

大気汚染　taikiosen
　air pollution

大気酸化　taikisanka
　atmospheric oxidation

大気酸素　taikisanso
　atmospheric oxygen

大量合成　tairyougousei
　large-scale synthesis

大量生産　tairyouseisan
　mass production, bulk production

大量生産品　tairyouseisanhin
　mass product

及ぶ　oyobu
　to attain, to come

丸底フラスコ　maruzokofurasuko
　round bottom flask

4 strokes｜元 予 互 太 天 内 毛

元素　genso
　element

元素分析　gensobunseki
　elementary analysis

元素周期律表　gensoshuukiritsuhyou
　periodic table of the elements

予備膨潤　yobiboujun
　preswelling

予備被覆　yobihifuku
　precoating

予備実験　yobijikken
　preliminary experiment

予備酸化　yobisanka
　preliminary oxidation

予備精製　yobiseisei
　prepurification

予防　yobou
　prevention; prophylaxis

予防手段　yoboushudan
　preventive measure

予混合　yokongou
　premix

予混合気体　yokongoukitai
　premixed gas

予混合燃焼　yokongounenshou
　premixed combustion

予燃　yonetsu
　preheating

予測値　yosokuchi
　predicted value

予測構造　yosokukouzou
　projected structure

互変異性　gohenisei
　tautomerism

互変異性体 *goheniseitai*
  tautomer

太白 *taihaku*
  refined sugar

太陽電池 *taiyoudenchi*
  solar cell, solar generator

太陽エネルギー *taiyouenerugi-*
  solar energy

太陽スペクトル *taiyousupekutoru*
  solar spectrum

天秤 *tenbin*
  balance

天井温度 *tenjouondo*
  ceiling temperature

天河石 *tenkaseki*
  amazonite $KAlSi_2O_3$

天然 *tennen*
  nature

天然アミノ酸 *tennenaminosan*
  natural amino acid

天然物 *tennenbutsu*
  natural product

天然物合成 *tennenbutsugousei*
  natural product synthesis

天然物化学 *tennenbutsukagaku*
  chemistry of natural compounds

天然同位体 *tennendouitai*
  natural isotope

天然エナンチオマ *tennenenanchioma*
  natural enantiomer

天然フラボノイド *tennenfurabonoido*
  natural flavonoid

天然ガソリン *tennengasorin*
  natural gasoline

天然ガス *tennengasu*
  natural gas

天然ガス液化 *tennengasuekika*
  natural gas liquefaction

天然ガス吸着 *tennengasukyuuchaku*
  natural gas adsorption

天然原料 *tennengenryou*
  natural raw material

天然ゴム *tennengomu*
  natural rubber

天然ゴム加硫物 *tennengomukaaryuubutsu*
  natural rubber vulcanizate

天然樹脂 *tennenjushi*
  natural resin

天然化合物 *tennenkagoubutsu*
  natural compound

天然基質 *tennenkishitsu*
  natural substrate

天然黒鉛 *tennenkokuen*
  natural graphite

天然混合物 *tennenkongoubutsu*
  natural mixture

天然高分子 *tennenkoubunshi*
  natural polymer

天然抗酸化剤 *tennenkousankazai*
  natural antioxidant

天然ポルフィリン *tennenporufirin*
  natural porphyrin

天然セメント *tennensemento*
  natural cement

天然繊維 *tennenseni*
  natural fiber

天然染料 *tennensenryou*
  natural dye

天然セスキテルペン *tennensesukiterupen*
  natural sesquiterpene

天然脂肪 *tennenshibou*
  natural fat

天然色 *tennenshoku*
  natural color

天然ソーダ *tennenso-da*
  natural soda

天然ウラン *tennenuran*
  natural uranium

天然油脂 *tennenyushi*
  natural fat and oil

天然有機物 *tennenyuukibutsu*
  natural organic substance

天然有機化合物　*tennenyuukikagoubutsu*
　　natural organic compound

天然有機炭素　*tennenyuukitanso*
　　natural organic carbon

天然材料　*tennenzairyou*
　　natural material

内圧　*naiatsu*
　　internal pressure

内部エネルギー　*naibuenerugi-*
　　internal energy

内部変換　*naibuhenkan*
　　internal conversion

内部表面　*naibuhyoumen*
　　inner surface

内部イオン対　*naibuiontsui*
　　internal ion pair

内部回転　*naibukaiten*
　　internal rotation

内部還元　*naibukangen*
　　internal reduction

内部摩擦　*naibumasatsu*
　　internal friction

内部二重結合　*naibunijuuketsugou*
　　internal double bond

内分泌性の　*naibunpitsuseino*
　　endocrine

内部酸化　*naibusanka*
　　internal oxidation

内部指示薬　*naibushijiyaku*
　　internal indicator

内部抵抗　*naibuteikou*
　　internal resistance

内胚葉　*naihaiyou*
　　endoderm

内因性アミノ酸　*naiinseiaminosan*
　　endogenous amino acid

内面　*naimen*
　　inside, inner surface

内挿　*naisou*
　　interpolation

内容物　*naiyoubutsu*
　　content, ingredient, constituent

毛焼き　*keyaki*
　　singeing

毛管　*moukan*
　　capillary

毛管圧力　*moukanatsuryoku*
　　capillary pressure

毛管分離　*moukanbunri*
　　capillary separation

毛管分析　*moukanbunseki*
　　capillary analysis

毛管ビュレット　*moukanbyuretto*
　　capillary buret

毛管カラム　*moukankaramu*
　　capillary column

毛管カラムガスクロマトグラフィー
　　moukankaramugasukuromatogurafi-
　　capillary column gas chromatography

毛管系　*moukankei*
　　capillary system

毛管効果　*moukankouka*
　　capillary effect

毛管クロマトグラフィー
　　moukankuromatogurafi-
　　capillary chromatography

毛管ピペット　*moukanpipetto*
　　capillary pipet

毛管力　*moukanryoku*
　　capillary force

毛管作用　*moukansayou*
　　capillary action

毛管水　*moukansui*
　　capillary water

毛細管　*mousaikan*
　　capillary

毛細管電極　*mousaikandenkyoku*
　　capillary electrode

毛細管構造　*mousaikankouzou*
　　capillary structure

毛細管吸引　*mousaikankyuuin*
　　capillary suction

毛細管流動　*mousaikanryuudou*
　　capillary flow

毛細管ゾーン　*mousaikanzo-n*
　capillary zone

毛細血管　*mousaikekkan*
　capillary tube, capillary

---

5 strokes | 包 凸 凹 必 斥 左 出
本 末 未 失 弁 甘 母

包接化合物　*housetsukagoubutsu*
　cage compound, clathrate

包摂化合物　*housetsukagoubutsu*
　inclusion commpound

包接水和物　*housetsusuiwabutsu*
　clathrate hydrate

包晶　*houshou*
　peritectic

包装　*housou*
　packaging

凸レンズ　*totsurenzu*
　convex lens

凹レンズ　*ourenzu*
　concave lens

必須アミノ酸　*hissuaminosan*
　essential amino acid

必須脂肪酸　*hissushibousan*
　essential fatty acid

斥力　*sekiryoku*
　repulsive force

左旋性の　*sasenseino*
　levorotatory

出口ガス圧力　*deguchigasuatsuryoku*
　outlet gas pressure

出来高　*dekidaka*
　yield, production

出願　*shutsugan*
　application

出熱　*shutsunetsu*
　heat output

出力　*shutsuryoku*
　output

出力電圧　*shutsuryokudenatsu*
　output voltage

出力電流　*shutsuryokudenryuu*
　output current

出力流れ　*shutsuryokunagare*
　output flow

本練り　*honneri*
　boiling off

末端　*mattan*
　terminal

末端アルキル鎖　*mattanarukirusa*
　terminal alkyl chain

末端分析　*mattanbunseki*
　terminal group analysis

末端置換基　*mattanchikanki*
　terminal substituent

末端エポキシド　*mattanepokishido*
　terminal epoxide

末端樹脂　*mattanjushi*
　terminated resin

末端官能基　*mattankannouki*
　terminal functional group

末端官能性　*mattankannousei*
　terminal functionality

末端結合　*mattanketsugou*
　terminal linkage

末端基　*mattanki*
　terminal group, end group

末端基分析　*mattankibunseki*
　end-group analysis

末端基効果　*mattankikouka*
　end group effect

末端二重結合　*mattannijuuketsugou*
　terminal double bond

末端オレフィン　*mattanorefin*
　terminal olefin

末端鎖　*mattansa*
　terminal chain

末端酸素　*mattansanso*
　terminal oxygen

未反応　*mihannou*
　unreacted

未加硫ゴム *mikaryuugomu*
unvulcanized rubber

未水和 *misuiwa*
unhydration

失活 *shikkatsu*
deactivation

失透 *shittou*
devitrification

弁理士 *benrishi*
patent agent, patent attorney

甘コウ *kankou*
calomel, mercurous chloride Hg_2Cl_2

母液 *boeki*
mother liquor

6 strokes │ 両 朱 再 曲

両面 *ryoumen*
double sided, double faced

両性物質 *ryouseibusshitsu*
amphoteric substance

両性電解質 *ryouseidenkaishitsu*
ampholyte

両性化合物 *ryouseikagoubutsu*
amphoteric substance

両性高分子電解質
ryouseikoubunshidenkaishitsu
polyampholyte

両性高分子膜 *ryouseikoubunshimaku*
amphoteric polymer film

両性セルロース誘導体
ryouseiseruro-suyuudoutai
ampholytic cellulose derivative

両性薬 *ryouseiyaku*
amphoteric drug

両親媒性 *ryoushinbaisei*
amphiphilic property

両親媒性アクリルアミド
ryoushinbaiseiakuriruamido
amphiphilic acrylamide

両親媒性分子 *ryoushinbaiseibunshi*
amphipatic molecule

両親媒性配位子 *ryoushinbaiseihaiishi*
amphiphilic ligand

両親媒性カチオン *ryoushinbaiseikachion*
amphiphilic cation

両親媒性系 *ryoushinbaiseikei*
amphiphatic system

両親媒性高分子 *ryoushinbaiseikoubunshi*
amphipathic polymer

両親媒性高分子電解質
ryoushinbaiseikoubunshidenkaishitsu
amphipathic polyelectrolyte

両親媒性共重合体
ryoushinbaiseikyoujuugoutai
amphiphilic copolymer

両親媒性ネットワーク
ryoushinbaiseinettowa-ku
amphiphilic network

両親媒性オリゴマ *ryoushinbaiseiorigoma*
amphiphilic oligomer

両親媒性ピロール *ryoushinbaiseipiro-ru*
amphiphilic pyrrole

両親媒性ポリエチレンオキシド
ryoushinbaiseiporiechirenokishido
amphiphilic polyethylene oxide

両親媒性ポルフィリン
ryoushinbaiseiprorufirin
amphipathic porphyrin

両親媒性リン脂質
ryoushinbaiseirinshishitsu
amphiphilic phospholipid

両親媒性シクロデキストリン
ryoushinbaiseishikurodekisutorin
amphiphile cyclodextrin

両親媒性層 *ryoushinbaiseisou*
amphiphilic layer

両親媒質 *ryoushinbaishitsu*
amphiphile

朱 *shu*
vermilion

再沈澱 *saichinden*
reprecipitation

再現性 *saigensei*
　reproducibility

再現像 *saigenzou*
　redevelopment

再配置 *saihaichi*
　rearrangement

再配向 *saihaikou*
　reorientation

再評価 *saihyouka*
　reassessment

再蒸留 *saijouryuu*
　redistillation

再開発 *saikaihatsu*
　redevelopment

再活性化 *saikasseika*
　reactivation

再検査 *saikensa*
　re-examination, verification

再結晶 *saikesshou*
　recrystallization

再結合 *saiketsugou*
　recombination

再吸収 *saikyuushuu*
　resorption, reabsorption

再生 *saisei*
　regeneration

再生ゴム *saiseigomu*
　regenerated rubber

再生可能エネルギー *saiseikanouenerugi-*
　renewable energy

再生器 *saiseiki*
　regenerator

再生繊維 *saiseiseni*
　regenerated fiber

再生油 *saiseiyu*
　reclaimed oil

再水和作用 *saisuiwasayou*
　rehydration

曲線 *kyokusen*
　curve

曲げ応力 *mageouryoku*
　bending stress

曲げ試験 *mageshiken*
　bending test

曲げ抵抗 *mageteikou*
　bending resistance

曲げ強さ *magetsuyosa*
　bending strength

7 strokes │ 束 寿

束縛電子 *sokubakudenshi*
　bound electron

束縛状態 *sokubakujoutai*
　bound state

束縛理論 *sokubakuriron*
　bonding theory

寿命 *jumyou*
　lifetime

8 strokes │ 長 表 画 果 毒 事

長波 *chouha*
　long wave

長時間作用 *choujikansayou*
　permanent effect

長期 *chouki*
　long-term, long-range

長期間安定 *choukikanantei*
　long term stability

長期間耐久性 *choukikantaikyuusei*
　long term durability

長期の *choukino*
　long-term, long-range

長期応力 *choukiouryoku*
　long term stress

長期接触 *choukisesshoku*
　long term contact

長期試験 *choukishiken*
　long-term experiment

長期照射 *choukishousha*
    long-term irradiation

長脚漏斗 *choukyakurouto*
    long-stem funnel, Bunsen funnel

長距離移動 *choukyoridou*
    long-distance transfer

長距離規則度 *choukyorikisokudo*
    long-range order

長鎖 *chousa*
    long chain

長鎖アルカン *chousaarukan*
    long chain alkane

長鎖アルキル基 *chousaarukiruki*
    long chain alkyl group

長鎖アルコール *chousaaruko-ru*
    long chain alcohol

長鎖塩基 *chousaenki*
    long chain base

長鎖飽和脂肪酸 *chousahouwashibousan*
    long-chain saturated fatty acid

長鎖の *chousano*
    long chain

長鎖パラフィン *chousaparafin*
    long chain paraffin

長鎖酸 *chousasan*
    long chain acid

長鎖脂肪酸 *chousashibousan*
    long chain fatty acid

長鎖疎水基 *chousasosuiki*
    long chain hydrophobic group

長鎖炭化水素 *chousatankasuiso*
    long chain hydrocarbon

長鎖トリグリセリド *chousatoriguriserido*
    long chain triglyceride

長石 *chouseki*
    feldspar

長アルキル鎖 *nagaarukirusa*
    long alkyl chain

長生き *nagaiki*
    longevity

長さ *nagasa*
    length

表す *arawasu*
    to show, to express

表 *hyou*
    table

表示 *hyouji*
    display

表面 *hyoumen*
    surface

表面圧 *hyoumenatsu*
    surface pressure

表面張力 *hyoumenchouryoku*
    surface tension

表面電位 *hyoumendeni*
    surface potential

表面エネルギー *hyoumenenerugi-*
    surface energy

表面反応 *hyoumenhannou*
    surface reaction

表面状態 *hyoumenjoutai*
    surface characteristics

表面準位 *hyoumenjuni*
    surface level

表面構造 *hyoumenkouzou*
    surface structure

表面粘性 *hyoumennensei*
    surface viscosity

表面積 *hyoumenseki*
    surface area

画分 *kakubun*
    fraction

果糖 *katou*
    D-fructose, fruit sugar, levulose $C_6H_{12}O_6$

毒 *doku*
    poison

毒物 *dokubutsu*
    poison

毒物分析 *dokubutsubunseki*
    toxicological analysis

毒物学 *dokubutsugaku*
    toxicology

毒ガス *dokugasu*
    poison gas

毒性 *dokusei*
toxicity

毒性学 *dokuseigaku*
toxicology

毒性元素 *dokuseigenso*
toxic element

毒素 *dokuso*
toxin

毒薬 *dokuyaku*
poison

事故防止 *jikoboushi*
accident prevention

事故確率 *jikokakuritsu*
accident probability

9 strokes | 発

発火 *hakka*
ignition

発火器 *hakkaki*
fuse

発火温度 *hakkaondo*
ignition temperature

発火点 *hakkaten*
ignition point, flash point, inflammation temperature

発見 *hakken*
discovery

発光 *hakkou*
emission

発酵 *hakkou*
fermentation

発光ダイオード *hakkoudaio-do*
light emitting diode, LED

発光現象 *hakkougenshou*
luminescence

発光放電 *hakkouhouden*
luminous discharge

発酵産物 *hakkousanbutsu*
fermentation product

発酵槽 *hakkousou*
fermenter

発光スペクトル *hakkousupekutoru*
emission spectrum

発光材料 *hakkouzairyou*
luminophore

発泡 *happou*
foaming

発泡ビニル *happoubiniru*
expanded vinyl

発泡ポリウレタン *happouporiuretan*
polyurethane foam

発泡プラスチック *happoupurasuchikku*
expanded plastic

発泡剤 *happouzai*
foaming agent

発泡材料 *happouzairyou*
expanded material

発散 *hassan*
divergence

発生 *hassei*
formation

発生器 *hasseiki*
generator

発振器 *hasshinki*
oscillator

発色団 *hasshokudan*
chromophore

発色反応 *hasshokuhannou*
color reaction

発色試薬 *hasshokushiyaku*
coloring reagent

発色剤 *hasshokuzai*
color coupler

発電機 *hatsudenki*
generator

発煙硫酸 *hatsuenryuusan*
fuming sulfuric acid, oleum H_2SO_4/SO_3

発煙硝酸 *hatsuenshousan*
fuming nitric acid HNO_3

発煙点 *hatsuenten*
smoke point

発癌物質 *hatsuganbusshitsu*
  carcinogen
発癌性化合物 *hatsuganseikagoubutsu*
  carcinogenic compound
発明 *hatsumei*
  invention
発明者 *hatsumeisha*
  inventor
発明的な *hatsumeitekina*
  inventive
発熱反応 *hatsunetsuhannou*
  exothermic reaction
発熱量 *hatsunetsuryou*
  heating value
発熱的な *hatsunetsutekina*
  exothermal, exothermic

---

10 strokes | 射 残

射出 *shashutsu*
  injection
射出成形 *shashutsuseikei*
  injection molding
残留分析 *zanryuubunseki*
  residue analysis

残留放射線 *zanryuuhoushasen*
  residual radiation
残留効果 *zanryuukouka*
  residual effect
残留溶剤 *zanryuuyouzai*
  residual solvent
残余窒素 *zanyochisso*
  residual nitrogen
残余電流 *zanyodenryuu*
  residual current

---

11 strokes | 疎 野

疎水基 *sosuiki*
  hydrophobic group
疎水性の *sosuiseino*
  hydrophobic
野崎檜山岸反応 *nozakihiyamakishihannou*
  Nozaki-Hiyama-Kishi reaction
野生型 *yaseigata*
  wild type
野生酵母 *yaseikoubo*
  wild yeast

9.2
kanji Based on Radicals

亻, 人 **2a** | 人
+ 0 strokes

人工ダイヤモンド *jinkoudaiyamondo*
  artificial diamond
人工元素 *jinkougenso*
  artificial element
人工肥料 *jinkouhiryou*
  artificial fertilizer, synthetic fertilizer
人工鉱物 *jinkoukoubutsu*
  artificial mineral

人工繊維 *jinkouseni*
  synthetic fiber, artificial fiber, chemical fiber
人跡未到, 人跡未踏 *jinsekimitou, jinsekimitou*
  unexplored
人造肥料 *jinzouhiryou*
  chemical fertilizer
人造宝玉 *jinzouhougyoku*
  artificial gemstones
人造宝石 *jinzouhouseki*
  artificial jewels

人造樹脂 *jinzoujushi*
synthetic resin

人造石油 *jinzousekiyu*
synthetic oil

人造繊維 *jinzouseni*
synthetic fiber

人形石 *ningyouseki*
ningyoite $(U,Ca,Ce)_2(PO_4)_2 \cdot 1\text{-}2H_2O$

亻, 人 2a 仕 代 付
+ 3 strokes

仕上げ *shiage*
finishing

仕事率 *shigotoritsu*
power

仕様書 *shiyousho*
specification

代替合成 *daitaigousei*
alternative synthesis

代替研究 *daitaikenkyuu*
alternative study

代替基質 *daitaikishitsu*
alternative substrate

代替触媒 *daitaishokubai*
alternative catalyst

代替溶剤 *daitaiyouzai*
alternative solvent

代用物 *daiyoubutsu*
substitute

代謝回転 *taishakaiten*
metabolic turnover

代謝生成物 *taishaseiseibutsu*
metabolite

付着 *fuchaku*
adhesion

付加 *fuka*
addition

付加反応 *fukahannou*
addition reaction

付加化合物 *fukakagoubutsu*
addition compound

付加環化 *fukakanka*
cycloaddition

付加錯体 *fukasakutai*
addition complex

付加生成物 *fukaseiseibutsu*
addition product

付属設備 *fuzokusetsubi*
accessory equipment

亻, 人 2a 伊 伝 仮 全 合 会
+ 4 strokes

伊利石 *iriseki*
illite $(K,H_3O)(Al,Mg,Fe)_2(Si,Al)_4 \cdot O_{10}[(OH)_2,(H_2O)]$

伊藤石 *itouseki*
itoite $Pb_3[GeO_2(OH)_2](SO_4)_2$

伝動 *dendou*
transmission

伝導 *dendou*
conduction

伝導分析 *dendoubunseki*
conductometric analysis

伝導度 *dendoudo*
conductivity

伝導度滴定 *dendoudotekitei*
conductometric titration

伝導率 *dendouritsu*
conductivity

伝導性高分子 *dendouseikoubunshi*
conducting polymer

伝導性膜 *dendouseimaku*
conductive membrane

伝導性酸化物 *dendouseisankabutsu*
conducting oxide

伝導測定 *dendousokutei*
conductometry

伝導測定セル　dendousokuteiseru
conductance cell

伝導体　dendoutai
conductor

伝導帯　dendoutai
conduction band

伝導滴定　dendoutekitei
conductometric titration

伝熱　dennetsu
heat transfer

伝熱係数　dennetsukeisuu
heat-transfer coefficient

伝熱面　dennetsumen
heating surface

伝熱速度　dennetsusokudo
heat transfer rate

仮説　kasetsu
hypothesis

仮定　katei
assumption

全圧力　zenatsuryoku
total pressure

全窒素　zenchisso
total nitrogen

全電荷　zendenka
total charge

全反応　zenhannou
complete reaction

全硫黄　zeniou
total sulfur

全硬度　zenkoudo
total hardness

全効率　zenkouritsu
overall efficiency

全酸素消費量　zensansoshouhiryou
total oxygen demand

全収率　zenshuuritsu
overall yield

全水分　zensuibun
total moisture, total water content

全炭素量　zentansoryou
total carbon content

全容積　zenyouseki
total volume

全有機炭素　zenyuukitanso
total organic carbon

合金　goukin
alloy

合成　gousei
synthesis

合成ガス　gouseigasu
synthesis gas

合成ゴム　gouseigomu
synthetic rubber

合成法　gouseihou
synthesis method

合成インジゴ　gouseiinjigo
synthetic indigo blue

合成潤滑油　gouseijunkatsuyu
synthetic lubricant

合成樹脂　gouseijushi
synthetic resin

合成化学　gouseikagaku
synthetic chemistry

合成香料　gouseikouryou
synthetic perfume

合成燃料　gouseinenryou
synthetic fuel

合成繊維　gouseiseni
artificial fiber

合成染料　gouseisenryou
synthetic dye

合成洗剤　gouseisenzai
synthetic detergent

合成色素　gouseishikiso
synthetic colorant

合成する　gouseisuru
to synthesize

合繊　gousen
synthetic fiber

合剤　gouzai
medical compound

会合　kaigou
association

イ, 人 2a	位 伸 体 作 低 余 含
+ 5 strokes	

位置エネルギー　*ichienerugi-*
potential energy

位相　*isou*
phase

伸び　*nobi*
elongation

伸長　*shinchou*
elongation

伸度　*shindo*
ductility

伸縮自在　*shinshukujizai*
flexible

伸展性　*shintensei*
extensibility

体積　*taiseki*
volume

体積分率　*taisekibunritsu*
volume fraction

体積磁化率　*taisekijikaritsu*
bulk susceptibility

体積粘性率　*taisekinenseiritsu*
bulk viscosity

体心格子　*taishinkoushi*
body-centered lattice

体心立方格子　*taishinrippoukoushi*
body-centered cubic lattice

作動　*sadou*
release

作業模型　*sagyoumokei*
working model

作業周期　*sagyoushuuki*
working cycle

作用　*sayou*
action

作用部位　*sayoubui*
site of action

作用物質　*sayoubusshitsu*
active ingredient

作用機構　*sayoukikou*
mechanism of action

作用点　*sayouten*
site of action, point of action

作用と反作用　*sayoutohansayou*
action and reaction

作用様式　*sayouyoushiki*
mode of action

低圧　*teiatsu*
low pressure

低圧成形　*teiatsuseikei*
low-pressure molding

低分子の　*teibunshino*
low-molecular

低分子量　*teibunshiryou*
low molecular weight

低分子量化合物　*teibunshiryoukagoubutsu*
low-molecular-weight compound

低分子量ポリマ　*teibunshiryouporima*
low molecular weight polymer

低分子量炭化水素　*teibunshiryoutankasuiso*
low molecular weight hydrocarbon

低窒素　*teichisso*
low nitrogen

低電位　*teideni*
low potential

低毒性　*teidokusei*
low toxicity

低エネルギー　*teienerugi-*
low energy

低エネルギー合成　*teienerugi-gousei*
low energy synthesis

低含水量　*teigansuiryou*
low water content

低発熱　*teihatsunetsu*
low heat

低硫黄　*teiiou*
low sulfur

低蒸気圧　*teijoukiatsu*
low vapor pressure

低コストプロセス　teikosutopurosesu
low cost process

低極性　teikyokusei
low polarity

低級アルケン　teikyuuaruken
light alkene

低級アルコール　teikyuuaruko-ru
lower alcohol

低級オレフィン　teikyuuorefin
lower olefin

低密度ポリエチレン　teimitsudoporiechiren
low density polyethylene

低燃焼性　teinenshousei
low flammability

低温　teion
low temperature

低温分光法　teionbunkouhou
low temperature spectroscopy

低温分離　teionbunri
low temperature separation

低温抽出　teionchuushutsu
low temperature extraction

低温液化　teionekika
low temperature liquefaction

低温反応　teionhannou
low temperature reaction

低温橋かけ　teionhashikake
low temperature curing

低温異性化　teioniseika
low temperature isomerization

低温乾留　teionkanryuu
low temperature carbonization

低温硬化　teionkouka
cold cure, low temperature vulcanization

低温酸化　teionsanka
low temperature oxidation

低酸素含量　teisansoganryou
low oxygen content

低周波　teishuuha
low frequency

低スピン錯体　teisupinsakutai
low-spin complex

低炭素　teitanso
low carbon

低抵抗　teiteikou
low resistance

低溶解度, 低溶解性　teiyoukaido, teiyoukaisei
low solubility

低融点　teiyuuten
low melting point

余熱　yonetsu
waste heat

含量　ganryou
content

含浸　ganshin
impregnation

含浸剤　ganshinzai
impregnant

含硝硫酸　ganshouryuusan
nitrose, nitrosylsulfuric acid NOHSO$_4$

含水フランクリン石　gansuifurankurinseki
chalcophanite $(Mn,Zn)O \cdot 2MnO_2 \cdot 2H_2O$

含水量　gansuiryou
moisture content, humidity

含水炭素　gansuitanso
carbohydrate

含糖量　gantouryou
sugar content

含有　tenka
content

含有する　tenkasuru
to contain

亻, 人　**2a**　使 価 例 供
+ 6 strokes

使用法　shiyouhou
instruction

使用する　shiyousuru
to use

価電子 *kadenshi*
 valence electron

価電子状態 *kadenshijoutai*
 valence state

価電子帯 *kadenshitai*
 valence band

例外 *reigai*
 exception

例えば *tatoeba*
 for example

供給 *kyoukyuu*
 supply

供給する *kyoukyuu shite*
 to supply

供与体 *kyouyotai*
 donor

亻, 人 2a | 命 信 保 係 侵
+ 7 strokes

命名法 *meimeihou*
 nomenclature

信号品質 *shingouhinshitsu*
 signal quality

信号損失 *shingousonshitsu*
 loss of signal

信号対雑音比 *shingoutaizatsuonhi*
 signal-to-noise ratio

信頼できる情報 *shinraidekirujouhou*
 reliable information

保護 *hogo*
 protection

保護アミン *hogoamin*
 protected amin

保護アミノ酸 *hogoaminosan*
 protected amino acid

保護アルコール *hogoaruko-ru*
 protected alcohol

保護被覆 *hogohifuku*
 protective coating

保護カルボン酸 *hogokarubonsan*
 protected carboxylic acid

保護基 *hogoki*
 protecting group

保護基法 *hogokihou*
 protecting group method

保護層 *hogokisou*
 protective layer

保護コロイド *hogokoroido*
 protective colloid

保護膜 *hogomaku*
 protective layer, protective coating

保護ペプチド *hogopepuchido*
 protected peptid

保護試薬 *hogoshiyaku*
 protecting reagent

保護側鎖 *hogosokusa*
 protected side chain

保護剤 *hogozai*
 protective material

保持 *hoji*
 retention

保持時間 *hojijikan*
 retention time

保持係数 *hojikeisuu*
 retention coefficient

保持率 *hojiritsu*
 retention

保持体 *hojitai*
 support

保持体積 *hojitaiseki*
 retention volume

保持容量 *hojiyouryou*
 retention capacity

保温 *hoon*
 heat insulation

保温材 *hoonzai*
 heat insulator

保全性 *hozensei*
 maintainability

保存 *hozon*
 preservation

保存温度 *hozonondo*
storage temperature

保存則 *hozonsoku*
conservation law

保存する *hozonsuru*
to preserve, to conserve

促進剤 *sokushinzai*
accelerator

係数 *keisuu*
coefficient

侵食 *shinshoku*
corrosion

イ, 人 2a
+ 8 to 11 strokes

修 値 偶 側
停 偏 備 傾

修正値 *shuuseichi*
adjusted value

修飾 *shuushoku*
modification

値 *atai*
value

偶数 *guusuu*
even number

偶然誤差 *guuzengosa*
accidental error

側鎖 *sokusa*
side chain

側鎖異性 *sokusaisei*
side-chain isomerism

停止反応 *teishihannou*
termination reaction

停止速度 *teishisokudo*
termination rate

偏光 *henkou*
polarization

偏向 *henkou*
deflection

偏光フィルター *henkoufiruta-*
polarizing filter, polarization filter

偏光計 *henkoukei*
polarimeter

偏光顕微鏡 *henkoukenbikyou*
polarization microscope

偏光器 *henkouki*
polariscope

偏光光度計 *henkoukoudokei*
polarization photometer

偏光吸収スペクトル
henkoukyuushuusupekutoru
polarized absorption spectrum

偏光面 *henkoumen*
polarization plane, plane of polarization

偏光子 *henkoushi*
polarizer

偏差 *hensa*
deviation

備中石 *bichuuseki*
bicchulite $Ca_2Al_2SiO_6(OH)_2$

備える *sonaeru*
to provide, to make preparations, to possess

傾寫 *keisha*
decantation

冫 2b
+ 4 strokes

次

次亜 *jia*
hypo- (pref.)

次亜塩素酸 *jiaensosan*
hypochlorous acid HClO

次亜塩素酸塩 *jiaensosanen*
hypochlorite MClO

次亜塩素酸カルシウム
jiaensosankarushiumu
calcium hypochlorite $Ca(ClO)_2$

次亜塩素酸ナトリウム
jiaensosannatoriumu
sodium hypochlorite NaOCl

次亜フッ素酸, 次亜弗素酸 *jiafussosan*
hypofluorous acid HF

次亜弗素酸　*jiafussosan*
　hypofluorous acid HF

次亜ハロゲン酸　*jiaharogensan*
　hypohalogenous acid HXO

次亜ホウ酸, 次亜硼酸　*jiahousan*
　hypoborous acid

次亜硼酸, 次亜ホウ酸　*jiahousan*
　hypoborous acid

次亜燐酸, 次亜リン酸　*jiarinsan*
　phosphonous acid H_3PO_2 (= $HP(OH)_2$)

次亜燐酸塩　*jiarinsanen*
　hypophosphite

次亜燐酸ナトリウム　*jiarinsannatoriumu*
　sodium hypophosphite NaH_2PO_2

次亜硝酸　*jiashousan*
　hyponitrous acid $H_2N_2O_2$

次亜臭素酸塩　*jiashuusosanen*
　hypobromite M^IOBr

次亜ヨウ素酸, 次亜沃素酸　*jiayousosan*
　hypoiodous acid HIO

次元　*jigen*
　dimension

次ジホスホン酸　*jijihosuhonsan*
　hypodiphosphonic acid $H_4P_2O_4$

次二亜ホスホン酸　*jiniahosuhonsan*
　hypodiphosphonous acid $H_4P_2O_2$

次燐酸　*jirinsan*
　hypodiphosphoric acid $H_4P_2O_6$

冫　2b
+ 5 strokes　　状 冷 冶 求

状態　*joutai*
　state, condition

状態図　*joutaizu*
　phase diagram

冷や　*hiya*
　cold water

冷媒　*reibai*
　refrigerant

冷媒混合物　*reibaikongoubutsu*
　refrigerant mixture

冷陰極放電ランプ　*reiinkyokukhoudenranpu*
　cold cathode lamp

冷間硬化接着剤　*reikankoukasecchakuzai*
　cold-setting adhesive

冷却　*reikyaku*
　cooling

冷却蛇管　*reikyakudakan*
　cooling coil

冷却液　*reikyakueki*
　cooling liquid

冷却表面　*reikyakuhyoumen*
　cooling surface

冷却器　*reikyakuki*
　condenser

冷却コイル　*reikyakukoiru*
　cooling coil

冷却マントル　*reikyakumantoru*
　cooling jacket

冷却温度　*reikyakuondo*
　cooling temperature

冷却試験　*reikyakushiken*
　cooling test

冷却装置　*reikyakusouchi*
　refrigerator

冷却水　*reikyakusui*
　cooling water

冷却する　*reikyakusuru*
　to cool, to chill, to refrigerate

冷却塔　*reikyakutou*
　cooling tower

冷却浴　*reikyakuyoku*
　cooling bath

冷却剤　*reikyakuzai*
　refrigerant

冷水　*reisui*
　cold water

冷水塔　*reisuitou*
　cooling tower

冷凍 *reitou*
    refrigeration

冷凍サイクル *reitousaikuru*
    refrigerating cycle

冷凍剤 *reitouzai*
    refrigerant

冷蔵庫 *reizouko*
    refrigerator, ice box

冶金 *yakin*
    metallurgy

冶金学 *yakingaku*
    science of metallurgy

冶金反応 *yakinhannou*
    metallurgical reaction

求電子置換 *kyuudenshichikan*
    electrophilic substitution

求電子性 *kyuudenshisei*
    electrophilicity

求電子試薬 *kyuudenshishiyaku*
    electrophilic reagent

求核置換 *kyuukakuchikan*
    nucleophilic substitution

求核反応 *kyuukakuhannou*
    nucleophilic reaction

求核性 *kyuukakusei*
    nucleophilicity

求核試薬 *kyuukakushiyaku*
    nucleophile

求核的の *kyuukakutekino*
    nucleophilic

冫 2b 凍 凝
+ 8 to 14 strokes

凍結 *touketsu*
    freezing

凍結安定性 *touketsuanteisei*
    frost resistance

凍結防止 *touketsuboushi*
    freezing prevention

凍結防止剤 *touketsuboushizai*
    antifreezing agent

凍結乾燥 *touketsukansou*
    lyophilization, freeze drying

凍結水溶液 *touketsusuiyoueki*
    frozen aqueous solution

凍結剤 *touketsuzai*
    freezing agent

凍石 *touseki*
    steatite $Mg_3(OH)_2[Si_4O_{10}]$

凝着 *gyouchaku*
    adhesion

凝塊 *gyoukai*
    coagulum

凝灰岩 *gyoukaigan*
    tuff

凝結 *gyouketsu*
    coagulation

凝結時間 *gyouketsujikan*
    setting time

凝固 *gyouko*
    coagulation, clotting, flocculation

凝固防止剤 *gyoukoboushizai*
    anticoagulant

凝固因子 *gyoukoinshi*
    coagulation factor

凝固する *gyoukosuru*
    to coagulate

凝固点 *gyoukoten*
    freezing point, solidifying point

凝固点降下 *gyoukotenkouka*
    depression of freezing point

凝固剤 *gyoukozai*
    coagulating agent

凝析 *gyouseki*
    coagulation, flocculation

凝縮 *gyoushuku*
    condensation

凝縮物 *gyoushukubutsu*
    condensate

凝縮液 *gyoushukueki*
    condensate

凝縮器　*gyoushukuki*
  condenser

凝縮熱　*gyoushukunetsu*
  heat of condensation

凝縮温度　*gyoushukuondo*
  condensing temperature

凝集エネルギー　*gyoushuuenerugi-*
  cohesive energy

凝集作用　*gyoushuusayou*
  agglomeration

凝集素　*gyoushuuso*
  agglutinin

凝集体　*gyoushuutai*
  aggregate

子, 子　2c　｜孔 存 孤
(all *kanji*)

孔雀石　*kujakuseki*
  malachite $CuCO_3 \cdot Cu(OH)_2$

存在　*sonzai*
  existence

孤立化　*koritsuka*
  isolation

阝　2d　｜防 阻 附 阿
+ 4/5 strokes

防爆形　*boubakugata*
  explosion proof

防虫剤　*bouchuuzai*
  insecticide

防毒面　*boudokumen*
  gas mask

防腐剤　*boufuzai*
  antiseptic, preservative

防護壁　*bougoheki*
  protective wall

防護マスク　*bougomasuku*
  protective mask

防護布　*bougopu*
  protective fabric

防護手袋　*bougotetai*
  protective glove

防火材　*bouhizai*
  fire insulation

防腐　*bouhu*
  preservative

防除　*boujo*
  prevention

防火加工　*boukakakou*
  fireproofing, antiflaming

防汚化合物　*bouokagoubutsu*
  antifouling compound

防汚塗料　*bouotoryou*
  antifouling paint

防止　*boushi*
  prevention

防止機構　*boushikikou*
  preventing mechanism

防湿　*boushitsu*
  moistureproof

防止剤　*boushizai*
  inhibitor

防止材料　*boushizairyou*
  prevention material

防食被覆　*boushokuhifuku*
  corrosion protective coating

防食システム　*boushokushisutemu*
  anti-corrosive system

防食剤　*boushokuzai*
  anticorrosive

防縮性　*boushukusei*
  shrinkproofing

防水加工　*bousuikakou*
  waterproofing

防水剤　*bousuizai*
  waterproofing agent

阻害　*sogai*
  inhibition

阻害物質　*sogaibusshitsu*
inhibitor

阻害剤　*sogaizai*
inhibitor

附着　*fuchaku*
adhesion, cohesion, agglutination

附着力　*fuchakuryoku*
adhesive force

附加　*fuka*
addition

附加反応　*fukahannou*
addition reaction

附加化合物　*fukakagoubutsu*
addition compound

阿片　*ahen*
opium

阿片アルカロイド　*ahenarukaroido*
opium alkaloid

阝 2d
+ 6 to 13 strokes

限 降 院 除 陶 階
隔 隕 際 障 隣

限外顕微鏡　*gengaikenbikyou*
ultramicroscope

限外濾過　*gengairoka*
ultrafiltration

限界値　*genkaichi*
threshold value

限界電流　*genkaidenryuu*
limiting current

限定　*gentei*
limit

限定要因　*genteiyouin*
limiting factor

降職　*koushoku*
degradation

降水　*kousui*
precipitation

院イオン交換樹脂　*inionkoukanjushi*
anion-exchange resin

除去　*jokyo*
remove, eliminate

除湿　*joshitsu*
dehumidification

除草剤　*josouzai*
herbicide

除草剤耐性　*josouzaitaisei*
herbicide resistant

除タンパク　*jotanpaku*
deproteinization

陶製三角架　*touseisankakuka*
clay triangle

陶石　*touseki*
pottery stone

階調度　*kaichoudo*
gradient

隔壁　*kakuheki*
septum

隔膜　*kakumaku*
diaphragm

隔離　*kakuri*
segregation

隔絶　*kakuzetsu*
isolation, separation

隕石　*inseki*
meteorite

際　*sai*
time, occasion, when

障害　*shougai*
barrier, hindrance

隣位　*rini*
adjacent position, consecutive position, vicinal position

隣接基　*rinsetsuki*
neighboring group

隣接基関与　*rinsetsukikanyo*
neighboring effect, neighboring group participation

隣接基効果　*rinsetsukikouka*
neighboring-group effect

隣接金属　*rinsetsukinzoku*
adjacent metal

隣接効果　*rinsetsukouka*
　adjacent effect

隣接プロトン　*rinsetsupuroton*
　vicinal proton

卩　2e　| 印 卵
(all *kanji*)

印刷　*insatsu*
　printing

卵割　*rankatsu*
　cleavage

刂, 刀　2f　| 切 別 制 副
(all *kanji*)

切断　*setsudan*
　cutting, breakage, cleavage

別途　*betto*
　special

制働質　*seidoushitsu*
　inhibitor

制御　*seigyo*
　control, monitoring

制御器　*seigyoki*
　controller

制御装置　*seigyosouchi*
　control equipment

刺激物　*shigekibutsu*
　stimulant

刺激臭　*shigekishuu*
　irritating odor

副反応　*fukuhannou*
　side reaction, secondary reaction

副産物　*fukusanbutsu*
　byproduct

副作用　*fukusayou*
　side effect, secondary effect

副生物　*fukuseibutsu*
　byproduct

力　2g　| 力 加
+ 0 to 3 strokes

力　*chikara*
　force, power, strength

力の定数　*chikaranoteisuu*
　force constant

力学　*rikigaku*
　dynamics, mechanics

力学的応力　*rikigakuouryoku*
　mechanical stress

力学試験　*rikigakushiken*
　mechanical test

力学的エネルギー　*rikigakutekienerugi-*
　mechanical energy

力学的キャラクタリゼーション
rikigakutekikyarakutarize-shon
　mechanical characterization

力学的性質　*rikigakutekiseishitsu*
　mechanical property

力価　*rikika*
　potency

加圧蒸留　*kaatsujouryuu*
　pressure distillation

加法　*kahou*
　summation

加熱　*kanetsu*
　heating

加熱減量　*kanetsugenryou*
　heat loss

加熱マントル　*kanetsumantoru*
　heating jacket

加熱試験　*kanetsushiken*
　heat test

加熱帯　*kanetsutai*
　heating zone

加硫　*karyuu*
　vulcanization, cure, curing

加硫遅延剤　*karyuuchienzai*
　retarder

加硫ゴム *karyuugomu*
  vulcanized rubber

加硫器, 加硫機 *karyuuki*
  vulcanizing apparatus

加硫促進剤 *karyuusokushinzai*
  vulcanization accelerator

加硫剤 *karyuuzai*
  vulcanizing agent, curing agent

加成反応 *kaseihannou*
  addition reaction

加成性 *kaseisei*
  additivity

加成体 *kaseitai*
  addition compound, addition product

加湿 *kashitsu*
  humidification

加速度 *kasokudo*
  acceleration

加速反応 *kasokuhannou*
  accelerated reaction

加速器 *kasokuki*
  electron accelerator

加水分解 *kasuibunkai*
  hydrolysis

加水分解度 *kasuibunkaido*
  degree of hydrolysis

加水分解酵素 *kasuibunkaikouso*
  hydrolase

加水分解する *kasuibunkaisuru*
  to hydrolyze

加水解離 *kasuikairi*
  hydrolytic dissociation

力 **2g**
+ 5 to 9 strokes

助 効 動

助酵素 *jokouso*
  coenzyme

励起 *reiki*
  excitation

励起分子 *reikibunshi*
  excited molecule

励起電位 *reikideni*
  excitation potential

励起エネルギー *reikienerugi-*
  excitation energy

励起原子 *reikigenshi*
  excited atom

励起イオン *reikiion*
  excited ion

励起状態 *reikijoutai*
  excited state

励起状態反応 *reikijoutaihannou*
  excited state reaction

励起関数 *reikikansuu*
  excitation function

励起錯体 *reikisakutai*
  excited complex

励起酸素 *reikisanso*
  exited oxygen

励起スペクトル *reikisupekutoru*
  excitation spectrum

効果 *kouka*
  effect

効能 *kounou*
  efficacy

効率 *kouritsu*
  efficiency

動物栄養学 *doubutsueiyougaku*
  animal nutrition

動物実験 *doubutsujikken*
  animal experiment

動物繊維 *doubutsuseni*
  animal fiber

動物試験 *doubutsushiken*
  animal experiment

動物油 *doubutsuyu*
  animal oil

動摩擦 *doumasatsu*
  kinetic friction

動力 *douriki*
  power

動力学 *dourikigaku*
  dynamics, kinetics

動力学方程式 *dourikigakuhouteishiki*
  kinetic equation

動力源 *douryokugen*
  power source

動力炉 *douryokuro*
  nuclear power reactor

動的吸着 *doutekikyuuchaku*
  dynamic adsorption

動的粘弾性 *doutekinendansei*
  dynamic viscoelasticity

動的測定法 *doutekisokuteihou*
  dynamic measurement

又 **2h** | 双 収 受
(all *kanji*)

双極電極 *soukyokudenkyoku*
  bipolar electrode

双極子 *soukyokushi*
  dipole

双極子モーメント *soukyokushimo-mento*
  dipole moment

双極子能率 *soukyokushinouritsu*
  dipole moment

双極子遷移 *soukyokushiseni*
  dipole transition

双極子相互作用 *soukyokushisougosayou*
  dipole interaction

双性イオン *souseiion*
  zwitter ion

収率 *shuuritsu*
  yield

収率の増加 *shuuritsunozouka*
  yield increase

収量 *shuuryou*
  yield

収差 *shuusa*
  aberration

収縮 *shuushuku*
  shrinkage, contraction

収束 *shuusoku*
  convergence

収縮性フィルム *shuushukuseifirumu*
  shrinkable film

受容体 *juyoutai*
  acceptor, receptor

冖 **2i** | 冗 玄
(all *kanji*)

冗長 *jouchou*
  redundancy

写真 *shashin*
  photography

写真測量 *shashinsokuryou*
  photometry

亠 **2j** | 玄 交 充 忘 対
+ 3 to 5 strokes

玄武岩 *genbugan*
  basalt

交沸石 *koufusseki*
  harmotome $(Ba,Na,K)_{1-2}(Si,Al)_8O_{16} \cdot 6H_2O$

交換 *koukan*
  exchange, substitution

交換反応 *koukanhannou*
  exchange reaction

交流 *kouryuu*
  alternating current

交叉反応 *kousahannou*
  cross reaction

交叉結合 *kousaketsugou*
  cross linkage

充填剤 *juutenzai*
  filler

忘硝　*boushou*
　　Glauber's salt $Na_2SO_4 \cdot 10H_2O$

対物レンズ　*taibutsurenzu*
　　objective

対応状態　*taioujoutai*
　　corresponding state

対流　*tairyuu*
　　convection

対流電流　*tairyuudenryuu*
　　convection current

対流管　*tairyuukan*
　　convection tube

対流熱　*tairyuunetsu*
　　convection heating

対称　*taishou*
　　symmetry

対照分析　*taishoubunseki*
　　control analysis

対称分子　*taishoubunshi*
　　symmetric molecule

対称度　*taishoudo*
　　degree of symmetry

対称軸　*taishoujiku*
　　symmetry axis

対称化合物　*taishoukagoubutsu*
　　symmetric compound

対称面　*taishoumen*
　　plane of symmetrie

対称の中心　*taishounochuushin*
　　symmetry center, center of symmetry

対照試験　*taishoushiken*
　　control experiment

対称心　*taishoushin*
　　center of symmetry

対称操作　*taishousousa*
　　symmetry operation

対称数　*taishousuu*
　　symmetry number

対称炭素原子　*taishoutansogenshi*
　　symmetric carbon atom

対称要素　*taishouyouso*
　　symmetry element, element of symmetry

対数　*taisuu*
　　logarithm

対数平均　*taisuuheikin*
　　logarithmic mean

対数関数　*taisuukansuu*
　　logarithmic function

対数目盛　*taisuumemori*
　　logarithmic scale

対数座標　*taisuuzahyou*
　　logarithmic coordinates

亠 2j | 変
+ 7 strokes

変圧器　*henatsuki*
　　transformer

変分法　*henbunhou*
　　variation method

変位　*heni*
　　displacement

変位の法則　*heninohousoku*
　　displacement law

変化　*henka*
　　modification, change, variation, alteration

変角振動　*henkakushindou*
　　deformation vibration

変換　*henkan*
　　conversion, transformation, change

変換反応　*henkanhannou*
　　transformation reaction

変換因子　*henkaninshi*
　　conversion factor

変換機　*henkanki*
　　converter

変形　*henkei*
　　deformation

変性　*hensei*
　　denaturation

変性アルコール　*henseiaruko-ru*
　　denatured alcohol

変性蛋白質　*henseitanpakushitsu*
　denatured protein

変性剤　*henseizai*
　denaturing agent

変旋光　*hensenkou*
　mutarotation

変質　*henshitsu*
　modification, alteration, degeneration

変数　*hensuu*
　variable

変体　*hentai*
　modification, variant, abnormality

亠　2j　　高 斎 裏
+ 8 to 11 strokes

高圧　*kouatsu*
　high pressure

高圧合成　*kouatsugousei*
　high-pressure synthesis

高圧重合　*kouatsujuugou*
　high-pressure polymerization

高分子　*koubunshi*
　polymer, macromolecule

高分子物質　*koubunshibusshitsu*
　macromolecular compound

高分子電解質　*koubunshidenkaishitsu*
　polyelectrolyte

高分子化学　*koubunshikagaku*
　polymer chemistry

高分子化合物　*koubunshikagoubutsu*
　high-molecular compound

高分子可塑剤　*koubunshikasozai*
　polymeric plasticiser

高分子吸収体　*koubunshikyuushuutai*
　high-polymer absorbent

高張液　*kouchoueki*
　hypertonic solution

高張の　*kouchouno*
　hypertonic, hyperosmotic

高電圧　*koudenatsu*
　high voltage, high tension

高栄養な　*koueiyouna*
　nutritious, nutritive, nutrient-rich

高エネルギー放射線　*kouenerugi-houshasen*
　high-energy radiation

高エネルギー結合　*kouenerugi-ketsugou*
　high-energy bond

高沸点の　*koufuttenno*
　high-boiling

高百分率の　*kouhyakubunritsuno*
　high-percentage

高次　*kouji*
　higher order

高次反応　*koujihannou*
　higher-order reaction

高次方程式　*koujihouteishiki*
　higher equation

高次化合物　*koujikagoubutsu*
　compound of higher order

高次構造　*koujikouzou*
　higher order structure

高重合体　*koujuugoutai*
　high polymer

高感度の　*koukandono*
　highly sensitive

高級アルコール　*koukyuuaruko-ru*
　higher alcohol

高級脂肪酸　*koukyuushibousan*
　long-chain fatty acid

高密度　*koumitsudo*
　high density

高粘性の　*kounenseino*
　highly viscous, high-viscous

高濃度の　*kounoudono*
　highly concentrated

高温　*kouon*
　high temperature

高温フレーム　*kouonfure-mu*
　hot flame

高温乾留 kouonkanryuu
  high-temperature carbonization
高温計 kouonkei
  pyrometer
高温細菌 kouonsaikin
  thermophilic bacteria
高温接着剤 kouonsecchakuzai
  high temperature adhesive
高炉 kouro
  blast furnace
高炉ガス kourogasu
  blast furnace gas
高炉用コークス kouroyouko-kusu
  blast furnace coke
高真空 koushinkuu
  high-vacuum
高真空蒸留 koushinkuujouryuu
  high-vacuum distillation
高周波 koushuuha
  high frequency
高速度 kousokudo
  high-speed
高速液体クロマトグラフィー
  kousokuekitaikuromatogurafi-
  high performance liquid chromatography (HPLC)
高速クロマトグラフィー
  kousokukuromatogurafi-
  high-speed chromatography
高速走査 kousokusousa
  rapid scanning
高スピン錯体 kousupinsakutai
  high-spin complex
高融点の kouyuutenno
  high-melting
斎戒 saikai
  purification
商品名 shouhinmei
  trade name, brand name
裏抜け uranuke
  strike through

十 2k 干 老
+ 1 to 4 strokes

干渉 kanshou
  interference
干渉フィルター kanshoufiruta-
  interference filter
干渉計 kanshoukei
  interferometer
干渉性放射 kanshouseihousha
  coherent radiation
干渉スペクトル kanshousupekutoru
  interference spectrum
老廃物 rouhaibutsu
  waste product
老化 rouka
  aging
老化防止剤 roukaboushizai
  antioxidant, oxidation inhibitor

十 2k 協 直
+ 6 strokes

協力効果 kyouryokukouka
  synergistic effect
協力作用 kyouryokusayou
  synergism
直径 chokkei
  diameter
直交 chokkou
  orthogonal
直流 chokuryuu
  direct current, cocurrent
直鎖 chokusa
  straight chain
直鎖アルカン chokusaarukan
  linear alkane
直鎖アルコール chokusaaruko-ru
  linear alcohol

直鎖状分子 *chokusajoubunshi*
　straight-chain molecule

直鎖状炭化水素 *chokusajoutankasuiso*
　straight-chain hydrocarbon

直鎖パラフィン *chokusaparafin*
　normal paraffin

直線 *chokusen*
　straight line

直線分子 *chokusenbunshi*
　linear molecule

直線偏光 *chokusenhenkou*
　linearly polarized light

直線形分子 *chokusenkeibunshi*
　linear molecule

直線勾配 *chokusenkoubai*
　linear gradient

直接アミノ化 *chokusetsuaminoka*
　direct amination

直接アリル化 *chokusetsuariruka*
　direct allylation

直接アルキル化 *chokusetsuarukiruka*
　direct alkylation

直接アセチル化 *chokusetsuasechiruka*
　direct acetylation

直接置換 *chokusetsuchikan*
　direct substitution

直接液化 *chokusetsuekika*
　direct liquefaction

直接エステル化 *chokusetsuesuteruka*
　direct esterification

直接付加 *chokusetsufuka*
　direct addition

直接フッ素化 *chokusetsufussoka*
　direct fluorination

直接合成 *chokusetsugousei*
　direct synthesis

直接グリコシル化 *chokusetsugurikoshiruka*
　direct glycosylation

直接反応 *chokusetsuhannou*
　direct reaction

直接反転 *chokusetsuhanten*
　direct inversion

直接ヒドロキシル化 *chokusetsuhidorokishiruka*
　direct hydroxylation

直接インジェクション *chokusetsuinjekushon*
　direct injection

直接イオン化 *chokusetsuionka*
　direct ionization

直接重合 *chokusetsujuugou*
　direct polymerization

直接開始 *chokusetsukaishi*
　direct initiation

直接還元 *chokusetsukangen*
　direct reduction

直接カルボニル化 *chokusetsukaruboniruka*
　direct carbonylation

直接蛍光標識 *chokusetsukeikouhyoushiki*
　direct fluorescence

直接結晶化 *chokusetsukesshouka*
　direct crystallization

直接交換 *chokusetsukoukan*
　direct exchange

直接メチル化 *chokusetsumechiruka*
　direct methylation

直接メタル化 *chokusetsumetaruka*
　direct metalation

直接燃焼 *chokusetsunenshou*
　direct combustion

直接熱分解 *chokusetsunetsubunkai*
　direct pyrolysis

直接リチウム化 *chokusetsurichiumuka*
　direct lithiation

直接濾過 *chokusetsuroka*
　direct filtration

直接酸化 *chokusetsusanka*
　direct oxidation

直接単離 *chokusetsutanri*
　direct isolation

直接定量 *chokusetsuteiryou*
　direct determination

直接滴定 *chokusetsutekitei*
　direct titration

直接不斉合成 *chousetsufuseigousei*
　direct asymmetric synthesis

十　2k　+ 7 to 11 strokes　｜　南　真　準

南石 *minamiseki*
　minamiite $(Na,Ca,K)Al_3(SO_4)_2(OH)_6$

南部石 *nanbuseki*
　nambulite $(Li,Na)Mn^{2+}_4[Si_5O_{14}(OH)]$

南極石 *nankyokuseki*
　antarcticite $CaCl_2 \cdot 6H_2O$

真鍮 *shinchuu*
　brass

真核生物 *shinkakuseibutsu*
　eukaryote

真空 *shinkuu*
　vacuum

真空蒸発器 *shinkuujouhatsuki*
　vacuum evaporator

真空蒸留 *shinkuujouryuu*
　vacuum distillation

真空管 *shinkuukan*
　vacuum tube, pressure tubing, vacuum valve

真空乾燥 *shinkuukansou*
　vacuum drying

真空乾燥器 *shinkuukansouki*
　vacuum drier, vacuum exsiccator, vacuum desiccator

真空計 *shinkuukei*
　vacuum gage

真空ポンプ *shinkuuponpu*
　vacuum pump

真空プレス *shinkuupuresu*
　vacuum press

真空濾過器 *shinkuurokaki*
　vacuum filter

準安定 *junantei*
　metastable

準安定状態 *junanteijoutai*
　metastable state

準備状況 *junbijoukyou*
　initial condition

ト　2m　+ 1 stroke　｜　上　下

上限 *jougen*
　upper limit

下水 *gesui*
　waste water, sewage

下水処理 *gesuishori*
　water treatment

下部構造 *kabukouzou*
　substructure

下面発酵 *kamenhakkou*
　bottom fermentation

ト　2m　+ 3 strokes　｜　外　正　比

外圧 *gaiatsu*
　external pressure

外部光電効果 *gaibukoudenkouka*
　photoemissive effect

外部温度 *gaibuondo*
　outside temperature

外界温度 *gaikaiondo*
　ambient temperature

外酵素 *gaikouso*
　exoenzyme

外膜 *gaimaku*
　outer membrane

外挿 *gaisou*
　extrapolation

外核電子 *gaikakudenji*
　outer-shell electron

正長石 *seichouseki*
orthoclase K[AlSi$_3$O$_8$]

正電極 *seidenkyoku*
positive electrode

正鉛酸 *seiensan*
orthoplumbic acid H$_4$PbO$_4$, Pb(OH)$_4$

正鉛酸塩 *seiensanen*
orthoplumbate(IV) M$^I_4$PbO$_4$

正八面体 *seihamentei*
octahedron

正イオン *seiion*
positive ion

正確さ *seikakusa*
accuracy

正珪酸 *seikeisan*
orthosilicic acid, silicic acid H$_4$SiO$_4$

正珪酸ジルコニウム *seikeisanjirukoniumu*
zirconium orthosilicate, zirconium silicate ZrSiO$_4$

正規直交系 *seikichokkoukei*
orthonormal system

正規化 *seikika*
normalization

正極 *seikyoku*
positive electrode

正の *seino*
positive

正燐酸, 正リン酸 *seirinsan*
orthophosphoric acid, phosphoric acid H$_3$PO$_4$

正スズ酸, 正錫酸 *seisuzusan*
orthostannic acid H$_4$SnO$_4$

正錫酸亜鉛 *seisuzusanaen*
zinc orthostannate Zn$_2$SnO$_4$

正錫酸マグネシウム *seisuzusanmaguneshiumu*
magnesium orthostannate Mg$_2$SnO$_4$

正多面体 *seitamentai*
regular polyhedron

比電荷 *hidenka*
specific charge

比電気伝導率 *hidenkidendouritsu*
specific conductivity

比導電率 *hidoudenritsu*
specific electric conductivity

比放射能 *hihoushanou*
specific radioactivity

比重 *hijuu*
specific gravity, specific weight

比較湿度 *hikakushitsudo*
percentage humidity

比活性 *hikassei*
specific activity

比酵素活性 *hikousokassei*
specific enzyme activity

比屈折 *hikussetsu*
specific refraction

比吸収係数 *hikyuushuukeisuu*
specific extinction coefficient

比粘度 *hinendo*
specific viscosity

比熱 *hinetsu*
specific heat

比熱容量 *hinetsuyouryou*
specific heat capacity

比例減力 *hireigenryoku*
proportional reduction

比旋光度 *hisenkoudo*
specific rotation

比色分析 *hishokubunseki*
colorimetry

比色法 *hishokuhou*
colorimetric method

比色計 *hishokukei*
colorimeter

比色測定 *hishokusokutei*
colorimetric measurement

比色定量 *hishokuteiryou*
colorimetry

比体積 *hitaiseki*
specific volume

比容 *hiyou*
specific volume

ト 2m	点
+ 7 strokes	

点 *ten*
　point

点電荷 *tendenka*
　point charge

点火 *tenka*
　ignition

点火器 *tenkaki*
　lighter, fuse

点火剤, 点火薬 *tenkazai, tenkayaku*
　igniting agent

点滴培養 *tentekibaiyou*
　drop culture

儿, 勹 2n	色 危
+ 4 strokes	

色 *iro*
　color

色フィルター *irofiruta-*
　color filter

色ガラス *irogarasu*
　colored glass

色消し *irokeshi*
　achromatism

色消しレンズ *irokeshirenzu*
　achromatic lens

色収差 *iroshuusa*
　chromatic aberration

色スペクトル *irosupekutoru*
　chromatic spectrum

色彩計 *shikisaikei*
　colorimeter

色彩強度 *shikisaikyoudo*
　color intensity

色素 *shikiso*
　pigment, dye

色素塩基 *shikisoenki*
　color base

色素化学 *shikisokagaku*
　color chemistry

色素酸 *shikisosan*
　color acid

色素指示薬 *shikisoshijiyaku*
　indicator dye, color indicator

色素体 *shikisotai*
　chromatophore

色素タンパク質 *shikisotanpakushitsu*
　chromoprotein

危険 *kiken*
　danger, hazard, risk

危険地帯 *kikenchitai*
　danger zone

儿, 勹 2n	角 免 負
+ 5 to 7 strokes	

角度 *kakudo*
　angle

角度計 *kakudokei*
　goniometer

角銀鉱 *kakuginkou*
　horn silver, argentum cornu, cerargyrite, chlorargyrite AgCl

角鉛鉱 *kakunamarikou*
　horn lead, phosgenite $Pb_2[Cl_2,CO_3]$

角振動数 *kakushindousuu*
　angular frequency

角速度 *kakusokudo*
　angular velocity

角運動量 *kakuundouryou*
　angular momentum

免疫学 *menekigaku*
　immunology

免疫学的測定法
　menekigakutekisokuteihou
　immunoassay

免疫グロブリン　*menekiguroburin*
  immunoglobulin

免疫反応　*menekihannou*
  immune reaction

免疫化学　*menekikagaku*
  immunological chemistry

免疫系　*menekikei*
  immune system

免疫細胞　*menekisaibou*
  immune cell

免疫性　*menekisei*
  immunity

免疫体　*menekitai*
  antibody

負電位　*fudeni*
  negative potential

負電荷　*fudenka*
  negative electric charge

負イオン　*fuion*
  negative ion

負極　*fukyoku*
  negative pole, negative electrode

急冷　*kyuurei*
  quenching

ヽヽ八　20　| 公 父 羊 並 前 剪 着
(all *kanji*)　　| 養 興

公害　*kougai*
  pollution

公差　*kousa*
  tolerance

父斉反応　*fuseihannou*
  asymmetric reaction

羊毛　*youmou*
  wool

羊毛繊維　*youmouseni*
  wool fiber

並発反応　*heihatsuhannou*
  parallel reaction

並列の　*heiretsuno*
  parallel

並流　*heiryuu*
  co-current

並進　*heishin*
  translation

並進エネルギー分布　*heishinenerugi-bunpu*
  translational energy distribution

並進エネルギー交換　*heishinenerugi-koukan*
  translational energy exchange

並進操作　*heishinsousa*
  translation operation

並進運動　*heishinundou*
  transitional motion

前処理　*maeshori*
  pretreatment

前期解離　*zenkikairi*
  predissociation

前駆物質　*zenkubusshitsu*
  precursor

前駆体　*zenkutai*
  precursor, progenitor

剪断剛性　*sendangousei*
  shear rigidity, shear stiffness, shearing stiffness

剪断応力　*sendanouryoku*
  shear stress

剪断力　*sendanryoku*
  shear force

瓶　*bin*
  bottle

着色　*chakushoku*
  coloration

着色顔料　*chakushokuganryou*
  colour pigment

着色反応　*chakushokuhannou*
  coloring reaction

着色効果　*chakushokukouka*
  coloring effect

着色力　*chakushokuryoku*
  colouring strength

着色料 *chakushokuryou*
dye, colorant

着色剤 *chakushokuzai*
dye, coloring matter

養生 *youjou*
cure

養生方法 *youjouhouhou*
curing technique

興奮 *koufun*
excitation

興奮性 *koufunsei*
excitability

厂 **2p** 反
+ 2 strokes

反跳 *hanchou*
recoil

反芳香族 *hanhoukouzoku*
antiaromatic

反磁性 *hanjisei*
diamagnetism

反磁性体 *hanjiseitai*
diamagnetic substance

反結合軌道 *hanketsugoukidou*
antibonding orbital

反強磁性 *hankyoujisei*
antiferromagnetism

反応 *hannou*
reaction

反応物 *hannoubutsu*
reactant

反応中間体 *hannouchuukantai*
reaction intermediate

反応中心 *hannouchuushin*
reaction center

反応電流 *hannoudenryuu*
reaction current

反応動力学 *hannoudourikigaku*
reaction kinetics

反応型 *hannougata*
reaction type

反応原系 *hannougenkei*
reactant

反応方程式 *hannouhouteishiki*
reaction equation, chemical equation

反応時間 *hannoujikan*
reaction time

反応次数 *hannoujisuu*
reaction order

反応器 *hannouki*
reactor

反応機構 *hannoukikou*
reaction mechanism

反応混合物 *hannoukongoubutsu*
reaction mixture

反応工学 *hannoukougaku*
reaction engineering

反応熱 *hannounetsu*
heat of reaction

反応連鎖 *hannourensa*
reaction chain

反応性 *hannousei*
reactivity

反応性指数 *hannouseishisuu*
reactivity index

反応式 *hannoushiki*
reaction equation

反応室 *hannoushitsu*
reaction chamber

反応速度 *hannousokudo*
reaction velocity, reaction rate

反応速度論 *hannousokudoron*
reaction kinetics, chemical kinetics

反応速度式 *hannousokudoshiki*
rate equation

反応促進剤 *hannousokushinzai*
reaction accelerator

反応体 *hannoutai*
reactant

反応定数 *hannouteisuu*
reaction constant

反応座標 *hannouzahyou*
　reaction coordinate

反復性 *hanpukusei*
　repeatability

反作用 *hansayou*
　counter effect, counteraction

反射 *hansha*
　reflection

反対称 *hantaishou*
　antisymmetric

反転 *hanten*
　inversion, negation

反転中心 *hantenchuushin*
　center of inversion

反転温度 *hantenondo*
　inversion temperature

反陽子 *hanyoushi*
　antiproton

厂　2p　｜圧
+ 3 strokes

圧迫性 *appakusei*
　compression

圧搾 *assaku*
　compression

圧搾ガス *assakugasu*
　compressed gas

圧搾空気 *assakukuuki*
　compressed air

圧縮 *asshuku*
　compression

圧縮ガス *asshukugasu*
　compressed gas

圧縮機 *asshukuki*
　compressor

圧縮機油 *asshukukiyu*
　compressor oil

圧縮空気 *asshukukuuki*
　compressed air

圧縮率 *asshukuritsu*
　compressibility

圧縮酸素 *asshukusanso*
　compressed oxygen

圧縮性 *asshukusei*
　compressibility

圧縮性流体 *asshukuseiryuutai*
　compressible fluid

圧電気 *atsudenki*
　piezoelectricity

圧濾器, 圧濾機 *atsuroki*
　filter press

圧力 *atsuryoku*
　pressure

圧力調節機 *atsuryokuchousetsuki*
　pressure regulator

圧力計 *atsuryokukei*
　manometer, pressure gauge

圧力係数 *atsuryokukeisuu*
　pressure coefficient

圧力差 *atsuryokusa*
　pressure difference

圧力制御 *atsuryokuseigyo*
　pressure control

厂　2p　｜灰 厚 農
+ 4 to 11 strokes

灰 *hai*
　ash

灰分 *haibun*
　ash

灰化 *haika*
　incineration

灰浴 *haiyoku*
　ash bath

厚膜 *atsumaku*
　thick film

農学 *nougaku*
　agriculture

農芸化学 *nougeikagaku*
 agricultural chemistry

農業 *nougyou*
 agriculture

農業試験場 *nougyoushikenjou*
 agricultural experiment station

農産 *nousan*
 agricultural chemicals, crop protection agent, pesticide

農産化学 *nousankagaku*
 agricultural chemistry

農産工業 *nousankougyou*
 agro-industry

農薬 *nouyaku*
 agricultural chemicals, crop protection agent, pesticide

農薬合成 *nouyakugousei*
 pesticide synthesis

農薬モニタリング *nouyakumonitaringu*
 pesticide monitoring

農薬定量 *nouyakuteiryou*
 pesticide determination

辶, 廴 f | 迅 近 延 退 逆 送
+ 3 to 6 strokes

迅速分析 *jinsokubunseki*
 rapid analysis

近代科学 *kindaikagaku*
 modern science

近似 *kinji*
 approximately, about, roughly

近似値 *kinjichi*
 approximate value, approximation

近似する *kinjisuru*
 to approximate

近赤外 *kinsekigai*
 near-infrared

近接 *kinsetsu*
 adjacent, vicinal, neighboring

近視紫外 *kinshigai*
 near-ultraviolet

延性 *ensei*
 ductility

退化 *taika*
 degeneration

退行 *taikou*
 regression

退行係数 *taikoukeisuu*
 regression coefficient

退色 *taishoku*
 fading

退色速度 *taishokusokudo*
 fading speed

逆アルドール縮合 *gyakuarudo-rushukugou*
 retro-aldol condensation

逆圧 *gyakuatsu*
 back pressure

逆抽出 *gyakuchuushutsu*
 back extraction

逆止弁 *gyakudomeben*
 check valve

逆動作 *gyakudousa*
 reverse action

逆反応 *gyakuhannou*
 reverse reaction, inverse reaction, back reaction

逆磁性 *gyakujisei*
 diamagnetic

逆混合 *gyakukongou*
 back mixing

逆効果 *gyakukouka*
 opposite effect

逆格子 *gyakukoushi*
 reciprocal lattice

逆作用 *gyakusayou*
 adverse effect; reaction

逆旋的な *gyakusentekina*
 disrotatory

逆浸透 *gyakushintou*
 reverse osmosis

逆相 *gyakusou*
   reverse phase

逆相クロマトグラフィー
   gyakusoukuromatogurafi-
   reversed phase chromatography

逆対称 *gyakutaishou*
   antisymmetric

逆滴定 *gyakutekitei*
   back titration

逆転 *gyakuten*
   inversion

逆転温度 *gyakutenondo*
   inversion temperature

逆転写酵素 *gyakutenshakouso*
   reverse transcriptase

送風機 *soufuuki*
   fan, ventilator, blower

辶, 廴   **2q** | 連 速 逐 透 造 通
+ 7 strokes |

連合 *rengou*
   association

連合体 *rengoutai*
   association complex

連結 *renketsu*
   combination, connection

連結反応器 *renketsuhannouki*
   coupled reactor

連結構造 *renketsukouzou*
   coupled structure

連鎖 *rensa*
   chain, linkage

連鎖分解 *rensabunkai*
   chain decomposition

連鎖分岐 *rensabunki*
   chain branching

連鎖群 *rensagun*
   linkage group

連鎖破壊 *rensahakai*
   chain breaking

連鎖反応 *rensahannou*
   chain reaction

連鎖反応機構 *rensahannoukikou*
   chain reaction mechanism

連鎖移動 *rensaidou*
   chain transfer

連鎖移動反応 *rensaidouhannou*
   chain transfer reaction

連鎖移動剤 *rensaidouzai*
   chain transfer agent

連鎖重合 *rensajuugou*
   chain polymerization

連鎖価 *rensaka*
   linkage value

連鎖開始 *rensakaishi*
   chain initiation

連鎖開始反応 *rensakaishihannou*
   chain initiation reaction

連鎖結合 *rensaketsugou*
   chain combination

連鎖構造 *rensakouzou*
   chain structure

連鎖球菌 *rensakyuukin*
   streptococcus

連鎖プロセス *rensapurosesu*
   chain process

連鎖領域 *rensaryouiki*
   linkage region

連鎖酸化反応 *rensasankahannou*
   chain oxidation reaction

連鎖停止 *rensataishi*
   chain termination

連鎖停止反応 *rensateishihannou*
   chain terminating reaction

連鎖停止剤 *rensateishizai*
   chain stopper

連接二重結合 *rensetsunijuuketsugou*
   cumulated double bond

連続 *renzoku*
   continuity

連続アルゴリズム *renzokuarugorizumu*
   continuation algorithm

連続アルキル化 renzokuarukiruka
continuous alkylation

連続番号 renzokubangou
consecutive number

連続分別 renzokubunbetsu
continuous fractionation

連続注入 renzokuchuunyuu
continuous injection

連続エステル化 renzokuesuteruka
continuous esterification

連続フロー分析 renzokufuro-bunseki
continuous flow assay

連続フロー法 renzokufuro-hou
continuous flow method

連続合成 renzokugousei
continuous synthesis

連続反応 renzokuhannou
consecutive reaction, successive reaction

連続イオン交換 renzokuionkoukan
continuous ion exchange

連続蒸留 renzokujouryuu
continuous distillation

連続重合 renzokujuugou
continuous polymerization

連続加熱 renzokukanetsu
continuous heating

連続乾燥 renzokukansou
continuous drying

連続加水分解 renzokukasuibunkai
continuous hydrolysis

連続クロマトグラフィー renzokukuromatogurafi-
continuous chromatography

連続ミキサ renzokumikisa
continuous mixer

連続流れ renzokunagare
continuous flow

連続乳化 renzokunyuuka
continuous emulsification

連続冷却 renzokureikyaku
continuous cooling

連続ろ過 renzokuroka
continuous filtration

連続処理 renzokushori
continuous treatment

連続相 renzokusou
continuous phase

連続スペクトル renzokusupekutoru
continuous spectrum

連続体法 renzokutaihou
continuum method

速度 sokudo
velocity

速度係数 sokudokeisuu
velocity rate

速度論 sokudoron
kinetics

速度支配 sokudoshihai
kinetic control

速度定数 sokudoteisuu
rate constant

逐次アルキル化 chikujiarukiruka
tandem alkylation

逐次分離 chikujibunri
sequential separation

逐次代入法 chikujidainyuuhou
iteration method

逐次段階 chikujidankai
consecutive step

逐次エステル交換 chikujiesuterukoukan
tandem transesterification

逐次合成 chikujigousei
sequential synthesis

逐次反応 chikujihannou
consecutive reaction, successive reaction

逐次変換 chikujihenkan
sequential transformation

逐次開始 chikujikaishi
sequential initiation

逐次還元 chikujikangen
stepwise reduction

逐次ペリ環化反応
  chikujiperikankahannou
  tandem pericyclic reaction

逐次ラジカル反応 *chikujirajikaruhannou*
  tandem radical reaction

逐次ラジカル環化 *chikujirajikarukanka*
  tandem radical cyclization

逐次ラジカル環化反応
  chikujirajikarukankahannou
  tandem radical cyclization reaction

逐次酸化 *chikujisanka*
  sequential oxidizing

逐次転位 *chikujiteni*
  tandem rearrangement

透磁率 *toujiritsu*
  permeability

透過 *touka*
  penetration, transmission

透過度 *toukado*
  diffusion rate

透過確率 *toukakakuritsu*
  penetration probability

透過係数 *toukakeisuu*
  transmission coefficient

透過器 *toukaki*
  permeator

透過機構 *toukakikou*
  transmission mechanism

透過パーセント *toukapa-sento*
  percent transmission

透過プロセス *toukapurosesu*
  permeation process

透過率 *toukaritsu*
  permeability, transmittance

透過性 *toukasei*
  permeability

透過性分析 *toukaseibunseki*
  permeability analysis

透過制御 *toukaseigyo*
  permeation control

透過性評価 *toukaseihyouka*
  permeability evaluation

透過性膜 *toukaseimaku*
  permeable membrane

透過速度論 *toukasokudoron*
  permeation kinetics

透過特性 *toukatokusei*
  permeation characteristic

透気度 *toukido*
  air permeability

透明 *toumei*
  transparency

透明度 *toumeido*
  transparency

透明液体 *toumeiekitai*
  clear liquid

透明フィルム *toumeifirumu*
  clear film

透明画 *toumeiga*
  transparency

透明ガラス *toumeigarasu*
  transparent glass

透明重合体 *toumeijuugoutai*
  transparent polymer

透明性 *toumeisei*
  transparency

透明紙 *toumeishi*
  cellophane

透明点 *toumeiten*
  clearing point

透明溶液 *toumeiyoueki*
  clear solution

透析 *touseki*
  dialysis

透析膜 *tousekimaku*
  permeable membrane

透析装置 *tousekisouchi*
  dialyzer

透視平面 *toushiheimen*
  perspective plane

透視投影 *toushitouei*
  perspective projection

透視図 *toushizu*
  perspective drawing

透水性 *tousuisei*
water permeability

造型 *zoukei*
molding

造形 *zoukei*
molding

造粒 *zouryuu*
granulation

造粒機 *zouryuuki*
granulator

通風乾燥 *tsuufuukansou*
draught drying

通風計 *tsuufuukei*
draft gage

通常分析 *tsuujoubunseki*
routine analysis

通常加熱 *tsuujoukanetsu*
conventional heating

通常流体 *tsuujouryuutai*
normal fluid

通常重力 *tsuujuuryoku*
normal gravity

通研 *tsuuken*
laboratory

通気 *tsuuki*
aeration

通気率 *tsuukiritsu*
permeability

通気性 *tsuukisei*
gas permeability

通性の *tsuuseino*
facultative

辶, 廴 2q 進遊部
+ 8 strokes

進化 *shinka*
evolution

進歩 *shinpo*
progress, development

進展 *shinten*
progress, development

進度測定 *shintensokutei*
progress measurement

遊離 *yuuri*
free

遊離アミノ基 *yuuriaminoki*
free amino base

遊離アルミニウム *yuuriaruminiumu*
free aluminum

遊離チオール *yuurichio-ru*
free thiol

遊離エネルギー *yuurienerugi-*
free energy

遊離塩基 *yuurienki*
free base

遊離塩素 *yuurienso*
free chlorine

遊離ガス *yuurigasu*
free gas

遊離型 *yuurigata*
free form

遊離原子 *yuurigenshi*
free atom

遊離原子価 *yuurigenshika*
free valence

遊離配位子 *yuurihaiishi*
free ligand

遊離硫黄 *yuuriiou*
free sulfur

遊離カルボン酸 *yuurikarubonsan*
free carboxylic acid

遊離ケイ素 *yuurikeiso*
free silicon

遊離基 *yuuriki*
free radical

遊離基中間体 *yuurikichuukantai*
free radical intermediate

遊離基反応 *yuurikihannou*
radical reaction

遊離基開始剤　*yuurikikaishizai*
 free radical initiator

遊離金属イオン　*yuurikinzokuion*
 free metal ion

遊離コレステロール　*yuurikoresutero-ru*
 free cholesterol

遊離マグネシウム　*yuurimaguneshiumu*
 free magnesium

遊離ポリアミン　*yuuriporiamin*
 free polyamine

遊離酸　*yuurisan*
 free acid

遊離石灰　*yuurisekkai*
 free lime

遊離シアン化物　*yuurishiankabutsu*
 free cyanide

遊離脂肪酸　*yuurishibousan*
 free fatty acid

遊離シリカ　*yuurishirika*
 free silica

遊離水　*yuurisui*
 free water

遊離ステロール　*yuurisutero-ru*
 free sterol

遊離鉄　*yuuritetsu*
 free iron

部分　*bubun*
 part, portion

部分群　*bubungun*
 subgroup, subspecies

部分平衡　*bubunheikou*
 partial equilibrium

部分加水分解　*bubunkasuibunkai*
 partial hydrolysis

辶, 廴　2q　運 遅 遠 適 遮
+ 9 to 11 strokes

運動エネルギー　*undouenerugi-*
 kinetic energy

運動学　*undougaku*
 kinematics

運動量　*undouryou*
 momentum

運動量保存の法則　*undouryouhozonnohousoku*
 law of conservation of momentum

遅延　*chien*
 retardation

遅延発光　*chienhakkou*
 delayed luminescence

遅延時間　*chienjikan*
 delay time, retardation time

遅延実験　*chienjikken*
 retardation experiment

遅延蛍光　*chienkeikou*
 delayed fluorescence

遅延作用　*chiensayou*
 delayed action

遅延剤　*chienzai*
 retardant

遠隔制御　*enkakuseigyo*
 remote control

遠赤外　*ensekigai*
 far-infrared

遠紫外　*enshigai*
 far-ultraviolet

遠心分離　*enshinbunri*
 centrifugation

遠心分離機, 遠心分離器　*enshinbunriki, enshinbunriki*
 centrifugal separator

遠心機　*enshinki*
 centrifuge

遠心効果　*enshinkouka*
 centrifugal effect

遠心の　*enshinno*
 centrifugal

遠心力　*enshinryoku*
 centrifugal force

遠心送風機　*enshinsoufuuki*
 centrifugal fan

適合性　*tekigousei*
　compatibility

適応　*tekiou*
　adaption

適定　*tekitei*
　titration

適定曲線　*tekiteikyokusen*
　titration curve

適用　*tekiyou*
　applying

適用業務　*tekiyougyoumu*
　application

適用性　*tekiyousei*
　applicability

遮蔽定数　*shaheiteisuu*
　screening constant

辶, 廴　2q
+ 12/13 strokes　|　遷 選 遺 還

遷移元素　*senigenso*
　transition element

遷移時間　*senijikan*
　transition time

遷移状態　*senijoutai*
　transition state

遷移確率　*senikakuritsu*
　transition probability

遷移モーメント　*senimo-mento*
　transition moment

遷移温度　*seniondo*
　transition temperature

遷移点　*seniten*
　transition point

選別　*senbetsu*
　selection

選択　*sentaku*
　selection

選択法　*sentakuhou*
　selection method

選択係数　*sentakukeisuu*
　selectivity coefficient

選択吸収　*sentakukyuushuu*
　selective absorption

選択マーカー　*sentakuma-ka-*
　selection marker

選択率　*sentakuritsu*
　selectivity

選択酸化　*sentakusanka*
　selective oxidation

選択性　*sentakusei*
　selectivity

選択則　*sentakusoku*
　selection rule

選択する　*sentakusuru*
　to select

選炭　*sentan*
　coal dressing

遺伝暗号　*idenangou*
　genetic code

遺伝物質　*idenbusshitsu*
　genetic material

遺伝病　*idenbyou*
　genetic disease

遺伝学　*idengaku*
　genetics

遺伝マーカー　*idenma-ka*
　genetic marker

遺伝子　*idenshi*
　gene

遺伝子分析　*idenshibunseki*
　gene analysis

遺伝子型　*idenshigata*
　genotype

遺伝子工学　*idenshikougaku*
　genetic engineering

遺伝子ライブラリー　*idenshiraiburari-*
　gene library

遺伝子多型　*idenshitagata*
　polymorphism

遺伝子突然変異　*idenshitotsuzenheni*
　gene mutation

遺伝子融合　*idenshiyuugou*
　gene fusion

還元　*kangen*
　reduction, deoxidation

還元電位　*kangendeni*
　reduction potential

還元炎　*kangenen*
　reducing flame

還元型, 還元形　*kangengata, kangengata*
　reduced form

還元方程式　*kangenhouteishiki*
　equation of reduction

還元酵素　*kangenkouso*
　reductase

還元力　*kangenryoku*
　reducing power

還元糖　*kangentou*
　reducing sugar

還元剤　*kangenzai*
　reducing agent

還流　*kanryuu*
　reflux

還流比　*kanryuuhi*
　reflux ratio

還流冷却器　*kanryuureikyakuki*
　reflux condenser

冂　2r　｜円 用 周 耐
(all *kanji*)

円偏光　*enhenkou*
　circulary polarized light

円偏光二色性　*enhenkounishokusei*
　circular dichroism

円二色性　*ennishokusei*
　circular dichroism

円筒　*entou*
　cylinder

円筒乾燥　*entoukansou*
　drum drying

用量　*youryou*
　dose

周波数　*shuuhasuu*
　frequency

周囲温度　*shuuiondo*
　ambient temperature

周囲湿度　*shuuishitsudo*
　ambient humidity

周波数計　*shuuisuukei*
　frequency meter

周期、周期性　*shuuki, shuukisei*
　period, cycle

周期反応　*shuukihannou*
　oscillatory reaction, periodic reaction

周期表　*shuukihyou*
　periodic table

周期関数　*shuukikansuu*
　periodic function

周期系　*shuukikei*
　periodic table of the elements

周期律　*shuukiritsu*
　periodic law

周期性　*shuukisei*
　periodicity

周期数　*shuukisuu*
　periodic number

耐アルカリ性　*taiarukarisei*
　alkali resistance

耐炎性　*taiensei*
　flame resistance

耐硫黄性　*taiiousei*
　sulfur tolerance

耐火粘土　*taikanendo*
　fire clay

耐火性　*taikasei*
　fire resistance, refractoriness

耐火材料　*taikazairyou*
　refractory material

耐菌試験 *taikinshiken*
　fungus test

耐光性 *taikousei*
　light resistance

耐久性 *taikyuusei*
　durability

耐摩耗層 *taimamousou*
　wear resistant layer

耐摩耗材料 *taimamouzairyou*
　wear-resistant material

耐然性 *tainensei*
　burning resistance

耐熱性 *tainetsusei*
　thermal resistance

耐熱性繊維 *tainetsuseiseni*
　heat-resistant fiber

耐酸の *taisanno*
　acid-proof

耐酸性 *taisansei*
　acid resistance

耐性 *taisei*
　resistance

耐性機構 *taiseikikou*
　resistant mechanism

耐性菌 *taiseikin*
　antibiotic-resistant bacteria

耐蝕性 *taishokusei*
　corrosion resistance

耐食性 *taishokusei*
　corrosion resistance

耐衝撃性 *taishougekisei*
　impact strength, shock resistance

耐水 *taisui*
　waterproof, water-resistant

耐水性 *taisuisei*
　water resistance

耐薬品性 *taiyakuhinsei*
　chemical resistance

耐溶剤性 *taiyouzaisei*
　solvent resistance

耐油性 *taiyusei*
　oil resistance

几　2s　風 段
(all *kanji*)

風信子石 *fushinjiseki*
　zircon $ZrSiO_4$

風化 *fuuka*
　weathering

風乾 *fuukan*
　air drying

風袋 *fuutai*
　tare

段階 *dankai*
　stage, step

段階分離 *dankaibunri*
　stepwise separation

段階的合成 *dankaitekigousei*
　stepwise synthesis

段階的変換 *dankaitekihenkan*
　stepwise conversion

段階的活性化 *dankaitekikasseika*
　stepwise activation

段階的固相合成 *dankaitekikosougousei*
　stepwise solid-phase synthesis

段階的熱分解 *dankaitekinetsubunkai*
　stepwise pyrolysis

段階的生成 *dankaitekiseisei*
　stepwise formation

段数 *dansuu*
　number of steps

段階重合 *denkaijuugou*
　step polymerization

段階的酸化 *denkaitekisanka*
　stepwise oxidation

殺虫剤 *sacchuuzai*
　insecticide

殺菌 *sakkin*
　sterilization

殺そ剤 *sassozai*
　rodenticide

殺菌剤 *satsukinzai*
　disinfectant, germicide

殺菌剤の　*satsukinzaino*
bactericidal

殺細菌作用　*satsusaikinsayou*
bactericidal action

殺線虫剤　*satsusenchuuzai*
nematicide

匚　2t　巨 医 臨
(all *kanji*)

巨大分子　*kyodaibunshi*
macromolecule

巨視的の　*kyoshitekino*
macroscopic

医化学　*ikagaku*
medicinal chemistry

臨界圧　*rinkaiatsu*
critical pressure

臨界分子量　*rinkaibunshiryou*
critical molecular weight

臨界直径　*rinkaichokkei*
critical diameter

臨界エネルギー　*rinkaienerugi-*
critical energy

臨界粘度　*rinkainendo*
critical viscosity

臨界温度　*rinkaiondo*
critical temperature

臨界湿度　*rinkaishitsudo*
critical humidity

臨界相　*rinkaisou*
critical phase

臨界点　*rinkaiten*
critical point

臨界溶解　*rinkaiyoukai*
critical solution

臨界容積　*rinkaiyouseki*
critical volume

臨床試験　*rinshoushiken*
clinical test

氵, 水　3a　｜　永 汚 沈
+ 1 to 4 strokes

永久硬度　*eikyuukoudo*
permanent hardness

永久双極モーメント
eikyuusoukyokumo-mento
permanent dipole moment

氷酢酸　*hyousakusan*
glacial acetic acid

氷晶　*hyoushou*
ice crystal

氷晶石　*hyoushouseki*
cryolite, ice stone Na_3AlF_6

氷点　*hyouten*
freezing point

氷点曲線　*hyoutenkyokusen*
freezing point curve

汚濁　*odaku*
pollution

汚泥　*odei*
sludge, slurry

汚染　*osen*
contamination, pollution

沈着　*chinchaku*
deposition

沈着効率　*chinchakukouritsu*
deposition efficiency

沈澱　*chinden*
precipitation

沈澱防止剤　*chindenboushizai*
suspending agent

沈殿物　*chindenbutsu*
precipitate

沈澱物　*chindenbutsu*
precipitate

沈澱する　*chindensuru*
to precipitate

沈澱滴定　*chindentekitei*
precipitation analysis

沈澱剤　*chindenzai*
precipitant

沈降 *chinkou*
sedimentation, deposition

沈降物 *chinkoubutsu*
sediment, precipitate, deposit

沈降平衡 *chinkouheikou*
sedimentation equilibrium

沈降法 *chinkouhou*
sedimentation method

沈降速度 *chinkousokudo*
sedimentation velocity

沈降槽 *chinkousou*
sedimentation tank

沈降タンク *chinkoutanku*
settling tank, settler

沈降炭酸カルシウム *chinkoutansankarushiumu*
precipitated calcium carbonate

沈積 *chinseki*
sedimentation

没食子酸 *bosshokushisan*
gallic acid $C_6H_2(OH)_3COOH$

⺡, 水 3a 沸 油 波 注 泡 法 泥
\+ 5 strokes

沸石 *fusseki*
zeolite

沸点 *futten*
boiling point

沸点上昇 *futtenjoushou*
boiling-point elevation

沸点上昇法 *futtenjoushouhou*
ebullioscopy

沸点曲線 *futtenkyokusen*
boiling-point curve

沸騰 *futtou*
boiling

沸騰フラスコ *futtoufurasuko*
distillation flask, boiling flask

沸騰熱 *futtounetsu*
boiling heat

沸騰温度 *futtouondo*
boiling temperature, temperature of ebullition

沸騰水 *futtousui*
boiling water

油 *abura*
oil

油分離器, 油分離機 *aburabunriki, aburabunriki*
oil separator

油吸収度 *aburakyuushuudo*
oil absorbency

油とり *aburatori*
degreasing agent

油圧 *yuatsu*
oil pressure

油長 *yuchou*
oil length

油品質 *yuhinshitsu*
oil quality

油状 *yujou*
oily, greasy

油潤滑 *yujunkatsu*
oil lubrication

油気熱交換器 *yukinetsukoukanki*
vapor heat exchanger

油混合物 *yukongoubutsu*
oil mixture

油膜 *yumaku*
oil film

油ポンプ *yuponpu*
oil pump

油留分 *yuryuubun*
oil fraction

油酸 *yusan*
oleic acid, oleinic acid
$CH_3(CH_2)_7CH=CH(CH_2)_7COOH$

油酸塩 *yusanen*
oleate $CH_3(CH_2)_7CH=CH(CH_2)_7COOR$

油性 *yusei*
oily, greasy

油性塗料 *yuseitoryou*
oil-based paint

油脂　*yushi*
　fats and oils

油脂安定度　*yushianteido*
　fat stability

油脂工業　*yushikoukyou*
　oil and fat industry

油脂精製　*yushiseisei*
　oil and fat refining

油槽　*yusou*
　oil tank

油水分離器　*yusuibunriki*
　oily water separator

油ワニス　*yuwanisu*
　oil varnish

油浴　*yuyoku*
　oil bath

油溶性　*yuyousei*
　fat soluble

油溶性の　*yuyouseino*
　oil-soluble

油剤　*yuzai*
　oil-based compound

波長　*hachou*
　wavelength

波長シフト　*hachoushifuto*
　wavelength shift

波動方程式　*hadouhouteishiki*
　wave equation

波動関数　*hadoukansuu*
　wave function

波状態　*hajoutai*
　wave condition

波数　*hasuu*
　wave number

注型　*chuukei*
　casting

注型溶媒　*chuukeiyoubai*
　casting solvent

注入　*chuunyuu*
　injection

注射　*chuusha*
　injection

注射器　*chuushaki*
　injection syringe

注射する　*chuushasuru*
　to inject

注油　*chuuyu*
　greasing, lubricating

泡　*awa*
　foam, froth, bubble

泡安定剤　*awaanteizai*
　foam stabilizer

泡立ち　*awadachi*
　foaming

泡立ち試験　*awadachishiken*
　foaming test

泡立て剤　*awadatezai*
　foaming agent

泡止め剤　*awadomezai*
　defoaming agent

泡ガラス　*awagarasu*
　foam glas

泡切り剤　*awakirizai*
　defoaming agent

法律　*houritsu*
　law

泥膏　*deikou*
　paste

泥炭　*deitan*
　peat

氵, 水　3a　浮 洗 活 浄 海
+ 6 strokes

浮選　*fusen*
　flotation

浮遊物　*fuyuubutsu*
　suspended matter

洗瓶　*senbin*
　washing bottle

洗浄　*senjou*
　washing

洗浄剤 *senjouzai*
cleaning agent

洗剤 *senzai*
detergent

活性 *kassei*
activity

活性分子 *kasseibunshi*
activated molecule, excited molecule

活性中心 *kasseichuushin*
active center

活性エステル *kasseiesuteru*
active ester

活性複合体 *kasseifukugoutai*
activated complex

活性因子 *kasseiinshi*
active factor

活性状態 *kasseijoutai*
active state

活性化 *kasseika*
activation

活性化エネルギー *kasseikaenerugi-*
activation energy

活性化エンタルピー *kasseikaentarupi-*
enthalpy of activation

活性化エントロピー *kasseikaentoropi-*
entropy of activation

活性化合物 *kasseikagoubutsu*
active compound

活性化状態 *kasseikajoutai*
active state

活性化温度 *kasseikaondo*
activation temperature

活性化酸素 *kasseikasanso*
active oxygen

活性化する *kasseikasuru*
to activate

活性化剤 *kasseikazai*
activator

活性サイト *kasseisaito*
active site

活性錯体 *kasseisakutai*
activated complex

活性酸素 *kasseisanso*
active oxygen

活性水素 *kasseisuiso*
active hydrogen

活性体 *kasseitai*
activator

活性炭 *kasseitan*
active carbon, active charcoal

活性剤 *kasseizai*
activator

活動電位 *katsudoudeni*
action potential

活量 *katsuryou*
activity

活量係数 *katsuryoukeisuu*
activity coefficient

浄水 *jousui*
clean water

浄水器 *jousuiki*
water puifier

海塩 *kaien*
sea salt

海泡石 *kaihouseki*
sepiolite $Mg_4Si_6O_{15}(OH)_2 \cdot 6H_2O$

海緑石 *kairyokuseki*
glauconite
$(K,Na)(Fe^{3+},Al,Mg)_2(Si,Al)_4O_{10}(OH)_2$

海水 *kaisui*
seawater, salt-water

氵, 水 3a 酒 消 浸 浴
+ 7 strokes

酒 *sake*
alcohol, sake

酒毒 *shudoku*
alcoholic poisoning

酒石 *shuseki*
tartar $HOOC(CHOH)_2COOK$

酒石酸 *shusekisan*
tartaric acid $HOOC(CHOH)_2COOH$

酒石酸アンモニウム
　shusekisananmoniumu
　ammonium tartrate
　$NH_4OOC(CHOH)_2COONH_4$

酒石酸塩　*shusekisanen*
　tartrate $MOOC(CHOH)_2COOM$

酒石酸カリウム　*shusekisankariumu*
　potassium tartrate $KOOC(CHOH)_2COOK$

酒石酸ナトリウム　*shusekisannatoriumu*
　sodium tartrate $NaOOC(CHOH)_2COONa$

酒石酸水素カリウム
　shusekisansuioskariumu
　potassium hydrogentartrate
　$HOOC(CHOH)_2COOK$

消毒　*shoudoku*
　disinfection

消毒剤　*shoudokuzai*
　disinfectant

消泡機　*shouhouki*
　foam breaker

消泡剤　*shouhouzai*
　defoaming agent

消化　*shouka*
　digestion

消火器　*shoukaki*
　fire extinguisher

消化酵素　*shoukakouso*
　digestive enzyme

消化率　*shoukaritsu*
　digestibility

消火剤　*shoukazai*
　fire extinguisher, fire-extinguishing agent

消石灰　*shousekkai*
　slaked lime $Ca(OH)_2$

浸し掛け　*hitashigake*
　dipping

浸漬電極　*shinsekidenkyoku*
　dipping electrode

浸漬タンク　*shinsekitanku*
　dipping tank

浸染　*shinsen*
　dyeing

浸出　*shinshutsu*
　leaching

浸透　*shintou*
　osmosis, penetration

浸透圧　*shintouatsu*
　osmotic pressure

浸透係数　*shintoukeisuu*
　osmotic coefficient

浸透率　*shintouritsu*
　permeability

浸透性　*shintousei*
　permeability

浸透剤　*shintouzai*
　permeate

浴　*yoku*
　bath

氵, 水　3a
+ 8 strokes　│　混 淡 清 添

混濁　*kondaku*
　turbidity, muddiness

混合　*kongou*
　mixing, mixture

混合, 混合物　*kongou, kongoubutsu*
　mixture

混合物　*kongoubutsu*
　mixture

混合できない　*kongoudekinai*
　immiscible, nonmiscible, unmixable

混合塩　*kongouen*
　mixed salt

混合エントロピー　*kongouentoropi-*
　entropy of mixing

混合エーテル　*kongoue-teru*
　mixed ether ROR′

混合ガス　*kongougasu*
　mixed gas

混合比　*kongouhi*
　mixing ratio

混合肥料 kongouhiryou
mixed fertilizer, compound fertilizer

混合ケトン kongouketon
mixed keton

混合機 kongouki
mixer

混合気 kongouki
gaseous mixture

混合無水酸 kongoumusuisan
mixed anhydride

混合熱 kongounetsu
heat of mixing

混合酸化物 kongousankabutsu
mixed oxide

混合触媒 kongoushokubai
mixed catalyst

混合する kongousuru
to mix, to blend

混合溶媒 kongouyoubai
solvent mixture

混合材 kongouzai
admixture

混入 konnyuu
mixing, contamination

混酸 konsan
mixed acid

混成物 konseibutsu
mixture, mix, composition, blend

混成エーテル konseie-teru
mixed ether ROR′

混成ガス konseigasu
mixed gas

混成軌道 konseikidou
hybrid orbial

混成無水物 konseimusuibutsu
mixed anhydride

混晶 konshou
mixed crystal

混和性 konwasei
miscibility

混融点 konyuuten
mixed melting point

混在物 konzaibutsu
inclusion

淡紅銀鉱 tankouginkou
proustite Ag_3AsS_3

淡水 tansui
fresh water

淡水化 tansuika
desalination

清め kiyome
purification

清澄 seichou
clarification

清澄器 seichouki
clarifier

清澄剤 seichouzai
clarifier

清浄 seijou
clarification

清浄器 seijouki
purifying apparatus

清浄浴 seijouyoku
cleaning bath

清浄剤 seijouzai
cleaning agent

清涼剤 seiryouzai
refrigerant

清水 seisui
pure water

添加 tenka
addition

添加アルコール tenkaaruko-ru
added alcohol

添加物 tenkabutsu
additive, admixture

添加塩 tenkaen
additive salt

添加ガス tenkagasu
additive gas

添加時間 tenkajikan
addition time

添加効果 tenkakouka
addition effect

添加速度　*tenkasokudo*
　addition rate

添加する　*tenkasuru*
　to add

添加体　*tenkatai*
　addition product

添加剤　*tenkazai*
　additive

添加剤効果　*tenkazaikouka*
　additive effect

添加剤濃度　*tenkazainoudo*
　additive concentration

氵, 水　3a
+ 9 strokes　｜測 湖 温 湿 湯
　　　　　　　滋 渡 減

測光　*sokkou*
　photometry

測定　*sokutei*
　measurement

測定範囲　*sokuteihani*
　measuring range

測定法　*sokuteihou*
　measuring method

測定器　*sokuteiki*
　measuring apparatus, measuring equipment

測定装置　*sokuteisouchi*
　measuring equipment

湖北石　*kohokuseki*
　hubeite $Ca_2MnFe[HSi_4O_{13}]\cdot 2H_2O$

温置　*onchi*
　incubation

温度　*ondo*
　temperature

温度調節　*ondochousetsu*
　temperature control

温度調節器　*ondochousetsuki*
　thermoregulator

温度変化　*ondohenka*
　temperature change, temperature variation

温度ジャンプ　*ondojanpu*
　sudden change of temperature

温度計　*ondokei*
　thermometer

温度係数　*ondokeisuu*
　temperature coefficient

温度勾配　*ondokoubai*
　temperature gradient

温度曲線　*ondokyokusen*
　temperature curve

温度目盛　*ondomemori*
　temperature scale

温度の影響　*ondonoeikyou*
　temperature influence

温度制御　*ondoseigyo*
　temperature control

温度シフト　*ondoshifuto*
　temperature shift

温度測定　*ondosokutei*
　temperature measurement

温石綿　*onsekimen*
　chrysotile $Mg_3Si_2O_5(OH)_4$

温泉華　*onsenka*
　tufa, sinter

湿気　*shikki*
　moisture, dampness, humidity

湿電池　*shitsudenchi*
　wet cell battery

湿度　*shitsudo*
　humidity

湿度計　*shitsudokei*
　hygrometer

湿潤　*shitsujun*
　wetting

湿潤力　*shitsujunryoku*
　wetting power

湿潤性　*shitsujunsei*
　wettability

湿潤する　*shitsujunsuru*
　to wet, to damp, to humidify, to moisten

湿潤剤　*shitsujunzai*
　wetting agent

湯浴　yua
　water bath

湯加減　yukagen
　water temperature

滋賀石　shigakenseki
　shigaite $Al_4Mn_7(SO_4)_2(OH)_{22} \cdot 8H_2O$

渡環反応　tokanhannou
　transannular reaction

渡辺鉱　watabekou
　watanabeite $Cu_4(As,Sb)_2S_5$

減圧　genatsu
　reduced pressure

減圧蒸留　genatsujouryuu
　vacuum distillation

減圧濃縮　genatsunoushuku
　vacuum concentration

減感　genkan
　desensitization

減感剤　genkanzai
　desensitizer

減極　genkyoku
　depolarization

減極剤　genkyokuzai
　depolarizer

減力　genryoku
　reduction

減力剤　genryokuzai
　reducer

減湿　genshitsu
　dehumidification

減衰　gensui
　attenuation

減衰係数　gensuikeisuu
　attenuation coefficient, damping coefficient

氵, 水　3a
+ 10 to 11 strokes　｜滑 溝 滞 溢 滅
　　　　　　　　　　漂 滴 滲 漏

滑石　kasseki
　talc

溝呂木ヘック反応　mizorokihekkuhannou
　Mizoroki-Heck reaction

滞留時間　tairyuujikan
　residence time, retention time

溢流　itsuryuu
　overflow

滅菌　mekkin
　sterilization

滅菌器　mekkinki
　sterilizer

漂白　hyouhaku
　bleaching

漂白剤　hyouhakuzai
　bleaching agent

滴瓶　tekibin
　dropping bottle

滴下漏斗　tekikarouto
　dropping funnel

滴下試験　tekikashiken
　drop test

滴下水銀電極　tekikasuigindenkyoku
　dropping mercury electrode

滴定分析　tekiteibunseki
　titration analysis

滴定液　tekiteieki
　titrant

滴定法　tekiteihou
　titration method

滴定曲線　tekiteikyokusen
　titration curve

滴定量　tekiteiryou
　titer

滴剤　tekizai
　drop

滲出　shinshutsu
　extraction, effusion

漏れ　more
　leakage

漏斗　routo
　funnel

漏斗台　*routodai*
　funnel stand

漏斗管　*routokan*
　funnel tube

氵, 水　**3a**
+ 12/13 strokes

潜　潤　澱　濃　濁

潜熱　*sennetsu*
　latent heat

潜伏期　*senpukuki*
　incubation period

潤滑油　*junkatsuyu*
　lubricating oil

潤滑剤　*junkatsuzai*
　lubricant, slip agent

澱粉　*denpun*
　starch $(C_6H_{10}O_5)_n$

澱粉価　*denpunka*
　starch value

澱粉添加剤　*denpuntenkazai*
　starch additives

澱粉糖　*denpuntou*
　starch sugar

澱粉プラスチック　*dnepunpurasuchikku*
　starch plastic

濃度　*noudo*
　concentration

濃度比　*noudohi*
　ration of concentration

濃度計　*noudokei*
　densimeter

濃度勾配　*noudokoubai*
　concentration gradient

濃度効果　*noudokouka*
　concentration effect

濃度測定　*noudosokutei*
　concentration measurement

濃塩酸　*nouensan*
　concentrated hydrochloric acid

濃厚　*noukou*
　density, concentration

濃厚化　*noukouka*
　thickening

濃硫酸　*nouryuusan*
　concentrated sulfuric acid

濃縮　*noushuku*
　enrichment, concentration

濃縮物　*noushukubutsu*
　concentrate

濃縮廃水　*noushukuhaisui*
　wastewater concentration

濃縮比　*noushukuhi*
　concentration ratio

濃縮係数　*noushukukeisuu*
　concentration factor

濃縮器　*noushukuki*
　concentrator

濃縮する　*noushukusuru*
　to concentrate

濃縮体　*noushukutai*
　concentrate

濃縮塔　*noushukutou*
　concentrating column, enriching column

濃縮ウラン　*noushukuuran*
　enriched uranium

濃縮溶液　*noushukuyoueki*
　concentrated solution

濃淡電池　*noutandenchi*
　concentration cell

濁度　*dakudo*
　turbidity

濁度計　*dakudokei*
　turbidimeter

濁度測定　*dakudosokutei*
　turbidimetry

濁沸石　*dakufutsuseki*
　laumontite

濁り度　*nigorido*
　turbidity, tarnish

濁り測定　*nigorisokutei*
turbidimetry, turbidity measurement, turbidimetric analysis, nephelometric analysis

氵, 水　3a
+ 15 to 17 strokes

濾 瀝 灌

濾板　*roban*
filter plate

濾液　*roeki*
filtrate

濾過　*roka*
filtration

濾過液　*rokaeki*
filtrate

濾過フラスコ　*rokafurasuko*
filtering flask

濾過ケーク　*rokake-ku*
filter cake

濾過器　*rokaki*
filter

濾過装置　*rokasouchi*
filter stand, filtering apparatus, filtration equipment

濾過する　*rokasuru*
to filter

濾過用木炭　*rokayoumokutan*
filter charcoal

濾過用漏斗　*rokayourouto*
filtering funnel

濾光器　*rokouki*
light filter

濾紙　*roshi*
filter paper

濾床　*roshou*
filter bed

瀝青ウラン鉱　*rekiseiurankou*
pitchblende UO_2

灌漑　*kangai*
irrigation

土　3b
+ 0 to 4 strokes

土 地 至 坑 均
走 赤

土壌　*dojou*
soil

土壌微量元素　*dojoubiryougenso*
soil trace element

土壌微生物学　*dojoubiseibutsugaku*
soil microbiology

土壌窒素　*dojouchisso*
soil nitrogen

土壌汚染　*dojouosen*
soil pollution

土壌炭酸塩　*dojoutansanen*
soil carbonate

土壌炭素　*dojoutanso*
soil carbon

土器　*doki*
earthenware, stoneware

地下水　*chikasui*
ground water

地球化学　*chikyuukagaku*
geochemistry

地金　*jigane*
ore, mineral, ground metal

至適条件　*shitekijouken*
optimum conditions

至適温度　*shitekiondo*
optimum temperature

先端技術　*sentangijutsu*
advanced technology, high technology

坑ヒスタミン剤, 坑ヒスタミン薬　*kouhisutaminzai, kouhisutaminyaku*
antihistamine

均一　*kinitsu*
homogeneity

均一系　*kinitsukei*
homogeneous system

均一系反応　*kinitsukeihannou*
homogeneous reaction

均染　*kinsen*
level dyeing

均質　*kinshitsu*
　homogeneity

均質沈澱　*kinshitsuchinden*
　homogeneous precipitation

均質化　*kinshitsuka*
　homogenization

均質系　*kinshitsukei*
　homogeneous system, one-phase system

均質体　*kinshitsutai*
　homogeneous substance

走化性　*soukasei*
　chemotaxis

赤ばん　*akaban*
　bieberite $CoSO_4 \cdot 7H_2O$

赤金鉱　*akaganekou*
　akaganeite $Fe^{3+}O(OH,Cl)$

赤針鉱　*akaharikou*
　pyrobelonite $PbMn(VO_4)(OH)$

赤縞メノウ　*akakoumenou*
　sardonyx

赤リン　*akarin*
　red phosphorus

赤外　*sekigai*
　infrared

赤外分光法　*sekigaibunkouhou*
　infrared spectroscopy

赤外吸収　*sekigaikyuushuu*
　infrared absorption

赤外炉　*sekigairo*
　infrared oven

赤外線　*sekigaisen*
　infrared radiation

赤外線加熱　*sekigaisenkanetsu*
　infrared heating

赤外線乾燥　*sekigaisenkansou*
　infrared drying

赤外スペクトル　*sekigaisupekutoru*
　infrared spectrum

赤血塩　*sekiketsuen*
　potassium hexacyanoferrate(III),
　potassium ferricyanide $K_3Fe(CN)_6$

赤熱, 赤熱温度　*sekinetsu, sekinetsuondou*
　red heat

赤燐　*sekirin*
　red phosphorus

赤色硫化水銀　*sekishokuryuukasuigin*
　red mercury(II) sulfide HgS

赤色酸化鉛　*sekishokusankaen*
　minium, lead(II,IV) oxide, read lead Pb_3O_4

赤色酸化水銀　*sekishokusankasuigin*
　red mercury(II) oxide HgO

赤色沃化水銀　*sekishokuyoukasuigin*
　red mercuric iodide HgI_2

赤鉄鉱　*sekitekkou*
　hematite, iron glance, red iron ore $\alpha\text{-}Fe_2O_3$

赤血球　*sekkekkyuu*
　erythrocyte

赤銅鉱　*shakudoukou*
　cuprite Cu_2O

土　3b + 6 and more strokes	型　培　堆　起　埋　型 境　塗　塑　塊　堅　塔 増　壊　壜

型穴　*kataana*
　impression

埋込み　*umekomi*
　imbedding

埋込クラスタ　*umekomikurasuta*
　embedded cluster

起爆する　*kibakusuru*
　to initiate, to start

起爆薬　*kibakuyaku*
　initiator

起源　*kigen*
　origin, source

起磁力　*kijiryoku*
　magnetomotive force

堆積物　*taisekibutsu*
　sediment, settling

堆積作用　*taisekisayou*
　sedimentation, deposition

培地 *baichi*
medium

培養 *baiyou*
cultivation, incubation

培養液 *baiyoueki*
nutrient liquid

培養基 *baiyouki*
culture medium, nutrient medium

基 *ki*
group, radical, residue

基本材料, 基本物質 *kihonzairyou, kihobusshitsu*
basic material, raw material

基準 *kijun*
basis

基準試料 *kijunshiryou*
authentic sample

基線 *kisen*
base line

基質 *kishitsu*
substrate

基質濃度 *kishitsunoudo*
substrate concentration

基質特異性 *kishitsutokuisei*
substrate specifity

基礎科学 *kisokagaku*
basic science

基礎研究 *kisokenkyuu*
basic research, fundamental research

基礎試験 *kisoshiken*
basic study, preclinical study

基底状態 *kiteijoutai*
ground state

基剤 *kizai*
base

基材 *kizai*
substrate, reactant

塔 *tou*
column, tower

堅牢度試験 *kenroudoshiken*
fastness test

塊炭 *kaitan*
lump coal

塑性物質 *soseibusshitsu*
plastic material

塑性不安定 *soseifuantei*
plastic instability

塑性加工 *soseikakou*
plasticity processing

塗工液 *tokoueki*
coating solution

塗工フィルム *tokoufirumu*
coated film

塗工層 *tokousou*
coating layer

塗工塗料 *tokoutoryou*
coating color

塗料 *toryou*
paint, coating material

塗料技術 *toryougijutsu*
paint technology

塗料樹脂 *toryoujushi*
coating resin

塗料添加剤 *toryoutenkazai*
coating additives

塗料溶媒 *toryouyoubai*
coating solvent

境界潤滑 *kyoukaijunkatsu*
boundary lubrication

境界層 *kyoukaisou*
boundary layer

境膜 *kyoumaku*
laminar film

増幅器 *joufukuki*
amplifier

増援 *zouen*
reinforcement

増感 *zoukan*
intensification, sensitization

増粘剤 *zounenzai*
thickener

増量剤 *zouryouzai*
filler

増湿装置　zoushitsusouchi
　humidifier

壊変　kaihen
　degradation, decomposition,
　disintegration, decay, fragmentation

壜　bin
　bottle

扌, 手　3c　｜　手 抜 抑 抗 技 投
+ 0 to 4 strokes

手段　shudan
　means, measures

抜き出す　nukidasu
　to extract, to leave out, to remove

抜き型　nukigata
　cutting dies

抑える　osaeru
　hold down, sppress, control

抑制　yokusei
　inhibition, suppression, restraint,
　retardation

抑制技術　yokuseigijutsu
　inhibition technique

抑制作用　yokuseisayou
　depressant action

抑制体　yokuseitai
　repressor

抑制剤　yokuseizai
　inhibitor, restrainer, retarder

抗毒素　koudokuso
　antitoxin

抗働質　koudoushitsu
　inhibitor

抗癌薬　kouganyaku
　anticancer drug

抗原　kougen
　antigen

抗原抗体反応　kougenkoutaihannou
　antigen-antibody reaction

抗凝固　kougyouko
　anticoagulation

抗ヒスタミン剤, 抗ヒスタミン薬
　kouhisutaminzai, kouhisutaminyaku
　antihistamine

抗菌性物質　koukinseibutsushitsu
　antibacterials

抗乳化度　kounyuukado
　demulsibility

抗酸化剤　kousankazai
　antioxidant, oxidation inhibitor

抗生　kousei
　antibiotic

抗生物質　kouseibusshitsu
　antibiotic

抗真菌剤　koushinkinzai
　fungicide

抗体　koutai
　antibody

技術　gijutsu
　technology

技術データ　gijutsude-ta
　engineering data

技術マニュアル　gijutsumanyuaru
　engineering manual

技術サービス　gijutsuse-bisu
　technical service

技術者　gijutsusha
　engineer, technician

技術仕様書　gijutsushiyousho
　technical specification

技術スタッフ　gijutsusutaffu
　technical staff

投影　touei
　projection

投影法　toueihou
　projective method

投影式　toueishiki
　projection formula

投射　tousha
　projection

扌, 手　3c
+ 5 strokes

抽 抱 担 拡

抽出 *chuushutsu*
　extraction

抽出分 *chuushutsubun*
　extract

抽出物 *chuushutsubutsu*
　extract

抽出フロー *chuushutsufuro-*
　extraction flow

抽出ガスクロマトグラフィー
　chuushutsugasukuromatogurafi-
　extraction gas chromatography

抽出技術 *chuushutsugijutsu*
　extraction technology

抽出発酵 *chuushutsuhakkou*
　extractive fermentation

抽出反応 *chuushutsuhannou*
　extraction reaction

抽出法 *chuushutsuhou*
　extraction method

抽出実験 *chuushutsujikken*
　extraction experiment

抽出蒸留 *chuushutsujouryuu*
　extractive distillation

抽出カラム *chuushutsukaramu*
　extraction column

抽出過程 *chuushutsukatei*
　extraction process

抽出系 *chuushutsukei*
　extraction system

抽出器 *chuushutsuki*
　extractor

抽出機 *chuushutsuki*
　extractor

抽出効率 *chuushutsukouritsu*
　extraction efficiency

抽出クロマトグラフィー
　chuushutsukuromatogurafi-
　extraction chromatography

抽出濃度 *chuushutsunoudo*
　extraction concentration

抽出性 *chuushutsusei*
　extractability

抽出試料 *chuushutsushiryou*
　extraction sample

抽出収率 *chuushutsushuuritsu*
　extraction yield

抽出速度 *chuushutsusokudo*
　extraction rate

抽出速度論 *chuushutsusokudoron*
　extraction kinetics

抽出装置 *chuushutsusouchi*
　extractor, contactor

抽出塔 *chuushutsutou*
　extraction column

抽出溶媒 *chuushutsuyoubai*
　extraction solvent

抽出油 *chuushutsuyu*
　extracted oil

抽出有機相 *chuushutsuyuukisou*
　extracted organic phase

押出し *oshidashi*
　extrusion

押出機 *oshidashiki*
　extruder

押出成形 *oshidashiseikei*
　extrusion molding

抱水クロラル *housuikuroraru*
　chloral hydrate $CCl_3CH(OH)_2$

抵抗 *teikou*
　resistance, stability, persistence

抵抗計 *teikoukei*
　ohm meter, resistance meter

抵抗係数 *teikoukeisuu*
　resistance coefficient

抵抗温度計 *teikouondokei*
　resistance thermometer

抵抗パラメータ *teikouparame-ta*
　resistance parameter

抵抗率 *teikouritsu*
　resistivity

抵抗炉 *teikouro*
　resistance furnace, resistance oven, resistor furnace

抵抗線 *teikousen*
　resistance wire

担子菌類 *tanshikinrui*
　basidiomycet

担体 *tantai*
　carrier, support

担体表面 *tantaihyoumen*
　carrier surface

担体材料 *tantaizairyou*
　carrier material

拡大 *kakudai*
　enlargement, expansion, extension, increase

拡散 *kakusan*
　diffusion

拡散電位 *kakusandeni*
　diffusion potential

拡散反射 *kakusanhansha*
　diffuse reflection

拡散係数 *kakusankeisuu*
　diffusion coefficient

拡散光 *kakusankou*
　diffused light

拡散ポンプ *kakusanponpu*
　diffusion pump

拡散速度 *kakusansokudo*
　diffusion rate, diffusion velocity

拡散定数 *kakusanteisuu*
　diffusion constant

指示薬溶液 *shijiyakuyoueki*
　indicator solution

指紋領域 *shimonryouiki*
　fingerprint region

指数 *shisuu*
　exponent, index

指数因子 *shisuuinshi*
　exponential factor

指数関数 *shisuukansuu*
　exponential function

振動 *shindou*
　oscillation, vibration

振動エネルギー *shindouenerugi-*
　oscillation energy, vibrational energy

振動反応 *shindouhannou*
　oscillatory reaction

振動緩和 *shindoukanwa*
　vibrational relaxation

振動面 *shindoumen*
　plane of vibration

振動モード *shindoumo-do*
　mode of oscillation

振動量子数 *shindouryoushisuu*
　vibrational quantum number

振動子 *shindoushi*
　oscillator

振動スペクトル *shindousupekutoru*
　vibrational spectrum

振動数 *shindousuu*
　frequency

振動定数 *shindouteisuu*
　oscillation constant

扌, 手　3c
+ 6/7 strokes　| 指 振

指示 *shiji*
　indication

指示薬 *shijiyaku*
　indicator

扌, 手　3c
+ 8 strokes　| 振 推 排 接 探

振幅 *shinpuku*
　amplitude

推進力 *suishinryoku*
　driving force

推進薬　suishinyaku
　propellant
排煙　haien
　flue gas, waste gas, stack gas
排煙脱硫　haiendatsuryuu
　flue gas desulfurization
排気　haiki
　exhaust
排気弁　haikiben
　exhaust valve
排気ガス　haikigasu
　exhaust gas
排泄　haisetsu
　excretion
排水　haisui
　waste water, sewage
排水リサイクル　haisuirisaikuru
　waste water recycle
排水処理　haisuishori
　waste water treatment
接着　secchaku
　adhesion, glueing
接着力試験　secchakukashiken
　adhesion strength test
接着力　secchakuryoku
　bond strength
接着性　secchakusei
　adhesive property
接着試験　secchakushiken
　adhesion test
接着強さ　secchakutsuyosa
　bonding strength, adhesion strength
接着剤　secchakuzai
　adhesive
接着剤フィルム　secchakuzaifirumu
　adhesive film
接触　sesshoku
　contact
接触分解　sesshokubunkai
　catalytic cracking
接触中毒剤　sesshokuchuudokuzai
　contact poison

接触反応　sesshokuhannou
　catalytic reaction
接触イオン対　sesshokuiontsui
　contact ion pair
接触重合　sesshokujuugou
　catalytic polymerization
接触還元　sesshokukangen
　catalytic reduction
接触面　sesshokumen
　contact surface, contact area
接触酸化　sesshokusanka
　catalytic oxidation
接触点　sesshokuten
　contact point
接触剤　sesshokuzai
　catalyst, contact substance
接種　sesshu
　inoculation
接眼レンズ　setsuganrenzu
　ocular
接続管　setsusokukan
　connecting pipe
接続器　setsuzokuki
　connector
探究者　tankyuusha
　researcher
探傷　tanshou
　flaw detection

扌, 手　3c + 9 to 20 strokes	揺　揮　換 操　擬　攪

揺変性　youhensei
　thixotropy
揮発　kihatsu
　volatilization, evaporation
揮発分　kihatsubun
　volatile component
揮発度　kihatsudo
　volatility

揮発性　*kihatsusei*
　volatility

揮発成分　*kihatsuseibun*
　volatile component

揮発性物質　*kihatsuseibusshitsu*
　volatile matter

揮発性溶剤　*kihatsuseiyouzai*
　volatile solvent

揮発油　*kihatsuyu*
　volatile oil

換気　*kanki*
　ventilation

換気器　*kankiki*
　ventilator

換気機　*kankiki*
　ventilator

換算圧力　*kansanatsuryoku*
　reduced pressure

換算温度　*kansanondo*
　reduced temperature

操作　*sousa*
　operation

擬ハロゲン　*giharogen*
　pseudohalogen

擬似芳香族性　*gijihoukouzokusei*
　pseudoaromaticity

擬似酸　*gijisan*
　pseudo-acid

攪拌　*kakuhan*
　agitation

攪拌器　*kakuhanki*
　stirrer, agitator

攪拌機　*kakuhanki*
　stirrer, agitator

可 3d + 2 strokes　　可 右

可逆電池　*kagyakudenchi*
　reversible cell

可逆反応　*kagyakuhannou*
　reversible reaction

可逆変化　*kagyakuhenka*
　reversible change

可逆過程　*kagyakukatei*
　reversible process

可逆光化学反応　*kagyakukoukagakuhannou*
　reversible reaction

可逆サイクル　*kagyakusaikuru*
　reversible cycle

可逆性　*kagyakusei*
　reversibility

可逆阻害　*kagyakusogai*
　reversible inhibition

可干渉性　*kakanshousei*
　coherence

可燃物　*kanenbutsu*
　combustibles

可燃性　*kanensei*
　flammability

可燃性ガス　*kanenseigasu*
　inflammable gas

可燃性廃棄物　*kanenseihaikibutsu*
　combustible waste

可視炎　*kashien*
　visible flame

可視光　*kashikou*
　visible light

可視光線　*kashikousen*
　visible radiation

可塑　*kaso*
　plastic

可塑性　*kasosei*
　plasticity

可塑剤　*kasozai*
　liquefier

可とう性　*katousei*
　flexibility

可溶化　*kayouka*
　solubilization

可溶化剤 *kayoukazai*
solubilizer

可溶性 *kayousei*
solubility

右旋性の *usenseino*
dextrorotatory

右旋性酒石酸 *usenseishusekisan*
dextrotartaric acid, (+)-tartaric acid
HOOC(CHOH)$_2$COOH

3d
+ 3 to 4 strokes

吸 向 呈 乱 豆

吸着 *kyuuchaku*
adsorption

吸着分析 *kyuuchakubunseki*
adsorption analysis

吸着器 *kyuuchakuki*
adsorber

吸着クロマトグラフィー
kyuuchakukuromatogurafi-
adsorption chromatography

吸着熱 *kyuuchakunetsu*
heat of adsorption

吸着質 *kyuuchakushitsu*
adsorbate

吸着速度 *kyuuchakusokudo*
rate of adsorption

吸着装置 *kyuuchakusouchi*
adsorber

吸着定数 *kyuuchakuteisuu*
adsorption constant

吸着剤 *kyuuchakuzai*
adsorbent

吸引瓶 *kyuuinbin*
filtering flask

吸引管 *kyuuinkan*
suction tube

吸引器 *kyuuinki*
aspirator

吸引ポンプ *kyuuinponpu*
suction pump

吸引濾過 *kyuuinroka*
suction filtration

吸引漏斗 *kyuuinrouto*
suction filter, suction funnel

吸塵器, 吸塵機 *kyuujinki*
dust separator

吸光 *kyuukou*
extinction, absorbance

吸光度 *kyuukoudo*
absorbance

吸光係数 *kyuukoukeisuu*
extinction coefficient

吸光率 *kyuukouritsu*
absorpitivity

吸熱反応 *kyuunetsuhannou*
endothermic reaction

吸湿度 *kyuushitsudo*
hygroscopic degree

吸湿性 *kyuushitsusei*
hygroscopicity

吸湿性の *kyuushitsuseino*
hygroscopic

吸収 *kyuushuu*
absorption

吸収バンド *kyuushuubando*
absorption band

吸収エネルギー *kyuushuuenerugi-*
absorbed energy

吸収因子 *kyuushuuinshi*
absorption factor

吸収係数 *kyuushuukeisuu*
absorption coefficient

吸収機 *kyuushuuki*
absorber

吸収極大 *kyuushuukyokudai*
absorption maximum

吸収曲線 *kyuushuukyokusen*
absorption curve

吸収熱 *kyuushuunetsu*
heat of absorption

吸収能　*kyuushuunou*
　　absorbing power

吸収線　*kyuushuusen*
　　absorption line

吸収セル　*kyuushuuseru*
　　absorption cell

吸収スペクトル　*kyuushuusupekutoru*
　　absorption spectrum

吸収体　*kyuushuutai*
　　absorber

吸収帯　*kyuushuutai*
　　absorption band

吸収塔　*kyuushuutou*
　　absorption tower

吸収剤　*kyuushuuzai*
　　absorbent

吸水　*kyuusui*
　　water absorption

吸込み　*suikomi*
　　suction

吸収率　*kyuushuuritsu*
　　absorption coefficient

向上　*koujou*
　　improvement

向流　*kouryuu*
　　counterflow, countercurrent

向流分配　*kouryuubunpai*
　　countercurrent distribution

向流クロマトグラフィー
　kouryuukuromatogurafi-
　　countercurrent chromatography

呈色　*teishoku*
　　coloration

呈色反応　*teishokuhannou*
　　color reaction

乱れ　*midare*
　　turbulence

乱流　*ranryuu*
　　turbulent flow

乱流反応性　*ranryuuhannousei*
　　turbulent reacting

乱流拡散　*ranryuukakusan*
　　turbulent diffusion

豆石　*mameseki*
　　pisolite

豆炭　*mametan*
　　briquet

‖ 3d + 5 to 12 strokes	呼　知　品　短 豊　噴　器

呼吸　*kokyuu*
　　respiration

呼吸活性　*kokyuukassei*
　　respiratory acitivity

呼吸酵素　*kokyuukouso*
　　respiratory enzyme

知識　*chishiki*
　　knowledge

品質　*hinshitsu*
　　quality, grade

品質分析　*hinshitsubunseki*
　　quality analysis

品質管理　*hinshitsukanri*
　　quality control

品質検査　*hinshitsukensa*
　　quality inspection

品質基準　*hinshitsukijun*
　　quality standard

品質規格　*hinshitsukikaku*
　　quality specification

品質試験　*hinshitsushiken*
　　quality test

品質仕様　*hinshitsushiyou*
　　quality specification

短時間安定性　*tanjikananteisei*
　　short-time stability

短カラム　*tankaramu*
　　short column

短カラム液体クロマトグラフィー
　tankaramuekitaikuromatogurafi-
　　short-column liquid chromatography

短期研究 tankikenkyuu
short-term research

短期試験 tankishiken
short-term test

短工程合成法 tankouteigouseihou
short-route synthesis

短鎖 tansa
short chain

短鎖アルカン tansaarukan
short-chain alkane

短鎖アシル tansaashiru
short-chain acyl

短鎖分岐 tansabunki
short-chain branch

短鎖脂肪酸 tansashibousan
short-chain fatty acid

短繊維 tanseni
short fibre

短接触 tansesshoku
short contact

豊羽鉱 toyohakou
toyohaite $Ag_2FeSn_3S_8$

噴霧 funmu
atomization, spraying

噴霧微粒 funmubiryuu
spray atomization

噴霧乾燥 funmukansou
spray drying

噴霧乾燥機 funmukansouki
spray dryer

噴霧器 funmuki
atomizer, sprayer

噴霧機 funmuki
sprayer, atomizer, spraying apparatus, vaporizer

噴霧粒子 funmuryuushi
spray particle

噴霧室 funmushitsu
spray chamber

噴流 funryuu
water jet

噴射 funsha
injection, jet, spray

器具 kigu
instrument, apparatus

器官 kikan
organ

器差 kisa
instrument error

女 3e | 好 始 要 媒 嫌
(all kanji)

好中球 kouchuukyuu
neutrophilic

好気性細菌 koukiseisaikin
aerobic bacterium

始期 shiki
beginning period, initial term

要因 youin
factor

要因解析 youinkaiseki
factor analysis

媒介アルドール反応 baikaiarudo-ruhannou
mediated aldol reaction

媒介アルキル化 baikaiarukiruka
mediated alkylation

媒介不斉合成 baikaifuseigousei
mediated asymmetric synthesis

媒介合成 baikaigousei
mediated synthesis

媒介グリコシル化 baikaigurikoshiruka
mediated glycosylation

媒介変換 baikaihenkan
mediated transformation

媒介カップリング反応 baikaikappuringuhannou
mediated coupling reaction

媒染染料 baisensenryou
mordant dye

媒染浴 *baisenyoku*
  mordant bath

媒染剤 *baisenzai*
  mordant

媒質 *baishitsu*
  medium

媒質表面 *baishitsuhyoumen*
  medium surface

媒質効果 *baishitsuseiyu*
  medium effect

媒体 *baitai*
  medium

嫌気性細菌 *kenkiseisaikin*
  anaerobic bacteria

巾 3f | 希 帯
(all *kanji*)

希土類元素 *kidoruigenso*
  rare earth elements

希塩酸 *kiensan*
  diluted hydrochloric acid

希ガス *kigasu*
  noble gas

希元素 *kigenso*
  rare element

希硫酸 *kiryuusan*
  diluted sulfuric acid

希釈 *kishaku*
  dilution

希釈率 *kishakuritsu*
  dilution rate

希釈溶液 *kishakuyoueki*
  diluted solution

希釈剤 *kishakuzai*
  diluent

帯電防止性 *taidenboushisei*
  antistatic property

帯電防止剤 *taidenboushizai*
  antistatic agent

帯スペクトル *taisupekutoru*
  band spectrum

犭,犬 3g | 独
(all *kanji*)

独立栄養 *dokuritsueiyou*
  autotrophy

弓 3h | 引 弱 張 強 弾
(all *kanji*)

引張力 *hippariryoku*
  tensile force

引張試験 *hipparishiken*
  tension test

引火性 *inkasei*
  flammability, ignitability

引火性液体 *inkaseiekitai*
  inflammable fluid

引火性ガス *inkaseigasu*
  inflammable gas

引火点 *inkaten*
  flash point

弱電解質 *jakudenkaishitsu*
  weak electrolyte

弱電流 *jakudenryuu*
  weak current

弱毒 *jakudoku*
  weak poison

弱塩基 *jakuenki*
  weak base

弱酸 *jakusan*
  weak acid

張力計 *chouryokukei*
  tensiometer

強度因子 *kyoudoinshi*
  intensity factor

強塩基 *kyouenki*
  strong base

強磁性 *kyoujisei*
ferromagnetism

強磁性の *kyoujiseino*
ferromagnetic

強磁性体 *kyoujiseitai*
ferromagnetic material

強化プラスチック *kyoukapurasuchikku*
reinforced plastic

強粘液 *kyouneneki*
viscous liquid

強熱減量 *kyounetsugenryou*
heat loss

強熱温度 *kyounetsuondo*
ignition temperature

強酸 *kyousan*
strong acid

強制換気 *kyouseikanki*
forced ventilation

強誘電性 *kyouyuudensei*
ferroelectricity

弾力 *danryoku*
elasticity, flexibility

弾性 *dansei*
elasticity

弾性フィルム *danseifirumu*
elastic film

弾性ゴム *danseigomu*
elastic rubber

弾性重合体 *danseijuugoutai*
elastic polymer

弾性係数 *danseikeisuu*
stiffness coefficient

弾性研究 *danseikenkyuu*
elasticity study

弾性光ファイバ *danseikoufaiba*
elastic optical fiber

弾性繊維 *danseiseni*
elastic fiber

弾性試験 *danseishiken*
elasticity test

弾性体 *danseitai*
elastic body

イ 3i (all *kanji*)	行 径 律 後 徐 従 循 微 衝

行列 *gyouretsu*
matrix

径 *kei*
diameter, size

律速段階 *rissokudankai*
rate-determing step

後加硫 *atokaryuu*
after-vulcanization

後処理 *atoshori*
aftertreatment

徐冷 *jorei*
annealing, tempering

従来 *kouhou*
official bulletin

循環 *junkan*
circulation

循環過程 *junkankatei*
cyclic process

微分方程式 *bibunhouteishiki*
differential equation

微調整 *bichousei*
fine adjustment

微粉機 *bifunki*
pulverizing mill

微粉砕機 *bifunsaiki*
pulverizing mill

微結晶 *bikesshou*
microcrystal

微構造 *bikouzou*
microstructure

微量分析 *biryoubunseki*
microanalysis

微量元素 *biryougenso*
trace element

微量化学 *biryoukagaku*
microchemistry

微量成分分析 *biryouseibunbunseki*
trace analysis

微粒化　*biryuuka*
  atomization

微粒金属　*biryuukinzoku*
  fine-grained metal

微細構造　*bisaikouzou*
  fine structure, microstructure

微細構造　*bisaikouzou*
  microstructure

微生物　*biseibutsu*
  microorganism

微生物分解　*biseibutsubunkai*
  biodegradation

微生物学　*biseibutsugaku*
  microbiology

微生物定量　*biseibutsuteiryou*
  microbiological assay

微小　*bishou*
  minute, microscopic

衝撃試験　*shougekishiken*
  impact test

衝突　*shoutotsu*
  collision

衝突確率　*shoutotsukakuritsu*
  probability of collision

艹　**3j**　｜　形 彫 影
(all *kanji*)

形状係数　*keijoukeisuu*
  shape factor

形態　*keitai*
  form, shape, configuration

形態学　*keitaigaku*
  morphology

彫刻　*choukoku*
  engraving

影響　*eikyou*
  influence

艹　**3k**　｜　共 芽 苦 苛
+ 3 to 7 strokes　｜　革 茶 華 荷

共沸蒸留　*kyoufutsujouryuu*
  azeotropic distillation

共沸混合物　*kyoufutsukongoubutsu*
  azeotrope, azeotropic mixture

共沸点　*kyoufutten*
  azeotropic point

共原子価　*kyougenshika*
  covalence

共平面性　*kyouheimensei*
  coplanarity

共重合　*kyoujuugou*
  copolymerization

共重合体　*kyoujuugoutai*
  mixed polymer

共鳴　*kyoumei*
  resonance

共鳴エネルギー　*kyoumeienerugi-*
  resonance energy

共鳴効果　*kyoumeikouka*
  resonance effect

共鳴スペクトル　*kyoumeisupekutoru*
  resonance spectrum

共生　*kyousei*
  symbiosis

共振回路　*kyoushinkairo*
  resonance circuit

共晶　*kyoushou*
  eutectic

共役　*kyouyaku*
  conjugation

共役塩基　*kyouyakuenki*
  conjugate base

共役系　*kyouyakukei*
  conjugated system

共役二重結合　*kyouyakunijuuketsugou*
  conjugated double bond

共役酸　*kyouyakusan*
  conjugate acid

共有原子価 *kyouyuugenshika*
  covalence

共有結合 *kyouyuuketsugou*
  covalent bond

共融混合物 *kyouyuukongoubutsu*
  eutectic mixture

共分散 *kyoubunsan*
  covariance

芽胞 *gahou*
  spore

苦灰石 *kukaiseki*
  dolomite $CaMg(CO_3)_2$

苛性ソーダ *kaseiso-da*
  caustic soda NaOH

革 *kawa*
  leather

茶 *cha*
  tea

茶ポリフェノール *chaporifeno-ru*
  tea polyphenol

茶素 *chaso*
  theine, coffeine $C_8H_{10}N_4O_2$

華氏 *kashi*
  degree Fahrenheit

荷電 *kaden*
  electric charge

荷電電流 *kadendenryuu*
  charging current

荷電粒子 *kadenryuushi*
  charged particle

荷重試験 *kajuushiken*
  load test

黄鉛 *ouen*
  chrome yellow $PbCrO_4$

黄玉石 *ougyokuseki, kougyokuseki*
  topaz $Al_2[F_2,SiO_4]$

黄リン, 黄燐 *ourin*
  white phosphorus, yellow phosphorus

黄鉄鉱 *outekkou*
  pyrite FeS_2

菱面体 *ryoumentai*
  rhombohedral

菌類 *kinrui*
  fungus, fungi

菌類学 *kinruigaku*
  mycology

菌糸体 *kinshitai*
  mycelium

菌体 *kintai*
  bacterial cell

菌体外の *kintaigaino*
  extracellular

落下試験 *rakkashiken*
  drop test

落球粘度計 *rakkyuunendokei*
  falling ball viscometer

葉 *ha*
  leaf

葉緑素 *youryokuso*
  chlorophyll

葉緑体 *youryokutai*
  chloroplast

葉酸 *yousan*
  folic acid $C_{19}H_{19}N_7O_6$

葉酸塩 *yousanen*
  folate

葉酸類似体 *yousanruijitai*
  folate analog

葉酸誘導体 *yousanyuudoutai*
  folate derivative

葡萄糖 *budoutou*
  dextrose, grape sugar, D-glucose $C_6H_{12}O_6$

++   **3k**
+ 8/9 strokes

黄 菱 菌 落
葉 葡 蓄

黄銅 *oudou*
  yellow brass

黄銅鉱 *oudoukou*
  chalcopyrite $CuFeS_2$

蓄電池 *chikudenchi*
accumulator, storage battery

蓄熱器 *chikunetsuki*
regenerator

蓄熱炉 *chikunetsuro*
regenerative furnace

蓄熱式 *chikunetsushiki*
regenerative system

蓄熱式バーナ *chikunetsushikiba-na*
regenerative burner

蓄熱式熱交換器
chikunetsushikinetsukoukanki
heat regenerator

蓄熱室 *chikunetsushitsu*
regenerator

蓄積プロセス *chikupurosesu*
storage process

++ 3k   蓚 蔗 薄 薬 藍
+ 10 to 13 strokes

蓚酸 *shuusan*
oxalic acid HOOCCOOH

蓚酸亜鉛 *shuusanaen*
zinc oxalate ZnC_2O_4

蓚酸塩 *shuusanen*
oxalate $M^I_2C_2O_4$

蓚酸銀 *shuusangin*
silver oxalate $Ag_2C_2O_4$

蓚酸カリウム *shuusankariumu*
potassium oxalate $K_2C_2O_4$

蓚酸カルシウム *shuusankarushiumu*
calcium oxalate CaC_2O_4

蓚酸ナトリウム *shuusannatoriumu*
sodium oxalate $Na_2C_2O_4$

蓚酸ウラニル *shuusanuraniru*
uranyl oxalate $UO_2(C_2O_4)$

蔗糖 *shotou*
sucrose $C_{12}H_{22}O_{11}$

薄膜 *hakumaku*
thin film

薄層板 *hakusouban*
thin-layer plate

薄層電極 *hakusoudenkyoku*
thin-layer electrode

薄層フィルム *hakusoufirumu*
thin-layer film

薄層クロマトグラフィー
hakusoukuromatogurafi-
thin-layer chromatography

薄層クロマトグラフ分析
hakusoukuromatogurafubunseki
thin-layer chromatographic analysis

薄層クロマトグラフ検出
hakusoukuromatogurafukenshutsu
thin-layer chromatographic detection

薄層クロマトグラム
hakusoukuromatoguramu
thin-layer chromatogram

薄層セル *hakusouseru*
thin-layer cell

薬 *kusuri*
drug

薬屋 *kusuriya*
pharmacy, drug store

薬効 *yakkou*
drug efficacy

薬局 *yakkyoku*
pharmacy, drug store, chemist

薬物 *yakubutsu*
drug, medicine, medicament

薬物分子 *yakubutsubunshi*
drug molecule

薬物エナンチオマ *yakubutsuenanchioma*
drug enantiomer

薬物学 *yakubutsugaku*
pharmacology

薬物化学 *yakubutsukagaku*
pharmaceutical chemistry

薬液 *yakueki*
chemical solution

薬害 *yakugai*
medicine side effects

薬品 *yakuhin*
　chemical

薬品回収 *yakuhinkaishuu*
　chemical recovery

薬包 *yakuhou*
　cartridge

薬名 *yakumei*
　drug name, medicine name

薬剤 *yakuzai*
　drug

薬剤耐性 *yakuzaitaisei*
　drug resistance

藍色 *aiiro*
　indigo blue

藍染 *aizome*
　indigo dye

藍 *ran*
　indigo

藍銅鉱 *randoukou*
　azurite $2CuCO_3 \cdot Cu(OH)_2$

宀 3m　　安 完
+ 3/4 strokes

安息香酸 *ansokukousan*
　benzoic acid C_6H_5COOH

安息香酸塩 *ansokukousanen*
　benzoate C_6H_5COOM

安息香酸エステル *ansokukousanesuteru*
　benzoic acid ester C_6H_5COOR

安息香酸カリウム *ansokukousankariumu*
　potassium benzoate C_6H_5COOK

安息香酸メチル *ansokukousanmechiru*
　methyl benzoate $C_6H_5COOCH_3$

安息香酸ナトリウム *ansokukousannatoriumu*
　sodium benzoate C_6H_5COONa

安息酸 *ansokusan*
　benzoic acid C_6H_5COOH

安定度試験 *anteidoshiken*
　stability test

安定度定数 *anteidoteisuu*
　stability constant

安定同位体 *anteidouitai*
　stable isotope

安定化 *anteika*
　stabilization

安定化剤 *anteikazai*
　stabilizer

安定性 *anteisei*
　stability

安定剤 *anteizai*
　stabilizer

安全 *anzen*
　safety, security

安全弁 *anzenben*
　safety valve

安全ガラス *anzengarasu*
　safety glass, shatterproof glass,
　nonshattering glass, protecting glass

安全解析 *anzenkaiseki*
　safety analysis

安全管理 *anzenkanri*
　safety control, safety management

安全制御 *anzenseigyo*
　safety control

安全装置 *anzensouchi*
　protective device

完成品 *kanseihin*
　end product, finished product

完全 *kanzen*
　complete, perfect

完全合成 *kanzengousei*
　total synthesis

完全混合 *kanzenkongou*
　complete mixing

完全燃焼 *kanzennenshou*
　complete combustion

完全流体 *kanzenryuutai*
　perfect fluid, ideal liquid

⼧ **3m**	宝 実 官 定 空 突
+ 5 strokes	

宝石　*houseki*
    jewel, gem stone

実験　*jikken*
    experiment, test, trial

実験値　*jikkenchi*
    experimental result, test findings

実験台　*jikkendai*
    laboratory bench

実験データ　*jikkende-ta*
    experimental data

実験動物　*jikkendoubutsu*
    laboratory animal, test animal

実験誤差　*jikkengosa*
    experimental error

実験化学　*jikkenkagaku*
    experimental chemistry

実験結果　*jikkenkekka*
    experimental result

実験器具　*jikkenkigu*
    laboratory instrument

実験者　*jikkensha*
    researcher

実験式　*jikkenshiki*
    empirical formula

実験室　*jikkenshitsu*
    laboratory

実験装置　*jikkensouchi*
    experimental equipment

実験的研究　*jikkentekikenkyuu*
    experimental study

実験材料　*jikkenzairyou*
    test substance, experimental material

実施　*jisshi*
    to carry out, to execute, to enforce

実在気体　*jitsuzaikitai*
    real gas

官能基　*kannouki*
    functional group

官能基分析　*kannoukibunseki*
    functional group analysis

定着　*teichaku*
    fixation

定着液　*teichakueki*
    fixer

定義　*teigi*
    definition

定常状態　*teijoujoutai*
    steady state

定常流　*teijouryuu*
    stationary flow

定期試験　*teikishiken*
    periodic examination

定温　*teion*
    fixed temperature

定量　*teiryou*
    determination

定量分析　*teiryoubunseki*
    quantitative analysis

定量薄層クロマトグラフィー
    teiryouhakusoukuromatogurafi-
    quantitative thin layer chromatography

定量的分離　*teiryoutekibunri*
    quantitative separation

定量的抽出　*teiryoutekichuushutsu*
    quantitative extraction

定量的元素分析　*teiryoutekigensobunseki*
    quantitative elemental analysis

定量的合成　*teiryoutekigousei*
    quantitative synthesis

定量的変化　*teiryoutekihenka*
    quantitative change

定量的移動　*teiryoutekiidou*
    quantitative transfer

定量的回収　*teiryoutekikaishuu*
    quantitative recovery

定量的キャラクタリゼーション
    teiryoutekikyarakutarize-shon
    quantitative characterization

定量的な　*teiryoutekina*
    quantitative

定量的酸化 *teiryoutekisanka*
quantitative oxidation

定量的処理 *teiryoutekishori*
quantitative treatment

定性分析 *teiseibunseki*
qualitative analysis

定性試験 *teiseishiken*
qualitative test

定数 *teisuu*
constant

定容 *teiyou*
constant volume

空胴ガラス *kuudougarasu*
hollow glass

空隙率 *kuugekiritsu*
porosity

空泡 *kuuhou*
air bubble

空間群 *kuukangun*
space group

空間格子 *kuukankoushi*
space lattice

空気 *kuuki*
air

空気圧 *kuukiatsu*
atmospheric pressure

空気分析 *kuukibunseki*
air analysis

空気調和 *kuukichouwa*
air conditioning

空気液化 *kuukiekika*
air liquefaction

空気液化機 *kuukiekikaki*
air liquefier

空気加熱炉 *kuukikanetsuro*
air heating furnace

空気乾燥 *kuukikansou*
air drying

空気汚染 *kuukiosen*
air pollution

空気冷却 *kuukireikyaku*
air cooling

空気冷却器 *kuukireikyakuki*
air condenser

空気濾過器, 空気濾過機
kuukirokaki, kuukirokaki
air filter

空気流 *kuukiryuu*
airflow

空気酸化 *kuukisanka*
air oxidation

空気清浄 *kuukiseijou*
air filtration, air purification

空気湿度 *kuukishitsudo*
air humidity, air moisture

空冷 *kuurei*
air cooling

空冷コンデンサー *kuureikondensa-*
air condenser

突然変異源性テスト
totsuzehenigenseitesuto
mutagenicity test

突然変異 *totsuzenheni*
mutation

| 宀 3m | 室 容 宿 密 |
| + 6 to 12 strokes | 寄 寒 審 |

室外温度 *shitsugaiondo*
external temperature

室温 *shitsuon*
room temperature

容量 *youryou*
capacity

容量分析法 *youryoubunsekihou*
volumetric analysis

容量係数 *youryoukeisuu*
volumetric coefficient

容積 *youseki*
volume

容積百分率 *yousekihyakubunritsu*
volume percent

容積効率 *yousekikouritsu*
volumetric efficiency

容積測定 *yousekisokutei*
volume measurement

宿主 *shukushu*
host

密着 *micchaku*
adherence

密着性 *micchakusei*
adhesion

密着試験 *micchakushiken*
adherence test, adhesion testing

密着剤 *micchakuzai*
adhesive

密封 *mippuu*
sealing

密封技術 *mippuugijutsu*
sealing technology

密封加熱 *mippuukanetsu*
sealed heating

密度 *mitsudo*
density

密度不均一性 *mitsudofukinitsusei*
density inhomogeneity

密度勾配 *mitsudokoubai*
density gradient

密度測定法 *mitsudosokuteihou*
density measurement

寄生虫 *kiseichuu*
parasite

寒暖計 *kandankei*
thermometer

寒天 *kanten*
agar, agar-agar

寒剤 *kanzai*
freezing mixture

審査 *shinsa*
examination, investigation

⺌, 小 **3n**	小 当 劣 乳
(all *kanji*)	栄 蛍 螢 常

小麦澱粉 *komugidenpun*
wheat starch

小胞体 *shouhoutai*
endoplasmic reticulum

小糖 *shoutou*
oligosaccharide

当量 *touryou*
equivalent

劣化 *rekka*
degradation, decomposition, decay, deterioration

乳鉢 *nyuubachi*
mortar

乳白度 *nyuuhakudo*
opacity

乳白ガラス *nyuuhakugarasu*
milk glass, opal glass

乳化 *nyuuka*
emulsification

乳化安定剤 *nyuukaanteizai*
emulsion stabilizer

乳化法 *nyuukahou*
emulsion method

乳化重合 *nyuukajuugou*
emulsion polymerization

乳化重合反応 *nyuukajuugouhannou*
emulsion polymerization reaction

乳化重合速度論 *nyuukajuugousokudoron*
emulsion polymerization kinetics

乳化性油 *nyuukaseiyu*
emulsifying oil

乳化試験 *nyuukashiken*
emulsification test

乳化剤 *nyuukazai*
emulsifier

乳光 *nyuukou*
opalescence

乳酸 *nyuusan*
　　lactic acid $CH_3CH(OH)COOH$

乳酸エチル *nyuusanechiru*
　　ethyl lactate $CH_3CH(OH)COOC_2H_5$

乳酸塩 *nyuusanen*
　　lactate $CH_3CH(OH)COOM$

乳酸エステル *nyuusanesuteru*
　　lactic acid ester $CH_3CH(OH)COOR$

乳酸銀 *nyuusangin*
　　silver lactate $CH_3CH(OH)COOAg$

乳酸発酵 *nyuusanhakkou*
　　lactic acid fermentation

乳酸カリウム *nyuusankariumu*
　　potassium lactate $CH_3CH(OH)COOK$

乳酸カルシウム *nyuusankarushiumu*
　　calcium lactate $Ca(C_3H_5O_2)_2$

乳酸菌 *nyuusankin*
　　lactic acid bacteria

乳酸メチル *nyuusanmechiru*
　　methyl lactate $CH_3CH(OH)COOCH_3$

乳酸ナトリウム *nyuusannatoriumu*
　　sodium lactate $CH_3CH(OH)COONa$

乳酸リチウム *nyuusanrichiumu*
　　lithium lactate $CH_3CH(OH)COOLi$

乳酸測定 *nyuusansokutei*
　　lactate measurement

乳清 *nyuusei*
　　whey

乳脂 *nyuushi*
　　milk fat

乳糖 *nyuutou*
　　lactose $C_{12}H_{22}O_{11}$

乳剤 *nyuuzai*
　　emulsion

栄養物質 *eiyoubusshitsu*
　　nutrient

栄養学 *eiyougaku*
　　nutritional science

栄養価 *eiyouka*
　　nutritive value

栄養素 *eiyouso*
　　nutrient

蛍石 *hotaruishi*
　　fluorite CaF_2

蛍光 *keikou*
　　fluorescence

蛍光分光学法 *keikoubunkougakuhou*
　　fluorescence spectroscopy

蛍光分光計 *keikoubunkoukei*
　　fluorescence spectrometer

蛍光分析 *keikoubunseki*
　　fluorometry

蛍光物質 *keikoubusshitsu*
　　fluorescent substance

蛍光顕微鏡 *keikoukenbikyou*
　　fluorescence microscope

蛍光検出器 *keikoukenshutsuki*
　　fluorescence detector

蛍光光度分析 *keikoukoudobunseki*
　　fluorophotometry

蛍光強度 *keikoukyoudo*
　　fluorescence intensity

蛍光指示薬 *keikoushijiyaku*
　　fluorescence indicator

蛍光色 *keikoushoku*
　　fluorescent color

蛍光塗料 *keikoutoryou*
　　fluorescence paint

螢光 *keikou*
　　fluorescence

螢光板 *keikouban*
　　fluorescent screen

螢光分析 *keikoubunseki*
　　fluorescence analysis

螢光写真 *keikoushashin*
　　fluorography

螢光指示薬 *keikoushijiyaku*
　　fluorescence indicator

螢光スクリーン *keikousukuri-n*
　　fluorescent screen

螢光スペクトル　*keikousupekutoru*
　fluorescence spectrum

螢光灯　*keikoutou*
　fluorescent lamp

螢石　*keiseki*
　fluorite, calcium fluoride, fluorspar CaF_2

常圧　*jouatsu*
　normal pressure

常磁性　*joujisei*
　paramagnetism

常磁性共鳴　*joujiseikyoumei*
　paramagnetic resonance

常磁性体　*joujiseitai*
　paramagnetic substance

常磁性体　*joujiseitai*
　paramagnetic substance

常温　*jouon*
　normal temperature

山　3o (all *kanji*)	山　岩　崩　嵩

山口ラクトン化反応
　yamaguchirakutankahannou
　Yamaguchi lactonization

岩塩　*ganen*
　rock salt, halite NaCl

岩石学　*gansekigaku*
　petrology

岩石圏　*gansekiken*
　lithosphere

崩壊　*houkai*
　decay, degradation, disintegration

崩壊系列　*houkaikeiretsu*
　decay series

崩壊定数　*houkaiteisuu*
　decay constant

嵩　*kasa*
　bulk, volume, quantitiy

士　3p (all *kanji*)	吉

吉草　*kissou*
　valerian

吉草酸　*kissousan*
　valeric acid, pentanoic acid
　$CH_3(CH_2)_3COOH$

吉草酸エチルエステル
　kissousanechiruesuteru
　ethyl valerate $CH_3(CH_2)_3COOC_2H_5$

广　3q (all *kanji*)	応　度　座　麻
	廃　腐　摩　魔

応力　*ouryoku*
　stress, tension

応力解析　*ouryokukaiseki*
　stress analysis

応力緩和　*ouryokukanwa*
　stress relaxation

応答　*outou*
　response

応答時間　*outoujikan*
　response time

応用化学　*ouyoukagaku*
　applied chemistry

応用力学　*ouyourikigaku*
　applied mechanics

度　*do*
　degree

度数　*dosuu*
　frequency

座標　*zahyou*
　coordinate

座標軸　*zahyoujiku*
　coordinate axis

麻美油　*asamiyu*
　hempseed oil

麻酔薬　*masuiyaku*
anesthetic

麻酔剤　*masuizai*
anesthetic

麻薬　*mayaku*
narcotic drug

廃物　*haibutsu*
waste, garbage

廃液　*haieki*
waste liquid

廃ガス　*haigasu*
exhaust gas, waste gas

廃ガス分析　*haigasubunseki*
waste gas analysis

廃品　*haihin*
waste, gargabe

廃棄物　*haikibutsu*
waste, garbage

廃棄物回収　*haikibutsukaishuu*
waste recovery

廃棄物処理　*haikibutsushori*
waste treatment

廃棄材料　*haikizairyou*
waste material

廃熱　*hainetsu*
waste heat

廃熱ボイラー　*hainetsuboira-*
waste-heat boiler

廃水　*haisui*
waste water, sewage

廃水分析　*haisuibunseki*
waste-water analysis

廃水処理　*haisuishori*
wastewater treatment

廃油　*haiyu*
used oil, waste oil

腐泥　*fudei*
digested sludge

腐土素　*fudoso*
humin

腐敗　*fuhai*
putrefaction

腐敗菌　*fuhaikin*
putrefactive bacteria

腐乱　*furan*
decomposition

腐蝕　*fushoku*
corrosion

腐食　*fushoku*
corrosion

腐食性大気　*fushokuseitaiki*
corrosive atmosphere

腐食性薬品　*fushokuseiyakuhin*
corrosive chemical

腐食試験　*fushokushiken*
corrosion test

腐植質　*fushokushitsu*
humus

摩耗　*mamou*
abrasion, wear

摩耗量　*mamouryou*
abrasion loss

摩耗試験　*mamoushiken*
wear test, abrasion test

摩砕機　*masaiki*
mill

摩擦　*masatsu*
friction

摩擦感度　*masatsukando*
friction sensitivity

摩擦係数　*masatsukeisuu*
coefficient of friction

摩擦熱　*masatsunetsu*
heat of friction

摩擦試験　*masatsushiken*
friction test

摩擦抵抗　*masatsuteikou*
friction resistance

魔法瓶　*mahoubin*
thermos flask, vacuum flask

尸 3r 尿局屈属層
(all kanji)

尿酸　*nyousan*
　uric acid $C_5H_4N_4O_3$

尿素　*nyouso*
　urea H_2NCONH_2

尿素分解　*nyousobunkai*
　urea decomposition

尿素分析　*nyousobunseki*
　urea analysis

尿素合成　*nyousogousei*
　urea synthesis

尿素反応　*nyousohannou*
　urea reaction

尿素ホルムアルデヒド樹脂　*nyousohorumuarudehidojushi*
　urea formaldehyde resin

尿素樹脂　*nyousojushi*
　urea resin

尿素回路　*nyousokairo*
　urea cycle

尿素加水分解　*nyousokasuibunkai*
　urea hydrolysis

尿素共重合体　*nyousokyoujuugoutai*
　urea copolymer

尿素サイクル　*nyousosaikuru*
　urea cycle

局部の　*kyokubuno*
　local

局在化　*kyokuzaika*
　localization

屈折　*kussetsu*
　refraction

屈折角　*kussetsukaku*
　fefracting angle

屈折計　*kussetsukei*
　refractometer

屈折計分析　*kussetsukeibunseki*
　refractometric analysis

屈折率　*kussetsuritsu*
　refractive index

屈折力　*kussetsuryoku*
　refraction power

属する　*zokusuru*
　to belong to

層　*sou*
　layer, film

層間化合物　*soukankagoubutsu*
　intercalation compound

層流熱伝達　*souryounetsudentatsu*
　laminar heat transfer

層流　*souryuu*
　laminar flow

ロ 3s 回因図国固面
(all kanji)

回帰分析　*kaikibunseki*
　regression analysis

回路　*kairo*
　circuit

回路計　*kairokei*
　circuit tester

回折　*kaisetsu*
　diffraction

回折角　*kaisetsukaku*
　diffraction angle

回折計　*kaisetsukei*
　diffractometer

回折格子　*kaisetsukoushi*
　diffraction grating

回折線　*kaisetsusen*
　diffraction line

回折スペクトル　*kaisetsusupekutoru*
　diffraction spectrum

回収　*kaishuu*
　recovery

回転　*kaiten*
　rotation, revolving, turning

回転電極　*kaitendenkyoku*
　rotating electrode

| 回転エネルギー | *kaitenenerugi-* |
| rotational energy | |

回転エネルギー *kaitenenerugi-*
　rotational energy

回転偏光 *kaitenhenkou*
　rotatory polarization

回転変流機 *kaitenhenryuuki*
　rotary converter

回転異性体 *kaiteniseitai*
　rotamer

回転軸 *kaitenjiku*
　axis of rotation

回転角 *kaitenkaku*
　angle of rotation

回転乾燥機 *kaitenkansouki*
　rotary evaporator

回転機 *kaitenki*
　rotator

回転粘度計 *kaitennendokei*
　rotation viscosimeter

回転ポンプ *kaitenponpu*
　rotary pump

回転流量計 *kaitenryuuryoukei*
　rotameter

回転子 *kaitenshi*
　rotor

回転振動スペクトル
kaitenshindousupekutoru
　rotation-vibration spectrum

回転スペクトル *kaitensupekutoru*
　rotation spectrum

回転対称 *kaitentaishou*
　rotational symmetry

因子 *inshi*
　factor

因子分析 *inshibunseki*
　factor analysis

因数分解 *insuubunkai*
　factorization

図表 *zuhyou*
　graph, diagram

国際単位 *kokusaitani*
　international unit

固着 *kochaku*
　anchoring

固着剤 *kochakuzai*
　sticking agent

固化 *koka*
　caking

固形 *kokei*
　solid

固形分 *kokeibun*
　solid content

固相 *kosou*
　solid phase

固相液相平衡 *kosouekisouheikou*
　solid-liquid phase equilibrium

固相反応 *kosouhannou*
　solid-phase reaction, solid-state reaction

固相法 *kosouhou*
　solid phase method

固体 *kotai*
　solid

固体電解質 *kotaidenkaishitsu*
　solid electrolyte

固体塩基 *kotaienki*
　solid base

固体燃料 *kotainenryou*
　solid fuel

固体酸 *kotaisan*
　solid acid

固定 *kotei*
　fixation

固定窒素 *koteichisso*
　fixed nitrogen

固定化 *koteika*
　immobilization

固定化酵素 *koteikakouso*
　immobilized enzyme

固定化触媒 *koteikashokubai*
　fixed catalyst

固定相 *koteisou*
　stationary phase

固定炭素 *koteitanso*
　fixed carbon

固溶体　*koyoutai*
solid solution

固有値　*koyuuchi*
eigenvalue

固有接着　*koyuusecchaku*
specific adhesion

固有性　*koyuusei*
characteristic property

面外振動　*mengaishindou*
out-of-plane vibration

面心格子　*menshinkoushi*
face-centered lattice

面心立方格子　*menshinrippoukoushi*
face-centered cubic lattice

木　4a	木 材 林 析 松 枝
+ 0 to 5 strokes	相 査 柔 染 架

木化　*mokuka*
lignification

木炭　*mokutan*
charcoal

木タール　*mokuta-ru*
wood tar

木材　*mokuzai*
wood

木材高分子　*mokuzaikoubunshi*
wood polymer

木材熱分解　*mokuzainetsubunkai*
wood pyrolysis

木材パルプ　*mokuzaiparupu*
wool pulp

木材リグニン　*mokuzairigunin*
wood lignin

木材粒子　*mokuzairyuushi*
wood particle

木材製品　*mokuzaiseihin*
wood product

木材繊維　*mokuzaiseni*
wood fiber

材料　*zairyou*
material

林檎酸　*ringosan*
malic acid HOOCCH(OH)CH$_2$COOH

析出　*sekishutsu*
separation, isolation

析出電位　*sekishutsudeni*
deposition potential

松根油　*shoukonyu*
pine oil

枝分れ　*edawakare*
branching

枝分れ分子　*edawakarebunshi*
branched molecule

枝分れ鎖　*edawakarekusari*
branched chain

枝分れ度　*edawakeredo*
degree of branching

相　*sou*
phase

相分離　*soubunri*
phase separation

相同　*soudou*
homology

相互インダクタンス　*sougoindakutansu*
mutual inductance

相互リアクタンス　*sougoriakutansu*
mutual reactance

相互作用　*sougosayou*
interaction

相反法則　*souhanhousoku*
reciprocity law

相平衡　*souheikou*
phase equilibrium

相補結合　*souhoketsugou*
complementary bond

相補的な　*souhotekina*
complementary

相似　*souji*
analogy

相乗効果　*soujoukouka*
synergistic effect

相乗作用 *soujousayou*
synergism

相加平均 *soukaheikin*
arithmetical mean

相加効果 *soukakouka*
additive effect

相関 *soukan*
correlation

相関分析 *soukanbunseki*
correlation analysis

相関関数 *soukankansuu*
correlation function

相関係数 *soukankeisuu*
correlation coefficient

相関図 *soukanzu*
correlation diagram

相対原子質量 *soutaigenshishitsuryou*
relative atomic mass

相対配置 *soutaihaichi*
relative configuration

相対密度 *soutaimitsudo*
relative density

相対粘度 *soutainendo*
relative viscosity

相対性理論 *soutaiseiriron*
theory of relativity

相対湿度 *soutaishitsudo*
relative humidity

相対速度 *soutaisokudo*
relative velocity

相転移 *souteni*
phase transition

査定 *satei*
evaluation

柔軟剤 *juunanzai*
softener, softening agent

染料 *senryou*
dye, dyestuff

染色 *senshoku*
dyeing

染色堅ろう度 *senshokukenroudo*
color fastness, color permanence

染色体 *senshokutai*
chromosome

染浴 *senyoku*
dye bath

架橋 *kakyou*
cross-linking

架橋度 *kakyoudo*
degree of cross-linking

架橋剤 *kakyouzai*
cross-linking agent, vulcanizing agent

木 4a + 6 to 8 strokes	根 桂 桜 格 椰 極 棒 検 植

根茎 *konkei*
rhizome

桂皮アルデヒド *keihiarudehido*
cinnamaldehyde $C_6H_5CH=CHCHO$

桂皮アルコール *keihiaruko-ru*
cinnamyl alcohol $C_6H_5CH=CHCH_2OH$

桂皮酸 *keihisan*
cinnamic acid $C_6H_5CH=CHCOOH$

桜井反応 *sakuraihannou*
Sakurai reaction

格子 *koushi*
lattice

格子分光計 *koushibunkoukei*
grating spectrometer

格子エネルギー *koushienerugi-*
lattice energy

格子間原子 *koushikangenshi*
interstitial atom

格子間イオン *koushikanion*
interstitial ion

格子欠陥 *koushikekkan*
crystal defect, lattice defect

格子振動 *koushishindou*
lattice vibration

格子定数 *koushiteisuu*
lattice constant

椰子油　*yashiyu*
   coconut oil

極　*kyoku*
   pole

極値　*kyokuchi*
   extreme value

極性　*kyokusei*
   polarity

極性分子　*kyokuseibunshi*
   polar molecule

極性化合物　*kyokuseikagoubutsu*
   polar compound

極性結合　*kyokuseiketsugou*
   polar bond, polar linkage

極性基　*kyokuseiki*
   polar group

極性効果　*kyokuseikouka*
   polar effect

極性溶媒　*kyokuseiyoubai*
   polar solvent

極紫外　*kyokushigai*
   extreme ultraviolet

棒状温度計　*boujouondokei*
   bar thermometer, stem thermometer

検圧器　*kenatsuki*
   manometer

検光子　*kenkoushi*
   analyser

検温　*kenon*
   temperature measurement

検流計　*kenryuukei*
   galvanometer

検査　*kensa*
   inspection

検出　*kenshutsu*
   detection

検出限界　*kenshutsugenkai*
   detection limit, identification limit

検出方法　*kenshutsuhouhou*
   detection method

検出器　*kenshutsuki*
   detector

検糖器　*kentouki*
   saccharimeter

植物アルカロイド　*shokubutsuarukaroido*
   plant alkaloid

植物毒素　*shokubutsudokuso*
   phytotoxin

植物エキス　*shokubutsuekisu*
   plant extract

植物保護　*shokubutsuhogo*
   plant protection, crop protection

植物ホルモン　*shokubutsuhorumon*
   phytohormon

植物化学　*shokubutsukagaku*
   plant chemistry

植物生長調節剤　*shokubutsuseichouchousetsuzai*
   plant growth regulator

植物生化学　*shokubutsuseikagaku*
   plant biochemistry

植物性染料　*shokubutsuseisenryou*
   plant pigment

植物性ステロール　*shokubutsuseisutero-ru*
   vegetable sterol

植物性蛋白　*shokubutsuseitanpaku*
   plant protein

植物性油脂　*shokubutsuseiyushi*
   vegetable oils and fats

植物繊維　*shokubutsuseni*
   vegetable fiber

植物油　*shokubutsuyu*
   vegetable oil

木　4a
+ 10 to 12 strokes　　構　模　標　樟　横　機　樹　橋

構成　*kousei*
   constitution

構造　*kouzou*
   structure, constitution

構造分析　*kouzoubunseki*
   structural analysis

構造因子 *kouzouinshi*
structural factor

構造異性 *kouzouisei*
constitutional isomerism

構造異性体 *kouzouiseitai*
structural isomer

構造解析 *kouzoukaiseki*
structural analysis

構造活性相関 *kouzoukasseisoukan*
structure-activity relationship

構造粘性 *kouzounensei*
structural viscosity

構造式 *kouzoushiki*
structural formula

模型 *mokei*
model

模型酵素 *mokeikouso*
model enzyme

標準 *hyoujun*
standard, norm

標準圧 *hyoujunatsu*
normal pressure, atmospheric pressure

標準大気 *hyoujundaiki*
normal pressure, atmospheric pressure

標準電池 *hyoujundenchi*
normal element, standard cell

標準電位 *hyoujundeni*
standard potential

標準電極 *hyoujundenkyoku*
normal electrode

標準電極電位 *hyoujundenkyokudeni*
standard electrode potential

標準液 *hyoujuneki*
normal solution, standard solution

標準偏差 *hyoujunhensa*
standard deviation

標準状態 *hyoujunjoutai*
standard condition

標準化 *hyoujunka*
standardization

標準化する *hyoujunkasuru*
to standardize

標準燃料 *hyoujunnenryou*
reference fuel

標準温度 *hyoujunondo*
normal temperature

標準生成エンタルピー *hyoujunseiseientarupi-*
standard enthalpy of formation

標準試薬 *hyoujunshiyaku*
standard reagent

標準溶液 *hyoujunyoueki*
normal solution, standard solution

標識化合物 *hyoushikikagoubutsu*
labelled compound

標識付け *hyoushikizuke*
labelling

樟脳 *shounou*
camphor $C_{10}H_{16}O$

樟脳酸 *shounousan*
camphoric acid $C_{10}H_{16}O_4$

樟脳油 *shounouyu*
camphor oil

横断面 *oudanmen*
cross section

横座標 *yokozahyou*
abscissa

機械 *kikai*
machine

機械油 *kikaiabura*
machine oil

機械工学 *kikaikougaku*
mechanical engineering

機械的エネルギー *kikaitekienerugi-*
mechanical energy

機械的な *kikaitekina*
mechanical, mechanically

機械的性質 *kikaitekiseishitsu*
mechanical property

機器 *kiki*
instrument

機器分析 *kikibunseki*
instrumental analysis

機構 *kikou*
　mechanism

機能 *kinou*
　function

樹脂 *jushi*
　resin

樹脂化 *jushika*
　resinification

樹脂加工 *jushikakou*
　resin treatment

樹脂酸 *jushisan*
　resin acid

樹脂ワニス *jushiwanisu*
　resin varnish

樹脂油 *jushiyu*
　resin oil

橋かけ度 *hashikakedo*
　degree of cross-linking

橋かけ重合体 *hashikakejuugoutai*
　crosslinked polymer

橋かけ構造 *hashikakekouzou*
　bridged structure

橋かけ炭化水素 *hashikaketankasuiso*
　bridged hydrocarbon

橋かけ剤 *hashikakezai*
　cross-linking agent

月　4b
+ 0 to 5 strokes ｜ 月 肝 肥 青 胞 胆 背

月長石 *gecchouseki*
　moonstone

肝油 *kanyu*
　liver oil

肥料 *hiryou*
　fertilizer

青黴 *aokabi*
　penicillium, blue mold

青銅 *seidou*
　bronze

青金石 *seikinseki*
　lazurite, lapis lazuri $Na_3Ca(Al_3Si_3O_{12})S$

青藍 *seiran*
　indigo blue

青酸 *seisan*
　hydrogen cyanide HCN

青酸中毒 *seisanchuudoku*
　cyanide poisoning

胞子 *houshi*
　spore

胆汁酸 *tanjuusan*
　bile acid

胆汁色素 *tanjuushikiso*
　bile pigment

背圧 *haiatsu*
　back pressure

月　4b
+ 6/7 strokes ｜ 脆 脂 骨 能 豚

脆化 *zeika*
　embrittlement

脆化温度 *zeikaondo*
　brittle temperature

脆化点 *zeikaten*
　brittle point

脂気 *aburake*
　greasiness, oiliness

脂肪 *shibou*
　fat

脂肪分解 *shiboubunkai*
　lipolysis

脂肪分解酵素 *shiboubunkaikouso*
　lipase

脂肪含量 *shibouganryou*
　fat content

脂肪変性 *shibouhensei*
　fat degeneration

脂肪硬化 *shiboukouka*
　fat hardening

脂肪光沢 *shiboukoutaku*
  greasy luster

脂肪細胞 *shibousaibou*
  fat cell

脂肪酸 *shibousan*
  fatty acid

脂肪酸エステル *shibousanesuteru*
  fatty acid ester

脂肪層 *shibousou*
  layer of fat

脂肪体 *shiboutai*
  fat body

脂肪油 *shibouyu*
  fatty oil

脂肪族 *shibouzoku*
  aliphatics

脂肪族アルコール *shibouzokuaruko-ru*
  fatty alcohol

脂肪族化合物 *shibouzokukagoubutsu*
  aliphatic compound

脂肪族の *shibouzokuno*
  aliphatic

脂肪族炭化水素 *shibouzokutankasuiso*
  aliphatic hydrocarbon

脂環式化合物 *shikanshikikagoubutsu*
  alicyclic compound

脂質 *shishitsu*
  lipid

脂質二重膜 *shishitsunijuumaku*
  lipid double membrane

脂質生成 *shishitsuseisei*
  lipogenesis

脂溶性の *shiyouseino*
  fat soluble

骨粉 *koppun*
  bone meal

骨脂 *kosshi*
  bone fat

骨油 *kotsuyu*
  bone oil

骨材 *kotsuzai*
  aggregate

骨髄腫 *kotsuzuishu*
  myeloma

能動輸送 *noudouyousou*
  active transport

能力 *nouryoku*
  competence, ability

豚脂 *tonshi*
  lard

月 4b	膜 静
+ 10 strokes	

膜 *maku*
  membrane, film

膜安定性 *makuanteisei*
  membrane stability

膜圧 *makuatsu*
  membrane pressure

膜分離 *makubunri*
  membrane separation

膜電位 *makudeni*
  membrane potential

膜電極 *makudenkyoku*
  membrane electrode

膜現象 *makugenshou*
  membrane phenomenon

膜凝縮 *makugyoushuku*
  film condensation

膜平衡 *makuheikou*
  membrane equilibrium

膜表面 *makuhyoumen*
  membrane surface

膜蒸気分離 *makujoukibunri*
  membrane vapor separation

膜カラム *makukaramu*
  membrane column

膜構造 *makukouzou*
  membrane structure

膜クロマトグラフィー *makukuromatogurafi-*
  membrane chromatography

膜濾過器　*makurokaki*
　membrane filter

膜接触　*makusesshoku*
　membrane contact

膜振動　*makushindou*
　membrane vibration

膜浸透　*makushintou*
　membrane permeation

膜質の　*makushitsuno*
　membraneous

膜相互作用　*makusougosayou*
　membrane interaction

膜抵抗　*makuteikou*
　membrane resistance

膜透過性　*makutoukasei*
　membrane permeability

静圧　*seiatsu*
　static pressure

静電荷　*seidenka*
　static charge

静電気防止　*seidenkiboushi*
　antistatic

静電ポテンシャル　*seidenpotensharu*
　electrostatic potential

静電的荷電　*seidentekikaden*
　electrostatic charge

静電的な　*seidentekina*
　electrostatic

静電塗装　*seidentosou*
　electrostatic coating

静電容量　*seidenyouryou*
　electric capacity

静平衡　*seiheikou*
　static balance, static equilibrium

静力学　*seirikigaku*
　statics

静力学の　*seirikigakuno*
　static

静止電極　*seishidenkyoku*
　stationary electrode

静止エネルギー　*seishienerugi-*
　rest energy

静止位置　*seishiichi*
　rest position

静止状態　*seishijoutai*
　resting state

静止摩擦　*seishimasatsu*
　static friction

静止質量　*seishishitsuryou*
　rest mass

静水圧力　*seisuiatsuryoku*
　hydrostatic pressure, hydraulic pressure

月　**4b**	膠 膨
+ 11/12 strokes	

膠原質　*kougenshitsu*
　collagen

膠状　*koujou*
　colloid

膠質　*koushitsu*
　colloidal

膠質化学　*koushitsukagaku*
　colloid chemistry

膠質溶液　*koushitsuyoueki*
　colloidal solution

膠　*nikawa*
　glue

膨張　*bouchou*
　expansion, swelling, growth

膨張係数　*bouchoukeisuu*
　expansion coefficient

膨張率　*bouchouritsu*
　expansion coefficient, rate of expansion

膨張性　*bouchousei*
　expansibility

膨張剤　*bouchouzai*
　swelling substance, swelling agent

膨潤　*boujun*
　swelling

膨潤度　*boujundo*
　degree of swelling

膨潤比 *boujunhi*
swelling ratio

膨潤過程 *boujunkatei*
swelling process

膨潤速度論 *boujunsokudoron*
swelling kinetics

膨潤測定 *boujunsokutei*
swelling measurement

膨化 *bouka*
swelling

膨化能 *boukanou*
swelling capacity

膨化剤 *boukazai*
swelling agent

日 4c   日 白 百 明 昇
+ 0 to 4 strokes

日光 *nikkou*
daylight

日光照射 *nikkoushousha*
daylight radiation

日本酸 *nipponsan*
Japanic acid $HOOC(CH_2)_{19}COOH$

白血球 *hakkekkyuu*
leucocyte

白血病 *hakketsubyou*
leukemia

白金 *hakkin*
platinum (Pt, element 78)

白金電極 *hakkindenkyoku*
platinum electrode

白金粉末 *hakkinmatsudo*
platinum powder

白亜 *hakua*
chalk $CaCO_3$

白土 *hakudo*
clay

白土濾過 *hakudoroka*
clay filtration

白鉛鉱 *hakuenkou*
white lead ore, cerussite $PbCO_3$

白燐弾 *hakurindan*
white phosphorous

白色顔料 *hakushokuganryou*
white pigment

白色光 *hakushokukou*
white light

白色セメント *hakushokusemento*
white cement

白雲母 *shirounmo*
muscovite $KAl_2[(OH,F)_2AlSi_3O_{10}]$

百分率 *hyakubunritsu*
percent, percentage

明度 *meido*
lightness, brightness

明反応 *meihannou*
light reaction

明礬 *myouban*
alum $KAl(SO_4)_2 \cdot 12H_2O$

明礬石 *myoubanseki*
alunite $KAl(SO_4)_2 \cdot Al(OH)_3$

昇華 *shouka*
sublimation

昇華物 *shoukabutsu*
sublimate

昇華温度 *shoukaondo*
sublimation temperature

昇華点 *shoukaten*
sublimation point

日 4c   昼 時 曹 匙 乾
+ 5 to 7 strokes

昼光 *chuukou*
daylight

昼光フィルタ *chuukoufiruta*
daylight filter

時 *ji*
hour

時間 *jikan*
  time

時間分解測定 *jikanbunkaisokutei*
  time-resolved measurement

時定数 *jiteisuu*
  time constant

曹長石 *souchouseki*
  albite, soda feldspar $Na[AlSi_3O_8]$

匙 *saji*
  spoon, spatula

乾電池 *kandenchi*
  dry cell

乾熱滅菌 *kannetsumekkin*
  hot-air sterilization

乾留 *kanryuu*
  dry distillation, carbonization

乾性 *kansei*
  dryness

乾性ガス *kanseigasu*
  dry gas

乾式法 *kanshikihou*
  dry process

乾式試金 *kanshikishikin*
  dry assay

乾燥 *kansou*
  drying, desiccation

乾燥物質 *kansoubusshitsu*
  dry substance, dry matter

乾燥減量 *kansougenryou*
  loss on drying

乾燥時間 *kansoujikan*
  drying time

乾燥重量 *kansoujuuryou*
  dry weight

乾燥管 *kansoukan*
  drying tube

乾燥器 *kansouki*
  dryer, desiccator

乾燥機 *kansouki*
  dryer, desiccator

乾燥酵母 *kansoukoubo*
  dry yeast

乾燥密度 *kansoumitsudo*
  dry density

乾燥炉 *kansouro*
  drying oven, drying kiln, drying stove

乾燥室 *kansoushitsu*
  drying chamber

乾燥装置 *kansousouchi*
  drying equipment

乾燥する *kansousuru*
  to dry, to desiccate

乾燥塔 *kansoutou*
  drying tower

乾燥剤 *kansouzai*
  drying agent, desiccant

| 日 4c | 晶 量 最 皓 |
| +8 to 12 strokes | 暗 曇 |

晶化 *shouka*
  crystallization

晶析 *shouseki*
  crystallization

晶析装置 *shousekisouchi*
  crystallizer

晶子 *shoushi*
  crystallite

晶質 *shoushitsu*
  crystalloid

量 *ryou*
  quantity, weight, amount, volume

量感 *ryoukan*
  volume

量産 *ryousan*
  mass production

量子 *ryoushi*
  quantum

量子物理学 *ryoushibutsurigaku*
  quantum physics

量子化　ryoushika
　quantization

量子化学　ryoushikagaku
　quantum chemistry

量子化学解析　ryoushikagakukaiseki
　quantum-chemical analysis

量子欠損　ryoushikesson
　quantum defect

量子効率　ryoushikouritsu
　quantum efficiency

量子力学　ryoushirikigaku
　quantum mechanics

量子論　ryoushiron
　quantum theory

量子サイズ　ryoushisaizu
　quantum size

量子収率　ryoushishuuritsu
　quantum yield

量子収量　ryoushishuuryou
　quantum yield

量子数　ryoushisuu
　quantum number

量子転移　ryoushiteni
　quantum transition

量子流体　ryoushiyuutai
　quantum fluid

量増　ryouzou
　weight increase

最も　itomo
　extremely

最高被占分子軌道　saikouhisenbunshikidou
　highest occupied molecular orbital (HOMO)

最高被占軌道　saikouhisenkidou
　highest occupied molecular orbital (HOMO)

最終産物　saishuusanbutsu
　end product

最低空軌道　saiteikuukidou
　lowest unoccupied molecular orbital (LUMO)

最適温度　saitekiondo
　optimum temperature

最適pH　saitekipH
　optimum pH-value

皓礬　kouban
　epsomite, bitter salt $ZnSO_4 \cdot 7H_2O$

暗反応　anhannou
　dark reaction

曇り　kumori
　haze, clouding

曇り点　kumoriten
　cloud point

曇り点曲線　kumoritenkyokusen
　cloud point curve

火, 灬　4d　火 灯 炉 炎 黒
+ 0 to 7 strokes

火花放電　hibanahouden
　spark discharge

火花点火　hibanatenka
　spark ignition

火入れ　hiire
　pasteurization

火金石　kakinseki
　pyroaurite $Mg_6Fe^{3+}{}_2(CO_3)(OH)_{16} \cdot 4H_2O$

火力　karyoku
　heating power

火山灰　kazanbai
　volcanic ash

火山岩　kazangan
　volcanic rock, vulcanite

灯用ガス　touyougasu
　illuminating gas

灯油　touyu
　kerosine

炉　ro
　furnace, kiln

炎光光度計　enkoukoudokei
　flame photometer

炎症　enshou
　　inflammation

黒度　kokudo
　　blackness

黒銅鉱　kokudoukou
　　melaconit, tenorite CuO

黒鉛　kokuen
　　graphite

黒鉛電極　kokuendenkyoku
　　graphite electrode

黒鉛化　kokuenka
　　graphitization

黒燐　kokurin
　　black phosphorus

黒色火薬　kokushokukayaku
　　black powder, blasting powder

黒色酸化銅　kokushokusankadou
　　black copper oxide CuO

黒体　kokutai
　　black body

黒金剛石　kurokongouseki
　　black diamond, carbonade, boort

黒辰砂　kuroshinsa
　　metacinnabar HgS

黒雲母　kurounmo
　　biotite $K(Mg,Fe,Mn)_3(OH,F)_2[AlSi_3O_{10}]$

火, 灬　4d
+ 8 to 15 strokes

煙 焼 焔 煮 煩
照 熟 爆

煙道ガス　endougasu
　　flue gas

焼結炉　shouketsuro
　　sintering furnace

焼成　shousei
　　firing, burning, baking

焼成度　shouseido
　　burning degree

焼成温度　shouseiondo
　　burning temperature

焼入れ　yakiire
　　hardening, tempering

焼戻し　yakimodoshi
　　annealing, tempering

焔色分析　enshokubunseki
　　flame analysis

煮立つ　nitatsu
　　boil

煮沸　shafutsu
　　boiling

煩瑣　hansa
　　complicated

煩雑　hanzatsu
　　complicated

照度計　shoudokei
　　illumination meter

照合電極　shougoudenkyoku
　　reference electrode

照射　shousha
　　irradiation

照射線量　shoushasenryou
　　exposure

照射する　shoushasuru
　　to irradiate

熟成　jukusei
　　aging

爆発　bakuhatsu
　　explosion

爆発物　bakuhatsubutsu
　　explosive substance

爆発限界　bakuhatsugenkai
　　explosion limit

爆発反応　bakuhatsuhannou
　　explosive reaction

爆発性混合物　bakuhatsuseikongoubutsu
　　explosive mixture

爆発性の　bakuhatsuseino
　　explosive

爆発室　bakuhatsushitsu
　　combustion chamber

爆燃　bakunen
　　deflagration

ネ,示 4e | 示 神 禁
(all kanji)

示差熱分析 *shisanetsubunseki*
  differential thermal analysis (DTA)

示差温度計 *shisaondokei*
  differential thermometer

示差走査熱量測定 *shisasousanetsuryousokutei*
  differential scanning calorimetry (DSC)

示差滴定 *shisatekitei*
  differential titration

神経 *shinkei*
  nerve

神経伝達物質 *shinkeidentatsubusshitsu*
  neurotransmitter

神経系 *shinkeikei*
  nervous system

神経細胞 *shinkeisaibou*
  nerve cell, neurone

禁制原理 *kinseigenri*
  exclusion principle

禁制線 *kinseisen*
  forbidden line

禁制遷移 *kinseiseni*
  forbidden transition

禁制スペクトル *kinseisupekutoru*
  forbidden spectrum

禁止物質 *kinshibusshitsu*
  prohibited substance

禁止剤 *kinshizai*
  inhibitor

王 4f | 王 理 球 現
+ 1 to 7 strokes

王水 *ousui*
  aqua regia, nitrohydrochloric acid

玉形弁 *tamagataben*
  globe valve

主値 *shuchi*
  principal value

主軸 *shujiku*
  principle axis

主量子数 *shuryoushisuu*
  principal quantum number

主鎖 *shusa*
  principal chain

主成分 *shuseibun*
  principal component

主製品 *shuseihin*
  main produc

主スイッチ *shuusuicchi*
  main switch

理化学 *rikagaku*
  physics and chemistry

理論 *riron*
  theory

理論化学 *rironkagaku*
  theoretical chemistry

理論収量 *rironshuuryou*
  theoretical yield

理論的基礎 *rirontekikiso*
  theoretical basis

理論的モデル化 *rirontekimoderuka*
  theoretically modelling

理論的特性評価 *rirontekitokuseihyouka*
  theoretical characterization

理想液体 *risouekitai*
  idealized liquid

理想状態 *risoujoutai*
  ideal state

理想結晶 *risoukesshou*
  perfect crystal, ideal crystal

理想気体 *risoukitai*
  ideal gas

理想混合物 *risoukongoubutsu*
  ideal mixture

理想高分子鎖 *risoukoubunshisa*
  ideal polymer chain

理想吸着 *risoukyuuchaku*
  ideal adsorption

理想流体　*risouryuutai*
　ideal fluid

理想溶液　*risouyoueki*
　ideal solution

球状蛋白質　*kyuujoutanpakushitsu*
　globular protein

現像核　*genzoukaku*
　development nucleus

現像剤　*genzouzai*
　developer

王　4f
+ 8 to 13 strokes ｜ 琥 瑪 環

琥珀　*kohaku*
　succinite, amber $C_{40}H_{64}O_4$

琥珀酸　*kohakusan*
　succinic acid $HOOCCH_2CH_2COOH$

瑪瑙　*menou*
　agate

環　*kan*
　ring

環分裂　*kanbunretsu*
　ring cleavage

環分析　*kanbunseki*
　ring analysis

環電流　*kandenryuu*
　ring current

環外　*kangai*
　exocyclic

環状アルコール　*kanjouaruko-ru*
　cyclic alcohol

環状分子　*kanjoubunshi*
　cyclic molecule

環状電子反応　*kanjoudenshihannou*
　electrocyclic reaction

環状電子転位　*kanjoudenshiteni*
　electrocyclic rearrangement

環状エステル　*kanjouesuteru*
　cyclic ester

環状化合物　*kanjoukagoubutsu*
　cyclic compound

環状構造　*kanjoukouzou*
　cyclic structure

環状の　*kanjouno*
　cyclic

環状炭化水素　*kanjoutankasuiso*
　cyclic hydrocarbon

環化　*kanka*
　cyclization, annelation

環化重合　*kankajuugou*
　cyclization polymerization

環形成　*kankeisei*
　ring formation, cyclization

環境　*kankyou*
　environment

環境分析　*kankyoubunseki*
　environmental analysis

環境保護　*kankyouhogo*
　environmental protection

環境化学　*kankyoukagaku*
　environmental chemistry

環境研究　*kankyoukenkyuu*
　environmental research

環境汚染　*kankyouosen*
　environmental pollution

環境適応　*kankyoutekiou*
　environmental adaption

環式化合物　*kanshikikagoubutsu*
　cyclic compound

牛　4g
+ 0/4 strokes ｜ 牛 物

牛脂　*gyuushi*
　beef tallow

物産　*bussan*
　product

物質　*busshitsu*
　substance, material, compound

物質移動 busshitsuidou
mass transfer

物質移動係数 busshitsuidoukeisuu
mass transfer coefficient

物質移動速度 busshitsuidousokudo
mass-transfer rate

物質交代 busshitsukoutai
metabolism

物質量 busshitsuryou
amount of substance

物質収支 busshitsushuushi
mass balance

物質代謝 busshitsutaisha
metabolism

物理学 butsurigaku
physics

物理変化 butsurihenka
physical change

物理化学 butsurikagaku
physical chemistry

物理吸着 butsurikyuuchaku
physical adsorption, physisorption

物理性 butsurisei
physical property

物理的性質 butsuritekiseishitsu
physical property

物理的測定 butsuritekisokutei
physical measurement

牛 4g
+ 6/9 strokes

特 解

特許 tokkyo
patent

特許庁 tokkyochou
patent office

特許技術 tokkyogijutsu
patented technology

特許法 tokkyohou
patent law

特許権 tokkyoken
patent rights

特許クレーム tokkyokure-mu
claim

特許請求 tokkyoseikyuu
patent claim

特許侵害 tokkyoshingai
patent infringement

特許出願 tokkyoshutsugan
patent application

特級試薬 tokkyuushiyaku
extra pure reagent, special grade reagent

特徴 tokuchou
distinctive feature, characteristic

特異反応 tokuihannou
specific reaction

特異結合 tokuiketsugou
specific binding

特異性 tokuisei
specificity

特異試薬 tokuishiyaku
specific reagent

特異的吸着 tokuitekikyuuchaku
specific adsorption

特異的酸化 tokuitekisanka
specific oxidation

特異的性質 tokuitekiseishitsu
characteristic property

特性実験 tokujikken
characterization experiment

特産物 tokusanbutsu
special product, specialty

特産品 tokusanhin
special product, specialty

特性 tokusei
distinguishing feature, characteristic, property

特性反応 tokuseihannou
characteristic reaction

特性方程式 tokuseihouteishiki
characteristic equation

特性係数　*tokuseikeisuu*
　characteristic coefficient, characterization factor

特性基　*tokuseiki*
　characteristic group

特性曲線　*tokuseikyokusen*
　characteristic curve

特性温度　*tokuseiondo*
　characteristic temperature

特性振動数　*tokuseishindousuu*
　characteristic frequency

特性周波数　*tokuseishuuhasuu*
　characteristic frequency

特殊ゴム　*tokushugomu*
　speciality rubber

特殊品　*tokushuhin*
　specialties

特殊繊維　*tokushuseni*
　special fibre

特定同位体　*tokuteidouitai*
　specific isotope

特定異性体　*tokuteiiseitai*
　specific isomer

特定温度　*tokuteiondo*
　specific temperature

解毒　*gedoku*
　detoxication

解毒剤, 解毒薬　*gedokuzai*
　antidote

解熱剤, 解熱薬　*genetsuzai, genetsuyaku*
　antipyretic

解重合　*kaijuugou*
　depolymerization

解重合剤　*kaijuugouzai*
　depolymerizing agent

解決　*kaiketsu*
　solution (of a problem)

解離　*kairi*
　dissociation, disaggregation

解離度　*kairido*
　degree of dissociation

解離エネルギー　*kairienerugi-*
　dissociation energy

解離熱　*kairinetsu*
　heat of dissociation

解離力　*kairiryoku*
　dissociation power

解離定数　*kairiteisuu*
　dissociation constant

解析　*kaiseki*
　analysis

解糖系　*kaitoukei*
　glycolysis

解糖作用　*kaitousayou*
　glycolysis

解像力　*kaizouryoku*
　resolving power

方　4h　方 旋
(all *kanji*)

方法　*houhou*
　method, procedure, technique

方解石　*houkaiseki*
　calcite, calcspar $CaCO_3$

方程式　*houteishiki*
　equation

旋光　*senkou*
　optical rotation

旋光分散　*senkoubunsan*
　optical rotatory dispersion, ORD

旋光分析　*senkoubunseki*
　polarimetric analysis

旋光角　*senkoukaku*
　angle of rotation

旋光計　*senkoukei*
　polarimeter

旋光性　*senkousei*
　optical activity

旋光性物質　*senkouseibusshitsu*
　optically acitve substance

欠 4i	
欠 4j	処 改 麦 致 散 数 整 欠
(all *kanji*)	

処理 *shori*
treatment

改良品 *kairyouhin*
improved product

改質 *kaishitsu*
reforming

麦芽 *bakuga*
malt

麦芽エキス *bakugaekisu*
malt extract

麦芽糖 *bakugatou*
maltose, maltobiose, malt sugar
$C_{12}H_{22}O_{11}$

致死の *chishino*
lethal, deadly

致死量 *chishiryoo*
lethal dose

散光 *sankou*
diffused light

散布 *sanpu*
dispersion, scattering

散布する *sanpusuru*
to disperse

散乱 *sanran*
scattering

散剤 *sanzai*
powder

数 *kazu*
number

数値解析 *suuchikaiseki*
numerical analysis

数学 *suugaku*
mathematics

数学の問題 *suugakunomondai*
mathematical problem

数学的確率 *suugakutekikakuritsu*
mathematical probability

数学的計算 *suugakutekikeisan*
mathematical calculation

整流 *seiryuu*
rectification

整流管 *seiryuukan*
rectifier tubes

整流器 *seiryuuki*
rectifier

整数 *seisuu*
integer

整数定数 *seisuuteisuu*
integer constant

欠陥 *kekkan*
defect

忄, 心 4k	
(all *kanji*)	性 恒 情 感 慣 懸

性ホルモン *seihorumon*
sex hormone

性能試験 *seinoushiken*
performance test

恒温 *kouon*
constant temperature

恒温槽 *kouonsou*
incubator, thermostat bath

恒量 *kouryou*
constant weight

恒湿器 *koushitsuki*
humidistat

恒数 *kousuu*
constant

情報 *jouhou*
information, data

感度 *kando*
sensitivity, sensibility

感度分析 *kandobunseki*
sensibility analysis

感受率 *kanjuritsu*
susceptibility

感光計 *kankoukei*
   actinometer

感光性 *kankousei*
   photosensitivity

感光性樹脂 *kankouseijushi*
   photosensitive resin

感光セル *kankouseru*
   photosensitive cell

感染 *kansen*
   infection

慣性 *kansei*
   inertness

慣性係数 *kanseikeisuu*
   inertia factor

慣性モーメント *kanseimo-mento*
   moment of inertia

懸濁物質 *kendakubusshitsu*
   suspended matter

懸濁液 *kendakueki*
   suspension

懸濁重合 *kendakujuugou*
   suspension polymerization

戈, 弋   4n | 成 式
(all *kanji*)

成分 *seibun*
   ingredient, component, constituent

成分分析 *seibunfunseki*
   component analysis

成長ホルモン *seichouhorumon*
   growth hormone

成長因子 *seichouinshi*
   growth factor

成長解析 *seichoukaiseki*
   growth analysis

成形 *seikei*
   molding

成形プレス *seikeipuresu*
   molding press

式 *shiki*
   formula

式量 *shikiryou*
   formula weight

石   5a | 研 砂
+ 4 strokes

研究 *kenkyuu*
   research

研究チーム *kenkyuuchi-mu*
   research team

研究費 *kenkyuuhi*
   research costs

研究方法 *kenkyuuhouhou*
   resarch method

研究員 *kenkyuuin*
   researcher

研究会 *kenkyuukai*
   research society

研究開発 *kenkyuukaihatsu*
   research and development

研究活動 *kenkyuukatsudou*
   research activity

研究結果 *kenkyuukekka*
   research result

研究論文 *kenkyuuronbun*
   research thesis

研究領域 *kenkyuuryouiki*
   research area, research field

研究生 *kenkyuusei*
   research student

研究者 *kenkyuusha*
   researcher

研究室 *kenkyuushitsu*
   research laboratory

研究室棟 *kenkyuushitsutou*
   research laboratory building

研究所 *kenkyuusho*
   research institute

研磨 kenma
  polishing, sanding, grinding

研磨剤 kenmazai
  abrasive

研削材 kensakuzai
  abrasive

砂岩 sagan
  sandstone

砂金 sakin
  gold dust

砂鉄 satetsu
  iron sand

砂糖 satou
  sugar

砂浴 sayoku
  sand bath, sand-heat

砂 suna
  sand

石 5a
+ 5 to 10 strokes

破 硬 磁 確

破断 hadan
  break, fracture

破断点 hadanten
  fracture point

破壊 hakai
  fracture, breakdown

破壊試験 hakaishiken
  destruction test

破壊点 hakaiten
  failure point

破裂 haretsu
  bursting, explosion

破裂圧力 haretsuatsuryoku
  bursting pressure

破裂試験 haretsushiken
  bursting test

破砕 hasai
  crushing

破砕圧力 hasaiatsuryoku
  fracture pressure

破砕機 hasaiki
  crusher

砥粒加工 shirsuukakou
  abrasive machining

硬い塩基 kataienki
  hard base

硬い酸 kataisan
  hard acid

硬さ katasa
  hardness

硬度 koudo
  hardness

硬化 kouka
  setting, hardening, curing

硬化時間 koukajikan
  setting time

硬化脂肪 koukashibou
  hardened fat

硬化浴 koukayoku
  hardening bath

硬化油 koukayu
  hardened oil, hydrogenated oil

硬化剤 koukazai
  hardening agent

硬質ガラス koushitsugarasu
  hard glass

硬質ゴム koushitsugomu
  hard rubber

硬質陶器 koushitsutouki
  ironstone china

硬水 kousui
  hard water

磁場 jiba
  magnetic field

磁化 jika
  magnetization

磁界 jikai
  magnetic field

磁器 jiki
  porcelain

磁気 *jiki*
    magnetism

磁気分析 *jikibunseki*
    magnetochemical analysis

磁気異方性 *jikiihousei*
    magnetic anisotropy

磁気化学 *jikikagaku*
    magnetochemistry

磁気回転比 *jikikaitenhi*
    gyromagnetic ratio

磁気モーメント *jikimo-mento*
    magnetic moment

磁気量子数 *jikiryoushisuu*
    magnetic quantum number, total angular momentum quantum number

磁流鉄鉱 *jiryuutekkou*
    pyrrhotite

磁性 *jisei*
    magnetism

磁製乳鉢 *jiseinyuubachi*
    porcelain mortar

磁製ロート *jiseiro-to*
    porcelain funnel

磁製漏斗 *jiseirouto*
    porcelain funnel

磁製皿 *jiseizara*
    porcelain dish

磁石 *jishaku*
    magnet

磁子 *jishi*
    magneton

磁束 *jisoku*
    magnetic flux

磁束密度 *jisokumitsudo*
    magnetic flux density, magnetic induction

磁鉄鉱 *jitekkou*
    magnetite, magnetic iron ore FeO·Fe$_2$O$_3$

確認 *kakunin*
    identification

確率論 *kakuritsuron*
    theory of probability

立 5b 音 竜 産 辞 親 競
(all *kanji*)

音波破壊 *onpahakai*
    ultrasonic disintegration

竜脳 *ryuunou*
    borneol C$_{10}$H$_{17}$OH

産業 *sangyou*
    industry

産業廃棄物 *sangyouhaikibutsu*
    industrial waste

産業廃水 *sangyouhaisui*
    industrial waste water, industrial sewage

産業化学製品 *sangyoukagakuseihin*
    industrial chemical

辞書 *jisho*
    dictionary

親液性 *shinekisei*
    lyophilic

親水基 *shinsuiki*
    hydrophilic group

親水性の *shinsuiseino*
    hydrophilic

親和係数 *shinwakeisuu*
    affinity coefficient

親和定数 *shinwakeisuu*
    affinity constant

親和力 *shinwaryoku*
    affinity

親和性 *shinwasei*
    affinity

親油基 *shinyuki*
    lipophilic group

競争 *kyousou*
    competition

競争技術 *kyousougijutsu*
    competing technology

競争反応 *kyousouhannou*
    competitive reaction

競争阻害 *kyousousogai*
    competitive inhibition

目 5c
+ 0/1 strokes

目盛 *memori*
  graduation

目視観測 *mokushikansoku*
  observation, examination

目視検査 *mokushikensa*
  inspection

自動ビュレット *jidoubyuretto*
  automatic buret

自動ピペット *jidoupipetto*
  automatic pipet

自動酸化 *jidousanka*
  autoxidation

自動的な *jidoutekina*
  automatic

自動滴定 *jidoutekitei*
  automatic titration

自発反応 *jihatsuhannou*
  spontaneous reaction

自壊 *jikai*
  disintegration, degradation, decomposition, breakdown

自己分解 *jikobunkai*
  autolysis

自己放電 *jikohouden*
  self-discharge

自己インダクタンス *jikoindakutansu*
  self-induction

自己会合 *jikokaigou*
  self-association

自己吸収 *jikokyuushuu*
  self-absorption

自己免疫 *jikomeneki*
  autoimmunity

自己失活 *jikoshikkatsu*
  self-deactivation, self-quenching

自触媒反応 *jishokubaihannou*
  autocatalytic reaction

自触媒作用 *jishokubaisayou*
  autocatalysis

自溶 *jiyou*
  autogenous welding

自由電子 *jiyuudenshi*
  free electron

自由度 *jiyuudo*
  degree of freedom

自由エネルギー *jiyuuenerugi-*
  free energy

自由エンタルピ *jiyuuentarupi*
  free enthalpy

自由表面 *jiyuuhyoumen*
  free surface

自由回転 *jiyuukaiten*
  free rotation

自由水 *jiyuusui*
  free water

自然物 *shizenbutsu*
  natural product

自然現象 *shizengenshou*
  natural phenomena

自然発火 *shizenhakka*
  spontaneous ignition

自然法 *shizenhou*
  natural law

自然科学 *shizenkagaku*
  natural science

自然界 *shizenkai*
  nature, the natural world

自然律 *shizenritsu*
  natural law

自然数 *shizensuu*
  natural number, nonnegative integer

自然対数 *shizentaisuu*
  natural logarithm

目 5c
+ 2 to 9 strokes

見本 *mihon*
  sample

見掛け気孔率 *mikakekikouritsu*
apparent porosity

見掛け密度 *mikakemitsudo*
apparent density

見掛け粘度 *mikakenendo*
apparent viscosity

省エネルギー *shouenerugi-*
energy conservation

規格 *kikaku*
specification

規定 *kitei*
instruction

規定度 *kiteido*
normality

規定液 *kiteieki*
normal solution, standard solution

規定濃度 *kiteinoudo*
normal concentration

導電性ゴム *doudenseigomu*
conductive rubber

導電性表面 *doudenseihyoumen*
conducting surface

導電性樹脂 *doudenseijushi*
conductive resin

導電性高分子 *doudenseikoubunshi*
conducting polymer

導電性繊維 *doudenseiseni*
conductive fiber

導熱率 *dounetsuritsu*
thermal conductivity

導体 *doutai*
conductor

導体材料 *doutaizairyou*
conductor material

禾 5d	秒 科 香 秩 移 程 稀 積
(all *kanji*)	

秒 *byou*
second

科学 *kagaku*
science

科学院 *kagakuin*
science institute

科学者 *kagakusha*
scientist

科学雑誌 *kagakuzasshi*
science magazine

香気 *kouki*
aroma, flavor

香料 *kouryou*
flavour

香辛料 *koushinryou*
spice

秩序 *chitsujo*
order

移動 *idou*
transfer

程度 *teido*
extend, degree, level

稀硫酸 *kiryuusan*
diluted sulfuric acid

稀酸 *kisan*
diluted acid

稀釈 *kishaku*
dilution

稀釈熱 *kishakunetsu*
dilution heat

稀釈する *kishakusuru*
to dilute

稀釈剤 *kishakuzai*
diluent

稀溶液 *kiyoueki*
diluted solution

積分 *sekibun*
integral

積分学 *sekibungaku*
integral calculus

積分方程式 *sekibunhouteishiki*
integral equation

ネ 5e	初 装 補
+ 2 to 7 strokes	

初充電　*shojuuden*
　initial charge

初期値　*shokichi*
　initial value

初期条件　*shokijouken*
　initial condition

初期状態　*shokijoutai*
　initial state

初期化　*shokika*
　initialization

初期設定　*shokisettei*
　initialization

初速度　*shosokudo*
　initial rate, initial velocity

装置　*souchi*
　apparatus, equipment

補外　*hogai*
　extrapolation

補助電極　*hojodenkyoku*
　auxiliary electrode

補助化合物　*hojokagoubutsu*
　auxiliary compound

補助基　*hojoki*
　auxiliary group

補助剤　*hojozai*
　adjuvant

補助材料　*hojozairyou*
　auxiliary material

補間　*hokan*
　interpolation

補酵素　*hokouso*
　coenzyme

補強性　*hokyoutai*
　reinforcement

補強材　*hokyouzei*
　reinforcing material

補力　*horyoku*
　reinforcement, intensification, strenghtening

補力剤　*horyokuzai*
　intensifier

補色　*hoshoku*
　complementary color

補償　*hoshou*
　compensation

補償法　*hoshouhou*
　compensation method

補償点　*hoshouten*
　compensation point

補体　*hotai*
　complement

補体活性化　*hotaikasseika*
　complement activation

補体レセプター　*hotaireseputa-*
　complement receptor

ネ 5e	製 複
+ 8/9 strokes	

製品　*seihin*
　product

製品安全性　*seihinanzensei*
　product safety

製品分析　*seihinbunseki*
　product analysis

製品品質　*seihinhinshitsu*
　product quality

製品開発　*seihinkaihatsu*
　product development

製品研究　*seihinkenkyuu*
　product research

製品仕様書　*seihinshiyousho*
　product specification

製品収量　*seihinshuuryou*
　product yield

製品特性 *seihintokusei*
 product characteristics

製法 *seihou*
 preparation

製作 *seisaku*
 production, manufacturing

製作費 *seisakukhi*
 production costs

製造 *seisou*
 production

製造物責任 *seizoubutsusekinin*
 product liability

製造方法 *seizouhouhou*
 synthesis method

製造工程 *seizoukoutei*
 production process

製造する *seizousuru*
 to produce

複塩 *fukuen*
 double salt

複合体 *fukugoutai*
 complex

複環式化合物 *fukukanshikikagoubutsu*
 bicyclic compound

複屈折 *fukukussetsu*
 double refraction

複酸化物 *fukusankabutsu*
 double oxide

複製 *fukusei*
 duplication, replication, reproduction

複製機 *fukuseiki*
 duplicator

複素環 *fukusokan*
 heterocycle

複素環式化合物 *fukusokanshikikagoubutsu*
 heterocyclic compound

褐炭 *kattan*
 brown coal

褐鉄鉱 *kattekkou*
 limonite, brown iron ore $Fe_2O_3 \cdot xH_2O$

田 5f
(all *kanji*) 界 卑 留 略 累 異

界面 *kaimen*
 boundary surface, interface

界面張力 *kaimenchouryoku*
 interface tension, surface tension

界面エネルギー *kaimenenerugi-*
 surface energy

界面現象 *kaimengenshou*
 interfacial phenomenon

界面反応 *kaimenhannou*
 interface reaction

界面化学 *kaimenkagaku*
 surface chemistry

界面活性 *kaimenkassei*
 surface activity

界面活性剤 *kaimenkasseizai*
 tenside, surfactant

卑金属 *hikinzoku*
 base metal

留分 *ryuubun*
 fraction

留出物, 留出液
 ryuushutsubutsu, *ryuushutsueki*
 distillate

略図 *ryakuzu*
 schematic diagram, schematic drawing

累算器 *ruisanki*
 accumulator

累算温度 *ruisanondo*
 cumulative temperature

累積二重結合 *ruisekinijuuketsugou*
 cumulated double bond

異方性 *ihousei*
 anisotropy

異方性液体 *ihouseiekitai*
 anisotropic liquid

異常反応 *ijouhannou*
 anomalous reaction

異化作用 *ikasayou*
catabolism

異極化合物 *ikyokukagoubutsu*
heteropolar compound

異極結合 *ikyokuketsugou*
heteropolar bond

異燐 *irin*
yellow phosphorus

異性 *isei*
isomerism

異性化 *iseika*
isomerization

異性化合物 *iseikagoubutsu*
isomeric compound

異性化酵素 *iseikakouso*
isomerase

異性体 *iseitai*
isomer

異性体シフト *iseitaishifuto*
isomer shift

異種環化合物 *ishukankagoubutsu*
heterocyclic compound

異種溶解 *ishuyoukai*
heterolysis

異常 *ijou*
abnormality

□ 5g 置
+ 8 strokes

置換 *chikan*
substitution, replacement

置換アミド *chikanamido*
substituted amide

置換アリール *chikanari-ru*
substituted aryl

置換アルデヒド *chikanarudehido*
substituted aldehyde

置換アルコール *chikanaruko-ru*
substituted alcohol

置換ベンゼン *chikanbenzen*
substituted benzene

置換分 *chikanbun*
substituent

置換フェノール *chikanfeno-ru*
substituted phenol

置換複素環 *chikanfukusokan*
substituted heterocycle

置換反応 *chikanhannou*
substitution reaction

置換反応速度論 *chikanhannousokudoron*
substitution kinetics

置換ヘテロ芳香族化合物 *chikanheterohoukouzokukagoubutsu*
substituted heteroaromatic compound

置換法 *chikanhou*
substitution method

置換位置 *chikanichi*
substitution site

置換異性 *chikanisei*
substitution isomerism, position isomerism

置換異性体 *chikaniseitai*
substituted isomer

置換実験 *chikanjikken*
displacement experiment

置換重合体 *chikanjuugoutai*
substituted polymer

置換化合物 *chikankagoubutsu*
substituted compound

置換カルボン酸 *chikankarubonsan*
substituted carboxylic acid

置換基 *chikanki*
substituent

置換基導入 *chikankidounyuu*
substituent introduction

置換機構 *chikankikou*
substitution mechanism

置換基効果 *chikankikouka*
substituent effect

置換基制御 *chikankiseigyo*
substituent control

置換基定数　*chikankiteisuu*
　substituent constant

置換オレフィン　*chikanorefin*
　substituted olefin

置換ピリジン　*chikanpirijin*
　substituted pyridine

置換生成物　*chikanseiseibutsu*
　substitution product

置換速度　*chikansokudo*
　substitution rate

置換体　*chikantai*
　substitution product

血小板　*kesshouban*
　thrombocyte

血漿蛋白　*kesshoutanpaku*
　plasma protein

血液型　*ketsuekigata*
　blood group

血液凝固　*ketsuekigyouko*
　blood coagulation

血液検査　*ketsuekikensa*
　blood test

血石　*ketsuseki*
　heliotrope, bloodstone SiO_2

血糖　*kettou*
　blood sugar

皿　5h　｜　血
+ 1 stroke

血管　*kekkan*
　blood vessel

血球　*kekkyuu*
　blood cell

血清　*kessei*
　blood serum

血清アルブミン　*kesseiarubumin*
　serum albumin

血清学　*kesseigaku*
　serology

血清グロブリン　*kesseiguroburin*
　serum globulin

血清反応　*kesseihannou*
　serum reaction

血清カゼイン　*kesseikazein*
　serum caseine

血清蛋白　*kesseitanpaku*
　serum protein

血色素　*kesshikiso*
　hemoglobin

血しょう　*kesshou*
　blood plasma

血漿　*kesshou*
　plasma

疒　5i　｜　病 痕 癌
(all *kanji*)

病原細菌　*byougensaikin*
　pathogenic bacteria

病原糸状菌　*byougenshijoukin*
　pathogenic fungi

痕跡　*konseki*
　trace

痕跡分析　*konsekibunseki*
　trace analysis

癌　*gan*
　cancer, carcinoma

癌細胞　*gansaibou*
　cancer cell

糸状菌　*shijoukin*
　mold

糸　6a　｜　糸 系 紡 純 素
+ 0 to 4 strokes

系　*kei*
　line, series, system, group

紡績　*bouseki*
　spinning

純度 *jundo*
　purity, degree of purity

純度制御器 *jundoseigyoki*
　purity control

純度試験 *jundoshiken*
　purity test

純銀 *jungin*
　pure silver

純化 *junka*
　purification

純金 *junkin*
　pure gold, solid gold

純水 *junsui*
　pure water, purified water

素反応 *sohannou*
　elementary reaction

素粒子 *soryuushi*
　elementary particle

素材 *sozai*
　raw material

糸 6a　細 紺 組 終
+ 5 strokes

細度 *komakasa*
　fineness

細胞 *saibou*
　cell

細胞培養 *saiboubaiyou*
　cell culture

細胞分裂 *saiboubunretsu*
　cell division

細胞外酵素 *saibougaikouso*
　exoenzyme

細胞外の *saibougaino*
　extracellular

細胞壁 *saibouheki*
　cell wall

細胞核 *saiboukaku*
　cell nucleus

細胞間物質 *saiboukanbusshitsu*
　intercellular substance

細胞間液 *saiboukaneki*
　intercellular fluid

細胞間の *saiboukanno*
　intercellular

細胞器官 *saiboukikan*
　organelle

細胞膜 *saiboumaku*
　cell membrane

細胞内酵素 *saibounaikouso*
　endoenzyme

細胞内の *saibounaino*
　intracellular

細胞質 *saiboushitsu*
　cytoplasm

細胞質膜 *saiboushitsumaku*
　cell membrane

細胞周期 *saiboushuuki*
　cell cycle

細胞融合 *saibouyuugou*
　cell fusion

細粉 *saifun*
　fine powder

細管粘度計 *saikannendokei*
　capillary viscosimeter

細菌 *saikin*
　bacterium

細菌培養 *saikinbaiyou*
　bacterial culture

細菌毒素 *saikindokuso*
　bacterial toxin

細菌学 *saikingaku*
　bacteriology

細菌細胞 *saikinsaibou*
　bacterial cell

細孔 *saikou*
　pore

紺青 *konjou*
　Berlin blue $Fe_4[Fe(CN)_6]_3$

組合せ *kumiawase*
　combination

組成 *sosei*
  composition

組織 *soshiki*
  tissue

組織培養 *soshikibaiyou*
  tissue culture

組織学 *soshikigaku*
  histology

組織細胞 *soshikisaibou*
  tissue cell

終濃度 *shuunoudo*
  final concentration

終点 *shuuten*
  end point

糸 6a + 6 strokes | 統 絶 紫

統計 *toukei*
  statistics

統計分布 *toukeibunpun*
  statistical distribution

統計力学 *toukeirikigaku*
  statistical mechanics

統計理論 *toukeiriron*
  statistical theory

統計的効果 *toukeitekikouka*
  statistical effect

絶縁 *zetsuen*
  insulation

絶縁物 *zetsuenbutsu*
  insulating material

絶縁破壊 *zetsuenhakai*
  dielectric breakdown

絶縁層 *zetsuensou*
  insulating layer

絶縁体 *zetsuentai*
  insulator

絶縁耐力 *zetsuentairyoku*
  dielectric strength

絶縁テープ *zetsuente-pu*
  insulationg tape

絶縁材 *zetsuenzai*
  insulating material

絶対圧力 *zettaiatsuryoku*
  absolute pressure

絶対値 *zettaichi*
  absolute value

絶対誤差 *zettaigosa*
  absolute error

絶対配置 *zettaihaichi*
  absolute configuration

絶対の *zettaino*
  absolute

絶対温度 *zettaiondo*
  absolute temperature

絶対湿度 *zettaishitsudo*
  absolute humidity

紫外 *shigai*
  ultraviolet, UV

紫外分光 *shigaibunkou*
  ultraviolet spectroscopy

紫外光 *shigaikou*
  ultraviolet radiation

紫外吸収 *shigaikyuushuu*
  ultraviolet absorption

紫外線 *shigaisen*
  ultraviolet radiation

紫外スペクトル *shigaisupekutoru*
  ultraviolet spectrum

糸 6a + 8 to 12 strokes | 維 練 綿 緑 総 網 線 緩 縦 繊 縮 織

維持 *iji*
  maintenance

練り *neri*
  scouring

練炭 *rentan*
  briquette

綿実油 *menjitsuyu*
  cotton seed oil

綿状沈澱 *menjouchinden*
  flocculent precipitate

綿 *wata*
  cotton

緑青 *rokushou*
  verdigris $Cu(CH_3COO)_2 \cdot xCu(OH)_2 \cdot nH_2O$

緑柱石 *ryokuchuuseki*
  beryl $3BeO \cdot Al_2O_3 \cdot 6SiO_2$

総括反応速度 *soukatsuhannousokudo*
  overall reaction rate

網目 *amime*
  network

線状分子 *senjoubunshi*
  linear molecule

線状重合体 *senjoujuugoutai*
  linear polymer

線状高分子 *senjoukoubunshi*
  linear macromolecule

線形 *senkei*
  line shape

線形系 *senkeikei*
  linear system

線量 *senryou*
  dose

線量計 *senryoukei*
  dosimeter

線量測定法 *senryousokuteihou*
  dosimetry

線スペクトル *sensupekutoru*
  line spectrum

線図 *senzu*
  diagram

緩衝液 *kanshoueki*
  buffer solution

緩衝塩 *kanshouen*
  buffer salt

緩衝剤 *kanshouzai*
  buffer

緩和時間 *kanwajikan*
  relaxation time

縦電流 *tatedenryuu*
  longitudinal current

繊維 *seni*
  fiber

繊維芽細胞 *senigasaibou*
  fibroblast

繊維状物質 *senijoubusshitsu*
  fibrous material

繊維構造 *senikouzou*
  fiber structure

繊維素 *seniso*
  cellulose $(C_6H_{10}O_5)_n$

縮合 *shukugou*
  condensation

縮合物 *shukugoubutsu*
  condensate

縮合重合 *shukugoujuugou*
  condensation polymerization

縮合核 *shukugoukaku*
  condensed nucleus

縮合環 *shukugoukan*
  condensed ring

織物繊維 *shokubutsueni*
  textile fiber

米 6b
+ 0/4 strokes | 米粉

米澱粉 *komedenpun*
  rice starch

粉じん *funjin*
  dust

粉末 *funmatsu*
  powder

粉末度 *funmatsudo*
  fineness

粉末原料 *funmatsugenryou*
  powder raw material

粉末反応 *funmatsuhannou*
　powder reaction

粉末法 *funmatsuhou*
　powder method

粉末ミクロ組織 *funmatsumikurososhiki*
　powder microstructure

粉末にする *funmatsunisuru*
　to pulverize

粉末塗料 *funmatsutoryou*
　powder coatings

粉末冶金 *funmatsuyakin*
　powder metallurgy

粉砕 *funsai*
　pulverization, grinding, crushing

粉砕エネルギー *funsaienerugi-*
　grinding energy

粉砕機 *funsaiki*
　grinding machine, mill

粉砕混合物 *funsaikongoubutsu*
　grinding mixture

粉砕ロール *funsairo-ru*
　grinding mill

粉砕性 *funsaisei*
　grindability, crushability

粉粒体 *funsaitai*
　fine particle, pulverized matter

粉体 *funtai*
　powder

粉体爆発 *funtaibakuhatsu*
　dust explosion

粉体表面 *funtaihyoumen*
　powder surface

粉体混合物 *funtaikongoubutsu*
　powder mixture

粉体流 *funtairyuu*
　powder flow

粉体材料 *funtaizairyou*
　powder material

粉炭 *funtan*
　fine coal, slack coal

粉末粒子 *runmatsuryuushi*
　powder particle

米　6b
+ 5 strokes

粒粗粘断

粒化 *ryuuka*
　granulation

粒剤 *ryuuzai*
　granule, pellet

粗鉱 *seikou*
　crude ore, raw ore

粗銅 *sodou*
　black copper, blister copper

粗粒, 粗粒子 *soryuu, soryuushi*
　coarse grain

粗製品 *soseihin*
　crude product, raw product

粗製金属 *soseikinzoku*
　crude metal

粗繊維 *soseni*
　crude fiber

粗炭 *sotan*
　raw coal

粗糖 *sotou*
　raw sugar

粘着物 *nenchakubutsu*
　adhesive

粘着性 *nenchakusei*
　adhesiveness

粘稠な *nenchouna*
　viscous, viscid, thickly liquid, semifluid

粘稠剤, 粘稠化剤 *nenchuuzai, nenchuukazai*
　thickener

粘弾性 *nendansei*
　viscoelasticity

粘弾性物質 *nendanseibusshitsu*
　viscoelastic substance

粘弾性ポリマ *nendanseiporima*
　viscoelastic polymer

粘弾性流体流 *nendanseiryuutairyuu*
　viscoelastic fluid flow

粘弾性溶液 *nendanseiyoueki*
　viscoelastic solution

粘土 *nendo*
  clay

粘度 *nendo*
  viscosity

粘土灰 *nendohai*
  Clay ash

粘度法 *nendohou*
  viscometric method

粘度法則 *nendohousoku*
  viscosity law

粘度計 *nendokei*
  viscosimeter

粘土製品 *nendoseihin*
  clay product

粘度指数 *nendoshisuu*
  viscosity index

粘土触媒 *nendoshokubai*
  clay catalyst

粘液 *neneki*
  slime, mucilage, mucus

粘液酸 *nenekisan*
  mucic acid HOOC(CHOH)$_4$COOH

粘液層 *nenekisou*
  mucous layer

粘液蛋白質 *nenekitanpakushitsu*
  mucoprotein

粘結性 *nenketsusei*
  caking

粘性 *nensei*
  viscosity

粘性係数 *nenseikeisuu*
  viscosity coefficient, viscosity number

粘性液 *nenseiryuueki*
  viscous liquid

粘性膜 *nenseiryuumaku*
  viscous film

粘性流体 *nenseiryuutai*
  viscous liquid

粘性流体流 *nenseiryuutairyuu*
  viscous fluid flow

粘性溶媒 *nenseiryuuyoubai*
  viscous solvent

粘性ニュートン流体 *nenseinyu-tonyuutai*
  viscous Newtonian fluid

断面 *danmen*
  cross section

断面積 *danmenseki*
  cross-section area

断熱 *dannetsu*
  heat insulation

断熱圧縮 *dannetsuasshuku*
  adiabatic compression

断熱膨張 *dannetsubouchou*
  adiabatic expansion

断熱フォーム *dannetsufo-mu*
  thermal insulating foam

断熱反応器 *dannetsuhannouki*
  adiabatic reactor

断熱平衡 *dannetsuheikou*
  adiabatic equilibrium

断熱変化 *dannetsuhenka*
  adiabatic change

断熱冷却 *dannetsureikyaku*
  adiabatic cooling

断熱線 *dannetsusen*
  adiabatic curve

断熱的な *dannetsutekina*
  adiabatic

断熱材 *dannetsuzai*
  heat insulator

米 6b
+ 8 to 12 strokes

精糖翻

精銅 *seidou*
  refined copper

精白糖 *seihakutou*
  refined sugar

精密機器, 精密機械 *seimitsukiki, seimitsukikai*
  precision instrument

精密天秤 *seimitsutenbin*
  precision balance

精練 *seiren*
  scouring

精錬 *seiren*
  refining, smelting

精練剤 *seirenzai*
  scouring agent

精留 *seiryuu*
  rectification

精留器 *seiryuuki*
  rectifier

精製 *seisei*
  refining, purification

精製法 *seiseihou*
  refining process

精製糖 *seiseitou*
  refined sugar

精製油 *seiseiyu*
  refined oil

精糖 *seitou*
  refined sugar

精油 *seiyu*
  refined oil

精油所 *seiyusho*
  oil refinery

糖 *tou*
  sugar

糖アルデヒド *touarudehido*
  sugar aldehyde

糖アルコール *touaruko-ru*
  sugar alcohol

糖原質 *tougenshitsu*
  glucoside, glycoside

糖配座 *touhaiza*
  sugar conformation

糖反応 *touhannou*
  sugar reaction

糖ヒドラゾン *touhidorazon*
  sugar hydrazone

糖ジカルボン酸 *toujikarubonsan*
  sugar dicarboxylic acid

糖受容体 *toujuyoutai*
  sugar receptor

糖化 *touka*
  saccharification

糖環 *toukan*
  sugar ring

糖無水物 *toumusuibutsu*
  sugar anhydride

糖ペプチド合成 *toupepuchidogousei*
  glycopeptide synthesis

糖ペプチド類似体 *toupepuchidoruijitai*
  glycopeptide analog

糖ラクトン *tourakkuton*
  saccharic lactone

糖類 *tourui*
  saccharides

糖量計 *touryoukei*
  saccharimeter, sugar pydrometer

糖酸 *tousan*
  saccaric acid HOOC(CHOH)$_4$COOH

糖残基 *tousanki*
  sugar residue

糖シロップ *toushiroppu*
  sugar syrup

糖脂質 *toushishitsu*
  glycolipid

糖水溶液 *tousuiyoueki*
  sugar aqueous solution

糖タンパク質 *toutanpakushitsu*
  glycoprotein

糖蛋白質 *toutanpakushitsu*
  glycoprotein

翻訳 *honyaku*
  translation

舟 6c | 舟 般

(all *kanji*)

舟型, 舟形 *funegata*
  boat form, boat conformation

般式 *hanshiki*
　general formula

虫 6d | 蛇 蛋
+ 5 strokes

蛇紋岩, 蛇紋石　*jamongan, jamonseki*
　serpentine, green marble
　$(Mg,Fe)_3Si_2O_5(OH)_4$

蛋白　*tanpaku*
　protein

蛋白分解酵素　*tanpakubunkaikouso*
　protease, proteolytic enzyme, proteinase

蛋白加水分解　*tanpakukasuibunkai*
　proteolysis

蛋白工学　*tanpakukougaku*
　protein engineering, protein design

蛋白生合成　*tanpakuseigousei*
　protein biosynthesis

蛋白石　*tanpakuseki*
　opal $SiO_2 \cdot nH_2O$

蛋白繊維　*tanpakuseni*
　protein fiber

蛋白質　*tanpakushitsu*
　protein

蛋白質バンド　*tanpakushitsubando*
　protein band

蛋白質分解物　*tanpakushitsubunkaibutsu*
　protein degradation product

蛋白質分析法　*tanpakushitsubunsekihou*
　protein assay

蛋白質分子　*tanpakushitsubunshi*
　protein molecule

蛋白質電荷　*tanpakushitsudenka*
　protein charge

蛋白質電気泳動　*tanpakushitsudenkieidou*
　protein electrophoresis

蛋白質液体クロマトグラフィー
　tanpakushitsuekitaikuromatogurafi-
　protein liquid chromatography

蛋白質フラグメント
　tanpakushitsufuragumento
　protein fragment

蛋白質配座　*tanpakushitsuhaiza*
　protein conformation

蛋白質表面　*tanpakushitsuhyoumen*
　protein surface

蛋白質標識　*tanpakushitsuhyoushiki*
　protein marker

蛋白質異性体　*tanpakushitsuiseitai*
　protein isomer

蛋白質重合体　*tanpakushitsujuugoutai*
　protein polymer

蛋白質活性部位　*tanpakushitsukasseibui*
　protein active site

蛋白質結晶構造
　tanpakushitsukesshoukouzou
　protein crystal structure

蛋白質結晶成長
　tanpakushitsukesshouseichou
　protein crystal growth

蛋白質固定　*tanpakushitsukotei*
　protein immobilization

蛋白質クロマトグラフィー
　tanpakushitsukuromatogurafi-
　protein chromatography

蛋白質メチル化　*tanpakushitsumechiruka*
　protein methylation

蛋白質二次構造　*tanpakushitsunijikouzou*
　protein secondary structure

蛋白質オリゴマ　*tanpakushitsuorigoma*
　protein oligomer

蛋白質試料　*tanpakushitsushiryou*
　protein sample

蛋白質触媒作用
　tanpakushitsushokubaisayou
　protein catalysis

蛋白質組成　*tanpakushitsusosei*
　protein composition

蛋白質層　*tanpakushitsusou*
　protein layer

蛋白質代謝 *tanpakushitsutaisha*
  protein metabolism

蛋白質単結晶 *tanpakushitsutankesshou*
  protein single crystal

虫 6d 触 融 蟻 蝋
+ 7 to 11 strokes

触媒 *shokubai*
  catalyst

触媒毒 *shokubaidoku*
  catalyst poison

触媒反応 *shokubaihannou*
  catalytic reaction

触媒重合 *shokubaijuugou*
  catalyzed polymerization

触媒化学 *shokubaikagaku*
  catalysis

触媒活性 *shokubaikassei*
  catalytic activity

触媒量 *shokubairyo*
  catalytic amount

触媒作用 *shokubaisayou*
  catalysis

触媒水素化 *shokubaisuisoka*
  catalytic hydrogenation

融合 *yuugou*
  fusion

融合細胞 *yuugousaibou*
  fused cell

融解 *yuukai*
  fusion

融解塩 *yuukaien*
  fused salt

融解塩化物 *yuukaienkabutsu*
  molten chloride

融解エンタルピー *yuukaientarupi-*
  melt enthalpy

融解法 *yuukaihou*
  fusion method

融解状態 *yuukaijoutai*
  melt state

融解結晶化 *yuukaikesshouka*
  melt crystallization

融解熱 *yuukainetsu*
  heat of fusion

融解炉 *yuukairo*
  melting furnace, melter

融解速度論 *yuukaisokudoron*
  melting kinetics

融解帯 *yuukaitai*
  melting zone

融解特性 *yuukaitokusei*
  melting characteristic

融解曲線 *yuukakyokusen*
  melting curve

融点 *yuuten*
  melting point, fusion point

融点降下 *yuutenkouka*
  melting point depression

融和性 *yuuwasei*
  compatibility

融剤 *yuuzai*
  flux

蟻酸 *gisan*
  formic acid HCOOH

蟻酸アンモニウム *gisananmoniumu*
  ammonium formate $HCOONH_4$

蟻酸エチル *gisanechiru*
  ethyl formate $HCOOC_2H_5$

蟻酸塩 *gisanen*
  formate HCOOR, HCOOM

蟻酸ナトリウム *gisannatoriumu*
  sodium formate HCOONa

螺旋状構造 *rasenjoukouzou*
  spiral structure

螺旋角度 *rasenkakudo*
  helix angle

蝋 *rou*
  wax

蝋石　rouseki
　agalmatolite $Al_2O_3 \cdot 4SiO_2 \cdot H_2O$

竹 6f (all kanji) | 算 箆 管 範 篩

算術平均　sanjutsuheikin
　arithmetical mean

箆　hera
　spatula

管形反応器　kangatahannouki
　tubular reactor

管状炉　kanjouro
　tube furnace

管　kann
　tube

範囲　hani
　extend, scope, range

篩　furui
　sieve

篩分析　furuibunseki
　sieve analysis

言 7a + 2 to 4 strokes | 計 記 許 設

計器　keiki
　measuring instrument, meter, gauge, gage

計量器　keiryouki
　measuring apparatus

計算化学　keisankagaku
　computational chemistry

計算機　keisanki
　calculator

計測　keisoku
　measurement

計測器　keisokuki
　measuring instrument

計数器　keisuuki
　counter

計数装置　keisuusouchi
　counter

討究　toukyuu
　research

討論　touron
　discussion

記号　kikou
　mark, symbol

記憶　kioku
　storage, storing

記憶容量　kiokuyouryou
　memory capacity, storage capacity

記録計　kirokukei
　recorder

許容範囲　kyoyouhani
　tolerance limit

許容差　kyoyousa
　tolerance

許容遷移　kyoyouseni
　allowed transition

設置　secchi
　establishment, founding, institution

設備　setsubi
　equipment, facility, plant

言 7a + 5/6 strokes | 評 証 診 詳 試

評価　hyouka
　evaluation

証拠　shouko
　evidence, proof

証明　shoumei
　evidence, proof

証明書　shoumeisho
　certificate

証明する　shoumeisuru
　to prove

診断　shindan
　diagnosis

診断用試薬　*shindanyoushiyaku*
  diagnostic reagents

詳しい　*kuwashii*
  detailed, familiar with (something)

詳細　*shousai*
  detail

試験　*shiken*
  experiment, test, assay

試験データ　*shikende-ta*
  test data

試験法　*shikenhou*
  test method

試験方法　*shikenhouhou*
  test method, testing method

試験条件　*shikenjouken*
  test condition

試験管　*shikenkan*
  test tube

試験管はさみ　*shikenkanhasami*
  test-tube holder

試験管内の　*shikenkannaino*
  in vitro

試験結果　*shikenkekka*
  experimental result

試験器　*shikenki*
  testing instrument

試験機　*shikenki*
  testing apparatus

試験薬　*shikenkusuri*
  experimental drug

試験紙　*shikenshi*
  test paper

試験室　*shikenshitsu*
  laboratory

試験装置　*shikensouchi*
  test equipment

試金　*shikin*
  assay

試料　*shiryou*
  sample

試料瓶　*shiryoubin*
  sample bottle

試料採取　*shiryousaishu*
  sampling

試料採取器　*shiryousaishuki*
  sampler

試料水　*shiryousui*
  water sample

試作　*shisaku*
  trial manufacture

試作品　*shisakuhin*
  trial product

試薬　*shiyaku*
  reagent

試薬瓶　*shiyakubin*
  reagent bottle

試用期間　*shiyoukikan*
  trial period

試し, 試す　*tameshi, tamesu*
  test

言　7a　誤誘認説
+ 7 strokes

誤差　*gosa*
  error

誤差限度　*gosagendo*
  error limit, tolerance

誤差範囲　*gosahani*
  error range

誤差検出　*gousakenshutsu*
  error detecting

誘電分光法　*yuudenbunkouhou*
  dielectric spectroscopy

誘電分極　*yuudenbunkyoku*
  dielectric polarization

誘電分析　*yuudenbunseki*
  dielectric analysis

誘電泳動　*yuudeneidou*
  dielectrophoresis

誘電緩和　*yuudenkanwa*
  dielectric relaxation

誘電緩和スペクトロスコピー　*yuudenkanwasupekutorosukopi-*
　dielectric relaxation spectroscopy

誘電緩和スペクトル　*yuudenkanwasupekutoru*
　dielectric relaxation spectrum

誘電緩和特性　*yuudenkanwatokusei*
　dielectric relaxation characteristic

誘電係数　*yuudenkeisuu*
　dielectric coefficient

誘電吸収　*yuudenkyuushuu*
　dielectric absorption

誘電摩擦　*yuudenmasatsu*
　dielectric friction

誘電パラメータ　*yuudenparame-ta*
　dielectric parameter

誘電率　*yuudenritsu*
　dielectric constant

誘電作用　*yuudensayou*
　dielectric effect

誘電測定　*yuudensokutei*
　dielectric measurement

誘電体　*yuudentai*
　dielectric

誘電滴定　*yuudentekitei*
　dielectric titration

誘導　*yuudou*
　induction

誘導分解　*yuudoubunkai*
　induced degradataion

誘導物質　*yuudoubusshitsu*
　inductor, inducer

誘導電気炉　*yuudoudenkiro*
　induction furnace

誘導電流　*yuudoudenryuu*
　induced current, induction current

誘導合成　*yuudougousei*
　induced synthesis

誘導グラフト重合　*yuudougurafutojuugou*
　induced graft polymerization

誘導反応　*yuudouhannou*
　induced reaction

誘導変換　*yuudouhenkan*
　induced conversion

誘導変性　*yuudouhensei*
　induced denaturation

誘導放出　*yuudouhoushutsu*
　induced emission, stimulated emission

誘導時間　*yuudoujikan*
　induction time

誘導重合　*yuudoujuugou*
　induced polymerization

誘導架橋　*yuudoukakyou*
　induced crosslinking

誘導環化　*yuudoukanka*
　induced cyclization

誘導加水分解　*yuudoukasuibunkai*
　induced hydrolysis

誘導期　*yuudouki*
　induction period

誘導酵素　*yuudoukouso*
　inducible enzyme

誘導ポリマ　*yuudouporima*
　derived polymer

誘導プロセス　*yuudoupurosesu*
　induced process

誘導率　*yuudouritsu*
　inductivity

誘導酸化　*yuudousanka*
　induced oxidation

誘導体　*yuudoutai*
　derivative

誘導体化　*yuudoutaika*
　derivatization

誘導体化反応　*yuudoutaikahannou*
　derivatization reaction

誘導転位　*yuudouteni*
　induced rearrangement

誘発分解　*yuuhatsubunkai*
　induced disintegration

誘発発光　*yuuhatsuhakkou*
　induced emission, stimulated emission

誘発反応　*yuuhatsuhannou*
　induced reaction

誘発放射　*yuuhatsuhousha*
　induced radiation

誘発相転移　*yuuhatsusouteni*
　induced phase transition

誘起脱ハロゲン化　*yuukidatsuharogenka*
　induced dehalogenation

誘起脱プロトン化　*yuukidatsupurotonka*
　induced deprotonation

誘起電子　*yuukidenshi*
　induced electron

誘起腐食　*yuukifushoku*
　induced corrosion

誘起イオン化　*yuukiionka*
　induced ionization

誘起カチオン重合　*yuukikachionjuugou*
　induced cationic polymerization

誘起解離　*yuukikairi*
　induced dissociation

誘起改質　*yuukikaishitsu*
　induced modification

誘起過程　*yuukikatei*
　stimulated process

誘起結晶　*yuukikesshou*
　induced crystal

誘起金属化　*yuukikinzokuka*
　induced metallization

誘起光電子　*yuukikoudenshi*
　induced photoelectron

誘起効果　*yuukikouka*
　inductive effect

誘起共重合　*yuukikyoujuugou*
　induced copolymerization

誘起求核置換　*yuukikyuukakuchikan*
　induced nucleophilic substitution

誘起酸化　*yuukisanka*
　induced oxidation

誘起相転移　*yuukisouteni*
　induced phase transformation

認証　*ninshou*
　certification, authentication

説　*setsu*
　theory

説明　*setsumei*
　explanation

言　7a
+ 8 to 12 strokes

課　請　論
調　講　識

課題　*kadai*
　subject, theme, topic, problem

請求　*seikyuu*
　demand, request, claim

論理的分析　*ronritekibunseki*
　logical analysis

論理的な　*ronritekina*
　logic, logical

調査部　*chousabu*
　research division

調査員、調査官　*chousain, chousakan*
　examiner, tester

調整　*chousei*
　adjustment

調製　*chousei*
　preparation

調製的合成　*chouseitekigousei*
　preparative synthesis

調節　*chousetsu*
　regulation, adjustment

調節弁　*chousetsuben*
　control valve, regulating valve

調節器　*chousetsuki*
　regulator, modifier, controller

調節機構　*chousetsukikou*
　regulatory mechanism

調節酵素　*chousetsukouso*
　regulatory enzyme

調節剤　*chousetsuzai*
　modifier

調色　*choushoku*
　toning

調色液　*choushokueki*
　toner

調和振動子 *chouwashindoushi*
  harmonic oscillator

講究 *koukyuu*
  research

識別 *shikibetsu*
  recognition

貝 7b | 貝 貯 費 貴 賤 質
(all *kanji*)

貝褐炭 *kaikattan*
  agstein

貯槽 *chosou*
  reservoir, storage tank

貯蔵安定性 *chouzouanteisei*
  storage stability

貯蔵 *chozou*
  storage

貯蔵槽 *chozousou*
  storage tank

貯蔵タンク *chozoutanku*
  storage tank

費用 *hiyou*
  costs, charges, expenses

貴ガス *kigasu*
  noble gas

貴金属 *kikinzoku*
  noble metal

賤金属 *senkinzoku*
  base metal

質量 *shitsuryou*
  mass

質量分光分析 *shitsuryoubunkoubunseki*
  mass spectrometry

質量分光測定 *shitsuryoubunkousokutei*
  mass spectrometry

質量分析 *shitsuryoubunseki*
  mass spectrometry

質量分析法 *shitsuryoubunsekihou*
  mass spectroscopy

質量分析計 *shitsuryoubunsekikei*
  mass spectrometer

質量分析器 *shitsuryoubunsekiki*
  mass spectrograph

質量中心 *shitsuryouchuushin*
  center of mass

質量保存の法則 *shitsuryouhozonnohousoku*
  law of conservation of mass

質量係数 *shitsuryoukeisuu*
  mass coefficient

質量作用 *shitsuryousayou*
  mass action

質量作用の法則 *shitsuryousayounohousoku*
  law of mass action

質量速度 *shitsuryousokudo*
  mass velocity

質量スペクトル *shitsuryousupekutoru*
  mass spectrum

質量数 *shitsuryousuu*
  mass number

車 7c | 軌 軟 転
+ 2/4 strokes

軌道 *kidou*
  orbital

軌道対称 *kidoutaishou*
  orbital symmetry

軟化点 *banakaten*
  softening point

軟化 *nanka*
  softening

軟化剤 *nankazai*
  softening agent

軟質ガラス *nanshitsugarasu*
  soft glass

軟質ゴム *nanshitsugomu*
  soft rubber

軟質高分子　*nanshitsukoubunshi*
　　flexible polymer

軟質炭　*nanshitsutan*
　　soft coal

軟水　*nansui*
　　soft water

軟水化　*nansuika*
　　water softening

軟鉄　*nantetsu*
　　soft iron

軟らかい塩基　*yawarakaienki*
　　soft base

軟らかい酸　*yawarakaisan*
　　soft acid

転位　*teni*
　　rearrangement, dislocation

転移　*teni*
　　transfer, transition, metastasis

転移エネルギー　*tenienerugi-*
　　transition energy

転移エントロピー　*tenientoropi-*
　　transition entropy

転位反応　*tenihannou*
　　rearrangement reaction

転移状態　*tenijoutai*
　　transition state

転位活性　*tenikassei*
　　dislocation activity

転移酵素　*tenikouso*
　　transferase

転移熱　*teninetsu*
　　transition heat

転移温度　*teniondo*
　　transition temperature, conversion temperature

転位生成物　*teniseiseibutsu*
　　rearrangement product

転移点　*teniten*
　　transition point

転化　*tenka*
　　inversion, conversion

転化器　*tenkaki*
　　converter

転化酵素　*tenkakouso*
　　invertase

転化率　*tenkaritsu*
　　conversion ratio

転化速度　*tenkasokudo*
　　conversion rate

転化点　*tenkaten*
　　inversion point

転化糖　*tenkatou*
　　invert sugar

転炉　*tenro*
　　converter

転写　*tensha*
　　transfer; transcription

転写酵素　*tenshakouso*
　　transcriptase

転写阻害　*tenshasogai*
　　transcription inhibitor

転相　*tensou*
　　phase reversal

転座　*tenza*
　　translocation

車　7c	軸　軽　較　輝　輻　輪
+ 5 to 9 strokes	

軸　*jiku*
　　axis

軸対称　*jikutaishou*
　　axial symmetry

軽石　*karuishi*
　　pumice

軽ベンジン　*keibenjin*
　　light petrol

軽合金　*keigoukin*
　　light alloy

軽金属　*keikinzoku*
　　light metal

軽質ナフサ　*keishitsunafusa*
  light naphtha

軽水　*keisui*
  light water

軽水原子炉　*keisuigenshiro*
  light water nuclear reactor

軽油　*keiyu*
  light oil

較正　*kousei*
  calibration

較正曲線　*kouseikyokusen*
  calibration curve

輝度　*kido*
  brightness, brilliance, luminance

輝度係数　*kidokeisuu*
  luminance coefficient

輝銀鉱　*kiginkou*
  argentite Ag_2S

輻射, 輻射線　*fukusha, fukushasen*
  radiation

輻射エネルギー　*fukushaenerugi-*
  radiation energy

輻射源　*fukushagen*
  radiant source

輻射法則　*fukushahousoku*
  radiation law

輻射熱　*fukushanetsu*
  radiant heat

輸血　*yuketsu*
  transfusion

輸率　*yuritsu*
  transference number

輸送現象　*yusougenshou*
  transport phenomenon

𧾷, 足　**7d**	距
(all *kanji*)	

距離　*kyori*
  length, distance, range

酉　**7e**	酵　酪　醸
(all *kanji*)	

酵母　*koubo*
  yeast

酵母抽出物　*koubochuushutsubutsu*
  yeast extract

酵素　*kouso*
  enzyme

酵素母質　*kousoboshitsu*
  proenzyme, zymogen

酵素分解　*kousobunkai*
  enzymatic splitting

酵素学　*kousogaku*
  enzymology

酵素学的分解　*kousogakutekibunkai*
  enzymatic degradation

酵素学的分析　*kousogakutekibunseki*
  enzymatic analysis

酵素原　*kousogen*
  proenzyme, zymogen

酵素反応　*kousohannou*
  enzymatic reaction

酵素変換　*kousohenkan*
  enzymatic conversion

酵素法　*kousohou*
  enzymatic method

酵素化学　*kousokagaku*
  enzyme chemistry

酵素活性　*kousokassei*
  enzyme activity

酵素加水分解　*kousokasuibunkai*
  enzymatic hydrolysis

酵素系　*kousokei*
  enzyme system

酵素基質　*kousokishitsu*
  enzyme substrate

酵素基質複合体　*kousokishitsufukugoutai*
  enzyme-substrate complex

酵素の　*kousono*
  enzymatic

酵素製剤　*kousoseizai*
enzyme preparation

酵素阻害　*kousosogai*
enzyme inhibiton

酵素特異性　*kousotokuisei*
enzyme specificity

酵素誘導　*kousoyuudou*
enzyme induction

酵素誘導物質　*kousoyuudoubusshitsu*
enzyme inducer

酵素前駆体　*kousozenkutai*
zymogen, proenzyme

酪酸　*rakusan*
butyric acid C_3H_7COOH

酪酸エチル　*rakusanechiru*
ethyl butyrate $C_3H_7COOC_2H_5$

酪酸塩　*rakusanen*
butyrate

酪酸菌　*rakusankin*
butyric acid bacteria

酪酸クロライド　*rakusankuroraido*
butyric acid chloride C_3H_7COCl

醸造　*jouzou*
brewing

醸造する　*jouzousuru*
to brew

金　8a
+ 2 to 6 strokes　　針　鈴　鉱　銑

針　*hari*
needle

針状結晶　*shinjoukesshou*
needle crystal

針入, 針入度　*shinnyuu, shinyuudo*
penetration

針入度試験　*shinnyuudoshiken*
penetration test

鈴木宮浦カップリング
suzukimiyaurakappuringu
Suzuki-Miyaura coupling

鉱物　*koubutsu*
mineral

鉱物学　*koubutsugaku*
mineralogy

鉱物化学　*koubutsukagaku*
mineral chemistry

鉱物繊維　*koubutsuseni*
mineral fiber

鉱物油　*koubutsuyu*
mineral oil, crude oil, petroleum

鉱化　*kouka*
mineralize

鉱化剤　*koukazai*
mineralizer

鉱酸　*kousan*
mineral acid

鉱産物　*kousanbutsu*
mineral

鉱石　*kouseki*
ore, mineral

鉱質　*koushitsu*
mineral

鉱床　*koushou*
ore deposit

鉱水　*kousui*
mineral water

鉱油　*kouyu*
mineral oil

鉱山　*kouzan*
mine

鉱山調査　*kouzanchousa*
mine examination

銑鉄　*sentetsu*
pig iron

金　8a
+ 7 to 11 strokes　　鋳　錯　錠　鋼　錆
　　　　　　　　　　　　鍛　鎖　鎮　鏡

鋳鉄　*chuutetsu*
cast iron

鋳造 *chuuzouo*
  casting

鋳型 *igata*
  pattern, template

鋳型分子 *igatabunshi*
  template molecule

鋳型反応 *igatahannou*
  template reaction

鋳型化 *igataka*
  molding

鋳込法 *ikomihou*
  casting

鋳込 *ikomki*
  cast

鋳物 *imono*
  metal casting

鋳造法 *tyuuzouhou*
  casting

錯塩 *sakuen*
  complex salt

錯イオン *sakuion*
  complex ion

錯化合物 *sakukagoubutsu*
  complex compound

錯化合物生成 *sakukagoubutsuseisei*
  complexation

錯基 *sakuki*
  complex radical

錯体 *sakutai*
  complex

錯体化学 *sakutaikagaku*
  complex chemistry

錯滴定 *sakutekitei*
  complexometric titration, complexometry

錠剤 *jouzai*
  tablet

鋼へら *haganehera*
  steel spatula

鋼玉 *kougyoku*
  corundum Al_2O_3

錆 *sabi*
  rust

錆止め *sabidome*
  rust protection, rust prevention

錆止め塗装 *sabidometosou*
  anti-corrosive coat

錆止め剤 *sabidomezai*
  rust inhibitor, rust preventive

鍛冶 *tanya*
  metallurgy

鎖 *kusari*
  chain

鎖長 *sachou*
  chain length

鎖状分子 *sajoubunshi*
  chain molecule

鎖状異性 *sajouisei*
  chain isomerism

鎖状重合 *sajoujuugou*
  chain polymerization

鎖状重合体 *sajoujuugoutai*
  chain polymer

鎖状化合物 *sajoukagoubutsu*
  chain compound

鎖状構造 *sajoukouzou*
  chain structure

鎖式化合物 *sashikikagoubutsu*
  chain compound

鎮静剤 *chinseizai*
  tranquilizer

鎮痛剤 *chintsuuzai*
  analgesics

鏡映 *kyouei*
  reflection

鏡面 *kyoumen*
  mirror plane, symmetry plane, mirror surface

食 **8b** | 食 飲 飽 飼
(all *kanji*)

食塩 *shokuen*
  common salt, table salt NaCl

食塩水　*shokuensui*
　brine, saline solution

食品　*shokuhin*
　food

食品化学　*shokuhinkagaku*
　food chemistry

食品香料　*shokuhinkouryou*
　food flavor

食品添加物　*shokuhintenkabutsu*
　food additive, nutritional supplement

食品用染料　*shokuhinyousenryou*
　food dye

食物連鎖　*shokumotsurensa*
　food chain

食細胞　*shokusaibou*
　phagocyte

食細胞作用　*shokusaibousayou*
　phagocytosis

食用色素　*shokuyoushikiso*
　food colouring

飲料水　*inryousui*
　drinking water

飽和　*houwa*
　saturation

飽和アルコール　*houwaaruko-ru*
　saturated alcohol

飽和圧　*houwaatsu*
　saturation pressure

飽和窒素　*houwachisso*
　saturated nitrogen

飽和窒素複素環化合物
　houwachissofukusokankagoubutsu
　saturated nitrogen heterocycle

飽和度　*houwado*
　degree of saturation

飽和液体　*houwaekitai*
　saturated liquid

飽和複素環　*houwafukusokan*
　saturated heterocycle

飽和フルオロカーボン
　houwafuruoroka-bon
　saturated fluorocarbon

飽和蒸気　*houwajouki*
　saturated vapor

飽和化合物　*houwakagoubutsu*
　saturated compound

飽和ケトン　*houwaketon*
　saturated ketone

飽和効果　*houwakouka*
　saturation effect

飽和曲線　*houwakyokusen*
　saturation curve

飽和温度　*houwaondo*
　saturation temperature

飽和率　*houwaritsu*
　saturation factor

飽和酸素　*houwasanso*
　saturated oxygen

飽和線　*houwasen*
　saturation line

飽和限度　*houwasgendo*
　saturation limit

飽和脂肪酸　*houwashibousan*
　saturated fatty acid

飽和水　*houwasui*
　saturated water

飽和水蒸気　*houwasuijouki*
　saturated vapor

飽和水溶液　*houwasuiyoueki*
　saturated aqueous solution

飽和する　*houwasuru*
　to saturate

飽和炭化水素　*houwatankasuiso*
　saturated hydrocarbon

飽和炭素　*houwatanso*
　saturated carbon

飽和点　*houwaten*
　saturation point

飽和溶液　*houwayoueki*
　saturated solution

飼料添加物　*shiryoutenkabutsu*
　animal feed additive

佳 8c	集焦難
(all *kanji*)	

集電器 *shuudenki*
  collector
集じん電極 *shuujindenkyoku*
  collecting electrode
集光器 *shuukouki*
  condenser
集積二重結合 *shuusekinijuuketsugou*
  cumulative double bond
焦点 *shouten*
  focus, focus point
難燃加工 *nannenkakou*
  fireproof
難燃化剤 *nannenkazai*
  flame retardant
離液性 *riekisei*
  lyotropic property
雑種 *zasshu*
  hybrid
雑種形成 *zasshukeisei*
  hybridization
雑種細胞 *zasshusaibou*
  hybrid cell

雨 8d	雪雲零霰露
(all *kanji*)	

雪花セッコウ *sekkasekkou*
  alabaster $CaSO_4 \cdot 2H_2O$
雪状炭酸 *yukijoutansan*
  dry ice, carbon dioxide snow
雲母 *unmo*
  mica
零 *rei, zero*
  zero
零次反応 *reijihannou*
  zero-order reaction
零点 *reiten*
  zero point
零点調節 *reitenchousetsu*
  zero point adjustment
零点移動 *reitenidou*
  zero drift
霰石 *arareishi*
  aragonite $CaCO_3$
露出過度 *ro...kado*
  overexposure
露光 *rokou*
  exposure
露光時間 *rokoujikan*
  exposure time
露光過度 *rokoukado*
  overexposure
露光計 *rokoukei*
  exposure meter
露光指数 *rokoushisuu*
  exposure index
露出 *roshutsu*
  exposition
露点 *roten*
  dew point
露点圧力 *rotenatsuryoku*
  dew-point pressure
露点降下 *rotenkouka*
  dew-point depression
露点曲線 *rotenkyokusen*
  dew-point curve

門, 鬥 8e	閃閉間開関
(all *kanji*)	

閃光 *senkou*
  flash
閉殻 *heikaku*
  closed shell
閉殻金属イオン *heikakukinzokuion*
  closed-shell metal ion

閉管 *heikan*
  closed tube

閉環複分解 *heikanfukubunkai*
  ring-closing metathesis

閉環反応 *heikanhannou*
  ring-closure reaction

閉環オレフィン複分解
heikanorefinfukubunkai
  ring-closing olefin metathesis

閉鎖 *heisa*
  occlusion

閉鎖循環系 *heisajunkankei*
  closed circulatory system

閉鎖化 *heisaka*
  closing

閉塞 *heisoku*
  block

閉じた系 *tojitakei*
  closed system

間接分析 *kansetsubunseki*
  indirect analysis

間接照明 *kansetsushoumei*
  indirect lighting

間接測定 *kansetsusokutei*
  indirect measurement

間接滴定 *kansetsutekitei*
  indirect titration

間違いの原因 *machigainogenin*
  source of error

開いた系 *hiraitakei*
  open system

開発 *kaihatsu*
  development

開発部門 *kaihatsubumon*
  development department

開発する *kaihatsusuru*
  to develop

開閉器 *kaiheiki*
  switch

開放系 *kaihoukei*
  open system

開殻 *kaikaku*
  open shell

開環 *kaikan*
  ring cleavage

開環重合 *kaikanjuugou*
  ring-opening polymerization

開烈 *kairetsu*
  cleavage

開鎖 *kaisa*
  open chain

開始 *kaishi*
  initiation

開始反応 *kaishihannou*
  initiation reaction

開始期 *kaishiki*
  initiation period

開始剤 *kaishizai*
  initiator

関する *kansuru*
  to be related to, to concern, to involve

頁 **9a** (all *kanji*)	頃 項 順 頑 顆 頻 類 顔 顕

頃 *kei, koro, goro*
  approximately, about (time)

項 *kou*
  term, item

項間交差 *koukankousa*
  intersystem crossing

順列 *junretsu*
  permutation

順相クロマトグラフィー
junsoukuromatogurafi-
  normal phase chromatography

頑火輝石 *gankakiseki*
  enstatite $Mg_2Si_2O_6$

顆粒 *karyuu*
  granulate, granular material

顆粒化 *karyuuka*
  granulation

頻度 *hindo*
frequency

頻度因子 *hindoinshi*
frequency factor

類縁元素 *ruiengenso*
analogous element

類縁化合物 *ruienkagoubutsu*
analogues

類似 *ruiji*
similarity

類似分子 *ruijibushi*
analog molecule

類似反応 *ruijihannou*
analogous reaction

類似化合物 *ruijikagoubutsu*
analogue compound

類似の *ruijino*
similar

類似性 *ruijisei*
similarity

類似体合成 *ruijitaigousei*
analog synthesis

類似体の *ruijitaino*
analog

顔料 *ganryou*
pigment

顕微鏡 *kenbikyou*
microscope

顕微鏡分析 *kenbikyoubunseki*
microscopical analysis

顕微鏡法 *kenbikyouhou*
microscopy

顕微鏡試験 *kenbikyoushiken*
microscopic test

顕微鏡的結晶の *kenbikyoutekikesshouno*
microcrystalline

顕微鏡的な *kenbikyoutekina*
microscopic

顕晶質 *kenshoushitsu*
phanerocrystalline

馬 10a | 馬 騎
(all *kanji*)

馬尿酸 *banyousan*
hippuric acid $C_6H_5CONHCH_2COOH$

騎電力 *kitenryoku*
electromotive force

魚 11a | 魚 鯨 鱗
(all *kanji*)

魚油 *gyoyu*
fish oil

鯨油 *geiyu*
whale oil

鯨油酸 *geiyusan*
cetoleic acid
$CH_3(CH_2)_9CH=CH(CH_2)_9COOH$

鱗繊石 *rinsenseki*
lipidocrocite γ-FeOOH

III
Appendices

10
Bibliography

The following list of literature contains sources that were utilized for the compilation of this book, as well as books and online URLs that provide information on further topics related to chemistry, science and Japanese. In addition to this list, some important online sources are mentioned in previous chapters, i.e. in Section 2.6.3 (General Online Support for Japanese Language), Table 2.21 and Table 2.22 [Internet sources for character and text analysis (in addition to sources listed in Table 2.21)], Section 2.6.4 (Online Support for Analysis of Chemical Literature) and Section 4.3 (Online Sources of Japanese Patent Information).

10.1
Character Dictionaries

- Mark Spahn, Wolfang Hadamitzky: *Japanese Character Dictionary*, Nichigai Associates Inc., Tokyo, **1989** (ISBN 4-8169-0828-5).
- Andrew N. Nelson: *The Modern Reader's Japanese-English Character Dictionary*, Charles E. Tuttle Co., Inc., Tokyo, **1970**.
- J.H. Haig, A.N. Nelson: *The New Nelson Japanese-English Character Dictionary*, Charles E. Tuttle Co., Inc., Tokyo, **1997** (ISBN 0-8048-2036-8).
- P.G. O'Neill: *Essential Kanji*, Weatherhill, New York, **1990** (ISBN 0-8348-0222-8).
- プログレッシブ和英中辞典 (*puroguresshibuwaeichuujiten*, *Progressive Japanese–English Dictionary*), 小学館 (*shougakukan*), Tokyo, **1986** (ISBN 4-09-510251-9).
- Jack Halpern: *NTC's New Japanese-English Character Dictionary*, NTC Publishing Group, Lincolnwood, USA, **1992** (ISBN 0-8442-8434-3).
- Jack Halpern: *The Kodansha Kanji Learner's Dictionary*, Kodansha, Tokyo, **2001** (ISBN 4-7700-2855-5).
- *A New Dictionary of Kanji Usage*, Gakken, Tokyo, **1982** (ISBN 4-05-151312-2).

10.2
Grammar and Related Topics

- 日本語基本文法辞典 (*nihongokihonbunpoujiten*), Seichi Makino, Michio Tsutsui: *A Dictionary of Basic Japanese Grammar*, The Japanese Times, Tokyo, **1989** (ISBN 4-7890-0454-6).
- 日本語文法辞典【中級編】 (*nihongobunpoujiten [chuukyuuhen]*), Seichi Makino, Michio Tsutsui: *A Dictionary of Intermediate Japanese Grammar*, The Japanese Times, Tokyo, **1995** (ISBN 4-7890-0775-8).

- Hideichi Ono: *Japanese Grammar*, Hokuseido Press, Tokyo, **1988** (ISBN 4-590-00399-6).
- Masahiro Tanimori: *Handbook of Japanese Grammar*, Charles E. Tuttle Company, Tokyo, **1994** (ISBN 0-8048-1940-8).
- Samuel E. Martin: *A Reference Grammar of Japanese*, Charles E. Tuttle Company, Rutland, **1988** (ISBN 0-8048-1550-x).
- Yoko McClain: *Handbook of Modern Japanese Grammar*, Hokuseido Press, Tokyo, **1981** (ISBN 4-590-00570-0).
- Naomi Hanaoka McGloin: *A Student's Guide to Japanese Grammar*, 大修館 (*taishukan* Publishing Company), Tokyo, **1991** (ISBN 4-469-22065-5).
- Carol Akiyama, Nobuo Akiyama: *Barron's Japanese Grammar*, Barrons, New York (ISBN 081-2046-439).
- Francis G. Drohan: *A Handbook of Japanese Usage*, Charles E. Tuttle Company, Rutland, **1992** (ISBN 0-8048-1610-7).
- Roland A. Lange: *501 Japanese Verbs*, Barron's Educational Series, Woodbury, New York, **1988** (ISBN 0-8120-3991-2).
- Naoko Chino: *All about Particles*, Kodansha International, Tokyo, **1991** (ISBN 4-7700-1501-1).
- Kakuko Shoji: *Basic Connections*, Kodansha International, Tokyo, **1997** (ISBN 4-7700-1968-8).
- P. Motwani: *A Dictionary of Loanwords Usage – Katakana-English*, Maruzen, Tokyo, **1991** (ISBN 4-621-03578-9).

10.3
General Japanese–English Dictionaries

- 新村出: 広辞苑 (*shinmura izuru: koujien*), Iwanami Shoten (岩波書店), Tokyo, **1998** (ISBN 4-00-080111-2, 4-00-080112-0; approx. 220 000 entries).
- 松村明: 大辞林 (*matsumura akira: daijirin*), Sanseidou (三省堂), Tokyo, **1995** (ISBN 4-385-13905-9, 4-385-13900-8, approx. 233 000 entries).
- グランドコンサイス和英辞典 (*gurandokonsaisuwaeijiten*, Grand Concise Japanese–English Dictionary), Sanseido (三省堂), Tokyo, **2002** (ISBN 4-385-10905-2, over 200 000 entries).
- 松村明: 大辞泉 (*matsumura akira: daijisen*), Shougakukan (小学館), Tokyo, **1998** (ISBN 4-09-501212-9, approx. 220 000 entries).
- 梅棹忠夫: 日本語大辞典 (*umesao tadao: nihongodaijiten*), Koudansha (講談社), Tokyo, **1995** (ISBN 4-06-121057-2, 4-06-125002-7, over 170 000 entries).
- 金田一春彦, 池田弥三郎: 学研国語大辞典 (*kindaichi haruhiko, ikeda yasaburou: gakken kokugo daijiten*), 学習研究社 (*gakushuu kenkyuusha*), Tokyo, **1988** (ISBN 4-05-103502-6, 4-05-101902-0, approx. 120 000 entries).
- 金田一京助: 新明解国語辞典 (*kindaichi kyousuke: shin meikai kokugo jiten*), Sanseidou (三省堂), Tokyo, **1997** (ISBN 4-385-12099-X, approx. 76 000 entries).
- 見坊豪紀: 三省堂国語辞典 (*kenbou hidetoshi: sanseidou kokugo jiten*), Sanseidou (三省堂), Tokyo, **1992** (ISBN 4-385-13188-0, approx. 73 000 entries).
- 西尾実: 岩波国語辞典 (*nisho minoru: iwanami kokugo jiten*), 岩波書店 Iwanami Shoten, Tokyo, **1996** (ISBN 4-00-080041-8, approx. 57 000 entries).
- 宮地裕, 甲斐睦郎: 明治書院精選国語辞典 (*miyaji yutaka, kai mutsurou: meiji shoin seisen kokugo jiten*), Meiji Shoin (明治書院), Tokyo, **1995** (ISBN 4-625-70006-X, approx. 50 000 entries).
- Masuda Koh: *Kenkyusha's New Japanese-English Dictionary*, 研究者 (*kenkyusha*), Tokyo, **2003** (ISBN 07-8597-1289).

- 小学館 プログレッシブ和英中辞典 (*shougakukan puroguresshibuwaeichuujiten*) *Progressive Japanese-English Dictionary*, 小学館 (*shogakukan*), Tokyo, **2001** (ISBN 4-09-510253-5).
- *The Oxford-Duden Pictorial Japanaese & English Dictionary*, Oxford University Press, Oxford, **1989** (ISBN 0-19-864327-6).
- 基礎日本語学習辞典, *The Japan Foundation Basic Japanese English Dictionary*, Bonjinsha, Tokyo, **1986** (ISBN 4-89358-004-3).
- James C. Hepburn: *A Japanese and English Dictionary with an English and Japanese Index*, Charles E. Tuttle Company, Rutland, **1988** (ISBN 0-8048-1441-4) (see also version by Kodansha Gakujutsu Bunsho, **1986**).

10.4
Scientific Books and Dictionaries

- 吉郎橋本: 英独羅日化学語大辞典 (*hashimoto kichirou: eidokuranikagakugodaijiten*, English–German–Latin–Japanese Dictionary of Chemical Terms); 三共出版株式会社 (*sankyou shuppan kabushiki kaisha*), Tokyo, **1975**.
- 学術用語集化学編 (*gakujutsuyougoshuukagakuhen*, Japanese Scientific Terms Chemistry), compiled by Ministry of Education, Culture, Sports, Science and Technology Japan, 日本科学会 (*nihonkagakukai*; The Chemical Society of Japan), Tokyo, **1986** (ISBN 4-524-40821-5).
- N. Yamazaki, Y. Tomita, Y. Hirabayashi, Y. Hatano: 科学技術日本語案内 (*kagakugijutsunihongoannai*, Handbook of Scientific and Technical Japanese), 創拓社 (*sotakusha*), Tokyo.
- Rolf Schmid, Saburo Fukui: *Dictionary of Biotechnology*, Springer-Verlag, Berlin, Heidelberg, **1986** (ISBN 3-540-15566-X).
- 山田ひろし: 日中英化学用語辞典 (*yamada hiroshi: nichuueikagakuyougojiten*, Sanyo's Tri-Lingual Glossary of Chemical Terms), 三洋出版貿易株式会社 (*sanyou shuppan boueki kabushiki kaisha*; Sanyo Shuppan Boeki Co. Ltd.), Tokyo, **1976**.
- 英日中化学用語辞典 (*einichuukagakuyougojiten*, English–Japanese–Chinese Chemical Dictionary), New Times Press, Tokyo, **1998** (ISBN 4-7693-7066-0).
- 今井淑夫: 化学大百科 (*konsei shukufu: kagakudaihyakka*), 朝倉書店 (*chousoushoten*), Tokyo **1997** (ISBN 4-254-14045-2).
- 富井篤: 科学技術和英大辞典 (*tomii atsushi: kagakugijutsuwaeidaijiten*), オーム社 (*o-musha*) Tokyo, **1999** (ISBN 4-274-02409-1).
- 大木道側: 化学大辞典 (*taiboku dousoku: kagakudaijiten*), 東京化学同人 (*toukyou kagaku doujin*), Tokyo, **1989** (ISBN 4-8079-0323-3).
- 大木道側: 化学大辞典 (*taiboku dousoku: kagakudaijiten*), 東京化学同人 (*toukyou kagaku doujin*), Tokyo, **1995** (ISBN 4-8079-0411-6).
- 高本進: 化合物の辞典 (*kouppon shin: kagoubutsu no jiten*), 朝倉書店 (*chousoushoten*), Tokyo, **1997** (ISBN 4-254-14043-6).
- JIS工業用語大辞典 (*kougyouyougodaijiten*, Glossary of Technical Terms in Japanese Industrial Standards), Japanese Standards Association, Tokyo, **2001** (ISBN 4-542-20128-7).
- 科学技術和英表現大辞典 (*kagakugijutsuwaeihyougendaijiten*, New Japanese–English Dictionary of Scientific & Technical Expressions), 小倉書店 (*shousoushoten*), Tokyo, **1982**.
- 有機合成化学協会: 有機化合物辞典 (*yuukigousaikagakukyoukai: yuukikagoubutsujiten*, The Society of Synthetic Organic Chemistry Japan: Dictionary of Organic Compounds), 講談社サイエンティフィク (*koudanshasaientifiku*), Tokyo, **1998** (ISBN 4-06-139639-0).
- ファインケミカル事典 (*fainkemikarujiten*), The Society of Synthetic Organic Chemistry, Tokyo, **1982**.

10.5
Further Literature and Information Sources

- Yoriko Morita: *Japanese Patent Translation Handbook*, Published by the American Translators Association, Alexandria, **1997** (available at http://www.atanet.org).
- Edward E. Daub, R. Byron Bird, Nobou Inoue: *Basic Technical Japanese*, The University of Wisconsin Press, Wisconsin, **1990** (ISBN 0-299-12730-3).
- Klaus Hinkelmann: *Gewerblicher Rechtsschutz in Japan*, Carl Heymann Verlag, Köln, **2004** (ISBN 3-452-24622-1).
- Jon Sigurdson: *Science and Technology in Japan*, Cartermill Publishing, London, **1995** (ISBN 1-86067-012-1).
- Samuel Coleman: *Japanese Science from the Inside*, Routledge, London **1999** (ISBN 0-415-20169-1).
- M. Low, S. Nakayama, H. Yoshioka: *Science, Technology and Society in Contemporary Japan*, Cambridge University Press, Cambridge, **1999**.

10.5.1
Online Sources of Japanese Chemical Societies

- The Chemical Society of Japan: http://www.csj.jp/index-e.html;
- The Society of Chemical Engineers Japan: http://www.scej.org/en_html/index-e.htm;
- The Society of Polymer Science Japan: http://www.spsj.or.jp/index-e_old.htm;
- The Japanese Society for Analytical Chemistry: http://wwwsoc.nii.ac.jp/jsac/indexeng.html;
- Japanese Society for Bioscience, Biotechnology and Agrochemistry: http://www.jsbba.or.jp/e/index_e.html;
- The Society for Biotechnology Japan: http://wwwsoc.nii.ac.jp/sfbj;
- Japan Oil Chemists' Society: http://wwwsoc.nii.ac.jp/jocs/;
- The Society of Materials Science, Japan: http://www.jsms.jp/e-index.html;
- Catalysis Society of Japan: http://www.shokubai.org/index_e.html;
- The Japan Society of Plasma Science and Nuclear Fusion Research: http://www.jspf.or.jp/index-e.html.

10.5.2
Online Sources of Authorities and Institutes in Japan

- Ministry of Education, Culture, Sports, Science and Technology: http://www.mext.go.jp/english;
- Japan Science and Technology Agency: http://www.jst.go.jp/EN;
- Japan Society for the Promotion of Science: http://www.jsps.go.jp/english;
- National Institute of advanced Industrial Science and Technology: http://www.aist.go.jp/index_en.html;
- Japan Chemical Innovation Institute: http://www.jcii.or.jp/menu_e.html;
- Science and Technology Advisory Council Japan: http://www.stacj.org;
- Technology Administration, USA: http://www.technology.gov/reports/p_japan-patent.htm.

11
Subject Index

a

nitrogen containing hydrocarbons   3.3.1
a 亜   3.1.3, 2.2.1
accelerated patent examination 早期審査
  4.2.1, 5.4.1
acid (nomenclature)   3.1.2, 3.1.3, 3.3.3
acid esters (organic acids)   3.3.4
acid halogenides (organic acids)   3.3.4
adjectives   1.1.4, 2.1.3
adverbs 副詞   1.1.2, 1.1.4
alkenyl derivatives   3.2.1
alkyl derivatives   3.2.1
annual patent fees   5.1.3
application number   4.2.3
arithmetic operations   2.3.2
aromatic acids   3.3.3
aru ある   2.1.3
attack on patents   5.5

b

butsu 物   2.1.2

c

carboxylic acids   3.3.1
chemical classes   3.3.1
chemical elements   3.1.1
Chinese characters 漢字   1.2.3, 1.2.1, 2.2
chokuon 直音   1.2.2
chouon 長音   1.2.2
claims   5.2.1, 5.6.5
clouded sounds 濁音   1.2.2
colors   2.3.4
comma   1.2.1
complex (nomenclature)   3.1.2
conjugation   1.1.4
conjunctions 接続詞   1.1.2, 2.4
connection (sentences)   2.4.1
copula   1.1.2, 1.1.4, 2.1.3
counter   2.3.1
cyclic alkyl derivatives   3.2.1

d

da (copula) だ   1.1.2, 1.1.4
dai 第   2.3.1, 3.1.2, 3.1.4
daimeishi 代名詞   1.1.2
dakuon 濁音   1.2.2
Delphion   4.3.2
Denshi Jisho Dictionary   2.6.3
Derwent world patent index   4.3.2
desu (copula) です   1.1.2, 1.1.4
di   2.3.1
diacids   3.3.3
dialects   1.1.1
dictionaries   2.6.1, 2.6.3, 10.1, 10.3, 10.4
dioxides (nomenclature)   3.1.2
direct object   1.1.3
direct patent filing   5.3.1
document codes   4.2.1
doushi 動詞   1.1.4, 2.1.3

e

EDICT   2.6.3
Eijiro   2.6.3
elements   3.1.1
embedded sentences   2.4.1
en 塩   3.1.4, 2.2.1
enforcement of patent rights   5.7
espacenet   4.3.1
esters   3.3.4
European patent office   4.3.1
examination   5.4
examined patent publication 広告   4.2.1

f

File forming terms   4.2.4, 4.2.6
File index   4.2.4, 4.2.5
filing   5.3
first-to-file   4.1.2, 5.1.2
FI-terms   4.2.4, 4.2.5
formic acid derivatives   3.3.3
F-terms   4.2.4, 4.2.6

Japanese-English Chemical Dictionary. Edited by Markus Gewehr
Copyright © 2008 WILEY-VCH Verlag GmbH & Co. KGaA, Weinheim
ISBN: 978-3-527-31293-1

fukushi 副詞   1.1.2, 1.1.4
functional groups organic chemistry   3.3.1
furigana 振り仮名   1.2.1

g

ga が   1.1.3
generic substituents   3.2.2
genso no shuukihyou 元素の周期表 see inside book cover
germate consonants   1.2.2
godan 五段   1.1.4
gojuuon 五十音   1.2.2
goo jisho 辞書   2.6.3
grammar   1.1.2, 10.2
granted patent publication 登録   4.2.1
Greek letters   2.3.4

h

halides (nomenclature)   3.1.2
halogenides of organic acids   3.3.4
handakuon 半濁音   1.2.2
hannigoriten 半濁り点   1.2.2
Hantzsch-Widman system   3.3.2
heisei 平成   4.2.3
Hepburn system   1.2.2
heterocycles   3.3.2
hiragana   1.2.2, 1.2.1
honorific speech   1.1.1
hydrocarbons   3.3.1
hydrogen salts (nomenclature)   3.1.4

i

hydroxy diacids   3.3.3
ichidan 一段   1.1.4
imperial years   4.2.3
Industrial property digital library (IPDL)   4.3.1
infringement   5.7
INID Codes   4.2.2
inorganic acids   3.1.3
inorganic salts   3.1.4
INPADOC   4.3.1
inventive step   5.1.2
inventor rights   5.6.2
isomeric alkyl derivatives   3.2.1
i-type adjectives 形容詞   1.1.4

j

Japanese patent office 日本国特許庁   4.1.3
JAPIO   4.3.2
jekai je海   2.6.3
ji 次   3.1.3, 2.2.1
joshi 助詞   1.1.3, 2.4.1

josuushi 助数詞   2.3.1, 1.1.2
jouhou teikyou 情報提供   5.5.1

k

ka 化   2.1.2, 3.1.2
ka 過   3.1.3, 2.2.1
kagi 鉤   1.2.1
kakko 括弧   1.2.1
kanji 漢字   1.2.3, 1.2.1, 2.2
kanji combination   1.2.3, 2.2.2
kanji dictionaries   2.6.1, 2.6.3, 10.1
kanji identification   2.6.2
KanjiDB   2.6.3
KanjiDic   2.6.3
karada 体   2.1.2
katakana   1.2.2, 1.2.1
keigo 敬語   1.1.1
keiyoudoushi 形容動詞   1.1.4
keiyoushi 形容詞   1.1.4
kenjougo 謙譲語   1.1.1
ki 機, 器   2.1.2
koto 事   2.1.2
kouhyou 公表   4.2.1
koukai 公開   4.2.1
koukoku 広告   4.2.1
kun-reading   1.1.1, 1.2.3
kusuri 薬   2.1.2

l

laid-open patent publication 公開   4.2.1
language characteristics   1.1.1
licensing of patent rights   5.6.3
Lifescience Dictionary Project   2.6.4
loanwords   1.1.1, 1.2.4
long consonants   1.2.2
long vowels   1.2.2
maintenance fees   5.1.3

m

maru 丸   1.2.1, 2.4.1
mathematical symbols   2.3.2
meishi 名詞   1.1.2, 2.1.2
meta   3.2.3
minerals (nomenclature)   3.1.2, 3.1.4
modal verbs   2.4.1
mono   2.3.1
mono 物   2.1.2
monoxides (nomenclature)   3.1.2
mukou shimpan 無効審判   5.5.2

n

nakaguro 中黒   1.2.1
nakaten 中点   1.2.1

national Japanese phase for PCT applications   5.3.2
na-type adjectives 形容動詞   1.1.4
NCIPI   4.3.1
nigoriten 濁り点   1.2.2
nitrogen oxides   3.1.2
no の   2.1.3
Nomenclator   2.6.4
nomenclature inorganic compounds   3.1.2
nomenclature organic compounds   3.2
nouns 名詞   1.1.2, 2.1.2
noun-type predicate   1.1.4
novelty as patentability criterion   5.1.2
numbers   2.3.1

o

object (sentence)   1.1.2, 1.1.3
object counter 助数詞   2.3.1, 1.1.2
online dictionaries   2.6.3, 2.6.4
on-reading   1.1.1, 1.2.3
opposition   4.1.2, 4.2.1
ores (nomenclature)   3.1.2, 3.1.4
organic acids   3.3.3
organic acids derivatives   3.3.4
ortho   3.2.3
oxidation number   3.1.2, 3.1.4
oxides (nomenclature)   3.1.2, 3.1.4

p

palatalized sounds 拗音   1.2.2
para   3.2.3
particles 助詞   1.1.3, 2.4.1
passive verbs   2.1.1
patent abstract of Japan   4.3.1
patent application examination   5.4
patent application fees   5.1.3
patent claims   5.2.1
patent document types   4.2.1
patent filing   5.3.1
patent information sources   4.3.1, 4.3.2
patent infringement   5.7
patent law   5
patent numbers   4.2.3
patent owner rights   5.6.2
patent right   5.6
patent search   4.3.1
patent system, development   4.1.1
patent term extension   5.6.1
patent translation   4.3.1
patentability criteria   5.1.2
patentable inventions   5.1.1, 5.1.2
patentee rights   5.6.2
patenting fees   5.1.3
Patolis   4.3.2

periodic table of elements – see inside book cover
personal pronouns   2.1.1
phonetic   1.2.4, 1.2.2
phosphorus containing hydrocarbons   3.3.1
phosphorus oxoacids   3.1.3
predicate   1.1.2, 1.1.4
prefixes   2.2.1, 3.1.3
pronouns 代名詞   1.1.2
publication number   4.2.3
punctuation marks   1.2.1

r

radicals   1.2.3, 2.6.2
reading of *kanji*   1.2.3
registration number   4.2.3
relationship (sentences)   2.4.2
rights, inventors and patentees   5.6.2
ro-maji ローマ字   1.2.1, 1.2.2
romanization   1.2.2
rui 類   3.2.1
ryou 料   2.1.2

s

saikouhyou 再公表   4.2.1
salts (nomenclature)   3.1.4
saturated monoacids   3.3.3
semiclouded sounds 半濁音   1.2.2
sentence analysis   2.4.1
sentence connection   2.4.1
sentence relationship   2.4.2
sentence structure   1.1.2, 2.4.1
senyou jisshiken 専用実施権   5.6.3
sesquioxides (nomenclature)   3.1.2
setsuzokushi 接続詞   2.4.1, 1.1.2
shimpan 審判   5.3.1, 5.5.1, 5.5.2
shinsa seikyuu 審査請求   5.4.1
shouwa 昭和   4.2.3
so 素   3.1.1
sokuon 促音   1.2.2
sonkeigo 尊敬語   1.1.1
souki shinsa 早期審査   5.4.1
sound system   1.2.2
sources for text analysis   2.6
STN   4.3.2
stroke counting   2.6.2
subject   1.1.1, 1.1.2
subordinate clauses   2.4.1, 2.4.3, 2.4.4
substituents   3.2.2
substituents position   3.2.3
suffixes   2.2.1
sulfur containing hydrocarbons   3.3.1
sulfur oxoacids   3.1.3
suru する   2.1.3

syllables   1.2.2, 2.6.1, 6.2
symbols   2.3.2, 2.3.4
teki 的   2.1.3

t

teineigo 丁寧語   1.1.1
topic (sentence)   1.1.2, 1.1.3
touroku 登録   4.2.1
transcription   1.2.4, 1.2.2
translation   2.4, 4.3.1
translation: examples   2.5
translation: patents   4.3.1
translator associations   2.6.3
tri   2.3.1
trilateral cooperation   4.1.3
tsuujou jisshiken 通常実施権   5.6.3

u

units   2.3.3
unpalatalized sounds 直音   1.2.2
unsaturated acids   3.3.3

v

verb 動詞   1.1.4, 2.1.3
verb base   1.1.4
verbal predicate   1.1.4
vocabulary   1.1.1, 2.1.1
voiced sounds 濁音   1.2.2
vowel extender mark 長音   1.2.2

w

wa は   1.1.3
WaDoku *jiten* 和独辞典   2.6.3
Wikipedia   2.6.4
word order   1.1.2
writing direction   1.2.1
WWWJDIC   2.6.3

y

youon 拗音   1.2.2
yuusen shinsa 優先審査   5.4.1
zai 剤   2.1.2
zai 材   2.1.2